Springer Series in Electrophysics
Volume 9 Edited by Günter Ecker

Springer Series in Electrophysics

Editors: Günter Ecker Walter Engl Leopold B. Felsen

A.F. Alexandrov
L.S. Bogdankevich A.A. Rukhadze

Principles of Plasma Electrodynamics

With 51 Figures

Springer-Verlag
Berlin Heidelberg GmbH 1984

Professor Dr. Andrey F. Alexandrov
Moscow State University, Leninskie Gory, SU-117234 Moscow, USSR

Professor Larisa S. Bogdankevich
Professor Dr. Anri A. Rukhadze
P.N. Lebedev Physical Institute of the USSR Academy of Sciences, Leninskij prospekt, 53
SU-117333 Moscow, USSR

Title of the original Russian edition: *Osnovy elektrodinamiki plazmy*

ISBN 978-3-642-69249-9 ISBN 978-3-642-69247-5 (eBook)
DOI 10.1007/978-3-642-69247-5

Library of Congress Cataloging in Publication Data. Aleksandrov, A. F. (Andreĭ Fedorovich), 1935- Principles of plasma electrodynamics. (Springer series in electrophysics ; v. 9) Translation of: Osnovy elektrodinamiki plazmy. Includes index. 1. Plasma electrodynamics. I. Bogdankevich, L. S. (Larisa Semenovna), 1929-. II. Rukhadze, A. A. (Anri Amvrosievich), 1930-. III. Title. IV. Series. QC718.5E4A413 1984 530.4'4 84-5313

Softcover reprint of the hardcover 1st edition 1984

Typesetting: Schwetzinger Verlagsdruckerei, 6830 Schwetzingen
Offset printing: Beltz Offsetdruck, 6944 Hemsbach/Bergstr.
Bookbinding: J. Schäffer OHG, 6718 Grünstadt.
2153/3130-543210

Preface

The manuscript tackles one of the most interesting branches of plasma physics, the electrodynamics of the plasma. 99% of matter in the universe occur in the plasma state, – e.g., stars, gaseous nebulae, interstellar gas. The plasma also widely occurs on earth. Thus, the ionosphere protects human beings from the destroying effects of the solar radiation and provides the long-distance radio communication. Plasmas also show up in metals and semiconductors, and it is difficult to overestimate their importance in our everyday life. But even more important is that the power engineering of the future is connected with plasmas since the plasma is the fuel for thermonuclear recations and a practically unlimited source of energy harmless to the environment.

For the description of a hot plasma a unique logically complete and consistent theoretical model has been developed on the basis of the Maxwell-Vlasov equations. We tried to carry this idea through the entire text, which aims to present an orderly exposition of electromagnetic properties of the plasma within the Maxwell-Vlasov model. Both linear and nonlinear electrodynamics of the plasma are presented.

The first part (Chap. 1–5) deals with the linear electromagnetic properties of the plasma in thermodynamic equilibrium. The basic equations of the Maxwell-Vlasov model are introduced and the properties of the plasma in equilibrium are studied in the linear approximation of the electromagnetic field. The second part (Chaps. 6–9) analyzes the properties of the non-equilibrium plasma. The general theory of electromagnetic instabilities of such a plasma is developed in this part of the book. Note that the real plasma is always far from thermodynamic equilibrium and is usually unstable. Finally the principles of the nonlinear plasma electrodynamics along with the most characteristic nonlinear effects in the plasma are described in the third part (Chaps. 10–12). Presently the nonlinear electrodynamics of the plasma develops rapidly, therefore in the future the third part may undergo significant changes. However, we have made an attempt to cover the most complete and exploited branches of the nonlinear electrodynamics of the plasma.

The description is mathematically strict and systematic; the text is written as a manual supplied with many exercises and their solutions. It is quite sufficient for studying the electrodynamics of the plasma but is intended for a comparatively well prepared reader, i.e., physics students and young scientists. It is also useful for all specialists in plasma physics.

Compared to the 1978 Russian edition the English one is significantly revised and complemented. We hope that this book will be of interest for the foreign reader, too.

We wish to express our appreciation to Professor G. Ecker for his helpful assistance in the revision and preparation of the manuscript and to Dr. H. Lotsch for his attention, benevolence and effective help in improving the English version of the book.

Moscow, January 1984

A. F. Alexandrov · L. S. Bogdankevich
A. A. Rukhadze

Contents

Part I Electromagnetic Properties of a Plasma in Thermodynamic Equilibrium

Part III Principles of Nonlinear Electrodynamics of Plasma

Part I

Electromagnetic Properties of a Plasma in Thermodynamic Equilibrium

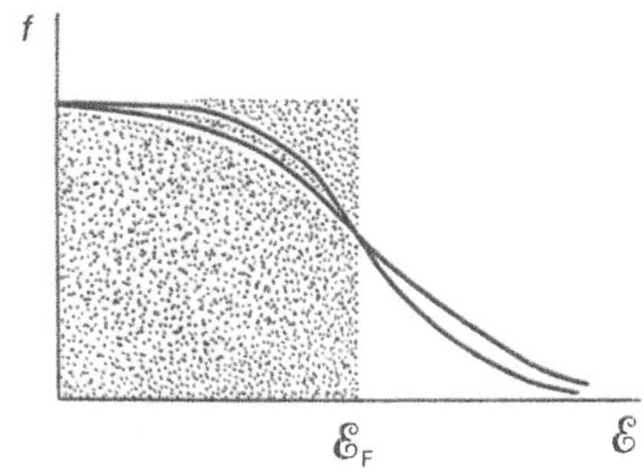

The first part of the manuscript presents the linear electrodynamics of a stable plasma in thermodynamic equilibrium. General equations of the electrodynamics of the plasma are formulated and their boundary conditions are derived. Various plasma models are discussed, with the primary focus being on the most general model, i.e., the kinetic equation with a self-consistent field. The expressions for the tensors of the complex conductivity and dielectric permittivity are obtained, both in the absence and in the presence of an external homogeneous magnetic field. The spectra of electromagnetic waves in a nondegenerate gaseous and degenerate solid-state plasma arc studied. Attention is also paid to the physical mechanisms of collisionless and collisional damping of electromagnetic waves in a plasma in thermodynamic equilibrium.

1. Basic Concepts of Plasma Physics

Basic plasma parameters, namely, the particle concentration, degree of ionization, temperature, particle distribution function, plasma frequency, Debye length and the plasma parameter are the primary focus of this chapter. Numerical estimations of the dependence of these parameters on specific conditions are presented. The applicability of the gas approximation to a description of the plasma is also discussed.

1.1 Definition of a Plasma

Before proceeding to an orderly exposition of the fundamentals of plasma electrodynamics, several questions must be answered: what is a plasma, where does it occur in nature, and what are its main characteristics?

Today the plasma state as a fourth state of matter is well known, along with the concepts of gaseous, liquid, and solid states. This is due to the wide occurrence of the plasma state in nature, and also to the prospects regarding the practical utilization of plasmas in various branches of modern technology.

The term "plasma" was introduced by I. Langmuir in 1923, when he was studying phenomena in electrical discharges in gases. Thus, the first definition of a plasma was connected with the concept of an ionized gas. Let us preliminarily define a plasma as an ionized gas comprising a large number of positively and negatively charged particles and sometimes also neutral atoms and molecules. It is the presence of the large number of charged particles in the plasma that results in those specific properties which allow us to define the plasma as a fourth state of matter and to distinguish it from ordinary gases.

The above definition of the plasma is neither exact nor complete. In fact, it is impossible to give a comprehensive definition of a plasma, since it would have to cover a great variety of phenomena under various conditions.

1.1.1 Plasmas in Nature

Practically everywhere one may encounter ionized gases. It is found, e.g., in upper layers of the earth's atmosphere, i.e., the ionosphere. While in lower layers of the atmosphere (up to an altitude of about 100 km) the number of charged particles is insignificantly small, their number becomes quite large in higher layers reaching a maximum at an altitude of 300–500 km. It is this layer of the ionosphere, the so-called F-layer, that allows the propagation of electromagnetic waves around the earth and stable radio communication. At very high altitudes the number of charged particles decreases, and the transition to the rarefied, interplanetary plasma occurs.

Another example is the plasma of the stellar range. Matter in the majority of objects in space (stars, nebulae, etc.) is ionized, i.e., in the plasma state. The fusion of light elements in thermonuclear reactions provides an immense energy output and plasma heating takes place in the plasma of stars, such as the sun. At present scientists from various countries are studying the possibilities of creating such a high-temperature plasma under terrestrial conditions in order to achieve controlled thermonuclear fusion and provide mankind with unlimited stores of energy. The development of this scientific enterprise is expanding and enriching our knowledge about plasmas and makes plasma science an independent and important branch of physics.

The plasma of gas-discharges presents another example of plasmas occurring in nature. Intense studies of these plasmas are associated with the necessity of developing classical and quantum electronics for which gas-discharge devices are of great importance.

Finally, we mention the solid-state plasmas, i.e., the electron plasmas of metals and the electron-hole plasmas of semiconductors.

Although there are plenty of other phenomena, e.g., magnetohydrodynamic or thermoionic converters, the above examples may suffice to illustrate the wide occurrence of the plasma state in nature and the great importance of studying its properties.

1.2 Plasma Parameters

As mentioned before, a plasma consists both of charged and neutral particles. Positively charged particles are ions (gaseous plasma) and holes (solid-state plasma); negatively charged particles are electrons and negative ions. Since the latter usually are insignificant in plasma phenomena, they may be ignored here.

The composition of a neutral plasma may be rather complex since, besides atoms and molecules in the ground state, it will contain excited atoms and molecules in addition. Since a plasma is a gas, we may use the same characteristics for its description as for a common gas. Let us then introduce the main plasma parameters using simple molecular-kinetic notions.

First we consider the *concentration (density)* of particles of different types N_α, the index α denoting the type of a particle. We shall mark all quantities for plasma electrons by the index e, for ions by i, and for neutrals by n. If several types of ions exist in the plasma, we define the concentration for ions of each type separately. The excitation of atoms and molecules will be of little interest to us. Therefore, N_n will mean the total number of neutral particles per unit volume without regard to their state.

Alternatively, the composition of the plasma can be described by the ratio of the electron density to that of the neutral particles, or the *degree of ionization* $r = N_e/N_n$. According to this, a plasma is *weakly ionized*, if $r < 10^{-2}$ to 10^{-3}, and *completely ionized*, if $r \to \infty$ holds. (For the "degree of ionization" one frequently uses also

$$r = N_e \Big/ \Big(N_n + \sum_i N_i\Big) .$$

For this definition $r = 1$ stands for complete ionization, and $r < 1$ for partial ionization).

Since plasmas consist of particles of different types, one must know their charge e_α and mass m_α. We have electrons with a charge $e_e \equiv e$, where $|e| = 4.8 \times 10^{-10}$ esu, and a mass $m_e = m = 9.1 \times 10^{-28}$ g, and ions with a charge $e_i = -Ze$ (Z is the multiplicity of ionization), and an approximate mass $m_i = M = A \cdot 1.66 \times 10^{-24}$ g, where A is the atomic mass of the corresponding gas. For neutral particles $e_n = 0$ and $m_n \approx m_i = M$ holds. In a solid-state plasma the effective mass of charge carriers (electrons and holes) differs from that of free electrons. Thus, to avoid misunderstandings we shall mark them as m_e^* and m_i^* where it is necessary. In metals we have $m_e^* \approx m_e$, in semiconductors usually $m_e^* \approx (0.01$ to $0.1)\, m_e$ and $m_i^* \approx m_i$. The charge of a negative carrier is equal to that of an electron, and the charge of a positive carrier is $e_i = -e$.

1.2.1 Plasmas in Thermodynamic Equilibrium and Quasiequilibrium. The Maxwell and Fermi Distribution Functions

The particles constituting a plasma are in random thermal motion. To describe this motion, one must introduce the concept of temperature for the plasma as a whole and for its components. We may use the concept of plasma temperatures T, T_α if we assume the *distribution functions* of particles of all types over the momenta p_α (or velocities) to be *Maxwellian*. If $T_\alpha \equiv T$, then the plasma is called *isothermal*.

More often the plasma is in partial thermodynamic equilibrium, and its components have different temperatures T_α

$$f_{M\alpha} = \frac{N_\alpha}{(2\pi m_\alpha k T_\alpha)^{3/2}} \exp\left(-\frac{p_\alpha^2}{2 m_\alpha k T_\alpha}\right) . \tag{1.2.1}$$

Such a plasma is *nonisothermal*.

If the distribution function is Maxwellian, the temperature T_α characterizes an average kinetic energy of the thermal motion

$$\frac{3}{2} kT_\alpha = \left(\overline{\frac{m_\alpha v_\alpha^2}{2}}\right), \tag{1.2.2}$$

where $k = 1.38 \times 10^{-16}$ erg/deg is the Boltzmann constant; T_α is the temperature of the component α [K]; v_α is the velocity of chaotic thermal motion of particles of type α [cm/s]. The upper line means averaging over all particles of type α.

The concept of plasma temperature can also be applied when the distribution functions differ from that of a Maxwellian. Then the temperature T_α means the quantity defined by (1.2.2).

We shall measure the temperature of plasma components on the absolute-temperature scale and use the energy units where the Boltzmann constant $k = 1$ and T_α is measured in electronvolts (1 eV = 11 600 K = 1.6×10^{-12} erg).

One more remark with respect to the distribution function of particles and the definition of plasma temperature: As is known from statistical physics, the Maxwellian distribution of particles is possible only for sufficiently high temperatures, when the *Fermi degeneracy* following from the Pauli exclusion principle is absent. The Fermi degeneracy concerning particles with half-integer spin (electrons, holes, and hydrogen ions) becomes essential when the Fermi energy exceeds the thermal

$$\mathscr{E}_{F\alpha} = \frac{p_{F\alpha}^2}{2m_\alpha} = \frac{(3\pi^2)^{2/3}\hbar^2 N_\alpha^{2/3}}{2m_\alpha} \gg kT_\alpha\,, \tag{1.2.3}$$

where $p_{F\alpha} = m_\alpha v_{F\alpha} = (3\pi^2)^{1/3}\hbar N_\alpha^{1/3}$ is the momentum at the Fermi boundary and $\hbar = h/2\pi = 1.05 \times 10^{-27}$ erg·s is Planck's constant. If (1.2.3) holds, the more general equilibrium distribution of fermions over momenta is defined by the expression

$$f_{F\alpha} = \frac{2}{(2\pi\hbar)^3}\left[\exp\left(\frac{p_\alpha^2}{2m_\alpha} - \mathscr{E}_{F\alpha}\right)\frac{1}{kT_\alpha} + 1\right]^{-1}, \tag{1.2.4}$$

called the *Fermi distribution function*.

The inequality (1.2.3) is satisfied at low temperatures and high concentrations $N_\alpha \gg 5 \times 10^{15}\, T_\alpha^{3/2} \cdot (m_\alpha/m)^{3/2}$. In such a *degenerate plasma* the concept of temperature as measure of the energy of random particle motion becomes senseless and is replaced by the Fermi energy $\mathscr{E}_{F\alpha} \approx 5 \times 10^{-27} N_\alpha^{2/3}\,(m/m_\alpha)$ erg, which is independent of the plasma temperature and grows with the particle concentration.

1.2.2 Characteristic Values of Plasma Parameters

Let us now consider the values of the main plasma parameters N_α and T_α for typical systems. In the *F-layer of the ionosphere* $N_e \approx N_i \approx 10^6\,\text{cm}^{-3}$, $N_n \lesssim 10^{10}\,\text{cm}^{-3}$, i.e., $r \gtrsim 10^{-4}$. The temperature is rather high, of the order of (3 to 5) $\times\ 10^3$ K. At *high altitudes* (greater than the earth's radius), i.e., in the interplanetary plasma, the concentration of charged particles varies within $10^{-2}\,\text{cm}^{-3} \lesssim N_e \approx N_i \lesssim 10\,\text{cm}^{-3}$ with $N_n \ll N_e$, and the temperature is $T \approx 10^4$ K. The concentration N and temperature T of the *stellar plasma* cover a very wide range; the concentration varies from 10^2 to $10^3\,\text{cm}^{-3}$ to 10^{22} to $10^{26}\,\text{cm}^{-3}$ and higher, the temperature from 10^4 to 10^5 K to 10^9 to 10^{10} K. Thus, for example, in the *solar corona* $N_e \approx N_i \approx 10^4$ to $10^8\,\text{cm}^{-3}$ and $T \approx 10^6$ to 10^8 K. Since *nuclear fusion* has a threshold, the plasma temperature in thermonuclear reactors should exceed $T > 10^8$ K and the concentration of charged particles, depending on the methods of plasma heating and confinement, has to be equal either to $N_e \approx N_i \approx 10^{14}$ to $10^{15}\,\text{cm}^{-3}$, or to $N_e \approx N_i \approx 10^{22}$ to $10^{23}\,\text{cm}^{-3}$. The *gas-discharge plasma*, in contrast to the thermonuclear one, is called a *low-temperature plasma*. Usually its temperature does not exceed 10^4 to 10^5 K and the concentration of charged particles is $N_e \approx N_i \approx 10^8$ to $10^{15}\,\text{cm}^{-3}$. Such a plasma is practically always weakly ionized, since $N_n \approx 10^{12}$ to $10^{17}\,\text{cm}^{-3}$. In a high-current discharge plasma, produced by an electric *explosion of metal wires*, $T \approx 10^4$ to 10^5 K and a concentration of charged particles $N_e \approx N_i \approx 10^8$ to $10^{20}\,\text{cm}^{-3}$ is found with practically complete plasma ionization. The electron *plasma in metals* has concentrations $N \approx 10^{21}$ to $10^{22}\,\text{cm}^{-3}$, while the effective mass of a charge carrier is of the order of a free-electron mass. Under these high concentrations free electrons in metals appear degenerate, and the condition of degeneracy (1.2.3) is satisfied up to the temperature $T \approx 10^4$ K. The *solid-state plasma in semiconductors*, containing negative (electrons) and positive (holes) charge carriers, can be either degenerate or nondegenerate. In semiconductors with a large number of light carriers (electrons), ($N_e \gtrsim 10^{16}$ to $10^{18}\,\text{cm}^{-3}$) and an effective mass $m_e^* \approx 10^{-2} m$, the degeneracy occurs at temperatures $T_\alpha < 10^2$ K. The degeneracy of heavy charge carriers (holes) occurs at lower temperatures, while in semiconductors with a small number of carriers ($N_e \lesssim 10^{14}$ to $10^{15}\,\text{cm}^{-3}$) the electron-hole plasma is usually nondegenerate.

The above-mentioned values of N_α and T_α are, of course, approximate and describe only the order of magnitude.

1.3 Quasi-Neutrality, Plasma Frequency and Debye Length

The above definition of a plasma is not complete. Not any ionized gas is a plasma. It must also possess the property of *quasi-neutrality*, i.e., on the average it must remain neutral for sufficiently long time and space intervals. The assumption of quasi-neutrality concerns the densities of electrons and ions in the form

$$\sum_{\alpha} e_{\alpha} N_{\alpha} = 0 , \qquad (1.3.1)$$

where $e_\alpha N_\alpha$ are, respectively, the charge and density of particles of type α.

For a plasma containing singly charged ions of only one type this condition reads

$$N_{\rm e} = N_{\rm i} , \qquad (1.3.2)$$

since an electron carries the charge $e = -e_i$.

Let us consider the time scale of charge separation. Imagine a plasma electron to deviate from the initial equilibrium position. A restoring force appears, defined in order of magnitude by the average interparticle force $F \approx e^2/r_{\rm av}^2$, where $r_{\rm av}$ is the average distance between the particles $r_{\rm av} = [3/(4\pi N_{\rm e})]^{1/3}$. As a result the electron oscillates with the frequency

$$\omega \approx \left(\frac{F}{m r_{\rm av}}\right)^{1/2} \approx \left(\frac{4\pi e^2 N_{\rm e}}{3m}\right)^{1/2} \sim \omega_{\rm p} , \qquad (1.3.3)$$

where the quantity

$$\omega_{\rm p} = \left(\frac{4\pi e^2 N_{\rm e}}{m}\right)^{1/2}$$

is called the *electron Langmuir frequency, electron plasma frequency*, or simply *plasma frequency*, and is a very important characteristic parameter. Naturally, one can regard a quantity reciprocal to the electron Langmuir frequency as a time scale of charge separation

$$\tau \sim 1/\omega_{\rm p} , \qquad (1.3.4)$$

since for time averages $t \gg \tau$ due to the particle oscillations the plasma as a whole behaves like a quasi-neutral system.

The plasma frequency, being independent of temperature, is the same both for the degenerate and nondegenerate plasma $\omega_{\rm p} \approx \sqrt{3 \cdot 10^9 N_{\rm e}}$, and for the plasma in semiconductors $\omega_{\rm p} \approx \sqrt{3 \cdot 10^9 N_{\rm e} m/m_{\rm e}}$. Accordingly we have for the ionospheric plasma $\omega_{\rm p} \approx 5 \times 10^7\,{\rm s}^{-1}$, for the thermonuclear and gas-discharge plasma $\omega_{\rm p} \approx 10^8$ to $10^{16}\,{\rm s}^{-1}$, and for the solid-state plasma $\omega_{\rm p} \approx 10^{13}$ to $5 \times 10^{15}\,{\rm s}^{-1}$.

Now let us consider the space scale of charge separation. Using simple physical reasoning, one can see that its value should be defined by the length on which density perturbations of charged particles may be shifted during the period of plasma oscillations as a result of the thermal motion of particles. Thus, the space scale of charge separation for the nondegenerate plasma is

$$d \approx \frac{v_{\mathrm{Te}}}{\omega_{\mathrm{p}}} = \left(\frac{kT_{\mathrm{e}}}{4\pi e^2 N_{\mathrm{e}}}\right)^{1/2} = r_{\mathrm{De}} , \tag{1.3.5}$$

where $v_{\mathrm{Te}} = (kT_{\mathrm{e}}/m)^{1/2}$ is the electron thermal velocity and

$$r_{\mathrm{De}} = \left(\frac{kT_{\mathrm{e}}}{4\pi e^2 N_{\mathrm{e}}}\right)^{1/2} \approx 7\left(\frac{T_{\mathrm{e}}[\mathrm{K}]}{N_{\mathrm{e}}}\right)^{1/2} \quad [\mathrm{cm}] \tag{1.3.6}$$

is called the *electron Debye length*, which plays a fundamental part in plasma physics.

In the degenerate plasma the characteristic energy of the chaotic electron motion is the Fermi energy, therefore in (1.3.6) kT_{e} must be replaced by $6\,\mathscr{E}_{\mathrm{Fe}}$

$$r_{\mathrm{De}} = \left(\frac{6\,\mathscr{E}_{\mathrm{Fe}}}{4\pi e^2 N_{\mathrm{e}}}\right)^{1/2} = \frac{\sqrt{3}\,v_{\mathrm{Fe}}}{\omega_{\mathrm{p}}} . \tag{1.3.7}$$

Thus, for plasma quasi-neutrality the characteristic dimensions L must be much greater than the Debye length

$$L \gg r_{\mathrm{De}} . \tag{1.3.8}$$

Only under these conditions may a system of charged particles be regarded as a plasma developing the typical collective effects. In the opposite case, we simply have a collection of separate charged particles which can be described by the electrodynamics of the vacuum.

For the ionospheric plasma we have $r_{\mathrm{De}} \approx 10^{-1}$ cm, for the thermonuclear and gas-discharge plasma $r_{\mathrm{De}} \approx 10^{-3}$ to 10^{-4} cm, for the solid-state plasma $r_{\mathrm{De}} \approx 10^{-5}$ to 10^{-7} cm. From these estimates it follows that under relevant conditions the Debye length is a very small quantity, and (1.3.8) is practically always satisfied with ample reserve.

1.4 Gas Approximation and Plasma Parameter

According to the given definition a plasma is regarded as a gaseous system containing charged particles. But we cannot consider any arbitrary system of particles to be a gas. A sum of charged particles constitutes a gas or, in other words, the *gas approximation* is valid, if the average potential energy of particles is smaller than their average kinetic energy. Only under this condition are gas particles almost free and interact weakly with each other. For the Coulomb interaction of particles we may write

$$\frac{e^2}{r_{\rm av}} \sim e^2 N^{1/3} \ll kT \,. \qquad (1.4.1)$$

An important plasma property is associated with this inequality

$$\eta = \frac{e^2}{r_{\rm av} kT} = \frac{e^2 N^{1/3}}{kT} \sim \frac{r_{\rm av}^2}{r_{\rm De}^2} \ll 1 \,. \qquad (1.4.2)$$

The quantity η is called the *plasma parameter* and the applicability condition for the gas approximation is $\eta \ll 1$. The inequality (1.4.2) implies the average distance between charged particles in the plasma to be smaller than the Debye length or, in other words, inside the Debye sphere (the sphere with the radius $r_{\rm D}$) there must be many particles.

In a real gaseous plasma the inequalities (1.4.1, 2) are usually satisfied with a wide margin. For the ionospheric plasma we have $\eta \lesssim 10^{-4}$ and for the thermonuclear and gas-discharge plasmas $\eta \lesssim 10^{-2}$.

For the degenerate plasma the plasma parameter η is the ratio of the average potential energy to the Fermi energy

$$\eta = \frac{e^2 N^{1/3}}{\mathscr{E}_{\rm Fe}} \sim \left(\frac{\hbar \omega_{\rm p}}{\mathscr{E}_{\rm Fe}}\right)^2 \sim \frac{r_{\rm av}^2}{r_{\rm De}^2} \ll 1 \,. \qquad (1.4.3)$$

Comparing (1.4.2, 3) we see that in the nondegenerate plasma an increase in density disfavours, in the degenerate case it favours the applicability of the gas approximation. In metals this approximation is valid only for concentrations $N_{\rm e} \gtrsim 10^{22}\,{\rm cm}^{-3}$, and in semiconductors with the effective mass $m^* = 10^{-2}$m for concentrations $N_{\rm e} \gtrsim 10^{16}$ to $10^{17}\,{\rm cm}^{-3}$. Thus, the condition (1.4.3) for real metals is only marginally fulfilled.

1.5 Exercises

1.5.1. Analyze the oscillations occurring in the homogeneous gaseous plasma due to a small displacement of the electrons with respect to the ions.

Solution. Denote the displacement vector of the electrons with respect to the ions by $\boldsymbol{S}$. The density of the uncompensated electron charge due to a displacement $\boldsymbol{S}$ is equal to

$$\varrho = {\rm div}\, e N_{\rm e} \boldsymbol{S} = e N_{\rm e}\, {\rm div}\, \boldsymbol{S} \,. \qquad (1.5.1)$$

This charge produces the electric field $\boldsymbol{E}$ which is determined through the Poisson equation

$$\operatorname{div} \boldsymbol{E} = 4\pi\varrho = 4\pi e N_e \operatorname{div} \boldsymbol{S} . \tag{1.5.2}$$

Noting that $\boldsymbol{E} = 0$ for $\boldsymbol{S} = 0$ it follows

$$\boldsymbol{E} = 4\pi e N_e \boldsymbol{S} . \tag{1.5.3}$$

Thus, the field $\boldsymbol{E}$ is parallel to the electron displacement $\boldsymbol{S}$ and acts on each electron with the force

$$\boldsymbol{F} = -e\boldsymbol{E} = -4\pi e^2 N_e \boldsymbol{S} \tag{1.5.4}$$

tending to return it into the initial equilibrium position. As a result we obtain the equation of electron motion in the form

$$m\frac{d^2\boldsymbol{S}}{dt^2} = -e\boldsymbol{E} = -4\pi e^2 N_e \boldsymbol{S} . \tag{1.5.5}$$

It describes plasma oscillations near equilibrium ($\boldsymbol{S} = 0$) with the frequency

$$\omega_0 = \omega_p = \left(\frac{4\pi e^2 N_e}{m}\right)^{1/2} . \tag{1.5.6}$$

1.5.2. Find the potential of a test charge q immersed in a spatially homogeneous gaseous plasma with the electron temperature T_e and the ion temperature T_i.

Solution. The charge q produces an electric field polarizing the plasma. As a result, besides the density of the external charge $q\,\delta(\boldsymbol{r})$, the induced charge density $\varrho(\Phi)$ appears in the plasma. Φ is the unknown potential and obeys the Poisson equation

$$\Delta\Phi = -4\pi\varrho(\Phi) - 4\pi q\,\delta(\boldsymbol{r}) . \tag{1.5.7}$$

The density of the induced charge is

$$\varrho = \sum_\alpha e_\alpha \tilde{N}_\alpha(\Phi) , \tag{1.5.8}$$

where $\tilde{N}_\alpha(\Phi)$ is the density of particles of type α, when the field Φ is present in the plasma.

In equilibrium the barometric formula

$$\tilde{N}_\alpha(\Phi) = N_\alpha \exp\left(-\frac{e_\alpha\Phi}{kT_\alpha}\right) \tag{1.5.9}$$

holds, where N_α is the unperturbed particle density in the absence of the charge q. Then the Poisson equation results in

$$\begin{aligned}\Delta\Phi &= -4\pi q\,\delta(\boldsymbol{r}) - 4\pi\sum_\alpha e_\alpha N_\alpha \exp\left(-\frac{e_\alpha\Phi}{kT_\alpha}\right)\\ &\approx -4\pi q\,\delta(\boldsymbol{r}) + \sum_\alpha \frac{4\pi e_\alpha^2 N_\alpha}{kT_\alpha}\,\Phi\,,\end{aligned} \tag{1.5.10}$$

where we expanded the exponentials in the range where $e_\alpha\Phi \ll kT_\alpha$ holds.

On developing the potential $\Phi(\boldsymbol{r})$ into the Fourier series

$$\Phi(\boldsymbol{r}) = \int d\boldsymbol{k}\, e^{i\boldsymbol{k}\cdot\boldsymbol{r}}\,\Phi(\boldsymbol{k}) \tag{1.5.11}$$

and using the relation

$$\delta(\boldsymbol{r}) = \frac{1}{(2\pi)^3}\int d\boldsymbol{k}\, e^{i\boldsymbol{k}\cdot\boldsymbol{r}}\,, \tag{1.5.12}$$

we obtain the solution of the Poisson equation in the form

$$\Phi(\boldsymbol{r}) = \frac{4\pi q}{(2\pi)^3}\int d\boldsymbol{k}\, e^{i\boldsymbol{k}\cdot\boldsymbol{r}}\left(k^2 + \frac{1}{r_D^2}\right)^{-1}, \quad \text{where} \tag{1.5.13}$$

$$\frac{1}{r_D^2} = \sum_\alpha \frac{4\pi e^2 N_\alpha}{kT_\alpha} = \frac{1}{r_{De}^2} + \frac{1}{r_{Di}^2} \tag{1.5.14}$$

(r_{De}, r_{Di} are the Debye lengths of electrons and ions, respectively).

If we integrate the expression (1.5.13) using the theory of analytic functions by closing the integration contour along a large circle in the upper half-plane of the complex $\boldsymbol{k}$ and evaluating the residue at $k = i/r_D$, we obtain

$$\Phi(r) = \frac{q}{r}\, e^{-r/r_D}\,.$$

Thus, the field of a test charge in a plasma differs from the field in vacuum since it is screened at larger distances $r > r_D$. This screening is a consequence of the displacement of the charged particles around the test charge. At distances $r < r_D$ the field of the test charge in the plasma practically does not differ from that in vacuum.

1.5.3. Show the regions of charge carrier degeneracy and applicability of the gas approximation in the diagram $N(T)$ of the electron plasma.

Solution. The degeneracy condition for the electron plasma reads

$$\mathscr{E}_F > kT\,.$$

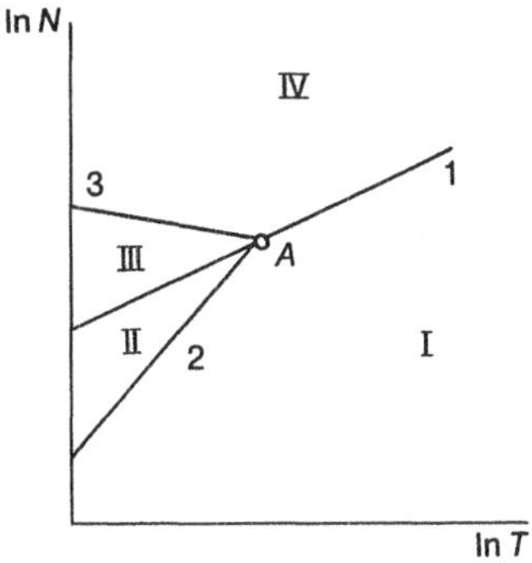

Fig. 1.1. $N(T)$ for an electron plasma

In the diagram of $\ln N$ versus $\ln T$ (Fig. 1.1) the condition

$$\mathscr{E}_F = kT = \frac{(3\pi^2)^{2/3}\hbar^2 N^{2/3}}{2m} \tag{1.5.15}$$

defines the straight line 1 separating the region of the degenerate plasma from the nondegenerate one.

The applicability condition of the gas approximation in the nondegenerate state is

$$\eta_{cl} = \frac{e^2 N^{1/3}}{kT} < 1 \,. \tag{1.5.16}$$

In the same diagram $\eta_{cl} = 1$ provides the straight line 2.

In the degenerate state for applicability of the gas approximation one must satisfy the condition

$$\eta_{qu} = \frac{e^2 N^{1/3}}{\mathscr{E}_F} < 1 \,. \tag{1.5.17}$$

As the Fermi energy depends only on N and is independent of T the condition $\eta_{qu} = 1$ gives the straight line 3 passing through the point A, where the three lines intersect, defined by

$$\mathscr{E}_F = kT = e^2 N^{1/3} \,.$$

In region I we have the nondegenerate plasma with weak interaction for which the gas approximation is valid. In region II there is the nondegenerate plasma with strong interaction, i.e., a classical fluid. In region III we have the degenerate plasma with strong interaction, i.e., a quantum fluid. In regions II and III the gas approximation is not valid. Finally, region IV belongs to the degenerate plasma with weak interaction for which the gas approximation is valid.

1.5.4. Estimate the plasma frequency, the Debye length of electrons and the plasma parameter for:
(a) the ionospheric plasma ($N_e \approx N_i \approx 10^7$ cm^{-3}, $T_e \approx T_i \approx 1$ eV);
(b) the plasma of a gas discharge ($N_e \approx N_i \approx 10^{12}$ cm^{-3}, $T_e \approx 10$ eV, $T_i \approx 0.1$ eV);
(c) the thermonuclear plasma ($N_e \approx N_i \approx 10^{15}$ cm^{-3}, $T_e \approx T_i \approx 10$ keV);
(d) the plasma in the magnetohydrodynamic converters ($N_e \approx N_i \approx 10^{19}$ cm^{-3}, $T_e \approx T_i \approx 0.3$ eV); and
(e) the electron plasma of metals under normal temperature ($N_e \approx N_i \approx 10^{22}$ to 10^{23} cm^{-3}).

Solution.
(a) $\omega_{pe} \approx 1.7 \times 10^8 \mathrm{s}^{-1}$, $r_{De} \approx 0.25$ cm, $\eta \approx 3 \times 10^{-5}$;

(b) $\omega_{pe} \approx 5.5 \times 10^{10} \mathrm{s}^{-1}$, $r_{De} \approx 2 \times 10^{-3}$ cm, $\eta \approx 1.5 \times 10^{-4}$;

(c) $\omega_{pe} \approx 1.7 \times 10^{12} \mathrm{s}^{-1}$, $r_{De} \approx 2 \times 10^{-3}$ cm, $\eta \approx 1.5 \times 10^{-6}$;

(d) $\omega_{pe} \approx 1.7 \times 10^{14} \mathrm{s}^{-1}$, $r_{De} \approx 1.5 \times 10^{-7}$ cm, $\eta \approx 1$;

(e) $\omega_{pe} \approx (0.5 \text{ to } 1.5) \times 10^{16} \mathrm{s}^{-1}$, $r_{De} \approx 10^{-7}$ to 10^{-6} cm, $\eta \approx 2$ to 1 .

2. Principles of Electrodynamics of Media with Dispersion in Space and Time

The electromagnetic field equations in the plasma and their boundary conditions are formulated, considering the plasma as a material medium with a frequency and a spatial dispersion. General properties of the complex conductivity and dielectric permittivity tensors and also the electromagnetic field energy in a medium are reported. Dispersion equations for the electromagnetic waves in the plasma and ways of solving them are given, i.e., the initial and boundary value problems are formulated.

2.1 Equations of the Electromagnetic Field in the Medium and Boundary Conditions

We shall formulate the basic equations treating the plasma as a material medium.

In the previous chapter the plasma was defined as a quasi-neutral system of charged particles. Specific plasma peculiarities appear only when the distribution of the charged particles in the plasma becomes inhomogeneous and creates macroscopic electromagnetic fields. In the plasma electromagnetic fields may also be produced by external sources. However, it is significant that these fields alter the distribution and motion of charged particles in the plasma by inducing charges and currents, which in turn again result in electromagnetic fields. This results in the *self-consistent* mutual interaction of charged particles and the field.

Thus, the Maxwell equations describing the electromagnetic field in the plasma must contain induced charges and currents. These equations are for the plasma as for any medium

$$
\begin{aligned}
\operatorname{curl}\boldsymbol{B} &= \frac{1}{c}\frac{\partial \boldsymbol{E}}{\partial t} + \frac{4\pi}{c}(j + j_0), \quad \operatorname{div}\boldsymbol{B} = 0\,, \\
\operatorname{curl}\boldsymbol{E} &= -\frac{1}{c}\frac{\partial \boldsymbol{B}}{\partial t}\,, \quad \operatorname{div}\boldsymbol{E} = 4\pi(\varrho + \varrho_0)\,,
\end{aligned}
\tag{2.1.1}
$$

where $\boldsymbol{B}$ is the magnetic induction and $\boldsymbol{E}$ the electric field; $\boldsymbol{j}$ and ϱ are the densities of the current and the space charge induced in the medium, respectively; $\boldsymbol{j}_0$ and ϱ_0 are current and charge densities of the external field sources.

The system (2.1.1) gains specific physical content if the physical meaning of the involved quantities is clear. The physical meaning of the electric field strength $\boldsymbol{E}$ and magnetic induction $\boldsymbol{B}$ is defined in vacuum as well as within media by the force on a test charge e moving with the velocity $\boldsymbol{v}$, the Lorentz force

$$\boldsymbol{F} = e\left(\boldsymbol{E} + \frac{1}{c}\,[\boldsymbol{v}, \boldsymbol{B}]\right) . \tag{2.1.2}$$

If we take the divergence of the first equation of the system (2.1.1), we find

$$\operatorname{div}\operatorname{curl}\boldsymbol{B} = \frac{1}{c}\frac{\partial}{\partial t}\operatorname{div}\boldsymbol{E} + \frac{4\pi}{c}\operatorname{div}(\boldsymbol{j} + \boldsymbol{j}_0) = 0 .$$

Inserting here the value of $\operatorname{div}\boldsymbol{E}$ from the last equation of (2.1.1), we find that the induced charges and currents satisfy the continuity equation, which expresses the conservation law of electric charge:

$$\frac{\partial(\varrho + \varrho_0)}{\partial t} + \operatorname{div}(\boldsymbol{j} + \boldsymbol{j}_0) = 0 . \tag{2.1.3}$$

Thus, in the system of field equations (2.1.1) the influence of the medium is characterized by the presence of one more vector quantity compared to the vacuum case. $\boldsymbol{j}$ is the density of the induced current in the medium. Due to the continuity equation (2.1.3) the density of the induced charge ϱ may be expressed by $\boldsymbol{j}$, since charge conservation holds for the external effects $(\varrho_0, \boldsymbol{j}_0)$ separately.

It is convenient to introduce the vector quantity $\boldsymbol{D}$, called the vector of electric displacement, through

$$\boldsymbol{D}(t, \boldsymbol{r}) = \boldsymbol{E}(t, \boldsymbol{r}) + 4\pi \int_{-\infty}^{t} dt' \boldsymbol{j}(t', \boldsymbol{r}) . \tag{2.1.4}$$

Using this definition and the continuity equation (2.1.3) one can eliminate the induced charge and current densities from the system of equations for the electromagnetic field in the medium and write

$$\begin{aligned} \operatorname{curl}\boldsymbol{B} &= \frac{1}{c}\frac{\partial \boldsymbol{D}}{\partial t} + \frac{4\pi}{c}\boldsymbol{j}_0 , \qquad \operatorname{div}\boldsymbol{B} = 0 , \\ \operatorname{curl}\boldsymbol{E} &= -\frac{1}{c}\frac{\partial \boldsymbol{B}}{\partial t} , \qquad \operatorname{div}\boldsymbol{D} = 4\pi\varrho_0 . \end{aligned} \tag{2.1.5}$$

2.1.1 Material Equations of Linear Electrodynamics

The system of equations (2.1.1 or 5) is not closed, unless the relation of the densities of the induced currents and charges to the electric field $\boldsymbol{E}$ (or the relation of $\boldsymbol{D}$ to $\boldsymbol{E}$) is given. The problem of any concrete model is to establish such a relation. In particular, we can obtain it for the plasma from the equations of motion of the particles. However, by general reasoning (independent of a specific model for the medium) one can state that in linear electrodynamics this relation must have the form

$$j_i(t,\boldsymbol{r}) = \int_{-\infty}^{t} dt' \int d\boldsymbol{r}' \sigma_{ij}(t,t',\boldsymbol{r},\boldsymbol{r}') E_j(t',\boldsymbol{r}') \,, \quad \text{and} \tag{2.1.6}$$

$$D_i(t,\boldsymbol{r}) = \int_{-\infty}^{t} dt' \int d\boldsymbol{r}' \varepsilon_{ij}(t,t',\boldsymbol{r},\boldsymbol{r}') E_j(t',\boldsymbol{r}') \,, \tag{2.1.7}$$

respectively. The indices i and j refer to the three space dimensions.

These relations, called the *material equations of linear electrodynamics*, account for the dependence of the currents and charges induced in the medium at a given moment t and a space point $\boldsymbol{r}$ upon the field values at all previous times and at any point of the medium. This is how time (frequency) and space (wavelength) dispersions in the medium appear. Physically the frequency dispersion is related to the inertia of particles and relaxation processes in the medium. Spatial dispersion concerns the propagation of the field action from one point of the medium to another due to transport processes or thermal motion of particles. The functions $\sigma_{ij}(t,t',\boldsymbol{r},\boldsymbol{r}')$ and $\varepsilon_{ij}(t,t',\boldsymbol{r},\boldsymbol{r}')$, called influence functions, are the kernels of the integrals (2.1.6, 7), characterizing the efficiency of field action transfer from one space-time point to another. The problem within concrete models of the medium is to find explicit expressions for these functions.

2.1.2 Derivation of Boundary Conditions

To solve problems in electrodynamics, the system of equations (2.1.1 or 5) must be supplemented with boundary conditions. These follow from the field equations by integrating them over an infinitely thin layer enclosing the boundary. Let us consider such an infinitely thin layer between media 1 and 2 and assume that the direction of the surface normal is $\boldsymbol{n}$ (Fig. 2.1). The second equation of the system (2.1.5) is

$$\operatorname{div}\boldsymbol{B} = 0 \,.$$

Applying Gauss theorem to a "pill box" of infinitesimal thickness, one finds

$$B_{1\mathrm{n}} = B_{2\mathrm{n}} \,. \tag{2.1.8}$$

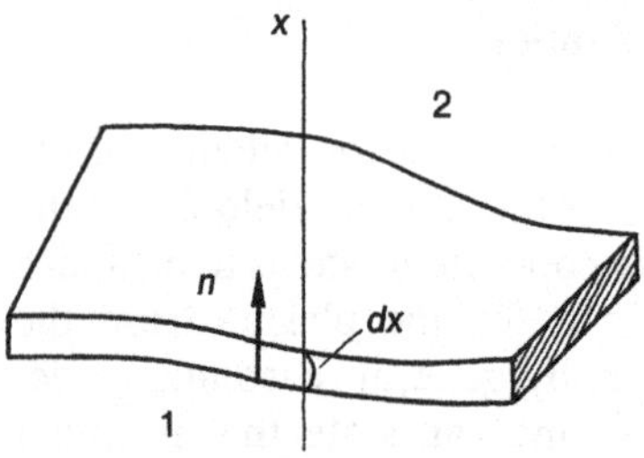

Fig. 2.1.

This is the continuity condition for the components of the magnetic vector normal to the surface of the media boundary.

We may obtain the continuity condition for the tangential components of the electric vector by analogously applying Stokes law to a loop using the third equation of (2.1.5). This results in

$$\boldsymbol{E}_{1t} = \boldsymbol{E}_{2t} \,. \tag{2.1.9}$$

Similar treatment of the last equation of (2.1.5) results in

$$D_{2n} - D_{1n} = 4\pi(\sigma + \sigma_0) \equiv 4\pi\sigma_0 - \int_1^2 [\boldsymbol{n}, [\boldsymbol{D}, \boldsymbol{n}]]\, dx \,, \tag{2.1.10}$$

where σ is a surface density of induced charges, and of the first in

$$[\boldsymbol{n}, \boldsymbol{B}_2 - \boldsymbol{B}_1] = \boldsymbol{B}_{2t} - \boldsymbol{B}_{1t} = \frac{4\pi}{c}(\boldsymbol{i} + \boldsymbol{i}_0) \,, \tag{2.1.11}$$

where $\boldsymbol{i}_0$ describes the surface current density of external sources and $\boldsymbol{i}$ is defined by

$$\boldsymbol{i} = -\frac{1}{4\pi}\int_1^2 \frac{\partial \boldsymbol{D}}{\partial t}\, dx = -\int_1^2 \left(\frac{1}{4\pi}\frac{\partial \boldsymbol{E}}{\partial t} + \boldsymbol{j}\right) dx = -\int_1^2 \boldsymbol{j}\, dx \,. \tag{2.1.12}$$

Equations (2.1.8–11) constitute a complete system of boundary conditions for the field equations (2.1.5), supplemented by the material equation (2.1.7).

2.2 Tensor of Complex Conductivity and Dielectric Permittivity

We now analyze the properties of the functional relations (2.1.6, 7) without referring to a specific plasma model.

We consider the medium homogeneous in space and time. In this case the kernels of the equations (2.1.6, 7) depend on $t - t'$ and $\boldsymbol{r} - \boldsymbol{r}'$ only. It follows

$$j_i(t, \boldsymbol{r}) = \int_{-\infty}^{t} dt' \int d\boldsymbol{r}' \sigma_{ij}(t-t', \boldsymbol{r}-\boldsymbol{r}') E_j(t', \boldsymbol{r}') , \tag{2.2.1}$$

$$D_i(t, \boldsymbol{r}) = \int_{-\infty}^{t} dt' \int d\boldsymbol{r}' \varepsilon_{ij}(t-t', \boldsymbol{r}-\boldsymbol{r}') E_j(t', \boldsymbol{r}') . \tag{2.2.2}$$

By applying the Fourier expansion, the electromagnetic field in the medium may be presented as the sum of plane monochromatic waves of the type

$$\boldsymbol{E}(t, \boldsymbol{r}) = \boldsymbol{E}(\omega, \boldsymbol{k}) \exp(-\mathrm{i}\omega t + \mathrm{i}\boldsymbol{k} \cdot \boldsymbol{r}) , \tag{2.2.3}$$

where ω is the frequency and $\boldsymbol{k}$ the wave vector. A corresponding dependency is valid for all other field quantities.

Then (2.2.1) reads

$$j_i(\omega, \boldsymbol{k}) = \int_{-\infty}^{t} dt' \int d\boldsymbol{r}' \, \sigma_{ij}(t-t', \boldsymbol{r}-\boldsymbol{r}') \times \exp[-\mathrm{i}\omega(t'-t) + \mathrm{i}\boldsymbol{k} \cdot (\boldsymbol{r}'-\boldsymbol{r})] E_j(\omega, \boldsymbol{k}) , \tag{2.2.4}$$

or

$$j_i(\omega, \boldsymbol{k}) = \int_{0}^{\infty} dt_1 \int d\boldsymbol{r}_1 \sigma_{ij}(t_1, \boldsymbol{r}_1) \exp(\mathrm{i}\omega t_1 - \mathrm{i}\boldsymbol{k} \cdot \boldsymbol{r}_1) E_j(\omega, \boldsymbol{k}) \tag{2.2.5}$$

with $t_1 = t - t'$, $\boldsymbol{r}_1 = \boldsymbol{r} - \boldsymbol{r}'$. Therefore, we immediately get the relation between the amplitudes $j_i(\omega, \boldsymbol{k})$ and $E_i(\omega, \boldsymbol{k})$:

$$j_i(\omega, \boldsymbol{k}) = \sigma_{ij}(\omega, \boldsymbol{k}) E_j(\omega, \boldsymbol{k}) , \quad \text{where} \tag{2.2.6}$$

$$\sigma_{ij}(\omega, \boldsymbol{k}) = \int_{0}^{\infty} dt_1 \int d\boldsymbol{r}_1 \sigma_{ij}(t_1, \boldsymbol{r}_1) \exp(\mathrm{i}\omega t_1 - \mathrm{i}\boldsymbol{k} \cdot \boldsymbol{r}_1) . \tag{2.2.7}$$

The quantity $\sigma_{ij}(\omega, \boldsymbol{k})$ is called the *tensor of complex conductivity* of the medium.

The *tensor of complex dielectric permittivity (dielectric tensor)* can be introduced in the same way. From (2.2.2) it follows

$$D_i(\omega, \boldsymbol{k}) = \varepsilon_{ij}(\omega, \boldsymbol{k}) E_j(\omega, \boldsymbol{k}) , \quad \text{where} \tag{2.2.8}$$

$$\varepsilon_{ij}(\omega, \boldsymbol{k}) = \int_{0}^{\infty} dt_1 \int d\boldsymbol{r}_1 \varepsilon_{ij}(t_1, \boldsymbol{r}_1) \exp(\mathrm{i}\omega t_1 - \mathrm{i}\boldsymbol{k} \cdot \boldsymbol{r}_1) \tag{2.2.9}$$

holds.

2.2.1 Dispersion in Time and Space

The dependency of these tensors on ω defines the frequency (time) dispersion and on the wave vector $\boldsymbol{k}$ the spatial dispersion of the electromagnetic field in the medium.

Taking into account (2.2.6 and 8), one can easily establish from the electric displacement (2.1.4) the following relation between the tensors $\sigma_{ij}(\omega, \boldsymbol{k})$ and $\varepsilon_{ij}(\omega, \boldsymbol{k})$:

$$\varepsilon_{ij}(\omega, \boldsymbol{k}) = \delta_{ij} + \frac{4\pi \mathrm{i}}{\omega}\sigma_{ij}(\omega, \boldsymbol{k}) . \tag{2.2.10}$$

δ_{ij} is the unit tensor. Naturally, we suppose $\omega \neq 0$.

It must be noted that the tensors $\sigma_{ij}(t, \boldsymbol{r})$ and $\varepsilon_{ij}(t, \boldsymbol{r})$ are real functions of their variables, since they connect the real quantities $\boldsymbol{E}(t, \boldsymbol{r})$, $\boldsymbol{j}(t, \boldsymbol{r})$ and $\boldsymbol{D}(t, \boldsymbol{r})$. The tensors $\sigma_{ij}(\omega, \boldsymbol{k})$ and $\varepsilon_{ij}(\omega, \boldsymbol{k})$, however, appear as complex functions even when their variables ω and $\boldsymbol{k}$ are real. The reality of $\sigma_{ij}(t, \boldsymbol{r})$ and $\varepsilon_{ij}(t, \boldsymbol{r})$ results in the following relations for $\sigma_{ij}(\omega, \boldsymbol{k})$ and $\varepsilon_{ij}(\omega, \boldsymbol{k})$:

$$\begin{aligned} &\varepsilon_{ij}(\omega, \boldsymbol{k}) = \varepsilon_{ij}^{*}(-\omega, -\boldsymbol{k}) , \\ &\mathrm{Re}\{\varepsilon_{ij}(\omega, \boldsymbol{k})\} = \mathrm{Re}\{\varepsilon_{ij}(-\omega, -\boldsymbol{k})\} , \\ &\mathrm{Im}\{\varepsilon_{ij}(\omega, \boldsymbol{k})\} = -\mathrm{Im}\{\varepsilon_{ij}(-\omega, -\boldsymbol{k})\} . \end{aligned} \tag{2.2.11}$$

Here $\mathrm{Re}\{\varepsilon_{ij}(\omega, \boldsymbol{k})\}$ and $\mathrm{Im}\{\varepsilon_{ij}(\omega, \boldsymbol{k})\}$ are the real and imaginary parts of the tensor $\varepsilon_{ij}(\omega, \boldsymbol{k})$, respectively.

2.2.2 The Case of the Isotropic Medium

For isotropic media the dielectric tensor simplifies further. In this case rotational invariance requires that the second rank tensor is composed of δ_{ij} and $k_i k_j$. Consequently for the isotropic medium the tensor $\varepsilon_{ij}(\omega, \boldsymbol{k})$ can be written as

$$\varepsilon_{ij}(\omega, \boldsymbol{k}) = \left(\delta_{ij} - \frac{k_i k_j}{k^2}\right)\varepsilon^{\mathrm{tr}}(\omega, k) + \frac{k_i k_j}{k^2}\varepsilon^{\mathrm{lo}}(\omega, k) , \tag{2.2.12}$$

meaning that among the nine tensor components $\varepsilon_{ij}(\omega, \boldsymbol{k})$ only two components $\varepsilon^{\mathrm{tr}}(\omega, k)$ and $\varepsilon^{\mathrm{lo}}(\omega, k)$ are independent. These are called the *transverse and longitudinal dielectric permittivity*, respectively, since the tensors $k_i k_j/k^2$ and $(\delta_{ij} - k_i k_j/k^2)$ are projectors into the directions parallel and perpendicular to the wave vector $\boldsymbol{k}$. Therefore, the quantity $\varepsilon^{\mathrm{lo}}(\omega, k)$ characterizes electromagnetic properties of the medium with respect to the longitudinal field and $\varepsilon^{\mathrm{tr}}(\omega, k)$ those with respect to the transverse field.

For the isotropic medium the tensor of complex conductivity takes the same form

$$\sigma_{ij}(\omega, \boldsymbol{k}) = \left(\delta_{ij} - \frac{k_i k_j}{k^2}\right)\sigma^{\mathrm{tr}}(\omega, \boldsymbol{k}) + \frac{k_i k_j}{k^2}\sigma^{\mathrm{lo}}(\omega, \boldsymbol{k}) . \tag{2.2.13}$$

Relation (2.2.10) holds for both components separately

$$\varepsilon^{\text{tr, lo}}(\omega, \boldsymbol{k}) = 1 + \frac{4\pi i}{\omega}\sigma^{\text{tr, lo}}(\omega, \boldsymbol{k}) \ . \tag{2.2.14}$$

The reality conditions (2.2.11) for the isotropic medium read:

$$\begin{aligned} \operatorname{Re}\{\varepsilon^{\text{tr, lo}}(\omega, \boldsymbol{k})\} &= \operatorname{Re}\{\varepsilon^{\text{tr, lo}}(-\omega, -\boldsymbol{k})\} \ , \\ \operatorname{Im}\{\varepsilon^{\text{tr, lo}}(\omega, \boldsymbol{k})\} &= -\operatorname{Im}\{\varepsilon^{\text{tr, lo}}(-\omega, -\boldsymbol{k})\} \ . \end{aligned} \tag{2.2.15}$$

It is obvious that the tensor $\sigma_{ij}(\omega, \boldsymbol{k})$ satisfies reality conditions of the same form.

2.2.3 The Kramers-Kronig Formulas

The dielectric tensor $\varepsilon_{ij}(\omega, \boldsymbol{k})$ is an analytical function of frequency in the upper half of the complex ω-plane since it is defined by the one-sided Fourier transformation (2.2.9). Therefore it satisfies the relation (by subtraction of δ_{ij} the approximate boundary conditions at infinity are ensured):

$$\varepsilon_{ij}(\omega, \boldsymbol{k}) - \delta_{ij} = \frac{1}{\pi i}\int_{-\infty}^{\infty} d\omega' \mathcal{P}\left\{\frac{\varepsilon_{ij}(\omega', \boldsymbol{k}) - \delta_{ij}}{\omega' - \omega}\right\} , \tag{2.2.16}$$

where $\mathcal{P}$ indicates that at $\omega' = \omega$ the principal value of the integral has to be taken.

On dividing this relation into the real and imaginary parts, we obtain the well-known *Kramers and Kronig formulas* which connect $\operatorname{Re}\{\varepsilon_{ij}(\omega, \boldsymbol{k})\}$ with $\operatorname{Im}\{\varepsilon_{ij}(\omega, \boldsymbol{k})\}$:

$$\begin{aligned} \operatorname{Re}\{\varepsilon_{ij}(\omega, \boldsymbol{k})\} - \delta_{ij} &= \frac{1}{\pi}\int_{-\infty}^{\infty} d\omega' \mathcal{P}\left\{\frac{\operatorname{Im}\{\varepsilon_{ij}(\omega', \boldsymbol{k})\}}{\omega' - \omega}\right\} , \\ \operatorname{Im}\{\varepsilon_{ij}(\omega, \boldsymbol{k})\} &= -\frac{1}{\pi}\int_{-\infty}^{\infty} d\omega' \mathcal{P}\left\{\frac{\operatorname{Re}\{\varepsilon_{ij}(\omega', \boldsymbol{k}) - \delta_{ij}\}}{\omega' - \omega}\right\} . \end{aligned} \tag{2.2.17}$$

In the isotropic medium the Kramers-Kronig formulas are valid for the longitudinal $\varepsilon^{\text{lo}}(\omega, \boldsymbol{k})$ and transverse $\varepsilon^{\text{tr}}(\omega, \boldsymbol{k})$ dielectric permittivities separately.

The Kramers-Kronig formulas for the tensor of complex conductivity read

$$\begin{aligned} \operatorname{Re}\{\sigma_{ij}(\omega, \boldsymbol{k})\} &= \frac{1}{\pi}\int_{-\infty}^{\infty} d\omega' \mathcal{P}\left\{\frac{\operatorname{Im}\{\sigma_{ij}(\omega', \boldsymbol{k})\}}{\omega' - \omega}\right\} , \\ \operatorname{Im}\{\sigma_{ij}(\omega, \boldsymbol{k})\} &= -\frac{1}{\pi}\int_{-\infty}^{\infty} d\omega' \mathcal{P}\left\{\frac{\operatorname{Re}\{\sigma_{ij}(\omega', \boldsymbol{k})\}}{\omega' - \omega}\right\} . \end{aligned} \tag{2.2.18}$$

In conclusion, it must be stressed that the problem of any concrete model of the plasma consists in calculating the tensors $\sigma_{ij}(\omega, \boldsymbol{k})$ and $\varepsilon_{ij}(\omega, \boldsymbol{k})$.

2.3 Energy of the Electromagnetic Field in the Medium

Proceeding with the study of the general properties of the tensor of dielectric permittivity in a material medium, we consider the problem of electromagnetic field energy. External forces producing a field in a medium naturally change the energy due to the interaction of the electromagnetic field with the sources. To calculate this work, we form the scalar product of the first equation of the system (2.1.5) with $\boldsymbol{E}$, of the third one with $\boldsymbol{B}$ and take the difference. As a result we obtain

$$\frac{1}{4\pi}\left(\boldsymbol{E}\cdot\frac{\partial \boldsymbol{D}}{\partial t}+\boldsymbol{B}\cdot\frac{\partial \boldsymbol{B}}{\partial t}\right)=-\frac{c}{4\pi}\operatorname{div}[\boldsymbol{E},\boldsymbol{B}]-\boldsymbol{E}\cdot\boldsymbol{j}_0\,. \tag{2.3.1}$$

Integrating this relation over a volume V, confined by the surface S, and using the Gauss theorem, we find

$$\frac{1}{4\pi}\int_V d\boldsymbol{r}\left(\boldsymbol{E}\cdot\frac{\partial \boldsymbol{D}}{\partial t}+\boldsymbol{B}\cdot\frac{\partial \boldsymbol{B}}{\partial t}\right)=-\frac{c}{4\pi}\oint d\boldsymbol{S}\,[\boldsymbol{E},\boldsymbol{B}]-\int_V d\boldsymbol{r}\,\boldsymbol{E}\cdot\boldsymbol{j}_0\,. \tag{2.3.2}$$

Passing to the limit of an unbounded medium and taking into account that the fields $\boldsymbol{E}$ and $\boldsymbol{B}$ vanish at infinity, we may neglect the first term on the right-hand side of (2.3.2). The second term describes the work A of the field against the external sources per unit time

$$\frac{dA}{dt}=\int_V d\boldsymbol{r}\,\boldsymbol{j}_0(\boldsymbol{r})\cdot\boldsymbol{E}(\boldsymbol{r})\,. \tag{2.3.3}$$

This work is balanced by a change in the field energy W. It follows

$$\frac{dW}{dt}=-\frac{dA}{dt}=-\int_V d\boldsymbol{r}\,\boldsymbol{j}_0\cdot\boldsymbol{E}=\frac{1}{4\pi}\int_V d\boldsymbol{r}\left(\boldsymbol{B}\cdot\frac{\partial \boldsymbol{B}}{\partial t}+\boldsymbol{E}\cdot\frac{\partial \boldsymbol{D}}{\partial t}\right). \tag{2.3.4}$$

Next we consider the case in which the fields $\boldsymbol{E}$, $\boldsymbol{B}$ and $\boldsymbol{D}$ are monochromatic plane waves $\exp(-i\omega t+i\boldsymbol{k}\cdot\boldsymbol{r})$. Taking into account the reality of the functions $\boldsymbol{E}(t,\boldsymbol{r})$, $\boldsymbol{B}(t,\boldsymbol{r})$ and $\boldsymbol{D}(t,\boldsymbol{r})$ we write

$$\boldsymbol{E}(t,\boldsymbol{r})=\boldsymbol{E}(\omega,\boldsymbol{k})\,\mathrm{e}^{-i\omega t+i\boldsymbol{k}\cdot\boldsymbol{r}}+\boldsymbol{E}^*(\omega,\boldsymbol{k})\,\mathrm{e}^{i\omega t-i\boldsymbol{k}\cdot\boldsymbol{r}}\,. \tag{2.3.5}$$

The quantities $\boldsymbol{B}$ and $\boldsymbol{D}$ are written analogously. Substituting these expressions into (2.3.4) and averaging over the time, we obtain

$$\frac{\overline{dW}}{dt} = \frac{i\omega}{4\pi} \int_V d\boldsymbol{r} (\boldsymbol{E} \cdot \boldsymbol{D}^* - \boldsymbol{E}^* \cdot \boldsymbol{D}) = \frac{i\omega}{4\pi} V (\boldsymbol{E} \cdot \boldsymbol{D}^* - \boldsymbol{E}^* \cdot \boldsymbol{D}) , \qquad (2.3.6)$$

where V contains the medium.

Note that by using (2.3.4) for the monochromatic fields of type (2.3.5) which do not vanish at infinity, strictly speaking, we are acting inconsistently. One can show, however, that the surface integral on the right-hand side of (2.3.2) does not increase as fast as the volume integral in the limit $V \to \infty$, so that it can be neglected.

Substituting (2.2.8) into (2.3.6), we obtain an expression for the average energy, dissipated in the medium per unit time

$$\begin{aligned} Q = \frac{\overline{dW}}{dt} &= \frac{i\omega}{4\pi} \int d\boldsymbol{r} [\varepsilon_{ij}^* (\omega, \boldsymbol{k}) - \varepsilon_{ji} (\omega, \boldsymbol{k})] E_i E_j^* \\ &= \frac{i\omega}{4\pi} V [\varepsilon_{ij}^* (\omega, \boldsymbol{k}) - \varepsilon_{ji} (\omega, \boldsymbol{k})] E_i E_j^* . \end{aligned} \qquad (2.3.7)$$

Deriving this relation, we have used the property of the dielectric tensor (2.2.11).

From (2.3.7) it follows directly that the quantity of heat delivered per unit volume and unit time is

$$\frac{Q}{V} = \frac{i\omega}{4\pi} [\varepsilon_{ij}^* (\omega, \boldsymbol{k}) - \varepsilon_{ji} (\omega, \boldsymbol{k})] E_i E_j^* . \qquad (2.3.8)$$

This equation allows us to draw an important conclusion: in a medium with a Hermitian dielectric tensor for real ω and $\boldsymbol{k}$, i.e., when $\varepsilon_{ij}^* (\omega, \boldsymbol{k}) = \varepsilon_{ji} (\omega, \boldsymbol{k})$ holds, heat is not generated. This implies that a plane monochromatic wave is not absorbed in such a medium. Thus, it follows that the absorption of the electromagnetic field in the medium is related to the anti-Hermitian part of the dielectric tensor.

Equation (2.3.8) takes a very simple form and allows further conclusions regarding the properties of the dielectric tensor for the isotropic medium when the tensor $\varepsilon_{ij} (\omega, \boldsymbol{k})$ is expressed by (2.2.12). In this case we obtain from (2.3.8)

$$\frac{Q}{V} = \frac{\omega}{2\pi k^2} (\mathrm{Im} \{\varepsilon^{\mathrm{lo}} (\omega, \boldsymbol{k})\} \, |\boldsymbol{k} \cdot \boldsymbol{E}|^2 + \mathrm{Im} \{\varepsilon^{\mathrm{tr}} (\omega, \boldsymbol{k})\} \, |[\boldsymbol{k}, \boldsymbol{E}]|^2) , \qquad (2.3.9)$$

where $\mathrm{Im} \{\varepsilon^{\mathrm{lo}} (\omega, \boldsymbol{k})\}$ and $\mathrm{Im} \{\varepsilon^{\mathrm{tr}} (\omega, \boldsymbol{k})\}$ are the imaginary parts of the components $\varepsilon^{\mathrm{lo}} (\omega, \boldsymbol{k})$ and $\varepsilon^{\mathrm{tr}} (\omega, \boldsymbol{k})$ with real ω and $\boldsymbol{k}$. The first term of this expression defines the absorption of the longitudinal field in the medium ($\boldsymbol{E} \parallel \boldsymbol{k}$), and the second one that of the transverse field ($\boldsymbol{E} \perp \boldsymbol{k}$).

2.3.1 The Dispersion of the Dielectric Permittivity Tensor

From (2.3.9) an important property of the tensor $\varepsilon_{ij}(\omega, \boldsymbol{k})$ for the isotropic medium in thermodynamic equilibrium follows. It is clear that in this case any electromagnetic wave must be absorbed. From $Q > 0$ it follows that in thermodynamic equilibrium

$$\mathrm{Im}\{\varepsilon^{\mathrm{lo}}(\omega, \boldsymbol{k})\} > 0\,, \quad \mathrm{Im}\{\varepsilon^{\mathrm{tr}}(\omega, \boldsymbol{k})\} > 0 \tag{2.3.10}$$

must be fulfilled for positive frequencies $\omega > 0$.

Because of the Kramers-Kronig formulas, the condition (2.3.10) restricts the values of the real part $\mathrm{Re}\{\varepsilon^{\mathrm{tr,\,lo}}(0, \boldsymbol{k})\}$, too. Actually, from (2.2.17) taking account of (2.2.15) we obtain

$$\varepsilon^{\mathrm{tr,\,lo}}(0, \boldsymbol{k}) - 1 = \frac{2}{\pi}\int_0^\infty d\omega' \frac{\mathrm{Im}\{\varepsilon^{\mathrm{tr,\,lo}}(\omega', \boldsymbol{k})\}}{\omega'} > 0\,.$$

Consequently, for the medium in thermodynamic equilibrium

$$\varepsilon^{\mathrm{tr,\,lo}}(0, \boldsymbol{k}) > 1 \tag{2.3.11}$$

holds.

The violation of any of the inequalities (2.3.10, 11) means a change in sign of Q, i.e., an energy transfer from the medium to the electromagnetic field. In such an active medium some random field fluctuation can increase with time thus decreasing the energy of the medium. Obviously, this is only possible if the medium is in a nonequilibrium state. We note here that the violation of any of the inequalities (2.3.10, 11) is not sufficient for the development of an instability and for the increase of the electromagnetic field energy in the medium. In the range where $\mathrm{Im}\{\varepsilon^{\mathrm{lo}}\} \leq 0$ and $\mathrm{Im}\{\varepsilon^{\mathrm{tr}}\} \leq 0$ holds, electromagnetic waves must be able to propagate in the medium.

With the "ansatz" (2.3.5), the electromagnetic field in the medium was assumed to be completely monochromatic. Actually, the field in the medium consists of a superposition of monochromatic components. When the frequencies do not vary much, one can use instead

$$\boldsymbol{E}(t, \boldsymbol{r}) = \boldsymbol{E}(\omega, \boldsymbol{k}, t)\,\mathrm{e}^{-\mathrm{i}\omega t + \mathrm{i}\boldsymbol{k}\cdot\boldsymbol{r}} + \boldsymbol{E}^*(\omega, \boldsymbol{k}, t)\,\mathrm{e}^{\mathrm{i}\omega t - \mathrm{i}\boldsymbol{k}\cdot\boldsymbol{r}}\,, \tag{2.3.12}$$

where $\boldsymbol{E}(\omega, \boldsymbol{k}, t)$ is a function slowly varying in time.

For the Fourier expansion of the field

$$\boldsymbol{E}(t, \boldsymbol{r}) = \int_{-\infty}^{\infty} d\omega'\,\mathrm{e}^{-\mathrm{i}\omega' t}\boldsymbol{E}(\omega', \boldsymbol{r}) \tag{2.3.13}$$

this means that the amplitude $\boldsymbol{E}(\omega', \boldsymbol{r})$ has a sharp maximum near $\omega' = \pm\,\omega$, implying

$$
\begin{aligned}
\boldsymbol{E}(\omega, \boldsymbol{k}, t) &= \int_0^\infty d\omega' \, \mathrm{e}^{\mathrm{i}(\omega-\omega')t} \boldsymbol{E}(\omega', \boldsymbol{k}) \,, \\
\boldsymbol{E}^*(\omega, \boldsymbol{k}, t) &= \int_{-\infty}^0 d\omega' \, \mathrm{e}^{-\mathrm{i}(\omega+\omega')t} \boldsymbol{E}^*(-\omega', \boldsymbol{k}) \,.
\end{aligned}
\tag{2.3.14}
$$

Due to the "ansatz" (2.3.12) adopting analogous relations for $\boldsymbol{B}(\omega, \boldsymbol{k}, t)$ and $\boldsymbol{D}(\omega, \boldsymbol{k}, t)$ we obtain from (2.3.4) by time averaging

$$
\begin{aligned}
\frac{1}{V}\frac{\overline{dW}}{dt} &= \frac{1}{4\pi}\left\{\frac{d}{dt}\left[B_i^*(\omega, \boldsymbol{k}, t)\, B_i(\omega, \boldsymbol{k}, t)\right] + E_i^*(\omega, \boldsymbol{k}, t)\,\frac{\partial E_j(\omega, \boldsymbol{k}, t)}{\partial t}\right. \\
&\quad \left.\times \frac{\partial}{\partial\omega}\,\omega\varepsilon_{ij}(\omega, \boldsymbol{k}) + E_j(\omega, \boldsymbol{k}, t)\,\frac{\partial E_i^*(\omega, \boldsymbol{k}, t)}{\partial t}\,\frac{\partial}{\partial\omega}\omega\varepsilon_{ji}^*(\omega, \boldsymbol{k})\right\} \\
&\quad + \frac{\mathrm{i}\omega}{4\pi}\left[\varepsilon_{ij}^*(\omega, \boldsymbol{k}) - \varepsilon_{ji}(\omega, \boldsymbol{k})\right] E_i(\omega, \boldsymbol{k}, t)\, E_j^*(\omega, \boldsymbol{k}, t)\,.
\end{aligned}
\tag{2.3.15}
$$

For a completely monochromatic field the quantities $\boldsymbol{E}(\omega, \boldsymbol{k}, t)$ and $\boldsymbol{B}(\omega, \boldsymbol{k}, t)$ are independent of t and (2.3.15) reduces to (2.3.7) giving the energy which is dissipated per unit time and volume in the medium. Only the last term on the right-hand side of (2.3.15) is nonzero in this case. If the field is not fully monochromatic and the medium is nonabsorbing, $\varepsilon_{ij}^*(\omega, \boldsymbol{k}) = \varepsilon_{ji}(\omega, \boldsymbol{k})$ holds and the last term in (2.3.15) can be neglected. Then we find

$$
\begin{aligned}
\frac{1}{V}\frac{\overline{dW}}{dt} = \frac{1}{V}\frac{dU}{dt} &= \frac{1}{4\pi}\frac{d}{dt}\left[B_i^*(\omega, \boldsymbol{k}, t)\, B_i(\omega, \boldsymbol{k}, t)\right. \\
&\quad \left. + E_i^*(\omega, \boldsymbol{k}, t)\, E_j(\omega, \boldsymbol{k}, t)\frac{\partial}{\partial\omega}\,\omega\varepsilon_{ij}(\omega, \boldsymbol{k})\right]\,.
\end{aligned}
\tag{2.3.16}
$$

Accordingly, the quantity

$$
\begin{aligned}
U = \frac{V}{4\pi}&\left[B_i^*(\omega, \boldsymbol{k}, t)\, B_i(\omega, \boldsymbol{k}, t)\right. \\
&\left. + E_i^*(\omega, \boldsymbol{k}, t)\, E_j(\omega, \boldsymbol{k}, t)\frac{\partial}{\partial\omega}\,\omega\varepsilon_{ij}(\omega, \boldsymbol{k})\right]
\end{aligned}
\tag{2.3.17}
$$

can be regarded as the average energy of the electromagnetic field in a nonabsorbing medium.

In the case of an isotropic nonabsorbing medium (2.3.17) further simplifies

$$
\begin{aligned}
U = \frac{V}{4\pi k^2}&\left\{|\boldsymbol{k}\boldsymbol{E}|^2\frac{\partial}{\partial\omega}\,\omega\varepsilon^{\mathrm{lo}}(\omega, \boldsymbol{k})\right. \\
&\left. + |[\boldsymbol{k}, \boldsymbol{E}]|^2\frac{\partial}{\partial\omega}\,\omega\left[\varepsilon^{\mathrm{tr}}(\omega, \boldsymbol{k}) - \frac{c^2k^2}{\omega^2}\right]\right\}\,.
\end{aligned}
\tag{2.3.18}
$$

Especially for thermodynamic equilibrium we obtain

$$\frac{\partial}{\partial\omega}\omega\varepsilon^{\mathrm{lo}}(\omega,\boldsymbol{k}) > 0, \qquad \frac{\partial}{\partial\omega}\omega\left[\varepsilon^{\mathrm{tr}}(\omega,\boldsymbol{k}) - \frac{c^2k^2}{\omega^2}\right] > 0\,. \tag{2.3.19}$$

When $U > 0$ holds, the field energy in the medium is positive. In the nonequilibrium case, U can become negative and the conditions (2.3.19) can be violated. Violation of any of the inequalities (2.3.19) indicates an electromagnetic instability of the medium. Then the corresponding wave is called a *negative energy wave*.

2.3.2 Average Force Affecting the Plasma in the Inhomogeneous High-Frequency Field

In conclusion, let us introduce one more energy quantity which is of great importance for the electrodynamics of material media especially of plasmas. The dielectric permittivity $\varepsilon_{ij}(\omega,\boldsymbol{k})$ is evidently a function of the density ϱ_{M}. Therefore the variation of the density $\delta\varrho_{\mathrm{M}}$ causes the variation of the dielectric permittivity

$$\delta\varepsilon_{ij}(\omega,\boldsymbol{k}) = \frac{\partial\varepsilon_{ij}(\omega,\boldsymbol{k})}{\partial\varrho_{\mathrm{M}}}\,\delta\varrho_{\mathrm{M}}\,, \tag{2.3.20}$$

which changes the energy of the electromagnetic field in the medium

$$\delta\overline{W} = \frac{V}{4\pi}\,\overline{\boldsymbol{E}\delta\boldsymbol{D}} = \frac{V}{4\pi}\left[E_iE_j^*\,\delta\varepsilon_{ij}^*(\omega,\boldsymbol{k}) + E_i^*E_j\,\delta\varepsilon_{ij}(\omega,\boldsymbol{k})\right]. \tag{2.3.21}$$

Supposing that the medium is nonabsorbing, i.e., the tensor $\varepsilon_{ij}(\omega,\boldsymbol{k})$ to be Hermitian, we obtain from (2.3.21) on account of (2.3.20)

$$\frac{\delta\overline{W}}{V} = \frac{1}{4\pi}\,E_i^*E_j\,\delta\varepsilon_{ij}(\omega,\boldsymbol{k}) = \frac{1}{4\pi}E_i^*E_j\frac{\partial\varepsilon_{ij}(\omega,\boldsymbol{k})}{\partial\varrho_{\mathrm{M}}}\,\delta\varrho_{\mathrm{M}}\,. \tag{2.3.22}$$

For $\delta\varrho_{\mathrm{M}} \to \varrho_{\mathrm{M}}$ this quantity is the potential energy (the sign inversed) per unit volume of the medium in the field of the electromagnetic wave. If the wave amplitude $\boldsymbol{E}(\omega,\boldsymbol{k},\boldsymbol{r})$ is spatially inhomogeneous the high-frequency electromagnetic field acts on the medium with an average force

$$\boldsymbol{F}_V = \frac{1}{4\pi}\varrho_{\mathrm{M}}\frac{\partial\varepsilon_{ij}(\omega,\boldsymbol{k})}{\partial\varrho_{\mathrm{M}}}\,\nabla(E_i^*E_j)\,. \tag{2.3.23}$$

In the case of an isotropic medium

$$\boldsymbol{F}_V = \frac{1}{4\pi}\left[\varrho_{\mathrm{M}}\frac{\partial}{\partial\varrho_{\mathrm{M}}}\varepsilon^{\mathrm{lo}}(\omega,\boldsymbol{k})\nabla|\boldsymbol{E}^{\mathrm{lo}}|^2 + \varrho_{\mathrm{M}}\frac{\partial}{\partial\varrho_{\mathrm{M}}}\varepsilon^{\mathrm{tr}}(\omega,\boldsymbol{k})\nabla|\boldsymbol{E}^{\mathrm{tr}}|^2\right] \tag{2.3.24}$$

holds. Here $\boldsymbol{E}^{\text{lo}} = \boldsymbol{k}(\boldsymbol{k}\cdot\boldsymbol{E})/k^2$ and $\boldsymbol{E}^{\text{tr}} = \boldsymbol{E} - \boldsymbol{E}^{\text{lo}}$ are the longitudinal and transverse components of the electromagnetic field with respect to the wave vector $\boldsymbol{k}$. If N_α is the density of particles of type α, then

$$\boldsymbol{F}_{V\alpha} = \frac{1}{4\pi} N_\alpha \frac{\partial \varepsilon_{ij}(\omega, \boldsymbol{k})}{\partial N_\alpha} \nabla (E_i^* E_j) \, . \tag{2.3.25}$$

The quantity $\boldsymbol{F}_{V\alpha}$ characterizes the average force influencing particles of type α.

2.4 Electromagnetic Waves in the Medium

From vacuum electrodynamics it is well known that monochromatic electromagnetic waves $\exp(-\mathrm{i}\omega t + \mathrm{i}\boldsymbol{k}\cdot\boldsymbol{r})$ can exist in the absence of external sources. In vacuum ω and $\boldsymbol{k}$ have real values, connected by the relation

$$\omega = kc \, . \tag{2.4.1}$$

An equation connecting the frequency with the wave vector $\omega(\boldsymbol{k})$ is called a *dispersion equation*. Equation (2.4.1) is an example valid for electromagnetic waves in vacuum.

Now we investigate the dispersion equation $\omega(\boldsymbol{k})$ for electromagnetic waves in media. If the medium is nonabsorbing, ω and $\boldsymbol{k}$ are real quantities (as known from the previous section). The tensor $\varepsilon_{ij}(\omega, \boldsymbol{k})$ is Hermitian for this case. For absorbing media they become complex.

Considering nontrivial solutions of the field equations (2.1.5) when the external sources are absent and assuming the field to depend on time and space coordinates in the form of a plane monochromatic wave $\exp(-\mathrm{i}\omega t + \mathrm{i}\boldsymbol{k}\cdot\boldsymbol{r})$, we can write these equations as follows

$$\begin{aligned} [\boldsymbol{k}, \boldsymbol{B}]_i &= -\frac{\omega}{c}\, \varepsilon_{ij}(\omega, \boldsymbol{k})\, E_j \, , \qquad \boldsymbol{k}\cdot\boldsymbol{B} = 0 \, , \\ [\boldsymbol{k}, \boldsymbol{E}] &= \frac{\omega}{c} \boldsymbol{B} \, , \qquad k_i \varepsilon_{ij}(\omega, \boldsymbol{k}) E_j = 0 \, . \end{aligned} \tag{2.4.2}$$

By elimination of $\boldsymbol{B}$ a system of three homogeneous equations for the components of the field $\boldsymbol{E}$ is easily derived, i.e.,

$$\left[k^2 \delta_{ij} - k_i k_j - \frac{\omega^2}{c^2}\, \varepsilon_{ij}(\omega, \boldsymbol{k}) \right] E_j = 0 \, . \tag{2.4.3}$$

Nontrivial solutions of this system of equations exist only if

$$\Lambda = \left| k^2 \delta_{ij} - k_i k_j - \frac{\omega^2}{c^2} \varepsilon_{ij}(\omega, \boldsymbol{k}) \right| = 0 \tag{2.4.4}$$

holds, where Λ is the determinant of the coefficient tensor in (2.4.3). Equation (2.4.4) is the dispersion equation, relating the frequency ω to the wave vector $\boldsymbol{k}$.

2.4.1 The Case of the Isotropic Medium

For the isotropic medium with a dielectric tensor of the form (2.2.12) the system (2.4.3) decomposes into two equations

$$\varepsilon^{\mathrm{lo}}(\omega, \boldsymbol{k}) \boldsymbol{E}^{\mathrm{lo}} = 0 , \qquad \left[k^2 - \frac{\omega^2}{c^2} \varepsilon^{\mathrm{tr}}(\omega, \boldsymbol{k}) \right] \boldsymbol{E}^{\mathrm{tr}} = 0 , \tag{2.4.5}$$

where $\boldsymbol{E}^{\mathrm{lo}} = \boldsymbol{k}(\boldsymbol{k} \cdot \boldsymbol{E})/k^2$ is the component of the electric field $\boldsymbol{E}$ parallel to the wave vector $\boldsymbol{k}$, i.e., the longitudinal field and $\boldsymbol{E}^{\mathrm{tr}} = \boldsymbol{E} - \boldsymbol{E}^{\mathrm{lo}}$ is the electric field component perpendicular to the vector $\boldsymbol{k}$, i.e., the transverse field.

The dispersion equation for the isotropic medium consequently consists of two separate equations

$$\varepsilon^{\mathrm{lo}}(\omega, \boldsymbol{k}) = 0 , \qquad k^2 - \frac{\omega^2}{c^2} \varepsilon^{\mathrm{tr}}(\omega, \boldsymbol{k}) = 0 , \tag{2.4.6}$$

giving the condition of existence for longitudinal and transverse waves, respectively.

2.4.2 Longitudinal Waves in an Anisotropic Medium

In the general case of anisotropic media (2.4.3) does not split into equations for longitudinal and transverse waves. Thus, the electromagnetic field generally is neither purely longitudinal nor purely transverse. However, at low frequencies the field $\boldsymbol{E}(\omega, \boldsymbol{k})$ turns out to be longitudinal with a high degree of accuracy. The condition under which the field in the anisotropic plasma is longitudinal will be discussed later in more detail. The longitudinal field derives from a scalar potential, giving for plane monochromatic waves

$$\boldsymbol{E}(\omega, \boldsymbol{k}) = -\mathrm{i}\boldsymbol{k}\, \Phi(\omega, \boldsymbol{k}) . \tag{2.4.7}$$

Then, from the last equation of the system (2.4.2) we obtain

$$k_i \varepsilon_{ij}(\omega, \boldsymbol{k})\, k_j\, \Phi(\omega, \boldsymbol{k}) = 0 , \tag{2.4.8}$$

leading to the dispersion equation for longitudinal (potential) waves in the anisotropic medium

$$\varepsilon(\omega, \boldsymbol{k}) = \frac{k_i k_j \varepsilon_{ij}(\omega, \boldsymbol{k})}{k^2} = 0 . \tag{2.4.9}$$

The quantity $\varepsilon(\omega, \boldsymbol{k})$ defined by (2.4.9), is called the *longitudinal dielectric permittivity of the anisotropic medium.* It is easy to show that the quantities $\varepsilon(\omega, \boldsymbol{k})$ and $\varepsilon^{lo}(\omega, \boldsymbol{k})$ coincide, if the tensor $\varepsilon_{ij}(\omega, \boldsymbol{k})$ has the form of (2.2.12).

Note that in the case of longitudinal (potential) waves the space charge is given by

$$\begin{aligned} \varrho(\omega, \boldsymbol{k}) &= \frac{\boldsymbol{k} \cdot \boldsymbol{j}(\omega, \boldsymbol{k})}{\omega} = \frac{k_i \sigma_{ij}(\omega, \boldsymbol{k}) E_j(\omega, \boldsymbol{k})}{\omega} \\ &= -\mathrm{i} \frac{k_i \sigma_{ij}(\omega, \boldsymbol{k}) k_j}{\omega} \Phi(\omega, \boldsymbol{k}) = -\frac{k^2}{4\pi} [\varepsilon(\omega, \boldsymbol{k}) - 1] \Phi(\omega, \boldsymbol{k}) . \end{aligned} \tag{2.4.10}$$

Substituting this relation into Poisson's equation and assuming $\varrho_0 = 0$, we get

$$k^2 \Phi(\omega, \boldsymbol{k}) = 4\pi\varrho(\omega, \boldsymbol{k}) = -k^2 [\varepsilon(\omega, \boldsymbol{k}) - 1] \Phi(\omega, \boldsymbol{k}) \tag{2.4.11}$$

leading to the relation

$$\begin{aligned} \varepsilon(\omega, \boldsymbol{k}) &= 1 - \frac{4\pi\varrho(\omega, \boldsymbol{k})}{k^2 \Phi(\omega, \boldsymbol{k})} \\ &= 1 + 4\pi\alpha(\omega, \boldsymbol{k}) . \end{aligned} \tag{2.4.12}$$

The quantity $\alpha(\omega, \boldsymbol{k})$ is called *polarizability of the medium.*

As already stated, the dispersion equation (2.4.4) establishes the relation between ω and $\boldsymbol{k}$ for the electromagnetic waves which can exist in the medium, the so-called eigenmodes. It determines the complex frequencies $\omega(\boldsymbol{k})$ for a given real value $\boldsymbol{k}$ or, in other words, the spectrum of natural oscillations of the medium. On the other hand, for any real value ω we can find a projection of the complex wave vector $\boldsymbol{k}(\omega)$ in the directions of space. These two approaches correspond to two different formulations of the problem of solving the integro-differential equations for the electromagnetic field. The first case corresponds to the initial value problem in which the time development is calculated from a known initial state of the electromagnetic field in the medium. The latter corresponds to the boundary value problem in which the spatial dependency of the field in the medium is determined from known values on some surface. In the following we consider these two problems in more detail.

2.5 Initial Value Problem

We consider a spatially homogeneous and infinite medium in which at the initial time $t = 0$ an electromagnetic field produced by external sources is given. At subsequent times $(t > 0)$ they are switched off. Here we shall consider the time variation of the resulting field in the medium. To solve this problem, it is not sufficient to know the initial values of $\boldsymbol{E}(0, \boldsymbol{r})$ and $\boldsymbol{B}(0, \boldsymbol{r})$ alone, one must also know the initial value of the electric displacement $\boldsymbol{D}(0, \boldsymbol{r})$. This means that the knowledge of the "history" of the field $\boldsymbol{E}(t, \boldsymbol{r})$ is necessary, since according to

$$D_i(0, \boldsymbol{r}) = \int_{-\infty}^{0} dt' \int d\boldsymbol{r}' \varepsilon_{ij}(0-t', \boldsymbol{r}-\boldsymbol{r}') E_j(t', \boldsymbol{r}') \tag{2.5.1}$$

the quantity $\boldsymbol{D}(0, \boldsymbol{r})$ is determined by the field $\boldsymbol{E}(t, \boldsymbol{r})$ at all time moments in the past $(t < 0)$. The physical reason of this necessity has to do with the time or frequency dispersion or, in other words, with inertia of the charge carriers and the relaxation processes in the medium.

Thus, in the initial value problem of electrodynamics the values of $\boldsymbol{B}(0, \boldsymbol{r})$ and $\boldsymbol{D}(0, \boldsymbol{r})$ or (which is the same) of $\boldsymbol{B}(0, \boldsymbol{r})$ and $\boldsymbol{E}(t, \boldsymbol{r})$ for all time moments $t \leq 0$ have to be prescribed. To solve this initial value problem, we shall use the one-sided Fourier transformation in time

$$\begin{aligned} \boldsymbol{E}(t, \boldsymbol{r}) &= \int_{-\infty + i\sigma}^{\infty + i\sigma} d\omega\, e^{-i\omega t} \int d\boldsymbol{k}\, e^{i\boldsymbol{k}\cdot\boldsymbol{r}} \boldsymbol{E}(\omega, \boldsymbol{k}), \qquad (\mathrm{Im}\{\omega\} = \sigma \geq 0)\,; \\ \boldsymbol{E}(\omega, \boldsymbol{k}) &= \frac{1}{(2\pi)^4} \int_0^{\infty} dt\, e^{i\omega t} \int d\boldsymbol{r}\, e^{-i\boldsymbol{k}\cdot\boldsymbol{r}} \boldsymbol{E}(t, \boldsymbol{r})\,. \end{aligned} \tag{2.5.2}$$

Let us analogously transform the electric displacement $\boldsymbol{D}(t, \boldsymbol{r})$ and the magnetic induction $\boldsymbol{B}(t, \boldsymbol{r})$. Then we obtain from (2.1.5)

$$\begin{aligned} \frac{\omega}{c} \boldsymbol{D}(\omega, \boldsymbol{k}) + [\boldsymbol{k}, \boldsymbol{B}(\omega, \boldsymbol{k})] &= \frac{i}{c} \boldsymbol{D}(t = 0, \boldsymbol{k}), \quad \boldsymbol{k} \cdot \boldsymbol{B}(\omega, \boldsymbol{k}) = 0\,, \\ \frac{\omega}{c} \boldsymbol{B}(\omega, \boldsymbol{k}) - [\boldsymbol{k}, \boldsymbol{E}(\omega, \boldsymbol{k})] &= \frac{i}{c} \boldsymbol{B}(t = 0, \boldsymbol{k}), \quad \boldsymbol{k} \cdot \boldsymbol{D}(\omega, \boldsymbol{k}) = 0\,. \end{aligned} \tag{2.5.3}$$

Here $\boldsymbol{D}(t = 0, \boldsymbol{k})$ and $\boldsymbol{B}(t = 0, \boldsymbol{k})$ are the spatial Fourier transforms of the initial values $\boldsymbol{D}(0, \boldsymbol{r})$ and $\boldsymbol{B}(0, \boldsymbol{r})$

$$\boldsymbol{B}(t = 0, \boldsymbol{k}) = \frac{1}{(2\pi)^3} \int d\boldsymbol{r}\, e^{-i\boldsymbol{k}\cdot\boldsymbol{r}} \boldsymbol{B}(t = 0, \boldsymbol{r})\,.$$

Taking account of (2.2.8) the system of equations (2.5.3) can easily be reduced to the following system of three inhomogeneous algebraic equations:

$$\begin{aligned} \Lambda_{ij}(\omega, \boldsymbol{k})\ E_j(\omega, \boldsymbol{k}) &\equiv \left[k^2\delta_{ij} - k_ik_j - \frac{\omega^2}{c^2}\,\varepsilon_{ij}(\omega, \boldsymbol{k})\right] E_j(\omega, \boldsymbol{k}) \\ &= \frac{\mathrm{i}\omega}{c^2}\, D_i(t=0, \boldsymbol{k}) \ + \frac{\mathrm{i}}{c}\,[\boldsymbol{k}, \boldsymbol{B}(t=0, \boldsymbol{k})]_i \ . \end{aligned} \tag{2.5.4}$$

The solution has the form

$$E_i(\omega, \boldsymbol{k}) = \frac{\Lambda_i(\omega, \boldsymbol{k})}{\Lambda(\omega, \boldsymbol{k})}\ , \tag{2.5.5}$$

where $\Lambda(\omega, \boldsymbol{k})$ is the determinant of the system (2.5.4)

$$\Lambda(\omega, \boldsymbol{k}) = \left|k^2\delta_{ij} - k_ik_j - \frac{\omega^2}{c^2}\,\varepsilon_{ij}(\omega, \boldsymbol{k})\right|\ , \tag{2.5.6}$$

and $\Lambda_i(\omega, \boldsymbol{k})$ is the signed minor corresponding to the right-hand part of (2.5.4), i.e., to the initial values of $\boldsymbol{D}(0, \boldsymbol{r})$ and $\boldsymbol{B}(0, \boldsymbol{r})$.

Substituting (2.5.5) into the Fourier representation (2.5.2), we obtain $\boldsymbol{E}(t, \boldsymbol{r})$ for arbitrary initial conditions. For a monochromatic initial perturbation with given $\boldsymbol{k}$ we have

$$E_i(t, \boldsymbol{k}) = \int\limits_{-\infty+\mathrm{i}\sigma}^{\infty+\mathrm{i}\sigma} d\omega\, \mathrm{e}^{-\mathrm{i}\omega t}\,\frac{\Lambda_i(\omega, \boldsymbol{k})}{\Lambda(\omega, \boldsymbol{k})}\ . \tag{2.5.7}$$

This integral is usually computed by means of contour integration in the plane of the complex variable ω. The integration contour in (2.5.7) lying above the real axis ($\sigma \geq 0$) is closed by the semi-circle with an infinite radius in the lower ω-plane, and the integral over the closed contour is evaluated by the sum of residues. Mostly these residues are the contributions of the poles of the integrand at the zeros of the determinant (2.5.6)[1]:

$$\Lambda(\omega, \boldsymbol{k}) = 0\ . \tag{2.5.8}$$

Denoting the roots of (2.5.8), which is the dispersion equation (2.4.4), by $\omega_n(\boldsymbol{k})$ we find the following dependence of the field on the time:

$$\boldsymbol{E}(t, \boldsymbol{k}) \sim \sum_n C_n \exp[-\mathrm{i}\omega_n(\boldsymbol{k})t]\ , \tag{2.5.9}$$

i.e., the field is a superposition of plane waves with frequencies defined by (2.5.8).

1 Integration of (2.5.7) in the presence of branch cuts and other singularities of the integrand needs a special discussion. Their contribution to the integral (2.5.7) does not result in a purely exponential time dependency.

The roots $\omega_n(\boldsymbol{k})$ of the dispersion equation generally are complex. The sign of the imaginary part shows whether the oscillations with the corresponding frequencies $\mathrm{Re}\{\omega_n\}$ increase or decrease with time. When, as for the stable system, all the imaginary parts $\mathrm{Im}\{\omega_n(\boldsymbol{k})\}$ are negative, then the terms of (2.5.9) decrease with time. The quantity $\delta_n = |\mathrm{Im}\{\omega_n(\boldsymbol{k})\}|$ is called the *damping decrement* of the corresponding monochromatic perturbation. Under these conditions after an initial interval of time the summand with the minimal damping decrement contributes dominantly to the field.

If among the roots of the dispersion equation there is a root with $\mathrm{Im}\{\omega_n(\boldsymbol{k})\} = 0$, then the corresponding term of (2.5.9) describes a long-lived natural oscillation of the medium.

Finally, if there is a root with $\mathrm{Im}\{\omega_n(\boldsymbol{k})\} > 0$, the corresponding mode will increase with time. This is possible only when the medium is unstable. The quantity $\delta_n = \mathrm{Im}\{\omega_n(\boldsymbol{k})\}$ then is called the *increment of oscillation*. Note that the appearance of roots with a positive imaginary part is a sufficient condition for the instability of the medium.

2.6 Boundary Value Problem

Equation (2.4.4) determines not only the time, but also the space variation of the electromagnetic field. In particular, the field penetration and propagation in the medium can be determined by the roots of the dispersion equation. Of course, the dielectric tensor $\varepsilon_{ij}(\omega, \boldsymbol{k})$ from (2.4.4), strictly speaking, is determined only for an unbounded and homogeneous medium. However, this seeming inconsistency should be permissible for medium dimensions significantly greater than the wavelength of the electromagnetic oscillations studied. Then (2.4.4) can correctly describe space variations of electromagnetic waves at distances from the boundary significantly exceeding the wavelength. At such distances the space variation of the field is not governed by the boundary conditions, but by medium properties.

Treating the initial-value problem we consider complex solutions of the dispersion equation $\omega_n(\boldsymbol{k})$ for real values of $\boldsymbol{k}$. In solving boundary-value problems, however, we usually solve for $\boldsymbol{k}(\omega)$ assuming ω and two orthogonal components of $\boldsymbol{k}(\omega)$ to a given direction to be real. Here the spatial variation of the field is determined by

$$\boldsymbol{E}(t, \boldsymbol{r}) \sim \sum_n C_n \exp[-\mathrm{i}\omega t + \mathrm{i}\boldsymbol{k}_n(\omega)\cdot\boldsymbol{r}] , \tag{2.6.1}$$

where $\boldsymbol{k}_n(\omega)$ satisfies (2.4.4).

In general, $\boldsymbol{k}_n(\omega)$ is a complex quantity. If $\mathrm{Im}\{k_{n\theta}(\omega)\} > 0$ holds, [θ is the angle between the given direction and the vector $\boldsymbol{k}_n(\omega)$ and $k_{n\theta}(\omega)$ is the component of the wave vector in this direction], a wave will be damped in this direction. In the opposite case it will grow. However, it is impossible to draw conclusions about the stability of the medium from the sign of $\operatorname{Im}\{k_{n\theta}(\omega)\}$. A strict solution and thorough analysis of the dispersion curves is necessary. On the basis of the initial value problem, i.e., the solution of (2.4.4) for $\omega(\boldsymbol{k})$, the analysis of the problem of stability is simplified. If the medium is known to be stable, i.e., $\operatorname{Im}\{\omega_n(\boldsymbol{k})\} < 0$, then $\operatorname{Im}\{k_{n\theta}(\omega)\}$ characterizes the spatial damping of this oscillation "mode" in the given direction. Similarly $\operatorname{Im}\{k_{n\theta}(\omega)\}$ is the characteristic of spatial wave amplification with the assigned frequency for the unstable medium.

In the general case of complex $\boldsymbol{k}(\omega)$ the wave $\boldsymbol{E} \sim \exp(-\mathrm{i}\omega t + \mathrm{i}\boldsymbol{k}\cdot\boldsymbol{r})$ can be called "planar" only conditionally, since the planes of constant phase (the planes perpendicular to the vector $\operatorname{Re}\{\boldsymbol{k}(\omega)\}$) do not coincide with the planes of constant amplitude (the planes perpendicular to the vector $\operatorname{Im}\{\boldsymbol{k}(\omega)\}$). Therefore, such waves are called inhomogeneous in contrast to homogeneous waves for which the planes mentioned coincide.

2.6.1 The Phase and Group Velocities of Waves

If the wave damping is weak, i.e., $|\operatorname{Im}\{\boldsymbol{k}(\omega)\}| \ll |\operatorname{Re}\{\boldsymbol{k}(\omega)\}|$ (i.e., in a weakly absorbing transparent medium), one can show that phase and group velocities coincide with those appearing in the absence of absorption (or damping). For the nonabsorbing transparent medium

$$\boldsymbol{v}_{\mathrm{ph}} = \frac{\omega \boldsymbol{k}}{k^2} \tag{2.6.2}$$

characterizes the propagation velocity of the constant phase level $[(-\omega t + \boldsymbol{k}\cdot\boldsymbol{r}) = \text{const}]$ and is consequently called the *phase velocity* of the wave. The quantity

$$\boldsymbol{v}_{\mathrm{g}} = \frac{d\omega}{d\boldsymbol{k}} = \nabla_{\boldsymbol{k}}\omega \tag{2.6.3}$$

characterizes the velocity of amplitude (and with that energy) displacement and is called the *group velocity*.

Group and phase velocity do not necessarily have the same direction. The angle between the vectors $\boldsymbol{v}_{\mathrm{g}}$, $\boldsymbol{v}_{\mathrm{ph}}$ may be acute or blunt. In the case of an acute angle one can speak of a wave with positive dispersion, or a forward wave; in the case of a blunt angle the wave is called backward with negative dispersion. In the latter case the direction of energy propagation is opposite to the phase velocity of the wave.

2.6.2 Correlation Between the Initial and Boundary Value Problems

In conclusion, we shall briefly discuss wave damping (growth) in weakly absorbing (amplifying) media and the relation between the initial and boundary value problems for such media.

In weakly absorbing media the anti-Hermitian part of the dielectric tensor, responsible for the wave absorption, see (2.3.8), is small compared to the Hermitian part. Consequently, in (2.4.4) the imaginary summands, compared to the real ones, are small, i.e., $\operatorname{Im}\{\Lambda(\omega,\boldsymbol{k})\} \ll \operatorname{Re}\{\Lambda(\omega,\boldsymbol{k})\}$. Solving for the behaviour of the wave field in time (initial value problem), we can get the approximate solutions of (2.4.4) in the form[2]

$$\omega \to \omega(\boldsymbol{k}) + \mathrm{i}\delta(\boldsymbol{k}) ,$$

where $\omega(\boldsymbol{k})$ are the real roots of the equation

$$\operatorname{Re}\{\Lambda(\omega,\boldsymbol{k})\} = 0 \tag{2.6.4}$$

defining the spectrum of oscillation frequencies and

$$\delta(\boldsymbol{k}) = -\frac{\operatorname{Im}\{\Lambda(\omega,\boldsymbol{k})\}}{\dfrac{\partial}{\partial\omega}\operatorname{Re}\{\Lambda(\omega,\boldsymbol{k})\}} \tag{2.6.5}$$

are the corresponding damping decrements (increments) of the oscillations.

When $\delta(\boldsymbol{k}) > 0$, the medium transfers its energy to the wave and oscillation buildup occurs. For $\delta(\boldsymbol{k}) < 0$ dissipation of the wave energy takes place and the energy of the medium increases.

To analyze the spatial damping of the waves in a given direction we must solve the dispersion equation (2.4.4) with respect to $k_\theta(\omega)$:

$$k_\theta(\omega) = \operatorname{Re}\{k_\theta(\omega)\} + \mathrm{i}\operatorname{Im}\{k_\theta(\omega)\} , \tag{2.6.6}$$

where $\operatorname{Re}\{k_\theta(\omega)\}$ are the real roots of (2.6.4) and the quantity

$$\operatorname{Im}\{k_\theta(\omega)\} = -\frac{\operatorname{Im}\{\Lambda(\omega,\boldsymbol{k})\}}{\dfrac{\partial}{\partial k_\theta}\operatorname{Re}\{\Lambda(\omega,\boldsymbol{k})\}} \tag{2.6.7}$$

characterizes wave damping (increase) in space.

From (2.6.5) and (2.6.7) it is easy to establish the following relation between wave damping in space and time:

$$\operatorname{Im}\{k_\theta(\omega)\} = -\frac{\delta(\theta)}{v_g(\theta)} , \tag{2.6.8}$$

2 To avoid complicated notations, we shall denote complex and real frequencies by the same letter ω. In cases when $\operatorname{Re}\{\omega\}$ and $\operatorname{Im}\{\omega\}$ appear separately, they will be denoted by ω and δ, respectively. Below we shall write this in the conventional form $\omega \rightleftarrows \omega + \mathrm{i}\delta$.

$$v_g(\theta) = -\frac{\dfrac{\partial}{\partial k_\theta}\,\mathrm{Re}\{\Lambda(\omega,\boldsymbol{k})\}}{\dfrac{\partial}{\partial \omega}\,\mathrm{Re}\{\Lambda(\omega,\boldsymbol{k})\}}, \tag{2.6.9}$$

where $v_g(\theta) = \partial\omega/\partial k_\theta$ is the group velocity of the wave in the given direction.

Equations (2.6.5–7) are especially instructive for a longitudinal wave in the isotropic medium when $\Lambda(\omega,\boldsymbol{k}) = \varepsilon^{\mathrm{lo}}(\omega,\boldsymbol{k})$ holds, see (2.4.6). If the medium is in equilibrium $\mathrm{Im}\{\varepsilon^{\mathrm{lo}}(\omega,\boldsymbol{k})\} > 0$ holds for $\omega > 0$ and the waves always damp both in space and time, i.e., $\delta(\boldsymbol{k}) < 0$ and $\mathrm{Im}\{k_\theta(\omega)\} > 0$. A change in sign of $\mathrm{Im}\{\varepsilon^{\mathrm{lo}}(\omega,\boldsymbol{k})\}$ may result in a change in sign of $\delta(\boldsymbol{k})$ and $\mathrm{Im}\{k_\theta(\omega)\}$ as well, i.e., in oscillation buildup or instability of the medium. Hence, we again come to the conclusion that $\mathrm{Im}\{\varepsilon^{\mathrm{lo}}(\omega,\boldsymbol{k})\} < 0$ for $\omega > 0$ indicates instability.

2.7 Electro- and Magnetostatics

So far we have dealt with media with time and spatial dispersion. Now let us consider the static limit, i.e., when $\partial/\partial t = 0$. Then the field equations have the form

$$\begin{aligned} \mathrm{curl}\,\boldsymbol{B} &= \frac{4\pi}{c}(\boldsymbol{j}+\boldsymbol{j}_0)\,, \qquad & \mathrm{div}\,\boldsymbol{B} &= 0\,, \\ \mathrm{curl}\,\boldsymbol{E} &= 0\,, & \mathrm{div}\,\boldsymbol{E} &= 4\pi(\varrho+\varrho_0) \end{aligned} \tag{2.7.1}$$

and the continuity equation reads

$$\mathrm{div}\,\boldsymbol{j} = 0\,. \tag{2.7.2}$$

Introducing the scalar and vector potentials

$$\boldsymbol{E} = -\,\mathrm{grad}\,\Phi\,, \qquad \boldsymbol{B} = \mathrm{curl}\,\boldsymbol{A} \tag{2.7.3}$$

the field equations (2.7.1) develop into

$$\Delta\Phi = -4\pi(\varrho+\varrho_0)\,, \qquad \Delta\boldsymbol{A} = -\frac{4\pi}{c}(\boldsymbol{j}+\boldsymbol{j}_0)\,. \tag{2.7.4}$$

Let us consider the material equations connecting the charge $\varrho(\boldsymbol{r})$ and current $\boldsymbol{j}(\boldsymbol{r})$ induced in the medium, to the electric $\boldsymbol{E}(\boldsymbol{r})$ and magnetic $\boldsymbol{B}(\boldsymbol{r})$ fields in spatially infinite media. Then in the linear approximation over fields

$\boldsymbol{E}$ and $\boldsymbol{B}$ for the Fourier transforms $\varrho(\boldsymbol{k})$ and $\boldsymbol{j}(\boldsymbol{k})$ we obtain the following material equations

$$\begin{aligned} \varrho(\boldsymbol{k}) &= -\,\alpha(\boldsymbol{k})k^2\Phi(\boldsymbol{k}) - \frac{1}{c}\tilde{T}_i(\boldsymbol{k})A_i(\boldsymbol{k}) \,, \\ j_i(\boldsymbol{k}) &= T_i(\boldsymbol{k})\Phi(\boldsymbol{k}) + \Pi_{ij}(\boldsymbol{k})A_j(\boldsymbol{k}) \,. \end{aligned} \tag{2.7.5}$$

To construct any specific model of the medium means defining the coupling factors $\alpha(\boldsymbol{k})$, $T_i(\boldsymbol{k})$, $\tilde{T}_i(\boldsymbol{k})$ and $\Pi_{ij}(\boldsymbol{k})$ in (2.7.5). They must be expressed in terms of the dielectric tensor $\varepsilon_{ij}(\omega,\boldsymbol{k})$ in the static limit. Thus we shall consider the dynamic limit $\omega \neq 0$, and then pass over to the static one $\omega \to 0$. In the dynamic limit we introduce the scalar and vector potentials

$$\boldsymbol{B} = \operatorname{curl}\boldsymbol{A}, \qquad \boldsymbol{E} = -\frac{1}{c}\frac{\partial \boldsymbol{A}}{\partial t} - \operatorname{grad}\Phi \tag{2.7.6}$$

with the calibration condition $\operatorname{div}\boldsymbol{A} = 0$ provided. Then for the Fourier transforms $\boldsymbol{E}(\omega,\boldsymbol{k})$ and $\boldsymbol{B}(\omega,\boldsymbol{k})$ we obtain

$$\boldsymbol{B} = \mathrm{i}[\boldsymbol{k},\boldsymbol{A}] \,, \qquad \boldsymbol{E} = \frac{\mathrm{i}\omega}{c}\boldsymbol{A} - \mathrm{i}\boldsymbol{k}\Phi \,. \tag{2.7.7}$$

Taking account of (2.2.6)

$$j_i = \sigma_{ij}(\omega,\boldsymbol{k})E_j \tag{2.7.8}$$

and the continuity equation

$$\omega\varrho + (\boldsymbol{k}\cdot\boldsymbol{j}) = 0 \tag{2.7.9}$$

from (2.7.5), we obtain the necessary coefficients in the static limit

$$\begin{aligned} \alpha(\boldsymbol{k}) &= \lim_{\omega\to 0}\frac{\mathrm{i}k_i}{\mathrm{k}^2\omega}\sigma_{ij}(\omega,\boldsymbol{k})k_j, \qquad & \Pi_{ij}(\boldsymbol{k}) &= \lim_{\omega\to 0}\frac{\mathrm{i}\omega}{c}\sigma_{ij}(\omega,\boldsymbol{k}) \,, \\ T_i(\boldsymbol{k}) &= \lim_{\omega\to 0}\frac{1}{\mathrm{i}}\sigma_{ij}(\omega,\boldsymbol{k})k_j, & \tilde{T}_i(\boldsymbol{k}) &= \lim_{\omega\to 0}\frac{1}{\mathrm{i}}k_j\sigma_{ij}(\omega,\boldsymbol{k}) \,. \end{aligned} \tag{2.7.10}$$

The field equations (2.7.4) along with (2.7.5 and 10) form the complete system of equations of electro- and magnetostatics of media with spatial dispersion. Note that the electric and magnetic fields correlate when $\tilde{T}_i \neq 0$ and $T_i \neq 0$, i.e., the static electric field produced in the medium by external charges induces the magnetic field and the static magnetic field produced by external currents induces the electric field. Such a situation occurs in the medium with the nonvanishing static conductivity $\sigma_{ij}(\omega,\boldsymbol{k})$.

In the isotropic medium as well as in the medium in thermodynamic equilibrium in the classical limit ($\hbar \to 0$) the electric and magnetic fields do not correlate in the static limit since $T_i(\boldsymbol{k}) = \tilde{T}_i(\boldsymbol{k}) = 0$. Moreover, then the conductivity tensor appears finite in the static limit (Exercise 2.8.3). Therefore $\Pi_{ij}(\boldsymbol{k}) = 0$ and $\boldsymbol{j}(\boldsymbol{k}) = 0$ hold. Thus, (2.7.4) for the vector potential is analogous to that in the vacuum case, i.e., the magnetic field produced by external currents $\boldsymbol{j}_0(\boldsymbol{r})$ is independent of the properties of the medium

$$\Delta A = -\frac{4\pi}{c}\boldsymbol{j}_0(\boldsymbol{r}) \,. \tag{2.7.11}$$

There is no difference between the magnetostatics of the classical medium in thermodynamic equilibrium and that in vacuo.

According to (2.7.5), Eq. (2.7.4) for the scalar potential in the case of the classical medium in thermodynamic equilibrium has the form

$$\int d\boldsymbol{r}'\, \varepsilon(\boldsymbol{r}-\boldsymbol{r}')\,\Delta\Phi(\boldsymbol{r}') = -4\pi\varrho_0(\boldsymbol{r}) \,. \tag{2.7.12}$$

Here $\varepsilon(\boldsymbol{r})$ is connected with the static longitudinal dielectric permittivity (2.4.12) $\varepsilon(0,\boldsymbol{k}) = 1 + 4\pi\alpha(\boldsymbol{k})$ by

$$\varepsilon(0,\boldsymbol{k}) = \int d\boldsymbol{r}\; \varepsilon(\boldsymbol{r})\, \mathrm{e}^{-\mathrm{i}\boldsymbol{k}\cdot\boldsymbol{r}} \,. \tag{2.7.13}$$

Note that in the thermodynamically nonequilibrium medium (or in the case of the medium in equilibrium in the quantum limit) the factors $T_i(\boldsymbol{k})$ and $\tilde{T}_i(\boldsymbol{k})$ are nonvanishing; therefore the static electric and magnetic fields correlate in the medium.

2.8 Exercises

2.8.1. Calculate the electrostatic field of a point charge q situated at $\boldsymbol{r} = \boldsymbol{r}_0$ in an anisotropic homogeneous medium.

Solution. The static charge density

$$\varrho_0(\boldsymbol{r}) = q\delta(\boldsymbol{r}-\boldsymbol{r}_0) \tag{2.8.1}$$

produces an electric potential field ($\mathrm{curl}\,\boldsymbol{E} = 0, \boldsymbol{E} = -\,\mathrm{grad}\,\Phi$) which is subject to

$$\mathrm{div}\,\boldsymbol{D} = 4\pi\varrho_0 = 4\pi q\delta(\boldsymbol{r}-\boldsymbol{r}_0) \,. \tag{2.8.2}$$

Expanding all the quantities into Fourier series

$$A(\boldsymbol{r}) = \int d\boldsymbol{k}\; \mathrm{e}^{\mathrm{i}\boldsymbol{k}\cdot\boldsymbol{r}} A(\boldsymbol{k}) \tag{2.8.3}$$

and taking the static limit as $\omega \to 0$, we obtain

$$
\begin{aligned}
\mathrm{i}k_i D_i(0, \boldsymbol{k}) &= \mathrm{i}k_i \varepsilon_{ij}(0, \boldsymbol{k}) E_j(0, \boldsymbol{k}) \\
&= k_i k_j \varepsilon_{ij}(0, \boldsymbol{k}) \Phi(\boldsymbol{k}) = 4\pi \varrho_0(\boldsymbol{k}) = \frac{4\pi q}{(2\pi)^3} \mathrm{e}^{-\mathrm{i}\boldsymbol{k}\cdot\boldsymbol{r}_0} .
\end{aligned}
\tag{2.8.4}
$$

It follows

$$
\Phi(\boldsymbol{k}) = \frac{4\pi q}{(2\pi)^3} \frac{\mathrm{e}^{-\mathrm{i}\boldsymbol{k}\cdot\boldsymbol{r}_0}}{k_i k_j \varepsilon_{ij}(0, \boldsymbol{k})} ,
\tag{2.8.5}
$$

and finally

$$
\Phi(\boldsymbol{r}) = \frac{q}{2\pi^2} \int d\boldsymbol{k} \frac{\exp[\mathrm{i}\boldsymbol{k}\cdot(\boldsymbol{r}-\boldsymbol{r}_0)]}{k_i k_j \varepsilon_{ij}(0, \boldsymbol{k})} .
\tag{2.8.6}
$$

For an isotropic medium the denominator has the form $k_i k_j \varepsilon_{ij}(0, \boldsymbol{k}) = k^2 \varepsilon^{\mathrm{lo}}(0, k)$, i.e., the static field of the charge is given by the longitudinal dielectric permittivity.

Note that (2.7.5, 6) can be generalized to the case of an oscillating charge $q \sim \exp(-\mathrm{i}\omega t)$, if the frequency is sufficiently small. Then the field can still be derived from a potential

$$
\begin{aligned}
\Phi(\omega, \boldsymbol{k}) &= \frac{4\pi q}{(2\pi)^3} \frac{\mathrm{e}^{\mathrm{i}\boldsymbol{k}\cdot\boldsymbol{r}_0}}{k_i k_j \varepsilon_{ij}(\omega, \boldsymbol{k})} , \\
\Phi(\omega, \boldsymbol{r}) &= \frac{q}{2\pi^2} \int d\boldsymbol{k} \frac{\exp[\mathrm{i}\boldsymbol{k}\cdot(\boldsymbol{r}-\boldsymbol{r}_0)]}{k_i k_j \varepsilon_{ij}(\omega, \boldsymbol{k})} .
\end{aligned}
\tag{2.8.7}
$$

For the isotropic medium when $k_i k_j \varepsilon_{ij}(\omega, \boldsymbol{k}) = k^2 \varepsilon^{\mathrm{lo}}(\omega, k)$ holds, these formulas are valid for arbitrary ω.

In the vacuum limit $\varepsilon_{ij}(0, \boldsymbol{k}) \to \delta_{ij}$ the potential from (2.8.6) reduces to the well-known Coulomb potential

$$
\Phi(\boldsymbol{r}) = \frac{q}{2\pi^2} \int d\boldsymbol{k} \frac{\mathrm{e}^{\mathrm{i}\boldsymbol{k}\cdot\boldsymbol{R}}}{k^2} = \frac{q}{R} , \quad \text{where} \quad \boldsymbol{R} = \boldsymbol{r} - \boldsymbol{r}_0 .
\tag{2.8.8}
$$

If the relation $k_i k_j \varepsilon_{ij}(0, \boldsymbol{k}) = k^2 + 1/r_{\mathrm{scr}}^2$ holds, we obtain the screened potential

$$
\Phi(\boldsymbol{r}) = \frac{q}{R} \exp(-R/r_{\mathrm{scr}}) ,
\tag{2.8.9}
$$

r_{scr} being the screening radius. For an electron-ion plasma the Debye length gives the distance of screening (Exercise 1.5.2).

If $k_i k_j \varepsilon_{ij}(0, k) = k^2 - k_0^2$ holds we get

$$\Phi(r) = \frac{q}{R} \cos k_0 R \,. \tag{2.8.10}$$

The field of a test charge in such a medium has a periodic character which indicates instability, as will be shown below.

2.8.2. Calculate the magnetic field produced by a straight stationary current filament in an isotropic and homogeneous medium.

Solution. By orienting the z-axis along the direction of the current, we have

$$j_0(r) = e_z j_0 \delta(x) \delta(y) \,, \qquad \varrho_0 = 0 \,, \tag{2.8.11}$$

where e_z is the unit vector along the z-axis.

Expanding all the quantities in a Fourier series

$$A(r) = \int dk \, e^{ik \cdot r} A(k)$$

and assuming for convenience $j_0(r) = \lim_{\omega \to 0} j_0(r) \exp(-i\omega t)$ we obtain from the field equations

$$[k, B(\omega, k)]_i + \frac{\omega}{c}\left[\left(\delta_{ij} - \frac{k_i k_j}{k^2}\right)\varepsilon^{tr} + \frac{k_i k_j}{k^2}\varepsilon^{lo}\right] E_j(\omega, k)$$

$$= -\frac{4\pi i}{c} j_{0i}(\omega, k) \tag{2.8.12}$$

$$k \cdot B(\omega, k) = 0 \,, \quad [k, E(\omega, k)] = \frac{\omega}{c} B(\omega, k) \,, \quad \varepsilon^{lo}(\omega, k) \, k \cdot E(\omega, k) = 0$$

the magnetic field in the form

$$B(r) = \frac{4\pi i}{c} \lim_{\omega \to 0} \int dk \frac{e^{ik \cdot r}[k, j_0(k)]}{k^2 - \frac{\omega^2}{c^2}\varepsilon^{tr}(\omega, k)} \,. \tag{2.8.13}$$

Here $j_0(k)$ is the Fourier transform of the current density $j_0(r)$, i.e.,

$$j_0(k) = \frac{j_0 e_z}{(2\pi)^2} \delta(k_z) \,.$$

In vacuo $\varepsilon^{tr} = 1$ holds and (2.8.13) reduces to

$$B(r) = \frac{2 j_0}{c r} e_\phi \,, \tag{2.8.14}$$

where $\boldsymbol{e}_\phi$ is the unit vector in the azimuthal direction. Equation (2.8.14) is valid for any medium with

$$\lim_{\omega\to 0} \frac{\omega^2}{c^2}\, \varepsilon^{\mathrm{tr}}(\omega, \boldsymbol{k}) = 0 \; . \tag{2.8.15}$$

This assumption holds true for classical media in thermodynamic equilibrium (Exercise 2.8.4). Violation of condition (2.8.15) indicates that the classical medium is in a state of nonequilibrium.

Thus, the magnetic field produced by a static current within a classical medium in equilibrium does not differ from that produced in vacuo. This result no longer holds for a time-varying current, of course, which can be seen from (2.7.12), valid for arbitrary frequencies ω of the current.

2.8.3. Derive how the standard notation of the field equations

$$\begin{aligned} \operatorname{curl} \boldsymbol{H} &= \frac{1}{c}\frac{\partial \boldsymbol{D}_0}{\partial t} + \frac{4\pi}{c}\boldsymbol{j}_0 \; , \qquad \operatorname{div} \boldsymbol{B} = 0 \; , \\ \operatorname{curl} \boldsymbol{E} &= -\frac{1}{c}\frac{\partial \boldsymbol{B}}{\partial t} \; , \qquad \operatorname{div} \boldsymbol{D}_0 = 4\pi\varrho_0 \end{aligned} \tag{2.8.16}$$

and the material equations for homogeneous and isotropic media

$$\begin{aligned} \boldsymbol{B}(t, \boldsymbol{r}) &= \int_{-\infty}^{t} dt' \int d\boldsymbol{r}' \,\mu(t-t', \boldsymbol{r}-\boldsymbol{r}')\, \boldsymbol{H}(t', \boldsymbol{r}') \; , \\ \boldsymbol{D}_0(t, \boldsymbol{r}) &= \int_{-\infty}^{t} dt' \int d\boldsymbol{r}' \,\varepsilon_0(t-t', \boldsymbol{r}-\boldsymbol{r}')\, \boldsymbol{E}(t', \boldsymbol{r}') \end{aligned} \tag{2.8.17}$$

is related to the notation given by (2.1.5) and (2.2.2) (for the isotropic and homogeneous medium).

Solution. Let us write down the system of equations (2.8.16) for fields of the type $\exp(-\mathrm{i}\omega t + \mathrm{i}\boldsymbol{k}\cdot\boldsymbol{r})$:

$$\frac{\mathrm{i}[\boldsymbol{k}, \boldsymbol{B}(\omega, \boldsymbol{k})]}{\mu(\omega, \boldsymbol{k})} = -\frac{\mathrm{i}\omega}{c}\,\varepsilon_0(\omega, \boldsymbol{k})\, \boldsymbol{E}(\omega, \boldsymbol{k}) + \frac{4\pi}{c}\boldsymbol{j}_0(\omega, \boldsymbol{k}) \; , \tag{2.8.18}$$

$$\boldsymbol{k}\cdot\boldsymbol{B}(\omega, \boldsymbol{k}) = 0 \; , \qquad [\boldsymbol{k}, \boldsymbol{E}(\omega, \boldsymbol{k})] = \frac{\omega}{c}\boldsymbol{B}(\omega, \boldsymbol{k}) \; ,$$

$$\mathrm{i}\boldsymbol{k}\cdot\boldsymbol{E}(\omega, \boldsymbol{k})\,\varepsilon_0(\omega, \boldsymbol{k}) = 4\pi\varrho_0(\omega, \boldsymbol{k}) \; ,$$

where the Fourier-expanded equations (2.8.17)

$$B(\omega, k) = \mu(\omega, k) H(\omega, k), \qquad D_0(\omega, k) = \varepsilon_0(\omega, k) E(\omega, k),$$

$$\begin{aligned} \mu(\omega, k) &= \int_0^\infty dt_1 \int dr_1 \mu(t_1, r_1) \exp(i\omega t_1 - ik \cdot r_1), \\ \varepsilon_0(\omega, k) &= \int_0^\infty dt_1 \int dr_1 \varepsilon_0(t_1, r_1) \exp(i\omega t_1 - ik \cdot r_1) \end{aligned} \tag{2.8.19}$$

have been used.

The quantity $\varepsilon_0(\omega, k)$ is called dielectric permittivity and $\mu(\omega, k)$ *magnetic permeability* of the medium.

On the other hand, (2.1.5) reads for fields of this type in the isotropic and homogeneous case as follows

$$\begin{aligned} i[k, B(\omega, k)]_i &= -\frac{i\omega}{c}\left[\left(\delta_{ij} - \frac{k_i k_j}{k^2}\right)\varepsilon^{tr}(\omega, k)\right. \\ &\quad \left. + \frac{k_i k_j}{k^2}\varepsilon^{lo}(\omega, k)\right] E_j(\omega, k) + \frac{4\pi}{c} j_{0i}(\omega, k), \end{aligned} \tag{2.8.20}$$

$$k \cdot B(\omega, k) = 0, \quad [k, E(\omega, k)] = \frac{\omega}{c} B(\omega, k),$$

$$ik \cdot E(\omega, k)\, \varepsilon^{lo}(\omega, k) = 4\pi\varrho_0(\omega, k).$$

Here the Fourier-transformed material equation (2.2.8) with the specialization of (2.2.12) for $\varepsilon_{ij}(\omega, k)$ is used. Comparison of (2.8.18) and (2.8.20) yields

$$\begin{aligned} \varepsilon_0(\omega, k) &= \varepsilon^{lo}(\omega, k), \\ 1 - \frac{1}{\mu(\omega, k)} &= \frac{\omega^2}{c^2 k^2}[\varepsilon^{tr}(\omega, k) - \varepsilon^{lo}(\omega, k)]. \end{aligned} \tag{2.8.21}$$

According to the Bohr-Van-Leeuwen theorem the static magnetic permeability of any classical medium in equilibrium is equal to one, i.e., $\mu(0, k) = 1$. This implies that the dielectric tensor $\varepsilon_{ij}(\omega, k)$ can have at most a pole of the first order for $\omega \to 0$ (within a classical model and for the equilibrium state).

Another definition of $\varepsilon_0(\omega, k)$ and $\mu(\omega, k)$ can be given by demanding that the ratio between the potential of a high-frequency point charge in the medium and that of the charge in vacuo should be $1/\varepsilon_0(\omega, k)$. The magnetic field of a high-frequency straight current also differs from the field in vacuo by the factor $\mu(\omega, k)$. Using (2.8.6, 12) we obtain

$$\varepsilon_0(\omega, k) = \varepsilon^{\mathrm{lo}}(\omega, k)\,,$$
$$1 - \frac{1}{\mu(\omega, k)} = \frac{\omega^2}{k^2c^2 - \omega^2}\,[\varepsilon^{\mathrm{tr}}(\omega, k) - 1]\,. \tag{2.8.22}$$

In the static limit ($\omega \to 0$), expectedly, (2.8.21, 22) coincide, which reflects the ambiguity in determining the value of $\mu(\omega, k)$. Physically, $\mu(\omega, k)$ is the magnetic permeability only in the static limit.

2.8.4. Calculate the energy loss which a fastly moving charged particle suffers in the isotropic and homogeneous medium.

Solution. The charge and current densities are

$$\varrho_0 = q\delta(\boldsymbol{r} - \boldsymbol{v}t)\,, \quad \boldsymbol{j}_0 = q\boldsymbol{v}\delta(\boldsymbol{r} - \boldsymbol{v}t)\,, \tag{2.8.23}$$

where $\boldsymbol{v}$ is the velocity of the charge q.

We expand all quantities in a Fourier series

$$A(t, \boldsymbol{r}) = \int d\omega \int d\boldsymbol{k}\, A(\omega, \boldsymbol{k}) \exp(-\mathrm{i}\omega t + \mathrm{i}\boldsymbol{k}\cdot\boldsymbol{r}) \tag{2.8.24}$$

and write the field equations

$$\begin{aligned} \Lambda_{ij}(\omega, \boldsymbol{k})\, E_j(\omega, \boldsymbol{k}) &= \left[k^2\delta_{ij} - k_jk_i - \frac{\omega^2}{c^2}\,\varepsilon_{ij}(\omega, \boldsymbol{k})\right] E_j(\omega, \boldsymbol{k}) \\ &= \frac{4\pi\mathrm{i}\omega}{c^2}\, j_{0i}(\omega, \boldsymbol{k})\,, \quad \text{where} \end{aligned} \tag{2.8.25}$$

$$\boldsymbol{j}_0(\omega, \boldsymbol{k}) = \frac{q\boldsymbol{v}}{(2\pi)^3}\,\delta(\omega - \boldsymbol{k}\cdot\boldsymbol{v}) \tag{2.8.26}$$

is the transform of the current density.

In the isotropic medium (2.8.25) takes the form

$$\begin{aligned} &\frac{\omega^2}{c^2}\,\varepsilon^{\mathrm{lo}}(\omega, k)\,\frac{\boldsymbol{k}\boldsymbol{k}\cdot\boldsymbol{E}}{k^2} - \left[k^2 - \frac{\omega^2}{c^2}\,\varepsilon^{\mathrm{tr}}(\omega, k)\right]\left(\boldsymbol{E} - \frac{\boldsymbol{k}\boldsymbol{k}\cdot\boldsymbol{E}}{k^2}\right) \\ &= -\frac{4\pi\mathrm{i}\omega}{c^2}\,\boldsymbol{j}_0(\omega, \boldsymbol{k})\,. \end{aligned} \tag{2.8.27}$$

Taking the scalar product with $\boldsymbol{k}$ we obtain

$$\boldsymbol{k}\cdot\boldsymbol{E}(\omega, \boldsymbol{k}) = -\frac{4\pi\mathrm{i}}{\omega\varepsilon^{\mathrm{lo}}(\omega, k)}\,\boldsymbol{k}\cdot\boldsymbol{j}_0(\omega, \boldsymbol{k})\,. \tag{2.8.28}$$

Inserting this expression in (2.8.27) we obtain

$$E_i(\omega, \boldsymbol{k}) = -\frac{4\pi i\omega}{k^2}\left(\frac{k_i k_j}{\omega^2 \varepsilon^{\mathrm{lo}}(\omega, k)} - \frac{k^2\left(\delta_{ij} - \dfrac{k_i k_j}{k^2}\right)}{c^2\left[k^2 - \dfrac{\omega^2}{c^2}\varepsilon^{\mathrm{tr}}(\omega, k)\right]}\right) j_{0j}(\omega, \boldsymbol{k}) \,. \tag{2.8.29}$$

Inversion of the Fourier transformation gives

$$\begin{aligned} \boldsymbol{E}(t, \boldsymbol{r}) = &-\frac{4\pi i q}{(2\pi)^3}\int d\boldsymbol{k}\frac{\exp\left[i\boldsymbol{k}\cdot(\boldsymbol{r}-\boldsymbol{v}t)\right]}{k^2}\left\{\frac{\boldsymbol{k}}{\varepsilon^{\mathrm{lo}}(\boldsymbol{k}\cdot\boldsymbol{v}, k)}\right. \\ &\left. -\frac{k^2\boldsymbol{k}\cdot\boldsymbol{v}}{c^2}\left(\boldsymbol{v} - \frac{\boldsymbol{k}\,\boldsymbol{k}\cdot\boldsymbol{v}}{k^2}\right)\left[k^2 - \frac{(\boldsymbol{k}\cdot\boldsymbol{v})^2}{c^2}\,\varepsilon^{\mathrm{tr}}(\boldsymbol{k}\cdot\boldsymbol{v}, k)\right]^{-1}\right\} \end{aligned} \tag{2.8.30}$$

where we applied (2.8.26) for the ω-integration.

The energy loss of the moving particle follows from the work done by the electromagnetic field, which it generates within the medium. According to (2.3.3) the work per unit length is equal to

$$\begin{aligned} W = q\,\frac{\boldsymbol{v}\cdot\boldsymbol{E}}{v}\bigg|_{\boldsymbol{r}=\boldsymbol{v}t} &= \frac{iq^2}{2\pi^2 v}\int d\boldsymbol{k}\frac{\boldsymbol{k}\cdot\boldsymbol{v}}{k^2} \\ &\times\left(\frac{1}{\varepsilon^{\mathrm{lo}}(\boldsymbol{k}\cdot\boldsymbol{v}, k)} - \frac{k^2v^2 - (\boldsymbol{k}\cdot\boldsymbol{v})^2}{k^2c^2 - (\boldsymbol{k}\cdot\boldsymbol{v})^2\,\varepsilon^{\mathrm{tr}}(\boldsymbol{k}\cdot\boldsymbol{v}, k)}\right). \end{aligned} \tag{2.8.31}$$

There are two contributions $W = W^{\mathrm{lo}} + W^{\mathrm{tr}}$, where

$$\begin{aligned} W^{\mathrm{lo}} &= -\frac{2q^2}{\pi v^2}\int_0^\infty d\omega\,\omega\int_0^\infty \frac{\xi d\xi}{k^2}\,\mathrm{Im}\left\{\frac{1}{\varepsilon^{\mathrm{lo}}(\omega, k)}\right\}, \\ W^{\mathrm{tr}} &= +\frac{2q^2}{\pi c^2}\int_0^\infty d\omega\,\omega\int_0^\infty \frac{\xi^3 d\xi}{k^2}\,\mathrm{Im}\left\{\frac{1}{k^2 - \dfrac{\omega^2}{c^2}\,\varepsilon^{\mathrm{tr}}(\omega, k)}\right\}. \end{aligned} \tag{2.8.32}$$

Here the notations $\omega = \boldsymbol{k}\cdot\boldsymbol{v}$, $k^2 = \xi^2 + \omega^2/v^2$ are introduced and (2.2.15) is accounted for. Note that contributions arise not only in the ranges of ω and k, where $\varepsilon^{\mathrm{lo}}(\omega, k)$ and $\varepsilon^{\mathrm{tr}}(\omega, k)$ differ from zero and significant field absorption occurs, but also in ranges where the imaginary parts of $\varepsilon^{\mathrm{lo}}(\omega, k)$ and $\varepsilon^{\mathrm{tr}}(\omega, k)$ as well as the real parts $\mathrm{Re}\{\varepsilon^{\mathrm{lo}}(\omega, k)\}$ and $\mathrm{Re}\{k^2 - \omega^2/c^2\,\varepsilon^{\mathrm{tr}}(\omega, k)\}$ are very small so that the integrands become very peaked. In these ranges the charge excites longitudinal and transverse electromagnetic waves, respectively. Under equilibrium conditions, where $\mathrm{Im}\{\varepsilon^{\mathrm{lo}}(\omega, k)\} > 0$ and

$\mathrm{Im}\{\varepsilon^{\mathrm{tr}}(\omega,k)\}>0$, these parts of the energy loss of a fast particle can be written as

$$\begin{aligned}\Delta W^{\mathrm{lo}} &= \frac{2q}{v^2}\int_0^\infty d\omega\,\omega\int_0^\infty \frac{\xi\,d\xi}{k^2}\,\delta[\varepsilon^{\mathrm{lo}}(\omega,k)]\,,\\ \Delta W^{\mathrm{tr}} &= \frac{2q^2}{c^2}\int_0^\infty d\omega\,\omega\int_0^\infty \frac{\xi^3 d\xi}{k^2}\,\delta\left[k^2-\frac{\omega^2}{c^2}\,\varepsilon^{\mathrm{tr}}(\omega,k)\right].\end{aligned} \tag{2.8.33}$$

Here we have taken the limit

$$\lim_{\delta\to+0}\mathrm{Im}\left\{\frac{1}{x+\mathrm{i}\delta}\right\} = -\pi\delta(x)\,.$$

ΔW^{lo} and ΔW^{tr} represent the energy losses of a fast particle on behalf of the excitation of longitudinal and transverse waves in the medium. These losses are called the *polarization* and *Cherenkov losses*, respectively.

Note that in the case of an isotropic plasma without spatial dispersion $\varepsilon^{\mathrm{lo}}=\varepsilon^{\mathrm{tr}}=1-\omega_{\mathrm{pe}}^2/\omega^2$ holds (valid for $\omega>kv_{\mathrm{te}}$, as shown in Chap. 4). Then

$$\begin{aligned}\Delta W^{\mathrm{tr}} &= 0\,,\\ \Delta W^{\mathrm{lo}} &= \frac{q^2\omega_{\mathrm{pe}}^2}{2v^2}\int_0^\infty \frac{d\xi^2}{\xi^2+(\omega_{\mathrm{pe}}^2/\omega^2)} \simeq \frac{q^2\omega_{\mathrm{pe}}^2}{2v^2}\ln\frac{v^2}{v_{\mathrm{te}}^2}\,,\end{aligned} \tag{2.8.34}$$

where the divergency has been cut off at $\xi_{\max}^2=\omega_{\mathrm{pe}}^2/v_{\mathrm{Te}}^2$.

2.8.5. Obtain the material equations and the corresponding dispersion relations assuming external charges and currents to be an initial cause of the influence on the medium and electric and magnetic fields to be a response to this influence. Study the case of the homogeneous medium.

Solution. In this case the material equation should be written as

$$E_i(\boldsymbol{r},t) = \int_{-\infty}^{t} dt'\int d\boldsymbol{r}'\,\Lambda_{ij}^{-1}(t-t',\boldsymbol{r}-\boldsymbol{r}')\,j_{0j}(t',\boldsymbol{r}')\,. \tag{2.8.35}$$

Hence

$$\begin{aligned}E_i(\omega,\boldsymbol{k}) &= \frac{4\pi\mathrm{i}\omega}{c^2}\Lambda_{ij}^{-1}(\omega,\boldsymbol{k})\,j_{0j}(\omega,\boldsymbol{k})\,,\\ \Lambda_{ij}^{-1}(\omega,\boldsymbol{k}) &= \int_0^\infty dt_1\int d\boldsymbol{r}_1\Lambda_{ij}^{-1}(t_1,\boldsymbol{r}_1)\,.\end{aligned} \tag{2.8.36}$$

$\Lambda_{ij}(\omega,\boldsymbol{k})$ coincides with (2.5.4) and satisfies the dispersion relation

$$\Lambda_{il}^{(0)}\Lambda_{lj}^{-1}(\omega, k) - \delta_{ij} = \frac{1}{\pi i}\int_{-\infty}^{\infty} d\omega' \mathscr{P} \frac{\Lambda_{il}^{(0)}\Lambda_{lj}^{-1}(\omega', \boldsymbol{k}) - \delta_{ij}}{\omega' - \omega},$$

$$\Lambda_{ij}^{(0)} = \left(k^2 - \frac{2}{c^2}\right)\delta_{ij} - k_i k_j \,. \tag{2.8.37}$$

For the isotropic medium for $\varepsilon_{ij}(\omega, \boldsymbol{k})$ determined by (2.2.12) we obtain the dispersion relation for the function $1/\varepsilon^l(\omega, k)$

$$\frac{1}{\varepsilon(\omega, \boldsymbol{k})} - 1 = \frac{1}{\pi i}\int_{-\infty}^{\infty} d\omega' \mathscr{P} \frac{\dfrac{1}{\varepsilon^l(\omega', \boldsymbol{k})} - 1}{\omega' - \omega} \,. \tag{2.8.38}$$

Then for $\omega \to 0$ there follows the inequality $1/\varepsilon^l(0, k) < 1$, which does not contradict (2.3.11) but is more general since it admits the existence of negative $\varepsilon^l(0, k)$.

3. Equations of Plasma Dynamics

The simplest plasma models are considered in this chapter, i.e., an independent particle model, a hydrodynamic model and a general model. The Vlasov kinetic equation for a collisionless plasma is analyzed, as are the Boltzmann-Landau equation, taking into account collisions of charged particles, and equations with the model BGK integral. Proceeding from the general kinetic model, the applicability conditions for the simplest plasma models are presented.

3.1 Simplest Plasma Models

3.1.1 The Model of Independent Particles

The methods of quantitative description of plasmas have developed from very simple models to advanced concepts. The first and simplest models represent the plasma as a system of noninteracting charged particles, moving independently in external fields. If the plasma consists of N_α particles of type α, which are in a chaotic thermal motion, it would be necessary to solve $\Sigma_\alpha N_\alpha$ equations of motion provided the initial conditions, i.e., the same number of particle coordinates and velocities, are known. Naturally this problem is impracticable because of the impossibility to solve such a large number of equations and to have exact assignment of the initial conditions. Significant simplification is achieved when the thermal spread of the particle velocities is ignored. This is equivalent to considering the motion of a so-called mean particle which is at rest in the absence of external forces. Such a simple *model of independent particles* ignores both the interparticle interaction and the thermal spread of the particle velocities.

The equations of motion for the "mean" particle of type α within this model are

$$\begin{aligned} \frac{d\boldsymbol{r}_\alpha}{dt} &= \boldsymbol{V}_\alpha \,, \\ \frac{d\boldsymbol{p}_\alpha}{dt} &= e_\alpha\left(\boldsymbol{E} + \frac{1}{c}\,[\boldsymbol{V}_\alpha, \boldsymbol{B}]\right) + \boldsymbol{g}_\alpha m_\alpha \,, \end{aligned} \tag{3.1.1}$$

where $\boldsymbol{p}_\alpha = m_\alpha \boldsymbol{V}_\alpha / (1 - V_\alpha^2/c^2)^{1/2}$ is the momentum of the particle of type α and $\boldsymbol{g}_\alpha$ is the acceleration due to nonelectromagnetic forces.

This force may be, for example, the gravitational force. In plasma problems $\boldsymbol{g}_\alpha$ often stands for the acceleration due to the curvature of the magnetic field which is analogous to the gravitational force. The free thermal motion of particles along magnetic field lines with nonvanishing curvature results in a centrifugal force perpendicular to the magnetic field with the equivalent acceleration $g_\alpha \approx v_{\mathrm{T}\alpha}^2/R_0$ ($v_{\mathrm{T}\alpha}$ is the average thermal velocity of particles; R_0 the radius of the curvature of the magnetic field lines).

Also the field $\boldsymbol{g}_\alpha$ can be used to incorporate the effect of collisions of charged particles of the given type with other particles into the model. The force produced by collisions can be approximated by the friction term

$$\boldsymbol{g}_\alpha = -\nu_{\alpha\beta}(\boldsymbol{V}_\alpha - \boldsymbol{V}_\beta) ,$$

where $\nu_{\alpha\beta}$ is an *effective collision frequency* for encounters between particles of type α and β.

In (3.1.1) the magnetic induction $\boldsymbol{B}$ and the electric field $\boldsymbol{E}$ include the fields due to the charge and current densities of the charge carriers. The complete problem is a self-consistent one. Equations (3.1.1) must be solved simultaneously with the field equations (2.1.1), in which the densities of the induced charge and current are

$$\varrho = \sum_\alpha N_\alpha e_\alpha , \quad j_\alpha = \sum_\alpha N_\alpha e_\alpha \boldsymbol{V}_\alpha . \tag{3.1.2}$$

$\boldsymbol{V}_\alpha$ satisfies (3.1.1) and the continuity equation

$$\frac{\partial N_\alpha}{\partial t} + \operatorname{div} N_\alpha \boldsymbol{V}_\alpha = 0 . \tag{3.1.3}$$

The validity of the model of independent particles is very limited since it does not account for the thermal motion and for correlations of the particle motion. It can be applied only to strongly rarefied plasmas with practically uncorrelated particle motion.

3.1.2 The Hydrodynamic Model

To the opposite extreme case of dense plasmas the *hydrodynamic model* is applicable. When the motion of the particles is so strongly correlated that there is no difference among electron, ion and neutral components then the plasma behaves like a neutral conducting fluid. The self-consistent system of equations of magnetohydrodynamics is used to describe such a medium

$$\frac{\partial \varrho_{\mathrm{m}}}{\partial t} + \operatorname{div} \varrho_{\mathrm{m}} \boldsymbol{V}_{\mathrm{m}} = 0 \,,$$

$$\begin{aligned} \varrho_{\mathrm{m}} \frac{d\boldsymbol{V}_{\mathrm{m}}}{dt} &= \varrho_{\mathrm{m}} \left[\frac{\partial \boldsymbol{V}_{\mathrm{m}}}{\partial t} + (\boldsymbol{V}_{\mathrm{m}} \cdot \nabla) \boldsymbol{V}_{\mathrm{m}} \right] = -\nabla p \\ &\quad + \frac{1}{4\pi} [\operatorname{curl} \boldsymbol{B}, \boldsymbol{B}] + \eta \Delta \boldsymbol{V}_{\mathrm{m}} + \left(\zeta + \frac{\eta}{3} \right) \operatorname{grad} \operatorname{div} \boldsymbol{V}_{\mathrm{m}} \,, \end{aligned} \tag{3.1.4}$$

$$\frac{\partial \boldsymbol{B}}{\partial t} = \operatorname{curl} [\boldsymbol{V}_{\mathrm{m}}, \boldsymbol{B}] - \frac{c^2}{4\pi} \operatorname{curl} \left(\frac{1}{\sigma} \operatorname{curl} \boldsymbol{B} \right), \quad \operatorname{div} \boldsymbol{B} = 0 \,.$$

Here ϱ_{m} is the mass density of the plasma with the conductivity σ, $\boldsymbol{V}_{\mathrm{m}}$ is the plasma velocity and ζ and η are viscosity coefficients.

This system must be complemented by the equations of state and heat transfer

$$p = p(\varrho_{\mathrm{m}}, T) \,,$$

$$\varrho_{\mathrm{m}} T \left[\frac{\partial S}{\partial t} + (\boldsymbol{V}_{\mathrm{m}} \cdot \nabla) S \right] = \sigma_{ij} \frac{\partial V_{\mathrm{m}i}}{\partial r_j} + \operatorname{div} (\chi \nabla T) + \frac{c^2}{(4\pi)^2 \sigma} (\operatorname{curl} \boldsymbol{B})^2 \,, \tag{3.1.5}$$

where $S(\varrho_{\mathrm{m}}, T)$ is the entropy per unit mass, χ the thermal conductivity and σ_{ij} the viscous stress tensor

$$\sigma_{ij} = \eta \left(\frac{\partial V_{\mathrm{m}i}}{\partial r_j} + \frac{\partial V_{\mathrm{m}j}}{\partial r_i} - \frac{2}{3} \delta_{ij} \operatorname{div} \boldsymbol{V}_{\mathrm{m}} \right) + \zeta \delta_{ij} \operatorname{div} \boldsymbol{V}_{\mathrm{m}} \,. \tag{3.1.6}$$

In magnetohydrodynamics the electric and thermal conductivities σ, χ of the plasma as well as the viscosity coefficients ζ and η are considered to be known (their calculation is not possible within the limits of validity of this model). The equation of state $p(\varrho_{\mathrm{m}}, T)$ and the entropy $S(\varrho_{\mathrm{m}}, T)$ are given in a completely ionized ideal electron-ion plasma by

$$p = \frac{(1+Z) \varrho_{\mathrm{m}} \varkappa T}{M} \,, \qquad S = \frac{3}{2} \frac{1+Z}{M} \ln \left(\frac{p}{\varrho_{\mathrm{m}}^{5/3}} \right) , \tag{3.1.7}$$

where Z is the multiplicity of ion charge and M is the mass of the plasma ions.

Magnetohydrodynamics uses a number of phenomenological values, namely the transfer coefficients σ, χ, ζ, η, which need an independent definition. Its validity for the description of real plasmas (gaseous or solid-state) is limited. Magnetohydrodynamics, as well as the model of independent particles, do not account for the thermal motion of the particles. The former considers the plasma so strongly correlated that it ignores the difference between the separate plasma components. Therefore various modified hydrodynamics also called *quasihydrodynamics* were developed. These can be two-fluid (for electron and ion fluids) or three-fluid (for electron, ion and neutral fluids) models, phenomenologically taking into account interparticle interactions and thermal motion.

Such models necessarily are approximate and to discuss their limitations more general models of the plasma are necessary. The most general model is the kinetic plasma description based on the statistical representation of the plasma as a system of a large number of particles. In the following we shall mainly apply the kinetic plasma model and only in some cases the simplified hydrodynamic model.

3.2 Kinetic Equation with a Self-Consistent Field

In the kinetic theory of gases the system is described by distribution functions, which are defined as the probability density for the event that a particle is in a definite state at a given time t and at a given space point $\boldsymbol{r}$. If the state of the particle of type α is characterized by the momentum $\boldsymbol{p}$ and its energy ε_α is uniquely defined by the momentum, then the distribution function depends on the coordinates $\boldsymbol{p}, \boldsymbol{r}$ and t. The quantity $f_\alpha(\boldsymbol{p}, \boldsymbol{r}, t)\, d\boldsymbol{p}\, d\boldsymbol{r}$ means the number of particles of type α at time t in the phase space interval $d\boldsymbol{p}\, d\boldsymbol{r}$. Consequently the density of particles at the point $\boldsymbol{r}$, t is given by

$$N_\alpha(\boldsymbol{r}, t) = \int d\boldsymbol{p}\, f_\alpha(\boldsymbol{p}, \boldsymbol{r}, t) \ . \tag{3.2.1}$$

This relation is a normalization condition of the distribution function.

If the distribution function is known, we may calculate the mean value of any physical quantity, e.g., the mean velocity or the mean energy of particles of type α:

$$\begin{aligned} \boldsymbol{V}_\alpha(\boldsymbol{r}, t) &= \frac{\int f_\alpha(\boldsymbol{p}, \boldsymbol{r}, t)\, \boldsymbol{v}\, d\boldsymbol{p}}{N_\alpha(\boldsymbol{r}, t)} \ , \\ \mathscr{E}_\alpha(\boldsymbol{r}, t) &= \frac{\int f_\alpha(\boldsymbol{p}, \boldsymbol{r}, t)\, \mathscr{E}_\alpha(\boldsymbol{p})\, d\boldsymbol{p}}{N_\alpha(\boldsymbol{r}, t)} \ . \end{aligned} \tag{3.2.2}$$

To obtain the distribution function, a kinetic equation has to be solved. For a highly rarefied gas we can neglect the interaction of particles in first-order approximation and consider them as completely independent. In this approximation the distribution of the particles in a phase space volume near the point $\boldsymbol{p}, \boldsymbol{r}$ can vary only due to particle inflow and outflow through the surface enclosing this volume. If particles are neither created nor lost the total time derivative of the distribution function vanishes

$$\frac{df_\alpha(\boldsymbol{p}, \boldsymbol{r}, t)}{dt} = \frac{\partial f_\alpha}{\partial t} + \frac{\partial f_\alpha}{\partial \boldsymbol{p}} \frac{d\boldsymbol{p}}{dt} + \frac{\partial f_\alpha}{\partial \boldsymbol{r}} \frac{d\boldsymbol{r}}{dt} = 0 \ . \tag{3.2.3}$$

This is the continuity equation in the phase space of the particles of type α analogous to the Liouville equation.

According to the equations of motion

$$\frac{d\boldsymbol{p}}{dt} = \boldsymbol{F}_\alpha , \qquad \frac{d\boldsymbol{r}}{dt} = \boldsymbol{v} \tag{3.2.4}$$

the kinetic equation (3.2.3) can be written as

$$\frac{\partial f_\alpha}{\partial t} + \boldsymbol{v}\frac{\partial f_\alpha}{\partial \boldsymbol{r}} + \boldsymbol{F}_\alpha \frac{\partial f_\alpha}{\partial \boldsymbol{p}} = 0 , \tag{3.2.5}$$

where $\boldsymbol{F}_\alpha$ is the force acting on the particles of type α. For charged particles this force is the Lorentz force

$$\boldsymbol{F}_\alpha = e_\alpha \left(\boldsymbol{E} + \frac{1}{c}[\boldsymbol{v}, \boldsymbol{B}]\right) . \tag{3.2.6}$$

Equation (3.2.5) for charged particles of type α then reads

$$\frac{\partial f_\alpha}{\partial t} + \boldsymbol{v}\frac{\partial f_\alpha}{\partial \boldsymbol{r}} + e_\alpha \left(\boldsymbol{E} + \frac{1}{c}[\boldsymbol{v}, \boldsymbol{B}]\right) \frac{\partial f_\alpha}{\partial \boldsymbol{p}} = 0 . \tag{3.2.7}$$

$\boldsymbol{E}(\boldsymbol{r}, t)$ and $\boldsymbol{B}(\boldsymbol{r}, t)$ denote the electric and magnetic field at the position of the particle. When deriving (3.2.6), we assumed these fields to be given. They are determined by (2.1.1), however, where the charge and current densities ϱ and $\boldsymbol{j}$ depend on the distribution functions

$$\varrho = \sum_\alpha e_\alpha \int f_\alpha(\boldsymbol{p}, \boldsymbol{r}, t)\, d\boldsymbol{p} , \qquad \boldsymbol{j} = \sum_\alpha e_\alpha \int \boldsymbol{v} f_\alpha(\boldsymbol{p}, \boldsymbol{r}, t)\, d\boldsymbol{p} . \tag{3.2.8}$$

The summation extends over all types of charged particles of the gas.

Obviously the solution of (3.2.7) is a self-consistent problem, since the particle distributions f_α determine the sources of the electromagnetic field supporting the phase space flow. Therefore, (3.2.7) is called the *kinetic equation with a self-consistent field* or *Vlasov's equation.*

3.3 Boltzmann Kinetic Equation

The Vlasov equation derived in the previous section is valid only if the interaction of particles can be ignored. In other words, it is the zero-order approximation in the interaction parameter η representing the ratio of the average potential energy of particle interaction to the kinetic energy of thermal motion. Taking account of correlations in particle motion gives rise to a nonzero right-hand side of (3.2.5)

$$\frac{\partial f_\alpha}{\partial t} + \boldsymbol{v}\frac{\partial f_\alpha}{\partial \boldsymbol{r}} + F_\alpha \frac{\partial f_\alpha}{\partial \boldsymbol{p}} = \left(\frac{\partial f_\alpha}{\partial t}\right)_{\text{col}}, \tag{3.3.1}$$

the *collision integral*. It describes the variation of the distribution function due to particle collisions. To first order in the interaction parameter η only binary collisions contribute. Interactions of three or more particles are of higher order in η. In neutral gases the parameter η is so small that it is sufficient to consider only binary collisions. Then the collision integral has the form

$$\left(\frac{\partial f_\alpha}{\partial t}\right)_{\text{col}} = \sum_\beta \left(\frac{\partial f_\alpha}{\partial t}\right)_{\text{col}}^{\alpha\beta}, \tag{3.3.2}$$

where the index α denotes the particle species considered and the summation index β runs over all collision partners. We shall consider only elastic collisions, which conserve energy and do not change the particle structure. Inelastic collisions, i.e., excitation and ionization processes, are neglected which is allowable in the energy range far from excitation or ionization energies.

Let $\boldsymbol{p}_\alpha$, $\boldsymbol{p}_\beta$ and $\boldsymbol{p}'_\alpha$, $\boldsymbol{p}'_\beta$ be the momenta of the colliding particles α and β before and after their interaction (Fig. 3.1). Using $W(\boldsymbol{p}_\alpha, \boldsymbol{p}_\beta, \boldsymbol{p}'_\alpha, \boldsymbol{p}'_\beta)$, the transition probability per unit time of the elastic scattering process $\boldsymbol{p}_\alpha + \boldsymbol{p}_\beta \rightleftarrows \boldsymbol{p}'_\alpha + \boldsymbol{p}'_\beta$, the variation of the distribution function of particles of type α due to collisions with particles of type β can be formulated as

$$\begin{aligned}\left(\frac{\partial f_\alpha}{\partial t}\right)_{\text{col}}^{\alpha\beta} &= -\int d\boldsymbol{p}_\beta d\boldsymbol{p}'_\alpha d\boldsymbol{p}'_\beta\, W(\boldsymbol{p}_\alpha, \boldsymbol{p}_\beta, \boldsymbol{p}'_\alpha, \boldsymbol{p}'_\beta) \\ &\quad \times [f_\alpha(\boldsymbol{p}_\alpha) f_\beta(\boldsymbol{p}_\beta) - f_\alpha(\boldsymbol{p}'_\alpha) f_\beta(\boldsymbol{p}'_\beta)] \,. \end{aligned} \tag{3.3.3}$$

With the account of the momentum and the energy conservation law this expression is known to be the *Boltzmann elastic collision integral*. The problem of formulating collision integrals consists in the calculation of the scattering probability $W(\boldsymbol{p}_\alpha, \boldsymbol{p}_\beta; \boldsymbol{p}'_\alpha, \boldsymbol{p}'_\beta)$. We shall do this quantum mechanically as it is more illustrative than the classical derivation. Suppose that the potential energy of interaction of two particles is a function of their distance only

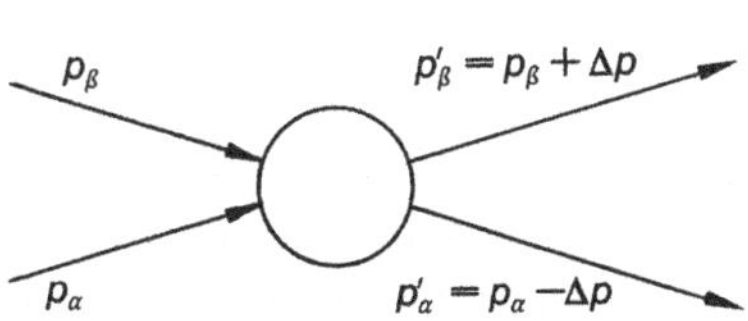

Fig. 3.1.

$$
\begin{aligned}
U(|\boldsymbol{r}_\alpha - \boldsymbol{r}_\beta|) &= \int d\boldsymbol{k}\, U(\boldsymbol{k}) \exp[\mathrm{i}\boldsymbol{k}\cdot(\boldsymbol{r}_\alpha - \boldsymbol{r}_\beta)] \,, \\
U(\boldsymbol{k}) &= \frac{1}{(2\pi)^3} \int d\boldsymbol{r}\, U(|\boldsymbol{r}|)\, \mathrm{e}^{-\mathrm{i}\boldsymbol{k}\cdot\boldsymbol{r}} \,.
\end{aligned}
\tag{3.3.4}
$$

The Fourier transform $U(\boldsymbol{k})$ of the interaction potential $U(\boldsymbol{r})$ then depends only on the modulus of the vector $\boldsymbol{k}$. According to quantum mechanics the probability of scattering of two particles can be calculated by evaluation of the elements of the interaction matrix using unperturbed initial and final states (when the interaction energy is small compared to the kinetic energy). This approximation, known as Born's approximation in the collision theory, is valid if $|U| r_0 \ll \hbar v$ holds, (r_0 is the interaction radius of the potential $U(\boldsymbol{r})$ and v the velocity of the scattered particle). In Born's approximation we have

$$
W = \frac{2\pi}{\hbar} |U_{\boldsymbol{p}_\alpha, \boldsymbol{p}_\beta, \boldsymbol{p}'_\alpha, \boldsymbol{p}'_\beta}|^2 \,\delta[(\mathscr{E}'_\alpha + \mathscr{E}'_\beta) - (\mathscr{E}_\alpha + \mathscr{E}_\beta)] \,, \tag{3.3.5}
$$

where $U_{\boldsymbol{p}_\alpha, \boldsymbol{p}_\beta, \boldsymbol{p}'_\alpha, \boldsymbol{p}'_\beta}$ is the matrix element of the interaction potential evaluated with the wave functions of free particles. The initial state (before scattering) is characterized by the momenta $\boldsymbol{p}_\alpha$ and $\boldsymbol{p}_\beta$, and the final state (after scattering) by $\boldsymbol{p}'_\alpha$ and $\boldsymbol{p}'_\beta$.

In a homogeneous medium the wave function of a free nonrelativistic particle with momentum $\boldsymbol{p}$ and energy $\mathscr{E} = p^2/2m$ is the plane wave

$$
\Psi_{\mathrm{p}} = A \exp\left(-\mathrm{i}\frac{\mathscr{E}}{\hbar} t + \mathrm{i}\frac{\boldsymbol{p}\cdot\boldsymbol{r}}{\hbar}\right), \tag{3.3.6}
$$

where $A = (2\pi\hbar)^{-2}$ is a normalization constant (note that the wave function of a free particle is normalized by the Dirac δ-function). Taking account of (3.3.4), we obtain the interaction matrix

$$
U_{\boldsymbol{p}_\alpha, \boldsymbol{p}_\beta, \boldsymbol{p}'_\alpha, \boldsymbol{p}'_\beta} = \int d\boldsymbol{k}\, U(\boldsymbol{k}) \langle \boldsymbol{p}'_\alpha | \mathrm{e}^{\mathrm{i}\boldsymbol{k}\cdot\boldsymbol{r}_\alpha} | \boldsymbol{p}_\alpha \rangle \langle \boldsymbol{p}'_\beta | \mathrm{e}^{-\mathrm{i}\boldsymbol{k}\cdot\boldsymbol{r}_\beta} | \boldsymbol{p}_\beta \rangle \,, \tag{3.3.7}
$$

where $\langle \boldsymbol{p}'_\alpha | \exp(\mathrm{i}\boldsymbol{k}\cdot\boldsymbol{r}_\alpha) | \boldsymbol{p}_\alpha \rangle$ is the matrix element of the operator $\exp(\mathrm{i}\boldsymbol{k}\cdot\boldsymbol{r}_\alpha)$ using the wave functions (3.3.6) of the initial and final states of the particle of type α. Evaluating this matrix element the free motion of particles must be accounted for. At the time t, for which we make estimates, the position of the particle is given by $\boldsymbol{r}_\alpha = \boldsymbol{r}_{\alpha 0} + \boldsymbol{v}_\alpha t$. One obtains

$$
\langle \boldsymbol{p}'_\alpha | \mathrm{e}^{\mathrm{i}\boldsymbol{k}\cdot\boldsymbol{r}_\alpha} | \boldsymbol{p}_\alpha \rangle = \delta(\boldsymbol{p}'_\alpha - \boldsymbol{p}_\alpha + \hbar\boldsymbol{k})\, \delta\left(\frac{p'^2_\alpha}{2m_\alpha} - \frac{p^2_\alpha}{2m_\alpha} + \hbar\boldsymbol{k}\cdot\boldsymbol{v}_\alpha\right). \tag{3.3.8}
$$

The matrix element $\langle \boldsymbol{p}'_\beta | \exp(-\mathrm{i}\boldsymbol{k}\cdot\boldsymbol{r}_\beta) | \boldsymbol{p}_\beta \rangle$ can be calculated analogously. Equation (3.3.8) accounts for the conservation of momentum and energy in the scattering process: the particle of type α with momentum $\boldsymbol{p}_\alpha$ emits an "interaction quantum" and passes over to the state with momentum $p'_\alpha = p_\alpha - \hbar k$ (Fig. 3.1). This quantum is absorbed by the particle of type β with momentum p_β which passes over to the state with momentum $p'_\beta = p_\beta + \hbar k$. The momentum of the "interaction quantum" equals $\Delta p = hk$ and its energy is $\hbar k \cdot v_\alpha = \hbar k \cdot v_\beta$.

Substituting (3.3.7) and (3.3.5) into (3.3.3) after simple calculations leads to

$$\left(\frac{\partial f_\alpha}{\partial t}\right)_{\text{col}}^{\alpha\beta} = \frac{1}{(2\pi)^2 \hbar} \int dp_\beta \, dk \, |U(k)|^2$$
$$\times \delta\left[\frac{(p_\alpha - \hbar k)^2}{2m} + \frac{(p_\beta + \hbar k)^2}{2m_\beta} - \frac{p_\alpha^2}{2m_\alpha} - \frac{p_\beta^2}{2m_\beta}\right]$$
$$\times \left[f_\alpha(p_\alpha) f_\beta(p_\beta) - f_\alpha(p_\alpha - \hbar k) f_\beta(p_\beta + \hbar k)\right] , \tag{3.3.9}$$

which obviously satisfies the conservation laws of momentum and energy.

Within the applicability of Born's approximation, i.e., for small values of the interaction energy in comparison with the kinetic energy of particles, (3.3.9) has a general character due to the absence of any restrictions concerning the law of particle interaction. The transition to the classical limit $\hbar k \ll p_\alpha, p_\beta$ simplifies (3.3.9). Supposing that the interaction quantum is small corresponds to the assumption of small values of the interaction energy of particles. In fact, such an interaction must result in insignificant momentum exchange. Expanding the integrand of (3.3.9) in powers of $\hbar k$ and keeping in mind that $U(k)$ depends only on the absolute value of the vector k, we obtain

$$\left(\frac{\partial f_\alpha}{\partial t}\right)_{\text{col}}^{\alpha\beta} = \frac{\partial}{\partial p_{\alpha i}} \int dp_\beta \, I_{ij}^{\alpha\beta}(p_\alpha, p_\beta) \left[\frac{\partial f_\alpha(p_\alpha)}{\partial p_{\alpha j}} f_\beta(p_\beta) - f_\alpha(p_\alpha) \frac{\partial f_\beta(p_\beta)}{\partial p_{\beta j}}\right] , \tag{3.3.10}$$

where

$$I_{ij}^{\alpha\beta} = (2\pi)^3 \pi \int dk \, |U(k)|^2 \, \delta(k \cdot u) \, k_i k_j$$
$$= (2\pi)^3 \frac{\pi}{2} \frac{u^2 \delta_{ij} - u_i u_j}{u^2} \int k^2 dk \, |U(k)|^2 \delta(k \cdot u) \tag{3.3.11}$$

and $u = v_\alpha - v_\beta$ is the relative velocity of the colliding particles.

3.3.1 The Fokker–Planck Equation

Using (3.3.10, 11) the kinetic equation (3.3.1) for the distribution function of the particles of type α taking account of close collisions with all the particles of type β (where β, in particular, can coincide with α) finally reads

$$\frac{\partial f_\alpha}{\partial t} + \boldsymbol{v}\frac{\partial f_\alpha}{\partial \boldsymbol{r}} + \boldsymbol{F}_\alpha \frac{\partial f_\alpha}{\partial \boldsymbol{p}_\alpha} = \sum_\beta \left(\frac{\partial f_\alpha}{\partial t}\right)^{\alpha\beta}_{\mathrm{col}}$$
$$= \frac{\partial}{\partial p_{\alpha i}}\left(D_{ij}\frac{\partial f_\alpha}{\partial p_{\alpha j}} - A_i f_\alpha\right), \quad \text{where} \tag{3.3.12}$$

$$D_{ij} = \sum_\beta \int d\boldsymbol{p}_\beta\, I_{ij}^{\alpha\beta}(\boldsymbol{p}_\alpha, \boldsymbol{p}_\beta) f_\beta(\boldsymbol{p}_\beta) \quad \text{and}$$
$$A_i = \sum_\beta \int d\boldsymbol{p}_\beta\, I_{ij}^{\alpha\beta}(\boldsymbol{p}_\alpha, \boldsymbol{p}_\beta)\frac{\partial f_\beta(\boldsymbol{p}_\beta)}{\partial p_{\beta j}} \tag{3.3.13}$$

are the diffusion and friction coefficients in the momentum space, respectively.

This kinetic equation is frequently called the *Fokker–Planck equation.*

The collision integral (3.3.10) and the kinetic equation (3.3.12) are valid quite generally since they do not specify the interaction law. In particular, they can be applied to plasmas with any degree of ionization if the law of binary interaction is known.

3.4 Collision Integral of Charged Particles

Now we want to exploit the derived general relations for a completely ionized plasma, where only collisions of charged particles are essential. To determine the energy of interaction of two charged particles in the plasma, one has to calculate the potential caused by a charged particle of type α moving with the velocity $\boldsymbol{v}_\alpha$, i.e., $\boldsymbol{r}_\alpha = \boldsymbol{r}_{0\alpha} + \boldsymbol{v}_\alpha t$. To obtain the potential induced by the charge, Poisson's equation is applied

$$\operatorname{div}\boldsymbol{D} = 4\pi e_\alpha \delta(\boldsymbol{r} - \boldsymbol{r}_{0\alpha} - \boldsymbol{v}_\alpha t) . \tag{3.4.1}$$

Using the Fourier representation of the space variables and the material equation (2.2.8), we obtain the potential of the moving particle as

$$\Phi(|\boldsymbol{r}_\alpha - \boldsymbol{r}|) = \frac{4\pi e_\alpha}{(2\pi)^3}\int d\boldsymbol{k}\frac{\exp[\mathrm{i}\boldsymbol{k}\cdot(\boldsymbol{r}_\alpha - \boldsymbol{r})]}{k_i k_j \varepsilon_{ij}(\boldsymbol{k}\cdot\boldsymbol{v}_\alpha, \boldsymbol{k})} , \tag{3.4.2}$$

where $\varepsilon_{ij}(\boldsymbol{k}\cdot\boldsymbol{v}_\alpha, \boldsymbol{k})$ is the dielectric tensor of the plasma at the frequency $\omega = \boldsymbol{k}\cdot\boldsymbol{v}_\alpha$.

Consequently the interaction energy of two particles of type α and β is

$$U(\boldsymbol{r}) = e_\beta \Phi_\alpha = \frac{4\pi e_\alpha e_\beta}{(2\pi)^3}\int d\boldsymbol{k}\frac{\mathrm{e}^{\mathrm{i}\boldsymbol{k}\cdot\boldsymbol{r}}}{k_i k_j \varepsilon_{ij}(\boldsymbol{k}\cdot\boldsymbol{v}_\alpha, \boldsymbol{k})} . \tag{3.4.3}$$

For the Fourier transform of the interaction energy we get

$$U(\boldsymbol{k}) = \frac{e_\alpha e_\beta}{2\pi^2} \frac{1}{k_i k_j \varepsilon_{ij}(\boldsymbol{k}\cdot\boldsymbol{v}_\alpha, \boldsymbol{k})} \,. \tag{3.4.4}$$

Substituting (3.4.4) into (3.3.11) results in

$$I_{ij}^{\alpha\beta} = 2\pi e_\alpha e_\beta \int d\boldsymbol{k} \frac{k_i k_j \,\delta(\boldsymbol{k}\cdot\boldsymbol{v}_\alpha - \boldsymbol{k}\cdot\boldsymbol{v}_\beta)}{|k_i k_j \varepsilon_{ij}(\boldsymbol{k}\cdot\boldsymbol{v}_\alpha, \boldsymbol{k})|^2} \,. \tag{3.4.5}$$

The collision term (3.3.10) with the dynamically screened kernel (3.4.5) is called the *Lenard–Balescu collision integral.*

Essentially identical results are obtained, if we assume that the particle interaction within the plasma is not different from the vacuum case, i.e., $\varepsilon_{ij} = \delta_{ij}$, and introduce appropriate cut-offs of the integration divergence at a minimum and maximum wave vector. This approach leads to the well-known *Landau formula*

$$I_{ij}^{\alpha\beta} = 2\pi e_\alpha^2 e_\beta^2 L \frac{u^2 \delta_{ij} - u_i u_j}{u^3} \,, \tag{3.4.6}$$

where $L = \ln|k_{\max}/k_{\min}|$ is the so-called *Coulomb logarithm*, introduced by Landau.

The cut-off at the lower limit of k is necessary since the particle interaction in the plasma is of Coulomb character only at small distances $r \leq r_{\max} = 1/k_{\min}$. At distances $r > r_{\max}$ the interaction is screened due to the plasma polarization. In thermodynamic equilibrium the screening radius coincides with the Debye length $r_{\max} = r_D \approx (T/4\pi e^2 N)^{1/2}$. The Lenard-Balescu collision term automatically takes account of this screening effect.

As the cut-off at the upper limit of k, we recall that at very small distances $r < r_{\min} = 1/k_{\max}$ the energy of the Coulomb interaction of particles exceeds their kinetic energy of free motion which violates the applicability condition of the perturbation expansion. Hence, $r_{\min}$ should be determined from $e^2/r_{\min} \approx T$. The resulting Coulomb logarithm $L \approx \ln(200\, T/N^{1/3})$ does not vary much and in wide ranges of temperature and density $L \approx 10$ to 20 holds. In the following we shall consider this L constant.

At large distances (small k) Eq. (3.4.5) which contains the plasma polarization does not diverge which justifies the Debye cut-off used in the Landau formula (3.4.6). A classical justification of the cut-off at small distances $r < r_{\min} \approx e^2/T$ can be given by a consideration of many particle collisions. In quantum mechanics, however, there appears another minimum interaction radius $r_{\min}$ through Heisenberg's uncertainty principle $rp \sim r v_T m > \hbar$, namely the de Broglie wavelength of a free particle $r_{\min} \sim \lambda_p = \hbar/m v_T$. In the general case one has to cut off at

$$r_{\min} = \max\left(\frac{e^2}{T}, \frac{\hbar}{m v_T}\right) .$$

The minimum interaction radius is determined by the de Broglie wavelength only at very high temperatures when $e^2 \ll \hbar v_T$.

Summarizing, the kinetic equation for a completely ionized plasma taking account of two-particle Coulomb collisions can be written as

$$\frac{\partial f_\alpha}{\partial t} + \boldsymbol{v}\frac{\partial f_\alpha}{\partial \boldsymbol{r}} + e_\alpha\left(\boldsymbol{E} + \frac{1}{c}[\boldsymbol{v},\boldsymbol{B}]\right)\frac{\partial f_\alpha}{\partial \boldsymbol{p}_\alpha} = 2\pi L e_\alpha^2 \sum_\beta e_\beta^2 \frac{\partial}{\partial p_{\alpha i}}$$

$$\times \int d\boldsymbol{p}_\beta \frac{u^2\delta_{ij} - u_i u_j}{u^3}\left[f_\beta(\boldsymbol{p}_\beta)\frac{\partial f_\alpha(\boldsymbol{p}_\alpha)}{\partial p_{\alpha j}} - f_\alpha(\boldsymbol{p}_\alpha)\frac{\partial f_\beta(\boldsymbol{p}_\beta)}{\partial p_{\beta j}}\right], \quad (3.4.7)$$

which is called the *Landau equation*.

3.4.1 The Case of the Degenerate Plasma

Terminating this section we may give the collision integral of charged particles in the degenerate plasma. Due to Pauli's exclusion principle, which states that at most two Fermions (electrons or holes) can occupy the same state (characterized by the momentum $\boldsymbol{p}$, here), the collision integral must be corrected. Since each state in phase space has the volume $(2\pi\hbar)^3$ the collision integral (3.3.9) of the degenerate plasma is

$$\left(\frac{\partial f_\alpha}{\partial t}\right)_{\text{col}}^{\alpha\beta} = (2\pi)^3 \frac{2\pi}{\hbar}\int d\boldsymbol{p}_\beta\, d\boldsymbol{k}\,|U(\boldsymbol{k})|^2$$

$$\times\,\delta\left(\frac{(\boldsymbol{p}_\alpha - h\boldsymbol{k})^2}{2m_\alpha} + \frac{(\boldsymbol{p}_\beta + \hbar\boldsymbol{k})^2}{2m_\beta} - \frac{p_\alpha^2}{2m_\alpha} - \frac{p_\beta^2}{2m_\beta}\right)$$

$$\times\left\{f_\alpha(\boldsymbol{p}_\alpha)f_\beta(\boldsymbol{p}_\beta)\left[1 - \frac{(2\pi\hbar)^3}{2}f_\alpha(\boldsymbol{p}_\alpha - \hbar\boldsymbol{k})\right]\right.$$

$$\times\left[1 - \frac{(2\pi\hbar)^3}{2}f_\beta(\boldsymbol{p}_\beta + \hbar\boldsymbol{k})\right] - f_\alpha(\boldsymbol{p}_\alpha - \hbar\boldsymbol{k})f_\beta(\boldsymbol{p}_\beta + \hbar\boldsymbol{k})$$

$$\left.\times\left[1 - \frac{(2\pi\hbar)^3}{2}f_\alpha(\boldsymbol{p}_\alpha)\right]\left[1 - \frac{(2\pi\hbar)^3}{2}f_\beta(\boldsymbol{p}_\beta)\right]\right\}. \quad (3.4.8)$$

Expanding this expression in powers of the momentum transfer due to collisions, i.e., for $p_\alpha, p_\beta \gg hk$, we finally obtain

$$\left(\frac{\partial f_\alpha}{\partial t}\right)_{\text{col}}^{\alpha\beta} = \int d\boldsymbol{p}_\beta I_{ij}^{\alpha\beta}(\boldsymbol{p}_\alpha, \boldsymbol{p}_\beta)\left\{2\frac{\partial f_\alpha}{\partial p_{\alpha i}}\frac{\partial f_\beta}{\partial p_{\beta j}}\left[1 - \frac{(2\pi\hbar)^3}{2}(f_\alpha + f_\beta)\right]\right.$$

$$\left. - \frac{\partial^2 f_\alpha}{\partial p_{\alpha i}\partial p_{\alpha j}} f_\beta\left[1 - \frac{(2\pi\hbar)^3}{2}f_\beta\right] - \frac{\partial^2 f_\beta}{\partial p_{\beta i}\partial p_{\beta j}} f_\alpha\left[1 - \frac{(2\pi\hbar)^3}{2}f_\alpha\right]\right\}, \quad (3.4.9)$$

where $I_{ij}^{\alpha\beta}(\boldsymbol{p}_\alpha, \boldsymbol{p}_\beta)$ is determined by (3.3.11) and for charged particles by (3.4.5, 6) with the obvious substitution $T \to \mathscr{E}_F$. In (3.4.8, 9) Pauli's principle is accounted for by the factors

$$1 - \frac{(2\pi\hbar)^3}{2} f_\alpha(\boldsymbol{p}_\alpha)$$

describing the probability of vacancies for a particle of type α in the state with momentum $\boldsymbol{p}$. The process $\boldsymbol{p}_\alpha, \boldsymbol{p}_\beta \rightleftarrows \boldsymbol{p}'_\alpha, \boldsymbol{p}'_\beta$ is possible only if the states with momenta $\boldsymbol{p}'_\alpha, \boldsymbol{p}'_\beta$ are vacant. In the nondegenerate plasma these factors are close to one, and (3.4.8, 9) reduce to (3.3.9, 10).

3.5 Model Integral for Elastic Particle Collisions

To simplify the Boltzmann collision integral we apply the Born approximation to calculate the scattering probability $W(\boldsymbol{p}_\alpha, \boldsymbol{p}_\beta; \boldsymbol{p}'_\alpha, \boldsymbol{p}'_\beta)$. The use of this approximation presumes that the Fourier transform of the interaction potential $U(\boldsymbol{r})$, i.e., the integral (3.3.4), exists. This integral diverges if $U(\boldsymbol{r})$ increases for small r faster than r^{-2} which is the case for the interaction of charged particles with neutral plasma atoms and molecules or with a crystal lattice (for a solid-state plasma). Therefore, to describe collisions of charged particles with neutrals we must apply the general Boltzmann collision integral and calculate the scattering probability without the use of Born's approximation. Such calculations are very complex mathematically. Furthermore, there exists no universal potential of the charged particle-neutral interaction. Moreover, the collision integral (3.3.10) (valid only in Born's approximation) does not describe the interaction of charged particles with neutrals correctly. This is the reason why there is no simple unified theory of the collision integral for charged particles in weakly ionized gaseous plasmas and in solid-state plasmas as well. In practical calculations we nevertheless apply the model collisional term of charged particles.

Now we briefly consider a collision model which has acquired great importance recently, the so-called *Bhatnagar–Gross–Krook model* (*BGK model*). It cannot be derived from the general Boltzmann integral by means of any approximation; therefore it is called a *model*. The BGK model can be constructed by general physical reasoning, however.

Any collision model of elastic scattering should fulfil the general conservation laws of particle number, momentum and energy. For two-body interactions between particles of the types α and β these laws are written as

$$\int dp \left(\frac{\partial f_\alpha}{\partial t}\right)_{\text{col}}^{\alpha\beta} = 0 \,,$$

$$m_\alpha \int dp\, \boldsymbol{v} \left(\frac{\partial f_\alpha}{\partial t}\right)_{\text{col}}^{\alpha\beta} + m_\beta \int dp\, \boldsymbol{v} \left(\frac{\partial f_\beta}{\partial t}\right)_{\text{col}}^{\beta\alpha} = 0 \,, \tag{3.5.1}$$

$$m_\alpha \int dp\, v^2 \left(\frac{\partial f_\alpha}{\partial t}\right)_{\text{col}}^{\alpha\beta} + m_\beta \int dp\, v^2 \left(\frac{\partial f_\beta}{\partial t}\right)_{\text{col}}^{\beta\alpha} = 0 \,.$$

These expressions follow from the kinetic equations (3.3.1) by multiplying them by 1, $m_\alpha \boldsymbol{v}$ and $m_\alpha v^2$, integrating over the momentum and summing up over the species index. Note that the collision integral must vanish when the equilibrium distribution (Fermi or Maxwellian distribution) is substituted. This results from the well-known Boltzmann H-theorem according to which collisions destroy any perturbation of the equilibrium distribution (the Maxwellian distribution for an nondegenerate gas or the Fermi distribution for a degenerate gas). The process of approaching the equilibrium state is called *relaxation*. One has to distinguish the relaxation times related to different processes. Thus, if in the perturbed state the particles of type α have a momentum different from zero, the latter must relax due to collisions. The characteristic time of this process is called the *momentum relaxation time* of particles of type α. Usually it coincides with the mean time of free flight since for the isotropization of momenta only one collision event is needed. The inverse momentum relaxation time is called the collision frequency.

In a nondegenerate weakly ionized plasma the conservation laws (3.5.1) are satisfied for the BGK integral

$$\left(\frac{\partial f_\alpha}{\partial t}\right)_{\text{col}}^{\alpha\beta} = -\nu_{\alpha\beta}(f_\alpha - N_\alpha \Phi_{\alpha\beta}) \,, \tag{3.5.2}$$

with a velocity-independent collision frequency $\nu_{\alpha\beta}$ describing the momentum relaxation of the particles of type α due to collisions with the particles of type β.

The function $\Phi_{\alpha\beta}$, called the local Maxwellian, is given by

$$\Phi_{\alpha\beta} = \frac{1}{(2\pi m_\alpha T_{\alpha\beta})^{3/2}} \exp\left(-\frac{m_\alpha(\boldsymbol{v} - V_\alpha)^2}{2T_{\alpha\beta}}\right) \quad \text{with} \tag{3.5.3}$$

$$N_\alpha = \int dp\, f_\alpha \,, \qquad V_\alpha = \frac{1}{N_\alpha} \int dp\, \boldsymbol{v} f_\alpha \,,$$

$$T_\alpha = \frac{m_\alpha}{3N_\alpha} \int dp\, (\boldsymbol{v} - V_\alpha)^2 f_\alpha \,, \quad T_{\alpha\beta} = \frac{m_\alpha T_\beta + m_\beta T_\alpha}{m_\alpha + m_\beta} \,. \tag{3.5.4}$$

Substituting (3.5.2) into (3.5.1) one can show that the conservation of momentum and energy requires $m_\alpha \nu_{\alpha\beta} N_\alpha = m_\beta \nu_{\beta\alpha} N_\beta$. Note that the frequencies $\nu_{\alpha\beta}$ are introduced externally into the BGK model. They must be determined from the exact Boltzmann integral including the interaction potential specific for each type of particle encounters. We can estimate $\nu_{\alpha\beta}$ using simple molecular-kinetic considerations for the scattering of charged particles by neutrals and write $\nu_{\alpha n} = v_{T\alpha}\sigma_0 N_n$, where σ_0, the effective scattering cross section, is by the order of magnitude $\sigma_0 = \pi a^2$. Here a denotes the radius of the neutral atom, $a \approx 10^{-8}$ cm, and N_n the density of neutrals in the plasma. The BGK description is satisfactory only for interactions between particles of different kind. Therefore, it can be applied to weakly ionized plasmas when the scattering of charged particles by neutrals is predominant. In spite of its simplicity it is impractical to apply the BGK integral to completely ionized plasmas.

Substituting (3.5.2) into (3.3.1), we obtain the BGK kinetic equation for the nondegenerate plasma

$$\frac{\partial f_\alpha}{\partial t} + \boldsymbol{v}\frac{\partial f_\alpha}{\partial \boldsymbol{r}} + e_\alpha\left(\boldsymbol{E} + \frac{1}{c}\,[\boldsymbol{v},\boldsymbol{B}]\right)\frac{\partial f_\alpha}{\partial \boldsymbol{p}} = -\sum_\beta \nu_{\alpha\beta}\,(f_\alpha - N_\alpha \Phi_{\alpha\beta})\,, \tag{3.5.5}$$

where the summation extends over the neutral species index.

3.5.1 The Case of the Degenerate Plasma

It is easy to extend the BGK collision integral to degenerate solid-state plasmas. One only has to take into account that the equilibrium distribution of the degenerate plasma is the Fermi distribution. Conveniently the description of collisions between the light charge carriers (electrons and holes) with the heavy particles of the lattice by the BGK model is quite accurate. Moreover, the solid-state plasma can be considered weakly ionized. In analogy to (3.5.2), the BGK integral for the degenerate plasma can be written as

$$\left(\frac{\partial f_\alpha}{\partial t}\right)^{\alpha\beta}_{\text{col}} = -\,\nu_{\alpha\beta}\,(f_\alpha - f_{0\alpha})\,, \tag{3.5.6}$$

where $f_{0\alpha}$ is the Fermi distribution function

$$f_{0\alpha} = \frac{1}{(2\pi\hbar)^3}\left\{\exp\left[\left(\frac{p_\alpha^2}{2m_\alpha} - \mathscr{E}_{F\alpha}\right)T_\alpha - 1\right] + 1\right\}^{-1}. \tag{3.5.7}$$

For $\mathscr{E}_{F\alpha} = p_{F\alpha}^2/2m_\alpha = (3\pi^2)^{2/3}\hbar^2 N_\alpha^{2/3}/2m_\alpha \ll T_\alpha$, the Fermi distribution function approaches the Maxwellian

$$f_{0\alpha} \to \frac{N_\alpha}{(2\pi m_\alpha T_\alpha)^{3/2}}\exp\left(-\frac{p_\alpha^2}{2m_\alpha T_\alpha}\right). \tag{3.5.8}$$

The function $N_\alpha \Phi_{\alpha\beta}$ takes the same form for $m_\alpha \ll m_\beta$ and $V_\beta = 0$ when $T_{\alpha\beta} \to T_\alpha$. Thus, in the nondegenerate limit the integrals (3.5.6) and (3.5.2) coincide.

3.6 Discussion of the Simplest Plasma Models

Knowing the general kinetic equation with a self-consistent field, which also accounts for close binary particle collisions, we can discuss the applicability of the simple plasma models, studied in Sect. 3.1. We shall confine our interest to the rarefied collisionless plasma described by (3.2.7):

$$\frac{\partial f_\alpha}{\partial t} + \boldsymbol{v}\frac{\partial f_\alpha}{\partial \boldsymbol{r}} + e_\alpha\left(\boldsymbol{E} + \frac{1}{c}[\boldsymbol{v},\boldsymbol{B}]\right)\frac{\partial f_\alpha}{\partial \boldsymbol{p}} = 0 . \tag{3.6.1}$$

For plasma processes with characteristic time (τ) and space (L) scales shorter than the mean free time and the mean free path, i.e.,

$$\frac{1}{\tau} \gg \sum_\beta \nu_{\alpha\beta}, \quad L \ll l_\alpha = \frac{v_{\mathrm{T}\alpha}}{\sum_\beta \nu_{\alpha\beta}}, \tag{3.6.2}$$

(3.6.1) gives a valid description.

Investigating wave processes the frequency $\omega \sim 1/\tau$ and the wave vector $k \sim 1/L$ are such parameters. Corrections due to particle collisions are small under the condition (3.6.2) and neglected in the following.

Note that even in the complete absence of particle collisions a hydrodynamic plasma description can be valid. To derive such a formulation the closed system of equations for the moments of the distribution functions must be set up. The first moments are the number density and flow velocity of each species:

$$N_\alpha(t,\boldsymbol{r}) = \int d\boldsymbol{p} f_\alpha , \quad N_\alpha(t,\boldsymbol{r})\boldsymbol{V}_\alpha(t,\boldsymbol{r}) = \int d\boldsymbol{p}\,\boldsymbol{v} f_\alpha . \tag{3.6.3}$$

Hydrodynamic equations have some advantages compared with kinetic equations provided that they can be received in a closed form. They are simpler than kinetics which follows from the fact, already, that hydrodynamic quantities like $N_\alpha(t,\boldsymbol{r})$ and $\boldsymbol{V}_\alpha(t,\boldsymbol{r})$ depend on the four variables $\boldsymbol{r}$ and t only whereas the distribution function $f_\alpha(t,\boldsymbol{p},\boldsymbol{r})$ is a function of seven variables.

Hydrodynamic equations for the collisionless plasma are obtained by taking moments of (3.6.1):

$$\begin{aligned} &\frac{\partial N_\alpha}{\partial t} + \operatorname{div} N_\alpha \boldsymbol{V}_\alpha = 0 , \\ &\frac{\partial N_\alpha p_{\alpha i}}{\partial t} + \frac{\partial}{\partial r_j}\Pi_{\alpha ij} = e_\alpha N_\alpha\left(\boldsymbol{E} + \frac{1}{c}[\boldsymbol{V}_\alpha,\boldsymbol{B}]\right) . \end{aligned} \tag{3.6.4}$$

The first equation is the continuity equation. The second one, the equation of motion, is not "hydrodynamically closed", since the tensor

$$\Pi_{aij} = \int d\boldsymbol{p}\, p_i v_j f_a \tag{3.6.5}$$

appears.

3.6.1 Two-Fluid Hydrodynamics of a Cold Collisionless Plasma

In general the calculation of this tensor is rather complex. In the collisionless plasma, however, there are two special cases which allow to achieve closure. One of these cases is the cold plasma limit. Investigating processes with a characteristic velocity much higher than the thermal velocity of the particles (electrons and ions), i.e.,

$$\frac{L}{\tau} \sim \frac{\omega}{k} \gg v_{\mathrm{T}a} \tag{3.6.6}$$

the plasma can be regarded as "cold". Neglecting the thermal motion of the particles the velocity distribution can be approximated by

$$f_a \sim \delta(\boldsymbol{v} - \boldsymbol{V}_a) \,. \tag{3.6.7}$$

From (3.6.1) we get

$$\Pi_{aij} = N_a m_a V_{ai} V_{aj} (1 - V_a^2/c^2)^{-1/2}$$

and the equation of motion (3.6.4) is easily reduced to the Euler equation for a charged fluid with zero pressure

$$\left(\frac{\partial}{\partial t} + \boldsymbol{V}_a \cdot \nabla\right) \frac{\boldsymbol{V}_a}{\sqrt{1 - \dfrac{V_a^2}{c^2}}} = \frac{e_a}{m_a}\left(\boldsymbol{E} + \frac{1}{c}\,[\boldsymbol{V}_a, \boldsymbol{B}]\right) . \tag{3.6.8}$$

These equations together with the continuity equations (3.6.4) for the electrons and the ions ($a = \mathrm{e, i}$) form the closed system of hydrodynamic equations for the cold collisionless plasma. The current and charge densities are connected with the particle densities and fluxes (3.6.3) through

$$\boldsymbol{j} = \sum_a e_a N_a \boldsymbol{V}_a \,, \qquad \varrho = \sum_a e_a N_a \,, \tag{3.6.9}$$

where the summation extends over all charged particle species.

Substituting these expressions into Maxwell's equations (2.1.1) yields the self-consistent system of hydrodynamic and field equations which establishes the so-called *two-fluid magnetohydrodynamics* of a cold plasma. Obviously, (3.6.8) coincides with the equation of motion (3.1.1), derived under the assumption of independent particles. The present derivation allows to define the range of validity of this simple model. It can be applied if the inequalities (3.6.2, 6) hold.

Note that (3.6.2, 6) define the range of validity of two-fluid hydrodynamics in the cold plasma case only if there is no external magnetic field. For magnetized plasmas it is necessary to satisfy additional conditions, namely

$$|\omega \pm \Omega_\alpha| \gg k_z v_{T\alpha}\,,\ \nu_\alpha\,; \qquad \frac{k_\perp v_{T\alpha}}{\Omega_\alpha} \ll 1\,, \tag{3.6.10}$$

where $\Omega_\alpha = e_\alpha B/m_\alpha c$ is the *Larmor frequency* of the particle species α; $k_z \sim 1/L_\parallel$ and $k_\perp \sim 1/L_\perp$ are the characteristic lengths along and across the magnetic field direction, respectively.

3.6.2 One-Fluid Hydrodynamics of the Nonisothermal Plasma

Another hydrodynamic description of the collisionless plasma – the so-called *one-fluid hydrodynamics* – is valid in a second range of parameters determined by

$$v_{\mathrm{Ti}} \ll \frac{L}{\tau} \sim \frac{\omega}{k} \ll v_{\mathrm{Te}}\,. \tag{3.6.11}$$

The first inequality allows that the cold ion component of such processes can be described by the equations (3.6.4, 8) in the nonrelativistic limit, naturally, since $|V_\mathrm{i}| \ll c$. The ion equations at the same time characterize the motion of the plasma mass. Due to the second inequality we may neglect the time derivative when applying (3.6.1) for the electron component. It is easy to see that in this case the Boltzmann distribution with account of the potential of an external field

$$f_\mathrm{e} = f_\mathrm{e}\left[\frac{m}{2T_\mathrm{e}}(v - V_\mathrm{e})^2 + \frac{e\Phi}{T_\mathrm{e}}\right] \tag{3.6.12}$$

is a solution. V_e is the directed velocity of the electrons. The temperature T_e must be assumed constant.

The electron current is determined by

$$j_\mathrm{e} = eN_\mathrm{e}V_\mathrm{e}\,, \tag{3.6.13}$$

with account of the barometric law for the electron density

$$N_\mathrm{e} = N_0 \exp\left(-\frac{e\Phi}{T_\mathrm{e}}\right) \tag{3.6.14}$$

which follows from (3.6.12). N_0 is the unperturbed density in the absence of the electric field.

Substituting (3.6.12) into (3.6.5) and taking account of the condition $|V_e| \ll v_{Te}$, the equation of motion (3.6.4) for the electrons reduces to

$$E = -\frac{T_e}{eN_e} \nabla N_e - \frac{1}{c} [V_e, B] . \tag{3.6.15}$$

Using this relation, we eliminate the electric field from the ion equation. Under the condition of quasineutrality $\sum_\alpha e_\alpha N_\alpha = 0$, a formulation containing the density of the total current

$$j = \sum_\alpha e_\alpha N_\alpha V_\alpha = -eN_e V_e + e_i N_i V_i \tag{3.6.16}$$

is finally obtained

$$\frac{\partial V_i}{\partial t} + (V_i \cdot \nabla) V_i = -\frac{v_s^2}{N_i} \nabla N_i + \frac{1}{MN_i c} [j, B] . \tag{3.6.17}$$

Here $v_s = \sqrt{ZT_e/M}$ is the *ion-acoustic velocity* and $Z = |e_i/e|$ denotes the charge number of the ions.

The Maxwell equations (2.1.1) also simplify at low frequencies since the displacement current can be neglected under the condition $\omega \sim 1/\tau \ll \omega_{pi}$. The electric field can be eliminated noting that only the electric field component perpendicular to the magnetic field contributes to the equation

$$\operatorname{curl} E = -\frac{1}{c} \frac{\partial B}{\partial t} . \tag{3.6.18}$$

Instead of (3.6.15) we can apply the simpler expression

$$E_\perp = -\frac{1}{c} [V_i, B] , \tag{3.6.19}$$

which follows directly from the Euler equation of motion for ions when $\omega \sim 1/\tau \ll \Omega_i$. Obviously $E_\perp$ coincides with the transverse projection of (3.6.15). Substituting (3.6.19) into (3.6.18) we obtain

$$\operatorname{curl} [V_i, B] = \frac{\partial B}{\partial t} . \tag{3.6.20}$$

Identifying the ion mass density with the total mass density $\varrho_m = MN_i$, we can write down the system of equations of one-fluid magnetohydrodynamics for the collisionless plasma:

$$\frac{\partial \boldsymbol{B}}{\partial t} = \operatorname{curl}[\boldsymbol{V}, \boldsymbol{B}] , \qquad \operatorname{div}\boldsymbol{B} = 0 ,$$

$$\varrho_{\mathrm{m}}\left[\frac{\partial \boldsymbol{V}}{\partial t} + (\boldsymbol{V}\cdot\nabla)\boldsymbol{V}\right] = -\,v_{\mathrm{s}}^2\nabla\varrho_{\mathrm{m}} - \frac{1}{4\pi}[\boldsymbol{B}, \operatorname{curl}\boldsymbol{B}] , \tag{3.6.21}$$

$$\frac{\partial \varrho_{\mathrm{m}}}{\partial t} + \operatorname{div}\varrho_{\mathrm{m}}\boldsymbol{V} = 0 .$$

It represents the limit of ideal conductivity $\sigma \to \infty$ and of negligible viscosity of the system of equations (3.1.4), the equation of state being

$$p = v_{\mathrm{s}}^2\varrho_{\mathrm{m}} = N_{\mathrm{e}}T_{\mathrm{e}} = ZN_{\mathrm{i}}T_{\mathrm{e}} . \tag{3.6.22}$$

In contrast to the general hydrodynamic description of a conducting fluid (Sect. 3.1), the temperature must be constant in time and space here, which turns out in (3.6.21). Further, according to (3.6.22), the total plasma pressure is determined by the electron temperature, which is possible only under the condition $T_{\mathrm{e}} \gg T_{\mathrm{i}}$.

The requirement of different temperatures of the plasma components can be argued directly from the basic assumption (3.6.11), too. As known from hydrodynamics, the typical velocity of a fluid element is of the order of the sound velocity $v_{\mathrm{s}} \sim \sqrt{p/\varrho_{\mathrm{m}}}$. According to (3.6.22), in the collisionless cold ion case $v_{\mathrm{s}} \sim \sqrt{T_{\mathrm{e}}/M}$ holds which is the reason why v_{s} is called the ion sound velocity. On the other hand, (3.6.11) requires that this velocity must exceed the thermal velocity of the ions, i.e., $v_{\mathrm{s}} \gg v_{\mathrm{Ti}}$, which is possible only in the nonisothermal plasma ($T_{\mathrm{e}} \gg T_{\mathrm{i}}$).

Summarizing, we recall and discuss the conditions under which the one-fluid magnetohydrodynamic (MHD) description of the collisionless nonisothermal plasma is valid:

$$v_{\mathrm{Ti}} \ll \frac{L}{\tau} \sim \frac{\omega}{k} \ll v_{\mathrm{Te}} , \qquad \omega \sim \frac{1}{\tau} \ll \Omega_{\mathrm{i}} \ll \omega_{\mathrm{pi}} . \tag{3.6.23}$$

Strictly speaking, in the presence of a strong magnetic field the first condition applies to the parallel projection of the wave vector, only. In other words, the parallel phase velocity must lie in the range between the thermal velocities (this will be elaborated in Chap. 5). The transverse projection must fulfil the inequality $k_{\perp}v_{\mathrm{T}\alpha}/\Omega_{\alpha} \sim v_{\mathrm{T}\alpha}/L_{\perp}\Omega_{\alpha} \ll 1$, i.e., the transverse wavelength must significantly exceed the Larmor radius. The condition $\Omega_{\mathrm{i}} \ll \omega_{\mathrm{pi}}$ is necessary to guarantee plasma quasineutrality for the low frequency processes $\omega \ll \Omega_{\mathrm{i}}$. However, it is still possible to pass over to the limit $\boldsymbol{B} \to 0$ in the MHD equations (3.6.21). In the absence of an external magnetic field the continuity equation and the Euler equation of the ions form a complete system of equations. Both (3.6.20) and the condition $\omega \ll \Omega_i$ are no longer needed. Plasma quasineutrality is ensured by the condition $\omega \sim 1/\tau \ll \omega_{pi}$, now.

Finally, the neglecting of particle collisions in our derivation of (3.6.21) must be argued. Although we were using the collisionless approach the condition $\omega \sim 1/\tau \gg \nu_e, \nu_i$ would be too strong. As for the cold ion kinetic equation one can, in fact, neglect collisions only for $\omega \gg \nu_i$, where ν_i is the total collision frequency of the ions. The electron kinetic equation, however, allows the neglecting of collisions in the range $\omega < \nu_e$ and $\nu_e \ll k_\parallel v_{Te} \sim v_{Te}/L_\parallel$, already. Thus, with respect to the collision frequencies the applicability conditions of the system (3.6.21) are

$$\omega \sim \frac{1}{\tau} \gg \nu_i \,, \qquad \omega, \quad \nu_e \ll k_\parallel v_{Te} \sim \frac{v_{Te}}{L_\parallel} \,. \tag{3.6.24}$$

Under certain conditions the hydrodynamic description is possible even for the rarefied collisionfree plasma. We only mention this limit since in this case the validity is most questionable. In the extreme case of a dense plasma with frequent particle collisions the validity of hydrodynamics is obvious.

3.6.3 The Hydrodynamic Description of a Degenerate Plasma

We can generalize (3.6.12) for the degenerate plasma and obtain the solutions (3.6.12) for an arbitrary degeneracy of the plasma

$$f_e = \frac{2}{(2\pi\hbar)^3} \left\{ \exp\left[\frac{m}{2T_e} (\boldsymbol{v} - \boldsymbol{V}_e)^2 - \frac{\mathscr{E}_{Fe}}{T_e} + \frac{e\phi}{T_e} \right] + 1 \right\}^{-1} . \tag{3.6.25}$$

At the same time the inequalities (3.6.11) ensure the constant temperature of the electrons $T_e = \text{const}$ if v_{Fe} is used instead of v_{Te}. Then from (3.6.25) we obtain

$$\frac{\partial f_e}{\partial \boldsymbol{r}} = \frac{\partial f_e}{\partial \mathscr{E}} e \frac{\partial \phi}{\partial \boldsymbol{r}} \,, \tag{3.6.26}$$

where $\mathscr{E} = m[\boldsymbol{v} - \boldsymbol{V}_e]^2/2$.

Now we can easily express the electric field $\boldsymbol{E} = -\nabla\phi - [\boldsymbol{V}_e, \boldsymbol{B}]/c$ in terms of the momenta of the distribution function of the electrons f_e. For the strongly degenerate plasma it can be written as

$$\boldsymbol{E} = -\frac{1}{5eN_e} \nabla(\mathscr{E}_{Fe} N_e) - \frac{1}{c} [\boldsymbol{V}_e, \boldsymbol{B}] \,. \tag{3.6.27}$$

When comparing (3.6.27) with (3.6.15) we summarize that the one-fluid hydrodynamic equations of the collisionless degenerate plasma take the form of (3.6.21) after the substitution

$$v_s^2 \nabla \varrho_M \rightarrow \nabla P . \tag{3.6.28}$$

Here $P = 1/5\, \mathscr{E}_{Fe} N_e$ is the pressure of the degenerate plasma, which coincides with the electron pressure and $N_e = ZN_i$. The applicability conditions of the hydrodynamic description of the degenerate plasma are analogous to (3.6.23, 24) under the substitution $v_{Te} \rightarrow v_{Fe}$. Moreover, the requirement of the plasma "nonisothermity" $\mathscr{E}_{Fe} \gg T_i$ remains valid and coincides with the requirement of its degeneracy.

3.7 Exercises

3.7.1. Verify that the Boltzmann integral for elastic collisions (3.3.3) vanishes if the distribution functions f_α and f_β are Maxwellian with $T_\alpha = T_\beta = T$.

Solution. The Maxwell distribution function for particles of type α with the temperature T has the form

$$f_\alpha(p_\alpha) = \frac{N_\alpha}{(2\pi m_\alpha T)^{3/2}} \exp\left(-\frac{p_\alpha^2}{2m_\alpha T}\right) . \tag{3.7.1}$$

Substituting this expression into (3.3.3) and taking account of the energy conservation law

$$\frac{p_\alpha^2}{2m_\alpha} + \frac{p_\beta^2}{2m_\beta} = \frac{p_\alpha'^2}{2m_\alpha} + \frac{p_\beta'^2}{2m_\beta} \tag{3.7.2}$$

we obtain

$$[f_\alpha(p_\alpha) f_\beta(p_\beta) - f_\alpha(p_\alpha') f_\beta(p_\beta')] = 0 . \tag{3.7.3}$$

Consequently, $(\partial f_\alpha/\partial t)_{\mathrm{col}}^{\alpha\beta} \equiv 0$.

3.7.2. Calculate the electron distribution function for the completely ionized stationary and homogeneous plasma in a constant external electric field. Use the Lorentz approximation. Evaluate the plasma conductivity for the Lorentz gas with infinitely heavy ions.

Solution. The gas where electrons do not interact with each other but interact with ions only is called the *Lorentz gas*. Such an approximation is valid for

$Z = |e_i/e| \gg 1$. The Landau kinetic equation for the Lorentz gas with heavy ions in an external electric field can be written as

$$e\boldsymbol{E} \cdot \frac{\partial f_e}{\partial \boldsymbol{p}} = \frac{\partial}{\partial p_i} 2\pi e^2 e_i^2 N_i L \frac{v^2 \delta_{ij} - v_i v_j}{v^3} \frac{\partial f_e}{\partial p_j} . \tag{3.7.4}$$

The solution of this equation can be expanded

$$f_e = f_M + \frac{\boldsymbol{v} \cdot \boldsymbol{f}_1(v)}{v} , \tag{3.7.5}$$

where f_M is the Maxwellian distribution function. For rather weak fields $|f_1| \ll |f_M|$; therefore

$$e\boldsymbol{E} \cdot \frac{\partial f_M}{\partial \boldsymbol{p}} = \frac{4\pi e^2 e_i^2 N_i L}{m^2} \frac{\boldsymbol{f}_1 \cdot \boldsymbol{v}}{v^4} \quad \text{or} \tag{3.7.6}$$

$$f_e = f_M + \frac{e\boldsymbol{E}}{m\nu(v)} \frac{\partial f_M}{\partial \boldsymbol{v}} , \quad \text{where} \tag{3.7.7}$$

$$\nu(v) = \frac{4\pi e^2 e_i^2 N_i L}{m^2 v^3} . \tag{3.7.8}$$

Hence we obtain the density of the electron current in the plasma:

$$\boldsymbol{j} = e \int d\boldsymbol{p}\, \boldsymbol{v} f_e = \frac{32}{3\pi} \frac{e^2 N_e}{m\nu_{\text{eff}}} \boldsymbol{E} = \sigma \boldsymbol{E} \tag{3.7.9}$$

and the plasma conductivity in the Lorentz gas model:

$$\sigma = \frac{32}{3\pi} \frac{e^2 N_e}{m\nu_{\text{eff}}} , \quad \text{where} \tag{3.7.10}$$

$$\nu_{\text{eff}} = \frac{4}{3} \sqrt{\frac{2\pi}{m}} \frac{e^2 e_i^2 N_i L}{T_e^{3/2}} .$$

Equation (3.7.10) is valid under the condition $\boldsymbol{u} = e\boldsymbol{E}/m\nu_{\text{eff}} \ll v_{Te}$, only. When $\boldsymbol{u} > v_{Te}$, the so-called *run-away electrons* appear. Therefore the concept of plasma conductivity can be used only up to some critical value of the field determined by the condition

$$\frac{eE_{cr}}{m\nu_{\text{eff}}} = v_{Te} \tag{3.7.11}$$

which is called the Dreicer field. For $E > E_{cr}$ there exists no stationary state of the plasma.

3.7.3. Using the Landau kinetic equation (3.4.7) investigate the process of temperature relaxation in a completely ionized plasma. Assume Maxwellian distributions for the electrons and ions with different temperatures T_e and T_i.

Solution. For Maxwellian velocity distributions the equal particle collision terms disappear:

$$\left(\frac{\partial f_e}{\partial t}\right)_{col}^{ee} = \left(\frac{\partial f_i}{\partial t}\right)_{col}^{ii} = 0 . \tag{3.7.12}$$

The process of temperature relaxation for T_e and T_i then is determined by electron-ion collisions

$$\frac{\partial f_e}{\partial t} = \left(\frac{\partial f_e}{\partial t}\right)_{col}^{ei} . \tag{3.7.13}$$

Inserting the Maxwellian distributions of the ions and the electrons, multiplying by $p_e^2/2\,\mathrm{m}$, and integrating over p_e we obtain

$$\frac{\partial T_e}{\partial t} = -\nu_T(T_e - T_i) , \quad \text{where} \tag{3.7.14}$$

$$\nu_T = \frac{e^2 e_i^2 L N_i}{6\pi^2 T_e T_i (m T_e M T_i)^{3/2}} \int dp_e\, dp_i \exp\left(-\frac{p_e^2}{2mT_e} - \frac{p_i^2}{2MT_i}\right) \times v_{ek} v_{em} \frac{u^2\delta_{km} - u_k u_m}{u^3} \tag{3.7.15}$$

with $\boldsymbol{u} = \boldsymbol{v}_e - \boldsymbol{v}_i$.

For $v_{Ti} \ll v_{Te}$ we can expand the integrand in powers of v_i/v_e which yields

$$v_{ek} v_{em} \frac{u^2\delta_{km} - u_k u_m}{u^3} \sim v_{ik} v_{im} \frac{v_e^2 \delta_{km} - v_{ek} v_{em}}{v_e^3} ,$$
$$\nu_T = 2\frac{m}{M}\nu_{eff} , \quad \nu_{eff} = \frac{4}{3}\sqrt{\frac{2\pi}{m}}\frac{e^2 e_i^2 N_i L}{T_e^{3/2}} . \tag{3.7.16}$$

From the conservation law of the total energy in the system it follows that $N_e T_e + N_i T_i = \text{const}$ or

$$N_e \frac{dT_e}{dt} + N_i \frac{dT_i}{dt} = 0 . \tag{3.7.17}$$

Consequently,

$$\frac{d}{dt}(T_e - T_i) = -\nu_T\left(1 + \left|\frac{e_i}{e}\right|\right)(T_e - T_i) . \tag{3.7.18}$$

Thus, the characteristic time of temperature relaxation is

$$\tau_{\rm T} \sim \frac{1}{\nu_{\rm T}} \approx \frac{M}{m} \frac{1}{\nu_{\rm eff}} .$$

3.7.4. Calculate the relaxation of the mean velocity for electrons distributed according to a shifted Maxwellian in a completely ionized plasma.

Solution. Assuming that the velocity distribution of the electrons

$$f_{\rm e} = \frac{N_{\rm e}}{(2\pi m T_{\rm e})^{3/2}} \exp\left(-\frac{m(\boldsymbol{v}-\boldsymbol{u})^2}{2T_{\rm e}}\right) \qquad (3.7.19)$$

does not change its form during the process of relaxation of the velocity $\boldsymbol{u}$, it is easy to show that the electron slowing down occurs only due to electron-ion collisions. Thus, after multiplying the equation

$$\frac{\partial f_{\rm e}}{\partial t} = \left(\frac{\partial f_{\rm e}}{\partial t}\right)_{\rm col}^{\rm ei} \qquad (3.7.20)$$

by $\boldsymbol{v}$ and integrating over the momentum, we obtain

$$\frac{\partial \boldsymbol{u}}{\partial t} = -\frac{4\pi e^2 e_{\rm i}^2 N_{\rm i} L}{m^2 N_{\rm e}} \int d\boldsymbol{p}\, f_{\rm e} \frac{\boldsymbol{v}}{v^3} . \qquad (3.7.21)$$

Deriving this relation, we have neglected the ion velocity compared to the electron velocity. Eq. (3.7.21) can be approximated by

$$\frac{\partial \boldsymbol{u}}{\partial t} = -\nu_{\rm p} \boldsymbol{u} , \quad \text{with} \qquad (3.7.22)$$

$$\nu_{\rm p} = \begin{cases} \nu_{\rm eff} & \text{for} \quad u \ll v_{\rm Te} , \\ 3\sqrt{\dfrac{\pi}{2}} \dfrac{v_{\rm Te}^3}{u^3} \nu_{\rm eff} & \text{for} \quad u \gg v_{\rm Te} . \end{cases} \qquad (3.7.23)$$

Thus, the relaxation time of the directed electron velocity equals $1/\nu_{\rm eff}$ for $u \ll v_{\rm Te}$ and increases proportional to $u^3/v_{\rm Te}^3$ for large u.

3.7.5. Analyze the relaxation process of a small anisotropy of the electron distribution function described by a two-temperature Maxwellian in a completely ionized plasma.

Solution. The electron distribution function is

$$f_e = \frac{N_e}{2\pi m T_\perp \sqrt{2\pi m T_\parallel}} \exp\left(-\frac{mv_\perp^2}{2T_\perp} - \frac{mv_\parallel^2}{2T_\parallel}\right) \tag{3.7.24}$$

and we assume $|T_\parallel - T_\perp| \ll T_\perp \approx T_\parallel \approx T$.

For the relaxation of the temperature anisotropy both electron-electron and electron-ion collisions are essential

$$\frac{\partial f_e}{\partial t} = \left(\frac{\partial f_e}{\partial t}\right)_{col}^{ee} + \left(\frac{\partial f_e}{\partial t}\right)_{col}^{ei}. \tag{3.7.25}$$

Multiplying this equation by $mv_\perp^2/2$ and by $mv_\parallel^2/2$, respectively, and integrating over the momentum, we obtain

$$\frac{dT_\perp}{dt} = -\frac{\nu_p}{2}(T_\perp - T_\parallel)\,, \qquad \frac{dT_\parallel}{dt} = -\nu_p(T_\parallel - T_\perp)\,,$$

$$\frac{d}{dt}(T_\perp - T_\parallel) = -\frac{3\nu_p}{2}(T_\perp - T_\parallel)\,, \quad \text{where} \tag{3.7.26}$$

$$\nu_p = \frac{4}{5}\nu_{eff}\left(1 + \frac{1}{\sqrt{2}}\left|\frac{e}{e_i}\right|\right); \qquad \nu_{eff} = \frac{4}{3}\sqrt{\frac{2\pi}{m}}\frac{e^2 e_i^2 N_i L}{T_e^{3/2}}. \tag{3.7.27}$$

Thus, the relaxation of a small temperature anisotropy as that of a small directed electron velocity occurs on the time scale $1/\nu_{eff}$.

3.7.6. Use the kinetic equation with the model BGK integral to calculate the electron distribution function and the plasma heating in an external electric field.

Solution. Taking into account only electron-neutral collisions, the kinetic equation with the BGK collision integral can be written as

$$eE \cdot \frac{\partial f_e}{\partial p} = -\nu_{en}(f_e - N_e \Phi_{en})\,, \quad \text{where} \tag{3.7.28}$$

$$\Phi_{en} = \frac{1}{(2\pi m T_{en})^{3/2}} \exp\left(-\frac{mv^2}{2T_{en}}\right), \quad \text{and} \tag{3.7.29}$$

$$T_{en} = \frac{mT_n + MT_e}{m + M}. \tag{3.7.30}$$

T_n is the temperature of the neutrals and M their mass.

The kinetic equation can be solved by the ansatz

$$f_e = f_0(v) + \frac{\boldsymbol{v} \cdot \boldsymbol{f}_1(v)}{v} , \tag{3.7.31}$$

assuming $|f_1| \ll f_0$. Averaging over the angles of the velocity with respect to $\boldsymbol{E}$ we obtain the two equations

$$\begin{aligned} \frac{e}{3v^2 m} \frac{\partial}{\partial v}(v^2 \boldsymbol{E} \cdot \boldsymbol{f}_1) &= -\nu_{en}(f_0 - N_e \Phi_{en}) , \\ \frac{e\boldsymbol{E}}{m} \frac{\partial f_0}{\partial v} &= -\nu_{en} \boldsymbol{f}_1 . \end{aligned} \tag{3.7.32}$$

Substituting f_1 from the second equation into the first one we obtain

$$\frac{e^2 E^2}{3m^2 v^2 \nu_{en}} \frac{\partial}{\partial v}\left(v^2 \frac{\partial f_0}{\partial v}\right) + \nu_{en}(f_0 - N_e \Phi_{en}) = 0 . \tag{3.7.33}$$

The solution of this equation is the Maxwellian distribution function with the temperature T_e given by

$$\frac{2e^2 E^2}{m^2 \nu_{en}} - \frac{3\nu_{en}}{m+M}(T_e - T_n) = 0 , \quad \text{or} \tag{3.7.34}$$

$$T_e \approx T_n + \frac{2M}{3m} \frac{e^2 E^2}{m \nu_{en}^2} . \tag{3.7.35}$$

The stationary value of the temperature is the result of the balance between the ohmic electron heating and the energy transfer from the electrons to the neutrals. Finally, the plasma conductivity is determined from

$$\boldsymbol{j} = e \int d\boldsymbol{p}\, \boldsymbol{v} \frac{\boldsymbol{v} \cdot \boldsymbol{f}_1}{v} = \frac{e^2 N_e}{m \nu_{en}} \boldsymbol{E} = \sigma \boldsymbol{E} , \tag{3.7.36}$$

which gives $\sigma = e^2 N_e / m \nu_{en}$.

3.7.7. Using the model of independent particles calculate the average force (*Miller's force*) acting on the electrons in an external electric high-frequency field with an inhomogeneous amplitude $\boldsymbol{E}(t, \boldsymbol{r}) = \boldsymbol{E}(\boldsymbol{r}) \sin \omega_0 t$. Consider the cases with and without an external homogeneous magnetic field superimposed. Neglect relativistic effects.

Solution. Let us write the equation of motion of the electrons in the form

$$\frac{\partial \boldsymbol{V}}{\partial t} + (\boldsymbol{V} \cdot \nabla)\boldsymbol{V} = \frac{e\boldsymbol{E}(t, \boldsymbol{r})}{m} + \frac{e}{mc}[\boldsymbol{V}, \boldsymbol{B}_0 + \boldsymbol{B}(t, \boldsymbol{r})] , \tag{3.7.37}$$

where $\boldsymbol{B}_0$ is the external homogeneous magnetic field and

$$\boldsymbol{B}(t,\boldsymbol{r}) = \boldsymbol{B}(\boldsymbol{r}) \cos \omega_0 t \,, \qquad \boldsymbol{B}(\boldsymbol{r}) = \frac{c}{\omega_0} \operatorname{curl} \boldsymbol{E}(\boldsymbol{r}) \,. \tag{3.7.38}$$

Assuming the fields $\boldsymbol{E}(t,\boldsymbol{r})$ and $\boldsymbol{B}(t,\boldsymbol{r})$ (and consequently the velocity $\boldsymbol{V}$) to be small, we obtain in the linear approximation

$$\frac{d\boldsymbol{V}_0}{dt} = \frac{e\boldsymbol{E}}{m}(t,\boldsymbol{r}) + \frac{\mathrm{e}}{\mathrm{mc}}[\boldsymbol{V}_0, \boldsymbol{B}_0] \,. \tag{3.7.39}$$

Substituting $\boldsymbol{V}_0(t)$ into the small nonlinear terms of (3.7.37) and averaging them over the time, we obtain the average force

$$\boldsymbol{F}_{\mathrm{av}} = -m\overline{(\boldsymbol{V}_0 \cdot \nabla)\boldsymbol{V}_0} + \frac{e}{c}\overline{[\boldsymbol{V}_0, \boldsymbol{B}]} \,. \tag{3.7.40}$$

In the absence of an external homogeneous magnetic field

$$\boldsymbol{V}_0 = -\frac{e\boldsymbol{E}(\boldsymbol{r})}{m\omega_0} \cos \omega_0 t \,, \qquad \boldsymbol{F}_{\mathrm{av}} = -\frac{e^2}{2m\omega_0^2}\{(\boldsymbol{E} \cdot \nabla)\boldsymbol{E} + [\boldsymbol{E}, \operatorname{curl} \boldsymbol{E}]\} = -\frac{e^2}{4m\omega_0^2} \nabla E^2(\boldsymbol{r}) \,. \tag{3.7.41}$$

The average force ejects electrons (and consequently the plasma) from the region of a strong high-frequency field[1].

In the presence of an external homogeneous magnetic field the opposite situation where the plasma is absorbed into the region of the strong high-frequency field is possible, too (see Exercise 5.8.12).

3.7.8. Calculate the average transverse force acting on the plasma electrons in a weakly inhomogeneous stationary magnetic field.

Solution. We first consider electrons without longitudinal (parallel to the magnetic field) velocity component which rotate around the magnetic field lines with the angular velocity $\Omega = eB(\boldsymbol{r})/mc$. Their position is $\boldsymbol{r} = \boldsymbol{r}_0(t) + \boldsymbol{\xi}(t)$, where $\boldsymbol{r}_0(t)$ denotes the coordinate of the centre of the Larmor rotation and $\boldsymbol{\xi}(t)$ is the position of the electron on the orbit; $\boldsymbol{r}_0(t)$ is in general large and slowly varying, $\boldsymbol{\xi}(t)$ is small and quickly varying. In the model of independent particles

1 The average force acting directly on the ions is M/m times smaller than the force (3.7.41). Thus, the ions are affected only by coupling to the electrons.

$$\dot{\xi} = \Omega \left[\xi, \frac{B}{B} \right], \quad \xi = \frac{1}{\Omega} \left[\frac{B}{B}, v_\perp \right], \tag{3.7.42}$$

where $v_\perp$ is the unperturbed velocity of the electron rotation.

Expanding $B(r)$ in powers of ξ and averaging the Lorentz force over the time, we find the average force perpendicular to the magnetic field acting on the plasma electrons:

$$\begin{aligned} F_{1\,\mathrm{av}} &= \frac{e}{c} [\xi, (\xi \cdot \nabla) B] = -\frac{m v_\perp^2}{2B} \left[\left(\frac{B}{B}, \nabla \right) B \right] \\ &= -\frac{m v_\perp^2}{2} \left(\frac{B \cdot \nabla}{B} \right) \frac{B}{B} = \frac{m v_\perp^2}{2R} n . \end{aligned} \tag{3.7.43}$$

R is the radius of curvature of the magnetic field lines and n is the unit vector directed from the centre of curvature to the position of the electron.

In the derivation of (3.7.43), Eq. (3.7.42) and the Maxwell equations for the stationary magnetic field $B(r)$ have been used:

$$\operatorname{curl} B = 0, \quad \operatorname{div} B = 0 . \tag{3.7.44}$$

The averages over the gyrophase involved are

$$\overline{\xi_i \xi_j} = \frac{1}{2} \xi^2 \delta_{ij} = \frac{v_\perp^2}{2\Omega^2} \delta_{ij} . \tag{3.7.45}$$

We now admit, along with the Larmor rotation, that the plasma electrons can move longitudinally with the velocity $v_\parallel$. Passing over to the coordinate system which rotates with the angular velocity $v_\parallel / R$ around the momentary centre of curvature of the magnetic field lines, we again have the case of electrons without a longitudinal velocity. In this system, however, there emerges an additional transverse inertial force, i.e., the centrifugal force equal to

$$F_{2\,\mathrm{av}} = \frac{m v_\parallel^2}{R} n . \tag{3.7.46}$$

The sum of (3.7.43) and (3.7.46) gives the total average force

$$F_{\mathrm{av}} = F_{1\,\mathrm{av}} + F_{2\,\mathrm{av}} = n \frac{m}{R} \left(v_\parallel^2 + \frac{v_\perp^2}{2} \right) . \tag{3.7.47}$$

This force on the electrons is equivalent to a gravitational acceleration

$$g = \frac{n}{R} \left(v_\parallel^2 + \frac{v_\perp^2}{2} \right) . \tag{3.7.48}$$

Such a force evidently affects the plasma ions as well and since it is independent of the sign of the charge both components experience forces into the same direction.

3.7.9. Using the model of independent particles show that the average force acting on the electron in the electric high-frequency $\boldsymbol{E}(\boldsymbol{r}, t) = \boldsymbol{E}(\boldsymbol{r}) \sin \omega_0 t$ and constant magnetic $\boldsymbol{B}_0$ fields is defined by

$$F_{\mathrm{av}} = -\frac{e^2}{4m\omega_0^2} \nabla \left\{ \frac{\omega_0}{\omega_0 + \Omega} E_\perp^{(+)2} + \frac{\omega_0}{\omega_0 - \Omega} E_\perp^{(-)2} + E_\parallel^2 \right\},$$

where $\Omega = eB_0/(mc)$; $E_\parallel$ and $E_\perp^{(\pm)}$ are longitudinal (i.e., along $\boldsymbol{B}_0$) and the right-hand and left-hand circular polarized transverse components of the electric field respectively.

4. Dielectric Permittivity and Oscillation Spectra of Unmagnetized Plasmas

On the basis of the kinetic equation with a self-consistent field, the expressions for the dielectric tensors of a homogeneous isotropic nondegenerate plasma with a Maxwellian particle distribution function and a degenerate plasma with a Fermi distribution function are obtained. Using the dielectric tensors the effect of interparticle collisions upon the character of propagation and the spectra of longitudinal and transverse waves are investigated.

4.1 Dielectric Permittivity of a Collisionless Homogeneous Isotropic Plasma

Knowing the system of kinetic equations for the charged particles we may proceed to study the electromagnetic properties of the plasma. It is natural to start with the simplest case of a spatially homogeneous isotropic plasma. Moreover, regarding collisions in the plasma as infrequent, we can neglect them and use the kinetic equation with a self-consistent field (Vlasov's equation) to obtain the dielectric permittivity. Such an approximation is valid for the description of processes occurring on a time scale short against the mean free time or possessing a space scale smaller than the mean free path. In this case the plasma is called collisionless.

In the absence of electromagnetic fields in a spatially homogeneous isotropic collisionless plasma, the distribution functions of the particles can be arbitrary functions of the momentum $|\boldsymbol{p}| = p$. Yet, in the nondegenerate plasma we assume the particle distribution to be Maxwellian with the temperature T_α and the density N_α for the particles of type α:

$$f_{0\alpha}(p) = f_{M\alpha}(p) = \frac{N_\alpha}{(2\pi m_\alpha T_\alpha)^{3/2}} \exp\left(-\frac{p_\alpha^2}{2m_\alpha T_\alpha}\right) . \tag{4.1.1}$$

For the degenerate plasma with the Fermi energy

$$\mathscr{E}_{F\alpha} = (2m_\alpha)^{-1} \times (3\pi^2)^{2/3} \hbar^2 N_\alpha^{2/3} > T_\alpha$$

the Fermi distribution is adopted:

$$f_{0\alpha}(p) = f_{F\alpha}(p) = \frac{2}{(2\pi\hbar)^3}\left[\exp\left(\frac{\mathscr{E}-\mathscr{E}_{F\alpha}}{T_\alpha}\right)+1\right]^{-1}, \tag{4.1.2}$$

where $\mathscr{E} = p_\alpha^2/2m_\alpha$. In the limit $T_\alpha \to 0$ it has the form

$$f_{0\alpha} = \begin{cases} \dfrac{2}{(2\pi\hbar)^3} & \text{if} \quad p \leq p_{F\alpha} = (3\pi^2)^{1/3}\hbar N_\alpha^{1/3}\,, \\ 0 & \text{if} \quad p > p_{F\alpha}\,. \end{cases}$$

In the opposite limiting case $T_\alpha \gg \mathscr{E}_{F\alpha}$ it coincides with the Maxwellian distribution (4.1.1).

To determine dielectric permittivity, we must admit a perturbation of the equilibrium distribution function $f_{0\alpha}(p)$. Such a deviation occurs due to small fluctuations of the electric and magnetic fields $\boldsymbol{E}(t,\boldsymbol{r})$ and $\boldsymbol{B}(t,\boldsymbol{r})$, which in turn are caused by the perturbation of the equilibrium. We write the perturbed distribution function

$$f_\alpha(\boldsymbol{p},\boldsymbol{r},t) = f_{0\alpha}(p) + \delta f_\alpha(\boldsymbol{p},\boldsymbol{r},t)\,. \tag{4.1.3}$$

and suppose the perturbation $\delta f_\alpha(\boldsymbol{p},\boldsymbol{r},t)$ and the values of the perturbed fields $\boldsymbol{E}$ and $\boldsymbol{B}$ to be small.

Substituting (4.1.3) into Vlasov's equation (3.2.7) and neglecting the terms of the second order, we obtain the linearized Vlasov equation for the perturbed distribution function[1]:

$$\frac{\partial \delta f_\alpha}{\partial t} + \boldsymbol{v}\frac{\partial \delta f_\alpha}{\partial \boldsymbol{r}} + e_\alpha \boldsymbol{E}\frac{\partial f_{0\alpha}(p)}{\partial \boldsymbol{p}} = 0\,. \tag{4.1.4}$$

In its unperturbed state the plasma is quasi-neutral, current and charge densities being absent. Due to the perturbing fields $\boldsymbol{E}$ and $\boldsymbol{B}$, induced charges and currents emerge in the plasma. According to (3.2.8) these charge and current densities are equal to

$$\begin{aligned} \varrho &= \sum_\alpha e_\alpha \int f_\alpha\, d\boldsymbol{p} = \sum_\alpha e_\alpha \int \delta f_\alpha\, d\boldsymbol{p}\,, \\ \boldsymbol{j} &= \sum_\alpha e_\alpha \int \boldsymbol{v} f_\alpha\, d\boldsymbol{p} = \sum_\alpha e_\alpha \int \boldsymbol{v}\,\delta f_\alpha\, d\boldsymbol{p}\,. \end{aligned} \tag{4.1.5}$$

The self-consistent fields $\boldsymbol{E}$ and $\boldsymbol{B}$, on the other hand, are determined by $\boldsymbol{j}$ and ϱ through the Maxwell equations (2.1.1).

---

1 Deriving (4.1.4) we have used $[\boldsymbol{v},\boldsymbol{B}]\cdot\partial f_{0\alpha}/\partial\boldsymbol{p} = 0$, valid for isotropic distribution functions.

Because of the linearity of (4.1.4) and of the field equations, there is no coupling between the Fourier expanded perturbations. Assuming the phasor $\exp(-\mathrm{i}\omega t + \mathrm{i}\boldsymbol{k}\cdot\boldsymbol{r})$ the solution of (4.1.4) is easily written

$$\delta f_\alpha = -\mathrm{i}e_\alpha \boldsymbol{E}\frac{\partial f_{0\alpha}}{\partial \boldsymbol{p}}(\omega - \boldsymbol{k}\cdot\boldsymbol{v})^{-1}. \tag{4.1.6}$$

Substituting this expression into (4.1.5) the density of the induced current

$$j_i = -\mathrm{i}\sum_\alpha e_\alpha^2 \int d\boldsymbol{p}\frac{v_i E_j \dfrac{\partial f_{0\alpha}}{\partial p_j}}{\omega - \boldsymbol{k}\cdot\boldsymbol{v}} \equiv \sigma_{ij}(\omega, \boldsymbol{k})\, E_j\,. \tag{4.1.7}$$

and the conductivity tensor

$$\sigma_{ij}(\omega\, \boldsymbol{k}) = -\mathrm{i}\sum_\alpha e_\alpha^2 \int d\boldsymbol{p}\,\frac{v_i}{\omega - \boldsymbol{k}\cdot\boldsymbol{v}}\,\frac{\partial f_{0\alpha}}{\partial p_j}\,. \tag{4.1.8}$$

follow.

Finally, using the relation between the complex dielectric permittivity and the complex conductivity (2.2.10) we can get the tensor $\varepsilon_{ij}(\omega, \boldsymbol{k})$

$$\begin{aligned}\varepsilon_{ij}(\omega, \boldsymbol{k}) &= \delta_{ij} + \frac{4\pi\mathrm{i}}{\omega}\sigma_{ij}(\omega, \boldsymbol{k})\\ &= \delta_{ij} + \sum_\alpha \frac{4\pi e_\alpha^2}{\omega}\int d\boldsymbol{p}\frac{v_i}{\omega - \boldsymbol{k}\cdot\boldsymbol{v}}\,\frac{\partial f_{0\alpha}}{\partial p_j}\,.\end{aligned} \tag{4.1.9}$$

In (4.1.7–9) the summation extends over the species index of the charged particles. In collisionless plasmas the neutrals apparently do not participate in electromagnetic phenomena. It must be underlined that the applicability of the “collisionless” approximation can be discussed only when the problem is solved with account taken of collisions.

4.1.1 Cherenkov Absorption and Radiation Emission of Waves

Our next concern is the treatment of the poles at $\omega = \boldsymbol{k}\cdot\boldsymbol{v}$ in the integrands of (4.1.7–9). At $\omega = \boldsymbol{k}\cdot\boldsymbol{v}$, these equations have no exact meaning since the integration result depends on the prescription of treatment of the singularity. To eliminate this ambiguity, one must take into account that the perturbation of the distribution function $\delta f_\alpha(\boldsymbol{p}, \boldsymbol{r}, t)$ disappears at $t \to -\infty$. For the given time-dependence $\delta f_\alpha \approx \exp(-\mathrm{i}\omega t)$ this means that we must have an infinitesimal positive imaginary part when $\boldsymbol{k}$ is real. On the other hand, when ω has a positive imaginary part, the poles of integrands in (4.1.7–9) do not lie on the

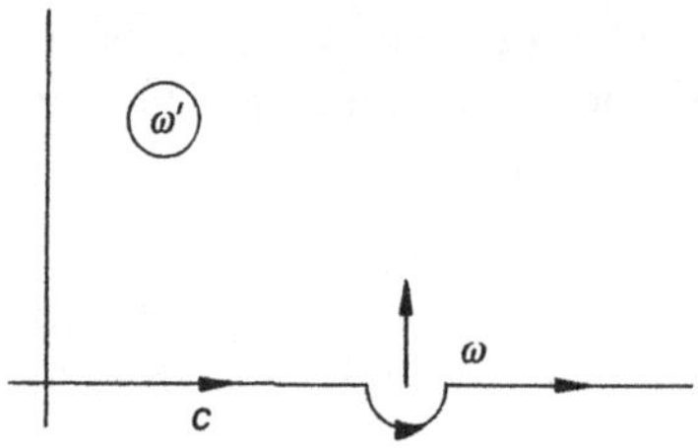

Fig. 4.1. Landau prescription

real axis which is the integration path but they are shifted into the upper half-plane ω' (Fig. 4.1). Thus, the prescription for the integration near $\omega = \boldsymbol{k}\cdot\boldsymbol{v}$ follows. One must avoid the singularity by integrating not over the real axis but over the contour C, shown in Fig. 4.1. In (4.1.7–9) the integration over such a contour is understood (*Landau prescription*).

We can use the well-known relation

$$\lim_{\nu\to 0}\frac{1}{x+\mathrm{i}\nu} = \mathcal{P}\frac{1}{x} - \mathrm{i}\pi\delta(x)\,, \tag{4.1.10}$$

where $\mathcal{P}$ denotes the prescription that at the singularity at $x = 0$ the principal value is to be taken, to write (4.1.9) in the form

$$\varepsilon_{ij}(\omega,\boldsymbol{k}) = \delta_{ij} + \sum_\alpha \frac{4\pi e_\alpha^2}{\omega} \times \int d\boldsymbol{p}\, v_i \frac{\partial f_{0\alpha}}{\partial p_j}\left[\mathcal{P}\frac{1}{\omega - \boldsymbol{k}\cdot\boldsymbol{v}} - \mathrm{i}\pi\delta(\omega - \boldsymbol{k}\cdot\boldsymbol{v})\right]. \tag{4.1.11}$$

In the integrand of (4.1.11) the first summand contributes to the Hermitian (real) part of the dielectric tensor and the second one to the anti-Hermitian (imaginary) part responsible for the wave absorption in the plasma. Only the resonant particles, which satisfy the condition $\omega = \boldsymbol{k}\cdot\boldsymbol{v}$ are responsible for the wave absorption. This condition can be written as $\omega/k = v_{\text{ph}} = v\cos\theta$, where θ is the angle between $\boldsymbol{k}$ and $\boldsymbol{v}$; v_{ph} is the phase velocity of the wave. This condition is nothing but the condition for *Cherenkov radiation emission* by a moving charged particle. The same condition governs the inverse process, namely the *Cherenkov wave absorption*. Thus, wave dissipation can occur in the absolute absence of particle collisions, namely by Cherenkov absorption by plasma particles. Because of the importance of this collisionless damping we want to study this effect in more detail.

Consider a particle moving with a velocity greater than the phase velocity of the wave. While "overtaking" the wave, this particle will be slowed down by the electric field of the wave. Thus, kinetic energy is transformed into wave energy (Cherenkov radiation). On the other hand, the potential barrier of the wave accelerates particles, which have velocities smaller than the

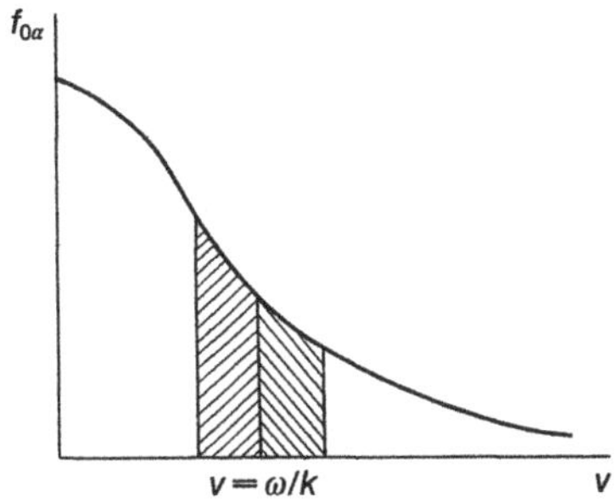

Fig. 4.2. Range of interaction of particles with a wave for a normal distribution over velocities

phase velocity of the wave. These particles take away energy from the wave (Cherenkov absorption). The velocity intervals of energy exchange between particles and waves are of equal width both for the particles exciting and absorbing waves. If the particle velocities are normally distributed (Fig. 4.2), i.e., if the distribution function decreases with increasing velocity, then there are less particles with $v > v_{ph}$ (giving away their energy to the wave) than particles with $v < v_{ph}$ (taking away energy) in intervals of equal width. Consequently, the total energy of the particles interacting with the wave increases and the wave is damped.

If, however, the distribution function $f_{0\alpha}(v)$ has a domain with a positive derivative, then electromagnetic waves can be amplified in the corresponding range of phase velocities.

4.1.2 The Longitudinal and Transverse Dielectric Permittivities of an Isotropic Plasma

For an isotropic plasma the tensor $\varepsilon_{ij}(\omega, \boldsymbol{k})$ can be written as

$$\varepsilon_{ij}(\omega, k) = \left(\delta_{ij} - \frac{k_i k_j}{k^2}\right)\varepsilon^{tr}(\omega, k) + \frac{k_i k_j}{k^2}\varepsilon^{lo}(\omega, k) , \tag{4.1.12}$$

where

$$\begin{aligned} \varepsilon^{lo}(\omega, k) &= 1 + \sum_\alpha \frac{4\pi e_\alpha^2}{\omega k^2} \int d\boldsymbol{p} \frac{(\boldsymbol{k}\cdot\boldsymbol{v})^2}{\omega - \boldsymbol{k}\cdot\boldsymbol{v}} \frac{\partial f_{0\alpha}}{\partial \mathscr{E}_\alpha} , \\ \varepsilon^{tr}(\omega, k) &= 1 + \sum_\alpha \frac{2\pi e_\alpha^2}{\omega k^2} \int d\boldsymbol{p} \frac{[\boldsymbol{k}, \boldsymbol{v}]^2}{\omega - \boldsymbol{k}\cdot\boldsymbol{v}} \frac{\partial f_{0\alpha}}{\partial \mathscr{E}_\alpha} \end{aligned} \tag{4.1.13}$$

are the longitudinal and transverse dielectric permittivities respectively, and $\mathscr{E}_\alpha(p)$ is the kinetic energy of the particle of type α.

For the nonrelativistic Maxwellian distribution (4.1.1), the integrals in (4.1.13) are known and

$$\varepsilon^{\mathrm{lo}}(\omega, k) = 1 + \sum_\alpha \frac{\omega_{p\alpha}^2}{k^2 v_{T\alpha}^2}\left[1 - I_+\left(\frac{\omega}{k v_{T\alpha}}\right)\right],$$

$$\varepsilon^{\mathrm{tr}}(\omega, k) = 1 - \sum_\alpha \frac{\omega_{p\alpha}^2}{\omega^2} I_+\left(\frac{\omega}{k v_{T\alpha}}\right) \quad \text{with} \tag{4.1.14}$$

$$I_+(x) = x e^{-x^2/2} \int_{i\infty}^{x} d\tau e^{\tau^2/2} = -\mathrm{i}\sqrt{\frac{\pi}{2}}\, x W\left(\frac{x}{\sqrt{2}}\right). \tag{4.1.15}$$

The function $W(x)$ has been studied in detail and tabulated.

The following asymptotic expansions of the function $I_+(x)$ are useful

$$I_+(x) = 1 + \frac{1}{x^2} + \frac{3}{x^4} + \ldots - \mathrm{i}\sqrt{\frac{\pi}{2}}\, x\, e^{-x^2/2} \tag{4.1.16}$$

for $|x| \gg 1, \quad |\mathrm{Re}\{x\}| \gg |\mathrm{Im}\{x\}|, \quad \mathrm{Im}\{x\} < 0$;

$$I_+(x) \approx -\mathrm{i}\sqrt{\frac{\pi}{2}}\, x \qquad \text{for} \quad |x| \ll 1;$$

$$I_+(x) = -\mathrm{i}\sqrt{2\pi}\, x\, e^{-x^2/2} \qquad \text{for} \quad |x| \gg 1, \quad |\mathrm{Im}\{x\}| \gg |\mathrm{Re}\{x\}|, \quad \mathrm{Im}\{x\} < 0.$$

The analytic investigation of oscillation spectra and of the character of propagation of the electromagnetic waves in the plasma must rely on these asymptotic expansions.

4.1.3 The Dielectric Permittivity of a Degenerate Plasma

There is no difficulty to calculate the longitudinal and transverse permittivities of the degenerate plasma with the Fermi distribution function (4.1.2). In case of zero temperature, taking into account

$$\frac{\partial f_{0\alpha}}{\partial \mathscr{E}_\alpha} = -\frac{2}{(2\pi\hbar)^3}\, \delta(\mathscr{E} - \mathscr{E}_{F\alpha}) \tag{4.1.17}$$

we obtain from (4.1.13)

$$\varepsilon^{\mathrm{lo}}(\omega, k) = 1 + \sum_\alpha \frac{3\omega_{p\alpha}^2}{k^2 v_{F\alpha}^2}\left[1 - \frac{\omega}{2 k v_{F\alpha}} \ln \frac{\omega + k v_{F\alpha}}{\omega - k v_{F\alpha}}\right],$$

$$\varepsilon^{\mathrm{tr}}(\omega, k) = 1 - \sum_\alpha \frac{3\omega_{p\alpha}^2}{2\omega^2}\left[1 + \left(\frac{\omega^2}{k^2 v_{F\alpha}^2} - 1\right)\right. \tag{4.1.18}$$

$$\left. + \left(-\frac{\omega}{2 k v_{F\alpha}} \ln \frac{\omega + k v_{F\alpha}}{\omega - k v_{F\alpha}}\right)\right],$$

where $v_{F\alpha} = p_{F\alpha}/m_\alpha = (3\pi^2)^{1/3}\hbar N_\alpha^{1/3}/m_\alpha$ is the velocity of particles of type α on the Fermi surface, i.e., with the momentum $p_{F\alpha}$.

Note that in (4.1.14, 18) the summation extends over all types of charged particles since the plasma was supposed to be either completely nondegenerate or completely degenerate. If one particle species is degenerate and another one is nondegenerate, then terms of the types (4.1.14, 18) appear in the dielectric permittivity.

4.2 Longitudinal Oscillations of a Collisionless Nondegenerate Plasma

We begin our study of electromagnetic waves in a homogeneous and isotropic plasma with the analysis of longitudinal waves in the collisionless case treated in the foregoing section. The dispersion equation (2.4.6) becomes in the case of a nondegenerate plasma

$$\varepsilon^{\mathrm{lo}}(\omega, k) = 1 + \sum_\alpha \frac{\omega_{\mathrm{p}\alpha}^2}{k^2 v_{\mathrm{T}\alpha}^2}\left[1 - I_+\left(\frac{\omega}{k v_{\mathrm{T}\alpha}}\right)\right] = 0\,. \tag{4.2.1}$$

The transcendental equation (4.2.1) has infinitely many complex solutions $\omega(k)$. We consider the most important of them, only, which correspond to weakly damped oscillations.

4.2.1 High-Frequency Plasma Waves

In the *fast wave range* where the phase velocity exceeds the thermal velocities

$$\frac{\omega}{k} \gg v_{\mathrm{Te}}, v_{\mathrm{Ti}} \tag{4.2.2}$$

we can use the asymptotic representation (4.1.16) for $I_+(x)$ and write the dispersion equation for the weakly damped waves $\mathrm{Re}\{\omega\} \gg \mathrm{Im}\{\omega\}$ in the form:

$$\begin{aligned}\varepsilon^{\mathrm{lo}}(\omega, k) \approx 1 &- \frac{\omega_{\mathrm{pe}}^2}{\omega^2}\left(1 + 3\frac{k^2 v_{\mathrm{Te}}^2}{\omega^2}\right)\\ &+ \mathrm{i}\sqrt{\frac{\pi}{2}}\frac{\omega\omega_{\mathrm{pe}}^2}{k^3 v_{\mathrm{Te}}^3}\exp\left(-\frac{\omega^2}{2k^2 v_{\mathrm{Te}}^2}\right) = 0\,.\end{aligned} \tag{4.2.3}$$

Here, we neglect the contribution of the ions since it is significant only if $T_{\mathrm{i}} > T_{\mathrm{e}}M^2/m^2$ holds, i.e., if the ion temperature is more than six orders of magnitude higher than the electron temperature. Plasmas with such a ratio of

T_i/T_e do not exist in nature, which follows from the values of the electron and the ion temperatures given for different plasmas in Chap. 1. Therefore, in the range of frequencies (4.2.2) such a plasma can be regarded as a pure electron plasma. The ions contribute only to neutralizing the electron charge (quasi-neutrality of the ground state).

Since $\omega \gg kv_{Te}$, the imaginary term in (4.2.3) is exponentially small, and to solve the dispersion equation we can apply the method described in Sect. 2.6 valid for weakly damped oscillations. We denote the imaginary part of the wave frequency by δ ($\omega \to \omega + i\delta$). The real part of ω is obtained from

$$\mathrm{Re}\{\varepsilon^{lo}(\omega, k)\} = 1 - \frac{\omega_{pe}^2}{\omega^2}\left(1 + \frac{3k^2 v_{Te}^2}{\omega^2}\right) = 0 , \tag{4.2.4}$$

which yields

$$\omega^2 = \omega_{pe}^2 (1 + 3k^2 r_{De}^2) . \tag{4.2.5}$$

We have taken into account that $\omega \approx \omega_{pe}$ and substituted ω_{pe} for ω in small terms.

Due to the condition (4.2.2), i.e., $k^2 r_{De}^2 \ll 1$, the second term in (4.2.5) is a small correction. Thus, weakly damped longitudinal oscillations with high phase velocity have a wavelength $\lambda \sim 1/k \gg r_{De}$. The longitudinal waves with the dispersion relation (4.2.5) are called *electron plasma waves* (electron Langmuir waves) or simply plasma waves. They form the *high frequency* ($\omega > \omega_{pe}$) *branch* of longitudinal oscillations in the isotropic plasma, which is also called the electron branch, since the ion contribution is negligible.

4.2.2 Landau Damping

The damping decrement δ of the plasma waves according to (2.6.5) follows in terms of $\mathrm{Im}\{\varepsilon^{lo}(\omega, k)\}$ when it is small against the real part:

$$\delta = \frac{\mathrm{Im}\{\varepsilon^{lo}(\omega, k)\}}{\partial \mathrm{Re}\{\varepsilon^{lo}(\omega, k)\}/\partial\omega} \approx -\sqrt{\frac{\pi}{8}}\frac{\omega_{pe}}{k^3 r_{De}^3}\exp\left(-\frac{1}{2k^2 r_{De}^2} - \frac{3}{2}\right) . \tag{4.2.6}$$

Hence, if $kr_{De} \ll 1$, the damping of the plasma waves is exponentially small. Actually, the phase velocities of the plasma waves exceed the thermal velocity of the electrons. Therefore, only fast electrons can absorb the plasma wave (Fig. 4.2). For a Maxwellian distribution the number of fast electrons in its tail decreases exponentially. This weak damping of plasma waves is called *Landau damping*.

The damping decrement (4.2.6) of plasma waves grows with increasing k, and $\delta \approx \omega$ for $kr_{De} \approx 1$. This expression for δ, however, cannot be used when it is large since we assumed $kr_{De} \ll 1$. It may only illustrate the tendency of

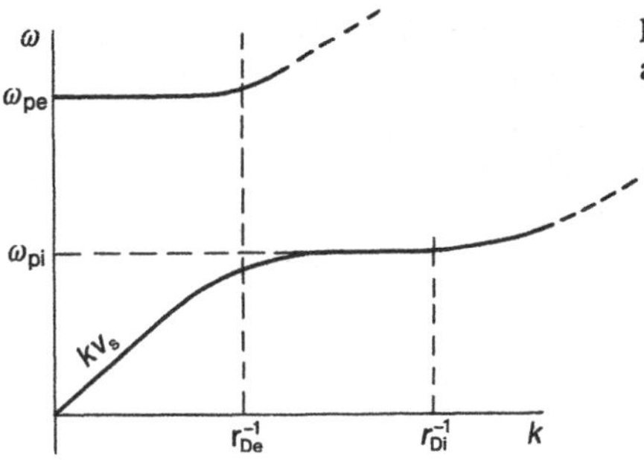

Fig. 4.3. Spectra of longitudinal oscillations in a nondegenerate isotropic plasma

growth with decreasing wavelength. The limiting case of strongly damped oscillations with $kr_{\mathrm{De}} \gg 1$ is treated in Exercise 4.7.3. The branch of longitudinal high-frequency plasma oscillations is shown in Fig. 4.3 (upper curve). The strongly damped part of the spectrum in the range $kr_{\mathrm{De}} > 1$ is shown by a dashed line.

4.2.3 Ion-Acoustic Waves in a Nonisothermal Plasma

In the *intermediate wave range* where the phase velocity satisfies the inequality

$$v_{\mathrm{Ti}} \ll \frac{\omega}{k} \ll v_{\mathrm{Te}} \,, \tag{4.2.7}$$

(4.2.1) takes the form

$$\begin{aligned} \varepsilon^{\mathrm{lo}}(\omega, k) &= 1 + \frac{\omega_{\mathrm{pe}}^2}{k^2 v_{\mathrm{Te}}^2}\left(1 + \mathrm{i}\sqrt{\frac{\pi}{2}}\,\frac{\omega}{k v_{\mathrm{Te}}}\right) \\ &\quad - \frac{\omega_{\mathrm{pi}}^2}{\omega^2}\left[1 + 3\,\frac{k^2 v_{\mathrm{Ti}}^2}{\omega^2} - \mathrm{i}\sqrt{\frac{\pi}{2}}\,\frac{\omega^3}{k^3 v_{\mathrm{Ti}}^3}\,\exp\left(-\frac{\omega^2}{2k^2 v_{\mathrm{Ti}}^2}\right)\right] = 0 \,, \end{aligned} \tag{4.2.8}$$

if we again assume weak damping, i.e. $\mathrm{Re}\{\omega\} \gg \mathrm{Im}\{\omega\}$. As the imaginary terms are small compared to the real terms we can easily apply the approximation (2.6.4, 5), which should be familiar, now. The resulting frequencies $\omega(k)$ and damping decrements $\delta(k)$ of the longitudinal waves in the intermediate range of phase velocities are

$$\begin{aligned} \omega^2 &= \omega_{\mathrm{pi}}^2\left[1 + 3k^2 r_{\mathrm{Di}}^2\left(1 + \frac{1}{k^2 r_{\mathrm{De}}^2}\right)\right]\left(1 + \frac{1}{k^2 r_{\mathrm{De}}^2}\right)^{-1} , \\ \delta &= -\sqrt{\frac{\pi}{8}}\,\frac{M\omega^4}{mZk^3 v_{\mathrm{Te}}^3}\left[1 + Z\sqrt{\frac{M}{m}}\left(\frac{T_{\mathrm{e}}}{T_{\mathrm{i}}}\right)^{3/2}\exp\left(-\frac{\omega^2}{2k^2 v_{\mathrm{Ti}}^2}\right)\right] . \end{aligned} \tag{4.2.9}$$

Here we account for the quasi-neutrality condition $N_e = ZN_i$, where the ion charge number $Z = |e_i/e|$. From the condition $\omega \gg kv_{Ti}$ it follows immediately that the mode (4.2.9) can exist only in the nonisothermal plasma ($T_e \gg T_i$) and in the range of wavelengths $k^2r_{Di}^2 \ll 1$. The spectrum of these oscillations is shown in Fig. 4.3 (lower curve). Since the frequency spectrum considerably depends on the ion plasma component, it is often called the *ion* or *low-frequency branch* of longitudinal oscillations in the plasma.

In the limit $kr_{De} \ll 1$ the spectrum (4.2.9) takes the especially simple form of the so-called ion-acoustic oscillations:

$$\omega^2 = k^2 Z \frac{T_e}{M}\left(1 + 3\frac{T_i}{ZT_e}\right) = k^2 v_s^2\,,$$
$$\delta = -\sqrt{\frac{\pi Zm}{8M}}\,\omega\left[1 + Z\sqrt{\frac{M}{m}}\left(\frac{T_e}{T_i}\right)^{3/2}\exp\left(-\frac{3}{2} - \frac{ZT_e}{2T_i}\right)\right]. \tag{4.2.10}$$

These oscillations of the nonisothermal plasma are called *ion-acoustic* because of the analogy between the relation (4.2.10) and the dispersion of ordinary acoustic oscillations in fluids. By analogy the phase velocity

$$v_s = \left[Z\frac{T_e}{M}\left(1 + \frac{3T_i}{ZT_e}\right)\right]^{1/2}$$

is called ion sound velocity.

From the condition $|\delta| \ll \omega$ we get a limit for the temperature ratio of the plasma components. One can easily see that

$$(T_e/T_i)^{3/2}\exp\left(-\frac{3}{2} - \frac{ZT_e}{2T_i}\right) \ll 1$$

must hold. Assuming $Z = 1$ we obtain $T_e/T_i > 6$.

In the range of short wavelengths when $k^2r_{De}^2 \gg 1$ but $k^2r_{Di}^2 \ll 1$, the spectrum (4.2.9) can be approximated

$$\omega^2 \approx \omega_{pi}^2\,,$$
$$\delta = -\sqrt{\frac{\pi Zm}{8M}}\frac{\omega_{pi}}{k^3r_{De}^3}\left[1 + Z\sqrt{\frac{M}{m}}\left(\frac{T_e}{T_i}\right)^{3/2}\exp\left(-\frac{\omega_{pi}^2}{2k^2v_{Ti}^2} - \frac{3}{2}\right)\right]. \tag{4.2.11}$$

The analysis of the decrement (4.2.11) shows that these oscillations are weakly damped for a highly nonisothermal plasma. They correspond to the horizontal part of the lower curve in Fig. 4.3. Physically, the emergence of oscillations in the nonisothermal plasma at the ion Langmuir frequency is quite natural. Under these conditions, the freely and fastly moving electrons produce a constant negative space charge. On this background the low-frequency ion oscillations can exist analogously to the electron Langmuir oscillations which need the positive charge background of the ions for existence.

4.2.4 The Low-Frequency Range, Debye Screening

In the range of *very slow low-frequency longitudinal waves*

$$\frac{\omega}{k} \ll v_{\mathrm{Ti}}, v_{\mathrm{Te}} , \tag{4.2.12}$$

according to (4.1.14), the longitudinal plasma permittivity goes over to the static limit:

$$\varepsilon^{\mathrm{lo}}(\omega, k) \approx \varepsilon^{\mathrm{lo}}(0, k) = 1 + \sum_{\alpha} \frac{\omega_{\mathrm{p}\alpha}^2}{k^2 v_{\mathrm{T}\alpha}^2} = 1 + \frac{1}{k^2 r_{\mathrm{D}}^2} , \quad \text{where} \tag{4.2.13}$$

$$r_{\mathrm{D}} = \left(\sum_{\alpha} r_{\mathrm{D}\alpha}^{-2}\right)^{-1/2}$$

is the Debye length of the nondegenerate plasma. In this frequency range the longitudinal field is screened within a radius equal to the Debye length (Exercise 2.8.2).

An analogous screening occurs when $kv_{\mathrm{Ti}} \ll \omega \ll kv_{\mathrm{Te}}$ in the frequency range $\omega > \omega_{\mathrm{pi}}$. Here the longitudinal permittivity (4.1.14) also does not depend on the frequency; it has the form of (4.2.12) but with the screening radius $r_{\mathrm{scr}} = r_{\mathrm{De}}$. This corresponds to the screening of a high-frequency field in the frequency range $\omega_{\mathrm{pi}} < \omega < \omega_{\mathrm{pe}}$.

4.3 Longitudinal Oscillations in the Collisionless Degenerate Plasma

As already mentioned, the electron plasma of metals and the electron-hole plasma of semiconductors are degenerate at low temperatures. According to (2.4.6 and 18), the dispersion equation for longitudinal oscillations in an isotropic degenerate plasma has the form

$$\varepsilon^{\mathrm{lo}}(\omega, k) = 1 + \sum_{\alpha} \frac{3\omega_{\mathrm{p}\alpha}^2}{k^2 v_{\mathrm{F}\alpha}^2} \left(1 - \frac{\omega}{2kv_{\mathrm{F}\alpha}} \ln \frac{\omega + kv_{\mathrm{F}\alpha}}{\omega - kv_{\mathrm{F}\alpha}}\right) = 0 . \tag{4.3.1}$$

4.3.1 High-Frequency Plasma Waves and Zero-Point Sound

In *the range of fast waves*, when the phase velocity exceeds the Fermi velocities of the electrons and the ions (holes)

$$\frac{\omega}{k} \gg v_{\mathrm{Fe}}\,,\, v_{\mathrm{Fi}}\,. \tag{4.3.2}$$

Equation (4.3.1) takes the form

$$\varepsilon^{\mathrm{lo}}(\omega, k) = 1 - \frac{\omega_{\mathrm{pe}}^2}{\omega^2}\left(1 + \frac{3}{5}\frac{k^2 v_{\mathrm{Fe}}^2}{\omega^2}\right) = 0\,. \tag{4.3.3}$$

Then the dispersion relation is

$$\omega^2 = \omega_{\mathrm{pe}}^2\left(1 + \frac{1}{5}\,k^2 r_{\mathrm{De}}^2\right), \tag{4.3.4}$$

where $r_{\mathrm{De}}^2 = 3\,v_{\mathrm{Fe}}^2/\omega_{\mathrm{pe}}^2$. This dispersion law is similar to that of the Langmuir oscillations of the nondegenerate plasma (4.2.5) and is limited to $k^2 r_{\mathrm{De}}^2 \ll 1$, too.

There is, however, an essential difference between the electron oscillations in the degenerate plasma and the Langmuir oscillations in the nondegenerate plasma. The high-frequency electron oscillations of the degenerate plasma remain undamped in the absence of particle collisions, whereas the Langmuir oscillations of the nondegenerate plasma extinguish by Landau damping though only exponentially weakly, as follows from (4.2.6). The reason is that according to the Fermi distribution there are no particles with velocities greater than the Fermi velocity which could contribute to the absorption.

In the limit $k^2 r_{\mathrm{De}}^2 \gg 1$ we obtain from (4.3.1)

$$\omega = k v_{\mathrm{Fe}}\left[1 + 2\exp\left(-\frac{2}{9}\,k^2 r_{\mathrm{De}}^2 - 2\right)\right]. \tag{4.3.5}$$

These oscillations, known as "zero-point sound", are the continuation of the Langmuir waves (4.3.4) into the range of short wavelengths (see Fig. 4.4, upper branch).

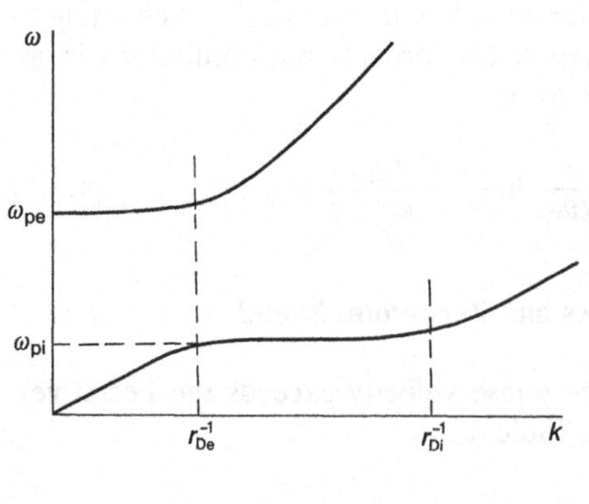

Fig. 4.4. Spectra of longitudinal oscillations in a degenerate isotropic plasma

4.3.2 Ion-Acoustic Waves in Degenerate Plasma

In *the range of intermediate phase velocities* $v_{\mathrm{Fi}} \ll \omega/k \ll v_{\mathrm{Fe}}$ Eq. (4.3.1) for the degenerate plasma is

$$\varepsilon^{\mathrm{lo}}(\omega, k) = 1 + \frac{3\,\omega_{\mathrm{pe}}^2}{k^2 v_{\mathrm{Fe}}^2}\left(1 + \mathrm{i}\,\frac{\pi\omega}{2\,k v_{\mathrm{Fe}}}\right) - \frac{\omega_{\mathrm{pi}}^2}{\omega^2} = 0\,. \tag{4.3.6}$$

In this case we obtain the following frequency and damping decrement

$$\omega^2 = \frac{\omega_{\mathrm{pi}}^2}{1 + 3\,\omega_{\mathrm{pe}}^2 k^{-2} r_{\mathrm{De}}^{-2}}\,, \qquad \delta = -\frac{3\,\pi M}{4\,Zm}\,\frac{\omega^4}{k^3 v_{\mathrm{Fe}}^3}\,. \tag{4.3.7}$$

Here the electrons play an active part in the absorption of oscillations since their random velocities greatly exceed the phase velocity. The ion velocities, however, are too small, so that the damping decrement is determined by the electrons alone.

These low-frequency oscillations, in analogy with the ion-acoustic oscillations (4.2.9) of the nondegenerate plasma, might be called ion-acoustic oscillations of the degenerate plasma, all the more as they exist not only in the completely degenerate plasma but also in the partially degenerate plasma where the electrons are degenerate but the ions nondegenerate. Here the frequency spectrum (4.3.7) is not modified. The damping decrement, however, changes since the plasma ions which are Maxwellian distributed now contribute to the absorption:

$$\delta = -\frac{3\,\pi M}{4\,Zm}\,\frac{\omega^4}{k^3 v_{\mathrm{Fe}}^3} - \sqrt{\frac{\pi}{8}}\,\frac{\omega^4}{k^3 v_{\mathrm{Ti}}^3}\exp\left(-\frac{\omega^2}{2\,k^2 v_{\mathrm{Ti}}^2}\right). \tag{4.3.8}$$

Note that in the completely degenerate electron-ion plasma ion-acoustic oscillations exist even for phase velocities $\omega/k \to v_{\mathrm{Fi}}$. In contrast, for the partially degenerate plasma these oscillations are possible in the limit $k v_{\mathrm{Fi}} \gg \omega_{\mathrm{pi}}$, only. In this limit their spectrum is given by

$$\omega = k v_{\mathrm{Fi}}\left[1 + 2\left(1 - \frac{\mathrm{i}\pi M}{m}\,\frac{v_{\mathrm{Fi}}^3}{v_{\mathrm{Fe}}^3}\right)\exp\left(-\frac{2\,k^2 v_{\mathrm{Fi}}^2}{3\,\omega_{\mathrm{pi}}^2} - 2\right)\right]. \tag{4.3.9}$$

We show the dispersion law of the low-frequency longitudinal waves in the degenerate plasma in Fig. 4.4 (lower branch).

4.3.3 Debye Screening in Degenerate Plasma

Finally, in *the range of very low frequencies* $\omega \ll k v_{\mathrm{Fe,\,i}}$, there remains the static screening of longitudinal fields since

$$\varepsilon^{\mathrm{lo}}(\omega, k) = 1 + \sum_{\alpha} \frac{3\,\omega_{\mathrm{p}\alpha}^2}{k^2 v_{\mathrm{F}\alpha}^2} = 1 + \frac{1}{k^2 r_{\mathrm{D}}^2} . \tag{4.3.10}$$

Here $r_{\mathrm{D}} = \sum_{\alpha} (3\,\omega_{\mathrm{p}\alpha}^2 / v_{\mathrm{F}\alpha}^2)^{-1/2}$ is the Debye length of the degenerate plasma which defines the screening distance for low-frequency fields.

A similar screening like in the nondegenerate plasma occurs for $v_{\mathrm{Fi}} \ll \omega/k \ll v_{\mathrm{Fe}}$ in the frequency range $\omega_{\mathrm{pe}} > \omega > \omega_{\mathrm{pi}}$. The screening radius equals r_{De} in this case.

4.4 Transverse Waves in Collisionless Isotropic Plasmas

Transverse electromagnetic waves in a homogeneous plasma are described by the dispersion equation (2.4.6)

$$k^2 - \frac{\omega^2}{c^2}\,\varepsilon^{\mathrm{tr}}(\omega, k) = 0 . \tag{4.4.1}$$

Here the transverse permittivity (4.1.14) of the nondegenerate plasma or (4.1.18) in the degenerate case has to be substituted.

As in the previous sections, we distinguish the limits of large and small phase velocities of the waves also called the limits of high and low frequencies.

4.4.1 Transverse Electromagnetic Waves

In *the range of fast waves*, when $\omega/k \gg v_{\mathrm{Te}}$ for the nondegenerate plasma and $\omega/k \gg v_{\mathrm{Fe}}$, respectively, for the degenerate case we substitute (4.1.14) or (4.1.18) into (4.4.1) yielding in this limit the approximate dispersion equation

$$k^2 - \frac{\omega^2}{c^2}\left(1 - \frac{\omega_{\mathrm{pe}}^2}{\omega^2}\right) = 0 . \tag{4.4.2}$$

Thus, the transverse wave spectrum of the isotropic plasma does not depend on the degree of degeneracy and is

$$\omega^2 = k^2 c^2 + \omega_{\mathrm{pe}}^2 . \tag{4.4.3}$$

In this frequency range the ion terms can be neglected and the plasma may be regarded as a pure electron plasma. Incidentally, the frequency spectrum of the transverse waves cannot depend on the thermal motion of the plasma particles, because the phase velocity of the wave is greater than the velocity of light and consequently considerably exceeds the thermal velocity

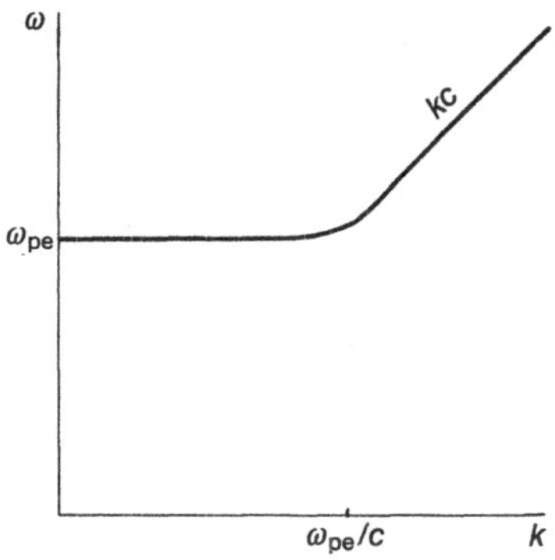

Fig. 4.5. Spectra of transverse waves in an isotropic plasma

of the plasma particles. (Note that we are studying the nonrelativistic equilibrium plasma.) As a result the absorption of these waves by plasma particles is impossible when collisions are neglected. Strictly speaking, (4.1.14) has a nonvanishing imaginary part at phase velocities greater than the velocity of light. This, however, results from the use of the Maxwellian distribution which is inaccurate for particle velocities $v \gtrsim c$. Therefore, this improper imaginary part cannot be applied. In the degenerate plasma, because of the low velocities, this problem does not exist.

Thus, fast high-frequency transverse waves are undamped in the collisionless isotropic plasma. We show their frequency spectrum in Fig. 4.5.

4.4.2 The Anomalous Skin-Effect

In *the range of transverse waves with small phase velocities*, $\omega \ll kv_{Te}$ for the nondegenerate plasma, and $\omega \ll kv_{Fe}$ for the degenerate. The contribution of the ions to the plasma dielectric permittivity may be practically always neglected, irrespective of the ratio between the phase velocity of the wave and the thermal velocity of the ions. For the nondegenerate plasma we get

$$\varepsilon^{tr}(\omega, k) = 1 + i\sqrt{\frac{\pi}{2}}\frac{\omega_{pe}^2}{\omega k v_{Te}} . \tag{4.4.4}$$

Substituting (4.4.4) into (4.4.1), we obtain the purely imaginary frequency

$$\omega = -i\sqrt{\frac{2}{\pi}}\frac{k^3 c^2 v_{Te}}{\omega_{pe}^2} . \tag{4.4.5}$$

Due to their strong collisionless absorption by the plasma electrons these low frequency transverse oscillations decay aperiodically. They should not be called waves in this frequency range. With respect to the boundary value problem, one usually speaks of fast absorption of the low-frequency field in the plasma. Solving (4.4.5) for $k(\omega)$, we obtain the space scale of damping, the so-called *penetration depth*

$$\lambda_{sk} = \frac{1}{\operatorname{Im}\{k\}} = 2\left(\frac{2}{\pi}\right)^{1/6}\left(\frac{c^2 v_{Te}}{\omega\,\omega_{pe}^2}\right)^{1/3}. \tag{4.4.6}$$

This penetration depth describes the shielding of a transverse field by the skin-effect, associated with energy dissipation due to Cherenkov absorption. We see that the frequency dependency of the collisionless penetration depth $\lambda_{sk} \sim \omega^{-1/3}$ differs from the skin-depth $\lambda_{sk} \sim \omega^{-1/2}$ due to particle collisions (Sect. 4.5). Therefore the collisionless *skin-effect* is called *anomalous*.

It is clear that in the degenerate plasma case where

$$\varepsilon^{tr}(\omega, k) \approx 1 + i\,\frac{3\pi\omega_{pe}^2}{4\,\omega k v_{Fe}} \tag{4.4.7}$$

an analogous skin-effect exists for the low-frequency transverse field. Substituting this expression into (4.4.1) we arrive at the aperiodically damped waves:

$$\omega = -i\,\frac{4}{3\pi}\,\frac{k^3 c^2 v_{Fe}}{\omega_{pe}^2}. \tag{4.4.8}$$

We obtain for the degenerate plasma the penetration depth

$$\lambda_{sk} = \frac{1}{\operatorname{Im}\{k\}} = 2\left(\frac{4}{3\pi}\,\frac{c^2 v_{Fe}}{\omega_{pe}^2\,\omega}\right)^{1/3}. \tag{4.4.9}$$

It should not be surprising that (4.4.7–9) differ from (4.4.4–6) by the substitution of v_{Te} for v_{Fe}. It follows from (4.4.6, 9) for the thickness of the skin-layer that $\lambda_{sk} \to \infty$ when $\omega \to 0$. The low-frequency transverse field can penetrate the collisionless plasma without limitation. The longitudinal field, however, is screened in the static limit ($\omega \to 0$), as shown above, the penetration depth being the Debye length.

4.5 Dielectric Permittivity and Oscillation Spectra of Weakly Ionized Plasmas with Account of Particle Collisions

Till now we have completely neglected particle collisions. Taking them into account allows firstly to determine the limits of applicability of the collisionless plasma approximation, used above. Secondly, the inclusion of ordinary collisional mechanisms of energy dissipation such as friction (i.e., momentum transfer from one particle to another), thermal conductivity, diffusion, and viscosity becomes possible. Finally, we can justify the stationary equilibrium distribution functions of charged particles, the Maxwellian or Fermian distribution, respectively, only in the presence of particle collisions.

We shall begin our analysis of collisional effects with the investigation of a weakly ionized nondegenerate plasma where elastic collisions of charged particles with neutrals dominate and the collision integral can be approximated by the BGK term (3.5.2). The neglect of collisions between charged particles simplifies the analysis and facilitates the study of the more complex case of a completely ionized plasma.

In a weakly ionized plasma the kinetic equation for the charged particles of type α with the BGK collision term is written in the form (Sect. 3.5)

$$\frac{\partial f_\alpha}{\partial t} + \boldsymbol{v}\frac{\partial f_\alpha}{\partial \boldsymbol{r}} + e_\alpha\left(\boldsymbol{E} + \frac{1}{c}[\boldsymbol{v},\boldsymbol{B}]\right)\frac{\partial f_\alpha}{\partial \boldsymbol{p}} = -\nu_{\alpha n}(f_\alpha - N_\alpha \Phi_{\alpha n}), \tag{4.5.1}$$

where $\nu_{\alpha n}$, the collision frequency with neutrals, is considered constant and

$$\begin{aligned} N_\alpha &= \int d\boldsymbol{p} f_\alpha, \qquad \Phi_{\alpha n} = \frac{1}{(2\pi m_\alpha T_{\alpha n})^{3/2}}\exp\left(-\frac{m_\alpha v^2}{2T_{\alpha n}}\right), \\ T_{\alpha n} &= \frac{m_\alpha T_n + M_n T_\alpha}{m_\alpha + M_n}. \end{aligned} \tag{4.5.2}$$

Note that the index "α" (α = e, i) corresponds to charged particles and "n" to neutrals. We further simplify by assuming equal masses and temperatures of the neutrals and the ions: $m_i = M_n = M$, $T_i = T_n$. This is the case if the plasma ions are generated by ionization of neutral atoms of the same substance. Neglecting terms of order m/M we have $T_{en} = T_e$. Using $T_{\alpha n} = T_\alpha$ in (4.5.2) the function $\Phi_{\alpha n}$ coincides with the Maxwellian normalized by one

$$\Phi_{\alpha n} = \frac{1}{(2\pi m_\alpha T_\alpha)^{3/2}}\exp\left(-\frac{m_\alpha v^2}{2T_\alpha}\right). \tag{4.5.3}$$

In the stationary state without external fields

$$f_{0\alpha} = N_{0\alpha}\Phi_{\alpha n}$$

is the only solution of (4.5.1). To investigate a small perturbation δf_α of the equilibrium caused by small fields $\boldsymbol{E}$ and $\boldsymbol{B}$, we can linearize (4.5.1) and obtain

$$\frac{\partial \delta f_\alpha}{\partial t} + \boldsymbol{v}\frac{\partial \delta f_\alpha}{\partial \boldsymbol{r}} + e_\alpha \boldsymbol{E}\frac{\partial f_{0\alpha}}{\partial \boldsymbol{p}} = -\nu_{\alpha n}\left(\delta f_\alpha - \Phi_{\alpha n}\int d\boldsymbol{p}\,\delta f_\alpha\right). \tag{4.5.4}$$

The solution of this linear kinetic equation for plane monochromatic waves $[\boldsymbol{E}, \delta f_\alpha \sim \exp(-\mathrm{i}\omega t + \mathrm{i}\boldsymbol{k}\cdot\boldsymbol{r})]$ can be written as

$$\delta f_\alpha = \mathrm{i}\frac{e_\alpha}{T_\alpha}\frac{(\boldsymbol{v}\cdot\boldsymbol{E})f_{0\alpha}}{\omega + \mathrm{i}\nu_{\alpha n} - \boldsymbol{k}\cdot\boldsymbol{v}} + \mathrm{i}\frac{\nu_{\alpha n}\eta_\alpha f_{0\alpha}}{\omega + \mathrm{i}\nu_{\alpha n} - \boldsymbol{k}\cdot\boldsymbol{v}}, \quad \text{where} \tag{4.5.5}$$

$$\eta_\alpha = \frac{1}{N_{0\alpha}}\int d\boldsymbol{p}\,\delta f_\alpha \tag{4.5.6}$$

is the perturbed particle density normalized by the equilibrium density.

Integrating (4.5.5) over the momentum or using the continuity equation for the particle species α we obtain for the density perturbation

$$\eta_\alpha = \frac{\boldsymbol{k}\cdot\boldsymbol{j}_\alpha}{e_\alpha N_{0\alpha}\omega}, \quad \boldsymbol{j}_\alpha = e_\alpha\int d\boldsymbol{p}\,\boldsymbol{v}\delta f_\alpha\,. \tag{4.5.7}$$

From (4.5.5, 7) we can determine the current density $\boldsymbol{j}_\alpha$ in terms of $\boldsymbol{E}$. By calculations similar to those in Sect. 4.1, we finally obtain for the longitudinal and transverse permittivities

$$\varepsilon^{\mathrm{lo}}(\omega,k) = 1 + \sum_\alpha \frac{\omega_{\mathrm{p}\alpha}^2}{k^2 v_{\mathrm{T}\alpha}^2}\frac{1 - I_+\left(\dfrac{\omega + \mathrm{i}\nu_{\alpha n}}{k v_{\mathrm{T}\alpha}}\right)}{1 - \dfrac{\mathrm{i}\nu_{\alpha n}}{\omega + \mathrm{i}\nu_{\alpha n}} I_+\left(\dfrac{\omega + \mathrm{i}\nu_{\alpha n}}{k v_{\mathrm{T}\alpha}}\right)},$$

$$\varepsilon^{\mathrm{tr}}(\omega,k) = 1 - \sum_\alpha \frac{\omega_{\mathrm{p}\alpha}^2}{\omega(\omega + \mathrm{i}\nu_{\alpha n})} I_+\left(\frac{\omega + \mathrm{i}\nu_{\alpha n}}{k v_{\mathrm{T}\alpha}}\right), \tag{4.5.8}$$

where the summation extends over the charged particle species only.

Before we come to the evaluation of the oscillation spectra in the isotropic plasma, taking account of particle collisions, we should note first that the static ($\omega \to 0$) longitudinal dielectric permittivity as in the collisionless plasma has the form, see (4.2.13),

$$\varepsilon^{\mathrm{lo}}(0,k) = 1 + \sum_\alpha \frac{\omega_{\mathrm{p}\alpha}^2}{k^2 v_{\mathrm{T}\alpha}^2} = 1 + \frac{1}{k^2 r_{\mathrm{D}}^2}, \tag{4.5.9}$$

leading to the Debye screening of longitudinal fields in the plasma. Therefore, particle collisions do not directly influence the electrostatic field in the plasma.

4.5.1 Collisional Damping of Longitudinal Waves

In the high-frequency range, for $\omega \gg k v_{\mathrm{T}\alpha}, \nu_{\alpha n}$, collisional effects are important, however. We have shown in Sect. 4.2 that the ion terms do not contribute to the collisionless longitudinal high-frequency oscillations. This remains valid when particle collisions are accounted for, but the dispersion equation is slightly modified:

$$\varepsilon^{\mathrm{lo}}(\omega, k) = 1 - \frac{\omega_{\mathrm{pe}}^2}{\omega^2}\left(1 + 3\frac{k^2 v_{\mathrm{Te}}^2}{\omega^2}\right)$$

$$+ \mathrm{i}\sqrt{\frac{\pi}{2}}\frac{\omega\,\omega_{\mathrm{pe}}^2}{k^3 v_{\mathrm{Te}}^3}\exp\left(-\frac{\omega^2}{2k^2 v_{\mathrm{Te}}^2}\right) + \mathrm{i}\frac{\omega_{\mathrm{pe}}^2\,\nu_{\mathrm{en}}}{\omega^3} = 0\,. \qquad (4.5.10)$$

Compared with (4.2.3) an additional imaginary term appears. The collisionless Cherenkov wave absorption is supplemented by the collisional energy dissipation of the field due to the transfer of electron momenta to neutral particles by collisions (electron friction). Within the BGK model collisionless and collisional dissipation mechanisms lead to additive contributions to the dispersion equation of longitudinal plasma oscillations. The nondissipative part of the dispersion equation and, consequently, the frequency spectrum of longitudinal oscillations (4.2.5) remain unchanged. In the expression for the damping decrement (4.2.6), however, there appears a collisional correction

$$\Delta\delta = -\nu_{\mathrm{en}}/2\,, \qquad (4.5.11)$$

which exceeds the collisionless Landau damping in the range of long wavelengths when

$$\nu_{\mathrm{en}} > \sqrt{\frac{\pi}{2}}\frac{\omega_{\mathrm{pe}}}{k^3 r_{\mathrm{De}}^3}\exp\left(-\frac{3}{2} - \frac{1}{2k^2 r_{\mathrm{De}}^2}\right). \qquad (4.5.12)$$

In the opposite limit of short wavelengths the collisional damping is negligibly small as compared to the collisionless Landau damping.

Next, we investigate the collisional modification of the low-frequency (slow) ion-acoustic oscillations which as in the isotropic collisionless plasma exist only in the strongly nonisothermal case $T_{\mathrm{e}} \gg T_{\mathrm{i}}$ with phase velocities in the range $v_{\mathrm{Ti}} \ll \omega/k \ll v_{\mathrm{Te}}$. When particle collisions are not too frequent, under the condition $\omega \gg \nu_{\mathrm{in}}$ and $|\omega + \mathrm{i}\nu_{\mathrm{e}}| \ll k v_{\mathrm{Te}}$ we obtain from (4.5.8) in the given range of phase velocities

$$\varepsilon^{\mathrm{lo}}(\omega, k) = 1 + \frac{\omega_{\mathrm{pe}}^2}{k^2 v_{\mathrm{Te}}^2}\left(1 + \mathrm{i}\sqrt{\frac{\pi}{2}}\frac{\omega}{k v_{\mathrm{Te}}}\right) - \frac{\omega_{\mathrm{pi}}^2}{\omega^2}\left(1 + 3\frac{k^2 v_{\mathrm{Ti}}^2}{\omega^2}\right)$$

$$+ \mathrm{i}\sqrt{\frac{\pi}{2}}\frac{\omega_{\mathrm{pi}}^2\,\omega}{k^3 v_{\mathrm{Ti}}^3}\exp\left(-\frac{\omega^2}{2k^2 v_{\mathrm{Ti}}^2}\right) + \mathrm{i}\frac{\omega_{\mathrm{pi}}^2\,\nu_{\mathrm{in}}}{\omega^3} = 0\,. \qquad (4.5.13)$$

The collisional correction of (4.2.8) consists in the last term which incorporates ion-neutral collisions (ion friction). The resulting correction of the damping decrement of ion-acoustic waves (4.2.9) is

$$\Delta\delta = -\nu_{\mathrm{in}}/2\,. \qquad (4.5.14)$$

Note that in the low-frequency range electron collisions do not contribute to the dissipation of longitudinal oscillations. This follows from the inequality $\nu_{en} \ll k v_{Te}$, according to which the wavelength is significantly smaller than the mean free path of the electrons. As a result, the electrons take energy from the wave dominantly by the Cherenkov absorption mechanism.

4.5.2 Damping of Transverse Waves

Finally, we consider the effect of particle collisions on transverse electromagnetic waves in weakly ionized plasmas. We have shown in the previous section that the high-frequency transverse waves with the dispersion law (4.5.3) which have phase velocities higher than the velocity of light are not absorbed in the collisionless limit. With account of infrequent collisions, for $\omega \gg \nu_{en}$, the transverse dielectric permittivity takes the form

$$\varepsilon^{tr}(\omega, k) = 1 - \frac{\omega_{pe}^2}{\omega^2}\left(1 - i\,\frac{\nu_{en}}{\omega}\right), \tag{4.5.15}$$

and the damping decrement due to electron friction becomes

$$\delta = -\frac{\nu_{en}}{2}\,\frac{\omega_{pe}^2}{\omega_{pe}^2 + k^2c^2}\,. \tag{4.5.16}$$

In the range of low frequencies (small phase velocities) the transverse waves are strongly damped in the collisionless plasma, already. The account of collisions does not change this situation.

4.5.3 Degenerate Plasma

There is no difficulty to study the oscillations of the weakly ionized degenerate plasma. We incorporate particle collisions by the model integral (3.5.6). Then the linearized kinetic equation for a small perturbation can be written as follows, see (4.5.4),

$$\frac{\partial \delta f_\alpha}{\partial t} + \boldsymbol{v}\cdot\frac{\partial \delta f_\alpha}{\partial \boldsymbol{r}} + e_\alpha(\boldsymbol{E}\cdot\boldsymbol{v})\,\frac{\partial f_{0\alpha}}{\partial \mathscr{E}} = -\nu_{\alpha n}\left(\delta f_\alpha + \frac{2\mathscr{E}_{F\alpha}}{3N_{0\alpha}}\int d\boldsymbol{p}\,\delta f_\alpha\right). \tag{4.5.17}$$

Using the "ansatz" $\delta f_\alpha \sim \exp(-i\omega t + i\boldsymbol{k}\cdot\boldsymbol{r})$ we have

$$\delta f_\alpha = -\frac{i}{\omega + i\nu_{\alpha n} - \boldsymbol{k}\cdot\boldsymbol{v}}\,\frac{\partial f_{0\alpha}}{\partial \mathscr{E}}\left(e_\alpha \boldsymbol{E}\cdot\boldsymbol{v} + \frac{2}{3}\,\nu_{\alpha n}\mathscr{E}_{F\alpha}\eta_\alpha\right). \tag{4.5.18}$$

By calculations similar to those given above for the longitudinal and transverse dielectric permittivities of the weakly ionized degenerate plasma we obtain

$$\varepsilon^{\mathrm{lo}}(\omega, k) = 1 + \sum_{\alpha} \frac{3\omega_{\mathrm{p}\alpha}^2}{k^2 v_{\mathrm{F}\alpha}^2}\left(1 - \frac{\omega + \mathrm{i}\nu_{\alpha n}}{2kv_{\mathrm{F}\alpha}} \ln \frac{\omega + \mathrm{i}\nu_{\alpha n} + kv_{\mathrm{F}\alpha}}{\omega + \mathrm{i}\nu_{\alpha n} - kv_{\mathrm{F}\alpha}}\right) \times \left(1 - \frac{\mathrm{i}\nu_{\alpha n}}{2kv_{\mathrm{F}\alpha}} \ln \frac{\omega + \mathrm{i}\nu_{\alpha n} + kv_{\mathrm{F}\alpha}}{\omega + \mathrm{i}\nu_{\alpha n} - kv_{\mathrm{F}\alpha}}\right)^{-1},$$

$$\varepsilon^{\mathrm{tr}}(\omega, k) = 1 - \sum_{\alpha} \frac{3\omega_{\mathrm{p}\alpha}^2}{2\omega(\omega + \mathrm{i}\nu_{\alpha n})}\left[1 + \left(\frac{(\omega + \mathrm{i}\nu_{\alpha n})^2}{k^2 v_{\mathrm{F}\alpha}^2} - 1\right)\right] \times \left(1 - \frac{\omega + \mathrm{i}\nu_{\alpha n}}{2kv_{\mathrm{F}\alpha}} \ln \frac{\omega + \mathrm{i}\nu_{\alpha n} + kv_{\mathrm{F}\alpha}}{\omega + \mathrm{i}\nu_{\alpha n} - kv_{\mathrm{F}\alpha}}\right). \quad (4.5.19)$$

We consider only the high-frequency limit, $|\omega + \mathrm{i}\nu_{\alpha n}| \gg kv_{\mathrm{F}\alpha}$, where the spatial dispersion of the dielectric tensor is extremely weak:

$$\varepsilon^{\mathrm{lo}}(\omega) = \varepsilon^{\mathrm{tr}}(\omega) = 1 - \frac{\omega_{\mathrm{pe}}^2}{\omega(\omega + \mathrm{i}\nu_{\mathrm{en}})}. \quad (4.5.20)$$

The resulting corrections to the damping decrement of transverse and longitudinal waves are the same as in the nondegenerate plasma, see (4.5.11, 16). In the degenerate plasma, however, there exists no damping of oscillations (both longitudinal and transverse) in the collisionless limit. The correction (4.5.14) is still valid for the damping decrement of ion-acoustic waves, which have in the collisionless limit the frequency spectrum (4.3.7).

4.6 Dielectric Permittivity and Oscillation Spectra of Fully Ionized Plasmas Taking Account of Particle Collisions

In fully ionized plasmas the collisions between charged particles prevail. First we shall analyze the oscillations of the nondegenerate plasma where the unperturbed distribution functions are Maxwellian. We admit different temperatures of the electron and ion components. We use the Landau collision integral to describe charged particle collisions. The Landau kinetic equation for the particles of type α is, see (3.4.7),

$$\frac{\partial f_\alpha}{\partial t} + \boldsymbol{v}\frac{\partial f_\alpha}{\partial \boldsymbol{r}} + e_\alpha\left(\boldsymbol{E} + \frac{1}{c}[\boldsymbol{v}, \boldsymbol{B}]\right)\frac{\partial f_\alpha}{\partial \boldsymbol{p}_\alpha} = \sum_\beta \left(\frac{\partial f_\alpha}{\partial t}\right)_{\mathrm{col}}^{\alpha\beta}$$

$$= \sum_\beta \frac{\partial}{\partial p_{\alpha i}} \int d\boldsymbol{p}_\beta I_{ij}^{\alpha\beta}\left[f_\beta(\boldsymbol{p}_\beta)\frac{\partial f_\alpha(\boldsymbol{p}_\alpha)}{\partial p_{\alpha j}} - f_\alpha(\boldsymbol{p}_\alpha)\frac{\partial f_\beta(\boldsymbol{p}_\beta)}{\partial p_{\beta j}}\right], \quad (4.6.1)$$

where

$$I_{ij}^{\alpha\beta} = 2\pi e_\alpha^2 e_\beta^2 \frac{u^2\delta_{ij} - u_i u_j}{u^3} L, \quad \boldsymbol{u} = \boldsymbol{v}_\alpha - \boldsymbol{v}_\beta.$$

In the stationary, homogeneous,. and field-free case the Maxwellian distribution functions with different temperatures T_e and T_i are not solutions of (4.6.1). The characteristic time of temperature equalization for the electron and ion components is of the order $M/(m\nu_{eff})$, however, and the plasma state with different temperature Maxwellian distributions may be regarded as a quasistationary equilibrium for processes occurring on a shorter time scale.

Small deviations from this equilibrium under the influence of small fields $\boldsymbol{E}$ and $\boldsymbol{B}$ lead to a perturbation δf_α which can be described by the linearization of (4.6.1):

$$\frac{\partial \delta f_\alpha}{\partial t} + \boldsymbol{v}\frac{\partial \delta f_\alpha}{\partial \boldsymbol{r}} + e_\alpha \boldsymbol{E}\frac{\partial f_{0\alpha}}{\partial \boldsymbol{p}_\alpha} = \sum_\beta \frac{\partial}{\partial p_{\alpha i}} \int d\boldsymbol{p}_\beta I_{ij}^{\alpha\beta}(\boldsymbol{p}_\alpha, \boldsymbol{p}_\beta) \left[\frac{\partial f_{0\alpha}}{\partial p_{\alpha j}} \delta f_\beta(\boldsymbol{p}_\beta) + \frac{\partial \delta f_\alpha}{\partial p_{\alpha j}} f_{0\beta}(\boldsymbol{p}_\beta) - f_{0\alpha}(\boldsymbol{p}_\alpha)\frac{\partial \delta f_\beta}{\partial p_{\beta j}} - \delta f_\alpha(\boldsymbol{p}_\alpha)\frac{\partial f_{0\beta}}{\partial p_{\beta j}}\right]. \tag{4.6.2}$$

It is impossible to solve this system of equations for the electrons and ions, which even after linearization remains a system of integral equations with rather complex kernels in general. Therefore, we cannot obtain the dielectric tensor for arbitrary frequencies ω and wave vectors $\boldsymbol{k}$ as it was possible for the weakly ionized plasma. The domain where we can solve (4.6.2) and obtain an analytical expression for $\varepsilon_{ij}(\omega, \boldsymbol{k})$ is rather wide, however, and covers practically all interesting cases.

In the range of high frequencies $\omega \gg \nu_\alpha$ or short wavelengths $kv_{T\alpha} \gg \nu_{\alpha n}$ the collision integral can be neglected to the zeroth order, and the solution of (4.6.2) can be written in the form of (4.1.6):

$$\delta f_\alpha^{(0)} = -\mathrm{i}\frac{e_\alpha \boldsymbol{E}}{\omega - \boldsymbol{k}\cdot\boldsymbol{v}}\frac{\partial f_{0\alpha}}{\partial \boldsymbol{p}_\alpha}. \tag{4.6.3}$$

We expand, as usual, all perturbed quantities in plane monochromatic waves $\exp(-\mathrm{i}\omega t + \mathrm{i}\boldsymbol{k}\cdot\boldsymbol{r})$.

To the first order we find the correction due to the collisions of the particles of the given type with all other particles (including particles of the same type):

$$\delta f_\alpha^{(1)} = \frac{\mathrm{i}}{\omega - \boldsymbol{k}\cdot\boldsymbol{v}} \sum_\beta \frac{\partial}{\partial p_{\alpha i}} \int d\boldsymbol{p}_\beta I_{ij}^{\alpha\beta}(\boldsymbol{p}_\alpha, \boldsymbol{p}_\beta) \left[\frac{\partial f_{0\alpha}}{\partial p_{\alpha j}} \delta f_\beta^{(0)}(\boldsymbol{p}_\beta) + f_{0\beta}(\boldsymbol{p}_\beta)\frac{\partial \delta f_\alpha^{(0)}}{\partial p_{\alpha j}} - f_{0\alpha}(\boldsymbol{p}_\alpha)\frac{\partial \delta f_\beta^{(0)}}{\partial p_{\beta j}} - \delta f_\alpha^{(0)}(\boldsymbol{p}_\alpha)\frac{\partial f_{0\beta}}{\partial p_{\beta j}}\right]. \tag{4.6.4}$$

The perturbation response (4.6.3) leads to the known expression for the dielectric tensor of the collisionless plasma, obtained in Sect. 4.1. The contribution (4.6.4) provides a correction to the induced charge and current densities in the plasma and consequently to the collisionless dielectric tensor.

Under the condition ω, $\nu_\alpha \ll kv_{T\alpha}$ for $\alpha = \mathrm{e, i}$ this correction may be neglected since the wavelength is much smaller than the mean free path of the particles. Then the dissipative and nondissipative collisional processes are insignificant.

In the opposite limit of high frequencies $\omega \gg \nu_\alpha$, $kv_{T\alpha}$ collisions must be accounted for, especially in the dissipative part of the dielectric tensor. This comes about since in the absence of collisions the dissipation is exponentially small at phase velocities greater than the thermal velocities. The calculations, though simple in idea, are rather complex and produce the following corrections to the collisionless dielectric permittivity

$$\delta\varepsilon^{\mathrm{lo}}(\omega, k) = \delta\varepsilon^{\mathrm{tr}}(\omega, k) = \mathrm{i}\frac{\omega_{\mathrm{pe}}^2 \nu_{\mathrm{eff}}}{\omega^3} \quad \text{for} \quad \omega \gg \nu_\mathrm{e}, kv_{\mathrm{Te}}\,, \quad \text{and} \tag{4.6.5}$$

$$\delta\varepsilon^{\mathrm{lo}}(\omega, k) = \mathrm{i}\frac{8}{5}\frac{\nu_{\mathrm{ii}}\,\omega_{\mathrm{pi}}^2 k^2 v_{\mathrm{Ti}}^2}{\omega^5}\,, \qquad \delta\varepsilon^{\mathrm{tr}}(\omega, k) = 0 \tag{4.6.6}$$

for $\omega \gg \nu_\mathrm{i}$, kv_{Ti} and ω, $\nu_\mathrm{e} \ll kv_{\mathrm{Te}}$.

In these expressions

$$\nu_{\mathrm{eff}} = \frac{4}{3}\sqrt{\frac{2\pi}{m}}\frac{e^2 e_\mathrm{i}^2 N_\mathrm{i} L}{T_\mathrm{e}^{3/2}}\,, \qquad \nu_{\mathrm{ii}} = \frac{4}{3}\sqrt{\frac{\pi}{M}}\frac{e_\mathrm{i}^4 N_\mathrm{i} L}{T_\mathrm{i}^{3/2}}\,. \tag{4.6.7}$$

4.6.1 Damping of Longitudinal High-Frequency Waves

Using (4.6.5, 6) we can calculate the collisional damping of the oscillation modes studied in the collisionless limit in Sects. 4.2 and 4.4. The high-frequency longitudinal oscillations with the frequency spectrum (4.2.5) experience, in addition to (4.2.6), a damping due to electron-ion collisions

$$\Delta\delta = -\nu_{\mathrm{eff}}/2\,, \tag{4.6.8}$$

which is similar to the correction (4.5.11) due to electron-neutral collisions. Comparing both dissipative effects and keeping in mind that they are additive, we can interpret the concept of the completely ionized plasma in more detail. Since the electron-ion collisions prevail under the condition

$$\nu_{\mathrm{eff}} \gg \nu_{\mathrm{en}}\,, \tag{4.6.9}$$

the plasma can be regarded as completely ionized in this range for the high-frequency plasma oscillations. In the opposite case, it must be regarded as weakly ionized. Note that (4.6.9) is satisfied for a degree of ionization $N_\mathrm{e}/N_\mathrm{n} \gtrsim (10^{-3}$ to $10^{-2})$ in real plasmas.

4.6.2 Collisional Damping of Ion-Acoustic Waves

For the low-frequency ion-acoustic branch of longitudinal oscillations of the nonisothermal plasma with $T_e \gg T_i$ the situation is slightly different. The account of ion-ion collisions, i.e., of high-frequency ion viscosity, provides the following correction to the collisionless damping decrement (4.2.9):

$$\Delta\delta = -\frac{4}{5}\nu_{ii}\frac{k^2 v_{Ti}^2}{\omega^2} . \tag{4.6.10}$$

The comparison of this correction with the contribution (4.5.14) due to ion-neutral collisions allows us to conclude that the plasma can be regarded as completely ionized for ion-acoustic oscillations if

$$\frac{4}{5}\nu_{ii}\frac{k^2 v_{Ti}^2}{\omega^2} \approx \frac{4}{5}\nu_{ii}\frac{T_i}{T_e} \gg \nu_{in} \tag{4.6.11}$$

and as weakly ionized in the opposite limit. This condition is satisfied for a degree of ionization $N_e/N_n \gtrsim (10^{-1}$ to $1)$ in the real plasma.

4.6.3 Damping of Transverse Waves

Finally, from the collisional correction to the transverse dielectric permittivity of the completely ionized plasma (4.6.5) we obtain the damping decrement of the high-frequency transverse waves with the spectrum (4.4.5), which are undamped in the collisionless limit:

$$\delta = -\frac{\nu_{eff}}{2}\frac{\omega_{pe}^2}{\omega_{pe}^2 + k^2c^2} . \tag{4.6.12}$$

Comparing this expression with the damping decrement (4.5.16) of transverse waves in the weakly ionized plasma shows that here as for the plasma oscillations the plasma may be regarded as completely ionized, when (4.6.9) is satisfied.

4.6.4 Degenerate Plasma

We conclude this section by the study of the effect of particle collisions on the oscillations of the completely ionized degenerate plasma. We confine our interest to the most interesting case when the electrons are degenerate and the ions nondegenerate. Further, we shall consider only high-frequency waves with $\omega > \nu_e$. When particle collisions are neglected these waves cannot be absorbed since they have phase velocities greater than the Fermi velocity or the velocity of light for the longitudinal or transverse high-frequency mode, respectively. Under these conditions the electron-ion collisions dominate. The calculation of the correction of the collisionless dielectric permittivity is similar to the above with the only difference that the collision integral (3.4.9) takes account of the degeneracy, now. Omitting details, we give the final result, see (4.6.5),

$$\delta\varepsilon^{\mathrm{lo}}(\omega, k) = \delta\varepsilon^{\mathrm{tr}}(\omega, k) = \mathrm{i}\frac{\omega_{\mathrm{pe}}^2 \nu_{\mathrm{Fe}}}{\omega^3}, \quad \text{where} \tag{4.6.13}$$

$$\nu_{\mathrm{Fe}} = 4\pi\sqrt{\frac{2\pi}{m}}\frac{e^2 e_{\mathrm{i}}^2 N_{\mathrm{i}} L}{\mathscr{E}_{\mathrm{Fe}}^{3/2}}. \tag{4.6.14}$$

Due to (4.6.13) the collisional damping decrements of high-frequency longitudinal and transverse waves are, see (4.6.8, 12),

$$\Delta\delta^{\mathrm{lo}} = -\frac{\nu_{\mathrm{Fe}}}{2}, \qquad \Delta\delta^{\mathrm{tr}} = -\frac{\nu_{\mathrm{Fe}}}{2}\frac{\omega_{\mathrm{Le}}^2}{\omega_{\mathrm{Le}}^2 + k^2c^2}. \tag{4.6.15}$$

4.7 Exercises

4.7.1. Derive the dielectric permittivity of the isotropic plasma and the average force affecting it in the SHF field in the model of independent particles.

Solution. In the model of independent particles the linearized system of equations may be written as (first-order quantities are V_α, δN_α, and $\boldsymbol{E}$)

$$\begin{aligned} &\frac{\partial \delta N_\alpha}{\partial t} + \operatorname{div} N_{0\alpha} V_\alpha = 0\,, \\ &\frac{\partial V_\alpha}{\partial t} = \frac{e_\alpha}{m_\alpha}\boldsymbol{E}\,. \end{aligned} \tag{4.7.1}$$

For perturbations of the type $\exp(-\mathrm{i}\omega t + \mathrm{i}\boldsymbol{k}\cdot\boldsymbol{r})$ we obtain the velocity of particles and the induced current density

$$V_\alpha = \frac{\mathrm{i}e_\alpha \boldsymbol{E}}{m_\alpha \omega}, \qquad j = \sum_\alpha e_\alpha N_{0\alpha} V_\alpha = \sum_\alpha \frac{\mathrm{i}e_\alpha^2 N_{0\alpha}}{m_\alpha \omega}\boldsymbol{E}. \tag{4.7.2}$$

Finally, for the longitudinal and transverse permittivities we obtain

$$\varepsilon^{\mathrm{lo}} = \varepsilon^{\mathrm{tr}} = 1 - \sum_\alpha \frac{\omega_{\mathrm{p}\alpha}^2}{\omega^2}. \tag{4.7.3}$$

These expressions represent the cold plasma limit $T_\alpha \to 0$ of (4.5.8).

According to (2.3.25) the average force affecting the unit volume of the plasma in the SHF field is

$$F_V = -\frac{e^2 N_{0e}}{m^2} \nabla |E|^2 . \tag{4.7.4}$$

This force "pushes" the plasma out of the strong field region.

4.7.2. Study small oscillations of the isotropic nonisothermal plasma on the basis of the one-fluid hydrodynamic equations (3.6.21).

Solution. For the isotropic plasma the linearized system of equations has the form of ($\varrho_M = \varrho_{0M} + \varrho_1, V$)

$$\varrho_{0M} \frac{\partial V}{\partial t} = - v_s^2 \nabla \varrho_1 , \quad \frac{\partial \varrho_1}{\partial t} + \operatorname{div} \varrho_{0M} V = 0 . \tag{4.7.5}$$

For plane wave perturbations we obtain for the system of algebraic equations

$$\omega \varrho_{0M} V - v_s^2 k \varrho_1 = 0 , \quad \omega \varrho_1 - \varrho_{0M} k \cdot V = 0 , \tag{4.7.6}$$

which yields the dispersion equation through the condition for the existence of nontrivial solutions:

$$\omega^2 = k^2 v_s^2 . \tag{4.7.7}$$

This dispersion equation coincides with (4.2.10) for the frequency spectrum of the long-wave ion-acoustic oscillations in the nonisothermal plasma.

4.7.3. Investigate the damping of the high-frequency Langmuir oscillations of the nondegenerate isotropic collisionless electron plasma in the limit of short wavelengths $k^2 r_{De}^2 \gg 1$.

Solution. Using the asymptotic expansion of the function $I_+(x)$ for $|x| \gg 1$ and $|\mathrm{Im}\{x\}| \gg |\mathrm{Re}\{x\}|$, $\mathrm{Im}\{x\} < 0$ in (4.2.1) we obtain the dispersion equation

$$1 + \mathrm{i}\sqrt{2\pi} \frac{\omega}{k v_{Te}} \frac{1}{k^2 r_{De}^2} \exp\left(-\frac{\omega^2}{2 k^2 v_{Te}^2}\right) = 0 . \tag{4.7.8}$$

Introducing the dimensionless variable ξ by

$$\omega = -\mathrm{i} k v_{Te} \xi \tag{4.7.9}$$

we have to solve

$$k^2 r_{De}^2 + \sqrt{2\pi}\, \xi \, \mathrm{e}^{\xi^2/2} = 0 \tag{4.7.10}$$

under the condition $|\mathrm{Re}\{\xi\}| \gg |\mathrm{Im}\{\xi\}|$. Splitting this equation into real and imaginary parts ($\xi = \xi' + \mathrm{i}\xi''$) we obtain

$$\sqrt{2\pi}\,\xi'\,\mathrm{e}^{\xi'^2/2}\cos(\xi'\xi'') - \sqrt{2\pi}\,\xi''\,\mathrm{e}^{\xi'^2/2}\sin(\xi'\xi'') + k^2r_{\mathrm{De}}^2 = 0\,,$$

$$\xi''\cos(\xi'\xi'') + \xi'\sin(\xi'\xi'') = 0\,. \tag{4.7.11}$$

From the second equation it follows that a solution lies in the sector $\xi' > 0$ and $\xi'' > 0$, when $\tan(\xi'\xi'') = -\xi''/\xi' \to -0$. The tangent is negative in the second and fourth quarters and approaches zero when its argument is about π or 2π. We take the first value of the argument which corresponds to the main branch of the tangent and write $\xi'\xi'' = \pi - z$ for $z \to +0$. Here $\tan(\xi'\xi'') = \tan(\pi - z) = -z$, i.e., $\xi''/\xi' = -z$, and therefore $\xi'^2 = (\pi - z)/z \to +\infty$ and $\cos(\xi'\xi'') = -1$. In view of this fact, the first equation of (4.7.11) results in

$$\sqrt{2\pi}\,\xi'\exp\left(\frac{\xi'^2}{2}\right) = k^2r_{\mathrm{De}}^2\,, \qquad \xi' \approx \sqrt{2\ln(k^2r_{\mathrm{De}}^2)} \gg 1\,. \tag{4.7.12}$$

Thus, the short-wave Langmuir oscillations are strongly damped with the damping decrement

$$\delta = \mathrm{Im}\{\omega\} = -kv_{\mathrm{Te}}\sqrt{2\ln(k^2r_{\mathrm{De}}^2)}\,. \tag{4.7.13}$$

The frequency $\mathrm{Re}\{\omega\}$ is easily found from the relation $\xi'\xi'' = \pi$. The complex solution in the short-wave limit is

$$\frac{\omega}{kv_{\mathrm{Te}}} = \frac{\pi}{\sqrt{\ln(k^2r_{\mathrm{De}}^2)}} - \mathrm{i}\sqrt{\ln(k^2r_{\mathrm{De}}^2)}\,. \tag{4.7.14}$$

Since $\mathrm{Re}\{\omega/k\} \ll v_{\mathrm{Te}}$, the bulk electrons participate in the Cherenkov absorption which explains the strong damping of these oscillations.

4.7.4. Use the Landau kinetic equation to derive the longitudinal and transverse permittivities of the isotropic fully ionized nondegenerate plasma under the condition of frequent collisions $\nu_\alpha \gg \omega$, $kv_{\mathrm{T}\alpha}$, $\alpha = \mathrm{e}, \mathrm{i}$.

Solution. Under these conditions the electrons contribute dominantly to the induced current whereas the ions remain unperturbed. The linearized electron kinetic equation has the form

$$\begin{aligned} e\boldsymbol{E}\frac{\partial f_{0\mathrm{e}}}{\partial \boldsymbol{p}} &= N_{0\mathrm{i}}\frac{\partial}{\partial p_i}I_{ij}^{\mathrm{ei}}(\boldsymbol{p})\frac{\partial \delta f_{\mathrm{e}}}{\partial p_j} + \frac{\partial}{\partial p_i}\int d\boldsymbol{p}'\, I_{ij}^{\mathrm{ee}}(\boldsymbol{p},\boldsymbol{p}') \\ &\times\left[\frac{\partial f_{0\mathrm{e}}}{\partial p_j}\delta f_{\mathrm{e}}(\boldsymbol{p}') + \frac{\partial \delta f_{\mathrm{e}}}{\partial p_j}f_{0\mathrm{e}}(\boldsymbol{p}') - f_{0\mathrm{e}}(\boldsymbol{p})\frac{\partial \delta f_{\mathrm{e}}}{\partial p_j'} - \delta f_{\mathrm{e}}(\boldsymbol{p})\frac{\partial f_{0\mathrm{e}}}{\partial p_j'}\right]. \end{aligned} \tag{4.7.15}$$

It is convenient to solve this equation by the Chapman-Enskog method, i.e., by expanding $\delta f_e(\boldsymbol{p})$ in Sonin polynomials. We confine ourselves to two terms of the expansion

$$\delta f_e(\boldsymbol{p}) = \frac{\boldsymbol{v}\cdot\boldsymbol{E}}{E}\left[a_0 + a_1\left(\frac{5}{2} - \frac{v^2}{2v_{Te}^2}\right)\right] f_{0e}\,, \tag{4.7.16}$$

which we substitute into (4.7.15). Multiplying the resulting equation by the polynomials 1 and $[5/2 - v^2/(2v_{Te}^2)]$ and integrating over the momentum, we obtain two equations for the expansion coefficients a_0 and a_1. For the plasma with singly charged ions $e_i = |e|$ these equations are

$$\frac{eE}{T_e} = -\nu_{eff}\left(a_0 + \frac{3}{2}a_1\right), \quad \frac{3}{2}a_0 + \frac{13 + 4\sqrt{2}}{4}a_1 = 0\,, \tag{4.7.17}$$

which yields

$$a_0 = -\frac{13 + 4\sqrt{2}}{6}a_1\,, \quad a_1 = -\frac{3}{2 + 2\sqrt{2}}\frac{eE}{\nu_{eff}T_e}\,. \tag{4.7.18}$$

Then, the induced electron current density can be obtained from (4.7.16)

$$\boldsymbol{j} \approx \boldsymbol{j}_e \approx 1.96\frac{e^2 N_{0e}}{m\nu_{eff}}\boldsymbol{E}\,. \tag{4.7.19}$$

Finally, the dielectric permittivity follows

$$\varepsilon^{lo} = \varepsilon^{tr} = 1 + \mathrm{i}\,1.96\frac{\omega_{pe}^2}{\omega\nu_{eff}}\,. \tag{4.7.20}$$

Strictly speaking, (4.7.20) is valid only under the condition $\omega\nu_{eff} \gg k^2 v_{Te}^2$. In the opposite case we must take account of small space- and time-dependent terms on the left-hand side of (4.7.15). The analysis shows that these terms are significant only for the longitudinal dielectric permittivity and do not contribute to the transverse dielectric permittivity.

4.7.5. Analyze the frequency dependency of the penetration depth of a transverse field.

Solution. The penetration depth of a transverse field with real frequency ω is determined by the roots $k(\omega)$ of the dispersion equation

$$k^2 = \frac{\omega^2}{c^2}\varepsilon^{tr}(\omega, k) \tag{4.7.21}$$

and given by

$$\lambda_{sk} = \frac{1}{\mathrm{Im}\{k(\omega)\}} . \tag{4.7.22}$$

For discussion we distinguish different frequency ranges:

a) In the range $\omega \gg \nu_e$, kv_0 (ν_e is the electron collision frequency, v_0 the average velocity of the electron random motion, $v_0 = v_{Te}$ for the nondegenerate plasma and $v_0 = v_{Fe}$ for the degenerate one) we have

$$\varepsilon^{tr} = 1 - \frac{\omega_{pe}^2}{\omega^2}\left(1 - i\frac{\nu_e}{\omega}\right) . \tag{4.7.23}$$

Here $\nu_e = \nu_{en}$ for the weakly ionized nondegenerate and degenerate plasma, $\nu_e = \nu_{eff}$ for the completely ionized nondegenerate plasma and $\nu_e = \nu_{Fe}$ for the degenerate one.

Substituting (4.7.23) into (4.7.21) yields

$$\lambda_{sk} = \begin{cases} \dfrac{2c\omega^2}{\omega_{pe}^2 \nu_e} & \text{for} \quad \omega \gg \omega_{pe} , \\[2ex] \dfrac{c}{\omega_{pe}} & \text{for} \quad \omega_{pe}\dfrac{v_0}{c} \ll \omega \ll \omega_{pe} . \end{cases} \tag{4.7.24}$$

b) In the range $kv_0 \gg \omega$, ν_e we have

$$\varepsilon^{tr} = 1 + i\alpha\frac{\omega_{pe}^2}{\omega k v_0} . \tag{4.7.25}$$

Here $\alpha = \sqrt{\pi/2}$ for the nondegenerate plasma and $\alpha = 3\pi/4$ for the degenerate one. In this frequency range we obtain from (4.7.21)

$$\lambda_{sk} = 2\left(\frac{c^2 v_0}{\alpha\omega_{pe}^2\omega}\right)^{1/3} , \tag{4.7.26}$$

for $\omega^* \ll \omega \ll \omega_{pe} v_0/c$, with $\omega^* = \nu_e^3 c^2/\omega_{pe}^2 v_0^2$.

Thus, the *anomalous skin-effect* given by (4.7.26) exists in the range $\nu_e \ll v_0\omega_{pe}/c$, only.

The existence condition $\omega > \omega^*$ is necessary for any ratio between ω and ν_e.

c) In the range $\nu_e \gg kv_0$, ω we have

$$\varepsilon^{tr} = 1 + i\alpha_1\frac{\omega_{pe}^2}{\omega\nu_e} \tag{4.7.27}$$

with $\alpha_1 = 1$ for weakly ionized and $\alpha_1 = 1.96$ for fully ionized plasmas. The skin-depth is

$$\lambda_{sk} = \left(\frac{2\nu_e c^2}{\alpha_1 \omega_{pe}^2 \omega}\right)^{1/2} \tag{4.7.28}$$

for $\omega \ll \omega^*, \nu_e$.

Thus, the *normal skin-effect* (4.7.28) exists in the frequency range $\omega \ll \omega^*$. The ration between ω^* and ν_e can be arbitrary.

The results are shown schematically in Fig. 4.6.

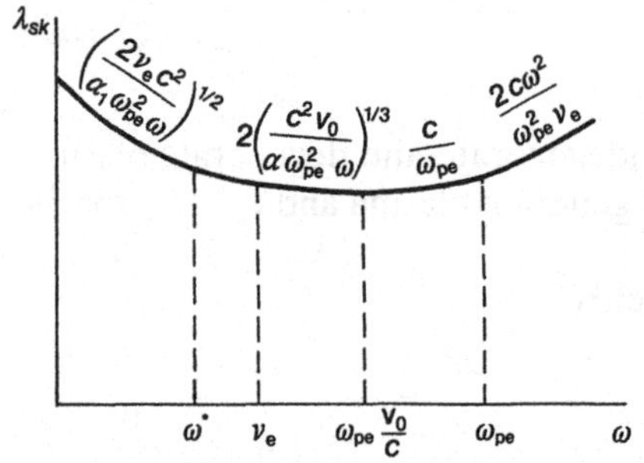

Fig. 4.6. Frequency dependence of the penetration depth of the transverse field

4.7.6. Verify that the ion-acoustic oscillations can exist in the frequency range $\nu_{en} \gg \omega, kv_{Te}$ in the highly collisional weakly ionized nonisothermal plasma with $T_e \gg T_i$ if $\omega > \nu_{in}$ and $\omega\nu_{en} \ll k^2 v_{Te}^2$. (In the completely ionized plasma such oscillations are impossible.)

Solution. According to (4.5.8) the ion-acoustic oscillations follow in the given frequency range from

$$\varepsilon^{lo}(\omega, k) = 1 + \frac{\omega_{pe}^2}{k^2 v_{Te}^2}\left(1 + i\frac{\omega\nu_{en}}{k^2 v_{Te}^2}\right) - \frac{\omega_{pi}^2}{\omega^2}\left(1 - i\frac{\nu_{in}}{\omega}\right) = 0\,. \tag{4.7.29}$$

We obtain the frequency ($\omega \to \omega + i\delta$) of weakly damped oscillations as

$$\omega^2 = \frac{\omega_{pi}^2}{1 + \omega_{pe}^2/(k^2 v_{Te}^2)}\,, \qquad \delta = -\frac{\omega^4 \nu_{en}}{2k^4 v_{Te}^2 v_s^2} - \frac{\nu_{in}}{2}\,. \tag{4.7.30}$$

Note that in the weakly ionized plasma the conditions

$$\nu_{en} \gg kv_{Te} \sim kv_s\sqrt{M/m} \approx \omega\sqrt{M/m} \gg \nu_{in}\sqrt{M/m}$$

are compatible since

$$\nu_{en} \approx v_{Te}\sigma_0 N_n \approx \sqrt{T_e M/(T_i m)}\, v_{Ti}\sigma_0 N_n \approx \sqrt{T_e M/(T_i m)}\,\nu_{in}\,.$$

For $T_e \gg T_i$ both inequalities can be satisfied.

For the completely ionized plasma we may write an equation similar to (4.7.29) and obtain a mode spectrum of the type (4.7.30). Because of the decrease of the collision frequencies of charged particles with increasing temperature, the assumed inequalities cannot be satisfied, however. Therefore, ion-acoustic oscillations cannot exist in the collision-dominated fully ionized plasma.

4.7.7. Derive the dielectric permittivity of the ultrarelativistic nondegenerate ($T_e \gg mc^2$) electron plasma and analyze the wave spectra (both longitudinal and transverse) in the high-frequency range.

Solution. Since ultrarelativistic particles have the energy $\mathscr{E} = cp$ the equilibrium distribution function is

$$f_{0e} = \frac{N_e}{8\pi}\left(\frac{c}{T_e}\right)^3 \exp\left(-\frac{cp}{T_e}\right). \tag{4.7.31}$$

The electron velocity is equal to the velocity of light: $v = \partial\mathscr{E}/\partial p = c$. In this sense the distribution function (4.7.31) is similar to the Fermi distribution. Substituting (4.7.31) into (4.1.13) we obtain

$$\begin{aligned} \varepsilon^{lo}(\omega, k) &= 1 + \frac{4\pi e^2 N_e}{k^2 T_e}\left(1 - \frac{\omega}{2ck}\ln\frac{\omega + ck}{\omega - ck}\right), \\ \varepsilon^{tr}(\omega, k) &= 1 - \frac{\pi e^2 N_e c}{\omega k T_e}\left[\frac{2\omega}{ck} - \left(\frac{\omega^2}{c^2k^2} - 1\right)\ln\frac{\omega + ck}{\omega - ck}\right]. \end{aligned} \tag{4.7.32}$$

These expressions are of similar structure to (4.1.18) for the degenerate plasma. It follows immediately that the imaginary parts of ε^{lo} and ε^{tr} vanish for $\omega/k > c$ and that wave absorption is impossible in this case. The frequency spectra are determined by the relations

$$\omega^2 = \begin{cases} \dfrac{4\pi e^2 N_e c^2}{3T_e} + \dfrac{6}{5}c^2k^2 & \text{for} \quad \omega \gg ck\,, \\[2ex] \dfrac{2\pi e^2 N_e c^2}{T_e} + c^2k^2 & \text{for} \quad \omega \to ck \end{cases} \tag{4.7.33}$$

for transverse waves, and

$$\begin{aligned} \omega^2 &= \frac{4\pi e^2 N_e c^2}{3T_e} + \frac{3}{5}c^2k^2 & \text{for} \quad \omega \gg ck\,, \\ \omega &= ck\left[1 + 2\exp\left(-2 - \frac{k^2 T_e}{2\pi e^2 N_e}\right)\right] & \text{for} \quad \omega \to ck \end{aligned} \tag{4.7.34}$$

for longitudinal waves.

In the range of small phase velocities (low frequencies) $\omega \ll ck$ the field is either strongly absorbed or screened. Indeed, for $\omega \ll ck$ we have

$$\varepsilon^{\rm lo} = 1 + \frac{4\pi e^2 N_{\rm e}}{k^2 T_{\rm e}}\left(1 + {\rm i}\frac{\pi}{2}\frac{\omega}{ck}\right), \qquad \varepsilon^{\rm tr} = 1 + {\rm i}\pi\frac{\pi e^2 N_{\rm e} c}{\omega k T_{\rm e}}. \tag{4.7.35}$$

The screening radius of the longitudinal field is equal to the Debye length $\lambda_{\rm sk}^{\rm lo} = [T_{\rm e}/(4\pi e^2 N_{\rm e})]^{1/2}$ and the penetration depth of the transverse field is determined by the anomalous skin effect

$$\lambda_{\rm sk}^{\rm tr} \approx \left(\frac{2}{\pi}\frac{T_{\rm e} c}{\pi e^2 N_{\rm e}\omega}\right)^{1/3}.$$

4.7.8. Analyze the relaxation process of a weak electron temperature anisotropy $(T_\perp - T_\parallel)$ in the completely ionized nonisothermal plasma $(T_{\rm e} \gg T_{\rm i})$ with account of polarization interaction of the electrons, i.e., interaction by means of the ion sound.

Solution. As shown in Exercise 3.7.5, both electron-electron and electron-ion collisions contribute to the electron temperature isotropization. The solution of Exercise 3.7.5 was obtained by neglecting the plasma polarization in the collision integral, i.e., the electron-wave interaction due to emission and absorption of plasma waves. This approximation is valid in nearly isothermal plasmas only, where ion-acoustic oscillations are impossible. In the nonisothermal plasma with $T_{\rm e} \gg T_{\rm i}$ the electrons easily exchange momentum with the ion sound. We take account of the interaction with this wave type only and analyze when it becomes decisive. The electron kinetic equation is

$$\frac{\partial f_{\rm e}}{\partial t} = \left(\frac{\partial f_{\rm e}}{\partial t}\right)_{\rm col}^{\rm ee} = \frac{\partial}{\partial p_{\rm i}}\int d\boldsymbol{p}'\, I_{ij}^{\rm ee}(\boldsymbol{p},\boldsymbol{p}') \times \left[\frac{\partial f_{\rm e}(\boldsymbol{p})}{\partial p_j} f_{\rm e}(\boldsymbol{p}') - f_{\rm e}(\boldsymbol{p})\frac{\partial f_{\rm e}(\boldsymbol{p}')}{\partial p_j'}\right], \tag{4.7.36}$$

where we take account of wave interaction in the electron-electron collision term by writing $I_{ij}^{\rm ee}(\boldsymbol{p},\boldsymbol{p}')$ in the form

$$I_{ij}^{\rm ee}(\boldsymbol{p},\boldsymbol{p}') = \pi\int\frac{d\boldsymbol{k}}{(2\pi)^3}\left(\frac{4\pi e^2}{k^2}\right)^2\frac{k_i k_j \delta(\boldsymbol{k}\cdot\boldsymbol{v} - \boldsymbol{k}\cdot\boldsymbol{v}')}{|\varepsilon(\boldsymbol{k}\cdot\boldsymbol{v},\boldsymbol{k})|^2} \approx \frac{\pi^2}{(2\pi)^3}\int_{kr_{\rm Di}<1} d\boldsymbol{k}\left(\frac{4\pi e^2}{k^2}\right)^2\frac{k_i k_j \delta(\boldsymbol{k}\cdot\boldsymbol{v} - \boldsymbol{k}\cdot\boldsymbol{v}')\,\delta[{\rm Re}\{\varepsilon(\boldsymbol{k}\cdot\boldsymbol{v},\boldsymbol{k})\}]}{|{\rm Im}\{\varepsilon(\boldsymbol{k}\cdot\boldsymbol{v},\boldsymbol{k})\}|}. \tag{4.7.37}$$

Restricting the integration to $kr_{Di} < 1$ we separate the effect of the interaction of the electrons with ion-acoustic oscillations, which exist in this domain only. The contribution of close collisions to the relaxation of the temperature anisotropy occurs in the range $kr_{Di} > 1$ and has been analyzed in Exercise 3.7.5 already. For ion-acoustic waves $\omega \ll kv_{Te}$, which allows the approximation

$$I_{ij}^{ee}(\boldsymbol{p},\boldsymbol{p}') = 2\pi e^4 \int_{-\infty}^{\infty} d\omega \int_{kr_{Di}<1} d\boldsymbol{k} \frac{k_i k_j}{k^4} \times \frac{\delta[\mathrm{Re}\{\varepsilon(\omega,\boldsymbol{k})\}]}{|\mathrm{Im}\{\varepsilon(\omega,\boldsymbol{k})\}|} \delta(\boldsymbol{k}\cdot\boldsymbol{v})\,\delta(\boldsymbol{k}\cdot\boldsymbol{v}') . \tag{4.7.38}$$

Regarding the temperature anisotropy as small, we obtain

$$\begin{aligned} \mathrm{Re}\{\varepsilon(\omega,\boldsymbol{k})\} &= 1 - \frac{\omega_{pi}^2}{\omega^2} + \frac{\omega_{pe}^2}{k^2 v_{Te}^2} , \\ \mathrm{Im}\{\varepsilon(\omega,\boldsymbol{k})\} &= \sqrt{\frac{\pi}{2}} \frac{\omega\,\omega_{pe}^2}{k^3 v_{Te}^3} + \sqrt{\frac{\pi}{2}} \frac{\omega\,\omega_{pi}^2}{k^3 v_{Te}^3} \exp\left(-\frac{\omega^2}{2k^2 v_{Te}^2}\right) . \end{aligned} \tag{4.7.39}$$

Substituting (4.7.39) into (4.7.38), we get by a lengthy but straightforward calculation

$$\begin{aligned} I_{ij}^{ee}(\boldsymbol{p},\boldsymbol{p}') &= 4\sqrt{2\pi}\, v_{Te} e^4 I \frac{[\boldsymbol{v},\boldsymbol{v}']_i [\boldsymbol{v},\boldsymbol{v}']_j}{|[\boldsymbol{v},\boldsymbol{v}']|^3} , \\ I &= \frac{1}{2}\left|\frac{e_i}{e}\right| \frac{T_e}{T_i} \left[\ln\left(\frac{e_i^2 M T_e^3}{e^2 m T_i^3}\right)\right]^{-1} . \end{aligned} \tag{4.7.40}$$

Hence, by order of magnitude $I_{ij}^{ee} \sim e^4 I / v_{Te}$, whereas the neglected contribution of close collisions is $I_{ij}^{ee} \sim e^4 L / v_{Te}$. Thus, the close collisions can be neglected compared to the polarization interaction under the condition $I \gg L$ or

$$\frac{T_e}{T_i}\left[\ln\left(\frac{e_i^2 M T_e^3}{e^2 m T_i^3}\right)\right]^{-1} \gg L = \ln\frac{r_{Di}}{r_{min}} . \tag{4.7.41}$$

For $T_e/T_i = 10^2$ to 10^3 this condition is satisfied.

Substituting (4.7.40) into (4.7.36) we obtain (as in Exercise 3.7.5)

$$\frac{d}{dt}(T_\perp - T_\parallel) = -\nu_{ee}(T_\perp - T_\parallel) , \quad \text{where} \tag{4.7.42}$$

$$\nu_{ee} = \frac{6}{5}\nu_{eff}\left|\frac{e}{e_i}\right|\frac{I}{L} , \quad \nu_{eff} = \frac{4}{3}\sqrt{\frac{2\pi}{m}}\frac{e^2 e_i^2 N_i L}{T_e^{3/2}} . \tag{4.7.43}$$

Together with the additive contribution of close collisions the final result is

$$\frac{d}{dt}(T_\perp - T_\parallel) = -\nu_f(T_\perp - T_\parallel) ,$$
$$\nu_f = \frac{6}{5}\nu_{eff}\left[1 + \left|\frac{e}{e_i}\right|\left(\frac{1}{\sqrt{2}} + \frac{I}{L}\right)\right] . \tag{4.7.44}$$

4.7.9. Using (4.5.8) for the longitudinal dielectric permittivity analyze the diffusion spread of a small inhomogeneity of the density of charged particles in a weakly ionized isotropic plasma.

Solution. The diffusion spread is a slow process with characteristic time $\tau \sim 1/|\omega| \gg 1/\nu_e, 1/\nu_i$. Therefore, for its description it is necessary to analyze the low-frequency limit of (4.5.8)

$$\varepsilon^{lo}(\omega, k) \approx 1 + \frac{i\omega_{pe}^2}{\omega\nu_{en} + ik^2v_{Te}^2} + \frac{i\omega_{pi}^2}{\omega\nu_{in} + ik^2v_{Ti}^2} . \tag{4.7.45}$$

Here $\omega \sim 1/\tau$ characterizes the time period of the spread of inhomogeneity with a space dimension $L \sim 1/k$. For low densities of charged particles the diffusion of electrons and ions occurs independently (a free diffusion) and is described by the poles in the electron and ion parts in (4.7.45)

$$\omega\nu_{\alpha n} + ik^2v_{T\alpha}^2 = 0 \quad \text{for} \quad \alpha = e, i .$$

Hence we obtain the free diffusion factors for each of the components

$$D_\alpha \approx v_{T\alpha}^2/\nu_{\alpha n} .$$

The zeros of $\varepsilon^{lo}(\omega, k)$ describe the spread of the inhomogeneity in plasmas taking account of self-consistent interaction of electrons and ions, i.e., the process of ambipolar diffusion. For the characteristic dimension of inhomogeneity greatly exceeding the Debye lengths of electrons and ions, taking account of $\nu_{in} \gg \nu_{en} m/M$,

$$\omega = -i\frac{k^2(v_{Ti}^2 + v_s^2)}{\nu_{in}} . \tag{4.7.46}$$

Thus the ambipolar diffusion factor is

$$D_a = (v_{Ti}^2 + v_s^2)/\nu_{in} .$$

4.7.10. Verify that the particular solution of (4.1.4) corresponding to undamped longitudinal waves (the *Van Campen modes*) is possible in an isotropic plasma.

Solution. Supposing (4.1.4) to be inhomogeneous, (4.1.6) should be supplemented by the solution of

$$(\omega - \boldsymbol{k} \cdot \boldsymbol{v})\,\delta f_a^{(1)} = 0\,, \quad \text{where}$$
$$\delta f_a^{(1)} = n_{a1}\, k\delta\,(\omega - \boldsymbol{k} \cdot \boldsymbol{v})\,. \tag{4.7.47}$$

Here n_{a1} is an arbitrary constant. The solution of (4.7.47) corresponds to the one-velocity (i.e., $v = \omega/k$) modulated beam of particles of type a with the density n_{a1}.

As a result, the solution of (4.1.6) takes the form

$$\delta f_a^{(0)} = -\frac{ie_a \boldsymbol{E}\dfrac{\partial f_{0a}}{\partial \boldsymbol{p}}}{\omega - \boldsymbol{k} \cdot \boldsymbol{v}} + n_{a1}\, k\delta\,(\omega - \boldsymbol{k} \cdot \boldsymbol{v})\,. \tag{4.7.48}$$

Introducing it into the Poisson equation we arrive at the dispersion equation for longitudinal oscillations of the isotropic electron plasma, cf. (4.1.13):

$$\varepsilon^{\mathrm{lo}}(\omega, k) = 1 + \frac{4\pi e^2}{\omega k^2} \int d\boldsymbol{p}\, \frac{(\boldsymbol{k} \cdot \boldsymbol{v})^2}{\omega - \boldsymbol{k} \cdot \boldsymbol{v}}\, \frac{\partial f_0}{\partial \mathscr{E}} + \frac{4\pi e}{k^3}\, d_{\mathrm{e}1} = 0\,. \tag{4.7.49}$$

Here $d_{\mathrm{e}1}$ is a new constant uniquely related to $n_{\mathrm{e}1}$.

The imaginary term in (4.7.49) can approach zero due to matching of the constant $n_{\mathrm{e}1}$ or $d_{\mathrm{e}1}$. This means that damping of such waves will be absent in the plasma. These waves are called the Van Campen modes after the Dutch physicist who was the first to demonstrate the completeness of the solutions of the form (4.7.48). This means that any perturbation in the plasma can be expanded into the set of functions of the form (4.7.48).

4.7.11. Define the degree of ionization under which the damping of plasma oscillation is governed by the collisions of charged particles, so that the plasma may be considered strongly ionized. Study the cases when (a) $T_{\mathrm{e}} \approx 1\,\mathrm{eV}$; and (b) $T_{\mathrm{e}} \approx 1\,\mathrm{keV}$.

Solution. The plasma may be considered strongly ionized when $\nu_{\mathrm{eff}}/\nu_{\mathrm{en}} > 1$. Assuming $\nu_{\mathrm{en}} = v_{\mathrm{Te}}\pi a^2 N_0$, where $a \approx 10^{-8}$ cm, we obtain the following conditions: (a) $N_{\mathrm{e}}/N_0 > 3 \cdot 10^{-4}$; (b) $N_{\mathrm{e}}/N_0 > 3 \cdot 10^{2}$.

4.7.12. See Exercise 4.7.11 for ion-acoustic waves if $T_{\mathrm{i}} \approx 0.1\,\mathrm{eV}$.

Solution. In this case the plasma is strongly ionized if

$$\frac{\nu_{\mathrm{ii}}}{\nu_{\mathrm{in}}} \cdot \frac{T_{\mathrm{i}}}{T_{\mathrm{e}}} > 1\,.$$

Thus (a) $N_i/N_0 > 5 \cdot 10^{-6}$; (b) $N_i/N_0 > 5 \cdot 10^{-2}$.

5. Dielectric Permittivity and Oscillation Spectra of Homogeneous Magneto-Active Plasmas

The spectra of electromagnetic oscillations of a plasma in thermodynamic equilibrium in an external constant homogeneous magnetic field are studied. The equilibrium particle distribution is assumed to be Maxwellian for the nondegenerate plasma and Fermian for the degenerate one. The dielectric tensor for the collisionless plasma is obtained on the basis of the Vlasov equation, using the kinetic equation with the collision BGK integral for a weakly ionized plasma and the Boltzmann-Landau equation for a completely ionized plasma. The analysis of the oscillation spectra of a cold magneto-active plasma is carried out. A special study is undertaken to demonstrate the effects of thermal motion and interparticle collisions upon the electromagnetic wave spectra.

5.1 Dielectric Tensor of the Homogeneous Collisionless Magneto-Active Plasma

We consider a plasma in thermodynamic equilibrium imbedded in an external, constant and homogeneous magnetic field $\boldsymbol{B}_0$. The equilibrium distributions of the charged particles are assumed to be Maxwellian (4.1.1) or Fermian (4.1.2). As in Chap. 4, we begin our analysis of the properties of such a plasma with the collisionless limit described by the Vlasov equation. The linearized Vlasov equation for a small deviation δf_α of the distribution function of the particle species α from the equilibrium reads

$$-\mathrm{i}(\omega - \boldsymbol{k}\cdot\boldsymbol{v})\,\delta f_\alpha + e_\alpha \boldsymbol{E}\cdot\frac{\partial f_{0\alpha}}{\partial \boldsymbol{p}_\alpha} + \frac{e_\alpha}{c}[\boldsymbol{v},\boldsymbol{B}_0]\frac{\partial \delta f_\alpha}{\partial \boldsymbol{p}_\alpha} = 0\,. \tag{5.1.1}$$

Taking the magnetic field oriented along the z-axis it is convenient to pass over to the cylindrical momentum coordinates $p_\perp$, ϕ, p_z defined by $p_x = p_\perp\cos\phi$, $p_y = p_\perp\sin\phi$. Note that we get $\boldsymbol{p} = m\gamma\boldsymbol{v}$, where $\gamma = (1-v^2/c^2)^{-1/2}$. Using these coordinates we obtain

$$-\mathrm{i}(\omega - \boldsymbol{k}\cdot\boldsymbol{v})\,\delta f_\alpha + e_\alpha \boldsymbol{E}\frac{\partial f_{0\alpha}}{\partial \boldsymbol{p}_\alpha} - \frac{\Omega_\alpha}{\gamma}\frac{\partial \delta f_\alpha}{\partial \phi} = 0\,, \tag{5.1.2}$$

where $\Omega_\alpha = e_\alpha B_0/(m_\alpha c)$. This linear inhomogeneous differential equation has the general solution

$$\delta f_\alpha = \frac{e_\alpha \gamma}{\Omega_\alpha} \int_c^\phi d\phi' \left(\boldsymbol{E} \cdot \frac{\partial f_{0\alpha}}{\partial \boldsymbol{p}_\alpha} \right)_{\phi'} \exp\left[\frac{\mathrm{i}\gamma}{\Omega_\alpha} \int_\phi^{\phi'} d\phi'' (\omega - \boldsymbol{k} \cdot \boldsymbol{v})_{\phi''} \right]. \tag{5.1.3}$$

The indices ϕ' and ϕ'' at the expressions in parentheses make explicit that these expressions are functions of ϕ' and ϕ''.

The integration constant C is determined by the requirement of periodicity $\delta f_\alpha(\phi + 2\pi) = \delta f_\alpha(\phi)$. It is easy to show that this requirement can be satisfied only if $|C| = \infty$. The integration constant has a clear physical meaning: since the time evolution of ϕ is caused by the cyclotron rotation of the particles around the z-axis we have $\phi = \Omega_\alpha t/\gamma$ and the variable ϕ is uniquely related to time. Then, from the requirement of a vanishing initial perturbation δf_α at $t \to -\infty$, i.e., the adiabatic switching on of the field in the infinite past, it follows that the lower limit of integration is $\phi = \pm\infty$. The sign corresponds to the sign of Ω_α for each type of particles α. Hence, (5.1.3) can be written as

$$\delta f_\alpha = \frac{e_\alpha \gamma}{\Omega_\alpha} \int_\infty^\phi d\phi' \left(\boldsymbol{E} \cdot \frac{\partial f_{0\alpha}}{\partial \boldsymbol{p}_\alpha} \right)_{\phi'} \exp\left[\frac{\mathrm{i}\gamma}{\Omega_\alpha} \int_\phi^{\phi'} d\phi'' (\omega - \boldsymbol{k} \cdot \boldsymbol{v})_{\phi''} \right]. \tag{5.1.4}$$

Using the relation (4.1.5) between the current density $\boldsymbol{j}$ and the perturbation δf_α we can determine the conductivity tensor $\sigma_{ij}(\omega, \boldsymbol{k})$ and, consequently, the dielectric tensor $\varepsilon_{ij}(\omega, \boldsymbol{k})$:

$$\begin{aligned} \varepsilon_{ij}(\omega, \boldsymbol{k}) = \delta_{ij} + \sum_\alpha \frac{4\pi \mathrm{i} e_\alpha^2}{\Omega_\alpha \omega} \int d\boldsymbol{p}_\alpha \gamma v_i \\ \times \int_\infty^\phi d\phi' \left(\frac{\partial f_{0\alpha}}{\partial p_j} \right)_{\phi'} \exp\left[\frac{\mathrm{i}\gamma}{\Omega_\alpha} \int_\phi^{\phi'} d\phi'' (\omega - \boldsymbol{k} \cdot \boldsymbol{v})_{\phi''} \right]. \end{aligned} \tag{5.1.5}$$

Due to the isotropy of the equilibrium distribution functions we may write $\partial f_0/\partial \boldsymbol{p} = \boldsymbol{v} \partial f_0/\partial \mathscr{E}$, where $\boldsymbol{p} = m\gamma \boldsymbol{v}$ and $\mathscr{E} = \sqrt{m^2c^4 + p^2c^2}$. Then, the dielectric tensor takes the form

$$\begin{aligned} \varepsilon_{ij}(\omega, \boldsymbol{k}) = \delta_{ij} + \mathrm{i} \sum_\alpha \frac{4\pi e_\alpha^2}{\Omega_\alpha \omega} \int d\boldsymbol{p}_\alpha \gamma \frac{\partial f_{0\alpha}}{\partial \mathscr{E}} v_i \\ \times \int_\infty^\phi d\phi' v_j(\phi') \exp\left[\frac{\mathrm{i}\gamma}{\Omega_\alpha} \int_\phi^{\phi'} d\phi'' (\omega - \boldsymbol{k} \cdot \boldsymbol{v})_{\phi''} \right]. \end{aligned} \tag{5.1.6}$$

Without loss of generality we may assume that the vector $\boldsymbol{k}$ lies in the xz-plane, i.e., $\boldsymbol{k} = (k_\perp, 0, k_z)$. Integrating over ϕ'' the exponential factor in (5.1.5) can be written as

$$\exp\left[\frac{i\gamma}{\Omega_\alpha}\int_{\phi}^{\phi'} d\phi''(\omega - \boldsymbol{k}\cdot\boldsymbol{v})_{\phi''}\right]$$

$$= \exp\left[-i\frac{\omega - k_z v_z}{\Omega_\alpha}\gamma(\phi - \phi')\right]\exp\left[-i\frac{k_\perp v_\perp \gamma}{\Omega_\alpha}(\sin\phi - \sin\phi')\right].$$

Using the expansion

$$\exp\left(\pm i b_\alpha \sin\phi\right) = \sum_{n=-\infty}^{\infty} \mathrm{J}_n(b_\alpha)\,\mathrm{e}^{in\phi},$$

where J_n is the Bessel function of the order n. With the argument $b_\alpha = \gamma k_\perp v_\perp/\Omega_\alpha$, (5.1.6) can be reduced to

$$\varepsilon_{ij}(\omega,\boldsymbol{k}) = \delta_{ij} + \sum_\alpha \frac{4\pi e_\alpha^2}{\omega}\int d\boldsymbol{p}_\alpha \frac{\partial f_{0\alpha}}{\partial \mathscr{E}}$$

$$\times \sum_{n=-\infty}^{\infty} \Pi_{ij}^{(n)}\left[\mathscr{P}\frac{1}{\omega - k_z v_z - n\Omega_\alpha/\gamma} - i\pi\delta(\omega - k_z v_z - n\Omega_\alpha/\gamma)\right]. \quad (5.1.7)$$

The symbol $\mathscr{P}$ denotes that the principal value of the integral has to be taken and the tensor $\Pi_{ij}^{(n)}$ is given by

$$\Pi_{ij}^{(n)} = \begin{pmatrix} \dfrac{v_\perp^2 n^2}{b_\alpha^2}\mathrm{J}_n^2 & i v_\perp^2 n \dfrac{\mathrm{J}_n \mathrm{J}_n'}{b_\alpha} & v_\perp v_z n \dfrac{\mathrm{J}_n^2}{b_\alpha} \\ -i v_\perp^2 n \dfrac{\mathrm{J}_n \mathrm{J}_n'}{b_\alpha} & v_\perp^2 \mathrm{J}_n'^2 & -i v_\perp v_z \mathrm{J}_n \mathrm{J}_n' \\ v_\perp v_z n \dfrac{\mathrm{J}_n^2}{b_\alpha} & i v_\perp v_z \mathrm{J}_n \mathrm{J}_n' & v_z^2 \mathrm{J}_n^2 \end{pmatrix}, \quad (5.1.8)$$

where $\mathrm{J}_n(b_\alpha)$ and $\mathrm{J}_n'(b_\alpha)$ are the Bessel function and its derivative with respect to the argument.

This representation of the dielectric tensor of the magneto-active plasma explicitly shows that cyclotron resonances are associated with the zeros of $\omega - k_z v_z - n\Omega_\alpha/\gamma$ in the denominator. The principal value integrals determine the Hermitian part of the tensor $\varepsilon_{ij}(\omega,\boldsymbol{k})$, and the terms containing the δ-function contribute to the anti-Hermitian part responsible for the wave absorption. Hence, in the collisionless magneto-active plasma only particles satisfying the condition

$$\omega - k_z v_z - n\frac{\Omega_\alpha}{\gamma} = 0 \quad (5.1.9)$$

can absorb waves. Equation (5.1.9) replaces the resonance condition $\omega = \boldsymbol{k}\cdot\boldsymbol{v}$ of the unmagnetized plasma.

The absorption mechanism can be easily understood. We decompose the motion of each charged particle into a force-free streaming along the magnetic field line and a perpendicular cyclotron rotation with the frequency Ω_α/γ. Then the charged particle can emit *cyclotron radiation* due to the rotational acceleration and Cherenkov waves because of the straight motion parallel to the field $\boldsymbol{B}_0$. The frequencies of the radiated waves can be obtained from the condition

$$\omega = n\frac{\Omega_\alpha}{\gamma} + k_z v_z\,, \qquad n = 0,\, \pm 1,\, \pm 2, \ldots\,.$$

For $n = 0$ the condition (5.1.9) corresponds to Cherenkov radiation, and for $n \neq 0$ to cyclotron radiation.

It must be emphasized that the anti-Hermitian part of the tensor $\varepsilon_{ij}(\omega, \boldsymbol{k})$ vanishes for $k_z = 0$ in the nonrelativistic plasma ($\gamma = 1$). Waves propagating strictly across the magnetic field remain undamped. In this case there appear no singularities in the integrand of (5.1.7), and the contributions from the δ-terms vanish. This statement is valid only for $\omega \lesssim \Omega_\alpha$, however. In the limit of high frequencies, $\omega \gg \Omega_\alpha$, the effect of the magnetic field is negligible. Then, absorption must also be possible for $k_z = 0$, since this limit corresponds to the isotropic plasma without an external magnetic field. This is physically obvious since the Larmor radius of the particles greatly exceeds the wavelength when $\Omega_\alpha \to 0$. The formal transition to the limit $\Omega_\alpha \to 0$ in (5.1.7) is nontrivial and connected with the problem of the asymptotic representation of the Bessel functions of high order at large arguments. For $\Omega_\alpha \to 0$, the arguments of the Bessel functions in (5.1.8) become large. Then, all terms with $|n| < n_{\max} \approx k_\perp v_\perp/\Omega_\alpha$ contribute to the same order, whereas the terms with $|n| > n_{\max}$ are exponentially small. Therefore, the summation in (5.1.7) must be extended up to $|n| = n_{\max}$. The denominator takes the form of $\omega - n_{\max}\Omega_\alpha = \omega - k_\perp v_\perp = \omega - \boldsymbol{k}\cdot\boldsymbol{v}$ and the collisionless wave absorption in the magnetic field free limit is reproduced. In the relativistic plasma waves with $k_z = 0$ can be absorbed. Due to the velocity-dependent factor $\gamma \neq 1$ in the δ-function the integral in (5.1.7) remains singular, which leads to wave absorption associated with relativistic effects.

5.1.1 Dielectric Tensor of the Quasi-Equilibrium Maxwellian Plasma

For Maxwellian distributions $f_{0\alpha}$ (4.1.1) the dielectric tensor $\varepsilon_{ij}(\omega, \boldsymbol{k})$ can be expressed in terms of tabulated functions. As a result we obtain:

$$\varepsilon_{xx} = 1 - \sum_\alpha \sum_n \frac{n^2 \omega_{p\alpha}^2}{\omega(\omega - n\Omega_\alpha)} \frac{A_n(z_\alpha)}{z_\alpha} I_+(\beta_{n\alpha}) ,$$

$$\varepsilon_{yy} = \varepsilon_{xx} + 2 \sum_\alpha \sum_n \frac{\omega_{p\alpha}^2 z_\alpha}{\omega(\omega - n\Omega_\alpha)} A_n'(z_\alpha) I_+(\beta_{n\alpha}) ,$$

$$\varepsilon_{xy} = -\varepsilon_{yx} = -\mathrm{i} \sum_\alpha \sum_n \frac{n \omega_{p\alpha}^2}{\omega(\omega - n\Omega_\alpha)} A_n'(z_\alpha) I_+(\beta_{n\alpha}) ,$$

$$\varepsilon_{xz} = \varepsilon_{zx} = - \sum_\alpha \sum_n \frac{\omega_{p\alpha}^2 n k_\perp}{\omega \Omega_\alpha k_z} \frac{A_n(z_\alpha)}{z_\alpha} I_+(\beta_{n\alpha}) , \tag{5.1.10}$$

$$\varepsilon_{yz} = -\varepsilon_{zy} = \mathrm{i} \sum_\alpha \sum_n \frac{\omega_{p\alpha}^2 k_\perp}{\omega \Omega_\alpha k_z} A_n'(z_\alpha) I_+(\beta_{n\alpha}) ,$$

$$\varepsilon_{zz} = 1 + \sum_\alpha \sum_n \frac{\omega_{p\alpha}^2}{k_z^2 v_{T\alpha}^2} \left[1 - \frac{n\Omega_\alpha}{\omega} A_n(z_\alpha) I_+(\beta_{n\alpha}) \right] , \quad \text{where}$$

$$A_n(z_\alpha) = \mathrm{e}^{-z_\alpha} \mathrm{J}_n(z_\alpha), \quad z_\alpha = \frac{k_\perp^2 v_{T\alpha}^2}{\Omega_\alpha^2}, \quad \beta_{n\alpha} = \frac{\omega - n\Omega_\alpha}{|k_z| v_{T\alpha}} . \tag{5.1.11}$$

Here $\mathrm{J}_n(z)$ is the Bessel function of an imaginary argument and $I_+(\beta)$ has been defined earlier, see (4.1.15). The asymptotic values of the function $I_+(\beta)$ are given by (4.1.16).

It is often sufficient to know the so-called longitudinal dielectric permittivity (2.5.9) which describes electrostatic oscillations in the anisotropic plasma. According to (2.5.9) and (5.1.7) this permittivity is defined by

$$\varepsilon(\omega, \boldsymbol{k}) = \frac{k_i k_j}{k^2} \varepsilon_{ij}(\omega, \boldsymbol{k}) = 1 - \sum_\alpha \frac{4\pi e_\alpha^2}{k^2} \times \int d\boldsymbol{p}_\alpha \frac{\partial f_{0\alpha}}{\partial \mathscr{E}} \left[1 - \sum_n \frac{\omega \mathrm{J}_n^2(b_\alpha)}{\omega - k_z v_z - n\Omega_\alpha/\gamma} \right] . \tag{5.1.12}$$

For Maxwellian distributions $f_{0\alpha}(\boldsymbol{p})$ it takes the form

$$\varepsilon(\omega, \boldsymbol{k}) = 1 + \sum_\alpha \frac{\omega_{p\alpha}^2}{k^2 v_{T\alpha}^2} \left[1 - \sum_n \frac{\omega}{\omega - n\Omega_\alpha} A_n(z_\alpha) I_+(\beta_{n\alpha}) \right] . \tag{5.1.13}$$

5.1.2 Dielectric Tensor of the Degenerate Plasma

For the degenerate plasma where the Fermi distribution characterizes the equilibrium an explicit expression of the dielectric tensor is available, too. We obtain from (5.1.7)

$$\varepsilon_{ij}(\omega, \boldsymbol{k}) = \delta_{ij} - \sum_{\alpha}\sum_{n}\frac{3\,\omega_{p\alpha}^2}{2\,\omega v_{F\alpha}^2}\int_0^{\pi} \sin\theta\, d\theta \frac{\Pi_{ij}^{(n)}(\theta)}{\omega - k_z v_{F\alpha}\cos\theta - n\Omega_\alpha}\,, \tag{5.1.14}$$

where $\Pi_{ij}^{(n)}(\theta)$ is the tensor

$$\Pi_{ij}^{(n)}(\theta) = \begin{pmatrix} \dfrac{n^2 v_{F\alpha}^2 \sin^2\theta J_n^2}{\xi_\alpha^2} & \dfrac{i v_{F\alpha}^2 \sin^2\theta J_n J_n'}{\xi_\alpha^2} & \dfrac{n v_{F\alpha}^2 \sin 2\theta J_n^2}{2\xi_\alpha^2} \\[2ex] -\dfrac{i v_{F\alpha}^2 \sin^2\theta J_n J_n^2}{\xi_\alpha} & v_{F\alpha}^2 \sin^2\theta J_n'^2 & -\dfrac{i v_{F\alpha}^2 \sin 2\theta J_n J_n'}{2} \\[2ex] \dfrac{n v_{F\alpha}^2 \sin 2\theta J_n^2}{2\xi_\alpha^2} & \dfrac{i v_{F\alpha}^2 \sin 2\theta J_n J_n'}{2} & v_{F\alpha}^2 \sin 2\theta J_n^2 \end{pmatrix} \tag{5.1.15}$$

and $J_n(\xi_\alpha)$ and $J_n'(\xi_\alpha)$ are the Bessel function and its derivative with respect to the argument $\xi_\alpha = k_\perp v_{F\alpha}\sin\theta/\Omega_\alpha$. Equation (5.1.14) leads to the following expression for the longitudinal permittivity of the degenerate plasma:

$$\begin{aligned} \varepsilon(\omega, \boldsymbol{k}) &= 1 + \sum_{\alpha}\frac{3\,\omega_{p\alpha}^2}{k^2 v_{F\alpha}^2}\left[1 - \frac{\omega}{2}\sum_{n}\int_0^{\pi}\sin\theta\, d\theta\right. \\ &\quad \left.\times J_n^2\left(\frac{k_\perp v_{F\alpha}\sin\theta}{\Omega_\alpha}\right)\frac{1}{\omega - k_z v_{F\alpha}\cos\theta - n\Omega_\alpha}\right]. \end{aligned} \tag{5.1.16}$$

5.2 Dielectric Permittivity and Oscillation Spectra of the Cold Collisionless Magneto-Active Plasma

The detailed analysis of the oscillations of a collisionless magneto-active plasma is very complex since the number of the solution branches is practically infinitely large compared to the number of modes existing in the unmagnetized plasma. Therefore, we shall deal with the most interesting and thoroughly studied oscillation types only. To begin with, we entirely neglect the thermal motion of the particles. In this cold plasma approximation the inequalities

$$\frac{k_\perp v_{0\alpha}}{\Omega_\alpha} \ll 1\,, \qquad \frac{k_z v_{0\alpha}}{\omega} \ll 1\,, \qquad \beta_{n\alpha} = \frac{\omega \pm n\Omega_\alpha}{k_z v_{0\alpha}} \gg 1 \tag{5.2.1}$$

must be satisfied, where $v_{0\alpha} = v_{T\alpha}$ for the nondegenerate plasma and $v_{0\alpha} = v_{F\alpha}$ for the degenerate plasma.

According to (5.1.10 and 14) the dielectric tensor has the form

$$\varepsilon_{ij} = \begin{pmatrix} \varepsilon_\perp & \mathrm{i}g & 0 \\ -\mathrm{i}g & \varepsilon_\perp & 0 \\ 0 & 0 & \varepsilon_\parallel \end{pmatrix} \tag{5.2.2}$$

This holds for the nondegenerate and the degenerate plasma when the thermal motion of the particles is negligible. The elements of ε_{ij} are

$$\begin{aligned} \varepsilon_\perp &= \varepsilon_{xx} = \varepsilon_{yy} = 1 - \sum_\alpha \frac{\omega_{p\alpha}^2}{\omega^2 - \Omega_\alpha^2}\,, \qquad \varepsilon_\parallel = \varepsilon_{zz} = 1 - \sum_\alpha \frac{\omega_{p\alpha}^2}{\omega^2}\,, \\ \varepsilon_{xy} &= -\varepsilon_{yx} = \mathrm{i}g = -\mathrm{i}\sum_\alpha \frac{\omega_{p\alpha}^2 \Omega_\alpha}{\omega(\omega^2 - \Omega_\alpha^2)}\,, \qquad \varepsilon_{xz} = \varepsilon_{zx} = \varepsilon_{yz} = \varepsilon_{zy} = 0\,. \end{aligned} \tag{5.2.3}$$

Since dissipative processes are neglected in the cold plasma approximation the tensor (5.2.2) is Hermitian.

Substituting (5.2.2, 3) into the wave equation

$$\left[k^2 \delta_{ij} - k_i k_j - \frac{\omega^2}{c^2}\varepsilon_{ij}(\omega)\right] E_j = 0 \tag{5.2.4}$$

we obtain

$$\begin{aligned} &\left(k_z^2 - \frac{\omega^2}{c^2}\varepsilon_\perp\right) E_x - \mathrm{i}\frac{\omega^2}{c^2} g E_y - k_\perp k_z E_z = 0\,, \\ &\mathrm{i}g\frac{\omega^2}{c^2} E_x + \left(k^2 - \frac{\omega^2}{c^2}\varepsilon_\perp\right) E_y = 0\,, \\ &-k_\perp k_z E_x + \left(k_\perp^2 - \frac{\omega^2}{c^2}\varepsilon_\parallel\right) E_z = 0\,. \end{aligned} \tag{5.2.5}$$

The condition for the existence of solutions of this system yields the dispersion equation for electromagnetic waves in the cold magneto-active plasma:

$$\begin{aligned} \Lambda(\omega, \boldsymbol{k}) = {} & k^2\left(\varepsilon_\perp k_\perp^2 + \varepsilon_\parallel k_z^2\right) - \frac{\omega^2}{c^2}\left[(\varepsilon_\perp^2 - g^2 - \varepsilon_\perp\varepsilon_\parallel) k_\perp^2 \right. \\ & \left. + 2k^2\varepsilon_\perp\varepsilon_\parallel\right] + \frac{\omega^4}{c^4}\varepsilon_\parallel(\varepsilon_\perp^2 - g^2) = 0\,. \end{aligned} \tag{5.2.6}$$

5.2.1 Wave Propagation Along the Magnetic Field

It follows from (5.2.5) that in contrast to the magnetic-field-free case the longitudinal and transverse waves are not independent in the magneto-active plasma. Equation (5.2.6) does not separate into independent equations for the transverse and longitudinal waves in general. Only for purely longitudinal wave propagation when the wave vector is directed along the magnetic field ($k_\perp = 0$) is the separation is complete. Then, the first two equations of the system (5.2.5) describe transverse waves obeying the dispersion equation

$$k^2c^2 = \omega^2(\varepsilon_\perp \pm g) . \tag{5.2.7}$$

The two signs on the right-hand side correspond to two independent modes with different polarization. The left-handed mode with

$$\frac{E_{y1}}{E_{x1}} = -\mathrm{i} \tag{5.2.8}$$

is called the *ordinary wave*, and the right-handed one with

$$\frac{E_{y2}}{E_{x2}} = \mathrm{i} \tag{5.2.9}$$

the *extraordinary wave*.

For purely longitudinal propagation ($k_\perp = 0$) a third solution described by the third equation of the system (5.2.5) exists. These longitudinal oscillations of the cold magneto-active plasma obey the dispersion equation

$$\varepsilon_\parallel = 0 \tag{5.2.10}$$

since $\boldsymbol{E} \parallel \boldsymbol{k} \parallel \boldsymbol{B}_0$. The frequency of these longitudinal oscillations is given by

$$\omega^2 = \omega_{\mathrm{pe}}^2 . \tag{5.2.11}$$

The solutions of (5.2.7) for the transverse waves with $k_\perp = 0$ are rather complicated in general. Simple analytic expressions for the frequency of these waves can be obtained by an asymptotic evaluation only

$$\begin{aligned}
\omega^2 &= k^2c^2 + \omega_{\mathrm{pe}}^2\left(1 \mp \frac{\Omega_{\mathrm{e}}}{\sqrt{k^2c^2 + \omega_{\mathrm{pe}}^2}}\right) \quad \text{for} \quad \omega \gg \Omega_{\mathrm{e}} , \\
\omega &= \frac{k^2c^2\Omega_{\mathrm{e}}}{\omega_{\mathrm{pe}}^2} \quad \text{for} \quad \Omega_{\mathrm{i}} \ll \omega \ll \Omega_{\mathrm{e}}, \omega_{\mathrm{pe}}^2/\Omega_{\mathrm{e}} , \\
\omega^2 &= \frac{k^2v_{\mathrm{A}}^2}{1 + v_{\mathrm{A}}^2/c^2}\left[1 \mp \frac{v_{\mathrm{A}}k}{\Omega_{\mathrm{i}}(1 + v_{\mathrm{A}}^2/c^2)^{3/2}}\right] \quad \text{for} \quad \omega \ll \Omega_{\mathrm{i}} ,
\end{aligned} \tag{5.2.12}$$

where $v_{\mathrm{A}} \equiv c\Omega_{\mathrm{i}}/\omega_{\mathrm{pi}} = B_0/\sqrt{4\pi N_{\mathrm{i}} M}$ is the Alfvén velocity.

Using these expressions and taking into account that due to (5.2.7) the solutions $\omega \to \Omega_{\mathrm{i}}$ and $\omega \to \Omega_{\mathrm{e}}$ exist for $k^2 \to \infty$, we can sketch the general course $\omega(k)$ of the solution branches.

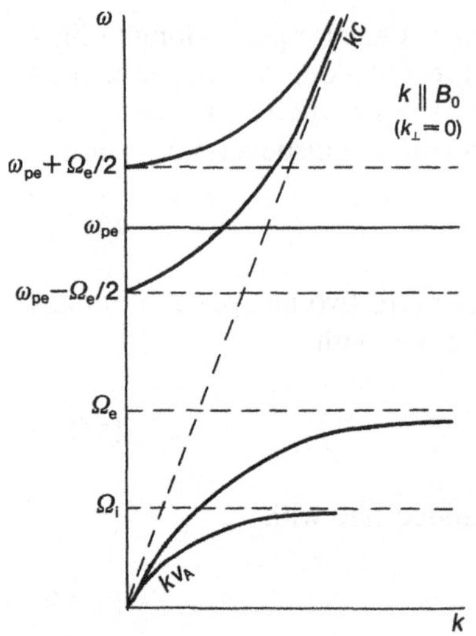

Fig. 5.1. Spectra of electromagnetic waves in a cold magneto-active plasma – longitudinal propagation

For $\omega_{pe} > \Omega_e$ the result is shown in Fig. 5.1. In the cold magneto-active plasma there exist five branches of oscillations propagating strictly along the external magnetic field. Four of them describe transverse waves; one branch represents longitudinal waves.

In the frequency range $\omega < \Omega_i$ the transverse waves are called *magnetohydrodynamic* (MHD) *waves*: the oscillation branch with the asymptotic behaviour $\omega \to \Omega_i$ for $k \to \infty$ is called the *Alfvén wave*. The oscillation branch with the limit $\omega \to \Omega_e$ for $k \to \infty$ is called the fast *magnetosonic wave* in the range $\omega < \Omega_i$ and the *spiral wave* or *helicon* in the range of intermediate frequencies $\Omega_i < \omega < \Omega_e$. Note that these oscillations are undamped in the collisionless limit.

5.2.2 Wave Propagation Across the Magnetic Field

In the case of strictly transverse wave propagation ($k_z = 0$) the analysis of (5.2.6) is equally simple. It separates into the two equations:

$$k^2 = \frac{\omega^2}{c^2}\varepsilon_\parallel \, , \qquad k^2 = \frac{\omega^2}{c^2}\frac{\varepsilon_\perp^2 - g^2}{\varepsilon_\perp} \tag{5.2.13}$$

describing ordinary and extraordinary waves, respectively, in the magneto-active plasma. The ordinary wave is purely transverse. Only the component E_z of the electric field is nonzero and the frequency spectrum is

$$\omega^2 = k^2c^2 + \omega_{pe}^2 \, . \tag{5.2.14}$$

The extraordinary wave is longitudinal-transverse with nonzero components E_y and E_x of the field. The frequency of this wave can be obtained analytically only in the limiting cases:

$$\omega^2 = \tfrac{1}{2}\left[\Omega_e^2 + 2\,\omega_{pe}^2 + k^2c^2 \pm \sqrt{4\,\omega_{pe}^2\,\Omega_e^2 + (k^2c^2 - \Omega_e^2)^2}\right] \quad \text{for} \quad \omega \gg \Omega_e\,,$$

$$\omega^2 = \frac{\omega_{pe}^2\,k^2c^2}{k^2c^2\,(1 + \omega_{pe}^2/\Omega_e^2) + \omega_{pe}^4/\Omega_e^2} \quad \text{for} \quad \Omega_i \ll \omega \ll \Omega_e\,, \tag{5.2.15}$$

$$\omega^2 = \frac{k^2\,v_A^2}{1 + v_A^2/c^2} \quad \text{for} \quad \omega \gg \Omega_i\,.$$

Using (5.2.14, 15) and noting that the second equation of (5.2.13) has the asymptotic values $\omega \to \sqrt{\omega_{pe}^2 + \Omega_e^2}$ and

$$\omega = \omega_{pi}/\sqrt{1 + \omega_{pe}^2/\Omega_e^2} \approx \sqrt{|\Omega_e\,\Omega_i|}$$

for $k^2 \to \infty$ it is easy to draw the dispersion curves $\omega(k)$. We show them in Fig. 5.2 for the plasma with $\omega_{pe} > |\Omega_e|$. In the cold collisionless plasma there are four branches of oscillations propagating strictly across the magnetic field. Three branches describe longitudinal-transverse waves and one branch, the one with $\omega = \omega_{pe}$ for $k \to 0$, is purely transverse. The branch existing in the frequency range $\omega < \sqrt{|\Omega_e\Omega_i|}$ is the MHD wave, propagating across the magnetic field $\boldsymbol{B}_0$. Oscillations corresponding to the Alfvén branch are absent.

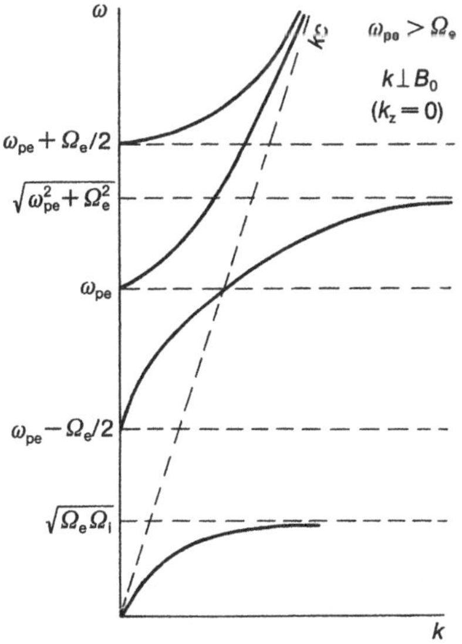

Fig. 5.2. Spectra of electromagnetic waves in a cold magneto-active plasma – transverse propagation

5.2.3 An Arbitrary Direction of Wave Propagation

In the general case of an arbitrary direction of wave propagation the analytic solution of (5.2.6) is rather difficult. Simple formulas for $\omega(k)$ can be obtained only for the Alfvén and MHD waves in the range of low frequencies

$$\omega^2 = \frac{k^2 v_A^2 \cos^2\theta}{1 + v_A^2/c^2}, \qquad \omega^2 = \frac{k^2 v_A^2}{1 + v_A^2/c^2}, \tag{5.2.16}$$

for helical waves (helicons) in the range of intermediate frequencies $\Omega_i \ll \omega \ll \Omega_e$, ω_{pe}^2/Ω_e

$$\omega = \frac{k^2 c^2}{\omega_{pe}^2} |\Omega_e \cos\theta| , \tag{5.2.17}$$

and for the ordinary and extraordinary waves which in the high frequency limit $\omega \gg \Omega_e$ approach

$$\omega^2 \rightarrow k^2 c^2 + \omega_{pe}^2 . \tag{5.2.18}$$

Here θ is the angle between the vectors $\boldsymbol{B}_0$ and $\boldsymbol{k}$.

The five oscillation branches $\omega(k)$ are shown in Fig. 5.3 for oblique propagation $\theta \neq 0$, $\pi/2$ assuming $\omega_{pe} > \Omega_e$.

5.2.4 Longitudinal Oscillations of the Magneto-Active Plasma

Concluding this section we consider the question when the waves in the cold magneto-active plasma are longitudinal and what their frequency spectrum

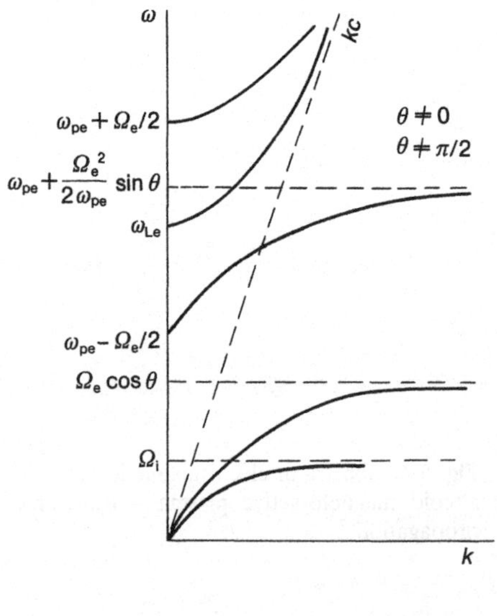

Fig. 5.3. Spectra of electromagnetic waves in a cold magneto-active plasma

is. We have already mentioned that only waves propagating along the external magnetic field are strictly longitudinal. However, if we formally derive the electric field of the wave from a potential, i.e., $\boldsymbol{E} = -\mathrm{i}\boldsymbol{k}\Phi$, then we obtain from (5.2.4) the condition for the existence of a longitudinal field in the cold magneto-active plasma:

$$k_i k_j \varepsilon_{ij}(\omega) = k_\perp^2 \varepsilon_\perp + k_z^2 \varepsilon_\parallel = 0 \,. \tag{5.2.19}$$

It is easy to solve this dispersion equation and to obtain the longitudinal wave spectra

$$\begin{aligned} \omega^2 &= \frac{\omega_{\rm pe}^2 + \Omega_{\rm e}^2}{2} \pm \frac{1}{2}\left[(\omega_{\rm pe}^2 + \Omega_{\rm e}^2)^2 - 4\,\omega_{\rm pe}^2 \Omega_{\rm e}^2 \cos^2\theta\right]^{1/2} , \\ \omega^2 &= \left(1 - \frac{m}{M} z \tan^2\theta\right) \Omega_{\rm i}^2 \,, \end{aligned} \tag{5.2.20}$$

where $z = |e_{\rm i}/e|$. These expressions are invalid near $\theta = \pi/2$, for $\cos^2\theta < m/M$. Instead the spectra of the longitudinal waves are in this case

$$\omega^2 = \omega_{\rm pe}^2 + \Omega_{\rm e}^2 \,, \qquad \omega^2 = \frac{\Omega_{\rm e}^2 \omega_{\rm pi}^2}{\omega_{\rm pe}^2 + \Omega_{\rm e}^2} \approx \Omega_{\rm e} \Omega_{\rm i} \,. \tag{5.2.21}$$

In the literature the oscillations corresponding to the first solution are called the *upper hybrid* and the oscillations

$$\omega^2 = \frac{\omega_{\rm Le}^2 \Omega_{\rm i}^2 \cos^2\theta}{\omega_{\rm Le}^2 + \Omega_{\rm e}^2}$$

corresponding to the second solution the *lower hybrid mode*. The θ-dependence of (5.2.19) is shown in Fig. 5.4 for $\omega_{\rm pe} > \Omega_{\rm e}$.

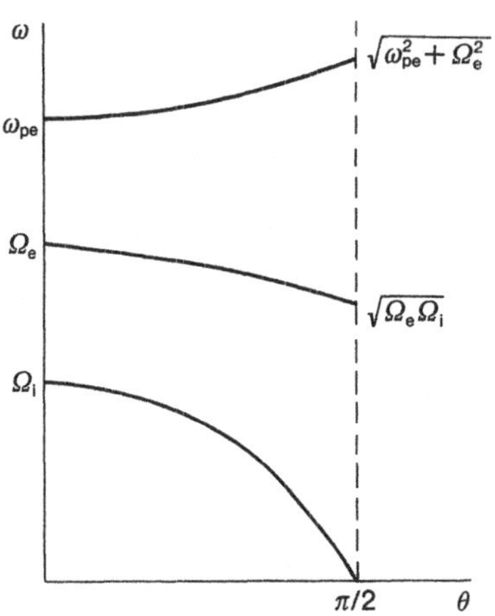

Fig. 5.4. Spectra of longitudinal oscillations in a cold magneto-active plasma

The answer to the question when longitudinal waves do exist in the magneto-active plasma is as follows. When the roots of (5.2.19) are close to the roots of (5.2.6), the waves corresponding to the latter can be considered longitudinal with a high degree of accuracy. It is easily seen that this is valid for arbitrary parameter values when

$$k \gg \omega/c \, . \tag{5.2.22}$$

Under this condition the waves in the magneto-active plasma are longitudinal. For $\theta \neq \pi/2$ there are three such branches and for $\theta = \pi/2$ only two.

5.3 Oscillations in Collisionless Magneto-Active Plasmas Taking Account of Thermal Effects

The account of the thermal particle motion leads to a damping of the five oscillation branches considered above. Besides, due to the thermal motion new oscillation branches appear in the magneto-active plasma, the most interesting being the low-frequency ion-acoustic oscillations and the cyclotron waves. We shall study the modification of the cold plasma branches first.

Using the general dielectric tensor (5.1.10 or 14) for the hot nondegenerate or degenerate plasma, the dispersion equation can be written as follows:

$$\begin{aligned} \Lambda(\omega, \boldsymbol{k}) &= \left| k^2 \delta_{ij} - k_i k_j - \frac{\omega^2}{c^2} \varepsilon_{ij}(\omega, \boldsymbol{k}) \right| \\ &= A k^4 + B \frac{\omega^2}{c^2} k^2 + C \frac{\omega^4}{c^4} = 0 \, , \quad \text{where} \end{aligned} \tag{5.3.1}$$

$$A = \frac{k_i k_j}{k^2} \varepsilon_{ij}(\omega, \boldsymbol{k}) = \frac{k_\perp^2}{k^2} \varepsilon_{xx} + \frac{k_z^2}{k^2} \varepsilon_{zz} + 2 \frac{k_\perp k_z}{k^2} \varepsilon_{xz} \, ,$$

$$\begin{aligned} B = \frac{k_i k_j}{k^2} \varepsilon_{i\mu} \varepsilon_{\mu j} - A \varepsilon_{ii}(\omega, \boldsymbol{k}) &= -\varepsilon_{xx} \varepsilon_{zx} + \varepsilon_{xz}^2 - \frac{k_z^2}{k^2} (\varepsilon_{yy} \varepsilon_{zz} + \varepsilon_{yz}^2) \\ &- \frac{k_\perp^2}{k^2} (\varepsilon_{xx} \varepsilon_{yy} + \varepsilon_{xy}^2) + 2 \frac{k_\perp k_z}{k^2} (\varepsilon_{xy} \varepsilon_{yz} - \varepsilon_{xz} \varepsilon_{zy}) \, , \end{aligned} \tag{5.3.2}$$

$$C = \operatorname{Det} |\varepsilon_{ij}(\omega, \boldsymbol{k})| = \varepsilon_{zz} (\varepsilon_{xx} \varepsilon_{yy} + \varepsilon_{xy}^2) + \varepsilon_{xx} \varepsilon_{yz}^2 - \varepsilon_{yy} \varepsilon_{xz}^2 + 2 \varepsilon_{xy} \varepsilon_{xz} \varepsilon_{yz} \, .$$

It is easy to see that $A(\omega, \boldsymbol{k})$ coincides with the "longitudinal" dielectric permittivity of the magneto-active plasma. Thus, we can obtain a condition for potential waves by comparing the solution of the exact dispersion equation (5.3.1) with the corresponding zero of the longitudinal dispersion equation

$$A(\omega, \boldsymbol{k}) = \varepsilon(\omega, \boldsymbol{k}) = \frac{k_i k_j}{k^2} \varepsilon_{ij}(\omega, \boldsymbol{k}) = 0 . \tag{5.3.3}$$

Taking into account that the coefficients A, B and C are linear, quadratic and cubic combinations of the components of the dielectric tensor and, consequently, of the plasma density, we arrive at the conclusion that all the components of the tensor are of the order one in the range of high frequencies ($\omega \gtrsim \omega_{\mathrm{pe}} \sim \Omega_{\mathrm{e}}$) and the condition that the oscillations are derivable from a scalar potential should be

$$\omega^2 \ll k^2 c^2 . \tag{5.3.4}$$

In the low-frequency range ($\omega \ll \Omega_{\mathrm{i}} \ll \omega_{\mathrm{pi}}$) the components of the tensor $\varepsilon_{ij} \sim \omega_{\mathrm{pi}}^2 / \Omega_{\mathrm{i}}^2 = c^2 / v_{\mathrm{A}}^2$ are large and this condition is more restrictive

$$\omega^2 \ll k^2 v_{\mathrm{A}}^2 . \tag{5.3.5}$$

The conditions (5.3.4, 5) are necessary but not sufficient in the respective frequency ranges. Oscillations not derivable from a potential are possible along with the potential oscillations. One potential branch is always present among the large number of oscillation branches, however.

5.3.1 Collisionless Damping of Waves in the Magneto-Active Plasma

We analyze the general dispersion equation (5.3.1) in the limit (5.2.1), when the thermal effects are weak, first. In the foregoing section such dissipative processes were completely ignored. The dielectric tensor (5.2.2) turned out to be Hermitian and, consequently, the plasma was nonabsorbing. The account of the Cherenkov and cyclotron dissipative mechanisms due to a weak thermal motion results in a small anti-Hermitian correction of the tensor (5.2.2) which is nonzero only for the nondegenerate (Maxwellian) plasma:

$$\delta\varepsilon_{ij} = \varepsilon_{ij}^{\mathrm{a}} = \begin{pmatrix} \varepsilon_{xx}^{\mathrm{a}} & \mathrm{i}g^{\mathrm{a}} & 0 \\ -\mathrm{i}g^{\mathrm{a}} & \varepsilon_{yy}^{\mathrm{a}} & 0 \\ 0 & 0 & \varepsilon_{\|}^{\mathrm{a}} \end{pmatrix} , \quad \text{where} \tag{5.3.6}$$

$$\varepsilon_{xx}^{\mathrm{a}} = \varepsilon_{\perp}^{\mathrm{a}} = \mathrm{i}\sqrt{\frac{\pi}{8}} \sum_{\alpha} \frac{\omega_{\mathrm{p}\alpha}^2}{\omega |k_z| v_{\mathrm{T}\alpha}} \left[\exp\left(-\frac{(\omega - \Omega_\alpha)^2}{2k_z^2 v_{\mathrm{T}\alpha}^2}\right) + \exp\left(-\frac{(\omega + \Omega_\alpha)^2}{2k_z^2 v_{\mathrm{T}\alpha}^2}\right)\right]$$

$$\varepsilon_{yy}^{\mathrm{a}} = \varepsilon_{\perp}^{\mathrm{a}} + \mathrm{i}\sqrt{2\pi} \sum_{\alpha} \frac{\omega_{\mathrm{p}\alpha}^2 k_{\perp}^2 v_{\mathrm{T}\alpha}}{\omega \Omega_\alpha^2 |k_z|} \exp\left(-\frac{\omega^2}{2k_z^2 v_{\mathrm{T}\alpha}^2}\right) ,$$

$$g^{\mathrm{a}} = \mathrm{i}\sqrt{\frac{\pi}{8}}\sum_{\alpha}\frac{\omega_{\mathrm{p}\alpha}^2}{|k_z|\omega v_{\mathrm{T}\alpha}}\left[\exp\left(-\frac{(\omega-\Omega_\alpha)^2}{2k_z^2 v_{\mathrm{T}\alpha}^2}\right) - \exp\left(-\frac{(\omega+\Omega_\alpha)^2}{2k_z^2 v_{\mathrm{T}\alpha}^2}\right)\right]$$

$$\varepsilon_{\|}^{\mathrm{a}} = \mathrm{i}\sqrt{\frac{\pi}{2}}\sum_{\alpha}\frac{\omega\omega_{\mathrm{p}\alpha}^2}{|k_z|^3 v_{\mathrm{T}\alpha}^3}\exp\left(-\frac{\omega^2}{2k_z^2 v_{\mathrm{T}\alpha}^2}\right). \tag{5.3.7}$$

The damping decrement resulting from this correction can be expressed by the general and very complex formula

$$\delta = -\frac{\mathrm{Im}\{\Lambda(\omega,\boldsymbol{k})\}}{\dfrac{\partial}{\partial\omega}\mathrm{Re}\{\Lambda(\omega,\boldsymbol{k})\}} = -\frac{k^4 A^{\mathrm{a}} + \dfrac{\omega^2}{c^2}k^2 B^{\mathrm{a}} + \dfrac{\omega^4}{c^4}C^{\mathrm{a}}}{k^4\dfrac{\partial A^{\mathrm{H}}}{\partial\omega} + k^2\dfrac{\partial}{\partial\omega}\dfrac{\omega^2}{c^2}B^{\mathrm{H}} + \dfrac{\partial}{\partial\omega}\dfrac{\omega^4}{c^4}C^{\mathrm{H}}}, \tag{5.3.8}$$

where

$$A^{\mathrm{a}} = \frac{k_\perp^2}{k^2}\varepsilon_\perp^{\mathrm{a}} + \frac{k_z^2}{k^2}\varepsilon_{\|}^{\mathrm{a}},$$

$$B^{\mathrm{a}} = -\left(1+\frac{k_z^2}{k^2}\right)\left(\varepsilon_\perp^{\mathrm{a}}\varepsilon_{\|}^{\mathrm{H}} + \varepsilon_{\|}^{\mathrm{a}}\varepsilon_\perp^{\mathrm{H}}\right) - 2\frac{k_\perp^2}{k^2}\left(\varepsilon_\perp^{\mathrm{a}}\varepsilon_\perp^{\mathrm{H}} - g^{\mathrm{a}}g^{\mathrm{H}}\right)$$

$$-\frac{k_\perp^2}{k^2}\left(\varepsilon_{yy}^{\mathrm{a}} - \varepsilon_\perp^{\mathrm{a}}\right)\varepsilon_\perp^{\mathrm{H}} - \frac{k_z^2}{k^2}\left(\varepsilon_{yy}^{2} - \varepsilon_\perp^{\mathrm{a}}\right)\varepsilon_{\|}^{\mathrm{H}},$$

$$C^{\mathrm{a}} = \varepsilon_{\|}^{\mathrm{a}}\left(\varepsilon_\perp^{\mathrm{H}2} - g^{\mathrm{H}2}\right) + 2\varepsilon_{\|}^{\mathrm{H}}\left(\varepsilon_\perp^{\mathrm{a}}\varepsilon_\perp^{\mathrm{H}} - g^{\mathrm{a}}g^{\mathrm{H}}\right) + \varepsilon_\perp^{\mathrm{H}}\varepsilon_{\|}^{\mathrm{H}}\left(\varepsilon_{yy}^{\mathrm{a}} - \varepsilon_\perp^{\mathrm{a}}\right) \tag{5.3.9}$$

$$A^{\mathrm{H}} = \frac{k_\perp^2}{k^2}\varepsilon_\perp^{\mathrm{H}} + \frac{k_z^2}{k^2}\varepsilon_{\|}^{\mathrm{H}},$$

$$B^{\mathrm{H}} = -\varepsilon_\perp^{\mathrm{H}}\varepsilon_{\|}^{\mathrm{H}}\left(1+\frac{k_z^2}{k^2}\right) - \frac{k_\perp^2}{k^2}\left(\varepsilon_\perp^{\mathrm{H}2} - g^{\mathrm{H}2}\right),$$

$$C^{\mathrm{H}} = \varepsilon_{\|}^{\mathrm{H}}\left(\varepsilon_\perp^{\mathrm{H}2} - g^{\mathrm{H}2}\right).$$

The components $\varepsilon_{ij}^{\mathrm{H}}$ and $\varepsilon_{ij}^{\mathrm{a}}$ are determined by (5.2.2) and (5.3.6), respectively.

In the range of high frequencies $\omega \gtrsim \Omega_{\mathrm{e}}$ the expression for the damping decrement δ as well as for the frequency $\omega(k)$ is rather involved. We therefore do not give the result and confine our interest to the range of low frequencies $\omega \ll \Omega_{\mathrm{e}}$, where the oscillation spectra are defined by (5.2.16) for the Alfvén and MHD oscillation branches and to the range of intermediate frequencies $\Omega_{\mathrm{i}} \ll \omega \ll \Omega_{\mathrm{e}}$, where the helicons (5.2.17) exist. The formulas for the corresponding damping decrements δ are

$$\delta = -\sqrt{\frac{\pi}{8}}\frac{\omega^6 k_\perp^2}{\Omega_i^2 |k_z|^5 v_{Te}^3}\exp\left(-\frac{\omega^2}{2k_z^2 v_{Te}^2}\right),$$

$$\delta = -\sqrt{\frac{\pi}{8}}\frac{m\omega k_\perp^2 v_{Te}}{Mk|k_z| v_A}\exp\left(-\frac{\omega^2}{2k_z^2 v_{Te}^2}\right), \tag{5.3.10}$$

$$\delta = -\sqrt{\frac{\pi}{8}}\frac{\omega^7 k_\perp^2}{2\Omega_e k|k_z|^6 v_{Te}^5}\exp\left(-\frac{\omega^2}{2k_z^2 v_{Te}^2}\right).$$

It is clear from the result that the Cherenkov absorption by the plasma electrons is responsible for the collisionless oscillation damping in these frequency ranges.

So far we have considered the nondegenerate Maxwellian plasma. As to the degenerate plasma with the Fermi distribution function, the anti-Hermitian part of the dielectric tensor vanishes exactly under the conditions (5.2.1), i.e., when the phase velocity exceeds the Fermi velocity. Consequently, there is no wave absorption in the collisionless degenerate plasma.

5.3.2 Spectra of Low-Frequency Slow Waves

We may naturally expect that new branches of oscillations appear in the hot plasma, in addition to the five branches of the cold magneto-active plasma. In fact, the low-frequency ion-acoustic oscillation branch existing in the unmagnetized nonisothermal plasma with $T_e \gg T_i$ in the range of phase velocities smaller than thermal velocity of the electrons occurs in the magneto-active plasma, too. To realize this statement we consider the frequency range

$$v_{Ti} \ll \frac{\omega}{k_z} \ll v_{Te}, \quad \omega \ll \Omega_i \ll \omega_{pi}. \tag{5.3.11}$$

Confining our interest to wavelengths longer than the Larmor radii, $k_\perp^2 v_{T\alpha}^2 \ll \Omega_\alpha^2$, the components (5.1.10) of the dielectric tensor of the nondegenerate plasma simplify. We obtain

$$\varepsilon_{ij}(\omega, \boldsymbol{k}) = \begin{pmatrix} \varepsilon_{xx} & 0 & 0 \\ 0 & \varepsilon_{yy} & \varepsilon_{yz} \\ 0 & \varepsilon_{zy} & \varepsilon_{zz} \end{pmatrix}, \quad \text{where} \tag{5.3.12}$$

$$\varepsilon_{xx} = \frac{\omega_{pi}^2}{\Omega_i^2} = \frac{c^2}{v_A^2},$$

$$\varepsilon_{yy} = \frac{c^2}{v_A^2} + i\sqrt{2\pi}\frac{\omega_{pe}^2 k_\perp^2 v_{Te}}{\Omega_e^2 \omega |k_z|},$$

$$\varepsilon_{yz} = -\varepsilon_{zy} = -\mathrm{i}\frac{\omega_{pe}^2 k_\perp}{\omega \Omega_e k_z}\left(1 + \mathrm{i}\sqrt{\frac{\pi}{2}}\frac{\omega}{|k_z| v_{Te}}\right),$$

$$\varepsilon_{zz} = -\frac{\omega_{pi}^2}{\omega^2} + \frac{\omega_{pe}^2}{k_z^2 v_{Te}^2}\left(1 + \mathrm{i}\sqrt{\frac{\pi}{2}}\frac{\omega}{|k_z| v_{Te}}\right). \tag{5.3.13}$$

Using (5.3.12) the dispersion equation (5.3.1) separates into two equations:

$$k_z^2 - \frac{\omega^2}{c^2}\varepsilon_{xx} = 0, \quad \left(k^2 - \frac{\omega^2}{c^2}\varepsilon_{yy}\right)\varepsilon_{zz} + \frac{\omega^2}{c^2}\varepsilon_{yz}^2 = 0. \tag{5.3.14}$$

The first one describes Alfvén waves which remain undamped in the considered approximation and have the spectrum

$$\omega^2 = k_z^2 v_A^2 = k^2 v_A^2 \cos^2\theta . \tag{5.3.15}$$

Thus, the branch of the fast Alfvén waves for $v_A \gg v_{Te}$ with the spectrum (5.2.16) extends into the range $v_{Ti} \ll v_A \ll v_{Te}$.

From the second equation we obtain the so-called *fast* and *slow magnetosonic waves*:

$$\omega_\pm^2 = \tfrac{1}{2}k^2\left[v_A^2 + v_s^2 \pm \sqrt{(v_A^2 + v_s^2)^2 - 4 v_A^2 v_s^2 \cos^2\theta}\,\right],$$
$$\delta_\pm = -\sqrt{\frac{\pi m}{8M}}\frac{k v_s}{2|\cos\theta|}\left(1 \pm \frac{(v_s^2\cos^2\theta - v_A^2)\cos 2\theta}{\sqrt{v_A^4 + v_s^4 - 2 v_s^2 v_A^2 \cos 2\theta}}\right), \tag{5.3.16}$$

where $v_s = \sqrt{zT_e/M}$ is the ion sound velocity and θ is the angle of wave propagation.

When the plasma pressure is low, $\beta = v_s^2/v_A^2 \ll 1$, the spectrum (5.3.16) takes an especially simple form. Then the fast magnetosonic waves become purely transverse ($\boldsymbol{E} \perp \boldsymbol{k}$) and obey

$$\omega_+^2 = k^2 v_A^2, \quad \delta_+ = -\sqrt{\frac{\pi m}{8M}}\frac{v_s \sin^2\theta}{v_A|\cos\theta|}\omega_+ . \tag{5.3.17}$$

We see that the fast magnetosonic waves extend the branch of the fast MHD waves with the spectrum (5.2.16) into the range of small phase velocities $\omega/k_z < v_{Te}$.

The slow magnetosonic wave becomes purely longitudinal in the low-pressure plasma $\beta \ll 1$ and its spectrum takes the form

$$\omega_-^2 = k^2 v_s^2 \cos^2\theta, \quad \delta_- = -\sqrt{\frac{\pi m}{8M}}\omega_- . \tag{5.3.18}$$

This is a new sixth oscillation branch of the magneto-active hot plasma analogous to the ion-acoustic waves of the isotropic plasma. It also exists only if the electrons are sufficiently hot compared with the ions, $T_e \gg T_i$.

5.3.3 Degenerate Plasma

Low-frequency slow waves can exist in the degenerate plasma, too. Under the conditions

$$v_{Fi} \ll \frac{\omega}{k_z} \ll v_{Fe}\ ; \qquad \omega^2 \ll \Omega_i^2 \ll \omega_{pi}^2\ ; \qquad \frac{k_\perp^2 v_{F\alpha}^2}{\Omega_\alpha^2} \ll 1 \tag{5.3.19}$$

the dielectric tensor (5.1.14) takes the form (5.3.12) with

$$\begin{aligned}
\varepsilon_{xx} &= \frac{\omega_{pi}^2}{\Omega_i^2} = \frac{c^2}{v_A^2}\,,\\
\varepsilon_{yy} &= \frac{c^2}{v_A^2} + i\frac{3\pi}{8}\frac{\omega_{pe}^2 k_\perp^2 v_{Fe}^2}{\omega|k_z|\Omega_e^2}\,,\\
\varepsilon_{yz} &= -\varepsilon_{zy} = -i\frac{\omega_{pe}^2 k_\perp}{\omega\Omega_e k_z}\left(1 + i\frac{3\pi}{8}\frac{\omega}{|k_z|\, v_{Fe}}\right).\\
\varepsilon_{zz} &= -\frac{\omega_{pi}^2}{\omega^2} + \frac{3\,\omega_{pe}^2}{k_z^2 v_{Fe}^2}\left(1 + i\frac{\pi}{2}\frac{\omega}{|k_z|\, v_{Fe}}\right).
\end{aligned} \tag{5.3.20}$$

Substituting these expressions into the dispersion equation (5.3.14) it follows that the spectrum (5.3.15) of the Alfvén waves remains unchanged. As to the fast and slow magnetosonic waves, their frequencies are still given by (5.3.16) in the degenerate plasma, but with $v_s = \sqrt{m/(3M)}\, v_{Fe}$. The damping decrement is

$$\delta_\pm = -\sqrt{\frac{\pi m}{8M}}\,\frac{k v_s}{2|\cos\theta|}\left(1 \pm \frac{(v_s^2\cos^2\theta - v_A^2)\cos 2\theta}{\sqrt{v_A^4 + v_s^4 - 2v_s^2 v_A^2\cos 2\theta}}\right). \tag{5.3.21}$$

Here also $v_s = \sqrt{m/(3M)}\, v_{Fe}$.

In the intermediate range of phase velocities $v_{0i} < \omega/k_z < v_{0e}$ the helical waves are possible under the conditions $\Omega_i \ll \omega \ll \Omega_e$, ω_{pe}^2/Ω_e. Their frequency is given as in the cold plasma by (5.2.17). The damping decrement due to thermal effects is defined by the expressions

$$\delta = -\frac{\sin^2\theta\,\omega^3}{\cos^2\theta\,\Omega_e k}\begin{cases}\sqrt{\dfrac{\pi}{8}}\,\dfrac{1}{v_{Te}} & \text{for the nondegenerate plasma,}\\[2ex] \dfrac{3\pi}{2}\,\dfrac{1}{v_{Fe}} & \text{for the degenerate plasma.}\end{cases} \tag{5.3.22}$$

Finally, we consider the zero-point oscillations treated in Sect. 4.3, which are undamped longitudinal oscillations of the degenerate plasma with phase velocities close to the Fermi velocity of the charge carriers. In the degenerate plasma they occur only in the presence of an external magnetic field. Since their phase velocity is much smaller than the velocity of light, they can be treated with a high degree of accuracy as potential oscillations. Since their frequency greatly exceeds the electron plasma frequency, the magnetic field practically does not influence the frequency spectrum (4.3.5) for $\omega_{pe} \gtrsim \Omega_e$. If $\Omega_e \gg \omega_{pe}$, $k_z v_{Fe}$, $k_\perp v_{Fe}$, however the spectrum of the zero-point sound is slightly modified

$$\omega = k_z v_{Fe}\left[1 + 2\exp\left(-\frac{2}{9}k^2 r_{De}^2 - 2\right)\right]. \tag{5.3.23}$$

The zero-point sound does not represent a new oscillation branch of the degenerate plasma, it rather continues the Langmuir oscillation branch into the range of short wavelengths $k^2 r_{De}^2 \gg 1$.

5.4 Cyclotron Waves

We have shown in Sect. 5.2 that the cold magneto-active plasma can sustain five different oscillation branches. Accounting for the electron thermal motion a sixth branch, the slow magnetosonic wave, appears, which transforms into the ion-acoustic wave (Sect. 5.3) in the low-pressure nondegenerate plasma under the conditions $\beta \ll 1$ and $T_e \gg T_i$. These branches do not cover all possible electromagnetic oscillations of the magneto-active plasma, however, since the dispersion equation of the hot magneto-active plasma is transcendental and the number of oscillation branches is infinitely large. The majority of these oscillations is strongly damped in time and space, however. Only in some frequency ranges and in particular extreme cases is the damping weak.

The cyclotron waves with frequencies near the cyclotron frequencies of the electrons or the ions $\omega = s\Omega_\alpha$, where $s = 1, 2, \ldots$, are especially interesting. They are of practical importance in the collisionless plasma because they can be used to heat the plasma when the Ohmic heating becomes ineffective. We shall not expand here the general theory of cyclotron oscillations, since we are interested in cyclotron waves propagating along the external magnetic field, only. For this special case the cyclotron resonance (cyclotron absorption) is easily obtained. It should be noted that resonances at higher multiples of the cyclotron frequency $s \geq 2$ do occur for $k_\perp \neq 0$ only and not for waves propagating strictly along the magnetic field.

5.4.1 Cyclotron Waves in the Nondegenerate Plasma

For the nondegenerate plasma and for propagation along the magnetic field (5.3.1) separates into three equations:

$$\varepsilon_{zz} = 1 + \sum_\alpha \frac{\omega_{p\alpha}^2}{k^2 v_{T\alpha}^2}\left[1 - I_+\left(\frac{\omega}{k v_{T\alpha}}\right)\right] = 0\,,$$
$$k^2c^2 - \omega^2(\varepsilon_{xx} \pm \mathrm{i}\varepsilon_{xy}) = k^2c^2 - \omega^2\left[1 - \sum_\alpha \frac{\omega_{L\alpha}}{\omega(\omega \mp \Omega_\alpha)} I_+\left(\frac{\omega \mp \Omega_\alpha}{k v_{T\alpha}}\right)\right] = 0\,. \tag{5.4.1}$$

The first equation of this system describes purely longitudinal oscillations. It exactly coincides with the dispersion equation for the longitudinal oscillations of the isotropic nondegenerate plasma analyzed in Sect. 4.2. The second and third equations describe the transverse ordinary (left-hand polarized) and extraordinary (right-hand polarized) wave propagating along the field, respectively. It is sufficient to discuss one of these equations since the other one is obtained by the substitution $\Omega_\alpha \rightleftarrows -\Omega_\alpha$.

We consider the cyclotron waves near the electron cyclotron frequency $\omega \approx \Omega_e$, only (more exactly, $|\omega - \Omega_e| \ll \Omega_e$). Further, we assume $|\omega - \Omega_\alpha| \gg k v_{Te}$, which means in optical terms that the frequency ω lies outside the resonance absorption line. The use of optical terms for the description of electromagnetic waves near resonance absorption frequencies of the plasma is very convenient. We shall determine the *refractive index* $n = c\,\mathrm{Re}\{k\}/\omega$ and the *absorption coefficient* $\chi = c\,\mathrm{Im}\{k\}/\omega$ from (5.4.1) instead of the natural oscillation frequency $\omega(\boldsymbol{k})$. Far from the electron absorption line we obtain

$$n^2 = 1 - \frac{\omega_{pe}^2}{\omega(\omega - \Omega_e)}\,,$$
$$\chi = \sqrt{\frac{\pi}{8}}\,\frac{\omega_{pe}^2 c}{n^2\omega^2 v_{Te}} \exp\left(-\frac{(\omega - \Omega_e)^2 c^2}{2n^2\omega^2 v_{Te}^2}\right). \tag{5.4.2}$$

Consequently, the absorption of electron cyclotron waves is exponentially small in the collisionless plasma outside the resonance, for $|\omega - \Omega_e|^3 \gg v_{Te}^2\,\omega_{pe}^2\,\Omega_e/c^2$. When the frequency ω approaches the cyclotron frequency Ω_e, the absorption grows and inside the absorption line, for $|\omega - \Omega_e|^3 < v_{Te}^2\,\omega_{pe}^2\,\Omega_e/c^2$, the wave becomes strongly damped

$$n + \mathrm{i}\chi = \frac{\mathrm{i} + \sqrt{3}}{2}\left(\sqrt{\frac{\pi}{2}}\,\frac{\omega_{pe}^2 c}{\omega^2 v_{Te}}\right)^{1/3}. \tag{5.4.3}$$

Comparing the penetration depth of the cyclotron wave with the depth of the anomalous skin-effect in the isotropic plasma (Sect. 4.4) we see that both coincide:

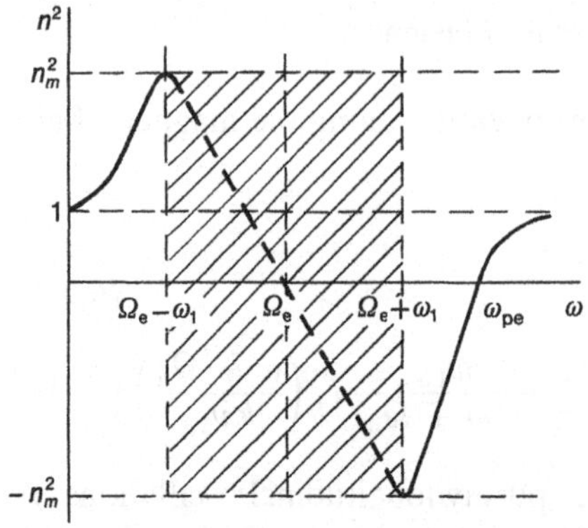

Fig. 5.5. Frequency dependence of the refractive index for the electron cyclotron wave

$$\lambda_{sk} = \frac{c}{\omega\chi} = 2\left(\sqrt{\frac{2}{\pi}}\frac{c^2 v_{Te}}{\omega\omega_{pe}^2}\right)^{1/3} . \tag{5.4.4}$$

This result is physically demonstrative. Since the electrons rotate with the Larmor frequency they can be regarded as oscillators with the natural frequency Ω_e and produce a field of the same frequency. Consequently, all the peculiarities of the isotropic plasma at the frequency ω must appear in the magneto-active plasma at the combination frequencies $\omega \pm \Omega_e$. In particular, the anomalous skin-effect of the transverse field which occurs in the isotropic plasma for $|\omega| \ll kv_{Te}$ is shifted to the range $|\omega - \Omega_e| \ll kv_{Te}$. We show in Fig. 5.5 the square of the refractive index $n^2(\omega)$ for the electron cyclotron wave. The hatched region corresponds to the region of the anomalous skin-effect. The width of the absorption range is characterized by $\omega_1 = (\omega_{pe}^2 \Omega_e v_{Te}^2/c^2)^{1/3}$. The plasma is "transparent" $[n^2(\omega) > 0]$ with respect to the electron cyclotron wave in the frequency range $\omega < \Omega_e$. The maximum of the refractive index is $n_m = (\omega_{pe}^2 c/\Omega_e^2 v_{Te})^{1/3}$.

5.4.2 Cyclotron Waves in the Degenerate Plasma

Cyclotron waves exist in the degenerate plasma, too. When the wave vector is parallel to the magnetic field, (5.3.1) separates into three equations:

$$\varepsilon_{zz} = 1 + \sum_\alpha \frac{3\omega_{p\alpha}^2}{k^2 v_{F\alpha}^2}\left(1 - \frac{\omega}{2kv_{F\alpha}}\ln\frac{\omega + kv_{F\alpha}}{\omega - kv_{F\alpha}}\right) = 0 ,$$

$$k^2c^2 - \omega^2(\varepsilon_{xx} \pm \mathrm{i}\varepsilon_{xy}) = k^2c^2 - \omega^2\left\{1 - \sum_\alpha \frac{3}{4}\frac{\omega_{p\alpha}^2}{\omega k v_{F\alpha}}\right. \tag{5.4.5}$$

$$\left.\times\left[\left(1 - \frac{(\omega \mp \Omega_\alpha)^2}{k^2 v_{F\alpha}^2}\right)\ln\frac{\omega \mp \Omega_\alpha + kv_{F\alpha}}{\omega \mp \Omega_\alpha - kv_{F\alpha}} + 2\frac{\omega \mp \Omega_\alpha}{kv_{F\alpha}}\right]\right\} = 0 .$$

As in the nondegenerate case, the first equation of this system describes purely longitudinal oscillations and exactly coincides with the dispersion equation for the longitudinal oscillations of the unmagnetized degenerate plasma, studied in Sect. 4.3. The second and third equations are symmetric under the substitution $\Omega_\alpha \rightleftarrows -\Omega_\alpha$. We analyze only one of them.

Confining our interest to electron cyclotron waves $\omega \approx \Omega_e$ we consider the second equation of (5.4.5) far from ($|\omega - \Omega_e| \gg kv_{Fe}$) and near ($|\omega - \Omega_e| \ll kv_{Fe}$), the resonance absorption line. A peculiarity of the degenerate plasma is the complete absence of the Cherenkov absorption of the cyclotron waves far from this line. Analogously to the nondegenerate plasma, the refractive index of the cyclotron wave is determined by (5.4.2) far from resonance for $|\omega - \Omega_e|^3 \gg \omega_{pe}^2 \Omega_e v_{Fe}^2/c^2$. For frequencies inside the absorption line, we obtain from (5.4.5), see (5.4.3),

$$n + i\chi = \frac{i + \sqrt{3}}{2}\left(\frac{3\pi}{4}\frac{\omega_{pe}^2 c}{\omega^2 v_{Fe}}\right)^{1/3} . \tag{5.4.6}$$

The penetration depth in the degenerate plasma is identical with the depth of the anomalous skin-effect in the unmagnetized degenerate plasma (Sect. 4.4):

$$\lambda_{sk} = 2\left(\frac{4}{3\pi}\frac{c^2 v_{Fe}}{\omega\omega_{pe}^2}\right)^{1/3} . \tag{5.4.7}$$

The general behaviour of the function $n^2(\omega)$ is the same as in the non-degenerate case shown in Fig. 5.5. The values of n_m and ω_1 are given by replacing the thermal velocity v_{Te} by v_{Fe}.

5.5 Dielectric Tensor of Weakly Ionized Magneto-Active Plasmas Taking Account of Particle Collisions

So far we have dealt with a collisionless plasma in an external magnetic field. Collisional effects are important for many reasons. Firstly, only the account of particle collisions in the magneto-active plasma justifies the use of the Maxwellian or Fermi distribution in the equilibrium. This is valid for time intervals greater than the mean free time since the latter characterizes the relaxation time of the velocity distribution function. For time intervals smaller than the mean free time, the distribution functions can be arbitrary. Secondly, dissipative processes such as friction, viscosity, diffusion, and thermal conductivity are effects caused by particle collisions in the plasma. These processes may become dominant when the collisions are frequent and they may even inhibit the collisionless Cherenkov and cyclotron dissipation.

5.5.1 Dielectric Tensor of the Quasi-Equilibrium Maxwellian Plasma

We begin to study the effect of collisions on electromagnetic waves in the magneto-active plasma with the weakly ionized nondegenerate system, since its kinetic model, the BGK collision model, allows a very transparent treatment of these effects. Moreover, it is possible to obtain the general dielectric tensor without any restrictions on the frequencies and the wavelengths from the BGK kinetic equation. The linearized kinetic equation for the particle component α is given in this model by

$$-\mathrm{i}(\omega - \boldsymbol{k}\cdot\boldsymbol{v})\,\delta f_\alpha + e_\alpha \frac{\boldsymbol{E}\,\partial f_{0\alpha}}{\partial \boldsymbol{p}_\alpha} - \Omega_\alpha \frac{\partial \delta f_\alpha}{\partial \phi} = -\,\nu_{\alpha n}(\delta f_\alpha - \eta_\alpha f_{0\alpha})\,, \qquad (5.5.1)$$

where $\eta_\alpha = \int d\boldsymbol{p}\;\delta f_\alpha / N_\alpha$. The equilibrium distribution function $f_{0\alpha}$ is assumed Maxwellian

$$f_{0\alpha} = \frac{N_\alpha}{(2\pi m_\alpha T_\alpha)^{3/2}} \exp\left(-\frac{m_\alpha v^2}{2T_\alpha}\right). \qquad (5.5.2)$$

This choice is justified by the fact that the magnetic field does not influence the stationary state of the system. Consequently, the argumentation used in Sect. 4.5 remains true for the magneto-active plasma.

The method of solution of (5.5.1) is the same as for the collisionless plasma (Sect. 5.1). We obtain the general solution

$$\begin{aligned}\delta f_\alpha &= \frac{1}{\Omega_\alpha}\int_\infty^\phi d\phi' \left(e_\alpha \boldsymbol{E}\frac{\partial f_{0\alpha}}{\partial \boldsymbol{p}_\alpha} - \nu_{\alpha n}\eta_\alpha f_{0\alpha}\right)_{\phi'} \exp\left[\frac{1}{\Omega_\alpha}\int_\phi^{\phi'} d\phi''\,(\omega - \boldsymbol{k}\cdot\boldsymbol{v} + \mathrm{i}\nu_{\alpha n})_{\phi''}\right] \\ &= \frac{\mathrm{i}e_\alpha}{T_\alpha} f_{0\alpha} \sum_n \frac{\exp\left(-\,in\phi + \mathrm{i}\dfrac{k_\perp v_\perp}{\Omega_\alpha}\sin\phi\right)}{\omega + \mathrm{i}\nu_{\alpha n} - k_z v_z - n\Omega_\alpha} \\ &\quad\times \left[\frac{n\Omega_\alpha}{k}\mathrm{J}_n E_x + \mathrm{i}v_\perp \mathrm{J}_n' E_y + \left(v_z E_z + \nu_{\alpha n}\eta_\alpha \frac{T_\alpha}{e_\alpha}\right)\mathrm{J}_n\right]. \end{aligned} \qquad (5.5.3)$$

Using (5.5.3) to calculate the induced current density of the particle species α

$$\boldsymbol{j}_\alpha = e_\alpha \int d\boldsymbol{p}\,\boldsymbol{v}\,\delta f_\alpha \qquad (5.5.4)$$

and eliminating η_α by means of the continuity equation

$$\eta_\alpha = \frac{\boldsymbol{k}\cdot\boldsymbol{j}_\alpha}{\omega e_\alpha N_\alpha} \qquad (5.5.5)$$

we arrive at the dielectric tensor of the weakly ionized magneto-active plasma

$$\varepsilon_{ij}(\omega, \boldsymbol{k}) = \delta_{ij} + \sum_{\alpha} \left[\delta_{i\mu} + \mathrm{i} \frac{\nu_{\alpha n} G_{\alpha i} k_{\mu}}{\omega - \mathrm{i}\nu_{\alpha n} \boldsymbol{k} \cdot \boldsymbol{G}_{\alpha}} \right] \times \frac{\omega + \mathrm{i}\nu_{\alpha n}}{\omega} \left[\varepsilon^{\alpha}_{\mu j}(\omega + \mathrm{i}\nu_{\alpha n}, \boldsymbol{k}) - \delta_{\mu j} \right] . \tag{5.5.6}$$

The tensor $\varepsilon^{\alpha}_{ij}(\omega + \mathrm{i}\nu_{\alpha n}, \boldsymbol{k})$ is the contribution of the particle species α to the dielectric tensor (5.1.10) of the collisionless magneto-active plasma at the shifted argument $\omega \to \omega + \mathrm{i}\nu_{\alpha n}$. The vector $\boldsymbol{G}_{\alpha}$ has the components

$$\boldsymbol{G}_{\alpha} = \left\{ \begin{array}{l} \dfrac{\Omega_{\alpha}}{k_{\perp}} \sum\limits_{n} \dfrac{n A_n(z_{\alpha})}{\omega + \mathrm{i}\nu_{\alpha n} - n\Omega_{\alpha}} I_{+}(\beta_{\alpha n}) \\ -\mathrm{i} \dfrac{\Omega_{\alpha}}{k_{\perp}} \sum\limits_{n} \dfrac{z_{\alpha} A'_n(z_{\alpha})}{\omega + \mathrm{i}\nu_{\alpha n} - n\Omega_{\alpha}} I_{+}(\beta_{\alpha n}) \\ -\dfrac{1}{k_z} \sum\limits_{n} A_n(z_{\alpha}) \left[1 - I_{+}(\beta_{\alpha n})\right] \end{array} \right\} , \tag{5.5.7}$$

where

$$z_{\alpha} = k_{\perp}^2 v_{\mathrm{T}\alpha}^2 / \Omega_{\alpha}, \; \beta_{\alpha n} = (\omega + \mathrm{i}\nu_{\alpha n} - n\Omega_{\alpha}) / k_z v_{\mathrm{T}\alpha} .$$

The summation in (5.5.6) extends over the charged particle components of the plasma.

The "longitudinal" dielectric permittivity of the nondegenerate weakly ionized plasma following from (5.5.6) is

$$\varepsilon(\omega, \boldsymbol{k}) = \frac{k_i k_j}{k^2} \varepsilon_{ij}(\omega, \boldsymbol{k}) = 1 + \sum_{\alpha} \frac{\omega_{\mathrm{p}\alpha}^2}{k^2 v_{\mathrm{T}\alpha}^2} \; \frac{1 - \sum\limits_{n} \dfrac{\omega + \mathrm{i}\nu_{\alpha n}}{\omega + \mathrm{i}\nu_{\alpha n} - n\Omega_{\alpha}} A_n(z_{\alpha}) I_{+}(\beta_{\alpha n})}{1 - \sum\limits_{n} \dfrac{\mathrm{i}\nu_{\alpha n}}{\omega + \mathrm{i}\nu_{\alpha n} - n\Omega_{\alpha}} A_n(z_{\alpha}) I_{+}(\beta_{\alpha n})} . \tag{5.5.8}$$

5.5.2 Degenerate Plasma

We can analogously calculate the dielectric tensor of the weakly ionized degenerate plasma. Writing the linearized kinetic equation with the BGK integral model in the form

$$-\mathrm{i}(\omega - \boldsymbol{k} \cdot \boldsymbol{v}) \delta f_{\alpha} + e_{\alpha} (\boldsymbol{E} \cdot \boldsymbol{v}) \frac{\partial f_{0\alpha}}{\partial \mathscr{E}_{\alpha}} - \Omega_{\alpha} \frac{\partial \delta f_{\alpha}}{\partial \phi} = -\nu_{\alpha n} \left(\delta f_{\alpha} + \frac{2}{3} \frac{\partial f_{0\alpha}}{\partial \mathscr{E}_{\alpha}} \mathscr{E}_{\mathrm{F}\alpha} \eta_{\alpha} \right) , \tag{5.5.9}$$

we obtain the solution

$$\delta f_\alpha = -\frac{1}{\Omega_\alpha}\int_\infty^\phi d\phi' \left[e_\alpha \left(\boldsymbol{E}\cdot\boldsymbol{v}\right)\frac{\partial f_{0\alpha}}{\partial \mathscr{E}_\alpha} + \frac{2}{3}\mathscr{E}_{\mathrm{F}\alpha}\nu_{\mathrm{an}}\eta_\alpha \frac{\partial f_{0\alpha}}{\partial \mathscr{E}_\alpha}\right]_{\phi'}$$

$$\times \exp\left[\frac{\mathrm{i}}{\Omega_\alpha}\int_\phi^{\phi'} d\phi''\left(\omega - \boldsymbol{k}\cdot\boldsymbol{v} + \mathrm{i}\nu_{\mathrm{an}}\right)_{\phi''}\right]. \tag{5.5.10}$$

Formulating the current density and using the continuity equation (5.5.5) we finally obtain the dielectric tensor of the degenerate plasma. It coincides in form with the tensor (5.5.6) where $\varepsilon_{ij}^\alpha(\omega + \mathrm{i}\nu_{\mathrm{an}}, \boldsymbol{k})$ is the contribution of the particle species α to the dielectric tensor (5.1.14) of the collisionless plasma at the argument $\omega \to \omega + \mathrm{i}\nu_{\mathrm{an}}$. The vector $\boldsymbol{G}_\alpha$ has the components

$$\boldsymbol{G}_\alpha = \left\{ \begin{array}{l} \dfrac{\Omega_\alpha}{2k_\perp}\sum\limits_n \int\limits_0^\pi \dfrac{d\theta \sin\theta\, n \mathrm{J}_n^2}{\omega + \mathrm{i}\nu_{\mathrm{an}} - k_z v_{\mathrm{F}\alpha}\cos\theta - n\Omega_\alpha} \\[2ex] -\mathrm{i}\dfrac{v_{\mathrm{F}\alpha}}{2}\sum\limits_n \int\limits_0^\pi \dfrac{d\theta \sin^2\theta\, \mathrm{J}_n'}{\omega + \mathrm{i}\nu_{\mathrm{an}} - k_z v_{\mathrm{F}\alpha}\cos\theta - n\Omega_\alpha} \\[2ex] \dfrac{v_{\mathrm{F}\alpha}}{2}\sum\limits_n \int\limits_0^\pi \dfrac{d\theta \sin\theta\cos\theta\, \mathrm{J}_n^2}{\omega + \mathrm{i}\nu_{\mathrm{an}} - k_z v_{\mathrm{F}\alpha}\cos\theta - n\Omega_\alpha} \end{array} \right\}. \tag{5.5.11}$$

For the "longitudinal" dielectric permittivity we get

$$\varepsilon(\omega, \boldsymbol{k}) = \frac{k_i k_j}{k^2}\varepsilon_{ij}(\omega, \boldsymbol{k}) \tag{5.5.12}$$

$$= 1 + \sum_\alpha \frac{3\omega_{\mathrm{p}\alpha}^2}{k^2 v_{\mathrm{F}\alpha}^2}\, \frac{1 - \sum\limits_n \dfrac{\omega + \mathrm{i}\nu_{\mathrm{an}}}{2}\int\limits_0^\pi \dfrac{d\theta\sin\theta\,\mathrm{J}_n^2}{\omega + \mathrm{i}\nu_{\mathrm{an}} - k_z v_{\mathrm{F}\alpha}\cos\theta - n\Omega_\alpha}}{1 - \sum\limits_n \dfrac{\mathrm{i}\nu_{\mathrm{an}}}{2}\int\limits_0^\pi \dfrac{d\theta\sin\theta\,\mathrm{J}_n^2}{\omega + \mathrm{i}\nu_{\mathrm{an}} - k_z v_{\mathrm{F}\alpha}\cos\theta - n\Omega_\alpha}}.$$

We want to stress that (5.5.6–12) are valid for any relation between the quantities ω, Ω_α, $k_\perp v_0$, $k_z v_0$, ν_{an}, where $v_0 = v_{\mathrm{T}}$ for the nondegenerate plasma and $v_0 = v_F$ for the degenerate one. In the limit $\nu_{\mathrm{an}} \to 0$ the dielectric tensor approaches the corresponding expression of the collisionless plasma (Sect. 5.1).

5.6 Dielectric Tensor of Completely Ionized Magneto-Active Plasmas Taking Account of Particle Collisions

To account for particle collisions in a completely ionized gas presents a much more difficult problem in solving the complex integro-differential equation with the Landau collision term. As in the case of the unmagnetized plasma (Sect. 4.6) we analyze some special cases only, which allow a rather simple solution.

The investigation of processes evolving fast on the time scale of temperature relaxation between electrons and ions ($\omega \gg \nu_{\text{eff}} m/M$) is simplified for the nondegenerate plasma by assuming Maxwellian zero-order distribution functions with different temperatures, i.e.,

$$f_{0\alpha} = \frac{N_\alpha}{(2\pi m_\alpha T_\alpha)^{3/2}} \exp\left(-\frac{m_\alpha v^2}{2T_\alpha}\right) . \tag{5.6.1}$$

In the isothermal plasma with $T_e = T_i$ (5.6.1) holds without the restriction to fast processes.

A small deviation of $\delta f_\alpha \sim \exp(-i\omega t + i\boldsymbol{k}\cdot\boldsymbol{r})$ from the equilibrium is described by the equation

$$-i(\omega - \boldsymbol{k}\cdot\boldsymbol{v})\,\delta f_\alpha - \Omega_\alpha \frac{\partial \delta f_\alpha}{\partial \phi} + e_\alpha \boldsymbol{E}\cdot\frac{\partial f_{0\alpha}}{\partial \boldsymbol{p}_\alpha} = \frac{\partial}{\partial p_{\alpha i}} \sum_\beta \int d\boldsymbol{p}_\beta I_{ij}^{\alpha\beta}(\boldsymbol{p}_\alpha, \boldsymbol{p}_\beta)$$

$$\times \left(\frac{\partial f_{0\alpha}}{\partial p_{\alpha j}}\delta f_\beta + \frac{\partial \delta f_\alpha}{\partial p_{\alpha j}} f_{0\beta} - f_{0\alpha}\frac{\partial \delta f_\beta}{\partial p_{\beta j}} - \delta f_\alpha \frac{\partial f_{0\beta}}{\partial p_{\beta j}}\right) . \tag{5.6.2}$$

Here

$$I_{ij}^{\alpha\beta} = 2\pi e_\alpha^2 e_\beta^2 L \frac{u^2\delta_{ij} - u_i u_j}{u^3} ,$$

$\boldsymbol{u} = \boldsymbol{v}_\alpha - \boldsymbol{v}_\beta$, and

$$L = \ln(r_D/r_{\min}) .$$

5.6.1 The High-Frequency Range

In general, the solution of (5.6.2) is very complex. It can easily be found, however, when the inequalities

$$\omega \gg \nu_\alpha ; \quad |\omega \pm \Omega_\alpha| \gg \nu_\alpha \tag{5.6.3}$$

are satisfied. In this limit the collision integral in (5.6.2) is a small term, and the method of successive approximations already used in Sect. 4.6 can be applied. This expansion in powers of the collision frequency gives

$$\delta f_\alpha = \delta f_\alpha^{(0)} + \delta f_\alpha^{(1)} , \tag{5.6.4}$$

where $\delta f_\alpha^{(0)}$ is the collisionless perturbation (5.1.4) and

$$\begin{aligned}\delta f_\alpha^{(1)} &= \sum_n \frac{\exp\left(-in\phi + \mathrm{i}\dfrac{k_\perp v_\perp \sin\phi}{\Omega_\alpha}\right)}{\omega - k_z v_z - n\Omega_\alpha} \\ &\times \frac{1}{2\pi}\int_{-\pi}^{\pi} d\phi' \exp\left(in\phi' - \mathrm{i}\frac{k_\perp v_\perp}{\Omega_\alpha}\sin\phi'\right) \sum_\beta \frac{\partial}{\partial p_{\alpha i}} \int d\boldsymbol{p}_\beta I_{ij}^{\alpha\beta}(\boldsymbol{p}_\alpha, \boldsymbol{p}_\beta) \\ &\times \left(\frac{\partial f_{0\alpha}}{\partial p_{\alpha j}}\delta f_\beta^{(0)} + f_{0\beta}\frac{\partial \delta f_\alpha^{(0)}}{\partial p_{\alpha j}} - f_{0\alpha}\frac{\partial \delta f_\beta^{(0)}}{\partial p_{\beta j}} - \delta f_\alpha^{(0)}\frac{\partial f_{0\beta}}{\partial p_{\beta j}}\right) .\end{aligned} \tag{5.6.5}$$

The calculation of the collisional correction of the induced current and consequently of the plasma dielectric permittivity is reduced to the evaluation of the integral

$$\delta \boldsymbol{j}_\alpha = e_\alpha \int d\boldsymbol{p}\, \boldsymbol{v} \delta f_\alpha^{(1)} , \tag{5.6.6}$$

which is rather complex, however, and possible in particular cases only.

The cold plasma limit (5.2.1) is such a special case and provides a purely anti-Hermitian collisional correction to the tensor (5.2.2)

$$\delta\varepsilon_{ij}^{\mathrm{a}}(\omega, \boldsymbol{k}) = \begin{pmatrix} \delta\varepsilon_\perp^{\mathrm{a}} & \mathrm{i}\delta g^{\mathrm{a}} & 0 \\ -\mathrm{i}\delta g^{\mathrm{a}} & \delta\varepsilon_\perp^{\mathrm{a}} & 0 \\ 0 & 0 & \delta\varepsilon_\parallel^{\mathrm{a}} \end{pmatrix} . \tag{5.6.7}$$

The components of (5.6.7) are

$$\begin{aligned}\delta\varepsilon_\perp^{\mathrm{a}} &= \mathrm{i}\frac{\omega_{\mathrm{pe}}^2 \nu_{\mathrm{eff}}}{\Omega_{\mathrm{e}}^2}\left[\left(\frac{\Omega_{\mathrm{e}}}{\omega^2 - \Omega_{\mathrm{e}}^2} - \frac{\Omega_{\mathrm{i}}}{\omega^2 - \Omega_{\mathrm{i}}^2}\right)^2 + \left(\frac{\omega}{\omega^2 - \Omega_{\mathrm{e}}^2} - \frac{\omega}{\omega^2 - \Omega_{\mathrm{e}}^2}\right)^2\right] , \\ \delta g^{\mathrm{a}} &= 2\mathrm{i}\frac{\omega_{\mathrm{pe}}^2 \nu_{\mathrm{eff}}\omega^2}{\Omega_{\mathrm{e}}^2}\left(\frac{\Omega_{\mathrm{e}}}{\omega^2 - \Omega_{\mathrm{e}}^2} - \frac{\Omega_{\mathrm{i}}}{\omega^2 - \Omega_{\mathrm{i}}^2}\right)\left(\frac{1}{\omega^2 - \Omega_{\mathrm{e}}^2} - \frac{1}{\omega^2 - \Omega_{\mathrm{i}}^2}\right) , \\ \delta\varepsilon_\parallel^{\mathrm{a}} &= \mathrm{i}\frac{\omega_{\mathrm{pe}}^2 \nu_{\mathrm{eff}}}{\omega^3}\end{aligned} \tag{5.6.8}$$

and ν_{eff} is defined by (4.6.7).

Thus, the collisional correction of the dielectric tensor is determined by collisions between particles of different types in the cold plasma.

5.6.2 The Range of Slow Waves

At small phase velocities when the condition (5.3.11) is satisfied, and the collisionless dielectric tensor takes the form of (5.3.12, 13), another simplification is possible. One can show that the electron-electron collisions are negligible in the given frequency range for $\nu_e \ll k_z v_{Te} \ll \Omega_e$, since the wavelength is then short compared to the mean free path of the electrons. The Cherenkov effect provides the main dissipative mechanism for the electron motion. In the frequency range $\omega \gg \nu_i$ ion-ion collisions contribute to the ion dissipation along with the Cherenkov mechanism. These collisions can be included easily when the solution of the ion kinetic equation by the method of successive approximations is justified. We obtain the following correction to the anti-Hermitian part of the dielectric tensor in this frequency range

$$\delta\varepsilon_{ij}^{a} = \begin{pmatrix} \delta\varepsilon_{xx}^{a} & 0 & 0 \\ 0 & \delta\varepsilon_{yy}^{a} & \delta\varepsilon_{yz}^{a} \\ 0 & -\delta\varepsilon_{yz}^{a} & \delta\varepsilon_{zz}^{a} \end{pmatrix} \quad \text{with} \tag{5.6.9}$$

$$\begin{aligned} \delta\varepsilon_{xx}^{a} &= \mathrm{i}\frac{7}{10}\frac{\omega_{pi}^2 \nu_{ii} k_\perp^2 v_{Ti}^2}{\omega \Omega_i^4} + \mathrm{i}\frac{\omega_{pe}^2 \nu_{eff}}{\omega \Omega_e}, \qquad \delta\varepsilon_{yy}^{a} = \mathrm{i}\frac{4}{5}\frac{\omega_{pi}^2 \nu_{ii} k_\perp^2 v_{Ti}^2}{\omega^3 \Omega_i^2}, \\ \delta\varepsilon_{zz}^{a} &= \mathrm{i}\frac{8}{5}\frac{\omega_{pi}^2 \nu_{ii} k_z^2 v_{Ti}^2}{\omega^5}, \qquad \delta\varepsilon_{yz}^{a} = -\delta\varepsilon_{zy}^{a} = -\frac{4}{5}\frac{\omega_{pi}^2 \nu_{ii} k_\perp k_z v_{Ti}^2}{\omega^4 \Omega_i}. \end{aligned} \tag{5.6.10}$$

The collision frequency ν_{ii} is defined by (4.6.7). Note that the ratio between ω and ν_{eff} may be arbitrary.

5.6.3 Degenerate Plasma

Concluding this section we investigate the role of particle collisions in a completely ionized, degenerate plasma at high frequencies when only electron processes are essential. In this limit we have to solve the electron-kinetic equation accounting for electron-electron and electron-ion collisions. We describe the equilibrium by the Fermi distribution function of the electrons:

$$f_{0e} = \begin{cases} \dfrac{2}{(2\pi\hbar)^3} & \text{if} \quad p < p_{Fe} = (3\pi^2)^{1/3}\hbar N_e^{1/3}, \\[2ex] 0 & \text{if} \quad p > p_{Fe}. \end{cases} \tag{5.6.11}$$

The solution of the linearized electron-kinetic equation (3.4.9) for a small perturbation $\delta f_e \sim \exp(\mathrm{i}\omega t + \mathrm{i}\boldsymbol{k}\cdot\boldsymbol{r})$ is straightforward in the limit (5.6.3). The method of successive approximations can then be applied to (3.4.9).

Omitting the calculation we give the final result for the cold plasma, which satisfies the conditions (5.2.1). We obtain the following anti-Hermitian correction of the dielectric tensor of the cold plasma (5.2.2):

$$\delta\varepsilon_{ij}^{\mathrm{a}} = \begin{pmatrix} \delta\varepsilon_{\perp}^{\mathrm{a}} & \mathrm{i}\delta g^{\mathrm{a}} & 0 \\ -\mathrm{i}\delta g^{\mathrm{a}} & \delta\varepsilon_{\perp}^{\mathrm{a}} & 0 \\ 0 & 0 & \delta\varepsilon_{\parallel}^{\mathrm{a}} \end{pmatrix} . \tag{5.6.12}$$

Here

$$\delta\varepsilon_{\perp}^{\mathrm{a}} = \mathrm{i}\frac{\omega_{\mathrm{pe}}^2 \nu_{\mathrm{Fe}} (\omega^2 + \Omega_{\mathrm{e}}^2)}{\omega (\omega^2 - \Omega_{\mathrm{e}}^2)^2} , \qquad \delta g^{\mathrm{a}} = 2\mathrm{i}\frac{\omega_{\mathrm{pe}} \nu_{\mathrm{Fe}} \Omega_{\mathrm{e}}}{(\omega^2 - \Omega_{\mathrm{e}}^2)^2} , \qquad \delta\varepsilon_{\parallel}^{\mathrm{a}} = \mathrm{i}\frac{\omega_{\mathrm{pe}}^2 \nu_{\mathrm{Fe}}}{\omega^3}$$

and ν_{Fe} is given by (4.6.14).

Note that (5.6.12, 13) are valid only in the high frequency range $\omega \gg \Omega_{\mathrm{i}}$, ω_{pi}, where the ion motion is negligible.

5.7 Electromagnetic Waves in Magneto-Active Plasmas Taking Account of Particle Collisions

Knowing the dielectric permittivity of the magneto-active plasma with particle collisions, we can solve the dispersion relation of the collisional system. We are interested in the effect of particle collisions on oscillation spectra when collisions are dominating and not only modifying the absorption of the waves in the plasma. These circumstances are given in the case of waves in the cold plasma, the MHD and Alfvén waves and also for the fast and slow magnetosonic waves in the hot plasma.

5.7.1 Damping of Waves in the Cold Magneto-Active Plasma

To begin with, we consider the cold plasma limit (5.2.1) in the frequency range (5.6.3). At high frequencies $\omega \gg \omega_{\mathrm{pi}}, \Omega_{\mathrm{i}}$ the electron component of the plasma only is affected. Then, as in the collisionless plasma, the Hermitian part of the dielectric tensor is defined by (5.2.2, 3). The anti-Hermitian part of the dielectric permittivity (5.3.6, 7), however, must be corrected by the contributions (5.5.6, 8), or (12) and can be written in the form

$$\delta\varepsilon_{ij}^{\mathrm{a}} = \begin{pmatrix} \delta\varepsilon_{\perp}^{\mathrm{a}} & \mathrm{i}\delta g^{\mathrm{a}} & 0 \\ -\mathrm{i}\delta g^{\mathrm{a}} & \delta\varepsilon_{\perp}^{\mathrm{a}} & 0 \\ 0 & 0 & \delta\varepsilon_{\parallel}^{\mathrm{a}} \end{pmatrix} \quad \text{with} \tag{5.7.1}$$

$$\delta\varepsilon_{\perp}^{\mathrm{a}} = \mathrm{i}\frac{\omega_{\mathrm{pe}}^2 \nu_{\mathrm{e}} (\omega^2 + \Omega_{\mathrm{e}}^2)}{\omega (\omega^2 - \Omega_{\mathrm{e}}^2)^2} , \quad \delta g^{\mathrm{a}} = \mathrm{i}\frac{2\,\omega_{\mathrm{pe}}^2 \Omega_{\mathrm{e}} \nu_{\mathrm{e}}}{(\omega^2 - \Omega_{\mathrm{e}}^2)^2} , \quad \delta\varepsilon_{\parallel}^{\mathrm{a}} = \mathrm{i}\frac{\omega_{\mathrm{pe}}^2 \nu_{\mathrm{e}}}{\omega^3} . \tag{5.7.2}$$

The collision frequencies are $\nu_e = \nu_{eff}$ for the completely ionized nondegenerate plasma, $\nu_e = \nu_{Fe}$ for the degenerate plasma, and $\nu_e = \nu_{en}$ for the weakly ionized plasma.

The collisional correction (5.7.1, 2) of the dielectric tensor determines the collisional damping decrement. The frequency spectra and the collisionless damping decrements are still given by the formulas of Sects. 5.2, 3, respectively. The collisional contribution to the damping decrement is

$$\Delta\delta = -\frac{k^4\delta A^a + k^2\frac{\omega^2}{c^2}\delta B^a + \frac{\omega^4}{c^4}\delta C^a}{k^4\frac{\partial A^H}{\partial\omega} + k^2\frac{\partial}{\partial\omega}\frac{\omega^2}{c^2}B^H + \frac{\partial}{\partial\omega}\frac{\omega^4}{c^4}C^H}, \tag{5.7.3}$$

where δA^a, δB^a, δC^a have the structure of (5.3.9) and may be derived by the substitution of $\delta\varepsilon_{ij}^a$ for ε_{ij}^a:

$$\delta A^a = \frac{k_\perp^2}{k^2}\delta\varepsilon_\perp^a + \frac{k_z^2}{k^2}\delta\varepsilon_\parallel^a ,$$

$$\delta B^a = -\left(1 + \frac{k_z^2}{k^2}\right)\left(\delta\varepsilon_\perp^a\varepsilon_\parallel^H + \delta\varepsilon_\parallel^a\varepsilon_\perp^H\right) - 2\frac{k_\perp^2}{k^2}(\delta\varepsilon_\perp^a\varepsilon_\perp^H - \delta g^a g^H) , \tag{5.7.4}$$

$$\delta C^a = \delta\varepsilon_\parallel^a(\varepsilon_\perp^{H2} - g^{H2}) + 2\varepsilon_\parallel^H(\delta\varepsilon_\perp^a\varepsilon_\perp^H - \delta g^a g^H) .$$

Together with the results of Sects. 5.2, 3 the solution of the problem of wave propagation at high frequencies in the cold magneto-active plasma is now complete. The formulas have a rather complicated form, however, and can be analyzed only numerically. A simple expression is available for the collisional damping decrement of the helical waves (5.2.17) only. We get from (5.7.3, 4)

$$\Delta\delta = -\frac{\nu_e}{2}\left(1 + \frac{k_z^2}{k^2}\right)\frac{k^2c^2}{\omega_{pe}^2} . \tag{5.7.5}$$

In contrast to the high-frequency electron oscillations where electron-electron collisions govern the collisional absorption the ion collisions become significant in the low-frequency range $\omega \ll \Omega_i$. The account of particle collisions does not change the frequencies (5.2.16) of the Alfvén and MHD waves belonging to this range but modifies their damping decrements

$$\Delta\delta = -\frac{1}{2}\frac{\nu_i\omega^2}{\Omega_i^2} . \tag{5.7.6}$$

We have $\nu_i = \nu_{eff}m/M$ for the completely ionized nondegenerate plasma, and $\nu_i = \nu_{in}$ for weak ionization. This formula follows from the general expression (5.7.3) by applying (5.6.7) and (5.5.6), and is valid in the cold plasma limit.

The contribution (5.7.6) is essential since the collisionless damping is exponentially small at low frequencies, see (5.3.10). The collisional damping is especially important in the case of nearly transverse wave propagation where the collisionless absorption can be ignored.

5.7.2 Collisional Damping of Low-Frequency Waves in the Hot Magneto-Active Plasma

Passing over to the hot plasma case in the low-frequency range $\omega \ll \Omega_\mathrm{i}$ we consider the oscillations with intermediate phase velocities $v_\mathrm{Ti} \ll \omega/k_z \ll v_\mathrm{Te}$. The corresponding waves are the Alfvén wave, the fast and the slow magnetosonic waves which have the spectra (5.3.15, 16) in the collisionless non-degenerate limit. The account of collisions changes the anti-Hermitian part of the tensor $\varepsilon_{ij}(\omega, \boldsymbol{k})$ only. The collisional correction is

$$\delta\varepsilon_{ij}^\mathrm{a} = \begin{pmatrix} \delta\varepsilon_{xx}^\mathrm{a} & 0 & 0 \\ 0 & \delta\varepsilon_{yy}^\mathrm{a} & \delta\varepsilon_{yz}^\mathrm{a} \\ 0 & \delta\varepsilon_{zy}^\mathrm{a} & \delta\varepsilon_{zz}^\mathrm{a} \end{pmatrix} . \tag{5.7.7}$$

For the completely ionized plasma the tensor components read

$$\begin{aligned} \delta\varepsilon_{xx}^\mathrm{a} &= \mathrm{i}\frac{7}{10}\frac{\omega_\mathrm{pi}^2 \nu_\mathrm{ii} k_\perp^2 v_\mathrm{Ti}^2}{\omega\Omega_\mathrm{i}^4} + \mathrm{i}\frac{\omega_\mathrm{pe}^2 \nu_\mathrm{eff}}{\omega\Omega_\mathrm{e}^2}, \qquad \delta\varepsilon_{yy}^\mathrm{a} = \mathrm{i}\frac{4}{5}\frac{\omega_\mathrm{pi}^2 \nu_\mathrm{ii} k_\perp^2 v_\mathrm{Ti}^2}{\omega^3\Omega_\mathrm{i}^2}, \\ \delta\varepsilon_{zz}^a &= \mathrm{i}\frac{8}{5}\frac{\omega_\mathrm{pi}^2 \nu_\mathrm{ii} k_z^2 v_\mathrm{Ti}^2}{\omega^5}, \qquad \delta\varepsilon_{yz}^\mathrm{a} = -\delta\varepsilon_{zy}^\mathrm{a} = -\frac{4}{5}\frac{\omega_\mathrm{pi}^2 \nu_\mathrm{ii} k_\perp k_z v_\mathrm{Ti}^2}{\omega^4\Omega_\mathrm{i}} \end{aligned} \tag{5.7.8}$$

and

$$\delta\varepsilon_{xx}^\mathrm{a} = \delta\varepsilon_{yy}^\mathrm{a} = \mathrm{i}\frac{\omega_\mathrm{pi}^2 \nu_\mathrm{in}}{\omega\Omega_\mathrm{i}^2}, \qquad \delta\varepsilon_{zz}^\mathrm{a} = \mathrm{i}\frac{\omega_\mathrm{pi}^2 \nu_\mathrm{in}}{\omega^3}, \qquad \delta\varepsilon_{yz}^\mathrm{a} = -\delta\varepsilon_{zy}^\mathrm{a} = 0 \tag{5.7.9}$$

for the weakly ionized case.

The Alfvén wave (5.3.15), undamped in the collisionless limit, is now damped according to

$$\delta = -\frac{1}{2}\left(\frac{m}{M}\nu_\mathrm{e} + \nu_\mathrm{i}\right), \quad \text{where} \tag{5.7.10}$$

$$\nu_\mathrm{e} = \nu_\mathrm{eff}, \quad \nu_\mathrm{i} = m\nu_\mathrm{eff}/M + 7\nu_\mathrm{ii} k_\perp^2 v_\mathrm{Ti}^2/(10\,\Omega_\mathrm{i}^2)$$

for complete and $\nu_a = \nu_{an}$ for weak ionization.

The collisionless damping decrement of the fast and the slow magnetosonic waves (5.3.16) must be corrected. We obtain

$$\Delta\delta_\pm = \frac{4}{5}\nu_{ii}\frac{v_{Ti}^2}{v_s^2}\left\{\frac{1}{2} + \frac{9}{8}\frac{v_s^2}{v_A^2}\sin^2\theta \mp \frac{1}{8}\left(1 + \frac{v_s^4}{v_A^4} - 2\frac{v_s^2}{v_A^2}\cos 2\theta\right)^{1/2}\right.$$

$$\times\left[\left(1 - \frac{v_s^2}{v_A^2}\cos 2\theta\right)\left(4 + 3\frac{v_s^2}{v_A^2}\sin^2\theta\right)\right.$$

$$\left.\left. + 2\sin^2\theta\left(2 + 3\frac{v_s^2}{v_A^2}(1+\cos^2\theta)\right)\right]\right\} \tag{5.7.11}$$

for the completely ionized plasma and

$$\Delta\delta_\pm = -\frac{\nu_{in}}{2}\frac{k^2 v_s^2\cos^2\theta}{\omega_\pm^4 - k^4 v_s^2 v_A^2\cos^2\theta}\left[\omega_\pm^2 - k^2 v_A^2 + \omega_\pm^2\left(\frac{\omega_\pm^2}{k^2 v_s^2\cos^2\theta} - 1\right)\right] \tag{5.7.12}$$

for the weakly ionized plasma.

These formulas take on an especially simple form for the low-pressure plasma with $\beta = v_s^2/v_A^2 \ll 1$, where the fast magnetosonic wave becomes transverse and the slow one longitudinal. Here, the collisional corrections of the damping decrement are

$$\Delta\delta_- = -\frac{1}{2}\begin{cases}\dfrac{8}{5}\nu_{ii}\dfrac{T_i}{T_e} & \text{for complete ionization ,}\\[2ex] \nu_{in} & \text{for weak ionization .}\end{cases} \tag{5.7.13}$$

The correction $\Delta\delta_+$ coincides with $\Delta\delta$ given in (5.7.6).

5.8 Exercises

5.8.1. Derive the dielectric tensor of the collisionless nonrelativistic magnetoactive plasma and the average force affecting the plasma in the SHF field in the model of independent particles.

Solution. The system of equations of the model of independent particles is

$$\begin{aligned}&\frac{\partial N_\alpha}{\partial t} + \operatorname{div} N_\alpha \boldsymbol{V}_\alpha = 0\ ,\\ &\frac{\partial \boldsymbol{V}_\alpha}{\partial t} + (\boldsymbol{V}_\alpha\cdot\nabla)\boldsymbol{V}_\alpha = \frac{e_\alpha}{m_\alpha}\left\{\boldsymbol{E} + \frac{1}{c}[\boldsymbol{V}_\alpha, \boldsymbol{B}]\right\}\ .\end{aligned} \tag{5.8.1}$$

In the unperturbed state we have $\boldsymbol{V}_{0\alpha} = 0$, $\boldsymbol{E}_0 = 0$. We orient the z-axis parallel to the field $B_0 \neq 0$, and consider perturbations of the form δN_α, $\delta\boldsymbol{V}_\alpha$, $\boldsymbol{E}$, $\boldsymbol{B} \sim \exp(-\mathrm{i}\omega t + \mathrm{i}\boldsymbol{k}\cdot\boldsymbol{r})$. The linearized equations of motion are

$$-\,\mathrm{i}\omega\delta V_\alpha = \frac{e_\alpha}{m_\alpha}E + \frac{e_\alpha}{m_\alpha c}[\delta V_\alpha, B_0] \, . \tag{5.8.2}$$

Hence

$$\delta V_\alpha = \frac{\mathrm{i}e_\alpha\omega}{m_\alpha(\omega^2-\Omega_\alpha^2)}\left(E - \frac{\Omega_\alpha^2}{\omega^2}\cdot\frac{B_0(E\cdot B_0)}{B_0^2}\right) - \frac{e_\alpha\Omega_\alpha[E,B_0]}{m_\alpha(\omega^2-\Omega_\alpha^2)B_0} \, . \tag{5.8.3}$$

The induced current density is

$$j_i = \sum_\alpha e_\alpha N_{0\alpha}\delta V_{\alpha i} = \sigma_{ij}E_j \, ,$$

giving the dielectric permittivity in the form

$$\varepsilon_{ij} = \begin{pmatrix} \varepsilon_\perp & \mathrm{i}g & 0 \\ -\mathrm{i}g & \varepsilon_\perp & 0 \\ 0 & 0 & \varepsilon_\parallel \end{pmatrix} \qquad \text{with}$$

$$\varepsilon_\perp = 1 - \sum_\alpha\frac{\omega_{\mathrm{p}\alpha}^2}{\omega^2-\Omega_\alpha^2} \, , \quad \varepsilon_\parallel = 1 - \sum_\alpha\frac{\omega_{\mathrm{p}\alpha}^2}{\omega^2} \, , \quad g = -\sum_\alpha\frac{\omega_{\mathrm{p}\alpha}^2\Omega_\alpha}{\omega(\omega^2-\Omega_\alpha^2)} \, . \tag{5.8.4}$$

These expressions coincide with (5.2.2, 3), obtained under the conditions (5.2.1), i.e., when the thermal motion of the particles is completely neglected.

To obtain the average force affecting the particles of type α we use (2.3.25)

$$\begin{aligned} F_{\mathrm{av}\alpha} &= \frac{1}{4\pi}\frac{\partial}{\partial N_\alpha}\varepsilon_{ij}(\omega,k)\nabla E_i^* E_j \\ &= \frac{1}{4\pi N_\alpha}(\varepsilon_\perp^{(\alpha)}-1)\nabla(|E_x|^2+|E_y|^2) \\ &\quad + g^{(\alpha)}\nabla(E_x^*E_y - E_xE_y^*) + (\varepsilon_\parallel^{(\alpha)}-1)\nabla|E_z|^2 \, . \end{aligned} \tag{5.8.5}$$

Here $\varepsilon_\perp^{(\alpha)}$, $\varepsilon_\parallel^{(\alpha)}$ and $g^{(\alpha)}$ are defined by (5.8.4) without the summation over α.

In contrast to the isotropic plasma, the average force (5.8.5) in the magneto-active plasma may lead to either expelling the particles of type α out of or forcing them into the region of the strong field. The direction of the effect depends on the SHF field polarization and the correlation between the frequency ω and the cyclotron frequency Ω_e. Thus, for linear polarization of the wave $E = (E_\perp, 0, 0)$ the force acting on the electrons of the magneto-active plasma is

$$F_{\mathrm{av\,e}} = -\frac{e^2}{m(\omega^2-\Omega_\mathrm{e}^2)}\nabla|E_\perp|^2 \, . \tag{5.8.6}$$

This field expels the electrons out of the strong field region when $\omega^2 > \Omega_e^2$ and it forces them inward when $\omega^2 < \Omega_e^2$.

5.8.2. Derive the oscillation modes of the nonisothermal magneto-active plasma using the model of one-fluid hydrodynamics. Neglect dissipative processes.

Solution. We linearize the system of equations

$$\frac{\partial \boldsymbol{B}}{\partial t} = \operatorname{curl}[\boldsymbol{V}, \boldsymbol{B}], \quad \operatorname{div} \boldsymbol{B} = 0,$$

$$\varrho_{\mathrm{M}}\left[\frac{\partial \boldsymbol{V}}{\partial t} + (\boldsymbol{V} \cdot \nabla)\boldsymbol{V}\right] = -v_{\mathrm{s}}^2 \nabla \varrho_{\mathrm{M}} - \frac{1}{4\pi}[\boldsymbol{B}, \operatorname{curl} \boldsymbol{B}], \tag{5.8.7}$$

$$\frac{\partial \varrho_{\mathrm{M}}}{\partial t} + \operatorname{div} \varrho_{\mathrm{M}} \boldsymbol{V} = 0$$

to describe small perturbations of the equilibrium $\varrho_{0\mathrm{M}}$, $\boldsymbol{V}_0 = 0$, $\boldsymbol{B}_0 \parallel 0z$. The perturbations $\delta\varrho_{\mathrm{M}}$, $\delta\boldsymbol{V}$, and $\boldsymbol{B}$ are assumed to depend on the time and the space coordinate in the form $\exp(-\mathrm{i}\omega t + \mathrm{i}\boldsymbol{k} \cdot \boldsymbol{r})$. As a result we obtain

$$\omega \boldsymbol{B} + [\boldsymbol{k}, (\delta \boldsymbol{V}, \boldsymbol{B}_0)] = 0, \quad \boldsymbol{k} \cdot \boldsymbol{B} = 0,$$

$$-\omega \delta \boldsymbol{V} + \frac{v_{\mathrm{s}}^2}{\varrho_{0\mathrm{M}}} \delta \varrho_{\mathrm{M}} \boldsymbol{k} + \frac{1}{4\pi \varrho_{0\mathrm{M}}}[\boldsymbol{B}_0, (\boldsymbol{k}, \boldsymbol{B})] = 0, \tag{5.8.8}$$

$$\omega \delta \varrho_{\mathrm{M}} - \varrho_{0\mathrm{M}} \boldsymbol{k} \cdot \delta \boldsymbol{V} = 0.$$

Here $v_{\mathrm{s}} = \sqrt{ZT_{\mathrm{e}}/M}$ is the velocity of the ion sound for the nonisothermal gaseous plasma and $v_{\mathrm{s}} = \sqrt{m/(3M)}\, v_{\mathrm{Fe}}$ for the degenerate solid-state one. Eliminating $\delta\varrho_{\mathrm{M}}$ from this system and decomposing it into Cartesian components we obtain two independent subsystems of equations:

$$\omega B_y + \delta V_y k_z B_0 = 0,$$

$$\omega \delta V_y + \frac{k_z B_0}{4\pi \varrho_{0\mathrm{M}}} B_y = 0 \tag{5.8.9}$$

$$\omega B_x + k_z B_0 \delta V_x = 0$$

$$\omega B_z + k_z B_0 \delta V_z - B_0(k_\perp \delta V_x + k_z \delta V_z) = 0,$$

$$\omega \delta V_x - \frac{v_{\mathrm{s}}^2}{\omega} k_\perp (k_\perp \delta V_x + k_z \delta V_z) + \frac{B_0}{4\pi \varrho_{0\mathrm{M}}}(B_x k_z - k_\perp B_z) = 0, \tag{5.8.10}$$

$$\omega \delta V_z - \frac{v_{\mathrm{s}}^2}{\omega} k_z (k_\perp \delta V_x + k_z \delta V_z) = 0.$$

The first subsystem consists of two equations for B_y and δV_y and describes the Alfvén wave with the spectrum

$$\omega^2 = k_z^2 v_A^2 . \tag{5.8.11}$$

The second one consists of four equations for B_x, B_z, δV_x, δV_z and describes the fast and slow magnetosonic waves with the common dispersion relation

$$\omega^4 - \omega^2 k^2 (v_A^2 + v_s^2) + k^4 v_A^2 v_s^2 \cos^2\theta = 0 , \tag{5.8.12}$$

where θ is the angle between $\boldsymbol{B}_0$ and $\boldsymbol{k}$. Hence

$$\omega_\pm^2 = \frac{k^2}{2}\left[v_A^2 + v_s^2 \pm \sqrt{(v_A^2 + v_s^2)^2 - 4 v_A^2 v_s^2 \cos^2\theta}\right] . \tag{5.8.13}$$

Equations (5.8.11, 13) coincide with those obtained in Sect. 5.3.

5.8.3. Study the longitudinal waves of the collisionless nondegenerate electron plasma which propagate strictly across the magnetic field (Bernstein modes).

Solution. According to (5.1.13) the dispersion equation of these waves is

$$1 = 2 \sum_{n=1}^{\infty} \frac{\omega_{pe}^2 n^2 \Omega_e^2}{k^2 v_{Te}^2 (\omega^2 - n^2 \Omega_e^2)} A_n\left(\frac{k^2 v_{Te}^2}{\Omega_e^2}\right) , \tag{5.8.14}$$

or in the long-wave range $k^2 v_{Te}^2 \ll \Omega_e^2$

$$1 - \frac{\omega_{pe}^2}{\omega^2 - \Omega_e^2} - \sum_{n=2}^{\infty} \frac{\omega_{pe}^2 n^2}{\omega^2 - n^2\Omega_e^2} \frac{1}{n!} \left(\frac{k^2 v_{Te}^2}{2\Omega_e^2}\right)^{n-1} = 0 . \tag{5.8.15}$$

In the limit $k \to 0$ the solutions of this equation are

$$\omega = \sqrt{\omega_{pe}^2 + \Omega_e^2} , \qquad \omega = n\Omega_e . \tag{5.8.16}$$

On the other hand, in the short-wave range $k^2 v_{Te}^2 \ll \Omega_e^2$ we have

$$1 = \sum_{n=1}^{\infty} \frac{\omega_{pe}^2 n^2}{\omega^2 - n^2 \Omega_e^2} \sqrt{\frac{2}{\pi}} \left(\frac{\Omega_e^2}{k^2 v_{Te}^2}\right)^{3/2} . \tag{5.8.17}$$

In the limit $k \to \infty$ the solutions of the dispersion equation are

$$\omega \to n\Omega_e .$$

The dispersion laws are shown in Fig. 5.6.

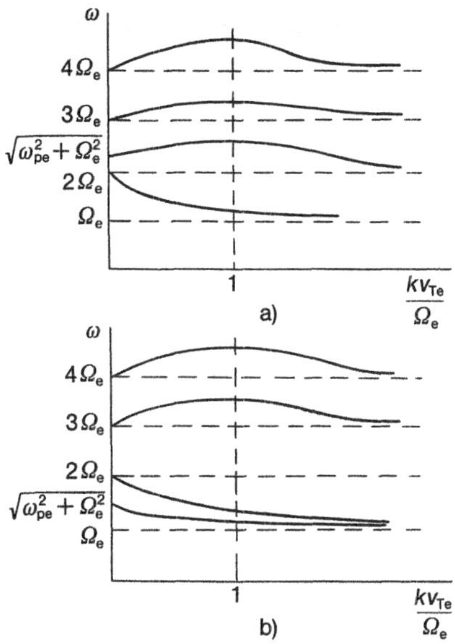

Fig. 5.6 a, b. Bernstein modes in a nondegenerate electron plasma: (**a**) $\omega_{pe} > \Omega_e$; (**b**) $\omega_{pe} < \Omega_e$

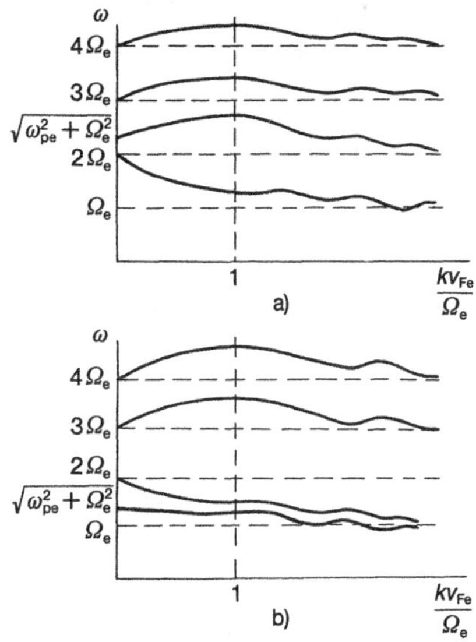

Fig. 5.7 a, b. Bernstein modes in a degenerate electron plasma: (**a**) $\omega_{pe} > \Omega_e$; (**b**) $\omega_{pe} < \Omega_e$

5.8.4. Analyze the Bernstein modes for the degenerate electron plasma.

Solution. Due to (5.1.16) the dispersion equation for longitudinal waves propagating across the magnetic field is

$$1 = \frac{3\,\omega_{pe}^2}{k^2 v_{Fe}^2} \sum_{n=1}^{\infty} \int_0^{\pi} d\theta \, \sin\theta \frac{n^2\Omega_e^2}{\omega^2 - n^2\Omega_e^2} \, J_n^2\left(\frac{kv_{Fe}\sin\theta}{\Omega_e}\right) . \tag{5.8.18}$$

As in the nondegenerate plasma (Exercise 5.8.3), the solutions of this equation are in the limit $k \to 0$

$$\omega = \sqrt{\omega_{pe}^2 + \Omega_e^2} , \qquad \omega = n\Omega_e . \tag{5.8.19}$$

In the limit $kv_{Fe} \gg \Omega_e$ we get from (5.8.18)

$$1 = \sum_{n=1}^{\infty} \frac{6\,\omega_{pe}^2\Omega_e^3}{\pi^2 k^3 v_{Fe}^3} \frac{n^2}{\omega^2 - n^2\Omega_e^2} \, J_0\left(\frac{kv_{Fe}}{\Omega_e}\right) . \tag{5.8.20}$$

Again we have $\omega \to n\Omega_e$ for $k \to \infty$ but the limit is approached in an oscillatory way (see Fig. 5.7).

5.8.5. Using (5.1.7, 8) estimate the collisionless absorption of the ordinary wave propagating across the magnetic field in the limit $k \to 0$. This absorption is determined by relativistic effects of the electron motion even in the nonrelativistic ($mc^2 \gg T_e$) Maxwellian plasma.

Solution. For the ordinary wave propagating across the magnetic field in an electron plasma we have

$$k^2 - \frac{\omega^2}{c^2}\varepsilon_{zz} = 0\,, \qquad (5.8.21)$$

$$\varepsilon_{zz} = 1 + \frac{4\pi e^2}{\omega}\int d\mathbf{p}\frac{\partial f_0}{\partial \mathscr{E}}\sum_{n=-\infty}^{\infty}\frac{v_z^2 J_n^2\left(\frac{kv_{Te}}{\Omega_e}\gamma\right)}{\omega - n\Omega_e/\gamma}\,.$$

In the limit $\Omega_e \gg kv_{Te}$, (5.8.21) yields

$$\varepsilon_{zz} = 1 - \frac{\omega_{pe}^2}{\omega^2} + \mathrm{i}\frac{2\pi^2 e^2 k^2}{15\,\omega^3 T_e}\int d\mathbf{p}\, f_{0e}\, v^4$$

$$\times\left[\delta\left(\omega - \Omega_e + \frac{\Omega_e v^2}{2c^2}\right) + \delta\left(\omega - \Omega_e - \frac{\Omega_e v^2}{2c^2}\right)\right]. \qquad (5.8.22)$$

Assuming $\Omega_e/\omega > 0$ we obtain

$$\varepsilon_{zz} = 1 - \frac{\omega_{pe}^2}{\omega^2} + \mathrm{i}\sqrt{\frac{\pi}{2}}\frac{32}{15}\frac{\omega_{pe}^2 k^2 c^7}{\omega^3\Omega_e v_{Te}^5}\left(1 - \frac{\omega}{\Omega_e}\right)^{5/2}\exp\left[-\frac{2c^2}{v_{Te}^2}\left(1 - \frac{\omega}{\Omega_e}\right)^2\right]. \qquad (5.8.23)$$

The substitution of (5.8.23) into (5.8.21) leads to the frequency and the damping decrement of the ordinary wave ($\omega \to \omega + \mathrm{i}\delta$):

$$\omega^2 = \omega_{Le}^2 + k^2c^2\,,$$

$$\delta = -\sqrt{\frac{\pi}{2}}\frac{16}{15}\frac{\omega_{pe}^2 k^2 c^7}{\omega^2\Omega_e v_{Te}^5}\left(1 - \frac{\omega}{\Omega_e}\right)^{5/2}\exp\left[-\frac{2c^2}{v_{Te}^2}\left(1 - \frac{\omega}{\Omega_e}\right)^2\right]. \qquad (5.8.24)$$

The ordinary wave is absorbed only near the cyclotron frequency $\omega \approx \Omega_e$.

5.8.6. Derive the refractive index and the damping coefficient of electromagnetic waves propagating along the external magnetic field for frequencies near the ion cyclotron frequency in the collisionless nondegenerate plasma.

Solution. Near the frequency $\omega \approx \Omega_i$ the general dispersion equation (5.4.1) reduces to

$$k^2c^2 = \omega^2\left[1 + \frac{\omega_{pe}^2}{\omega\Omega_e} - \frac{\omega_{pi}^2}{\omega(\omega - \Omega_i)}I_+\left(\frac{\omega - \Omega_i}{kv_{Ti}}\right)\right]. \qquad (5.8.25)$$

Far from the resonance absorption line $\omega \gg |\omega - \Omega_i| \gg kv_{Ti}$, the contribution of the electron term may be ignored in (5.8.25). Under these conditions the refractive index and the damping coefficient of the ion cyclotron wave are

$$n^2 = -\frac{\omega_{pi}^2}{\omega(\omega - \Omega_i)}, \qquad \chi = \sqrt{\frac{\pi}{8}}\frac{\omega_{pi}^2 c}{n^2\omega^2 v_{Ti}^2}\exp\left(-\frac{c^2(\omega - \Omega_i)^2}{2n^2\omega^2 v_{Ti}^2}\right). \quad (5.8.26)$$

Inside the absorption line, for $|\omega - \Omega_i| \ll kv_{Ti}$ the electron term is significant. Hence, for a high-pressure plasma with $v_{Ti}^2 \gg v_A^2$ we obtain from (5.8.25) the following weakly damped electron wave

$$n^2 \approx \frac{\omega_{pe}^2}{\omega\Omega_e}, \qquad \chi = \sqrt{\frac{\pi}{8}}\frac{\omega_{pi}^2 c}{n^2\omega^2 v_{Ti}}. \quad (5.8.27)$$

For a low-pressure plasma with $v_A^2 \gg v_{Ti}^2$ the electron term in (5.8.25) may be ignored and we obtain the highly damped ion cyclotron wave

$$n + i\chi = \frac{i + \sqrt{3}}{2}\left(\frac{2}{\sqrt{\pi}}\frac{\omega_{pi}^2 c}{\omega^2 v_{Ti}}\right)^{1/3}. \quad (5.8.28)$$

The quantity

$$\lambda_{sk} = \frac{c}{\omega\chi} = 2\left[\sqrt{\frac{2}{\pi}}c^2 v_{Ti}/(\omega\omega_{pi}^2)\right]^{1/3}$$

characterizes the penetration depth of this mode.

5.8.7. Study the ion-acoustic oscillations in the nonisothermal magnetized plasma with $T_e \gg T_i \approx 0$ under the conditions $\Omega_e^2 \gg \omega_{pi}^2 \gg \Omega_i^2$.

Solution. Under these conditions the ion-acoustic waves exist both for $\omega < \Omega_i$ and for $\omega > \Omega_i$. The dispersion equation is

$$1 - \frac{k_\perp^2}{k^2}\frac{\omega_{pi}^2}{\omega^2 - \Omega_i^2} - \frac{k_z^2\omega_{pi}^2}{k^2\omega^2} + \frac{\omega_{pe}^2}{k^2 v_{Te}^2}\left(1 + i\sqrt{\frac{\pi}{2}}\frac{\omega}{|k_z| v_{Te}}\right). \quad (5.8.29)$$

Neglecting the small imaginary part, we find the equation for the frequency spectrum

$$\omega^4\left(1 + \frac{\omega_{pi}^2}{k^2 v_s^2}\right) - \omega^2\left[\omega_{pi}^2 + \Omega_i^2\left(1 + \frac{\omega_{pi}^2}{k^2 v_s^2}\right)\right] + \omega_{pi}^2\Omega_i^2\cos^2\theta = 0. \quad (5.8.30)$$

For $\omega_{pi}^2 \gg \Omega_i^2$ and $\theta \neq 0$ we get the approximate roots

$$\omega_+ = \frac{\omega_{pi}^2 + \Omega_i^2(1 + \omega_{pi}^2/k^2v_s^2)}{1 + \omega_{pi}^2/k^2v_s^2},$$
$$\omega_-^2 = \frac{\omega_{pi}^2\Omega_i^2\cos^2\theta}{\omega_{pi}^2 + \Omega_i^2(1 + \omega_{pi}^2/k^2v_s^2)}. \quad (5.8.31)$$

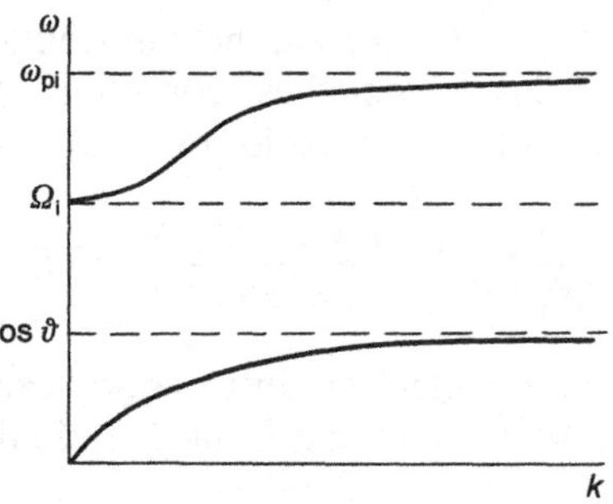

Fig. 5.8. Spectra of the ion-acoustic oscillations in a nonisothermal magnetized plasma

These spectra are shown in Fig. 5.8.

For $\theta \to 0$ we obtain from (5.8.29)

$$\omega_+^2 \to \Omega_i^2 \, , \quad \omega_-^2 = \frac{\omega_{pi}^2}{1 + \omega_{pi}^2 / k^2 v_s^2} \, . \tag{5.8.32}$$

Taking into account the small imaginary part of (5.8.29) we obtain the damping decrements

$$\delta_\pm = -\sqrt{\frac{\pi}{8}} \frac{\omega_\pm^4}{|k_z^3| v_s^2 v_{Te}} \frac{(\omega_\pm^2 - \Omega_i^2)^2}{(\omega_\pm^2 - \Omega_i^2)^2 + \omega_\pm^4 \tan^2 \theta} \, . \tag{5.8.33}$$

5.8.8. Derive the frequency spectrum and the damping decrement of the ion-acoustic waves of the weakly ionized degenerate electron-hole plasma in the frequency range $\Omega_e \gg \omega \gg \Omega_i$ and for $\Omega_e \gg \nu_{en} \gg k_z v_{Fe}$, $\omega \gg \nu_{in}$.

Solution. The ion-acoustic waves exist for phase velocities in the range $v_{Fe} \gg \omega / k_z \gg v_{Fi}$. Being longitudinal they satisfy the equation

$$1 + \frac{3\,\omega_{pe}^2}{k^2 v_{Fe}^2} \frac{k^2 D}{\omega} \frac{1}{\dfrac{k^2 D}{\omega} - 3\,\mathrm{i}} - \frac{\omega_{pi}^2}{\omega^2}\left(1 - \mathrm{i}\frac{\nu_{in}}{\omega}\right) = 0 \quad \text{with} \tag{5.8.34}$$

$$D = \frac{k_\perp^2}{k^2} D_\perp + \frac{k_z^2}{k^2} D_\parallel \, . \tag{5.8.35}$$

Here $D_\parallel = v_{Fe}^2 / \nu_{en}$ and $D_\perp = v_{Fe}^2 \nu_{en} / 2\,\Omega_e^2$ are the longitudinal and transverse *electron diffusion coefficients* in the strong magnetic field, respectively.

It follows from (5.8.34) that the ion-acoustic waves are weakly damped in the frequency range $\omega \ll k^2 D$. Under this condition their frequency and damping decrement are given by $(\omega \to \omega + \mathrm{i}\delta)$

$$\omega^2 = \frac{\omega_{pi}^2}{1 + \dfrac{3\,\omega_{pe}^2}{k^2 v_{Fe}^2}} \, , \qquad \delta = -\frac{\nu_{in}}{2} - \frac{3}{2} \frac{\omega^2}{k^2 D} \left(1 + \frac{k^2 v_{Fe}^2}{3\,\omega_{pe}^2}\right)^{-1} . \tag{5.8.36}$$

Obviously, the damping is caused by the ion friction and the electron diffusion.

5.8.9. Study the propagation of the ordinary wave across the magnetic field in the collisionless degenerate electron plasma for $\omega_{pe}^2 \gg \Omega_e^2 \gg \omega^2$.

Solution. Under these conditions the dispersion equation of the ordinary wave

$$k^2c^2 - \omega^2\varepsilon_{zz} = 0 \tag{5.8.37}$$

simplifies since the summand with $n = 0$ in ε_{zz}, see (5.1.14, 15), contributes dominantly:

$$\varepsilon_{zz} = 1 - \frac{3\,\omega_{pe}^2}{2\,\omega^2}\int_0^{\pi} d\theta \sin\theta \cos^2\theta\, J_0^2\left(\frac{kv_{Fe}}{\Omega_e}\sin\theta\right) . \tag{5.8.38}$$

Taking into account the asymptotic behaviour of the function $J_0(kv_{Fe}\sin\theta/\Omega_e)$ in the limit $kv_{Fe} \gg \Omega_e$ we obtain

$$\varepsilon_{zz} = 1 - \frac{3\,\omega_{pe}^2}{\omega^2}\frac{\Omega_e^2}{k^2v_{Fe}^2} J_1\left(\frac{kv_{Fe}}{\Omega_e}\right) . \tag{5.8.39}$$

For $\omega \ll \omega_{pe}$ we obtain from (5.8.37, 39) the condition for wave propagation

$$k^2c^2 + \frac{3\,\omega_{pe}^2\Omega_e^2}{k^2v_{Fe}^2} J_1\left(\frac{kv_{Fe}}{\Omega_e}\right) = 0 . \tag{5.8.40}$$

Solutions with $k^2 > 0$ (region of plasma transparency) exist only if $J_1(kv_{Fe}/\Omega_e) \approx 0$, i.e., $kv_{Fe}/\Omega_e \approx \mu_{1s}$, where μ_{1s} are the zeros of the Bessel function, $J_1(\mu_{1s}) = 0$. This implies that the degenerate high-density plasma is transparent for low-frequency oscillations when the wavelength is a multiple of the electron Larmor radius.

5.8.10. Study the cyclotron waves in the electron plasma at the higher harmonics $\omega \approx s\Omega_e$, $s \geq 2$ for oblique propagation $\theta \neq 0$.

Solution. Cyclotron waves propagating at an angle $\theta \neq 0$ in the plasma are absorbed also at the higher harmonics $\omega \approx s\Omega_e$. Assuming

$$|\omega - s\Omega_e| \ll k_{\|}v_{0e} , \quad k_{\perp}v_{0e} \ll \Omega_e ,$$

where v_{0e} is the random electron velocity (the thermal velocity for the non-degenerate plasma and the Fermi velocity for the degenerate plasma) we obtain

$$\varepsilon_{ij}(\omega, \boldsymbol{k}) = \begin{pmatrix} \varepsilon_{\perp} & ig & 0 \\ -ig & \varepsilon_{\perp} & 0 \\ 0 & 0 & \varepsilon_{\|} \end{pmatrix} . \tag{5.8.41}$$

The Hermitian part of the dielectric tensor is independent of the degree of degeneracy. Its components are

$$\varepsilon_\perp^{\mathrm{H}} = 1 - \frac{\omega_{\mathrm{pe}}^2}{\Omega_{\mathrm{e}}^2}\frac{1}{s^2-1}\,, \quad g^{\mathrm{H}} = -\frac{\omega_{\mathrm{pe}}^2}{\Omega_{\mathrm{e}}^2 s\,(s^2-1)}\,, \quad \varepsilon_\parallel^{\mathrm{H}} = 1 - \frac{\omega_{\mathrm{pe}}^2}{s^2\Omega_{\mathrm{e}}^2}\,. \tag{5.8.42}$$

The components of the anti-Hermitian part of $\varepsilon_{ij}(\omega, \boldsymbol{k})$ are

$$\varepsilon_\perp^{\mathrm{a}} = g^{\mathrm{a}} = \mathrm{i}\sqrt{\frac{\pi}{2}}\frac{s^2\omega_{\mathrm{pe}}^2}{\omega|k_\parallel|v_{\mathrm{Te}}}\left(\frac{k_\perp^2 v_{\mathrm{Te}}^2}{\Omega_{\mathrm{e}}^2}\right)^{s-1}\frac{1}{2^s s!}\,, \quad \varepsilon_\parallel^{\mathrm{a}} \approx 0 \tag{5.8.43}$$

for the nondegenerate plasma, and

$$\varepsilon_\perp^{\mathrm{a}} = g^{\mathrm{a}} = \mathrm{i}\frac{3}{2}\frac{\pi^2\omega_{\mathrm{pe}}^2}{\omega|k_\parallel|v_{\mathrm{Fe}}}\left(\frac{1}{2^s s!}\right)^2\left(\frac{k_\perp^2 v_{\mathrm{Fe}}^2}{\Omega_{\mathrm{e}}^2}\right)^{s-1}\,, \quad \varepsilon_\parallel^{\mathrm{a}} \approx 0 \tag{5.8.44}$$

for the degenerate plasma.

Evidently, the components of the anti-Hermitian part are small against the components of the Hermitian part. Therefore, the absorption is weak even inside the resonance line of the higher cyclotron harmonics. This allows to derive the refractive index and the damping coefficient

$$n_{1,2}^2 = \frac{(\varepsilon_\perp^{\mathrm{H}2} - g^{\mathrm{H}2} - \varepsilon_\perp^{\mathrm{H}}\varepsilon_\parallel^{\mathrm{H}})\sin^2\theta \pm \sqrt{(\varepsilon_\perp^{\mathrm{H}2} - g^{\mathrm{H}2} - \varepsilon_\perp^{\mathrm{H}}\varepsilon_\parallel^{\mathrm{H}})\sin^4\theta + 4g^{\mathrm{H}2}\varepsilon_\parallel^{\mathrm{H}2}\cos^2\theta}}{2(\varepsilon_\perp^{\mathrm{H}}\sin^2\theta + \varepsilon_\parallel^{\mathrm{H}}\cos^2\theta)}\,,$$

$$\begin{aligned}\chi_{1,2} = {} & \frac{-\mathrm{i}\varepsilon_\perp^{\mathrm{a}}}{4n_{1,2}(\varepsilon_\perp^{\mathrm{H}}\sin^2\theta + \varepsilon_\parallel^{\mathrm{H}}\cos^2\theta)} \\ & \times \{[\sin^4\theta\,(\varepsilon_\perp^{\mathrm{H}} - g^{\mathrm{H}})^2 + 2\sin^2\theta\cos^2\theta\,\varepsilon_\parallel^{\mathrm{H}}(\varepsilon_\perp^{\mathrm{H}} - g^{\mathrm{H}}) \\ & + \cos^2\theta\,(1+\cos^2\theta)\,\varepsilon_\parallel^{\mathrm{H}2}]\sqrt{(\varepsilon_\perp^{\mathrm{H}2} - g^{\mathrm{H}2} - \varepsilon_\perp^{\mathrm{H}}\varepsilon_\parallel^{\mathrm{H}})^2\sin^4\theta + 4g^{\mathrm{H}2}\varepsilon_\parallel^{\mathrm{H}2}\cos^2\theta} \\ & \pm [\sin^4\theta\,(\varepsilon_\perp^{\mathrm{H}2} - g^{\mathrm{H}2} - \varepsilon_\parallel^{\mathrm{H}}\varepsilon_\perp^{\mathrm{H}})(\sin^2\theta\,(\varepsilon_\perp^{\mathrm{H}} - g^{\mathrm{H}})^2 + \cos^2\theta\,\varepsilon_\parallel^{\mathrm{H}}(2\varepsilon_\perp^{\mathrm{H}} - 2g^{\mathrm{H}} - \varepsilon_\parallel^{\mathrm{H}}) \\ & + 4g^{\mathrm{H}}\varepsilon_\parallel^{\mathrm{H}2}\cos^2\theta\,(\varepsilon_\perp^{\mathrm{H}}\sin^2\theta - g^{\mathrm{H}}\sin^2\theta + \varepsilon_\parallel^{\mathrm{H}}\cos^2\theta)]\}\,.\end{aligned}$$

Note that $\varepsilon_\perp^{\mathrm{a}}$ is different for the nondegenerate and the degenerate plasma according to (5.8.43, 44).

5.8.11. Using (5.1.13) show that in the nonisotermal plasma with $T_{\mathrm{i}} > T_{\mathrm{e}}$ short-wave ($k_\perp\varrho_{\mathrm{i}} \gg 1$) low-frequency ($\omega \ll \Omega_{\mathrm{i}}$) potential oscillations with the spectrum

$$\omega^2 \approx k_z^2\frac{T_{\mathrm{i}}}{m}\,,$$

i.e., the spectrum of electron-acoustic waves, are possible. Such oscillations are also possible in the degenerate plasma if $T_{\mathrm{i}} > \mathscr{E}_{\mathrm{Fe}}$.

5.8.12. Using (5.5.8) for the longitudinal dielectric permittivity analyze the diffusion spread of a small inhomogeneity of the density of charged particles in the weakly ionized magnetized plasma.

Solution. Let us consider two limit cases: (a) plasma electrons are strongly magnetized ($\Omega_e \gg \nu_{en}$) and ions are unmagnetized ($\Omega_i \ll \nu_{in}$), and (b) both electrons and ions are strongly magnetized ($\Omega_\alpha \gg \nu_{\alpha n}$). Then

$$\varepsilon(\omega, k) = 1 + \frac{\omega_{pe}^2}{k^2 v_{Te}^2} \frac{ik^2 D_e}{\omega + ik^2 D_e} + \frac{\omega_{pi}^2}{k^2 v_{Ti}^2} \frac{ik^2 D_i}{\omega + ik^2 D_i} . \tag{5.8.45}$$

Hence for magnetized electrons

$$D_e = \frac{k_\perp^2}{k^2} \frac{v_{Te}^2 \nu_{en}}{\Omega_e^2} + \frac{k_\parallel^2}{k^2} \frac{v_{Te}^2}{\nu_{en}} .$$

For ions in case (a) $D_i = v_{Ti}^2/\nu_{in}$ and in case (b)

$$D_o = \frac{k_\perp^2}{k^2} \frac{v_{Ti}^2 \nu_{in}}{\Omega_i^2} + \frac{k_\parallel^2}{k^2} \frac{v_{Ti}^2}{\nu_{in}} .$$

Here $k^2 = k_\perp^2 + k_\parallel^2$ and $1/k_\perp = L_\perp$ is the dimension of the plasma inhomogeneity across the external magnetic field and $1/k_\parallel = L_\parallel$ along the external magnetic field.

Equalizing (5.8.45) to 0 and assuming the characteristic dimension of inhomogeneity to be much greater than the Debye length of electrons and ions yields

$$\omega = -i \frac{k^2 D_i (1 + T_e/T_i)}{1 + \dfrac{T_e}{T_i} \dfrac{D_i}{D_e}} . \tag{5.8.46}$$

Thus, for strongly magnetized electrons and ions the diffusion of the plasma across the magnetic field ($k_\parallel = 1/L_\parallel \to 0$) is always specified by electrons since $D_i \gg D_e$:

$$\omega = -ik_\perp^2 \frac{v_{Te}^2 \nu_{en}}{\Omega_e^2} \left(1 + \frac{T_i}{T_e}\right) \equiv -ik_\perp^2 D_\perp . \tag{5.8.47}$$

For unmagnetized ions the diffusion across the magnetic field results in

$$\omega = -ik_\perp^2 \frac{v_{Ti}^2}{\nu_{in}} \left(1 + \frac{T_e}{T_i}\right) \left(1 + \frac{\Omega_e \Omega_i}{\nu_{in} \nu_{en}}\right)^{-1} \equiv -ik_\perp^2 D_{eff} . \tag{5.8.48}$$

Part II

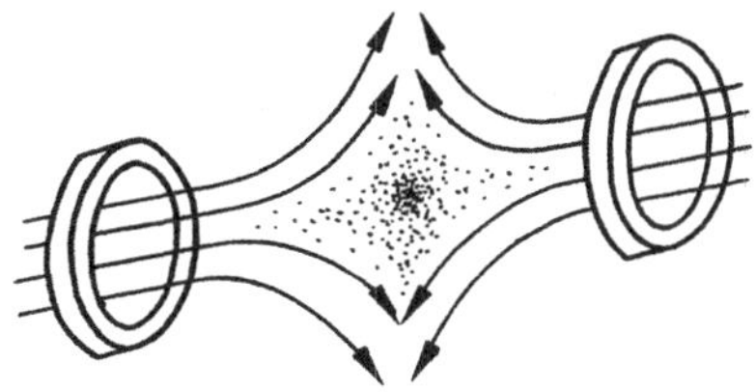

Electromagnetic Properties of Nonequilibrium Plasmas

The second part concerns the theory of the electromagnetic stability of a spatially inhomogeneous plasma which is not in thermodynamic equilibrium. The properties of a plasma carrying current in external electric and magnetic fields, and its properties in strong SHF fields and with a particle beam are analyzed. The geometrical optics approximation for a spatially inhomogeneous plasma is given. The analysis also pertains to the drift instabilities of a plasma confined by an external magnetic field. Here the properties of the spatially confined plasma are considered and the surface waves in a plasma waveguide are also studied.

6. Interaction of Charged Beams with the Plasma

Properties of a plasma not in thermodynamic equilibrium are investigated using the anisotropic particle distribution function for the magneto-active plasma; the dielectric tensor is obtained for such a plasma. On the basis of this tensor, the instability of a plasma with anisotropic particle temperature, the Cherenkov instability of a straight electron beam in a plasma, and the cyclotron instability of a rotating electron beam (flux of oscillators) in the plasma are studied.

6.1 Dielectric Tensor of the Homogeneous Anisotropic Nonequilibrium Plasma

In the previous chapters we have studied the electromagnetic properties of the plasma in thermodynamic equilibrium where the particles are distributed according to the Fermi or Maxwellian velocity distributions. The real plasma usually is far from the equilibrium state, however. The nonequilibrium state of the plasma depends on the method of its generation, e.g., by an electric discharge in a gas or on the surface of a solid, by the injection of charged particles into a neutral gas and its ionization, or by the injection of charge carriers into solids. Due to particle collisions the initial nonequilibrium state of the plasma approaches equilibrium. We have analyzed in Chap. 3 some of these problems and determined the collisional relaxation time of the most characteristic nonequilibrium states of the plasma (anisotropy of the momentum distribution, temperature anisotropy, etc.).

We shall see below that the relaxation of the nonequilibrium plasma due to collisionless effects can occur on a time scale short compared to the collisional relaxation time. Since the nonequilibrium plasma is unstable with respect to various small perturbations, which always exist due to fluctuations, large electromagnetic fields can develop in the plasma and destroy the initial nonequilibrium state.

The final stage of the nonequilibrium plasma due to instabilities can be determined on the basis of the nonlinear theory only. The initial stage of development of instabilities, however, presents a problem of linear electrodynamics according to which the growth rate of small perturbations is determined by (2.4.4)

$$\left| k^2\delta_{ij} - k_ik_j - \frac{\omega^2}{c^2}\varepsilon_{ij}(\omega, \boldsymbol{k}) \right| = 0 , \tag{6.1.1}$$

where $\varepsilon_{ij}(\omega, \boldsymbol{k})$ is the dielectric tensor of the spatially homogeneous nonequilibrium plasma. The roots $\omega_n(\boldsymbol{k})$ of this equation determine the development of small perturbations of the plasma

$$\phi_k(t) \sim \sum_n \phi_{nk}(0)\, \mathrm{e}^{-\mathrm{i}\omega_n(k)t} .$$

If only one root of the set $\omega_n(\boldsymbol{k})$ has a positive imaginary part, $\mathrm{Im}\{\omega_n(\boldsymbol{k})\} > 0$, the corresponding perturbation of the plasma increases with time and the plasma is unstable.

We consider in this chapter the instabilities of the nonequilibrium plasma with anisotropic distribution functions of the particles of the type α of the form

$$f_{0\alpha} = f_{0\alpha}(p_\perp, p_z) . \tag{6.1.2}$$

The external homogeneous magnetic field $\boldsymbol{B}_0$ is oriented along the z-axis. Examples of nonequilibrium plasmas of this type are the plasma with an anisotropic temperature, the plasma with a beam of charged particles moving parallel to the magnetic field or rotating around it.

Note that the distribution (6.1.2) satisfies the stationary and spatially homogeneous kinetic equation

$$[\boldsymbol{v}, \boldsymbol{B}_0] \cdot \frac{\partial f_{0\alpha}}{\partial \boldsymbol{p}} = 0 , \tag{6.1.3}$$

when electric and variable magnetic fields are absent and when particle collisions can be neglected.

The equation for a small deviation $\delta f_\alpha(p, t)$, caused by the variable fields $\boldsymbol{E}(\boldsymbol{r}, t)$ and $\boldsymbol{B}(\boldsymbol{r}, t)$, is

$$\frac{\partial \delta f_\alpha}{\partial t} + \boldsymbol{v} \cdot \frac{\partial \delta f_\alpha}{\partial \boldsymbol{r}} + \frac{e_\alpha}{c}[\boldsymbol{v}, \boldsymbol{B}_0] \cdot \frac{\partial \delta f_\alpha}{\partial \boldsymbol{p}} = -e_\alpha \left\{ \boldsymbol{E} + \frac{1}{c}[\boldsymbol{v}, \boldsymbol{B}] \right\} \cdot \frac{\partial f_{0\alpha}}{\partial \boldsymbol{p}} . \tag{6.1.4}$$

As in Sect. 5.1, we assume that the perturbed quantities depend on the time and space coordinates in the form of $\exp(-\mathrm{i}\omega t + \mathrm{i}\boldsymbol{k} \cdot \boldsymbol{r})$, directing the x-axis along the vector $\boldsymbol{k}_\perp$. Substituting the resulting perturbation $\delta f_\alpha(\boldsymbol{p}, t)$ into the expression for the induced current density

$$\boldsymbol{j} = \sum_\alpha \boldsymbol{j}_\alpha = \sum_\alpha e_\alpha \int d\boldsymbol{p}\, \boldsymbol{v} \delta f_\alpha(\boldsymbol{p}, t) , \tag{6.1.5}$$

we finally obtain the dielectric tensor of the nonequilibrium plasma state specified above

$$\varepsilon_{ij}(\omega, \boldsymbol{k}) = \delta_{ij} + \sum_{\alpha} \frac{4\pi i e_\alpha^2}{\omega \Omega_\alpha} \int d\boldsymbol{p}\, v_i \gamma \int_{\infty}^{\phi} d\phi' \frac{\partial f_{0\alpha}}{\partial p_l} \left(\frac{\omega - \boldsymbol{k}\cdot\boldsymbol{v}}{\omega} \delta_{lj} + \frac{k_l v_j}{\omega} \right)$$
$$\times \exp\left\{ \frac{i\gamma}{\Omega_\alpha} \left[(\omega - k_z v_z)(\phi' - \phi) - k_\perp v_\perp (\sin\phi' - \sin\phi) \right] \right\}, \qquad (6.1.6)$$

where

$$\gamma = (1 - v^2/c^2)^{1/2}, \qquad \Omega_\alpha = e_\alpha B_0 / m_\alpha c .$$

Taking account of the form of the unperturbed distribution function (6.1.2) and integrating (6.1.6) over the angles, we obtain

$$\varepsilon_{ij}(\omega, \boldsymbol{k}) = \delta_{ij} + \sum_{\alpha} \frac{4\pi e_\alpha^2}{\omega^2} \int d\boldsymbol{p} \sum_n \frac{(\omega - k_z v_z) \dfrac{\partial f_{0\alpha}}{\partial p_\perp} + k_z v_z \dfrac{\partial f_{0\alpha}}{\partial p_z}}{\left(\omega - k_z v_z - \dfrac{n\Omega_\alpha}{\gamma} \right) v_\perp} \Pi_{ij}^{(n)}, \qquad (6.1.7)$$

where

$$\Pi_{ij}^{(n)} = F_{\alpha i} F_{\alpha j}^{*} = \begin{pmatrix} v_\perp^2 \left(\dfrac{n J_n(x)}{x} \right)^2 & i v_\perp^2 \dfrac{n J_n(x) J_n'(x)}{x} & v_\perp v_z \dfrac{n J_n^2(x)}{x} \\ - i v_\perp^2 \dfrac{n J_n(x) J_n'(x)}{x} & v_\perp^2 J_n'^2(x) & - i v_z v_\perp J_n(x) J_n'(x) \\ v_\perp v_z \dfrac{n J_n^2(x)}{x} & i v_\perp v_z J_n(x) J_n'(x) & v_z^2 J_n^2(x) \end{pmatrix} \qquad (6.1.8)$$

$$\boldsymbol{F}_\alpha = \left[\frac{n\Omega_\alpha}{k_\perp \gamma} J_n(x); - i v_\perp J_n'(x); v_z J_n(x) \right], \qquad x = \frac{k_\perp v_\perp}{\Omega_\alpha} \gamma .$$

To calculate the dielectric permittivity explicitly, we need the exact form of the function $f_{0\alpha}(p_\perp, p_z)$. We assume

$$f_{0\alpha}(p_\perp, p_\parallel) = \frac{N_{0\alpha} \exp\left(- \dfrac{(p_\perp - p_{\perp 0})^2}{2 m_\alpha T_{\perp\alpha}} - \dfrac{(p_\parallel - p_{\parallel 0})^2}{2 m_\alpha T_{\parallel\alpha}} \right)}{2\pi m_\alpha p_{\perp 0} \sqrt{T_{\perp\alpha} T_{\parallel\alpha}}} \qquad (6.1.9)$$

describing a system of particles with average longitudinal $p_{\parallel 0}$ and transverse $p_{\perp 0} \gg \sqrt{m_\alpha T_{\perp\alpha}}$ momenta with respect to the direction of $\boldsymbol{B}_0$. The thermal spread of the momenta is defined by the temperatures $T_{\parallel\alpha}$ and $T_{\perp\alpha}$ assuming $T_{\parallel\alpha} \ll m_\alpha c^2$ and $T_{\perp\alpha} \ll m_\alpha c^2$.

In the absence of a mean directed motion ($p_{\perp 0} = p_{\parallel 0} = 0$) Eq. (6.1.9) simplifies to

$$f_{0\alpha} = \frac{N_{0\alpha} \exp\left(-\dfrac{p_\perp^2}{2m_\alpha T_\alpha} - \dfrac{p_\parallel^2}{2m_\alpha T_\alpha}\right)}{2\pi m_\alpha T_{\perp\alpha} \sqrt{2\pi m_\alpha T_{\parallel\alpha}}}, \tag{6.1.10}$$

corresponding to the plasma with an anisotropic temperature.

In the cold plasma limit $T_{\perp\alpha} = T_{\parallel\alpha} = 0$ we have

$$f_{0\alpha} = \frac{N_{0\alpha}}{2\pi p_{\perp 0}} \delta(p_\perp - p_{\perp 0})\, \delta(p_\parallel - p_{\parallel 0}) . \tag{6.1.11}$$

This distribution function describes a system of monoenergetic particles rotating perpendicularly to the external homogeneous magnetic field with the momentum $p_{\perp 0}$ and moving parallel to it with the momentum $p_{\parallel 0}$. It describes the real system when the thermal spread of the momenta may be neglected, i.e., for

$$p_{\perp 0} \gg \sqrt{m_\alpha T_{\perp 0}} \qquad \text{and} \qquad p_{\parallel 0} \gg \sqrt{m_\alpha T_{\parallel 0}} .$$

6.1.1 The Lorentz Transform of the Dielectric Tensor

The general calculation of the dielectric tensor (6.1.7) using the distribution (6.1.9) is very complex. We do not give it here since the study of concrete problems is much simpler. Considering such problems we can establish another method to find the dielectric tensor of the nonequilibrium plasma. It is rather general although the absence of particle rotation across the magnetic field, i.e., $p_{\perp 0} = 0$, must be assumed. Then we can introduce the inertial system moving along with the particles of the type α so that the longitudinal momentum $p_{\parallel 0}$ disappears. The momentum distribution of the particles of this type evidently has the form (6.1.10) in the moving frame. For $T_{\perp\alpha} = T_{\parallel\alpha}$ this distribution coincides with the Maxwellian. In this case the contribution of the particle species α to the dielectric tensor was calculated in the preceding chapter already, see (5.1.10). For the anisotropic plasma with $T_{\perp\alpha} \neq T_{\parallel\alpha}$ the calculation presents no difficulty and is given in the next section, see (6.2.4). Using these expressions we can obtain the dielectric tensor of the multicomponent plasma in the laboratory frame without solving the kinetic equation by application of the Lorentz transformation formulas. Actually, the total induced current in the plasma is the sum of the currents of its charged particle components

$$j_i(\omega, \boldsymbol{k}) = \sum_\alpha j_{\alpha i} = \sum_\alpha \sigma_{ij}^\alpha(\omega, \boldsymbol{k})\, E_j , \tag{6.1.12}$$

where $\sigma_{ij}^\alpha(\omega, \boldsymbol{k})$ is the contribution of the particle species α to the conductivity tensor in the laboratory frame.

To determine the tensor $\sigma_{ij}^{a}(\omega, \boldsymbol{k})$ we pass over to the frame moving with the velocity u_a given by[1]

$$p_{\|0} = m_a \gamma_a u_a \,, \qquad \gamma_a = (1 - u_a^2/c^2)^{1/2}. \tag{6.1.13}$$

In this frame we have

$$j'_{ia}(\omega'_a, \boldsymbol{k}'_a) = \sigma_{ij}^{(a)}(\omega'_a, \boldsymbol{k}'_a)\, E'_{ja}(\omega'_a, \boldsymbol{k}'_a) \,, \tag{6.1.14}$$

where ω'_a and $\boldsymbol{k}'_a$ are the Lorentz transformed frequency ω and the wave vector $\boldsymbol{k}$

$$\begin{aligned} \omega'_a &= (\omega - \boldsymbol{k} \cdot \boldsymbol{u}_a)\gamma_a \,, \\ \boldsymbol{k}'_a &= \boldsymbol{k} + \boldsymbol{u}_a \gamma_a \left[\frac{\boldsymbol{k} \cdot \boldsymbol{u}_a}{u_a^2}\left(1 - \frac{1}{\gamma_a}\right) - \frac{\omega}{c^2} \right], \end{aligned} \tag{6.1.15}$$

$\sigma_{ij}^{a}(\omega'_a, \boldsymbol{k}'_a)$ is the conductivity tensor of the particle component a in its respective moving frame; $\boldsymbol{E}'_a$ and $\boldsymbol{j}'_a$ are the electric field and the current densities in this frame related to $\boldsymbol{E}$ and $\boldsymbol{j}_a$ by the Lorentz transform

$$j_{ai} = \alpha_{ij}(\boldsymbol{u}_a)\, j'_{aj} \,, \qquad E'_{ai} = \beta_{ij}(\boldsymbol{u}_a)\, E_j \,, \qquad \text{where} \tag{6.1.16}$$

$$\begin{aligned} \alpha_{ij}(\boldsymbol{u}_a) &= \delta_{ij} + \gamma_a \left[\frac{u_{ai} u_{aj}}{u_a^2}\left(1 - \frac{1}{\gamma_a}\right) + \frac{k'_{aj} u_{ai}}{\omega'_a} \right], \\ \beta_{ij}(\boldsymbol{u}_a) &= \frac{\omega'_a}{\omega}\alpha_{ji}(\boldsymbol{u}_a) = \frac{\omega'_a}{\omega}\delta_{ij} + \gamma_a \left[\frac{u_{ai} u_{aj}}{u_a^2}\left(\frac{1}{\gamma_a} - 1\right) + \frac{k_i u_{aj}}{\omega} \right]. \end{aligned} \tag{6.1.17}$$

It is easy to derive from (6.1.12–17) the dielectric tensor of the multicomponent plasma in the laboratory frame:

$$\begin{aligned} \varepsilon_{ij}(\omega, \boldsymbol{k}) &= \delta_{ij} + \frac{4\pi \mathrm{i}}{\omega} \sum_a \alpha_{i\mu}(\boldsymbol{u}_a)\, \sigma_{\mu\nu}^{a}(\omega'_a, \boldsymbol{k}'_a)\, \beta_{\nu j}(\boldsymbol{u}_a) \\ &= \delta_{ij} + \sum_a \frac{\omega'_a}{\omega} \alpha_{i\mu}(\boldsymbol{u}_a)[\varepsilon_{\mu\nu}^{(a)}(\omega'_a, \boldsymbol{k}'_a) - \delta_{\mu\nu}]\, \beta_{\nu j}(\boldsymbol{u}_a) \,. \end{aligned} \tag{6.1.18}$$

---

1 If the distribution function f_{0a} is of the form (6.1.10) in this coordinate system, then it takes the form

$$f_{0a} = \frac{N_{0a}\gamma_a}{2\pi m_a T_{\perp a}\sqrt{2\pi m_a T_{\|a}}}\left(1 - \frac{u_a v_\|}{c^2}\right) \exp\left(-\frac{p_\perp^2}{2m_a T_a}\right) \exp\left(-\frac{\gamma_a^2 (p_\| - u_a\sqrt{p_\|^2/c^2 + \mathrm{m}_a^2})}{2m_a T_{\|a}}\right)$$

in the laboratory frame. N_{0a} is the particle density in the laboratory frame, and $T_{\perp a}$ and $T_{\|a}$ are the temperatures in the intrinsic frame.

Here $\varepsilon_{ij}^{\alpha}(\omega'_\alpha, \boldsymbol{k}'_\alpha)$ denotes the dielectric permittivity of the particles of the type α in their respective moving frame which is assumed to be known, see (5.1.107, 6.2.4).

Using (6.1.18), one should keep in mind that not only the frequency and the wave vector but also the particle density (due to the volume contraction) and the mass are transformed in the expression for $\varepsilon_{ij}^{\alpha}(\omega'_\alpha, \boldsymbol{k}'_\alpha)$. Hence, the plasma frequency is $\omega'_{p\alpha} = \omega_{p\alpha}\gamma_\alpha^{-1/2}$, where $\omega_{p\alpha} = \sqrt{4\pi e^2 N_{0\alpha}/m_\alpha}$ remains invariant and the Larmor frequency is transformed: $\Omega'_\alpha = \Omega_\alpha = e_\alpha B_0/m_\alpha c$. $N_{0\alpha}$ is the density of the particles of the type α in the laboratory frame, and m_α is their rest mass.

To conclude it should be noted that (6.1.16, 17) generalize the well-known *Minkowski material relations* for moving isotropic media to the case of anisotropic media with account taken of frequency or spatial dispersion. These relations become much simpler in the nonrelativistic limit $u_\alpha \ll c$

$$\omega'_\alpha = \omega - \boldsymbol{k}\cdot\boldsymbol{u}_\alpha\,, \qquad \boldsymbol{k}'_\alpha = \boldsymbol{k}\,,$$
$$\alpha_{ij}(\boldsymbol{u}_\alpha) = \delta_{ij} + \frac{k_j u_{i\alpha}}{\omega - \boldsymbol{k}\cdot\boldsymbol{u}_\alpha}\,, \qquad \beta_{ij}(\boldsymbol{u}_\alpha) = \frac{\omega - \boldsymbol{k}\cdot\boldsymbol{u}_\alpha}{\omega}\delta_{ij} + \frac{k_i u_{\alpha j}}{\omega}\,. \tag{6.1.19}$$

6.2 Instability of the Plasma with Anisotropic Temperature of the Particle Components

As a first example of the instabilities of nonequilibrium plasmas, we consider a plasma with anisotropic temperature of the particle components, i.e., with the distribution functions (6.1.10). Confining our interest to the nonrelativistic case and introducing the notations

$$\mathscr{E}_\parallel = \frac{mv_z^2}{2} = \frac{p_z^2}{2m}\,, \qquad \mathscr{E}_\perp = \frac{mv_\perp^2}{2} = \frac{p_\perp^2}{2m} \tag{6.2.1}$$

we can rewrite (6.1.7) in the form

$$\varepsilon_{ij}(\omega,\boldsymbol{k}) = \delta_{ij} - \left(\delta_{ij} - \frac{B_{0i}B_{0j}}{B_0^2}\right)\sum_\alpha \frac{2\pi e_\alpha^2}{\omega^2}\int d\boldsymbol{p}\, v^2 \frac{\partial f_{0\alpha}}{\partial \mathscr{E}_\perp} - \frac{B_{0i}B_{0j}}{B_0^2}\sum_\alpha \frac{4\pi e_\alpha}{\omega^2}$$
$$\times \int d\boldsymbol{p}\, v_z^2 \frac{\partial f_{0\alpha}}{\partial \mathscr{E}_\parallel} + \sum_\alpha\sum_n \frac{4\pi e_\alpha^2}{\omega^2}\int d\boldsymbol{p}\, \frac{F_{\alpha i}F_{\alpha j}^*\left(k_z v_z \frac{\partial f_{0\alpha}}{\partial \mathscr{E}_\parallel} + n\Omega_\alpha \frac{\partial f_{0\alpha}}{\partial \mathscr{E}_\perp}\right)}{\omega - n\Omega_\alpha - k_z v_z}\,, \tag{6.2.2}$$

where

$$F_\alpha = \left[\frac{n\Omega_\alpha}{k_\perp}\mathrm{J}_n\left(\frac{k_\perp v_\perp}{\Omega_\alpha}\right),\ -\mathrm{i}v_\perp \mathrm{J}_n'\left(\frac{k_\perp v_\perp}{\Omega_\alpha}\right),\ v_z \mathrm{J}_n\left(\frac{k_\perp v_\perp}{\Omega_\alpha}\right)\right]. \tag{6.2.3}$$

Substituting the distribution function (6.1.10) of the plasma with anisotropic temperature of the particle components into (6.2.2) we can write the tensor components

$$\begin{aligned}
\varepsilon_{xx} &= 1 - \sum_\alpha \sum_n \frac{n^2 \omega_{\mathrm{p}\alpha}^2 A_n(z_\alpha)}{\omega^2 z_\alpha} \\
&\quad \times \left[1 - \frac{T_{\perp\alpha}}{T_{\|\alpha}} + \frac{\omega}{\omega - n\Omega_\alpha} \frac{T_{\perp\alpha}}{T_{\|\alpha}} I_+(\beta_{n\alpha}) \zeta_{n\alpha} \right], \\
\varepsilon_{xy} &= -\varepsilon_{yx} = \mathrm{i} \sum_\alpha \sum_n \frac{n\omega_{\mathrm{p}\alpha}^2}{\omega^2} A_n'(z_\alpha) \\
&\quad \times \left[1 - \frac{T_{\perp\alpha}}{T_{\|\alpha}} - \frac{\omega}{\omega - n\Omega_\alpha} \frac{T_{\perp\alpha}}{T_{\|\alpha}} I_+(\beta_{n\alpha}) \zeta_{n\alpha} \right] \\
\varepsilon_{yy} &= \varepsilon_{xx} - 2 \sum_\alpha \sum_n \frac{n\omega_{\mathrm{p}\alpha}^2}{\omega^2} z_\alpha A_n'(z_\alpha) \\
&\quad \times \left[1 - \frac{T_{\perp\alpha}}{T_{\|\alpha}} - \frac{\omega}{\omega - n\Omega_\alpha} \frac{T_{\perp\alpha}}{T_{\|\alpha}} I_+(\beta_{n\alpha}) \zeta_{n\alpha} \right], \qquad (6.2.4) \\
\varepsilon_{xz} &= \varepsilon_{zx} = \sum_\alpha \sum_n \frac{n\omega_{\mathrm{p}\alpha}^2}{\omega\Omega_\alpha} \frac{T_{\perp\alpha}}{T_{\|\alpha}} \frac{A_n(z_\alpha)}{z_\alpha} \frac{k_\perp}{k_\|} [1 - I_+(\beta_{n\alpha})] \zeta_{n\alpha}, \\
\varepsilon_{yz} &= -\varepsilon_{zy} = -\mathrm{i} \sum_\alpha \sum_n \frac{\omega_{\mathrm{p}\alpha}^2}{\omega\Omega_\alpha} \frac{T_{\perp\alpha}}{T_{\|\alpha}} A_n'(z_\alpha) \frac{k_\perp}{k_\|} [1 - I_+(\beta_{n\alpha})] \zeta_{n\alpha}, \\
\varepsilon_{zz} &= 1 + \sum_\alpha \sum_n \frac{\omega_{\mathrm{p}\alpha}^2 (\omega - n\Omega_\alpha)}{k_z^2 v_{\mathrm{T}\|\alpha}^2 \omega} A_n(z_\alpha) [1 - I_+(\beta_{n\alpha})] \zeta_{n\alpha},
\end{aligned}$$

where

$$z_\alpha = \frac{k_\perp^2 v_{\mathrm{T}\perp\alpha}^2}{\Omega_\alpha^2}, \quad \beta_{n\alpha} = \frac{\omega - n\Omega_\alpha}{k_z v_{\mathrm{T}\|\alpha}}, \quad v_{\mathrm{T}\perp,\|\alpha} = \sqrt{\frac{T_{\perp,\|\alpha}}{m_\alpha}},$$

$$\zeta_{n\alpha} = 1 + \frac{n\Omega_\alpha}{\omega} \left(\frac{T_{\|\alpha}}{T_{\perp\alpha}} - 1 \right).$$

6.2.1 Plasma Instability with Anisotropic Temperature in the Absence of Magnetic Field

One can obtain from (6.2.4), in the limit $\boldsymbol{B}_0 \to 0$, the components of the dielectric tensor of the anisotropic plasma with $T_{\perp\alpha} \neq T_{\parallel\alpha}$ in the absence of an external magnetic field. As mentioned in Chap. 5, it is mathematically difficult to take this limit since it is associated with the summation of Bessel functions of large values of the argument. In order to derive the tensor $\varepsilon_{ij}(\omega, \boldsymbol{k})$ of the anisotropic unmagnetized plasma, it is easier to use the expression

$$\varepsilon_{ij}(\omega, \boldsymbol{k}) = \delta_{ij} + \sum_{\alpha} \frac{4\pi e_\alpha^2}{\omega} \int \frac{d\boldsymbol{p}}{\omega - \boldsymbol{k}\cdot\boldsymbol{v}} v_i \frac{\partial f_{0\alpha}}{\partial p_l} \left[\left(1 - \frac{\boldsymbol{k}\cdot\boldsymbol{v}}{\omega}\right)\delta_{lj} + \frac{k_l v_j}{\omega}\right] \tag{6.2.5}$$

which generalizes (4.1.9) to the case of arbitrary anisotropic distribution functions $f_{0\alpha}(\boldsymbol{p})$. We can introduce the following system of coordinates: the z-axis is oriented parallel to the vector $\boldsymbol{k}$ and the x-axis in such a way that the vector $\boldsymbol{n}_\parallel$ of the direction with the temperature $T_{\parallel\alpha}$ lies in the xz-plane. We denote the angle included between $\boldsymbol{n}_\parallel$ and $\boldsymbol{k}$ by θ; the plane perpendicular to the vector $\boldsymbol{n}_\parallel$ defines the directions where the temperature is equal to $T_{\perp\alpha}$ (Fig. 6.1).

As a result, we obtain for the dielectric tensor of the anisotropic plasma

$$\varepsilon_{ij}(\omega, \boldsymbol{k}) = \begin{pmatrix} \varepsilon_{11} & 0 & \varepsilon_{13} \\ 0 & \varepsilon_{22} & 0 \\ \varepsilon_{31} & 0 & \varepsilon_{33} \end{pmatrix}, \quad \text{where} \tag{6.2.6}$$

$$\varepsilon_{11} = 1 - \sum_\alpha \frac{\omega_{p\alpha}^2}{\omega^2}\left(1 + \frac{1}{T_\alpha^{*2}}\left\{\left[T_{\parallel\alpha}T_{\perp\alpha} - \frac{\sin^2 2\theta}{2}(T_{\parallel\alpha} - T_{\perp\alpha})^2\right]\right.\right.$$
$$\left.\left.\times\,[I_+(\beta_\alpha^*) - 1] + \frac{1}{4}\sin^2 2\theta\,(T_{\parallel\alpha} - T_{\perp\alpha})^2[\beta_\alpha^{*2} I_+(\beta_\alpha^*) - \beta_\alpha^{*2} - 1]\right\}\right)$$

$$\varepsilon_{13} = \varepsilon_{31} = \sum_\alpha \frac{\omega_{p\alpha}^2}{\omega^2}\frac{\sin 2\theta}{2T_\alpha^*}(T_{\perp\alpha} - T_{\parallel\alpha})[(\beta_\alpha^{*2} - 1)I_+(\beta_\alpha^*) - \beta_\alpha^{*2}], \tag{6.2.7}$$

$$\varepsilon_{22} = 1 - \sum_\alpha \frac{\omega_{p\alpha}^2}{\omega^2}\left[\left(1 - \frac{T_{\perp\alpha}}{T_\alpha^*}\right) + \frac{T_{\perp\alpha}}{T_\alpha^*} I_+(\beta_\alpha^*)\right],$$

$$\varepsilon_{33} = 1 + \sum_\alpha \frac{\omega_{p\alpha}^2}{\omega^2}\beta_\alpha^{*2}[1 - I_+(\beta_\alpha^*)].$$

Here

$$T_\alpha^* = T_{\parallel\alpha}\cos^2\theta + T_{\perp\alpha}\sin^2\theta, \qquad \beta_\alpha^* = \frac{\omega}{k}\sqrt{\frac{m_\alpha}{T_\alpha^*}}.$$

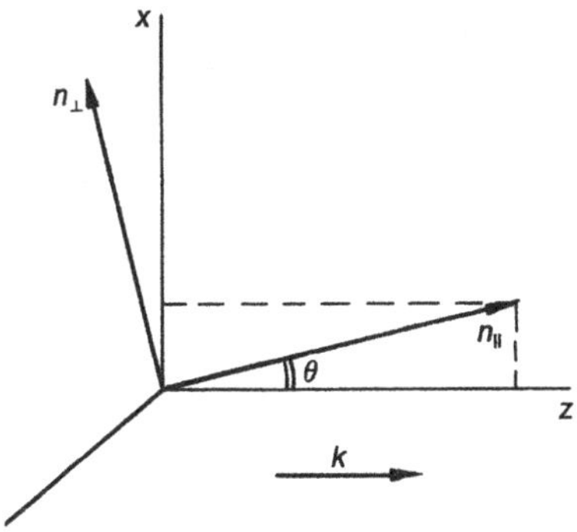

Fig. 6.1.

Equations (6.2.4, 7) show that, in contrast to the isotropic Maxwellian plasma, the anti-Hermitian components of the dielectric tensor, given by the balance of Cherenkov wave absorption and radiation, can change their signs in the anisotropic plasma. As stated above, this indicates a possible instability. However, the change in sign of the anti-Hermitian part of the tensor $\varepsilon_{ij}(\omega, \boldsymbol{k})$ is not a sufficient condition for an instability since oscillations must be present in the plasma for the development of an instability. To obtain a sufficient condition for instability we must analyze (6.1.1), which will be done below.

To begin with, we analyze the stability of an anisotropic plasma in the absence of an external magnetic field. Substituting (6.2.6) into (6.1.1) we obtain two independent equations:

$$k^2 - \frac{\omega^2}{c^2}\varepsilon_{22} = 0 , \tag{6.2.8}$$

$$\left(k^2 - \frac{\omega^2}{c^2}\varepsilon_{11}\right)\varepsilon_{33} + \frac{\omega^2}{c^2}\varepsilon_{13}^2 = 0 . \tag{6.2.9}$$

The first equation describes a purely transverse wave $\boldsymbol{k} \perp \boldsymbol{E} \parallel Oy$; the second one represents longitudinal-transverse waves with the field $\boldsymbol{E}$ in the xz-plane. Only for $\theta = 0$ or $\theta = \pi/2$ does this equation separate into two equations for purely longitudinal ($\boldsymbol{E} \parallel Oz \parallel \boldsymbol{k}$) and purely transverse ($\boldsymbol{k} \perp \boldsymbol{E} \parallel Ox$) modes:

$$\varepsilon_{33} = 0 , \tag{6.2.10}$$

$$k^2 - \frac{\omega^2}{c^2}\varepsilon_{11} = 0 . \tag{6.2.11}$$

We have $T_\alpha^* = T_{\parallel\alpha}$ for $\theta = 0$ and $T_\alpha^* = T_{\perp\alpha}$ for $\theta = \pi/2$.

It is easy to see that the *longitudinal oscillations are always stable in a plasma with anisotropic pressure.* Therefore, we shall study the transverse waves only, described by (6.2.8, 11). For simplicity, we assume $\theta = 0$ or $\theta = \pi/2$. For $\theta = 0$ we have $\varepsilon_{11} = \varepsilon_{22}$ so that (6.2.8 and 11) are identical. For

$\theta = \pi/2$ the components ε_{11} and ε_{22} are identical, too. However, in the latter case (6.2.8) is similar to the equation for transverse waves in the isotropic plasma with a temperature $T_\alpha = T_{\perp\alpha}$ and has only stable solutions. Thus, to obtain stability conditions of the anisotropic plasma it is sufficient to study (6.2.11), which has the form

$$k^2c^2 = \omega^2\left(1 - \sum_\alpha \frac{\omega_{p\alpha}^2}{\omega^2}\left\{1 + \frac{T_{\perp\alpha}}{T_{\|\alpha}}\left[I_+(\beta_{\|\alpha}) - 1\right]\right\}\right) \tag{6.2.12}$$

for $\theta = 0$, and

$$k^2c^2 = \omega^2\left(1 - \sum_\alpha \frac{\omega_{p\alpha}^2}{\omega^2}\left\{1 + \frac{T_{\|\alpha}}{T_{\perp\alpha}}\left[I_+(\beta_{\perp\alpha}) - 1\right]\right\}\right) \tag{6.2.12a}$$

for $\theta = \pi/2$. Hence, all conclusions derived for $\theta = 0$ are valid for $\theta = \pi/2$ under the substitution $T_{\|\alpha} \rightleftarrows T_{\perp\alpha}$. Therefore, we shall consider wave propagation at the angle $\theta = 0$ only.

For simplicity, we assume that the ion distribution is isotropic and that the electron component of the plasma possesses the anisotropic temperature distribution. Such a plasma is always unstable. In fact, we obtain from (6.2.12) under the condition

$$\beta_{\|e} \gg 1 \quad (\text{i.e., } \omega \gg kv_{\|e}) ,$$

$$k^2c^2 = \omega^2\left[1 - \frac{\omega_{pe}^2}{\omega^2}\left(1 + \frac{k^2v_{\perp e}^2}{\omega^2}\right)\right] . \tag{6.2.13}$$

The roots of this biquadratic equation are with high accuracy

$$\omega_1^2 = k^2c^2 + \omega_{pe}^2 , \qquad \omega_2^2 = -\frac{\omega_{pe}^2 k^2 v_{\perp e}^2}{\omega_{pe}^2 + k^2c^2} \lesssim -k^2v_{\perp e}^2 . \tag{6.2.14}$$

Since $\omega_2^2 < 0$ the second root is purely imaginary and describes the *aperiodic hydrodynamic instability of the anisotropic plasma.* It follows from the condition $\omega_2 \gg kv_{\|e}$ that this instability occurs for a sufficiently high anisotropy of the electron temperature $T_{\perp e} \gg T_{\|e}$ only. The analogous conclusion follows from (6.2.12a) for $T_{\|e} \gg T_{\perp e}$.

It is not true, however, that the anisotropy of the electron temperature distribution must be large for the development of a plasma instability. An instability can develop for a very small anisotropy, too. To realize this, we analyze (6.2.12) for $\beta_{\|e} \ll 1$, $\beta_i \gg 1$ (i.e., $kv_{Ti} \ll \omega \ll kv_{\|e}$), giving

$$k^2c^2 + \omega_{pe}^2\left(1 - \frac{T_{\perp e}}{T_{\|e}}\right) - i\sqrt{\frac{\pi}{2}}\frac{\omega\omega_{pe}^2 T_{\perp e}}{kv_{\|e}T_{\|e}} + \omega_{pi}^2 = 0 . \tag{6.2.15}$$

We obtain

$$\omega = -\mathrm{i}\frac{kv_{\parallel e} T_{\parallel e}}{\omega_{pe}^2 T_{\perp e}} \sqrt{\frac{\pi}{2}} \left[k^2c^2 + \omega_{pi}^2 + \omega_{pe}^2 \left(1 - \frac{T_{\perp e}}{T_{\parallel e}}\right)\right] . \tag{6.2.16}$$

In the infinite plasma where the wave vector $\boldsymbol{k}$ can be arbitrarily small the quantity $\delta = \mathrm{Im}\{\omega\}$ can become positive for $T_{\perp e} \gg T_{\parallel e}$. The corresponding oscillations are unstable. This instability is aperiodic, as the one discussed before, but in contrast to the former it is characterized by a change in sign of the imaginary part of ε_{11}, meaning that it is associated with the Cherenkov dissipation mechanism. Therefore it is called *kinetic*. The analogous analysis of (6.2.12 a) shows that this instability exists for $T_{\parallel e} > T_{\perp e}$, too. Its increment follows from (6.2.16) by the substitution $T_{\perp e} \rightleftarrows T_{\parallel e}$.

6.2.2 Instability of the Magneto-Active Plasma with Anisotropic Temperature

Also in the presence of an external magnetic field the plasma with an anisotropic temperature remains unstable. This is illustrated by the example of waves propagating strictly along the magnetic field ($k = k_\parallel$, $k_\perp = 0$). In this case (6.1.1) separates into two equations

$$k^2c^2 = \omega^2(\varepsilon_{xx} \pm \mathrm{i}\,\varepsilon_{xy}) , \tag{6.2.17}$$

$$\varepsilon_{zz} = 0 . \tag{6.2.18}$$

Equation (6.2.17) describes transverse (ordinary and extraordinary) waves, and (6.2.18), which coincides with (6.2.10), describes a purely longitudinal wave. The longitudinal oscillations as in the absence of a magnetic field are always stable in the anisotropic magneto-active plasma for $k_\perp = 0$. Substituting (6.2.4) into (6.2.17) for the transverse waves we obtain

$$k^2c^2 = \omega^2 \left(1 - \sum_\alpha \frac{\omega_{p\alpha}^2}{\omega^2} \left\{\frac{\omega}{\omega \mp \Omega_\alpha} I_+ \left(\frac{\omega \mp \Omega_\alpha}{kv_{\parallel\alpha}}\right) \right.\right.$$
$$\left.\left. \times \left[\frac{T_{\perp\alpha}}{T_{\parallel\alpha}} \pm \frac{\Omega_\alpha}{\omega}\left(1 - \frac{T_{\perp\alpha}}{T_{\parallel\alpha}}\right)\right] + \left(1 - \frac{T_{\perp\alpha}}{T_{\parallel\alpha}}\right)\right\}\right) . \tag{6.2.19}$$

Again, we assume the electron temperature to be only anisotropic. Naturally, one should consider this mode in the high-frequency range where the ions cannot participate. According to (6.2.19) the dispersion equation of the electron oscillations in the frequency range $|\omega - \Omega_e| \gg kv_{\parallel e}$ is written as

$$k^2c^2 = \omega^2 \left\{1 - \frac{\omega_{pe}^2}{\omega(\omega \mp \Omega_e)} + \mathrm{i}\sqrt{\frac{\pi}{2}} \frac{\omega_{pe}^2}{\omega k v_{\parallel e}} \left[\frac{T_{\perp e}}{T_{\parallel e}} \pm \frac{\Omega_e}{\omega}\left(1 - \frac{T_{\perp e}}{T_{\parallel e}}\right)\right] \right.$$
$$\left. \times \exp\left(-\frac{1}{2}\frac{(\omega \mp \Omega_e)^2}{k^2 v_{\parallel e}^2}\right)\right\} . \tag{6.2.20}$$

Hence, we obtain for the ordinary wave with $\omega \ll \Omega_e$ (in this frequency range it is called the *helical wave*)

$$\omega = \frac{k^2c^2}{\omega_{pe}^2}\Omega_e \, ,$$
$$\delta = -\sqrt{\frac{\pi}{2}}\frac{kc^2\Omega_e^2}{v_{\|e}\omega_{pe}^2}\left[\frac{T_{\perp e}}{T_{\|e}} + \frac{\Omega_e}{\omega}\left(1 - \frac{T_{\perp e}}{T_{\|e}}\right)\right] \exp\left(-\frac{(\omega - \Omega_e)^2}{2k^2v_{\|e}^2}\right) . \tag{6.2.21}$$

This wave is unstable for $T_{\perp e} \gg T\|_e$. Since $\omega \ll \Omega_e$ the anisotropy may be arbitrarily small for instability, however. Under the condition

$$k^2c^2 \approx \omega_{pe}^2\left(1 - \frac{T_{\|e}}{T_{\perp e}}\right) ,$$

it has a maximum increment and (6.2.21) reduces to

$$\omega = \Omega_e\left(1 - \frac{T_{\|e}}{T_{\perp e}}\right) ,$$
$$\delta_{max} = \sqrt{\frac{\pi}{8}}\omega_{pe}\left(\frac{T_{\perp e} - T_{\|e}}{T_{\perp e}}\right)^{3/2}\frac{T_{\perp e}v_{\|e}^2}{T_{\|e}c^2} \exp\left(-\frac{B_0^2T_{\perp e}}{8\pi N_{0e}T_{\|e}(T_{\perp e} - T_{\|e})}\right) . \tag{6.2.22}$$

The helical wave instability occurring in the frequency range $\omega \ll \Omega_e$ for $T_{\perp e} > T_{\|e}$ is caused by a change in sign of the anti-Hermitian part of the dielectric tensor. In this sense it is a kinetic instability. In the frequency range $\omega \gg \Omega_e$ the magnetic field can be ignored in (6.2.19), which then goes over to (6.2.12). Consequently, the hydrodynamic instability with the spectrum (6.2.14) is possible in the anisotropic plasma with $T_{\perp e} \gg T_{\|e}$, too.

Having solved the problem of the development of instabilities in collisionless plasmas with anisotropic temperature, we can compare the increments with the inverse time of the temperature relaxation due to particle collisions. Thus, the limit of validity of the derived formulas is obtained. As shown in Exercise 3.7.5, the relaxation of the electron temperature anisotropy is determined by electron-electron and electron-ion collisions in the completely ionized plasma and by electron-neutral collisions in the weakly ionized plasma. Therefore, the range of validity of the derived formulas is defined by the inequalities

$$\mathrm{Im}\,\{\omega\} \gg \nu_{eff} \, , \nu_{en} \tag{6.2.23}$$

for the completely and weakly ionized plasmas, respectively. From these conditions the minimal anisotropy of the electron temperature, necessary for instability, follows.

6.3 Interaction of a Straight Electron Beam with the Plasma. The Cherenkov Instability

Another example, widely used in practical applications, is a plasma with a small group of electrons of sufficiently high-ordered velocity moving through a medium of "resting" particles. In other words, the system under consideration is a plasma with a relativistic electron beam injected into it. The beam density is assumed to be smaller than that of the plasma. If the electron beam moves strictly parallel to the external field the equilibrium distribution function of the beam electrons has the form of (6.1.2). Moreover, we suppose that they have a Maxwellian distribution with nonrelativistic temperature in their intrinsic frame. The bulk of the "resting" particles is also assumed to be Maxwellian distributed, however, in the laboratory frame. To calculate the dielectric permittivity of the plasma, one can apply the general transformation formulas (6.1.18) where the tensor $\varepsilon_{ij}^{(\alpha)}(\omega_\alpha', \boldsymbol{k}_\alpha')$ is determined by (5.1.10) for the particles of type α.

We begin the analysis of the plasma-beam systems with a straight monoenergetic electron beam penetrating a cold plasma. It is evident that such a beam must be directed strictly parallel to the magnetic field. The validity of this model is restricted to fast processes with characteristic velocities greatly exceeding the thermal velocities of the beam and plasma particles which, consistently, may be completely ignored.

We can apply (5.2.2, 3) to the monoenergetic electron beam in its intrinsic frame and to the cold plasma in the laboratory frame. These equations, taking account of (6.1.17, 18), lead to the following components of the dielectric tensor:

$$\begin{aligned}
\varepsilon_{xx} &= \varepsilon_{yy} = 1 - \sum_\alpha \frac{\omega_{p\alpha}^2 \omega_\alpha'^2 \gamma_\alpha^{-1}}{\omega^2(\omega_\alpha'^2 - \Omega_\alpha^2)}\,, \\
\varepsilon_{xy} &= -\varepsilon_{yx} = -\mathrm{i}\sum_\alpha \frac{\omega_{p\alpha}^2 \omega_\alpha' \Omega_\alpha \gamma_\alpha^{-1}}{\omega^2(\omega_\alpha'^2 - \Omega_\alpha^2)}\,, \\
\varepsilon_{xz} &= \varepsilon_{zx} = -\sum_\alpha \frac{\omega_{p\alpha}^2 \omega_\alpha' k_\perp u_\alpha}{\omega^2(\omega_\alpha'^2 - \Omega_\alpha^2)}\,, \qquad (6.3.1) \\
\varepsilon_{yz} &= -\varepsilon_{zy} = \mathrm{i}\sum_\alpha \frac{\omega_{p\alpha}^2 \Omega_\alpha k_\perp u_\alpha}{\omega^2(\omega_\alpha'^2 - \Omega_\alpha^2)}\,, \\
\varepsilon_{zz} &= 1 - \sum_\alpha \left(\frac{\omega_{p\alpha}^2 \gamma_\alpha^{-1}}{\omega_\alpha'^2} + \frac{\omega_{p\alpha}^2 \gamma_\alpha k_\perp^2 u_\alpha^2}{\omega^2(\omega_\alpha'^2 - \Omega_\alpha^2)} \right).
\end{aligned}$$

Here $\omega_{p\alpha} = \sqrt{4\pi e_\alpha^2 N_{0\alpha}/m_\alpha}$ is the Langmuir frequency of the particles of the type α, $N_{0\alpha}$ being their density in the laboratory frame; $\omega_\alpha' = (\omega - \boldsymbol{k}\cdot\boldsymbol{u}_\alpha)\gamma_\alpha$

with $\gamma_\alpha = (1 - u_\alpha^2/c^2)^{-1/2}$; $\boldsymbol{u}_\alpha$ is the directed velocity of the particle species α and $\Omega_\alpha = e_\alpha B_0/(m_\alpha c)$ is the Larmor frequency.

Substituting (6.3.1) into (6.1.1) the oscillation spectrum of any multi-beam plasma with negligible thermal motion of the particles can be derived, especially that of a monoenergetic electron beam of small density penetrating the cold plasma. Note that the injected electron beam induces charges and currents in the plasma which neutralize the charge and the current of the beam (Exercise 6.5.1). The directed velocity of the plasma electrons forming the return current which neutralizes the beam current is small: $u_\mathrm{p} \approx u_\mathrm{b} N_\mathrm{b}/N_\mathrm{p} \ll u_\mathrm{b} \equiv u$. It can be neglected when the plasma density N_p greatly exceeds the electron beam density N_b. Therefore, (6.3.1) is applicable.

Substituting (6.3.1) into (6.1.1) and neglecting the beam contributions reproduces the dispersion equation of the cold magneto-active plasma, thoroughly analyzed in Sect. 5.2. The account of the beam terms provides small corrections to the spectra obtained there, when the frequency is far from the following resonances. It is easily seen from (6.3.1) that these contributions to the dielectric tensor become infinite under the conditions

$$\omega = k_z u \ , \quad \omega = k_z u \mp \Omega_\mathrm{e}/\gamma \ . \tag{6.3.2}$$

When the first equality called the *condition for the Cherenkov resonance* is satisfied, the beam contributes a second-order pole, and under the second condition called the *condition for the cyclotron (Doppler) resonance* it contributes a first-order pole. Thus, the interaction of a straight beam with the plasma is strongest when the condition for the Cherenkov resonance is satisfied.

6.3.1 Interaction of a Straight Electron Beam with Cold Isotropic Plasma

We begin the analysis of the interaction between a monoenergetic straight electron beam and the cold plasma with the simplest case without external magnetic field. For $\omega_\alpha' \gg \Omega_\alpha$ (6.2.1) separates into two equations:

$$\begin{aligned} &k^2c^2 - \omega^2\left(1 - \frac{\omega_\mathrm{pe}^2}{\omega^2} - \frac{\omega_\mathrm{b}^2}{\gamma\omega^2}\right) = 0 \ , \\ &(k^2c^2 - \omega^2 + \omega_\mathrm{pe}^2 + \omega_\mathrm{b}^2\gamma^{-1}) \\ &\times\left(1 - \frac{\omega_\mathrm{pe}^2}{\omega^2} - \frac{\omega_\mathrm{b}^2}{\gamma^3(\omega - k_z u)^2}\right) - \frac{k_\perp^2 u^2}{\omega^2}\,\frac{\omega_\mathrm{pe}^2\,\omega_\mathrm{b}^2}{\gamma(\omega - k_z u)^2} = 0 \ . \end{aligned} \tag{6.3.3}$$

Here ω_pe and ω_b are the Langmuir frequencies of the bulk electrons and of the beam electrons, respectively. We have neglected the ion terms in the derivation of (6.3.3), i.e., we consider only the interaction of the electron beam with the high-frequency plasma oscillations.

The first equation of (6.3.3) describes stable oscillations with the frequency

$$\omega^2 = \omega_{\mathrm{pe}}^2 + \omega_{\mathrm{b}}^2\gamma^{-1} + k^2c^2 . \tag{6.3.4}$$

The contribution of the beam electrons is negligibly small due to their low density. One can easily understand why these oscillations are stable. The electric field of these longitudinal waves is oriented along the y-axis. Thus the exchange of energy with the beam electrons is impossible:

$$\boldsymbol{E} \cdot \boldsymbol{u} = 0 . \tag{6.3.5}$$

The second equation of (6.3.3) describes a longitudinal-transverse wave with nonzero field components E_x and E_z. Here $\boldsymbol{E} \cdot \boldsymbol{u} \neq 0$ and the field can affect the beam electrons. The beam can be decelerated by the field, thus transferring part of its energy to the wave. The wave emerging from an initial fluctuation increases with time and the system appears unstable. As already noted the beam terms are most significant in the frequency range of the Cherenkov resonance. Therefore, the corresponding solution of (6.3.3) will be

$$\omega = k_z u + \delta ,$$

with $|\delta| \ll \omega$. As a result we have for $\omega^2 \approx k_z^2u^2 \neq \omega_{\mathrm{pe}}^2$

$$\delta^2 = \frac{\omega_{\mathrm{b}}^2\gamma^{-3}}{1 - \dfrac{\omega_{\mathrm{pe}}^2}{k_z^2u^2}} \left(1 + \frac{k_\perp^2}{k_z^2}\,\frac{\omega_{\mathrm{pe}}^2\gamma^2}{k^2c^2 - k_z^2u^2 + \omega_{\mathrm{pe}}^2}\right) \tag{6.3.6}$$

and for $\omega^2 \approx k_z^2u^2 \approx \omega_{\mathrm{pe}}^2$:

$$\begin{aligned} \delta_{1,2} &= \frac{-1 \pm \mathrm{i}\sqrt{3}}{2}\,\omega_{\mathrm{pe}}\left(\frac{N_{\mathrm{b}}}{2N_{\mathrm{p}}}\,\frac{k_\perp^2 + k_z^2\gamma^{-2}}{k^2\gamma}\right)^{1/3} , \\ \delta_3 &= \omega_{\mathrm{pe}}\left(\frac{N_{\mathrm{b}}}{2N_{\mathrm{p}}}\,\frac{k_\perp^2 + k_z^2\gamma^{-2}}{k^2\gamma}\right)^{1/3} . \end{aligned} \tag{6.3.7}$$

Equations (6.3.6, 7) show that the oscillations with the frequency $\omega \approx k_z u$ are unstable ($\mathrm{Im}\,\{\omega\} = \mathrm{Im}\,\{\delta\} > 0$). They increase with time if $k_z u \ll \omega_{\mathrm{pe}}$. Far from the plasma frequency, for $\omega \approx k_z u \neq \omega_{\mathrm{pe}}$, the increment is $\mathrm{Im}\,\{\delta\} \sim \omega_{\mathrm{pe}}(N_{\mathrm{b}}/N_{\mathrm{p}})^{1/2}$. If, however, $\omega \approx k_z u = \omega_{\mathrm{pe}}$ the increment is much larger: $\mathrm{Im}\,\{\delta\} \sim \omega_{\mathrm{pe}}(N_{\mathrm{b}}/N_{\mathrm{p}})^{1/3}$. This is understandable since a resonance occurs when the beam velocity coincides with the phase velocity of the plasma oscillations. Actually, the relativistic beam excites longitudinal waves

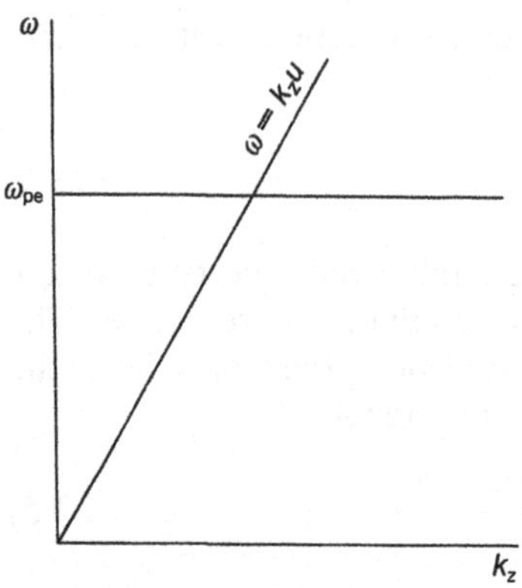

Fig. 6.2. Cherenkov resonance interaction of an electron beam with the longitudinal oscillations of an isotropic plasma

in the resonance case. The dispersion equation for longitudinal waves in the absence of a magnetic field can be written in this case as:

$$\varepsilon(\omega, \boldsymbol{k}) = \frac{k_i k_j}{k^2} \varepsilon_{ij}(\omega, \boldsymbol{k}) = 1 - \frac{\omega_{pe}^2}{\omega^2} - \frac{\omega_b^2 \gamma^{-3}}{(\omega - k_z u)^2} \frac{k_z^2 + k_\perp^2 \gamma}{k^2} = 0 . \quad (6.3.8)$$

When the beam is missing ($\omega_b = 0$) the electron plasma oscillations obey this equation. The electron beam and the plasma strongly interact at the intersection of the plasma oscillation spectrum with the frequency corresponding to the Cherenkov resonance condition (Fig. 6.2). Under this condition the solutions of (6.3.8) coincide with (6.3.7).

Thus, due to the interaction of the low-density electron beam with the isotropic plasma longitudinal waves are excited in the plasma under the condition of the Cherenkov resonance. Transverse electromagnetic waves cannot be excited in the isotropic plasma since their phase velocity is always greater than the velocity of light. The beam velocity is too small for resonant interaction (Chap. 4). Note that, according to (6.3.7), $\mathrm{Re}\{\delta_1\} < 0$ for an unstable root of the dispersion equation implying $u > \omega/k_z$. The beam electrons overtake the wave and transfer part of their kinetic energy to the field. The field amplitude increases with growing energy since the longitudinal wave is a wave with positive energy in the isotropic plasma:

$$|E^{lo}|^2 \frac{\partial}{\partial \omega} |\omega \varepsilon^{lo}(\omega)| = |E^{lo}|^2 \frac{\partial}{\partial \omega} \left[\omega \left(1 - \frac{\omega_{pe}^2}{\omega^2} \right) \right] > 0 . \quad (6.3.9)$$

6.3.2 Cherenkov Instability of the Electron Beam in the Cold Magneto-Active Plasma

The *Cherenkov instability of the electron beam* interacting with the plasma also appears in the presence of an external aligned magnetic field. If the magnetic field is weak, $|\delta| > \Omega_e \gamma^{-1}$, it can be neglected in (6.3.1). Then all the derived formulas remain valid. If the field is strong and $\Omega_e \gg \omega_{pe}$ we may take in (6.3.1) the limit of an infinitely strong magnetic field and obtain the two equations

$$k^2c^2 - \omega^2 = 0\,,$$
$$k_\perp^2 c^2 + (k_z^2c^2 - \omega^2)\left(1 - \frac{\omega_{pe}^2}{\omega^2} - \frac{\omega_b^2\gamma^{-3}}{(\omega - k_z u)^2}\right) = 0\,. \tag{6.3.10}$$

The first equation describes a purely transverse wave ($\boldsymbol{E} \parallel 0y$) with the phase velocity $\omega/k_z > c$; the beam does not interact with it. The second one concerns longitudinal-transverse waves with nonzero field components E_x and E_z. These modes are interacting with the beam. Without the beam term this equation gives two branches of oscillations

$$\omega_{1,2}^2 = \tfrac{1}{2}\left[\omega_{pe}^2 + k^2c^2 \pm \sqrt{(\omega_{pe}^2 + k^2c^2)^2 - 4\,\omega_{pe}^2 k_z^2 c^2}\,\right] \tag{6.3.11}$$

corresponding to fast and slow waves. The electron beam can resonantly interact only with the slow wave (Fig. 6.3) and only under the condition

$$\omega_{pe}^2 > k_\perp^2 u^2 \gamma^2\,. \tag{6.3.12}$$

The solution of the second equation of (6.3.10) corresponding to time-increasing oscillations has the form

$$\omega = k_z u + \delta = \sqrt{\omega_{pe}^2 - k_\perp^2 u^2\gamma^2} + \delta\,,$$

$$\delta = \begin{cases} \dfrac{-1 + \mathrm{i}\sqrt{3}}{2}\,\dfrac{\omega}{\gamma}\left(\dfrac{N_b}{2N_p}\right)^{1/3}\left(1 + \dfrac{k_\perp^2 u^2\gamma^2(\gamma^1 - 1)}{\omega_{pe}^2}\right)^{-1/3} & \\ \qquad\qquad \text{for} \quad \left|\dfrac{\delta}{\omega}\right| \ll \dfrac{1}{2(\gamma^2 - 1)}\,, & \\ \mathrm{i}\dfrac{\omega}{\gamma}\left(\dfrac{N_b(\gamma^2 - 1)}{N_p\gamma}\right)^{1/2}\left(1 + \dfrac{k_\perp^2 u^2\gamma^2(\gamma^2 - 1)}{\omega_{pe}^2}\right)^{-1/2} & \\ \qquad\qquad \text{for} \quad \left|\dfrac{\delta}{\omega}\right| \gg \dfrac{1}{2(\gamma^2 - 1)}\,. & \end{cases} \tag{6.3.13}$$

Since $\mathrm{Re}\,\{\delta\} \leq 0$ we have $u \gtrsim \omega/k_z$. Consequently, the beam electrons are faster than the wave and transfer a part of their energy to it. When the beam is nonrelativistic, for $\gamma \to 1$, the excited wave is quasilongitudinal to a high degree of accuracy. With increasing γ, however, its component not derivable from a potential becomes large. The energy transfer of the directed electron

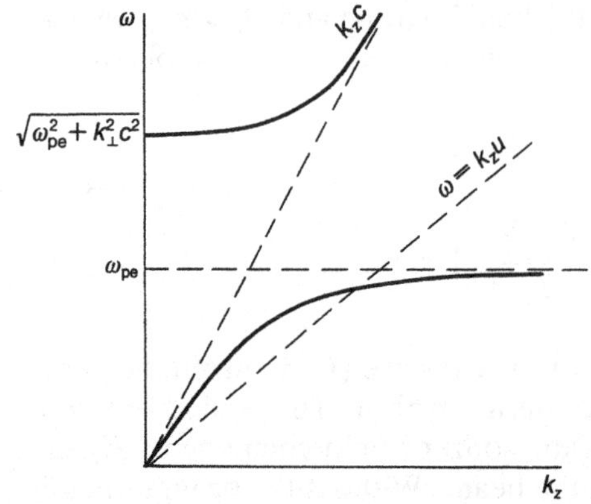

Fig. 6.3. Cherenkov resonance interaction of an electron beam with the oscillations of a cold magneto-active plasma

velocity to the wave can be easily understood. According to (6.3.3, 11) the group velocity of the slow wave propagating along the z-axis is

$$v_{\mathrm{gr}\,2,z} = \frac{\partial \omega_2}{\partial k_z} = \frac{2}{3}\,u\,, \tag{6.3.14}$$

i.e., it is always smaller than the beam velocity u.

Note that this velocity is positive. The wave energy is flowing in the direction of the beam motion. This is the reason why the wave is commonly called convective and the instability a convective instability. The perturbations do not grow in space but are washed away by the electron beam. In the reverse case, when it is growing in space and time, the instability is called absolute.

6.3.3 The Resonance Cherenkov Amplification of Waves

So far we have treated the problem of the electrodynamic stability of the nonequilibrium medium consisting of a cold plasma and a monoenergetic straight beam. By solving the initial value problem, we have determined the excitation conditions of initial fluctuations, their frequency spectra, and increments. Electromagnetic perturbations applied from the outside must also increase in such a medium. To prove this, we must solve the boundary value problem posed in Sect. 2.6, i.e., we must determine the complex components of the wave vector of a given direction of propagation from the dispersion equation. Dealing with the case of the strongly magnetized plasma with a beam the complex value of k_z is determined (when the Cherenkov resonance condition holds) from the second equation of (6.2.10). Thus, the problem considered is the amplification of waves propagating along the external magnetic field. We get

$$k_z = \frac{\omega}{u} + \delta k_z = \frac{1}{u}\sqrt{\omega_{pe}^2 - k_\perp^2 u^2 \gamma^2} + \delta k_z$$

$$\delta k_z = \begin{cases} \dfrac{1-\mathrm{i}\sqrt{3}}{2}\dfrac{\omega}{u}\left(\dfrac{\omega_b^2}{2\gamma^7 k_\perp^2 u^2}\right)^{1/3} & \text{for } \left|\dfrac{\delta k_z u}{\omega}\right| \ll (2\gamma^2)^{-1}\,, \\ -\mathrm{i}\dfrac{\omega}{k}\left(\dfrac{\omega_b^2}{\gamma^5 k_\perp^2 u^2}\right)^{1/2} & \text{for } \left|\dfrac{\delta k_z u}{\omega}\right| \gg (2\gamma^2)^{-1}\,. \end{cases} \tag{6.3.15}$$

The quantity $\mathrm{Im}\,\{\delta k_z\}$ is the coefficient of wave amplification in the direction $z > 0$.

Finally, we investigate the role of the thermal motion of the beam and the plasma electrons in the development of the Cherenkov instability which was completely ignored above. Thus, these results are valid under circumstances only when the phase velocities of the waves exceed the thermal velocity of the plasma electrons in the laboratory frame and the velocity spread of beam electrons in their intrinsic frame

$$\frac{\omega}{k_z} \geqslant u \gg v_{Te}\,, \quad \frac{\omega'}{k_z'} \gg v_{Tb}\,. \tag{6.3.16}$$

Here v_{Te} is the thermal velocity of the plasma electrons and V_{tb} is the non-relativistic velocity spread of the beam electrons in the intrinsic frame.

6.3.4 The Effect of Thermal Motion on the Cherenkov Instability of the Electron Beam

In order to specify the conditions (6.3.16) in more detail and to find out how the Cherenkov instability of the beam is modified when these conditions are violated, we now study the interaction between the motion of the particles. For simplicity, we confine our consideration to longitudinal waves propagating strictly along the directed velocity of the beam aligned with the magnetic field. The dispersion equation for the oscillations of this electron plasma-beam system is of the form:

$$\varepsilon(\omega, \boldsymbol{k}) = \frac{k_i k_j}{k^2}\varepsilon_{ij}(\omega, \boldsymbol{k}) = 1 - \frac{\omega_{pe}^2}{k^2 v_{Te}^2}\left[1 - I_+\left(\frac{\omega}{k v_{Te}}\right)\right]$$
$$- \frac{\omega_b^2 \gamma^{-1}}{k'^2 v_{Tb}^2}\left[1 - I_+\left(\frac{\omega'}{k' v_{Tb}}\right)\right] = 0\,. \tag{6.3.17}$$

Hence, the conditions (6.3.16) are actually the conditions for the negligibility of the thermal motion in the system with the Cherenkov instability developed (here $k' \approx k\gamma^{-1}$). Under these conditions (6.3.17) takes the form

$$1 - \frac{\omega_{pe}^2}{\omega^2} - \frac{\omega_b^2 \gamma^{-3}}{(\omega - \boldsymbol{k} \cdot \boldsymbol{u})^2} = 0 \tag{6.3.18}$$

and coincides with the dispersion equation for the longitudinal waves (6.3.8) with $k_\perp = 0$.

When the thermal motion of the particles is neglected, the Cherenkov instability is often called a *hydrodynamic* instability, which stresses that it is nondissipative and can be described by the hydrodynamics of a cold plasma. As shown above, this instability develops for $\boldsymbol{k} \cdot \boldsymbol{u} < \omega_{pe}$. We will show that the Cherenkov instability also develops for $\boldsymbol{k} \cdot \boldsymbol{u} > \omega_{pe}$ if the thermal motion of the particles is accounted for. In contrast to the hydrodynamic beam instability, the instability for $\boldsymbol{k} \cdot \boldsymbol{u} > \omega_{pe}$ is dissipative. Then it is caused by a change in sign of the Cherenkov wave damping and therefore called *kinetic*. Under these conditions the beam contribution to the real part of the longitudinal dielectric permittivity can be neglected and (6.3.17) may be written

$$1 - \frac{\omega_{pe}^2}{\omega^2}\left(1 + 3\,\frac{k^2 v_{Te}^2}{\omega^2}\right) + \mathrm{i}\sqrt{\frac{\pi}{2}}\,\frac{\omega_{pe}^2\,\omega}{k^3 v_{Te}^3}\exp\left(-\frac{\omega^2}{2k^2 v_{Te}^2}\right)$$
$$+\,\mathrm{i}\sqrt{\frac{\pi}{2}}\,\frac{\omega_b^2(\omega - \boldsymbol{k}\cdot\boldsymbol{u})\gamma^3}{k^3 v_{Tb}^3}\exp\left[-\frac{(\omega - \boldsymbol{k}\cdot\boldsymbol{u})^2\gamma^4}{2k^2 v_{Tb}^2}\right] = 0\,. \tag{6.3.19}$$

The solution of this equation is ($\omega \to \omega + \mathrm{i}\delta$)

$$\omega^2 = \omega_{pe}^2 + 3k^2 v_{Te}^2\,,$$

$$\delta = -\sqrt{\frac{\pi}{8}}\,\omega_{pe}^2\left[\frac{\omega_b^2\gamma^3}{k^3 v_{Tb}^3}\left(1 - \frac{\boldsymbol{k}\cdot\boldsymbol{u}}{\omega_{pe}^2}\right)\exp\left(-\frac{(\omega - \boldsymbol{k}\cdot\boldsymbol{u})^2\gamma^4}{2k^2 v_{Te}^2}\right)\right.$$
$$\left. +\,(\omega_{pe}^2/k^3 v_{Tb}^3)\exp\left(-\frac{\omega_{pe}^2}{2k^2 v_{Te}^2} - \frac{3}{2}\right)\right]. \tag{6.3.20}$$

If $\mathrm{Re}\,\{\delta\} > 0$ the system is unstable which is possible in the range $\boldsymbol{k}\cdot\boldsymbol{u} > \omega_{pe}$. The instability is caused by a change in sign of the Landau damping decrement of the beam term when the Cherenkov condition $u > \omega/k$ is satisfied. Note that the kinetic beam instability has a maximum increment near the frequency of the Cherenkov resonance $\omega \approx \omega_{pe} \approx \boldsymbol{k}\cdot\boldsymbol{u}$ as was the case with the hydrodynamic instability. The growth rate decreases with growing distance from the resonance frequency.

Fig. 6.4 shows the increment of the Cherenkov beam instability in dependence on $\boldsymbol{k}\cdot\boldsymbol{u}$. The continuous transition from the hydrodynamic instability (increasing branch) to the kinetic instability (decreasing branch) can be seen.

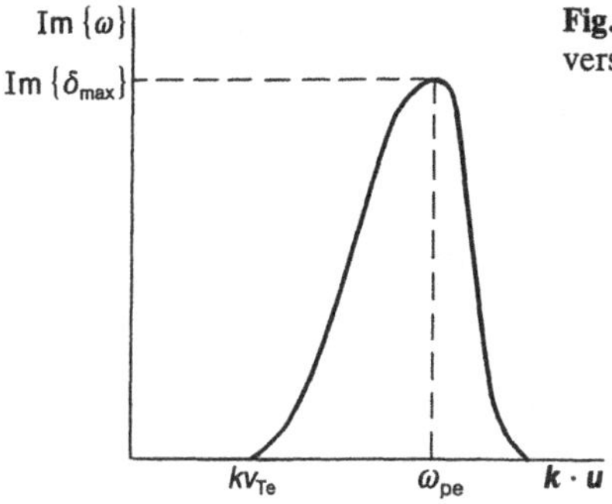

Fig. 6.4. Increment of the Cherenkov beam instability versus $\boldsymbol{k} \cdot \boldsymbol{u}$

6.4 Interaction of a Rotating Electron Beam (Beam of Oscillators) with the Plasma. Cyclotron Instability

As mentioned in Sect. 6.3, the interaction of an electron beam with the plasma is strong when the Cherenkov or the cyclotron resonance condition is fulfilled. In the case of a straight beam the Cherenkov interaction prevails since the second-order poles correspond to it whereas the cyclotron interaction is represented by first-order poles. However, the cyclotron interaction appears to be as strong as the Cherenkov interaction, when the beam electrons have a directed velocity component perpendicular to the external magnetic field, in addition to the longitudinal one.

It follows from the resonance condition (6.3.2) that the cyclotron interaction determines no limit for the phase velocity of the waves; it may be both smaller and larger than the beam velocity. Only the ratio between the cyclotron frequency of the electrons in the external magnetic field and the frequency of the electromagnetic wave is important, here. Therefore, to get the mechanism of the cyclotron interaction as clear as possible we consider the perturbations in the "rotating beam-plasma system" supposing $k_z = 0$, which excludes the Cherenkov resonance. The perturbations are assumed to propagate strictly across the magnetic field.

For simplicity, we further restrict the investigation to a monoenergetic rotating beam interacting with a cold electron plasma. The dielectric permittivity of this system is of the form

$$\varepsilon_{ij}(\omega, \boldsymbol{k}) = \varepsilon_{ij}^{(0)}(\omega) + \delta\varepsilon_{ij}(\omega, \boldsymbol{k}) \,, \tag{6.4.1}$$

where $\varepsilon_{ij}^{(0)}(\omega)$ is the dielectric permittivity of the cold electron plasma (Sect. 5.2):

$$\varepsilon_{ij}^{(0)}(\omega) = \begin{pmatrix} \varepsilon_{\perp 0} & \mathrm{i}g_0 & 0 \\ -\mathrm{i}g_0 & \varepsilon_{\perp 0} & 0 \\ 0 & 0 & \varepsilon_{\parallel 0} \end{pmatrix} \,, \quad \text{with} \tag{6.4.2}$$

$$\varepsilon_{\perp 0} = 1 - \frac{\omega_{pe}^2}{\omega^2 - \Omega_e^2}, \quad g_0 = -\frac{\omega_{pe}^2 \Omega_e}{\omega(\omega^2 - \Omega_e^2)}, \quad \varepsilon_{\parallel 0} = 1 - \frac{\omega_{pe}^2}{\omega^2}$$

and $\delta\varepsilon_{ij}(\omega, \boldsymbol{k})$ is a correction caused by the electron beam.

To calculate $\delta\varepsilon_{ij}(\omega, \boldsymbol{k})$ one cannot apply the transformation formulas (6.1.17, 18) since the beam electrons rotate around the magnetic field in addition to their parallel motion. It is not convenient to introduce a rotating coordinate system. Instead we calculate $\delta\varepsilon_{ij}(\omega, \boldsymbol{k})$ using (6.1.7, 8) and substituting the equilibrium distribution function (6.1.11) of the beam electrons. Simple calculations for perturbations with $k_z \neq 0$ yield

$$\begin{aligned}
\delta\varepsilon_{xx} &= \sum_n \left[\frac{2}{z} n \mathrm{J}_n(z) \mathrm{J}_n'(z) P_n + n^2 \mathrm{J}_n^2(z) Q_n\right], \\
\delta\varepsilon_{yy} &= \sum_n \left\{\frac{1}{z} [z^2 \mathrm{J}_n'^2(z)]' P_n + z^2 \mathrm{J}_n'^2(z) Q_n\right\}, \\
\delta\varepsilon_{zz} &= -\frac{\omega_b^2}{\omega^2}\frac{1}{\gamma^2} + \sum_n \left\{\left[\frac{2 k_z u_\parallel}{\omega - k_z u_\parallel} \mathrm{J}_n^2(z) + \frac{u_\parallel^2}{u_\perp^2} 2 z \mathrm{J}_n(z) \mathrm{J}_n'(z)\right] P_n \right. \\
&\quad \left. + \frac{u_\parallel^2}{u_\perp^2} z^2 \mathrm{J}_n^2(z) Q_n\right\}, \\
\delta\varepsilon_{xy} &= -\delta\varepsilon_{yx} = -\mathrm{i} \sum_n \left\{\frac{n}{z} [z \mathrm{J}_n(z) \mathrm{J}_n'(z)]' P_n + n z \mathrm{J}_n(z) \mathrm{J}_n'(z) Q_n\right\}, \\
\delta\varepsilon_{xz} &= \delta\varepsilon_{zx} = \sum_n \left\{\left[\frac{n \Omega_e k_z}{k_\perp (\omega - k_z u_\parallel)} \mathrm{J}_n^2(z) + \frac{u_\parallel}{u_\perp} 2 n \mathrm{J}_n(z) \mathrm{J}_n'(z)\right] P_n \right. \\
&\quad \left. + \frac{u_\parallel}{u_\perp} n z \mathrm{J}_n^2(z) Q_n\right\}, \\
\delta\varepsilon_{yz} &= -\delta\varepsilon_{zy} = \mathrm{i} \sum_n \left(\left\{\frac{\Omega_e k_z z \mathrm{J}_n(z) \mathrm{J}_n'(z)}{k_\perp (\omega - k_z u_\parallel)} + \frac{u_\parallel}{u_\perp} [z \mathrm{J}_n(z) \mathrm{J}_n'(z)]'\right\} P_n \right. \\
&\quad \left. + \frac{u_\parallel}{u_\perp} z^2 \mathrm{J}_n(z) \mathrm{J}_n'(z) Q_n\right).
\end{aligned} \tag{6.4.3}$$

Here we introduced the notations

$$\begin{aligned}
P_n &\equiv \frac{\omega_b^2 (\omega - k_z u_\parallel) \gamma^{-1}}{\omega^2 (\omega - k_z u_\parallel - n\Omega_e/\gamma)}, \\
Q_n &\equiv \frac{\omega_b^2 \gamma^{-3} \Omega_e^2 (\omega^2 - k_z^2 c^2)}{\omega^2 c^2 k_\perp^2 (\omega - k_z n_\parallel - u\Omega_e/\gamma)^2}, \\
z &= \frac{k_\perp u_\perp}{\Omega_e} \gamma, \quad \gamma = \left(1 - \frac{u_\parallel^2 + u_\perp^2}{c^2}\right)^{-1/2}.
\end{aligned} \tag{6.4.4}$$

Equation (6.4.4) shows that the beam terms of the dielectric tensor contribute the second-order poles at $\omega = k_z u_\parallel + n\Omega_e/\gamma$ if $u_\perp \neq 0$. Consequently, the beam strongly interacts with an electromagnetic wave in this frequency range. To prove this statement we analyze the dispersion equation (6.1.1) for electromagnetic waves propagating across the magnetic field, i.e., $k_z = 0$, $k_\perp = k$. In this case we obtain two separate equations, describing extraordinary and ordinary waves, respectively,

$$k^2 = \frac{\omega^2}{c^2} \frac{\varepsilon_{xx}\varepsilon_{yy} + \varepsilon_{xy}^2}{\varepsilon_{xx}}, \qquad k^2 = \frac{\omega^2}{c^2}\varepsilon_{zz} . \tag{6.4.5}$$

Here

$$\begin{aligned} \varepsilon_{xx} &= \varepsilon_{\perp 0} + \delta\varepsilon_{xx} , \qquad \varepsilon_{xy} = \mathrm{i}g_0 + \delta\varepsilon_{xy} , \\ \varepsilon_{yy} &= \varepsilon_{\perp 0} + \delta\varepsilon_{yy} , \qquad \varepsilon_{zz} = \varepsilon_{\parallel 0} + \delta\varepsilon_{zz} . \end{aligned} \tag{6.4.6}$$

The electric field of the extraordinary wave is perpendicular and that of the ordinary wave is parallel to the external magnetic field. Therefore, the extraordinary wave can interact with the transverse momentum component of the beam electrons, and the ordinary with the longitudinal component, only.

Neglecting the beam contribution, (6.4.5) describes the oscillations of the cold magneto-active plasma which propagate across the magnetic field. In Sect. 5.2 the spectrum of these oscillations has been analyzed. For electron oscillations we obtained

$$\begin{aligned} \omega_{1,2}^2 &= \tfrac{1}{2}\left[\Omega_e^2 + 2\,\omega_{pe}^2 + k^2c^2 \pm \sqrt{4\,\omega_{pe}^2\,\Omega_e^2 + (k^2c^2 - \Omega_e^2)^2}\,\right] , \\ \omega_3^2 &= k^2c^2 + \omega_{pe}^2 . \end{aligned} \tag{6.4.7}$$

6.4.1 Conditions for Resonance Cyclotron Interaction of a Rotating Beam with Electromagnetic Waves in the Plasma

The frequencies ω_1 and ω_2 correspond to two branches of the extraordinary wave and ω_3 corresponds to the ordinary wave. The intersections of the curves $\omega_\alpha(k)$ ($\alpha = 1, 2, 3$) with the line $\omega = n\Omega_e/\gamma$ define the ranges of the resonant cyclotron interaction of the rotating electron beam with these oscillations (Fig. 6.5 a). We see that the cyclotron resonance is impossible under the condition

$$\frac{n\Omega_e}{\gamma} < \omega_2(0) . \tag{6.4.8 a}$$

Under the condition

$$\omega_2(0) < \frac{n\Omega_e}{\gamma} < \omega_{pe} \tag{6.4.8 b}$$

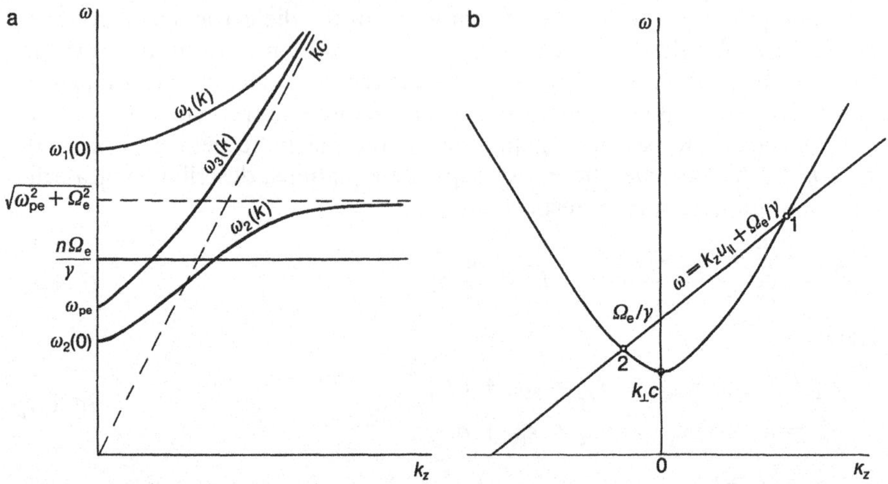

Fig. 6.5. (**a**) Ranges of the resonance cyclotron interaction of an electron beam with the oscillations of a cold magneto-active plasma – transverse propagation. (**b**) Ranges of the Cherenkov and cyclotron interactions of an electron beam with the longitudinal electron oscillations of a magneto-active plasma

the interaction between the beam and the lower branch of the extraordinary wave is possible. Under the condition

$$\omega_{\mathrm{pe}} < \frac{n\Omega_{\mathrm{e}}}{\gamma} < \sqrt{\omega_{\mathrm{pe}}^2 + \Omega_{\mathrm{e}}^2} \tag{6.4.8 c}$$

the resonance with the ordinary wave appears additionally. Under the condition

$$\sqrt{\omega_{\mathrm{pe}}^2 + \Omega_{\mathrm{e}}^2} < \frac{n\Omega_{\mathrm{e}}}{\gamma} < \omega_1(0) \tag{6.4.8 d}$$

the beam can interact with the ordinary wave, only. Finally, under the condition

$$\frac{n\Omega_{\mathrm{e}}}{\gamma} > \omega_1(0) \tag{6.4.8 e}$$

the resonant interaction of the beam with the ordinary wave and with the upper branch of the extraordinary wave as well is possible.

These conditions determine the possible resonance between the rotating beam and electromagnetic eigenmodes of the cold plasma. We must know the increments to determine what waves are really excited by the beam. To obtain them, one has to solve (6.4.5) taking account of the beam terms and of the conditions for the resonant interaction, $\omega = n\Omega_{\mathrm{e}}/\gamma$. We simplify the task by analyzing the limit of the rarefied plasma $\Omega_{\mathrm{e}} \gg \omega_{\mathrm{pe}}$. In such a plasma the

Cherenkov excitation of waves by a straight beam is possible when the plasma density exceeds some critical value which can be, however, arbitrarily small for the infinite plasma, see (6.3.12). This critical density is specified by the requirement that the wave can exist in the plasma with a phase velocity smaller than the velocity of light. Then it can be excited by the Cherenkov resonance with the beam electrons. As to the cyclotron resonance there exists no such requirement hence the cyclotron resonance is possible even in the complete absence of the plasma. Then (6.4.5) significantly simplifies:

$$\begin{aligned} k^2c^2 &= \omega^2 + \omega_b^2 \frac{u_\perp^2}{c^2} J_n'^2 \left(\frac{k u_\perp \gamma}{\Omega_e}\right) \left(1 - \frac{\omega\gamma}{n\Omega_e}\right)^{-2}, \\ k^2c^2 &= \omega^2 + \frac{\omega_b^2 u_\parallel^2}{\gamma c^2} J_n^2 \left(\frac{k u_\perp \gamma}{\Omega_e}\right) \left(1 - \frac{\omega\gamma}{n\Omega_e}\right)^{-2}. \end{aligned} \tag{6.4.9}$$

Here the plasma terms are completely neglected in the components of the dielectric tensor. Further, we have taken into account only the second-order pole of the beam terms (for the cyclotron resonance).

It is easy to solve (6.4.9) for the frequencies and the increments of the electromagnetic waves excited by the beam:

$$\omega \to \omega + \delta = \frac{n\Omega_e}{\gamma} + \delta . \tag{6.4.10}$$

In the approximation used the frequencies of the extraordinary and the ordinary waves coincide:

$$\omega = \frac{n\Omega_e}{\gamma} = kc . \tag{6.4.11}$$

Their increments are, respectively:

$$\begin{aligned} \frac{\delta_{\text{extraord}}}{\omega} &= \frac{1 + \mathrm{i}\sqrt{3}}{2} \left[\frac{\omega_b^2 \gamma}{2n^2 \Omega_e^2} \frac{u_\perp^2}{c^2} J_n'^2 \left(\frac{n u_\perp}{c}\right)\right]^{1/3}, \\ \frac{\delta_{\text{ord}}}{\omega} &= \frac{1 + \mathrm{i}\sqrt{3}}{2} \left[\frac{\omega_b^2 \gamma}{2n^2 \Omega_e^2} \frac{u_\parallel^2}{c^2} J_n^2 \left(\frac{n u_\perp}{c}\right)\right]^{1/3}. \end{aligned} \tag{6.4.12}$$

The increments of the extraordinary and the ordinary waves are nonzero in the presence of a transverse component of the electron velocity $u_\perp \neq 0$, only. They increase with increasing $u_\perp$. The presence of a longitudinal component of the electron velocity $u_\parallel \neq 0$ is necessary for the excitation of the ordinary wave. The maximum of the increments is obtained at low cyclotron harmonics n, especially for relativistic values of the transverse component of the electron velocity, when $u_\perp \ll c$.

6.4.2 Convective and Absolute Cyclotron Instabilities

Since we were assuming $k_z = 0$, so far, we cannot answer the question whether the cyclotron instability is convective or not. To answer this question we admit $k_z \neq 0$, confining ourselves to the case of small transverse electron velocities, $u_\perp \ll c$, however. Assuming that the beam is a small perturbation and taking account of the second-order poles of the dielectric permittivity (6.4.4) only, we obtain near the cyclotron resonance (6.3.2) the following dispersion equations for the extraordinary and ordinary waves, respectively:

$$k_\perp^2 + k_z^2 = \frac{\omega^2}{c^2}(1 + \delta\varepsilon_{yy}), \quad k_\perp^2 + k_z^2 = \frac{\omega^2}{c^2}(1 + \delta\varepsilon_{zz}). \tag{6.4.13}$$

Here

$$\begin{aligned} \delta\varepsilon_{yy} &= -\frac{\omega_b^2 u_\perp^2}{4\gamma\omega^2}\left(k_z^2 - \frac{\omega^2}{c^2}\right)\frac{1}{(\omega - k_z u_\parallel - \Omega_e/\gamma)^2}, \\ \delta\varepsilon_{zz} &= -\frac{\omega_b^2 u_\perp^2 k_\perp^2}{2\omega^4}\cdot\frac{(\omega - \Omega_e/\gamma)^2 - \omega^2 u_\parallel^2/c^2}{(\omega - k_z u_\parallel - \Omega_e/\gamma)^2}. \end{aligned} \tag{6.4.14}$$

In the limit $u_\perp \ll c$ only the fundamental harmonic $n = 1$ of the cyclotron resonances is accounted for.

To derive the solutions of (6.4.13) we write

$$\omega \to \omega + \delta = kc + \delta = \frac{\Omega_e}{\gamma} + k_z u_\parallel + \delta. \tag{6.4.15}$$

As a result we obtain for the increments of the extraordinary and the ordinary waves

$$\begin{aligned} \frac{\delta_{\text{extraord}}}{\omega} &= \frac{1 + i\sqrt{3}}{2}\left(\frac{\omega_b^2 k_\perp^2 u_\perp^2}{8\gamma\omega^4}\right)^{1/3}, \\ \frac{\delta_{\text{ord}}}{\omega} &= \frac{1 + i\sqrt{3}}{2}\left(\frac{\omega_b^2 k_\perp^4 u_\perp^2 u_\parallel^2}{4\omega^6}\right)^{1/3}. \end{aligned} \tag{6.4.16}$$

Now we can answer the question of the nature of the cyclotron beam instability of high- and low-frequency electromagnetic waves (Fig. 6.5 b)

$$\omega_{1,2}(k_z) = c\sqrt{k_z^2 + k_\perp^2} = \frac{\gamma_\parallel^2 \Omega_e}{\gamma}\left(1 \pm \frac{u_\parallel}{c}\sqrt{1 - \frac{k_\perp^2 c^2 \gamma^2}{\gamma_\parallel^2 \Omega_e^2}}\right). \tag{6.4.17}$$

Due to (6.4.15) their group velocity and their longitudinal wave number are, respectively,

$$v_{\mathrm{gr}1,2z} = \frac{\partial \omega_{1,2}}{\partial k_z} = c\frac{k_{z1,2}c}{\omega_{1,2}}, \quad k_{z1,2} = \frac{1}{u_\|}\left(\omega_{1,2} - \frac{\Omega_\mathrm{e}}{\gamma}\right). \tag{6.4.18}$$

For $\gamma_\| > k_\perp c\gamma/\Omega_\mathrm{e} > 1$ we get $k_{z1,2} > 0$ and the group velocities of both excited waves are positive (in this case the intersection point 2 of Fig. 6.5 b lies in the right half-plane). The wave energy flows in the direction of the beam, and the cyclotron instability is of the convective type. For $k_\perp c\gamma/\Omega_\mathrm{e} < 1$ we obtain $k_{z1} > 0$, $k_{z2} < 0$. Thus, the beam-excited low-frequency wave transfers its energy in the opposite direction of the beam motion, whereas the high-frequency wave, as usual, carries the energy parallel to the beam motion. In this case the low-frequency wave is not convected by the beam and the instability is absolute. In the limit $k_z = 0$ the cyclotron instability should also be regarded as absolute, since the longitudinal group velocities of the ordinary and extraordinary waves vanish and the electromagnetic field perturbations are not convected by the electron beam.

6.4.3 Screening of the Cyclotron Radiation in the Plasma

Closing this section we discuss the effect of the plasma on the cyclotron instability of the rotating beam. The presence of the plasma inhibits the development of this instability. For the densities obeying

$$\omega_{\mathrm{pe}}^2 > \frac{n^2\Omega_\mathrm{e}^2}{\gamma^2}\left(1 + \frac{\gamma}{n}\right) \tag{6.4.19}$$

it becomes impossible in the case $k_z = 0$. This is due to the fact that the cyclotron resonance $n\Omega_\mathrm{e}/\gamma$ is screened by the plasma, i.e., it is no longer a natural frequency of the system. Finally, note that the cyclotron instability is hydrodynamic. It is not associated with dissipation processes in the system and is excited by beams without thermal spread of the velocity. The account of a velocity spread can give rise to the appearance of the weak kinetic cyclotron instability (Exercise 6.5.6).

6.5 Exercises

6.5.1. Consider a relativistic electron beam of the radius r_0 and the velocity $\boldsymbol{u} \| 0z$ injected into a spatially infinite isotropic plasma in the plane $z = 0$ at the time $t = 0$. Derive the charge and current densities induced in the plasma which is assumed collisionless.

Solution. The charge and current densities of the beam are

$$\varrho_0 = eN_\mathrm{b}\eta(r_0 - r)[\eta(z) - \eta(z - ut)], \quad j_0 = \varrho_0 \boldsymbol{u}, \tag{6.5.1}$$

where $\eta(x) = 1$ for $x > 0$ and $\eta(x) = 0$ for $x < 0$. They satisfy the continuity equation. Since no electrons are created in the plasma volume the densities of the charge ϱ and the current j induced in the plasma must satisfy the continuity equation as well as

$$\frac{\partial \varrho_0}{\partial t} + \operatorname{div} j_0 = \frac{\partial \varrho}{\partial t} + \operatorname{div} j = 0 . \tag{6.5.2}$$

To solve the system of the field equations

$$\left.\begin{aligned} \operatorname{curl} B &= \frac{1}{c}\frac{\partial E}{\partial t} + \frac{4\pi}{c}(j + j_0) , \quad \operatorname{div} B = 0 , \\ \operatorname{curl} E &= -\frac{1}{c}\frac{\partial B}{\partial t} , \quad \operatorname{div} E = 4\pi(\varrho + \varrho_0) \end{aligned}\right\} \tag{6.5.3}$$

we apply the Fourier-Laplace transformation

$$\begin{aligned} A(r, t) &= \int_{-\infty + i\delta}^{\infty + i\delta} d\omega e^{-i\omega t} \int dk e^{ik \cdot r} A(k, \omega) , \\ A(k, \omega) &= \frac{1}{(2\pi)^4} \int_0^{\infty} dt e^{i\omega t} \int dr e^{-ik \cdot r} A(r, t) \end{aligned} \tag{6.5.4}$$

and the material equations for the cold isotropic plasma

$$\begin{aligned} j(\omega, k) &= \sigma(\omega) E(\omega, k) = \frac{[\varepsilon(\omega) - 1]\omega}{4\pi i} E(\omega, k) , \\ \varepsilon(\omega) &= 1 - \frac{\omega_{pe}^2}{\omega^2} . \end{aligned} \tag{6.5.5}$$

It is easy to obtain from (6.5.3) the Fourier transform of the induced electric field:

$$E(\omega, k) = \frac{4\pi i \left\{ \frac{\omega^2}{c^2} \varepsilon(\omega) j_0(\omega, k) - k(k \cdot j_0(\omega, k)) \right\}}{\omega \left\{ \varepsilon(\omega) \left[k^2 - \frac{\omega^2}{c^2} \varepsilon(\omega) \right] \right\}} \tag{6.5.6}$$

where $j_0(\omega, k)$ is the Fourier transform of the beam current:

$$j_0(\omega, k) = -\frac{e N_b r_0 u J_1(k_\perp r_0)}{(2\pi)^2 k_\perp k_z (\omega - k \cdot u)} . \tag{6.5.7}$$

The solution of the problem follows by substituting (6.5.5, 6) into the inverse Fourier transformation formula (6.5.4). Thereby the induced charge and current densities are determined for arbitrary time and at any space point in the plasma. We discuss these integral expressions in the asymptotic limit of large values of the time t and the coordinate z. Then all the transition processes have damped and the functions $\varrho(\boldsymbol{r}, t)$ and $\boldsymbol{j}(\boldsymbol{r}, t)$ become functions of $t' = t - z/u$. According to the Laplace theorem, we get for the asymptotic value

$$A_\infty(\boldsymbol{r}, t) = \lim_{\substack{t\to\infty \\ z\to\infty}} A(\boldsymbol{r}, t) = -\mathrm{i}\int d\boldsymbol{k} \lim_{\omega\to\boldsymbol{k}\cdot\boldsymbol{u}} (\omega - \boldsymbol{k}\cdot\boldsymbol{u})A(\omega, \boldsymbol{k})\,\mathrm{e}^{-\mathrm{i}\omega t + \mathrm{i}\boldsymbol{k}\cdot\boldsymbol{r}}. \tag{6.5.8}$$

Taking this into account, we obtain

$$\varrho_\infty(\boldsymbol{r}, t) = \frac{\mathrm{i}eN_\mathrm{b}r_0}{2\pi}\int_0^\infty dk_\perp \mathrm{J}_0(k_\perp r_0)\,\mathrm{J}_1(k_\perp r_0)\int_{-\infty}^{+\infty} dk_z \frac{\exp[\mathrm{i}k_z(z-ut)]}{k_z\,\varepsilon(\boldsymbol{k}\cdot\boldsymbol{u})} - \varrho_0$$

$$\boldsymbol{j}_\infty(\boldsymbol{r}, t) = \frac{\mathrm{i}\omega_\mathrm{pe}^2 r_0 eN_\mathrm{b}}{2\pi}\int_0^\infty dk_\perp \mathrm{J}_1(k_\perp r_0)\int_{-\infty}^{+\infty} dk_z \frac{\exp[\mathrm{i}k_z(z-ut)]}{k_z u\varepsilon(\boldsymbol{k}\cdot\boldsymbol{u})\left[k^2 - \frac{(\boldsymbol{k}\cdot\boldsymbol{u})^2}{c^2}\varepsilon(\boldsymbol{k}\cdot\boldsymbol{u})\right]}$$

$$\times\left\{\mathrm{i}\frac{k_\perp}{k}\mathrm{J}_1(k_\perp r)\,\boldsymbol{e}_r + \left[1 - \frac{u^2}{c^2}\varepsilon(\boldsymbol{k}\cdot\boldsymbol{u})\right]\mathrm{J}_0(k_\perp r)\,\boldsymbol{e}_z\right\}. \tag{6.5.9}$$

The integrals (6.5.9) are determined by the poles of the integrands

$$k_{z1,2} = \pm\frac{\omega_\mathrm{pe}}{u}, \qquad k_{z3,4} = \pm\,\mathrm{i}\gamma\sqrt{k_\perp^2 + \frac{\omega_\mathrm{pe}^2}{c^2}}\,. \tag{6.5.10}$$

The first two roots correspond to longitudinal oscillations of the isotropic plasma with $\omega = k_z u$ and the other ones represent transverse oscillations.

Naturally, the values of ϱ_∞ and $\boldsymbol{j}_\infty$ at a sufficiently large distance $z' = z - ut \gg c/\gamma\omega_\mathrm{pe}$ from the beam front are of the greatest interest. Then

$$\varrho_\infty = eN_\mathrm{b}\eta(r_0 - r)\left[\cos\left(\frac{\omega_\mathrm{pe}z'}{u}\right) - 1\right],$$

$$j_{z\infty} = eN_\mathrm{b}u\frac{\omega_\mathrm{pe}^2 r_0}{u^2}\left[\Psi_{00}(u)\cos\frac{\omega_\mathrm{pe}z'}{u} - \frac{u^2}{c^2}T_{00}(c)\right], \tag{6.5.11}$$

$$j_{r\infty} = -\,eN_\mathrm{b}u\frac{\omega_\mathrm{pe}r_0}{u}\Psi_{11}(u)\sin\left(\frac{\omega_\mathrm{pe}z'}{u}\right), \quad \text{where}$$

$$\Psi_{nm}(u) = \int_0^\infty dk_\perp k_\perp^m \frac{\mathrm{J}_n(k_\perp r)\,\mathrm{J}_0(k_\perp r_0)}{k_\perp^2 + \omega_\mathrm{pe}^2/u^2},$$

$$T_{00}(c) = \int_0^\infty dk_\perp \frac{\mathrm{J}_0(k_\perp r)\,\mathrm{J}_1(k_\perp r_0)}{k_\perp^2 + \omega_\mathrm{pe}^2/c^2}. \tag{6.5.12}$$

We see from (6.5.11) that the current $j_{r\infty}$ is a quickly oscillating function of z' and that it vanishes by averaging. The averaged induced charge exactly equals in magnitude the beam charge. Since it has the inverse sign the beam charge is neutralized in the average. At small plasma densities, for $\alpha = \omega_{pe}^2 r_0^2 / c^2 \ll 1$, the induced current $j_{z\infty}$ is small ($\sim \alpha$) compared to the beam current. In the case of a dense plasma with $\alpha \gg 1$, the current $j_{z\infty}$ is equal to the beam current by order of magnitude. Its averaged part has the inverse sign and up to an error $\sim 1/\alpha$ the magnitude of the beam current. Thus, in contrast to the charge neutralization occurring for plasma densities greater than the beam density, the current neutralization takes place only in the dense plasma with $\omega_{pe}^2 \gg c^2/r_0^2$.

6.5.2. Proceeding from (6.3.3) show that the *interchange instability* with $k_z = 0$ corresponding to the lamination of the electron beam into separate current-carrying filaments is possible in the plasma-beam system in the absence of an external magnetic field.

Solution. For interchange modes with $k_z = 0$ we obtain from (6.3.3)

$$(k^2c^2 - \omega^2 + \omega_{pe}^2)\left(1 - \frac{\omega_{pe}^2}{\omega^2}\right) - \frac{k^2u^2}{\omega^2}\frac{\omega_{pe}^2\omega_b^2}{\gamma\omega^2} = 0 \,. \tag{6.5.13}$$

Hence, in the limit $\omega^2 \ll \omega_{pe}^2$ we have

$$\omega^2 = -\frac{k^2u^2}{k^2c^2 + \omega_{pe}^2}\frac{\omega_b^2}{\gamma} \,. \tag{6.5.14}$$

Since $\omega^2 < 0$ the instability is aperiodic. It corresponds to the filamentation of the electron beam. The modes with $k > \omega_{pe}/c$ have the maximum increment. This implies that the lamination of the beam into filaments with radii $r_0 < c/\omega_{pe}$ is most probable. Note that (6.3.10) has no unstable solutions with $k_z = 0$, meaning that a strong magnetic field stabilizes the interchange stability.

6.5.3. Study the interaction of a low-density relativistic straight electron beam with the high-frequency electrostatic oscillations of the cold magnetoactive plasma.

Solution. The dispersion equation pertaining to the described system is written as

$$\frac{k_ik_j}{k^2}\varepsilon_{ij}(\omega, \boldsymbol{k}) = \frac{k_z^2}{k^2}\left(1 - \frac{\omega_{pe}^2}{\omega^2}\right) + \frac{k_\perp^2}{k^2}\left(1 - \frac{\omega_{pe}^2}{\omega^2 - \Omega_e^2}\right)$$
$$- \frac{k_z^2}{k^2}\frac{\omega_b^2\gamma^{-3}}{(\omega - k_zu)^2} - \frac{k_\perp^2\omega_b^2\gamma^{-1}}{(\omega - k_zu)^2 - \Omega_e^2\gamma^{-2}} = 0 \,. \tag{6.5.15}$$

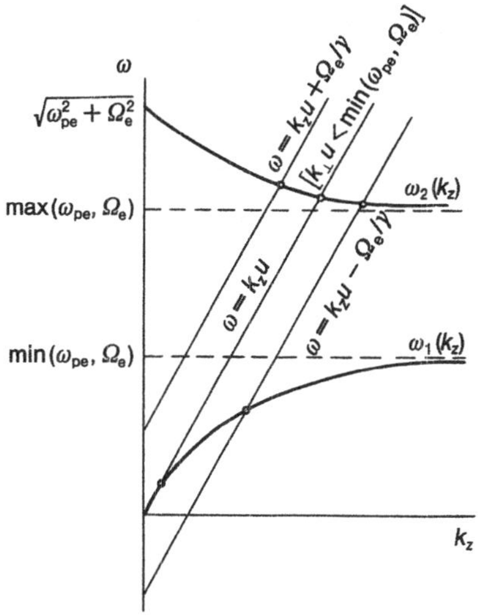

Fig. 6.6

In the absence of the beam we obtain the following longitudinal oscillations of the plasma:

$$\omega_{1,2} = \frac{1}{2}\left[\omega_{pe}^2 + \Omega_e^2 \pm \sqrt{(\omega_{pe}^2 + \Omega_e^2)^2 - 4\,\frac{k_z^2}{k_\perp^2 + k_z^2}\,\omega_{pe}^2\Omega_e^2}\,\right]. \quad (6.5.16)$$

Figure 6.6 shows the spectra $\omega_{1,2}(k_z)$ and the straight lines

$$\omega = k_z u\,, \qquad \omega = k_z u \pm \frac{\Omega_e}{\gamma} \quad (6.5.17)$$

corresponding to the Cherenkov and cyclotron resonance (with the normal and anomalous Doppler effects) of the beam interaction with the plasma oscillations. The strongest interaction arises at the intersections of these straight lines with the curves $\omega_{1,2}(k_z)$ where the development of the Cherenkov and cyclotron instabilities is possible. It is shown in Fig. 6.6 that the straight line corresponding to the anomalous Doppler effect $\omega = k_z u - \Omega_e/\gamma$ intersects both branches of the longitudinal waves. The Cherenkov line $\omega = k_z u$ always intersects the upper branch; for $k_\perp u < \min\{\omega_{pe}\Omega_e\}$ it also intersects the lower branch. The straight line corresponding to the normal Doppler effect $\omega = k_z u + \Omega_e/\gamma$ intersects the upper branch only. Further, Fig. 6.6 shows that the group velocity of the upper oscillation branch is always negative whereas that of the lower one is positive. Therefore the instability of the upper mode is always absolute and that of the lower one convective.

In the limiting cases of dense ($\omega_{pe}^2 \gg \Omega_e^2$) and rarefied ($\Omega_e^2 \gg \omega_{pe}^2$) plasmas one can obtain simple formulas for the oscillation frequencies and the increments.

In the dense plasma ($\omega_{pe}^2 \gg \Omega_e^2$) the Cherenkov instability $\omega = k_z u + \delta$ predominantly excites the upper Langmuir oscillations with the spectrum

$$\omega \approx \omega_{pe} , \quad \frac{\delta}{\omega} = \frac{-1 + i\sqrt{3}}{2\gamma} \left(\frac{N_b}{2N_p}\right)^{1/3} . \tag{6.5.18}$$

Under these conditions the cyclotron instability is possible with the anomalous Doppler effect ($\omega = k_z u - \Omega/\gamma + \delta$) only and it primarily excites the upper oscillation branch

$$\omega \approx \omega_{pe} , \quad \frac{\delta}{\omega} \simeq \frac{i}{2} \left(\frac{N_b}{N_p} \frac{\omega_{pe}}{\Omega_e}\right)^{1/2} . \tag{6.5.19}$$

We see that the increment of the Cherenkov instability is $\mathrm{Im}\{\delta\} \sim \omega\gamma^{-1}(N_b/N_p)^{1/3}$, compared with the increment of the cyclotron instability $\mathrm{Im}\{\delta\} \sim \omega(N_b/N_p)^{1/2}$. Nevertheless, the cyclotron instability can prevail if

$$\frac{1}{\gamma^{1/3}} > \left(\frac{\omega_{pe}}{\Omega_e}\right)^{1/2} \left(\frac{N_b}{N_p}\right)^{1/6} > \frac{1}{\gamma} . \tag{6.5.20}$$

This is possible in the case of ultrarelativistic electron beams only.

In the rarefied plasma ($\omega_{pe}^2 \ll \Omega_e^2$) the Cherenkov instability predominantly excites the lower branch of the Langmuir oscillations with the spectrum

$$\omega \approx \sqrt{\omega_{pe}^2 - k_\perp^2 u^2} , \quad \frac{\delta}{\omega} \approx \frac{-1 + i\sqrt{3}}{2\gamma} \left(\frac{N_b}{2N_p}\right)^{1/3} . \tag{6.5.21}$$

For the cyclotron instability with the anomalous Doppler effect we have

$$\omega \approx \omega_{pe} , \quad \frac{\delta}{\omega} \approx \frac{1}{3} \left(\frac{N_b}{6N_p}\right)^{1/2} \left(\frac{\omega_{pe}}{\Omega_e}\right)^{1/2} . \tag{6.5.22}$$

This increment is always much less than the increment (6.5.21). Therefore the cyclotron instability cannot develop in the rarefied plasma except for the case $k_\perp u > \omega_{pe}$ where the Cherenkov instability is impossible.

6.5.4. Study the interaction between two counter-streaming identical plasma beams moving parallel to the external magnetic field with velocities much smaller than the thermal velocity of the electrons.

Solution. Since the velocities of the beams are much smaller than the velocity of light, we can confine our analysis to electrostatic perturbations. The electrostatic dispersion equation for the system of two colliding plasma beams is written as

$$1+\sum_{e,i}\frac{\omega_{p\alpha}^2}{k^2v_{T\alpha}^2}\left[2-\sum_s\frac{\omega-\boldsymbol{k}\cdot\boldsymbol{u}}{\omega-\boldsymbol{k}\cdot\boldsymbol{u}-s\Omega_\alpha}A_s\left(\frac{k_\perp^2v_{T\alpha}^2}{\Omega_\alpha^2}\right)I_+\left(\frac{\omega-\boldsymbol{k}\cdot\boldsymbol{u}-s\Omega_\alpha}{k_zv_{T\alpha}}\right)\right.$$

$$\left.-\sum_s\frac{\omega+\boldsymbol{k}\cdot\boldsymbol{u}}{\omega+\boldsymbol{k}\cdot\boldsymbol{u}-s\Omega_\alpha}A_s\left(\frac{k_\perp^2v_{T\alpha}^2}{\Omega_\alpha^2}\right)I_+\left(\frac{\omega+\boldsymbol{k}\cdot\boldsymbol{u}-s\Omega_\alpha}{k_zv_{T\alpha}}\right)\right]=0\,. \tag{6.5.23}$$

We investigate the two limiting cases without an external magnetic field and with an infinitely strong one.

When there is no magnetic field (6.5.23) simplifies to

$$1+\sum_{e,i}\frac{\omega_{p\alpha}^2}{k^2v_{T\alpha}^2}\left[2-I_+\left(\frac{\omega-\boldsymbol{k}\cdot\boldsymbol{u}}{kv_{T\alpha}}\right)-I_+\left(\frac{\omega+\boldsymbol{k}\cdot\boldsymbol{u}}{kv_{T\alpha}}\right)\right]=0\,. \tag{6.5.24}$$

Under the condition $u \ll v_{Ti}$ the oscillations obeying this equation are stable, moreover, they are damping with time. This implies that there is no interaction between the beams.

For $v_{Ti} \ll u \ll v_{Te}$ we have from (6.5.24)

$$1+2\frac{\omega_{pe}^2}{k^2v_{Te}^2}\left(1-\mathrm{i}\sqrt{\frac{\pi}{2}}\frac{\omega}{kv_{Te}}\right)$$

$$-\frac{\omega_{pi}^2}{(\omega-\boldsymbol{k}\cdot\boldsymbol{u})^2}-\frac{\omega_{pi}^2}{(\omega+\boldsymbol{k}\cdot\boldsymbol{u})^2}=0\,. \tag{6.5.25}$$

If the small imaginary term describing the Cherenkov absorption by the electrons is neglected we obtain the spectrum

$$\omega_{1,2}^2=\alpha^{-1}\left\{\omega_{pi}^2+(\boldsymbol{k}\cdot\boldsymbol{u})^2\alpha\right.$$

$$\left.\pm\sqrt{[\omega_{pi}^2+(\boldsymbol{k}\cdot\boldsymbol{u})^2\alpha]^2+4(\boldsymbol{k}\cdot\boldsymbol{u})^2\alpha\,[2\,\omega_{pi}^2-(\boldsymbol{k}\cdot\boldsymbol{u})^2\alpha]}\right\}\,, \tag{6.5.26}$$

where $\alpha = 1 + 2\,\omega_{pe}^2/k^2v_{Te}^2$.

Under the condition

$$2\,\omega_{pi}^2>(\boldsymbol{k}\cdot\boldsymbol{u})^2\alpha \tag{6.5.27}$$

the root $\omega_2^2 \approx -(\boldsymbol{k}\cdot\boldsymbol{u})^2 < 0$ corresponding to aperiodically unstable oscillations appears. This indicates a strong interaction of the colliding beams. According to (6.5.27), this interaction is possible for velocities $u < v_s = \sqrt{T_e/M}$. It occurs only in the nonisothermal plasma with $T_e \gg T_i$, however, since $u \gg v_{Ti}$ was assumed.

This hydrodynamic instability also exists in the presence of a strong longitudinal magnetic field. In the limit $B_0 \rightarrow \infty$, assuming $v_{Ti} \ll u \ll v_{Te}$, we obtain from (6.5.23)

$$1+2\frac{\omega_{pe}^2}{k^2 v_{Te}^2}\left(1-i\sqrt{\frac{\pi}{2}}\frac{\omega}{|k_z|\, v_{Te}}\right)$$
$$-\frac{k_z^2}{k^2}\left(\frac{\omega_{pi}^2}{(\omega-\boldsymbol{k}\cdot\boldsymbol{u})^2}+\frac{\omega_{pi}^2}{(\omega+\boldsymbol{k}\cdot\boldsymbol{u})^2}\right)=0\,. \qquad (6.5.28)$$

Comparing with (6.5.25) we conclude that the instability is of the same nature as in case of the nonmagnetized plasma. The modifications are slight. The condition (6.5.27) becomes

$$2\,\omega_{pi}^2 > k^2u^2\alpha$$

and the instability increment becomes larger than $\omega_2^2 \approx -k^2u^2$.

6.5.5. Show that ions injected into an electron beam can be captured and accelerated under the condition $\omega \approx k_z u_e - \Omega_e/\gamma \approx k_z u_i$, i.e., when a longitudinal electron cyclotron wave with the anomalous Doppler effect and an ion Cherenkov wave are resonant. The directed electron and ion velocities are denoted by u_e and u_i, respectively.

Solution. Since the ion velocity is small compared to the velocity of light we can confine our analysis of the interaction between the electron and the ion beam to the case of longitudinal waves obeying the dispersion equation

$$\frac{k_\perp^2}{k^2}\left(1-\frac{\omega_{pe}^2\gamma^{-1}}{(\omega-k_z u_e)^2-\Omega_e^2/\gamma^2}\right)$$
$$+\frac{k_z^2}{k^2}\left(1-\frac{\omega_{pe}^2\gamma^{-3}}{(\omega-k_z u_e)^2}\right)-\frac{\omega_{pi}^2}{(\omega-k_z u_i)^2}=0\,. \qquad (6.5.29)$$

Here we assumed $\omega_{pi}^2 \gg \Omega_i^2$, i.e., nonmagnetized ions. The electrons, however, are assumed strongly magnetized, $\Omega_e^2 \gg \omega_{pe}^2\gamma$. Under these conditions the solution of (6.5.29) can be written as

$$\omega=\omega_0+\delta=k_z u_i+\delta\,, \qquad (6.5.30)$$

where ω_0 is the solution of (6.5.29) without the ion beam contribution

$$\omega_0 = k_z u_e-\frac{\Omega_e}{\gamma}-\frac{k_\perp^2}{k^2}\frac{\omega_{pe}^2}{2\,\Omega_e}\,. \qquad (6.5.31)$$

Finally, we obtain

$$\delta^3=-\frac{k_\perp^2}{k^2}\frac{\omega_{pe}^2\omega_{pi}^2}{2\,\Omega_e}\,. \qquad (6.5.32)$$

The following solution

$$\delta = \frac{1+\mathrm{i}\sqrt{3}}{2}\left(\frac{k_\perp^2}{k^2}\,\frac{\omega_{\mathrm{pe}}^2\omega_{\mathrm{pi}}^2}{2\,\Omega_{\mathrm{e}}}\right)^{1/3} . \tag{6.5.33}$$

corresponds to unstable oscillations.

Since $\mathrm{Re}\{\delta\} > 0$, the phase velocity of this mode is greater than the ion velocity. The wave accelerates the ions by overtaking them. The ions gain kinetic energy at the expense of the wave energy. Nevertheless, the wave amplitude grows with the increment $\mathrm{Im}\{\delta\} > 0$. Energy conservation requires, however, that the wave energy decreases. In fact, under the given conditions we have $\omega = \omega_0 < k_z u_{\mathrm{e}}$ and the energy of the cyclotron wave in the beam is negative:

$$|E|^2 \frac{\partial}{\partial\omega}(\omega\varepsilon(\omega))|_{\omega=\omega_0} \simeq 2|E|^2 \frac{k_\perp^2}{k^2}\,\frac{\omega_0(\omega_0 - k_z u_{\mathrm{e}})\,\omega_{\mathrm{pe}}^2\gamma^{-1}}{[(\omega - k_z u_{\mathrm{e}})^2 - \Omega_{\mathrm{e}}^2/\gamma^2]^2} < 0 . \tag{6.5.34}$$

6.5.6. Analyze the excitation of an ordinary electromagnetic wave propagating across the magnetic field. What is the effect of the thermal velocity spread of the electrons?

Solution. The dispersion equation for the ordinary wave propagating across the magnetic field has the form, see (6.4.5),

$$k^2c^2 - \omega^2\varepsilon_{zz} = 0 . \tag{6.5.35}$$

The contribution of the rotating beam taking account of the thermal velocity spread, i.e., of the beam with the distribution function (6.1.9), is calculated according to (6.1.7, 8):

$$\delta\varepsilon_{zz} = -\frac{\omega_{\mathrm{b}}^2}{\omega^2\gamma}\left[\frac{1}{\gamma_\|^2} - \sum_{n=0}\left(\alpha_{zz} - \beta_{zz}\frac{\partial}{\partial\kappa}\right)\frac{I_+(\kappa/a)}{\kappa}\right] , \tag{6.5.36}$$

where

$$\begin{aligned} \alpha_{zz} &= \left(2 + z\frac{d}{dz}\right)\left[\frac{n}{z}\mathrm{J}_n(z)\right]^2, \quad \beta_{zz} = \frac{u_\|^2}{c^2}\mathrm{J}_n^2 , \\ \kappa &= 1 - \frac{\omega\gamma}{n\Omega_{\mathrm{e}}} , \quad a = \frac{v_{\mathrm{Tb}}u}{\gamma c^2} , \quad z = \frac{k u_\perp \gamma}{\Omega_{\mathrm{e}}} . \end{aligned} \tag{6.5.37}$$

When deriving (6.5.36) we assumed $T_{\perp\mathrm{b}} = T_{\|\mathrm{b}} = T_{\mathrm{b}}$ and $v_{\mathrm{Tb}}^2 = T_{\mathrm{b}}/m$.

Using (6.5.36) it is easy to formulate (6.5.35) in the frequency range of the cyclotron resonance

$$k^2c^2 = \omega^2\left[1 - \frac{\omega_{\mathrm{pe}}^2}{\omega^2} - \frac{\omega_{\mathrm{b}}^2\gamma^{-1}u_\|^2}{\omega^2c^2}\,\mathrm{J}_n^2\left(\frac{k u_\perp\gamma}{\Omega_{\mathrm{e}}}\right)\frac{\partial}{\partial\kappa}\,\frac{I_+(\kappa/a)}{\kappa}\right] . \tag{6.5.38}$$

In the absence of the plasma, taking the limit $a \to 0$ of a monoenergetic beam this equation goes over to the second equation of (6.4.9) which describes the hydrodynamic cyclotron instability of the ordinary wave. Thus, the validity condition for (6.4.12) may be written as

$$|\kappa| \gg a = \frac{v_{\mathrm{Tb}} u}{\gamma c^2} . \tag{6.5.39}$$

This condition must be satisfied, in addition to (6.4.15) (the absence of screening of the cyclotron plasma radiation).

Another solution of (6.5.38) is the kinetic instability, existing for a finite beam temperature, i.e., for a finite value of a. This instability develops when the condition (6.5.39) is fulfilled. The spectrum of the unstable oscillations ($\omega \to \omega + \mathrm{i}\delta$) is

$$\begin{aligned} &\omega^2 = k^2c^2 + \omega_{\mathrm{pe}}^2 , \\ &\frac{\delta}{\omega} = \sqrt{\frac{\pi}{8}} \frac{\omega_{\mathrm{b}}^2 u_{\parallel}^2 \kappa}{\gamma \omega^2 c^2 a^2} \mathrm{J}_n^2\left(\frac{k u_{\perp}\gamma}{\Omega_{\mathrm{e}}}\right) \exp\left(-\frac{\kappa^2}{a^2}\right) . \end{aligned} \tag{6.5.40}$$

In contrast to the hydrodynamic instability which is subject to the condition $\omega > n\Omega_{\mathrm{e}}/\gamma$, the kinetic instability is possible for $\omega < n\Omega_{\mathrm{e}}/\gamma$ (i.e., $\kappa < 0$).

6.5.7. Study the excitation of halfcyclotron waves in a system of counter-streaming rotating monoenergetic electron beams of low density $\Omega_{\mathrm{e}} \gg \omega_{\mathrm{b}}\gamma^{1/2}$, and small transverse velocity $u_{\perp}^2 \ll c^2$.

Solution. In the system of counter-streaming beams the excitation of *half-cyclotron waves* is possible when the conditions of the Cherenkov and the cyclotron resonance are both given:

$$\omega \approx k_z u_{\parallel} \approx \frac{\Omega_{\mathrm{e}}}{\gamma} - k_z u_{\parallel} \approx \frac{\Omega_{\mathrm{e}}}{2\gamma} . \tag{6.5.41}$$

To analyze this double resonance, it is necessary to know the dielectric tensor for $k_z \neq 0$. In the limit $u_{\perp}^2 \ll c^2$ it is of the form, see (6.4.3):

$$\begin{aligned} \varepsilon_{xx} = \varepsilon_{yy} = 1 - \frac{\omega_{\mathrm{b}}^2}{2\gamma\omega^2} \sum_{s=\pm 1} \Bigg(& \frac{2(\omega - sk_z u_{\parallel})}{(\omega - sk_z u_{\parallel})^2 - \Omega_{\mathrm{e}}^2/\gamma^2} \\ & + u_{\perp}^2 \left(k_z^2 - \frac{\omega^2}{c^2}\right) \frac{(\omega - k_z u_{\parallel} s)^2 + \Omega_{\mathrm{e}}^2/\gamma^2}{[(\omega - sk_z u_{\parallel})^2 - \Omega_{\mathrm{e}}^2/\gamma^2]^2} \Bigg) , \\ \varepsilon_{xy} = -\varepsilon_{yx} = \mathrm{i}\frac{\omega_{\mathrm{b}}^2}{\gamma\omega^2} \sum_{s=\pm 1} & \frac{(\omega - sk_z u_{\parallel})\,\Omega_{\mathrm{e}}}{\gamma} \\ & \times \left(\frac{1}{(\omega - sk_z u_{\parallel})^2 - \Omega_{\mathrm{e}}^2/\gamma^2} + \frac{u_{\perp}^2 (k_z^2 - \omega^2/c^2)}{[(\omega - sk_z u_{\parallel})^2 - \Omega_{\mathrm{e}}^2/\gamma^2]^2} \right) , \end{aligned} \tag{6.5.42}$$

$$\varepsilon_{zz} \equiv 1 - \sum_{s=\pm 1} \frac{\omega_b^2 \gamma^{-1} \gamma_\parallel^{-2}}{(\omega - sk_z u_\parallel)^2} - \frac{2\omega_b^2}{\gamma\omega^2} \sum_{s=\pm 1} \left(\frac{k_\perp^2 u_\parallel^2}{(\omega - sk_z u_\parallel)^2 - \Omega_e^2/\gamma^2} \right.$$

$$\left. + \frac{k_\perp^2 u_\perp^2 \Omega_e}{\gamma\omega^2} \frac{\{sk_z u_\parallel [(\omega - sk_z u_\parallel)\omega + \Omega_e^2/\gamma^2] + \omega^2(\omega - sk_z u_\parallel) u_\parallel^2/c^2\}}{[(\omega - sk_z u_\parallel)^2 - \Omega_e^2/\gamma^2]^2} \right)$$

$$\varepsilon_{xz} = \varepsilon_{zx} = \frac{\omega_b^2 k_\perp}{\omega^2 \gamma} \sum_{s=\pm 1} s u_\parallel (\omega - sk_z u_\parallel)$$

$$\times \left(\frac{1}{(\omega - sk_z u_\parallel)^2 - \Omega_e^2/\gamma^2} + \frac{u_\perp^2 (k_z^2 - \omega^2/c^2)}{[(\omega - sk_z u_\parallel)^2 - \Omega_e^2/\gamma^2]^2} \right),$$

$$\varepsilon_{yz} = -\varepsilon_{zy} = \mathrm{i} \frac{\omega_b^2 k_\perp}{\gamma\omega^2} \sum_{s=\pm 1} \frac{s u_\parallel \Omega_e}{\gamma}$$

$$\times \left(\frac{1}{(\omega - sk_z u_\parallel)^2 - \Omega_e^2/\gamma^2} + \frac{u_\perp^2 (k_z^2 - \omega^2/c^2)}{[(\omega - sk_z u_\parallel)^2 - \Omega_e^2/\gamma^2]^2} \right).$$

Substituting (6.5.42) into the general dispersion equation (6.1.1) and taking into account the terms linear in the density only, which contain the second-order poles of the Cherenkov and cyclotron resonance (6.5.41), we obtain

$$1 - \frac{\omega_b^2}{\gamma\gamma_\parallel^2 (\omega - k_z u_\parallel)^2 (1 + \gamma_\parallel^2 k_\perp^2/k_z^2)}$$
$$- \frac{\gamma^{-1}\gamma_\parallel^{-2} \omega_b^2 (2\gamma_\parallel^2 - 1) u_\perp^2 k_\perp^2 / u_\parallel^2 k_z^2}{(\omega + k_z u_\parallel - \Omega_e/\gamma)^2 (1 + \gamma_\parallel^2 k_\perp^2/k_z^2)} = 0 . \tag{6.5.43}$$

Under the condition $u_\perp^2 \ll u_\parallel^2$, we get the solution

$$\omega = \omega_0 + \delta = \Omega_e/\gamma - k_z u_\parallel + \delta , \quad \text{where} \tag{6.5.44}$$

$$\omega_0 = k_z u_\parallel \pm \sqrt{\frac{\omega_b^2 \gamma^{-1} \gamma_\parallel^{-2}}{1 + \gamma_\parallel^2 k_\perp^2/k_z^2}} \approx k_z u_\parallel \approx \frac{\Omega_e}{2\gamma} \tag{6.5.45}$$

is the solution of (6.5.43) for $u_\perp = 0$ and the increment of the halfcyclotron wave is

$$\frac{\delta^3}{\omega_0^3} = \frac{2\gamma_\parallel^2 - 1}{\gamma^{3/2}\gamma_\parallel^3} \left(\frac{k_z^2}{k_\perp^2 \gamma_\parallel^2 + k_z^2} \right)^{3/2} \frac{\omega_b^3 k_\perp^2 u_\perp^2}{2\omega_0^5} . \tag{6.5.46}$$

6.5.8. Analyze the stability of a monoenergetic electron beam with Gaussian distributed azimuthal angles of the electron velocity with respect to the excitation of cyclotron waves. (Such a beam is created when a straight monoenergetic electron beam penetrates a thin foil).

Solution. The distribution function of the electron beam is

$$f_{0e} = A\delta(\mathscr{E} - \mathscr{E}_0) \exp\left(\frac{\theta^2}{\theta_0^2}\right), \tag{6.5.47}$$

where A is the normalization constant, $\mathscr{E}_0$ the electron energy, and θ_0 the spread of the azimuthal angle ($\theta_0^2 \ll 1$). The normalization gives

$$\int f_{0e}\, d\boldsymbol{p} = N_e\,, \qquad A = \frac{N_e c^3}{2\pi\theta_0^2 \mathscr{E}\sqrt{\mathscr{E}_0^2 - m^2c^4}}\,. \tag{6.5.48}$$

The distribution function (6.5.47) can be written in the form

$$f_{0e} = A\delta(\mathscr{E} - \mathscr{E}_0) \exp\left(-\frac{p_\perp^2}{p^2\theta_0^2}\right) \tag{6.5.49}$$

which is convenient to calculate the dielectric tensor according to (6.1.7, 8).

We consider the extraordinary cyclotron wave propagating across the magnetic field. When the density of the beam is low, i.e., for $\Omega_e \gg \omega_b\sqrt{\gamma}$, we should account for the terms linear in the beam density, only:

$$k^2c^2 - \omega^2\varepsilon_{yy} = 0\,, \qquad \varepsilon_{yy} \simeq 1 + \frac{1}{4}\frac{\omega_b^2\theta_0^2}{(\omega - \Omega_e/\gamma)^2}\frac{(\gamma^2-1)}{\gamma^3}\,, \tag{6.5.50}$$

where $\gamma = \mathscr{E}_0/mc^2$. Calculating ε_{yy} we have assumed a small velocity spread $\theta_0^2 \ll 1$. Therefore we are confined to the first cyclotron resonance $\omega \approx \Omega_e/\gamma$.

We obtain the solution ($\omega \to \omega + \delta$)

$$\omega^2 \approx k^2c^2 \approx \Omega_e^2/\gamma^2\,, \qquad \frac{\delta}{\omega} = \frac{-1 + \mathrm{i}\sqrt{3}}{2}\left(\frac{\omega_b^2}{8\omega^2}\frac{\gamma^2-1}{\gamma^3}\theta_0^2\right)^{1/3}. \tag{6.5.51}$$

6.5.9. With the example of the isotropic plasma, show that the Cherenkov beam instability can develop also when its increment is smaller than the collision frequency of the plasma electrons *(dissipative beam instability)*.

Solution. As shown in Sect. 6.3, longitudinal waves are excited by the Cherenkov instability in the isotropic plasma. Taking account of electron collisions we have instead of (6.3.7)

$$1 - \frac{\omega_{pe}^2}{\omega^2}\left(1 - \mathrm{i}\frac{\nu_e}{\omega}\right) - \frac{\omega_b^2\gamma^{-3}}{(\omega - k_z u)^2}\frac{k_z^2 + k_\perp^2\gamma^2}{k^2} = 0\,. \tag{6.5.52}$$

In the absence of the beam this equation describes weakly damped plasma oscillations with the spectrum ($\omega \to \omega + i\delta_0$)

$$\omega \approx \omega_{pe} , \qquad \delta_0 = -\frac{\nu_e}{2} . \tag{6.5.53}$$

In the presence of the beam these oscillations can be increasing. In fact, under the conditions for the Cherenkov resonance ($\omega = k_z u + i\delta \approx \omega_{pe} + i\delta$) we obtain from (6.5.53)

$$\frac{\delta}{\omega_{pe}} = \begin{cases} \dfrac{i + \sqrt{3}}{2} \left(\dfrac{N_b}{2N_p} \dfrac{1}{\gamma} \dfrac{k_\perp^2 + k_z^2 \gamma^{-2}}{k^2} \right)^{1/3} & \text{if} \quad |\delta| \gg \nu_e , \\ \dfrac{i + 1}{\sqrt{2}} \left(\dfrac{N_b}{N_p} \dfrac{\omega_{Le}}{\gamma \nu_e} \dfrac{k_\perp^2 + k_z^2 \gamma^{-2}}{k^2} \right)^{1/2} & \text{if} \quad |\delta| < \nu_e . \end{cases} \tag{6.5.54}$$

The first limiting case corresponds to the collisionless instability (6.3.7) and the second one to the dissipative instability. Obviously the beam instability can be realized only when $|\delta| \gg \nu_b$, where ν_b^{-1} is the relaxation time of the ordered velocity of the beam electrons (Exercise 3.7.4).

6.5.10. Study the excitation of the zero-point sound in a degenerate isotropic plasma by a low-density nonrelativistic monoenergetic electron beam.

Solution. The dispersion equation for the longitudinal waves of the system can be written as

$$1 + \frac{3\omega_{pe}^2}{k^2 v_{Fe}^2} \left(1 - \frac{\omega}{2kv_{Fe}} \ln \frac{\omega + kv_{Fe}}{\omega - kv_{Fe}} \right) - \frac{\omega_b^2}{(\omega - \boldsymbol{k} \cdot \boldsymbol{u})^2} = 0 . \tag{6.5.55}$$

Assuming the beam to be a small perturbation, we obtain

$$\begin{aligned} \omega &= \omega_0 + \delta = kv_{Fe} + \delta \approx \boldsymbol{k} \cdot \boldsymbol{u} + \delta , \\ \omega_0 &= kv_{Fe} \left[1 + 2 \exp\left(-\frac{2}{3} \frac{\omega_0^2}{\omega_{pe}^2} - 2 \right) \right] , \\ \frac{\delta}{\omega_0} &= \frac{-1 + i\sqrt{3}}{2} \left[\frac{4}{3} \frac{N_b}{N_p} \exp\left(-\frac{2}{3} \frac{\omega_0^2}{\omega_{pe}^2} - 2 \right) \right]^{1/3} . \end{aligned} \tag{6.5.56}$$

These formulas are applicable under the condition

$$\frac{N_b}{6N_p} \ll \exp\left[-4 \left(1 + \frac{\omega_0^2}{3\omega_{pe}^2} \right) \right] . \tag{6.5.57}$$

Along with the inequality $\mathrm{Im}\{\delta\} > \nu_{\mathrm{eff}}$, where ν_{eff} is the electron collision frequency in the degenerate plasma, this condition is easily satisfied for metals and degenerate semiconductors.

6.5.11. Proceeding from (6.1.18) show that the dispersion relation for electromagnetic waves in the system comprising two types of plasma, plasma 1 isotropic in the intrinsic frame and moving in the laboratory frame, and plasma 2, resting in the laboratory frame, splits into

$$k^2 - \frac{\omega^2}{c^2}\left\{\varepsilon^{(2)\mathrm{tr}}(\omega, k) + \frac{\omega'^2}{\omega^2}\left[\varepsilon^{(1)\mathrm{tr}}(\omega', k') - 1\right]\right\} = 0\,, \tag{6.5.58}$$

$$\begin{aligned} &\left\{k^2 - \frac{\omega^2}{c^2}\left[\varepsilon^{(2)\mathrm{tr}}(\omega, k) + \frac{\omega'^2}{\omega^2}\left(\varepsilon^{(1)\mathrm{tr}}(\omega', k') - 1\right)\right]\right\} \\ &\quad \times \left[\varepsilon^{(2)\mathrm{lo}}(\omega, k) + \varepsilon^{(1)\mathrm{lo}}(\omega', k') - 1\right] \\ &\quad - \frac{k^2u^2 - (\boldsymbol{k}\cdot\boldsymbol{u})^2}{c^2(1 - u^2/c^2)}\left\{\varepsilon^{(1)\mathrm{lo}}(\omega', k') + \frac{\omega'^2}{c^2k'^2}\left[\varepsilon^{(1)\mathrm{tr}}(\omega', k') - \varepsilon^{(1)\mathrm{lo}}(\omega', k')\right]\right\} \\ &\quad \times \left\{\varepsilon^{(2)\mathrm{lo}}(\omega, k) - 1 + \frac{\omega^2}{c^2}\left[\varepsilon^{(2)\mathrm{tr}}(\omega, k) - \varepsilon^{(2)\mathrm{lo}}(\omega, k)\right]\right\} = 0\,. \end{aligned} \tag{6.5.59}$$

Here $\boldsymbol{u}$ is the velocity of plasma 1 with respect to plasma 2; ω' and k' are the frequency and the wave vector transformed according to (6.1.15). Equation (6.5.58) describes the purely transverse wave and (6.5.59) the mixed longitudinal-transverse wave.

7. Plasmas in an External Homogeneous Electric Field

The instability of a plasma embedded in external constant and variable electric fields is studied both in the absence and in the presence of an external magnetic field. Special attention is paid to the dispersion equation for non-degenerate and degenerate plasmas accounting for interparticle collisions. The plasma stability in variable fields is considered both when the frequency of the external field exceeds all characteristic plasma frequencies and when the frequency is near one of the natural plasma oscillation frequencies, i.e., when a parametric interaction between high-frequency fields and the plasma occurs.

7.1 The Distribution Function of the Charged Particles in an External Electric Field

Another wide-spread example of a plasma not in thermodynamic equilibrium is an unbounded and homogeneous plasma in an external field with an anisotropic velocity distribution function of the particles.

One can hardly imagine any real plasma without an external electric field. The field is used both for the creation of the plasma (gas discharge in constant or high-frequency electric fields, optical discharge) and for its heating, confinement and acceleration (Ohmic heating in constant or high-frequency fields, heating by laser radiation, radiation confinement and acceleration of plasmas in variable fields, etc.). Moreover, the dispersion of electromagnetic waves with finite amplitude is strongly affected by the plasma behaviour in the electric field. Therefore, the analysis of the plasma characteristics in an external constant or variable electric fields is of great practical importance.

7.1.1 Plasma in a Strong Constant Electric Field

First, we shall study the stationary state of the plasma in a stationary and homogeneous external electric field. We begin with the configuration where the constant electric field $\boldsymbol{E}_0$ is parallel to the external magnetic field $\boldsymbol{B}_0$. Then the individual charged particles are accelerated with time:

$$p_0 = \frac{m_a u_a}{\sqrt{1 - u_a^2/c^2}} = e_a E_0 t \, . \tag{7.1.1}$$

This acceleration, however, is terminated when the particle collides with another one and loses the momentum gained in the field. It lasts only for a time $t < \nu_a^{-1}$, where ν_a is the collision frequency of momentum transfer of this particle. Because of the great difference in the electron and ion masses the effect of the electric field on the ions may be ignored usually and only its effect on the light charged particles (electrons) must be taken into account. Therefore we need to consider the electrons only, ν_a being the frequency of electron collisions, i.e., $a = \mathrm{e}$ and $\nu_a = \nu_\mathrm{e}$.

Thus, for $t < \nu_\mathrm{e}^{-1}$ the plasma may be regarded as collisionless. In this interval all the electrons are accelerated by the electric field according to the law (7.1.1) whereas the ions remain immobile. If the electron momentum distribution was of the form of $f_0(\boldsymbol{p})$ before the switching on of the field, i.e., if it was the equilibrium Maxwellian or Fermi distribution, then at the time t it is shifted by the value $\boldsymbol{p}_0(t) = e\boldsymbol{E}_0, t$. For the example of an initially nonrelativistic Maxwellian electron distribution it reads at the moment t:

$$f_0(\boldsymbol{p}, t) = \frac{N_\mathrm{e}}{(2\pi m T_\mathrm{e})^{3/2}} \exp\left(-\frac{[\boldsymbol{p} - \boldsymbol{p}_0(t)]^2}{2mT_\mathrm{e}}\right) . \tag{7.1.2}$$

In the degenerate plasma the distribution function becomes at the moment t the shifted Fermi distribution function:

$$f_0(\boldsymbol{p}, t) = \begin{cases} \dfrac{2}{(2\pi\hbar)^3} & \text{for} \quad |\boldsymbol{p} - \boldsymbol{p}_0| < p_\mathrm{F} \, , \\ 0 & \text{for} \quad |\boldsymbol{p} - \boldsymbol{p}_0| > p_\mathrm{F} \, , \end{cases} \tag{7.1.3}$$

where $p_\mathrm{F} = (3\pi^2 N_\mathrm{e})^{1/3}\hbar$.

The form (7.1.2) of the shifted nonstationary distribution function can be easily derived from Vlasov's kinetic equation

$$\frac{\partial f_0}{\partial t} + e\boldsymbol{E}_0 \frac{\partial f_0}{\partial \boldsymbol{p}} = 0 \, , \tag{7.1.4}$$

assuming $f_0(\boldsymbol{p})$ to be the initial distribution (without the field). A simple substitution of the function $f_0[\boldsymbol{p} - \boldsymbol{p}_0(t)]$ into (7.1.4) confirms this.

7.1.2 Runaway Electrons

In the nondegenerate completely ionized plasma the nonstationary solution (7.1.2) for the electrons appears to be valid for an arbitrary time (for $t > \nu_\mathrm{ei}^{-1}$ also) if

$$E_0 \gg E_{cr} = \frac{m\nu_{ei}}{e} v_{Te} , \quad \text{where} \tag{7.1.5}$$

$$\nu_{ei} = (4/3)\sqrt{2\pi/m}\, e^2 e_i^2 N_i L / T_e^{3/2}$$

is the frequency of electron-ion collisions defining the friction force between the electrons and the ions $\boldsymbol{F}_{fr} = -m\nu_{ei}\boldsymbol{u}_e$. This condition has a demonstrative physical meaning. For $E_0 > E_{cr}$ the electron acquires during the interval between the collisions an energy greater than its thermal energy T_e. As a result, the frequency of electron-ion collisions sharply decreases ($\nu_{ei} \sim 1/u^3$, see Exercise 4.7.3) and together with this frequency the friction force between the electrons and the ions also tends to zero; it cannot balance the accelerating force $e\boldsymbol{E}_0$ and all the plasma electrons pass over to the domain of constant acceleration or to the *"runaway domain"* described by (7.1.2). With time the electron velocity becomes relativistic, and the distribution function $f_0[\boldsymbol{p} - \boldsymbol{p}_0(t)]$ should be written in the form of the relativistic Maxwellian distribution function.

7.1.3 The Stationary Distribution Function of Electrons in a Weak Constant Electric Field

Under a condition opposite to (7.1.5), when the field $\boldsymbol{E}_0$ is weak, quite another picture of the completely ionized nondegenerate plasma can be given. In this case an electron acquires an energy smaller than the thermal energy during the mean free time, and the friction force finally compensates the accelerating force of the electric field. Then the distribution function of the electrons becomes stationary. To determine this function one must solve the equation

$$\frac{e\boldsymbol{E}_0}{m}\frac{\partial f_{0e}}{\partial \boldsymbol{v}} = \left(\frac{\partial f_{0e}}{\partial t}\right)_{col}^{ei} + \left(\frac{\partial f_{0e}}{\partial t}\right)_{col}^{ee} . \tag{7.1.6}$$

Here $(\partial f_{0e}/\partial t)_{col}^{ee}$ and $(\partial f_{0e}/\partial t)_{col}^{ei}$ are the electron-electron and the electron-ion collision integrals, respectively (Sect. 3.4).

For weak fields $\boldsymbol{E}_0$ the function f_{0e} only slightly differs from the isotropic Maxwellian distribution function f_{00}. Therefore (7.1.6) can be linearized assuming

$$f_{0e} = f_{00} + \delta f_{0e} , \quad \text{where} \tag{7.1.7}$$

$$f_{00} = \frac{N_e}{(2\pi m T_e)^{3/2}} \exp\left(-\frac{mv^2}{2T_e}\right) . \tag{7.1.8}$$

The perturbation δf_{0e} satisfies the linearized equation:

$$\frac{e\boldsymbol{E}_0}{m}\frac{\partial f_{00}}{\partial \boldsymbol{v}} = N_i \frac{\partial}{\partial p_i} I_{ij}^{ei}(\boldsymbol{p}) \frac{\partial \delta f_{0e}}{\partial p_j} + \frac{\partial}{\partial p_i} \int d\boldsymbol{p}' \, I_{ij}^{ee}(\boldsymbol{p},\boldsymbol{p}') \times \left[\frac{\partial f_{00}}{\partial p_j}\delta f_{0e}(\boldsymbol{p}') + \frac{\partial \delta f_{0e}}{\partial p_j} f_{00}(\boldsymbol{p}') - f_{00}(\boldsymbol{p})\frac{\partial \delta f_{0e}}{\partial p_j'} - \delta f_{0e}(\boldsymbol{p})\frac{\partial f_{00}}{\partial p_j'}\right]. \tag{7.1.9}$$

Equation (7.1.9) can be solved easily by the Chapman-Enskog method, i.e., by expanding $\delta f_{0e}(\boldsymbol{p})$ in Sonin polynomials. Considering only two terms of the expansion, we may write

$$\delta f_{0e}(\boldsymbol{p}) = \frac{\boldsymbol{v}\cdot\boldsymbol{E}_0}{E_0}\left[a_0 + a_1\left(\frac{5}{2} - \frac{v^2}{2v_{Te}^2}\right)\right] f_{00}\,. \tag{7.1.10}$$

Substituting this expression into (7.1.9), multiplying by the polynomials 1 and $[5/2 - v^2/(2v_{Te}^2)]$ and integrating over the velocity gives

$$\frac{eE_0}{T_e} = -\nu_{ei}\left(a_0 + \frac{3}{2}a_1\right), \quad \frac{3}{2}a_0 + \frac{13+4\sqrt{2}}{4}a_1 = 0\,, \tag{7.1.11}$$

when the ion charge is $e_i = |e|$. Solving for a_0 and a_1 the unknown function f_{0e} is

$$f_{0e} = \frac{N_e}{(2\pi m T_e)^{3/2}} \exp\left(-\frac{mv^2}{2T_e}\right)\left[1 + \frac{1}{1+\sqrt{2}}\frac{\boldsymbol{u}_e\cdot\boldsymbol{v}}{v_{Te}^2}\left(\frac{2\sqrt{2}-1}{2} + \frac{3}{4}\frac{v^2}{v_{Te}^2}\right)\right], \tag{7.1.12}$$

where $\boldsymbol{u}_e = e\boldsymbol{E}_0/m\nu_{ei}$.

The mean directed velocity of the electrons, determined from (7.1.12),

$$\langle \boldsymbol{v}\rangle = N_e^{-1}\int \boldsymbol{v} f_{0e}\, d\boldsymbol{p} \approx 1.96\,\boldsymbol{u}_e \tag{7.1.13}$$

is much smaller than the thermal velocity, i.e., $u_e \ll v_{Te}$, and fulfills the condition of applicability which is the inequality opposite to (7.1.5).

Equation (7.1.12) slightly differs from the Maxwellian distribution function with the mean velocity $\boldsymbol{u}_e$ (7.1.2) expanded in a power series. Using (7.1.2) instead of (7.1.12) we obtain small quantitative differences in various specific results. However, these results qualitatively coincide. For simplicity we use (7.1.2) with $\boldsymbol{p}_0 = m\boldsymbol{u}_e = e\boldsymbol{E}_0/\nu_{ei}$ also for the case of a weak field.

In a weakly ionized nondegenerate or degenerate plasma, immersed in an external electric field, a stationary state can exist also if

$$E_0 < E_{cr} = \frac{m\nu_{en}}{e} v_0\,, \tag{7.1.14}$$

v_0 being the random velocity of the electron motion (the thermal or the Fermi velocity). This condition implies that an electron acquires a velocity smaller than its random velocity during the mean free time in the field.

To determine the stationary distribution function of the electrons in a weakly ionized nondegenerate plasma we use the kinetic equation with the model BGK integral (Sect. 3.5):

$$\frac{e\boldsymbol{E}_0}{m}\frac{\partial f_{0\mathrm{e}}}{\partial \boldsymbol{v}} = -\nu_{\mathrm{en}}(f_{0\mathrm{e}} - N_\mathrm{e}\Phi_{\mathrm{en}}) \ . \tag{7.1.15}$$

As a result we obtain

$$\begin{aligned} f_{0\mathrm{e}} &= \frac{N_\mathrm{e}}{(2\pi m T_\mathrm{e})^{3/2}}\exp\left(-\frac{mv^2}{2T_\mathrm{e}}\right)\left(1+\frac{\boldsymbol{u}_\mathrm{e}\cdot\boldsymbol{v}}{v_{\mathrm{Te}}^2}\right) \\ &\approx \frac{N_\mathrm{e}}{(2\pi m T_\mathrm{e})^{3/2}}\exp\left[-\frac{m(\boldsymbol{v}-\boldsymbol{u}_\mathrm{e})^2}{2T_\mathrm{e}}\right] , \end{aligned} \tag{7.1.16}$$

where $\boldsymbol{u}_\mathrm{e} = e\boldsymbol{E}_0/m\nu_{\mathrm{en}}$.

The electron distribution function for the weakly ionized degenerate plasma in the external electric field is obtained analogously. Here it is necessary to solve the equation (Sect. 3.5)

$$e\boldsymbol{E}_0\frac{\partial f_{0\mathrm{e}}}{\partial \boldsymbol{p}} = -\nu_{\mathrm{en}}[f_{0\mathrm{e}} - f_{00}(p)] \ , \tag{7.1.17}$$

where $f_{00}(p)$ is the isotropic Fermi distribution function of the degenerate electron gas. Then we obtain

$$f_{0\mathrm{e}}(\boldsymbol{p}) = f_{00}(p) - \frac{e\boldsymbol{E}_0}{\nu_{\mathrm{en}}}\frac{\partial f_{00}}{\partial \boldsymbol{p}} \approx f_{00}(|\boldsymbol{p}-\boldsymbol{p}_0|) \ , \tag{7.1.18}$$

where $f_{00}(|\boldsymbol{p}-\boldsymbol{p}_0|)$ is determined by (7.1.3).

7.1.4 Plasma in a High-Frequency Electric Field

The derived relations are valid for a constant field. It is not difficult to generalize them to time-varying fields and, most interesting, to periodically time-varying high-frequency fields

$$\boldsymbol{E}_0(t) = \boldsymbol{E}_0 \sin\omega_0 t \ . \tag{7.1.19}$$

For $\omega_0 \gg \nu_e$ the electron collisions can be ignored and we obtain from Vlasov's equation

$$\frac{\partial f_{0e}}{\partial t} + e\boldsymbol{E}_0 \sin \omega_0 t \frac{\partial f_{0e}}{\partial \boldsymbol{p}} = 0 \tag{7.1.20}$$

the solution in the form of the shifted Maxwellian or Fermi distribution function:

$$f_{0e} = f_{0e}(|\boldsymbol{p} - \boldsymbol{p}_0(t)|) , \quad \text{where} \tag{7.1.21}$$

$$\boldsymbol{p}_0(t) = e \int^t \boldsymbol{E}_0(t)\, dt = -\frac{e\boldsymbol{E}_0}{\omega_0} \cos \omega_0 t = \frac{m\boldsymbol{u}_e(t)}{\sqrt{1 - u_e^2/c^2}} . \tag{7.1.22}$$

For $\omega_0 < \nu_e$ the collision integral becomes the main term of the kinetic equation and (7.1.6–18) appear to be valid under the substitution $\boldsymbol{E}_0 \to \boldsymbol{E}_0(t) = \boldsymbol{E}_0 \sin \omega_0 t$.

7.2 Stability of the Nonmagnetized Plasma in an External Constant Electric Field

Under the condition (7.1.5) which is already satisfied for relatively weak fields in the real plasma the electrons are accelerated up to large velocities significantly exceeding their thermal velocity during the mean free time. The resulting state is a state in which all the electrons are moving with respect to the motionless ions. Below such an electron beam will be shown to be unstable, and the increment of the instability is rather large, of the order of the Langmuir ion frequency or even larger. In spite of the nonstationarity of the electron distribution function (7.1.2), the value of the directed velocity does not vary significantly during such a short time. Therefore, to analyze the stability of this system in the first-order approximation, the velocity $\boldsymbol{u}_e$ may be regarded as constant. (In the theory of the stability of a plasma in an external electric field the equivalent approximation is known as the *adiabatic approximation*). For simplicity particle collisions will also be ignored, which is justified when the plasma is sufficiently dilute and thus collisionless.

Under the given restrictions it is allowed to analyze the plasma stability on the basis of the dispersion equation

$$\left| k^2\delta_{ij} - k_i k_j - \frac{\omega^2}{c^2} \varepsilon_{ij}(\omega, \boldsymbol{k}) \right| = 0 , \tag{7.2.1}$$

where $\varepsilon_{ij}(\omega, \boldsymbol{k})$ is calculated by means of the transformation formulas derived in Chap. 6.

7.2.1 Buneman Instability of Nonmagnetized Plasma in Strong Electric Field

Under the influence of a sufficiently strong electric field the electrons move relative to the motionless ions with a velocity much greater than their thermal velocity. In the nonmagnetized plasma (7.2.1) is of the form

$$\left(k^2 - \frac{\omega^2}{c^2} + \frac{\omega_{pe}^2\gamma^{-1} + \omega_{pi}^2}{c^2}\right)\left(1 - \frac{\omega_{pe}^2\gamma^{-3}}{(\omega - \boldsymbol{k}\cdot\boldsymbol{u})^2} - \frac{\omega_{pi}^2}{\omega^2}\right) - \frac{\omega_{pe}^2\gamma^{-1}\omega_{pi}^2 k_\perp^2 u^2}{\omega^2 c^2 (\omega - \boldsymbol{k}\cdot\boldsymbol{u})^2} = 0\,. \tag{7.2.2}$$

Here $\gamma = (1 - u^2/c^2)^{-1/2}$, $\boldsymbol{u}$ is the velocity of the electric drift of the electrons and $k_\perp$ is the component of the wave vector of the perturbation which is perpendicular to the velocity $\boldsymbol{u}$.

Under the condition

$$\omega_{pe}^2 \geqslant (\boldsymbol{k}\cdot\boldsymbol{u})^2\gamma^3 \tag{7.2.3}$$

Eq. (7.2.2) has unstable solutions, and the maximum increment is determined by

$$\omega = \frac{1 + i\sqrt{3}}{2}\left(\frac{m}{2M}\right)^{1/3}\gamma\left(1 + \frac{k_\perp^2\gamma^2u^2}{c^2(k_\perp^2 + k_z^2\gamma^2)}\right)^{1/3}\boldsymbol{k}\cdot\boldsymbol{u}\,. \tag{7.2.4}$$

It is achieved under such circumstances that (7.2.3) holds with the equality sign, this is called the resonance case and the instability is known as the *Buneman instability*[1].

As mentioned above, this quickly growing aperiodic instability occurs for $u \gg v_{Te}$. We now show that this instability is also possible for $u < v_{Te}$. At these small velocities of the electron drift the oscillations excited in the system are longitudinal with a high degree of accuracy and, consequently, obey the dispersion equation

$$\varepsilon(\omega, \boldsymbol{k}) = \frac{k_i k_j}{k^2}\varepsilon_{ij}(\omega, \boldsymbol{k}) = 1 + \frac{\omega_{pe}^2}{k^2 v_{Te}^2}\left[1 - I_+\left(\frac{\omega - \boldsymbol{k}\cdot\boldsymbol{u}}{k v_{Te}}\right)\right] + \frac{\omega_{pi}^2}{k^2 v_{Ti}^2}\left[1 - I_+\left(\frac{\omega}{k v_{Ti}}\right)\right] = 0\,. \tag{7.2.5}$$

---

1 It is easily seen from (7.2.2) that in the nonresonant case when (7.2.3) is satisfied excluding equality, the increment of the Buneman instability is significantly smaller: $\mathrm{Im}\{\omega\} \sim (m/M)^{1/2}\,\boldsymbol{k}\cdot\boldsymbol{u}$.

At high velocities of the electric drift of the electrons, when $u \gg v_{Te}$, this equation coincides with (7.2.2) in the limit $u \ll c$ and describes the quickly growing Buneman instability.

7.2.2 Ion-Acoustic Instability of Plasma with a Current

For $u \ll v_{Te}$ we obtain from (7.2.5) in the frequency range $kv_{Ti} \ll \omega \ll kv_{Te}$

$$1 + \frac{\omega_{pe}^2}{k^2 v_{Te}^2}\left(1 + i\sqrt{\frac{\pi}{2}}\frac{\omega - \boldsymbol{k}\cdot\boldsymbol{u}}{kv_{Te}}\right) - \frac{\omega_{pi}^2}{\omega^2} + i\sqrt{\frac{\pi}{2}}\frac{\omega_{pi}^2\omega}{k^3 v_{Ti}^3}\exp\left(-\frac{\omega^2}{2k^2 v_{Ti}^2}\right) = 0 \,. \tag{7.2.6}$$

The imaginary terms due to the Cherenkov dissipation of the wave by the plasma electrons and ions are small compared to the real terms. Therefore the solution can be obtained in the form of $\omega \to \omega + i\delta$, where $|\delta| \ll \omega$. As a result we find

$$\omega^2 = \frac{\omega_{pi}^2}{1 + \omega_{pe}^2/(k^2 v_{Te}^2)} \,,$$
$$\frac{\delta}{\omega} = -\sqrt{\frac{\pi}{8}}\frac{M}{m}\frac{\omega^3}{k^3 v_{Te}^3}\left(1 - \frac{u}{v_{ph}}\cos\theta\right) - \sqrt{\frac{\pi}{8}}\frac{\omega^3}{k^3 v_{Ti}^3}\exp\left(-\frac{\omega^2}{2k^2 v_{Ti}^2}\right) . \tag{7.2.7}$$

Here $v_{ph} = \omega/k$ is the phase velocity of the wave and θ the angle included between the vectors $\boldsymbol{u}$ and $\boldsymbol{k}$.

For $\boldsymbol{u} = 0$ the spectrum (7.2.7) is that of the ion-acoustic oscillations, occurring in the nonisothermal plasma with $T_e \gg T_i$. Equation (7.2.7) shows that the damping decrement is reduced for a nonzero velocity of the electron drift $\boldsymbol{u}$ and for $u > u_{cr}$ when $\delta > 0$ these oscillations become unstable. It is obvious that this instability is possible for $u > v_{ph}/\cos \gg v_{Ti}$ only, i.e., when the velocity of the electron drift is larger than the phase velocity of the ion-acoustic oscillations. Thus, this instability leading to the buildup of ion-acoustic oscillations is a pure Cherenkov instability. Therefore it is often called the *ion-acoustic instability of the plasma with a current.*

We have shown in Sect. 4.2 for the plasma without an electron drift, i.e., for $\boldsymbol{u} = 0$, that the ion-acoustic oscillations are possible for a sufficiently nonisothermal plasma only, i.e., for $T_e > 6\,T_i$. In the system with an electron drift this condition is weakened. When the drift velocity approaches the critical value of marginal instability where the imaginary part of the longitudinal dielectric permittivity tends to zero, the inequality $T_e > 3\,T_i$ which follows from the applicability of the expansion of Re $\{I_+(\omega/k_z v_{Ti})\}$ in powers of $k_z v_{Ti}/\omega$ already ensures the existence of ion-acoustic oscillations. The requirement of a small imaginary part of the dielectric permittivity, leading to the more strict condition in the case of the isotropic plasma, is automatically satisfied here.

7.2.3 The Critical Velocity

We get from (7.2.7) for the most interesting ion-acoustic oscillations with a long wavelength

$$\omega^2 = k^2 v_s^2 ,$$
$$\frac{\delta}{\omega} = -\sqrt{\frac{\pi m}{8M}}\left(1 - \frac{u}{v_s}\cos\theta\right) - \sqrt{\frac{\pi}{8}}\left(\frac{T_e}{T_i}\right)^{3/2} \exp\left(-\frac{T_e}{2T_i}\right) . \quad (7.2.8)$$

Then we obtain for the critical drift velocity

$$u_{cr} = \frac{v_s}{\cos\theta}\left[1 + \left(\frac{M}{m}\right)^{1/2}\left(\frac{T_e}{T_i}\right)^{3/2} \exp\left(-\frac{T_e}{2T_i}\right)\right] , \quad (7.2.9)$$

i.e., $u_{cr} > v_s$.

When the oscillation frequency grows, the velocity u_{cr} decreases and becomes smaller than v_s. Short-wave oscillations with a wavelength smaller than the Debye length of the electrons possess the smallest critical velocity. For these oscillations

$$\omega = \omega_{pi} , \quad (7.2.10)$$
$$\frac{\delta}{\omega} = -\sqrt{\frac{\pi}{8}\frac{M}{m}}\frac{\omega_{pi}^3}{k^3 v_{Te}^3}\left(1 - \frac{u\cos\theta}{v_{ph}}\right) - \sqrt{\frac{\pi}{8}}\frac{\omega_{pi}^3}{k^3 v_{Ti}^3} \exp\left(-\frac{\omega_{pi}^2}{2k^2 v_{Ti}^2}\right) .$$

Finally, note that (7.2.5) has stable solutions only for very small velocities of the electron drift, when $u < v_{Ti}$. Under this condition the current-carrying plasma in the external electric field is stable.

Figure 7.1 shows the increment of the unstable oscillations of the plasma in the constant electric field as a function of $\boldsymbol{k} \cdot \boldsymbol{u}$. The increment of the

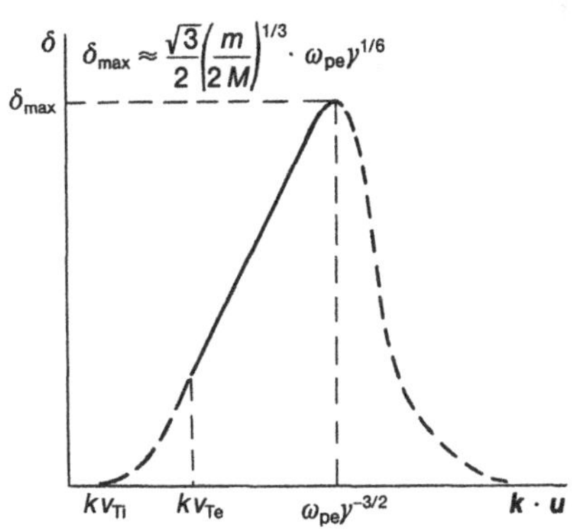

Fig. 7.1. Increment of the plasma oscillations in a constant electric field versus $\boldsymbol{k} \cdot \boldsymbol{u}$

kinetic instability is shown by the dashed line and the increment of the hydrodynamic instability by the solid line. The maximum increment is

$$|\delta|_{\max} \approx \sqrt{\frac{3}{4}}\left(\frac{m}{2M}\right)^{1/3} \omega_{\mathrm{pe}}\gamma^{1/6}.$$

7.2.4 The Effect of Collisions on the Development of Instabilities

Till now we have neglected particle collisions in the plasma. Thus, strictly speaking, the formulas derived are valid for $|\delta| \gg \nu_{\mathrm{e}}$ only, i.e., when the time of growth of all the oscillation processes is much smaller than the time between the particle collisions. This requirement is substantial for the high-frequency hydrodynamic instability developing for $u > v_{\mathrm{Te}}$, since the electrons can acquire this velocity only during a time smaller than their mean free time for $E_0 > E_{\mathrm{cr}}$. Explicitly, we have

$$\omega_{\mathrm{pe}}\left(\frac{m}{M}\right)^{1/3} \gamma^{1/6} \gg \nu_{\mathrm{e}}\,, \tag{7.2.11}$$

where ν_{e} corresponds to ν_{ei} for the completely ionized plasma and to ν_{en} for the weakly ionized plasma.

The ion-acoustic instability of the nonisothermal plasma in an electric field can develop both for $E_0 > E_{\mathrm{cr}}$ and for $E_0 < E_{\mathrm{cr}}$. Here the drift velocity of the electrons must reach the critical value for the buildup of ion-acoustic oscillations. When the plasma is sufficiently rarefied and the critical velocity is reached during a time smaller than the mean free time of the electrons, then their distribution function has the form of (7.1.2). If the plasma density is high and the electrons undergo many collisions during the acceleration in the field $\boldsymbol{E}_0$, then their stationary distribution function has the form of (7.1.12) for the completely ionized plasma or of (7.1.16) for the weakly ionized plasma. The distribution function (7.1.16) is a shifted Maxwellian and (7.1.12) is not Maxwellian. As mentioned above, this function may also be replaced by a Maxwellian, for simplicity. This permits us to apply the general theory, based on the Lorentz transformation formulas which we discussed previously. For the case of the nondegenerate weakly ionized plasma with collisions the application of these formulas (Sect. 5.4) gives us the following dispersion equation for the longitudinal electromagnetic waves, which generalizes (7.2.5):

$$1 + \frac{\omega_{\mathrm{pe}}^2}{k^2 v_{\mathrm{Te}}^2}\,\frac{1 - I_+\left(\dfrac{\omega - \boldsymbol{k}\cdot\boldsymbol{u} + \mathrm{i}\nu_{\mathrm{e}}}{k v_{\mathrm{Te}}}\right)}{1 - \dfrac{\mathrm{i}\nu_{\mathrm{e}}}{\omega - \boldsymbol{k}\cdot\boldsymbol{u} + \mathrm{i}\nu_{\mathrm{e}}}\, I_+\left(\dfrac{\omega - \boldsymbol{k}\cdot\boldsymbol{u} + \mathrm{i}\nu_{\mathrm{e}}}{k v_{\mathrm{Te}}}\right)}$$

$$+ \frac{\omega_{\mathrm{pi}}^2}{k^2 v_{\mathrm{Ti}}^2}\,\frac{1 - I_+\left(\dfrac{\omega + \mathrm{i}\nu_{\mathrm{i}}}{k v_{\mathrm{Ti}}}\right)}{1 - \dfrac{\mathrm{i}\nu_{\mathrm{i}}}{\omega + \mathrm{i}\nu_{\mathrm{i}}}\, I_+\left(\dfrac{\omega + \mathrm{i}\nu_{\mathrm{i}}}{k v_{\mathrm{Ti}}}\right)} = 0\,. \tag{7.2.12}$$

Here ν_e and ν_i are the frequencies of electron-neutral and ion-neutral collisions.

In the frequency range $\omega, \nu_e \ll k v_{Te}$ and $\omega \gg \nu_i, k v_{Ti}$ of the ion-acoustic oscillations (7.2.12) is of the form

$$1 + \frac{\omega_{pe}^2}{k^2 v_{Te}^2}\left(1 + i\sqrt{\frac{\pi}{2}}\frac{\omega - \boldsymbol{k}\cdot\boldsymbol{u}}{k v_{Te}}\right) - \frac{\omega_{pi}^2}{\omega^2}\left(1 - i\frac{\nu_i}{\omega}\right)$$
$$+ i\sqrt{\frac{\pi}{2}}\frac{\omega_{pi}^2\omega}{k^3 v_{Ti}^3}\exp\left(-\frac{\omega^2}{2k^2 v_{Ti}^2}\right) = 0 . \tag{7.2.13}$$

It differs from (7.2.6) by a small imaginary term, which accounts for the ion-neutral collisions. Consequently, the spectrum (7.2.7) is altered. In the expression for δ there appears an additional summand

$$\Delta\delta = -\nu_i/2 . \tag{7.2.14}$$

The same correction appears in (7.2.8, 10).

Thus, the account of ion collisions leads to a higher critical velocity for the buildup of ion-acoustic oscillations in the weakly ionized plasma. For long-wavelength oscillations we have instead of (7.2.9)

$$u_{cr} = \frac{v_s}{\cos\theta}\left[1 + \frac{\omega_{pi}^2}{\omega^2}\left(\frac{M}{m}\right)^{1/2}\left(\frac{T_e}{T_i}\right)^{3/2}\exp\left(-\frac{T_e}{2T_i}\right) + \sqrt{\frac{2M}{\pi m}}\frac{\nu_i}{\omega}\right] . \tag{7.2.15}$$

For the completely ionized plasma with the electron distribution function (7.1.2) the results (7.2.13 to 15) remain valid if we substitute

$$\nu_i \rightarrow \frac{8}{5}\nu_{ii}\frac{k^2 v_{Ti}^2}{\omega^2} , \quad \text{where} \quad \nu_{ii} = \frac{4}{3}\sqrt{\frac{\pi}{M}}\, e_i^4 L N_i T_i^{-3/2} \tag{7.2.16}$$

is the frequency of the ion-ion collisions. This is easily seen if we remember that the collisional correction to the longitudinal dielectric permittivity of the completely ionized plasma is determined in the range of the ion-acoustic frequencies by (4.6.6).

7.2.5 The Case of the Degenerate Plasma

Finally, we consider the stability of the degenerate plasma in an external electric field when the electron velocity distribution function has the form of the shifted Fermi distribution (7.1.3). Here we deal with the stability of the longitudinal waves only, regarding the electron drift velocity as small compared to the velocity of light. Using the transformation formulas (6.1.18) and the explicit form of the longitudinal dielectric permittivity in the degenerate plasma (4.5.19), we obtain

$$1+\frac{3\omega_{\mathrm{pe}}^2}{k^2v_{\mathrm{Fe}}^2}\left(1-\frac{1}{2}\frac{\omega-\boldsymbol{k}\cdot\boldsymbol{u}+\mathrm{i}\nu_{\mathrm{e}}}{kv_{\mathrm{Fe}}}\ln\frac{\omega-\boldsymbol{k}\cdot\boldsymbol{u}+\mathrm{i}\nu_{\mathrm{e}}+kv_{\mathrm{Fe}}}{\omega-\boldsymbol{k}\cdot\boldsymbol{u}+\mathrm{i}\nu_{\mathrm{e}}-kv_{\mathrm{Fe}}}\right)$$

$$\times\left(1-\frac{\mathrm{i}\nu_{\mathrm{e}}}{2kv_{\mathrm{Fe}}}\ln\frac{\omega-\boldsymbol{k}\cdot\boldsymbol{u}+\mathrm{i}\nu_{\mathrm{e}}+kv_{\mathrm{Fe}}}{\omega-\boldsymbol{k}\cdot\boldsymbol{u}+\mathrm{i}\nu_{\mathrm{e}}-kv_{\mathrm{Fe}}}\right)^{-1} \tag{7.2.17}$$

$$-\frac{\omega_{\mathrm{pi}}^2}{k^2v_{\mathrm{Ti}}^2}\left[1-I_+\left(\frac{\omega+\mathrm{i}\nu_{\mathrm{i}}}{kv_{\mathrm{Ti}}}\right)\right]\left[1-\frac{\mathrm{i}\nu_{\mathrm{i}}}{\omega+\mathrm{i}\nu_{\mathrm{i}}}I_+\left(\frac{\omega+\mathrm{i}\nu_{\mathrm{i}}}{kv_{\mathrm{Ti}}}\right)\right]^{-1}=0\,.$$

The ions are assumed nondegenerate in the derivation of this equation.

In the range of the ion-acoustic oscillation frequencies ν_{i}, $kv_{\mathrm{Ti}} \ll \omega$, $\nu_{\mathrm{e}} \ll kv_{\mathrm{Fe}}$ we obtain from (7.2.17)

$$1+\frac{3\omega_{\mathrm{pe}}^2}{k^2v_{\mathrm{Fe}}^2}\left(1+\mathrm{i}\frac{\pi}{2}\frac{\omega-\boldsymbol{k}\cdot\boldsymbol{u}}{kv_{\mathrm{Fe}}}\right)-\frac{\omega_{\mathrm{pi}}^2}{\omega^2}\left(1-\mathrm{i}\frac{\nu_{\mathrm{i}}}{\omega}\right)$$

$$+\,\mathrm{i}\sqrt{\frac{\pi}{2}}\frac{\omega_{\mathrm{pi}}^2\omega}{k^3v_{\mathrm{Ti}}^3}\exp\left(-\frac{\omega^2}{2k^2v_{\mathrm{Ti}}^2}\right)=0\,, \tag{7.2.18}$$

which is analogous to (7.2.13). Therefore the given analysis is also applicable here and yields formulas similar to (7.2.7–10, 15) with the substitutions $v_{\mathrm{Te}} \to v_{\mathrm{Fe}}$ and $T_{\mathrm{e}} \to \mathscr{E}_{\mathrm{Fe}}$.

7.3 Stability of the Magnetized Plasma in an External Constant Electric Field

A strong magnetic field affects the stability of a plasma in an external constant electric field; thus we have to generalize our results regarding the magneto-active plasma. We consider $\boldsymbol{B}_0 \parallel \boldsymbol{E}_0$ only. As before, the electron distribution function is assumed to be the shifted Maxwellian (7.1.2) or the shifted Fermi (7.1.3) distribution. We suppose that the oscillations developing do not alter the electron streaming velocity (adiabatic approximation). For simplicity, we confine our analysis to sufficiently strong magnetic fields where the electrons are magnetized ($\Omega_{\mathrm{e}}^2 \gg \omega_{\mathrm{pe}}^2$) and the ions, on the contrary, not magnetized ($\Omega_{\mathrm{i}}^2 \ll \omega_{\mathrm{pi}}^2$).

7.3.1 The Buneman Instability of the Magneto-Active Plasma

Under these restrictions the analysis of (7.2.1) is significantly simplified. Thus, in the limit of a strong magnetic field, for the case when the electric drift velocity of the electrons greatly exceeds their random velocity and the latter can be neglected, (7.2.1) takes the form

$$k_\perp \varepsilon_{xx} + \left(k_z^2 - \frac{\omega^2}{c^2}\varepsilon_{xx}\right)\varepsilon_{zz} = k_\perp^2\left(1 - \frac{\omega_{pi}^2}{\omega^2}\right)$$

$$+ \left[k_z^2 - \frac{\omega^2}{c^2}\left(1 - \frac{\omega_{pi}^2}{\omega^2}\right)\right]\left[1 - \frac{\omega_{pe}^2\gamma^{-3}}{(\omega - k_z u)^2} - \frac{\omega_{pi}^2}{\omega^2}\right] = 0 . \tag{7.3.1}$$

We obtain unstable solutions under the condition

$$\omega_{pe}^2 \gg k^2u^2\gamma^3 . \tag{7.3.2}$$

The instability is aperiodic since $\omega \ll k_z u$, and its increment becomes maximum for the resonance, when (7.3.2) holds with the equality sign.

$$\omega = \frac{1 + \mathrm{i}\sqrt{3}}{2}\left(\frac{mk^2}{2Mk_z^2}\right)^{1/3} k_z u\gamma . \tag{7.3.3}$$

By comparison of (7.2.3, 4) with (7.3.3) it follows that the external magnetic field hinders the development of the high-frequency Buneman instability in a strong electric field. The range (7.3.2) where the instability exists is narrower than the range (7.2.3) for the waves propagating at an angle $\theta \neq 0$ with respect to the drift velocity. The increment of the instability in the magneto-active plasma (7.3.3) is smaller than in the case without magnetic field (7.2.4).

As in the nonmagnetized plasma, the high-frequency instability occurs only for high electric drift velocities of the electrons exceeding their random velocity. Equally as in the absence of a magnetic field, instabilities are possible in the magnetized plasma at insignificant drift velocities which are smaller than the thermal electron velocity but greater than the thermal ion velocity. However, the account of the thermal motion of particles, i.e., the explicit form of the distribution function of the charged particles, is important in the analysis of these instabilities. This rather complicates the analysis of (7.2.1). At small drift velocities we may study the longitudinal instead of the general dispersion equation, which simplifies our task.

7.3.2 The Ion-Acoustic Instability of Plasma with a Current in a Magnetic Field

For the nondegenerate electron-ion plasma in a constant electric field parallel to the magnetic field the equation for the longitudinal oscillations has the form

$$
\begin{aligned}
\varepsilon(\omega, \boldsymbol{k}) &= \frac{k_i k_j}{k^2} \varepsilon_{ij}(\omega, \boldsymbol{k}) \\
&= 1 + \frac{\omega_{\mathrm{pe}}^2}{k^2 v_{\mathrm{Te}}^2} \left[1 - \sum_n \frac{\omega - \boldsymbol{k} \cdot \boldsymbol{u}}{\omega - \boldsymbol{k} \cdot \boldsymbol{u} - n\Omega_{\mathrm{e}}} A_n \left(\frac{k_\perp^2 v_{\mathrm{Te}}^2}{\Omega_{\mathrm{e}}^2} \right) \right] \\
&\quad \times I_+ \left(\frac{\omega - \boldsymbol{k} \cdot \boldsymbol{u} - n\Omega_{\mathrm{e}}}{k_z v_{\mathrm{Te}}} \right) \\
&\quad + \frac{\omega_{\mathrm{pi}}^2}{k^2 v_{\mathrm{Ti}}^2} \left[1 - \sum_n \frac{\omega}{\omega - n\Omega_{\mathrm{i}}} A_n \left(\frac{k_\perp^2 v_{\mathrm{Ti}}^2}{\Omega_{\mathrm{i}}^2} \right) I_+ \left(\frac{\omega - n\Omega_{\mathrm{i}}}{k_z v_{\mathrm{Ti}}} \right) \right] = 0 \,.
\end{aligned}
\tag{7.3.4}
$$

At small velocities of the electron drift $u \ll v_{\mathrm{Te}}$ unstable solutions of (7.3.4) can be expected in the frequency range $k_z v_{\mathrm{Ti}} \ll \omega \ll k_z v_{\mathrm{Te}}$ where we can approximate

$$
1 + \frac{\omega_{\mathrm{pe}}^2}{k^2 v_{\mathrm{Te}}^2} \left(1 + \mathrm{i} \sqrt{\frac{\pi}{2}} \frac{\omega - \boldsymbol{k} \cdot \boldsymbol{u}}{|k_z| \, v_{\mathrm{Te}}} \right) - \frac{\omega_{\mathrm{pi}}^2}{\omega^2} + \mathrm{i} \sqrt{\frac{\pi}{2}} \frac{\omega_{\mathrm{pi}}^2 \omega}{k^3 v_{\mathrm{Ti}}^3} \exp\left(- \frac{\omega^2}{2 k^2 v_{\mathrm{Ti}}^2} \right) = 0
\tag{7.3.5}
$$

As before, the ions are considered nonmagnetized and the electrons strongly magnetized in the derivations. Equation (7.3.5) differs from the analogous equation (7.2.6) only by a small imaginary term originating from the Cherenkov effect of the electrons. Thus, the analysis of (7.2.6) remains valid and the resulting expressions (7.2.7–10) must be modified by the following substitution only

$$
1 - \frac{u}{v_{\mathrm{ph}}} \cos\theta \rightarrow \frac{1}{\cos\theta} \left(1 - \frac{u}{v_{\mathrm{ph}}} \cos\theta \right) .
\tag{7.3.6}
$$

As a result, the critical velocity of the electron drift becomes smaller and the increment of the instability larger than in the absence of the magnetic field. Consequently the external magnetic field facilitates the development of the ion-acoustic instability in an external electric field.

Figure 7.2 shows this increment as a function of $\boldsymbol{k} \cdot \boldsymbol{u}$. The dependence is qualitatively the same as in the absence of the magnetic field.

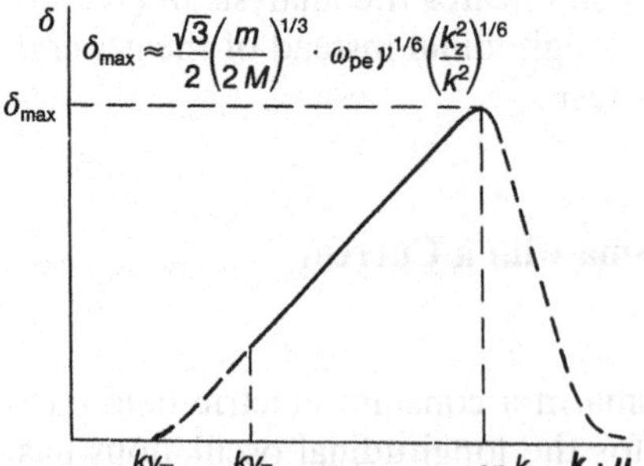

Fig. 7.2. Increment of the oscillations of a nondegenerate plasma in external constant electric and magnetic fields versus $\boldsymbol{k} \cdot \boldsymbol{u}$

7.3.3 Effect of Collisions on the Development of Instabilities

Coming to the discussion of the effect of collisions on the instabilities investigated in this section we note first that the validity of the formulas derived so far is restricted. Regarding the high-frequency oscillations excited when $u > v_{Te}$ we must demand that the electrons should be accelerated to a velocity exceeding their thermal velocity during their mean free time. This gives the condition $E_0 > E_{cr}$, which means that all processes must develop in a time interval shorter than the mean free time of the electrons. The requirement $|\delta| > \nu_e$ then leads to the condition (7.2.11). The slow ion-acoustic instability can develop both for $E_0 > E_{cr}$ and for $E_0 < E_{cr}$. In this case the electrons must reach a mean drift velocity greater than the ion thermal velocity. As shown in the foregoing section, this is possible not only in the dilute collisionless plasma, but also in the collision-dominated plasma. Ion collisions play an important part in the buildup of ion-acoustic oscillations and as in the magnetic field free case the buildup occurs for $\omega > \nu_i$ only.

Using the results of Sects. 5.5, 6, we can write the dispersion equation for ion-acoustic oscillations of the magneto-active plasma with the electron distribution function (7.1.2). In the frequency range $\Omega_e \gg \omega \gg \Omega_i$ where the electrons are magnetized and the ions not it is of the form

$$1 + \frac{\omega_{pe}^2}{k^2 v_{Te}^2}\left(1 + i\sqrt{\frac{\pi}{2}}\frac{\omega - \boldsymbol{k}\cdot\boldsymbol{u}}{|k_z|\, v_{Te}}\right) - \frac{\omega_{pi}^2}{\omega^2}\left(1 - i\frac{\nu_i}{\omega}\right)$$

$$+ i\sqrt{\frac{\pi}{2}}\frac{\omega_{pi}^2\omega}{k^3 v_{Ti}^3}\exp\left(-\frac{\omega^2}{2k_z^2 v_{Ti}^2}\right) = 0\,, \qquad (7.3.7)$$

where $\nu_i = \nu_{in}$ or $\nu_i = 8\nu_{ii}k^2 v_{Ti}^2/5\omega^2$ for the weakly or completely ionized plasma, respectively. This equation differs from (7.3.5) by the presence of a small imaginary term accounting for ion collisions only. The correction for the damping decrement, see (7.2.7, 8, 10), when the substitution (7.3.6) is made becomes

$$\Delta\delta = -\frac{\nu_i}{2}\,. \qquad (7.3.8)$$

Consequently the ion collisions have a stabilizing effect on the electric drift excitation of the ion-acoustic instability in the nonisothermal plasma.

7.3.4 The Case of Degenerate Plasma

Finally, we briefly discuss the stability of the degenerate plasma in the external electric and magnetic field. As before, only the electrons are considered degenerate and their electric drift velocity is assumed smaller than the Fermi velocity but greater than the ion thermal velocity. Under these conditions the general dispersion equation

$$\varepsilon(\omega,k) = \frac{k_i k_j}{k^2}\varepsilon_{ij}(\omega,k) = 1 + \frac{\omega_{pi}^2}{k^2 v_{Ti}^2}\frac{1 - I_+\left(\dfrac{\omega + i\nu_i}{k v_{Ti}}\right)}{1 - \dfrac{i\nu_i}{\omega + i\nu_i} I_+\left(\dfrac{\omega + i\nu_i}{k v_{Ti}}\right)}$$

$$+ 3\frac{\omega_{pe}^2}{k^2 v_{Fe}}\left(1 - \sum_n \frac{\omega - k_z u + i\nu_e}{2}\int_0^\pi \frac{\sin\theta d\theta\, J_n^2(k_\perp v_{Fe}\sin\theta/\Omega_e)}{\omega - k_z u + i\nu_e - k_z v_{Fe}\cos\theta - n\Omega_e}\right)$$

$$\times\left(1 - \sum_n \frac{i\nu_e}{2}\int_0^\pi \frac{\sin\theta d\theta\, J_n^2(k_\perp v_{Fe}\sin\theta/\Omega_e)}{\omega - k_z u + i\nu_e - k_z v_{Fe}\cos\theta - n\Omega_e}\right)^{-1} = 0 \quad (7.3.9)$$

can be simplified if the frequency range $k_z v_{Ti} \ll \omega \ll k_z v_{Fe}$ is considered. As a result we obtain the equation

$$1 + \frac{3\omega_{pe}^2}{k^2 v_{Fe}^2}\left(1 + i\frac{\pi}{2}\frac{\omega - \boldsymbol{k}\cdot\boldsymbol{u}}{|k_z|\, v_{Fe}}\right) - \frac{\omega_{pi}^2}{\omega^2}\left(1 - i\frac{\nu_i}{\omega}\right)$$
$$+ i\sqrt{\frac{\pi}{2}}\frac{\omega_{pi}^2\omega}{k^3 v_{Ti}^3}\exp\left(-\frac{\omega^2}{2k^2 v_{Ti}^2}\right) = 0\,, \quad (7.3.10)$$

which differs from (7.2.18) by the small imaginary term due to the Cherenkov effect of the electrons. Therefore, all the conclusions with respect to (7.2.18) remain valid when the substitution (7.3.6) is accounted for. There results a destabilizing effect of the magnetic field on the current instability of the degenerate plasma.

7.4 The Plasma in a Superhigh-Frequency Electric Field

The problems of the interaction of superhigh-frequency (SHF) electric fields with plasmas are associated with a number of applied problems of plasma physics. Among these the SHF discharge, i.e., the plasma creation by SHF fields, is of the greatest importance. The creation of a high-temperature thermonuclear plasma by SHF heating is also related to these problems. Plasma confinement by SHF fields is of great interest for scientists engaged in thermonuclear physics. Recently, the possibility to accelerate plasma clusters by means of the radiation pressure of SHF fields was intensively investigated. The theory of the interaction of SHF fields with plasmas has been widely developed and covers an increasing range of plasma phenomena. A full presentation of this theory falls outside the framework of this book. Therefore we confine our interest to the general principles and to the description of the phenomena, studied in detail both theoretically and experimentally.

We have derived in Sect. 7.1 the distribution functions, see (7.1.21, 22), of the charged particles for the homogeneous plasma in an external high-frequency electric field, parallel to the external magnetic field,

$$\boldsymbol{E}_0(t) = \boldsymbol{E}_0 \sin \omega_0 t \,. \tag{7.4.1}$$

The stability of such a plasma will be analyzed below by considering small perturbations of these distributions. To begin with, we analyze SHF fields with frequencies higher than all characteristic plasma frequencies

$$\omega_0 \gg \omega_{p\alpha}, \Omega_\alpha, \nu_\alpha \,. \tag{7.4.2}$$

In the first-order approximation the plasma in a SHF field may be regarded as isotropic. Consequently, the field $E_0(t)$ obeys the dispersion equation for transverse waves (Chap. 4):

$$k_0 = \frac{\omega_0}{c}\sqrt{1 - \frac{\omega_{pe}^2}{\omega_0^2}} \approx \frac{\omega_0}{c} \,. \tag{7.4.3}$$

The wavelength $k_0^{-1} = c/\omega_0$ characterizes the inhomogeneity of the SHF field neglected in (7.4.1). Therefore the SHF field can be considered homogeneous for processes with a characteristic length of the inhomogeneity $1/k$ much smaller than $1/k_0$ only, i.e., for $k \gg \omega_0/c$. Assuming this condition in the following, we are allowed to analyze quasilongitudinal oscillations only, when studying the stability of the plasma in the SHF field. Actually, the oscillation frequencies are of the order of the characteristic plasma frequencies, i.e., $\omega \sim (\omega_{p\alpha}, \Omega_\alpha)$. Taking account of (7.4.2) we come to the condition $\omega \ll \omega_0 \approx k_0 c \ll kc$, which is the condition of validity for the quasilongitudinal approximation.

Since the effect of a high-frequency electric field on the plasma ions can be neglected compared to its effect on the electrons, the ion velocity distribution function can be regarded as isotropic (usually Maxwellian) whereas the electron distribution function is given by (7.1.22, 23):

$$\begin{aligned} f_{0e}(\boldsymbol{p}) &= f_{0e}[\boldsymbol{p} - \boldsymbol{p}_0(t)] \,, \\ \boldsymbol{p}_0(t) &= e\int^t \boldsymbol{E}_0(t)\,dt = -\frac{e\boldsymbol{E}_0}{\omega_0}\cos\omega_0 t = \frac{m\boldsymbol{u}_0(t)}{\sqrt{1 - u_0^2/c^2}} \,. \end{aligned} \tag{7.4.4}$$

For simplicity we consider the nonrelativistic case $u_0(t) < c$ only.

Linearizing the Vlasov equation of the electrons and the ions describing the collisionless plasma in an external SHF field parallel to the constant magnetic field, we obtain for small deviations from the zero-order distribution functions

$$\frac{\partial \delta f_{\mathrm{e}}}{\partial t} + \mathrm{i}\boldsymbol{k} \cdot \boldsymbol{v} \delta f_{\mathrm{e}} + \frac{e E_0 \sin \omega_0 t}{m} \frac{\partial \delta f_{\mathrm{e}}}{\partial \boldsymbol{v}} + \frac{e}{mc} [\boldsymbol{v}, \boldsymbol{B}_0] \frac{\partial \delta f_{\mathrm{e}}}{\partial \boldsymbol{v}}$$

$$+ \frac{e\boldsymbol{E}}{m} \frac{\partial f_{0\mathrm{e}}(\boldsymbol{p} - \boldsymbol{p}_0)}{\partial \boldsymbol{v}} = 0 , \tag{7.4.5}$$

$$\frac{\partial \delta f_{\mathrm{i}}}{\partial t} + \mathrm{i}\boldsymbol{k} \cdot \boldsymbol{v} \delta f_{\mathrm{i}} + \frac{e_{\mathrm{i}}}{Mc} [\boldsymbol{v}, \boldsymbol{B}_0] \frac{\partial \delta f_{\mathrm{i}}}{\partial \boldsymbol{v}} + \frac{e_{\mathrm{i}} \boldsymbol{E}}{M} \frac{\partial f_{0\mathrm{i}}(\boldsymbol{p})}{\partial \boldsymbol{v}} = 0 .$$

Here $f_{0\mathrm{e}}(\boldsymbol{p} - \boldsymbol{p}_0)$ and $f_{0\mathrm{i}}(\boldsymbol{p})$ are the zero-order distribution functions of the electrons and the ions; δf_{e} and δf_{i} are the small perturbations which depend on the coordinates in the form of $\exp(\mathrm{i}\boldsymbol{k} \cdot \boldsymbol{r})$ because of the assumption of homogeneity; $\boldsymbol{E} = -\nabla \Phi$ is the electric field due to the perturbations satisfying the Poisson equation

$$k^2 \Phi = 4\pi e \int d\boldsymbol{p}\, \delta f_{\mathrm{e}} + 4\pi e_{\mathrm{i}} \int d\boldsymbol{p}\, \delta f_{\mathrm{i}} . \tag{7.4.6}$$

In order to solve (7.4.5, 6) it is convenient to introduce a new function

$$\Psi_{\mathrm{e}}(\boldsymbol{p}, t) = \exp\left(-\mathrm{i} \frac{e}{m} \frac{\boldsymbol{k} \cdot \boldsymbol{E}_0 \sin \omega_0 t}{\omega_0^2}\right) \delta f_{\mathrm{e}}\left(\boldsymbol{p} + \frac{e\boldsymbol{E}_0}{\omega_0} \cos \omega_0 t\right) . \tag{7.4.7}$$

The system of equations then takes the form

$$\frac{\partial \Psi_{\mathrm{e}}}{\partial t} + \mathrm{i}\boldsymbol{k} \cdot \boldsymbol{v} \Psi_{\mathrm{e}} - \Omega_{\mathrm{e}} \frac{\partial \Psi_{\mathrm{e}}}{\partial \phi} - \mathrm{i}\boldsymbol{k} \frac{\partial f_{0\mathrm{e}}(\boldsymbol{p})}{\partial \boldsymbol{v}} \frac{4\pi e}{mk^2}$$

$$\times \left[e \int d\boldsymbol{p}\, \Psi_{\mathrm{e}}(\boldsymbol{p}) + e_{\mathrm{i}} \int d\boldsymbol{p}\, \delta f_{\mathrm{i}}(\boldsymbol{p}) \exp\left(-\mathrm{i} \frac{e}{m} \frac{\boldsymbol{k} \cdot \mathrm{E}_0}{\omega_0^2} \sin \omega_0 t\right)\right] = 0 , \tag{7.4.8}$$

$$\frac{\partial \delta f_{\mathrm{i}}}{\partial t} + \mathrm{i}\boldsymbol{k} \cdot \boldsymbol{v} \delta f_{\mathrm{i}} - \Omega_{\mathrm{i}} \frac{\partial \delta f_{\mathrm{i}}}{\partial \phi} - \mathrm{i}\boldsymbol{k} \frac{\partial f_{0\mathrm{i}}(\boldsymbol{p})}{\partial \boldsymbol{v}} \frac{4\pi e_{\mathrm{i}}}{Mk^2}$$

$$\times \left[e_{\mathrm{i}} \int d\boldsymbol{p}\, \delta f_{\mathrm{i}} + e \int d\boldsymbol{p}\, \Psi_{\mathrm{e}} \exp\left(-\frac{\mathrm{i}e}{m} \frac{\boldsymbol{k} \cdot \boldsymbol{E}_0}{\omega_0^2} \sin \omega_0 t\right)\right] = 0 ,$$

where $f_{0\mathrm{e}}(\boldsymbol{p})$ and $f_{0\mathrm{i}}(\boldsymbol{p})$ are the isotropic zero-order distribution functions in the intrinsic frame of the electrons and the ions.

We apply the expansion

$$\exp(\pm \mathrm{i}\boldsymbol{k} \cdot \boldsymbol{r}_{\mathrm{E}} \sin \omega_0 t) = \sum_{l=-\infty}^{\infty} \mathrm{e}^{\pm \mathrm{i} l \omega_0 t} \mathrm{J}_l(\boldsymbol{k} \cdot \boldsymbol{r}_{\mathrm{E}}) , \tag{7.4.9}$$

where $\boldsymbol{r}_{\mathrm{E}} = e\boldsymbol{E}_0 / m\omega_0^2$ is the amplitude of the oscillations of the electrons in the external SHF electric field. Note that (7.4.8) constitutes a system with periodic coefficients. Introducing the decomposition

$$(\Psi_e, \delta f_i) = e^{-i\omega t} \sum_{n=-\infty}^{\infty} e^{-in\omega_0 t} (\Psi_{en}, \delta f_{in}) , \tag{7.4.10}$$

we find from (7.4.8)

$$\begin{aligned} &-i(\omega + n\omega_0)\Psi_{en} + ik \cdot v\Psi_{en} - \Omega_e \frac{\partial \Psi_{en}}{\partial \phi} - ik \frac{\partial f_{0e}}{\partial v} \frac{4\pi e}{mk^2} \\ &\quad \times \left[e \int dp\, \Psi_{en} + e_i \sum_l J_{l-n}(k \cdot r_E) \int dp \delta f_{il} \right] = 0 , \\ &- i(\omega + n\omega_0)\delta f_{in} + ik \cdot v\delta f_{in} - \Omega_i \frac{\partial \delta f_{in}}{\partial \phi} \\ &\quad - ik \frac{\partial f_{0i}}{\partial v} \frac{4\pi e_i}{Mk^2} \left[e_i \int dp\, \delta f_{in} + e \sum_l J_{n-l}(k \cdot r_E) \int dp\, \Psi_{en} \right] = 0 . \end{aligned} \tag{7.4.11}$$

Using the notations

$$u_{en} = e \int dp\, \Psi_{en} , \quad u_{in} = e_i \int dp\, \delta f_{in} \tag{7.4.12}$$

we can write the formal solution of (7.4.11) in the form of coupled equations as

$$\begin{aligned} u_{en} &= - \delta\varepsilon_e(\omega + n\omega_0, k)\left[u_{en} + \sum_l J_{l-n}(k \cdot r_E)\, u_{il} \right] , \\ u_{in} &= - \delta\varepsilon_i(\omega + n\omega_0, k)\left[u_{in} + \sum_l J_{n-l}(k \cdot r_E)\, u_{el} \right] . \end{aligned} \tag{7.4.13}$$

Here $\delta\varepsilon_e(\omega, k)$ and $\delta\varepsilon_i(\omega, k)$ are the partial contributions of the electrons and the ions to the longitudinal dielectric permittivity.

7.4.1 The Dispersion Equation for Oscillations in the Plasma in a SHF Field

The condition for the existence of solutions of this infinite system of coupled equations is the dispersion equation for small longitudinal oscillations in the external SHF electric field. Since it is a determinant of infinite order, it is impossible to analyze it in general. However, for the most interesting limiting cases this determinant can be essentially simplified and the oscillation spectrum can be found. For example, all the values $\delta\varepsilon_{e,i}(\omega + n\omega_0, k)$ with $n \neq 0$ are small compared to 1 in the limit of very high frequencies, when the external field satisfies (7.4.2). Assuming that only u_{e0} and u_{i0} are nonzero in (7.4.13) we obtain

$$\begin{aligned} &[1 + \delta\varepsilon_e(\omega, k)]\, u_{e0} + \delta\varepsilon_e(\omega, k)\, J_0(k \cdot r_E)\, u_{i0} = 0 , \\ &\delta\varepsilon_i(\omega, k)\, J_0(k \cdot r_E)\, u_{e0} + [1 + \delta\varepsilon_i(\omega, k)]\, u_{i0} = 0 , \end{aligned} \tag{7.4.14}$$

which gives the dispersion equation

$$1 + \delta\varepsilon_e(\omega, \boldsymbol{k}) + \delta\varepsilon_i(\omega, \boldsymbol{k}) + [1 - J_0^2(\boldsymbol{k}\cdot\boldsymbol{r}_E)]\,\delta\varepsilon_e(\omega, \boldsymbol{k})\,\delta\varepsilon_i(\omega, \boldsymbol{k}) = 0 \ . \qquad (7.4.15)$$

Deriving (7.4.15) we fully neglected particle collisions. However, this equation is also valid for the collisional plasma if the oscillating velocity component of the particles is small compared to the thermal velocity. The external SHF field has no effect on the collision process, i.e., on the cross sections of scattering. When particle collisions are accounted for there appear right-hand sides in the system of equations (7.4.11), namely the linearized collision integrals of the electrons and the ions. These terms have an analogous structure to those valid in the absence of the SHF electric field. Consequently, the solutions of these equations still have the form (7.4.13). Particle collisions should be accounted for in the expressions for $\delta\varepsilon_e(\omega + n\omega_0, \boldsymbol{k})$ and $\delta\varepsilon_i(\omega + n\omega_0, \boldsymbol{k})$, however (Sects. 4.5, 6 and 5.5, 6).

We first analyze (7.4.15) for the isotropic and nondegenerate gaseous plasma when $\delta\varepsilon_e(\omega, \boldsymbol{k})$ and $\delta\varepsilon_i(\omega, \boldsymbol{k})$ are given by (Sect. 4.1)

$$\delta\varepsilon_\alpha(\omega, k) = \frac{\omega_{p\alpha}^2}{k^2 v_{T\alpha}^2}\left[1 - I_+\left(\frac{\omega}{k v_{T\alpha}}\right)\right] . \qquad (7.4.16)$$

7.4.2 High-Frequency Electro-Acoustic Oscillations

In the high-frequency limit $\omega \gg k v_{T\alpha}$ (the cold plasma limit) some exponentially small imaginary terms can be neglected and we obtain from (7.4.15)

$$1 - \frac{\omega_{pe}^2}{\omega^2} - \frac{\omega_{pi}^2}{\omega^2} - \frac{\omega_{pe}^2\,\omega_{pi}^2}{\omega^4}[1 - J_0(\boldsymbol{k}\cdot\boldsymbol{r}_E)] = 0 \ . \qquad (7.4.17)$$

For this case we find the oscillation spectra

$$\omega_1^2 = \omega_{pe}^2 + \omega_{pi}^2 J_0^2(\boldsymbol{k}\cdot\boldsymbol{r}_E) \ , \quad \omega_2^2 = \omega_{pi}^2[1 - J_0^2(\boldsymbol{k}\cdot\boldsymbol{r}_E)] \ . \qquad (7.4.18)$$

The first mode is the well-known high-frequency Langmuir mode, modified by the external SHF field. The second mode is a new one. It is analogous to the ion-acoustic mode. Physically, the mean energy of the oscillatory motion of the electrons in a high-frequency field acts like an electron temperature (Fig. 7.3). This becomes obvious in the long-wavelength limit $\boldsymbol{k}\cdot\boldsymbol{r}_E \ll 1$, where we have

$$\omega_2^2 = \omega_{pi}^2\frac{(\boldsymbol{k}\cdot\boldsymbol{r}_E)^2}{2} \equiv (\boldsymbol{k}\cdot\boldsymbol{W}_s)^2 \ , \qquad (7.4.19)$$

with $\boldsymbol{W}_s = (\omega_{pi}/\omega_0)(\boldsymbol{v}_E/\sqrt{2})$. Here $\boldsymbol{v}_E = \omega_0\boldsymbol{r}_E$ is the amplitude of the electron velocity oscillations in the external field.

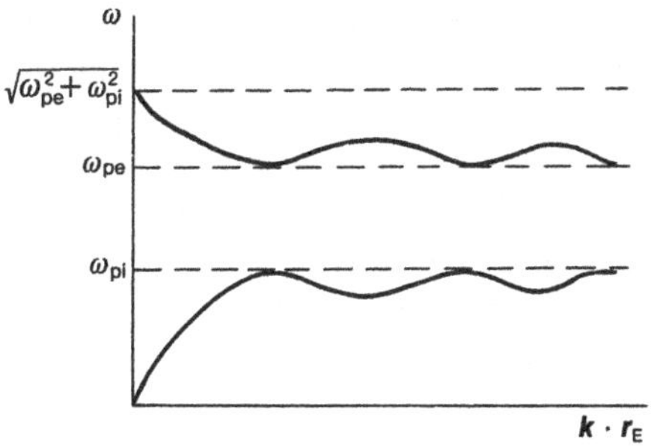

Fig. 7.3. Oscillation spectra of a cold plasma in an external SHF electric field

Interpreting W_s as the velocity of sound, one may speak of anisotropic electron acoustic oscillations of the plasma immersed in a SHF electric field. Note that these acoustic oscillations occur for sufficiently strong fields only. We must require $W_s \gg v_{Te}$ leading to the condition $v_E \gg (\omega_0/\omega_{pi})\, v_{Te} \gg v_{Te}\sqrt{M/m}$ for the electron velocity oscillations. In the short-wavelength limit $\boldsymbol{k}\cdot\boldsymbol{r}_E \gg 1$ the frequencies ω_2 and ω_1 are oscillating functions of k. Asymptotically ω_2 tends to ω_{pi} and ω_1 to ω_{pe} (Fig. 7.3).

7.4.3 Ion-Acoustic Oscillations of the Plasma in a SHF Field

Coming to the study of (7.4.15) in the frequency range $kv_{Ti} \ll \omega \ll kv_{Te}$ we want to note first that in the absence of the external high-frequency field the ion sound oscillations of the plasma exist in this range if the plasma is nonisothermal, $T_e \gg T_i$. In this frequency range the dispersion relation (7.4.15) accounting for a SHF field is written as

$$1 + \frac{\omega_{pe}^2}{k^2 v_{Te}^2}\left(1 + i\sqrt{\frac{\pi}{2}}\frac{\omega}{kv_{Te}}\right) - \frac{\omega_{pi}^2}{\omega^2} - \frac{\omega_{pi}^2}{\omega^2}\left[1 - J_0^2(\boldsymbol{k}\cdot\boldsymbol{r}_E)\right]$$
$$\times \frac{\omega_{pe}^2}{k^2 v_{Te}^2}\left(1 + i\sqrt{\frac{\pi}{2}}\frac{\omega}{kv_{Te}}\right) = 0\,. \tag{7.4.20}$$

Exponentially small ion absorption is neglected here. We obtain the following spectrum

$$\omega^2 = \omega_{pi}^2\left(1 - \frac{J_0^2(\boldsymbol{k}\cdot\boldsymbol{r}_E)}{1 + k^2 r_{De}^2}\right), \qquad \delta = -\sqrt{\frac{\pi}{8}}\frac{\omega_{pe}^2\omega_{pi}^2}{k^3 v_{Te}^3}\frac{J_0^2(\boldsymbol{k}\cdot\boldsymbol{r}_E)}{[1 + (kr_{De})^{-2}]^2}\,. \tag{7.4.21}$$

In the long-wavelength limit $\boldsymbol{k}\cdot\boldsymbol{r}_E \ll 1$ and $k^2 r_{De}^2 \ll 1$ it takes the simple form

$$\omega^2 = (\boldsymbol{k}\cdot\boldsymbol{W}_s)^2 + k^2 v_s^2\,, \qquad \delta = -\sqrt{\frac{\pi}{8}\frac{m}{M}}\,kv_s\,. \tag{7.4.22}$$

For $v_s > W_s$ the spectrum of the ion-acoustic oscillations is distorted slightly by the high-frequency field and, as in the absence of the field, this mode exists in the nonisothermal plasma only. For $W_s > v_s$ (or $v_E \gg v_{Te}$) a specific kind of sound appears in the plasma. In this limit the spectrum (7.4.22) represents the continuation of the spectrum (7.4.19) into the range of low phase velocities or high-frequency fields of small intensity. Note that oscillations can exist in this range both for the nonisothermal and the isothermal plasma.

7.4.4 Spectra of Oscillations of the Magneto-Active Plasma in a SHF Field

In order to study the effect of an external magnetic field on the oscillation spectra of the plasma in a SHF field, we have to analyze (7.4.15) with $\delta\varepsilon_e(\omega, \boldsymbol{k})$ and $\delta\varepsilon_i(\omega, \boldsymbol{k})$ given by (see Sect. 5.1)

$$\delta\varepsilon_\alpha(\omega, \boldsymbol{k}) = \frac{\omega_{p\alpha}^2}{k^2 v_{T\alpha}^2}\left[1 - \sum_n \frac{\omega}{\omega - n\Omega_\alpha} A_n\left(\frac{k_\perp^2 v_{T\alpha}^2}{\Omega_\alpha^2}\right) I_+\left(\frac{\omega - n\Omega_\alpha}{k_z v_\alpha}\right)\right]. \quad (7.4.23)$$

For $k_\perp = 0$ and $\omega \gg \Omega_\alpha$ (7.4.23) coincides with (7.4.16) since the former becomes independent of the magnetic field under these conditions. Since the oscillation spectra are the same we have to consider oscillations with $k_\perp \neq 0$ only. We begin the analysis of (7.4.15) in the cold plasma limit $k_\perp v_{T\alpha} \ll \Omega_\alpha$, $|\omega - \Omega_\alpha| \gg k_z v_{T\alpha}$ where it is of the form

$$1 - \frac{\omega_{pe}^2 + \omega_{pi}^2}{\omega^2}\frac{k_z^2}{k^2} - \frac{\omega_{pe}^2\omega_{pi}^2}{\omega^4}\frac{k_\perp^2}{k^2}\left(\frac{\omega_{pe}^2}{\omega^2 - \Omega_e^2} + \frac{\omega_{pi}^2}{\omega^2 - \Omega_i^2}\right)[1 - \mathrm{J}_0^2(\boldsymbol{k}\cdot\boldsymbol{r}_E)]$$
$$\times\left(\frac{k_z^2}{k^2} + \frac{k_\perp^2}{k^2}\frac{\omega^2}{\omega^2 - \Omega_e^2}\right)\left(\frac{k_z^2}{k^2} + \frac{k_\perp^2}{k^2}\frac{\omega^2}{\omega^2 - \Omega_i^2}\right) = 0\,. \quad (7.4.24)$$

In the frequency range $\omega \ll \Omega_i$ it is reduced to a quadratic equation in the unknown ω^2. Its roots are approximately equal to

$$\omega_1^2 = \frac{\omega_{pe}^2 k_z^2}{k^2}\left\{1 + \frac{\omega_{pi}^2}{\Omega_i^2}\frac{k_\perp^2}{k^2}[1 - \mathrm{J}_0^2(\boldsymbol{k}\cdot\boldsymbol{r}_E)]\right\}$$
$$\times\left(1 + \frac{\omega_{pi}^2}{\Omega_i^2}\frac{k_\perp^2}{k^2}\left\{1 + \frac{\omega_{pe}^2}{\Omega_e^2}\frac{k_\perp^2}{k^2}[1 - \mathrm{J}_0^2(\boldsymbol{k}\cdot\boldsymbol{r}_E)]\right\}\right)^{-1}, \quad (7.4.25)$$

$$\omega_2^2 = \frac{k_z^2}{k^2}\omega_{pi}^2[1 - \mathrm{J}_0^2(\boldsymbol{k}\cdot\boldsymbol{r}_E)]\left\{1 + \frac{\omega_{pi}^2 k_\perp^2}{\Omega_i^2 k^2}[1 - \mathrm{J}_0^2(\boldsymbol{k}\cdot\boldsymbol{r}_E)]\right\}^{-1}.$$

With decreasing high-frequency electric field, the first mode goes over to the low-frequency mode of the cold magneto-active plasma (Sect. 4.2). The second mode, however, is a new one. Its frequency ω_2 tends to zero in the absence of the high-frequency field, i.e., the oscillations disappear. Physically, these oscillations are analogous to the acoustic oscillations with a high frequency oscillatory motion of the electrons, see (7.4.18, 19). As in the case of the isotropic plasma, they occur when the velocity v_E is sufficiently high $v_E > v_{Te}\omega_0/\omega_{pi} \gg v_{Te}\sqrt{M/m}$.

In the range of intermediate phase velocities $v_{Ti} \ll \omega/k_z \ll v_{Te}$ and intermediate frequencies $\Omega_i < \omega < \Omega_e$, i.e., in the ion-acoustic range, we obtain from (7.4.15)

$$1 + \frac{\omega_{pe}^2}{k^2 v_{Te}^2}\left(1 + \mathrm{i}\sqrt{\frac{\pi}{2}}\frac{\omega}{|k_z|\, v_{Te}}\right) - \frac{\omega_{pi}^2}{\omega^2} - \frac{\omega_{pi}^2\omega_{pe}^2}{\omega^2 k^2 v_{Te}^2}\,[1 - \mathrm{J}_0^2(\boldsymbol{k}\cdot\boldsymbol{r}_E)]$$

$$\times\left(1 + \mathrm{i}\sqrt{\frac{\pi}{2}}\frac{\omega}{|k_z|\, v_{Te}}\right) = 0 \qquad (7.4.26)$$

for the magneto-active plasma in the high-frequency electric field, if we ignore the ion thermal motion for simplicity. Taking into account that the Cherenkov dissipation by the electrons is small, we get the following oscillation spectrum ($\omega \to \omega + \mathrm{i}\delta$)

$$\omega^2 = \omega_{pi}^2\left(1 - \frac{\mathrm{J}_0^2(\boldsymbol{k}\cdot\boldsymbol{r}_E)}{1 + k^2 r_{De}^2}\right), \qquad \delta = -\sqrt{\frac{\pi}{8}}\frac{\omega_{pe}^2\omega_{pi}^2}{k^2|k_z|v_{Te}^3}\,\frac{\mathrm{J}_0^2(\boldsymbol{k}\cdot\boldsymbol{r}_E)}{[1 + (kr_{De})^{-2}]^2}\,. \qquad (7.4.27)$$

Since it is completely analogous to the spectrum (7.4.21) the analysis given there is also valid here. A small difference in the Cherenkov terms leads to an increase of the damping decrement in the magnetized plasma, however.

7.4.5 Oscillations of the Degenerate Plasma in a SHF Field

Finally we study the longitudinal oscillations of the degenerate plasma in the SHF electric field. Only the case of the nonmagnetized plasma is considered. In the high-frequency limit $\omega \gg kv_{Fe,i}$ the degree of degeneracy is insignificant if the thermal motion is neglected, and (7.4.17–19) are valid. In the intermediate frequency range $kv_{Fi} \ll \omega \ll kv_{Fe}$ we can use (4.1.18) and we obtain from (7.4.15)

$$1 + \frac{3\,\omega_{pe}^2}{k^2 v_{Fe}^2}\left(1 + \mathrm{i}\frac{\pi}{2}\frac{\omega}{kv_{Fe}}\right) - \frac{\omega_{pi}^2}{\omega^2} - \frac{3\,\omega_{pe}^2\omega_{pi}^2}{\omega^2 k^2 v_{Fe}^2}$$

$$\times\,[1 - \mathrm{J}_0^2(\boldsymbol{k}\cdot\boldsymbol{r}_E)]\left(1 + \mathrm{i}\frac{\pi}{2}\frac{\omega}{kv_{Te}}\right) = 0\,. \qquad (7.4.28)$$

This dispersion relation is analogous to (7.4.20) and yields an oscillation spectrum similar to (7.4.21):

$$\begin{aligned} \omega^2 &= \omega_{pi}^2\left(1-\frac{J_0^2(\boldsymbol{k}\cdot\boldsymbol{r}_E)}{1+k^2v_{Fe}^2/3\,\omega_{pe}^2}\right), \\ \delta &= -\frac{3\pi}{4}\,\frac{\omega_{pe}^2\,\omega_{pi}^2}{k^3v_{Fe}^3}\cdot\frac{J_0^2(\boldsymbol{k}\cdot\boldsymbol{r}_E)}{[1+(3\,\omega_{pe}^2/k^2v_{Fe}^2)^2]^2}\,. \end{aligned} \tag{7.4.29}$$

We see that the kind of acoustic waves caused by a high-frequency electric field can exist both in the degenerate and the nondegenerate plasma. Thus, in the limit $k^2v_{Fe}^2 \ll \omega_{pe}^2$ and $\boldsymbol{k}\cdot\boldsymbol{r}_E \ll 1$ we have, cf. (7.4.22),

$$\omega^2 = k^2v_s^2 + (\boldsymbol{k}\cdot\boldsymbol{W}_s)^2\,, \quad \delta = -\frac{\pi}{4}\sqrt{\frac{m}{3M}}\,kv_s\,, \tag{7.4.30}$$

where $v_s^2 = mv_{Fe}^2/3\,M$ and $\boldsymbol{W}_s = \boldsymbol{v}_E\omega_{pi}/(\sqrt{2}\,\omega_0)$.
However, for the real occurrence of this kind of sound $v_E \gg v_{Fe}$ is a necessary condition.

7.5 Parametric Interaction of SHF Electric Fields with a Plasma

We have shown in the preceding section that there appear a number of new oscillation modes in the plasma if it is exposed to a SHF electric field of very high or high frequency. The modes of the plasma not exposed to a high-frequency field also exist and their spectra are distorted when the frequency of the field significantly exceeds all the characteristic frequencies of the plasma. It is essential that these oscillations are stable, i.e., that their amplitudes do not grow with time. When the frequency of the external field approaches the characteristic oscillation frequencies of the plasma we have a quite different situation. In this case the high-frequency field and the plasma strongly interact by parametric resonance, and even in a weak field growing oscillations can appear in the plasma.

To investigate these parametric instabilities we consider the system of equations (7.4.13), assuming now that the frequency of the external field ω_0 is of the order of the natural electron frequencies of the plasma and thus much greater than the ion frequencies. Especially for the nonmagnetized plasma we have $\omega_0 \approx \omega_{pe}$ and therefore $\omega_0 \gg \omega_{pi}$. The terms $\delta\varepsilon_i(\omega + n\omega_0)$ of (7.4.13) are small for all $n \neq 0$ in this case which allows us to neglect all quantities except u_{i0} and to write the dispersion relation of this system as

$$1 = \frac{\delta\varepsilon_i(\omega,\boldsymbol{k})}{1+\delta\varepsilon_i(\omega,\boldsymbol{k})}\sum_n J_n^2(\boldsymbol{k}\cdot\boldsymbol{r}_E)\frac{\delta\varepsilon_e(\omega+n\omega_0,\boldsymbol{k})}{1+\delta\varepsilon_e(\omega+n\omega_0,\boldsymbol{k})}\,. \tag{7.5.1}$$

Comparing with (7.4.15) valid for $\omega_0 \gg \omega_{pe}$, we see that both relations coincide if it is justified to confine oneself to the contribution of the term $n = 0$.

7.5.1 Resonant Parametric Excitation of the High-Frequency Longitudinal Oscillations of the Plasma by SHF Fields

We analyze (7.5.1) in the cold plasma limit when $\omega \gg kv_{\mathrm{Ti}}$, $|\omega \pm n\omega_0| \gg kv_{\mathrm{Te}}$ holds, assuming that the k-dependence of the partial permittivities $\delta\varepsilon_{\mathrm{e}}(\omega + n\omega_0, \boldsymbol{k})$ and $\delta\varepsilon_{\mathrm{i}}(\omega, \boldsymbol{k})$ can be neglected. For simplicity we further assume a collisionless plasma and nonmagnetized ions. These assumptions allow to write (7.5.1) in the form

$$\frac{\omega^2}{\omega_{\mathrm{pi}}^2} = 1 - \sum_n \mathrm{J}_n^2(\boldsymbol{k}\cdot\boldsymbol{r}_{\mathrm{E}})\frac{\delta\varepsilon_{\mathrm{e}}(\omega + n\omega_0)}{1 + \delta\varepsilon_{\mathrm{e}}(\omega + n\omega_0)} \,. \tag{7.5.2}$$

Evidently, unstable hydrodynamic modes with $\omega^2 < 0$ can occur in the frequency range $\omega_0 \gg |\omega|$ only. The frequency must be a multiple $n \neq 0$ of the frequency ω_0 and it must fulfil

$$|1 + \delta\varepsilon_{\mathrm{e}}(n\omega_0)| \ll 1 \,. \tag{7.5.3}$$

This implies that a strong instability can develop if the frequency of the external SHF field or its harmonics are close to the natural frequencies of the longitudinal electron oscillations of the plasma. Replacing ω by zero in the right-hand side of (7.5.2) we obtain in the first-order approximation

$$\omega^2 = \frac{2\mathrm{J}_n^2(\boldsymbol{k}\cdot\boldsymbol{r}_{\mathrm{E}})\,\omega_{\mathrm{pi}}^2}{1 + \delta\varepsilon_{\mathrm{e}}(n\omega_0)} \,. \tag{7.5.4}$$

When the external magnetic field is weak, for $\omega_0 \gg \Omega_{\mathrm{e}}$, we have

$$\omega^2 = 2\frac{m}{M}\omega_{\mathrm{pe}}^2\mathrm{J}_n^2(\boldsymbol{k}\cdot\boldsymbol{r}_{\mathrm{E}})\left(1 - \frac{\omega_{\mathrm{pe}}^2}{n^2\omega_0^2}\right)^{-1} . \tag{7.5.5}$$

We see that the plasma is unstable ($\omega^2 < 0$) when it is opaque with respect to the harmonics of the SHF field

$$n^2\omega_0^2 \lesssim \omega_{\mathrm{pe}}^2 \,. \tag{7.5.6}$$

The increment of the instability following from (7.5.5) becomes infinite when (7.5.6) is satisfied. This indicates that one cannot assume $\omega = 0$ in the right-hand side of (7.5.2). We have to be more careful and to keep terms of the order $\omega^2/(n^2\omega_0^2)$. Doing so it follows from (7.5.2)

$$\frac{\omega^2}{\omega_{\mathrm{pi}}^2} = 1 - \sum_n \mathrm{J}_n^2(\boldsymbol{k}\cdot\boldsymbol{r}_{\mathrm{E}})\frac{\omega_{\mathrm{pe}}^2}{\omega_{\mathrm{pe}}^2 - (n\omega_0 + \omega)^2} \,. \tag{7.5.7}$$

When the conditions

$$\omega^2 \gg \omega_{pi}^2\,, \qquad 1 \gg |\Delta_n| \equiv \left|\frac{\omega_{pe}^2}{n^2\omega_0^2} - 1\right| \gg \sqrt{\frac{m}{M}} \tag{7.5.8}$$

are given, this equation has the form

$$4\frac{\omega^4}{\omega_{pe}^4} - \frac{\omega^2}{\omega_{pe}^2}\Delta_n^2 - 2\,\mathrm{J}_n^2(\boldsymbol{k}\cdot\boldsymbol{r}_\mathrm{E})\,\frac{m}{M}\Delta_n = 0\,. \tag{7.5.9}$$

The quantity Δ_n is called the *frequency shift* of the resonant parametric interaction of the SHF field with the plasma. According to (7.5.9), the growth rate (increment) of the oscillations is a function of Δ_n. For large frequency shifts

$$\Delta_n^2 \gg \frac{\omega^2}{\omega_{pe}^2} = -2\,\frac{m}{M}\,\frac{\mathrm{J}_n^2(\boldsymbol{k}\cdot\boldsymbol{r}_\mathrm{E})}{\Delta_n} \tag{7.5.10}$$

the solution (7.5.5) is valid, giving for the order of the increment $\mathrm{Im}\{\omega\} \sim \omega_{pe}(m/M)^{1/2}$. For smaller frequency shifts the increment grows until, under the condition

$$\frac{\omega^2}{\omega_{pe}^2}\Delta_n + \frac{m}{M}\mathrm{J}_n^2(\boldsymbol{k}\cdot\boldsymbol{r}_\mathrm{E}) = 0 \tag{7.5.11}$$

it becomes maximum:

$$\omega_{max}^2 = -\,\omega_{pe}\left[\frac{m}{2M}\,\mathrm{J}_n^2(\boldsymbol{k}\cdot\boldsymbol{r}_\mathrm{E})\right]^{2/3}. \tag{7.5.12}$$

Thus, the maximum increment $\mathrm{Im}\{\omega_{max}\} \sim (m/M)^{1/3}\,\omega_{pe}$ also falls into the range of opacity, when $\Delta_n > 0$ (i.e., $n\omega_0 < \omega_{pe}$). The distance from resonance in the maximum is

$$(\Delta_n)_{min} \approx \left[4\frac{m}{M}\mathrm{J}_n^2(\boldsymbol{k}\cdot\boldsymbol{r}_\mathrm{E})\right]^{1/3}. \tag{7.5.13}$$

It is not true, however, that an instability could occur in the range of opacity with respect to the SHF field only; the plasma can be unstable in the range of transparency $n\omega_0 > \omega_{pe}$, too. There the instability becomes kinetic and has a small increment, only (Exercise 7.7.9). We show in Fig. 7.4 the increment of the hydrodynamic parametric instability in the nonmagnetized plasma as a function of the ratio of ω_0/ω_{pe}.

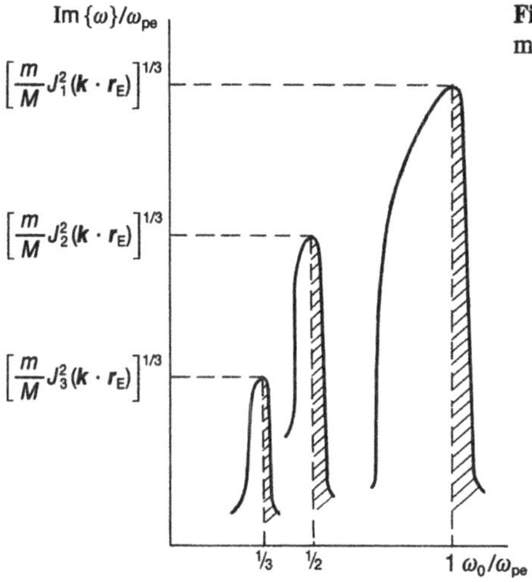

Fig. 7.4. Increment of the hydrodynamic parametric instability in a plasma versus ω_0/ω_{pe}

7.5.2 The Effect of a Magnetic Field on the Development of the Parametric Instability of the Plasma in a SHF Field

An external magnetic field parallel to the SHF electric field does not affect this instability as long as the field is weak and if $\omega_0 > \Omega_e$ holds. A qualitatively new character of the parametric interaction of the external SHF field with the plasma arises in strong fields for $\Omega_e > \omega_0$. These new modes of interaction are associated with the two branches of longitudinal electron oscillations in the magneto-active plasma which are called the upper and the lower hybrid oscillations and which are defined by (Sect. 5.2)

$$\omega_{1,2}^2 = \tfrac{1}{2}(\omega_{pe}^2 + \Omega_e^2) \pm \tfrac{1}{2}\sqrt{(\omega_{pe}^2 + \Omega_e^2)^2 - 4\,\omega_{pe}^2\Omega_e^2\cos^2\theta}\,, \tag{7.5.14}$$

where θ is the angle between the wave vector and the magnetic field. The magneto-active plasma becomes parametrically unstable when the harmonic frequencies of the SHF field are approximately equal to one of the natural frequencies (7.5.14), i.e., for

$$n^2\omega_0^2 \approx \omega_\alpha^2\,; \quad \alpha = 1,2\,. \tag{7.5.15}$$

The parametric instabilities of the magneto-active plasma are analyzed in the same way as in the absence of the magnetic field. Introducing the frequency difference to the resonance frequencies (7.5.15)

$$\Delta_{n\alpha} = \cos^2\theta\frac{\omega_{pe}^2}{n^2\omega_0^2 - \Omega_e^2} + \sin^2\theta\frac{\omega_{pe}^2}{n^2\omega_0^2} - 1 \tag{7.5.16}$$

we can write (7.5.12) under the conditions (7.5.8) as

$$A_a^2 \frac{\omega^4}{\omega_{pe}^4} - \frac{\omega^2}{\omega_{pe}^2} - 2\frac{m}{M} \Delta_{na} J_n^2 (\boldsymbol{k} \cdot \boldsymbol{r}_E) = 0 \, , \quad \text{where} \tag{7.5.17}$$

$$A_a = -\frac{\partial \Delta_{na}}{\partial n\omega_0} = -\frac{\partial \delta\varepsilon_e (n\omega_0)}{\partial n\omega_0} \, . \tag{7.5.18}$$

Equation (7.5.17) has the same form as (7.5.9) and for large frequency shifts, when

$$\frac{\Delta_{na}^2}{A_a^2} \gg \frac{\omega^2}{\omega_{pe}^2} = -2 \frac{m}{M} \frac{J_n^2 (\boldsymbol{k} \cdot \boldsymbol{r}_E)}{\Delta_n} \, , \tag{7.5.19}$$

it also gives an increment $\mathrm{Im}\{\omega\} \sim \omega_{pe} \sqrt{m/M}$. With decreasing frequency shift the increment increases and, under the condition

$$(\Delta_{na})_{\min} \approx \left[\frac{m}{M} A_a^2 J_n^2 (\boldsymbol{k} \cdot \boldsymbol{r}_E) \right]^{1/3} \tag{7.5.20}$$

it becomes maximum

$$\omega_{\max}^2 = -\omega_{pe}^2 \left[\frac{1}{|A_a|} \frac{m}{M} J_n^2 (\boldsymbol{k} \cdot \boldsymbol{r}_E) \right]^{2/3} \, . \tag{7.5.21}$$

The order of the maximum increment is $\mathrm{Im}\{\omega_{\max}\} \sim \omega_{pe} (m/M)^{1/3}$.

In the limit of a weak magnetic field, $\boldsymbol{B}_0 \to 0$, the two natural frequencies ω_a coalesce into the one frequency ω_{pe}. We get $A_a \to 2$, and (7.5.15–21) pass over into (7.5.6–12). The parametric interaction of the strong SHF field with the plasma is not modified much in the case of a finite magnetic field, for oscillations propagating strictly along the magnetic field. Finally, in the limit of a very strong magnetic field, $\boldsymbol{B}_0 \to \infty$, (7.5.15–21) become identical with (7.5.6–12) by the simple substitution $\omega_{pe} \to \omega_{pe} \cos\theta$. The interaction between the SHF field and the plasma has the same character as in the case without magnetic field.

Note that, physically, the parametric instability of the plasma in a SHF field is analogous to the aperiodic hydrodynamic instability of the plasma in a constant electric field occurring when the drift velocity of the electrons exceeds their thermal velocity (Sect. 7.2). As in the latter case the parametric instability is caused by the relative motion of the electrons with respect to the ions. For the parametric instability the motion is oscillatory, however, which leads to the specific dependence of the increments on the frequency of the SHF field. The instability develops when the harmonics of the frequency are close to the natural frequencies of the longitudinal electron oscillations of the plasma. Both the hydrodynamic instability of the current-carrying plasma and the parametric instability can exist only in strong SHF fields. The drift velocity of the electrons must exceed the thermal velocity in both cases.

However, external SHF fields of low intensity can also excite plasma oscillations parametrically. The oscillating velocity of the electrons is much smaller than the thermal velocity in this case. To investigate these instabilities we consider (7.5.1) in the limit of low intensity of the SHF field, $\boldsymbol{k}\cdot\boldsymbol{r}_E \ll 1$. It is sufficient to analyze the three first terms with $n = 0, \pm 1$ of the expansion in Bessel functions, only. As a result (7.5.1) takes the form

$$\frac{1+\delta\varepsilon_i(\omega,\boldsymbol{k})+\delta\varepsilon_e(\omega,\boldsymbol{k})}{\delta\varepsilon_i(\omega,\boldsymbol{k})[1+\delta\varepsilon_e(\omega,\boldsymbol{k})]} + \frac{(\boldsymbol{k}\cdot\boldsymbol{r}_E)^2}{4}\left[\frac{1}{\varepsilon(\omega+\omega_0,\boldsymbol{k})}+\frac{1}{\varepsilon(\omega-\omega_0,\boldsymbol{k})}\right] = 0\,, \tag{7.5.22}$$

if the condition $\omega_{pi} \ll \omega_0$ is given. When ω_0 is close to the natural frequencies of the longitudinal electron modes we have $|\varepsilon(\omega \pm \omega_0)| \ll 1$.

To begin with, we analyze (7.5.22) for the collisionless nonmagnetized plasma in an external SHF field and in particular consider the low-frequency oscillations $\omega \ll kv_{Ti}$. Assuming that the wavelength is much longer than the Debye length of the electrons, we obtain

$$\frac{k^2v_{Ti}^2}{\omega_{pi}^2}\left(1-i\sqrt{\frac{\pi}{2}}\frac{\omega}{kv_{Ti}}\right)+\frac{k^2v_{Te}^2}{\omega_{pe}^2}\left(1-i\sqrt{\frac{\pi}{2}}\frac{\omega}{kv_{Te}}\right) + \frac{\boldsymbol{k}\cdot\boldsymbol{r}_E}{2}\left(\frac{\omega_0^2(\omega_0^2-\omega_{pe}^2)}{(\omega_0^2-\omega_{pe}^2)^2-4\,\omega_0^2\omega^2}\right) = 0\,. \tag{7.5.23}$$

Since the imaginary terms describing the Cherenkov dissipation are small, we get the simplification

$$\omega_0^2\Delta^2 - 4\,\omega^2 - \eta\omega_0^2\Delta\left(1+i\sqrt{\frac{\pi}{2}}\frac{\omega}{kv_{Ti}}\frac{T_i}{T_e+T_i}\right) = 0\,. \tag{7.5.24}$$

Here $\eta = E_0^2\cos^2\theta\,[8\pi N(T_e+T_i)]^{-1}$ characterizes the ratio of the energy density of the SHF field causing the instability to the internal energy of the plasma; θ is the angle between the direction of the external field and the wave vector; $\Delta = (\omega_{pe}^2/\omega_0^2) - 1$ is the frequency difference of the SHF field. The roots of this equation in the low-frequency range are approximately

$$\omega = \begin{cases} i\dfrac{\eta-\Delta}{\eta}kv_{Ti}\sqrt{\dfrac{2}{\pi}}\dfrac{T_e+T_i}{T_i} & \text{for } \dfrac{\omega^2}{\omega_0^2} \ll \dfrac{\Delta^2-\eta\Delta}{4}\,, \\[2ex] \pm\sqrt{\dfrac{\Delta^2-\eta\Delta}{4}}\,\omega_0 & \text{for } \dfrac{\omega^2}{\omega_0^2} \approx \dfrac{\Delta^2-\eta\Delta}{4}\,. \end{cases} \tag{7.5.25}$$

We see that these oscillations can arise for $\Delta > 0$, only, consequently the domain of existence is the frequency range $\omega_0^2 < \omega_{pe}^2$ where the plasma is opaque with respect to the high-frequency field. The oscillations are aperiodically unstable for $\eta \geqslant \Delta$ and the smaller the frequency difference Δ is, the lower is the minimum necessary value of η. The critical high-frequency field causing the parametric instability can be of small intensity, then. The applicability condition of the collisionless approximation, the condition justifying neglecting collisions in the expressions for $\delta\varepsilon(\omega \pm \omega_0)$, requires that $|\Delta| > \nu_e/\omega_0$, where ν_e is the frequency of the electron-ion or the electron-neutral collisions. Consequently, the condition

$$\eta = \frac{E_0^2 \cos^2\theta}{8\pi N(T_e + T_i)} > 4\frac{\nu_e}{\omega_0} \ll 1 \tag{7.5.26}$$

is necessary for the development of an aperiodic instability. The numerical factor 4 follows from a more exact account of particle collisions (Exercise 7.7.10). The ratio of the velocity of the high-frequency electron oscillations to their themal velocity is a small value, $v_E^2/v_{Te}^2 \sim \eta \ll 1$, in this case.

7.5.3 The Ion-Acoustic Parametric Instability of the Nonisothermal Plasma

For even lower intensities of the SHF fields the parametric instability appears in nonisothermal plasmas with $T_e \gg T_i$, in which the ion-acoustic oscillations can exist. They are excited by the high-frequency electric field in the ion-acoustic frequency range $kv_{Ti} \ll \omega \ll kv_{Te}$ for $k^2v_{Te}^2 \ll \omega_{pe}^2$ where (7.5.22) takes the form

$$-\frac{\omega^2}{\omega_{pi}^2} + \frac{k^2v_{Te}^2}{\omega_{pe}^2}\left(1 - \mathrm{i}\sqrt{\frac{\pi}{2}}\frac{\omega}{kv_{Te}}\right) - \frac{(\boldsymbol{k}\cdot\boldsymbol{r}_E)}{2}\frac{\Delta}{\Delta^2 - 4\omega^2/\omega_0^2} = 0\,. \tag{7.5.27}$$

It is essential that the ion-acoustic oscillations can be excited parametrically in the range of plasma transparency with respect to the SHF field $\omega_0^2 > \omega_{pe}^2$ where $\Delta < 0$. Actually, under the condition of resonance

$$\omega^2 = k^2v_s^2 = (\omega_0 - \omega_{pe})^2\,, \quad \text{or} \tag{7.5.28}$$

$$\omega_0 = \omega_{pe} + kv_s\,, \tag{7.5.28 a}$$

i.e., when the frequency of the SHF field is equal to the sum of the electron Langmuir frequency and the ion-acoustic frequency, we obtain from (7.5.27) the increment of the parametric instability[2] $(\omega \to \omega + \mathrm{i}\delta)$

---

2 The parametric instability can be interpreted as a decay of a nonlinear wave in the nonisothermal plasma into the electron Langmuir and ion-acoustic waves since this instability is accompanied by an increase of the amplitudes of these waves in the plasma (Chap. 12).

$$\delta = \frac{\omega_0^2 - \omega_{pe}^2}{16\,\omega^2}\sqrt{\frac{2}{\pi}}\frac{(\boldsymbol{k}\cdot\boldsymbol{r}_E)^2}{k^2 r_{De}^2} k v_{Te}\,. \tag{7.5.29}$$

The range of existence of this instability evidently is the range of plasma transparency $\omega_0^2 > \omega_{pe}^2$. Here, from the validity condition for neglecting the electron collisions in the dielectric permittivity $\varepsilon(\omega \pm \omega_0)$, it follows that $\mathrm{Im}\{\omega\} = \delta > \nu_e$, or

$$\eta = \frac{E_0^2 \cos^2\theta}{8\pi N T_e} > \sqrt{8\pi \frac{m}{M}\frac{\nu_e}{\omega_0}}\,. \tag{7.5.30}$$

It is easy to see that the threshold defined by (7.5.30) is much lower than the threshold (7.5.26).

7.5.4 The Effect of the Magnetic Field on the Development of the Low-Frequency Parametric Instabilities

Continuing our analysis we study the effect of an external aligned magnetic field on the parametric instability of the plasma in a weak SHF field. First of all note that the formulas (7.5.23–30) derived for wave propagation along the magnetic field, i.e., for $\theta = 0$, are still valid for the magneto-active plasma. When the waves are propagating at the angle θ with respect to the magnetic field there appears, as in the case of the aperiodic parametric instability in strong SHF fields, the possibility of parametric interaction of the wave field with the plasma at the upper and lower electron hybrid frequencies, cf. (7.5.15):

$$\omega_0^2 \approx \omega_\alpha^2\,. \tag{7.5.31}$$

The frequencies ω_α, $\alpha = 1{,}2$, are defined by (7.5.14).

We shall not give here a thorough analysis of the parametric instabilities of the magneto-active plasma in a weak SHF field (Exercise 7.7.11). We only note that both the aperiodic and the ion-acoustic parametric instabilities can be excited at both branches of the hybrid electron oscillations. Their increments and excitation thresholds are by an order of magnitude the same as in the absence of the magnetic field.

7.5.5 The Case of the Degenerate Plasma

Concluding this section we shall briefly discuss the existence condition of the parametric instability in the degenerate solid-state plasma. For simplicity the plasma electrons are considered to be degenerate and the ions to be non-degenerate. Since it is difficult to get a velocity of the electron oscillations higher than the Fermi velocity, we confine our interest to weak SHF fields yielding small oscillation velocities. In other words, we analyze (7.5.22) only for the degenerate nonmagnetized plasma. Since $\mathscr{E}_{Fe} \gg T_e \gtrsim T_i$ in the plasma with degenerate electrons, the low-frequency ion-acoustic oscillations can be excited here. The parametric instability occurring in the frequency range $kv_{Ti} \ll \omega \ll kv_{Fe}$ has the lowest threshold. If, moreover, $k^2 v_{Fe}^2 \ll \omega_{pe}^2$ we obtain from (7.5.22) in the resonance case $\omega_0 \approx \omega_{pe} \gg \omega$

$$-\frac{\omega^2}{\omega_{pi}^2} + \frac{k^2 v_{Fe}^2}{3\,\omega_{pe}^2}\left(1 - \mathrm{i}\frac{\pi}{2}\frac{\omega}{kv_{Fe}}\right) - \frac{(\boldsymbol{k}\cdot\boldsymbol{r}_E)^2}{2}\frac{\Delta}{\Delta^2 - 4\,\omega^2/\omega_0^2} = 0\,, \qquad (7.5.32)$$

where $\Delta = (\omega_{pe}^2/\omega_0^2) - 1$. Deriving this equation we assumed that $\Delta \gg \nu_e/\omega_0$ and $\mathrm{Im}\,\{\omega\} \gg \nu_e$.

Equation (7.5.32) has the same form as (7.5.27). Therefore, introducing the notation $v_s^2 = 3\,v_{Fe}^2\,m/M$ we have the same resonance conditions as for the nondegenerate plasma, i.e., (7.5.28). The increment of the parametric instability, cf. (7.5.22),

$$\delta = \frac{\omega_0^2 - \omega_{pe}^2}{24\,\pi\omega^2}\frac{(\boldsymbol{k}\cdot\boldsymbol{r}_E)^2}{k^2 r_{De}^2} kv_{Fe} \qquad (7.5.33)$$

and its threshold, cf. (7.5.30),

$$\eta = \frac{E_0^2\cos^2\theta}{8\,\pi N\mathscr{E}_{Fe}} > \frac{3}{4}\frac{\nu_e}{\omega_0} \qquad (7.5.34)$$

are essentially the same. Here ν_e is the effective frequency of the electron-electron collisions (the inverse of the momentum relaxation time) in the degenerate plasma.

7.6 Plasma Parametric Instability with Respect to Nonpotential Perturbations

We have dealt above with the plasma parametric instability with respect to perturbations derivable from a scalar potential, $\boldsymbol{E} = -\nabla\Phi$. As the dispersion equation (7.5.1) for these perturbations shows, the external electric field becomes effective only when the angle between the wave vector and the field $\boldsymbol{E}_0$ differs from $\pi/2$. If this angle equals $\pi/2$, the field $\boldsymbol{E}_0$ does not appear in (7.5.1) and the plasma is stable with respect to longitudinal perturbations. However, this assumption is no longer valid for oscillations not derivable from a potential. Moreover, the plasma in an external high-frequency field is anisotropic in the absence of a magnetic field. The oscillations of such a plasma cannot be strictly longitudinal. Nevertheless, the deviation of (7.5.1) for $\boldsymbol{k}\cdot\boldsymbol{r}_{\rm E}\neq 0$ from the exact dispersion equation for arbitrary oscillations is very small. This means that these oscillations are longitudinal to a high degree of accuracy. Contributions not derivable from a potential are essential for $\boldsymbol{k}\cdot\boldsymbol{r}_{\rm E}=0$, only, i.e., for waves which propagate strictly across the field $\boldsymbol{E}_0$.

To analyze the plasma stability with respect to perturbations possessing transverse components, the kinetic equations for the electrons and the ions, cf. (7.4.5),

$$\begin{aligned}&\frac{\partial\delta f_{\rm e}}{\partial t}+{\rm i}\boldsymbol{k}\cdot\boldsymbol{v}\delta f_{\rm e}+\frac{e\boldsymbol{E}_0}{m}\sin\omega_0 t\frac{\partial\delta f_{\rm e}}{\partial v}+\frac{e}{m}\left\{\boldsymbol{E}+\frac{1}{c}[\boldsymbol{v},\boldsymbol{B}]\right\}\frac{\partial f_{0\rm e}(\boldsymbol{p}-\boldsymbol{p}_0)}{\partial\boldsymbol{v}}=0\\&\frac{\partial\delta f_{\rm i}}{\partial t}+{\rm i}\boldsymbol{k}\cdot\boldsymbol{v}\delta f_{\rm i}+\frac{e_{\rm i}}{M}\left\{\boldsymbol{E}+\frac{1}{c}[\boldsymbol{v},\boldsymbol{B}]\right\}\frac{\partial f_{0\rm i}(\boldsymbol{p})}{\partial\boldsymbol{v}}=0\end{aligned}\tag{7.6.1}$$

must be supplemented by the complete system of the Maxwell equations

$$\begin{aligned}&{\rm curl}\,\boldsymbol{E}=-\frac{1}{c}\frac{\partial\boldsymbol{B}}{\partial t}\,,\qquad {\rm curl}\,\boldsymbol{B}=\frac{1}{c}\frac{\partial\boldsymbol{E}}{\partial t}+\frac{4\pi}{c}\sum_{\alpha={\rm e,i}}e_\alpha\int d\boldsymbol{p}\,\boldsymbol{v}\delta f_\alpha\,,\\&{\rm div}\,\boldsymbol{B}=0\,,\qquad {\rm div}\,\boldsymbol{E}=4\pi\sum_{\alpha={\rm e,i}}e_\alpha\int d\boldsymbol{p}\,\delta f_\alpha\,.\end{aligned}\tag{7.6.2}$$

As in Sects. 7.4, 5, we assume here that the velocity of the electron oscillations in the high-frequency electric field is nonrelativistic,

$$|\boldsymbol{p}_0|=m|\boldsymbol{u}_0|+|(e\boldsymbol{E}_0)/(\omega_0)\cos\omega_0 t|\ll mc\,.$$

Further, we confine ourselves to the case of the nonmagnetized plasma for simplicity.

The evaluation of (7.6.1, 2) is completely analogous to that given in Sect. 7.4 for longitudinal oscillations. We introduce the variables (7.4.7) and expand all the perturbed quantities ($\Psi_{\rm e}$, $\delta f_{\rm i}$, $\boldsymbol{E}$, and $\boldsymbol{B}$) in the series (7.4.10). As a result we obtain a rather complicated system of coupled equations which, however, is easily solved when the frequency ω_0 of the field is of the order of the electron plasma frequency but larger than the ion plasma frequency, i.e., $\omega_0\sim\omega_{\rm pe}\gg\omega_{\rm pi}$. If we further assume $\boldsymbol{k}\cdot\boldsymbol{r}_{\rm E}\ll 1$ but $k\gg\omega_0/c$ which is the homogeneity requirement for the high-frequency field valid under the condition $\boldsymbol{v}_{\rm E}=\omega_0\boldsymbol{r}_{\rm E}\ll c$, then we obtain the following dispersion equation

$$\begin{aligned}&\frac{\varepsilon^{\rm lo}(\omega,\boldsymbol{k})}{\delta\varepsilon_{\rm i}^{\rm lo}(\omega,\boldsymbol{k})[1+\delta\varepsilon_{\rm e}^{\rm lo}(\omega,\boldsymbol{k})]}+\frac{1}{2}\frac{(\boldsymbol{k}\cdot\boldsymbol{v}_{\rm E})^2}{\omega_0^2-\omega_{\rm pe}^2}\\&-\frac{[\boldsymbol{k},\boldsymbol{v}_{\rm E}]^2}{2k^2c^2}\frac{\delta\varepsilon_{\rm e}^{\rm lo}(\omega,\boldsymbol{k})}{1+\delta\varepsilon_{\rm e}^{\rm lo}(\omega,\boldsymbol{k})}=0\,.\end{aligned}\tag{7.6.3}$$

It generalizes (7.5.1) in the limit $\boldsymbol{k} \cdot \boldsymbol{r}_{\mathrm{E}} \ll 1$ for oscillations not derivable from a potential and differs from it by the presence of the last term. The latter is essential for $\boldsymbol{k} \cdot \boldsymbol{v}_{\mathrm{E}} = 0$, only. The waves must propagate strictly across the external high-frequency field in order that their components not derivable from a potential show up.

For the nondegenerate gaseous plasma with Maxwellian distributed particle velocities we obtain the approximate solution of (7.6.3):

$$\frac{\omega^2}{\omega_{\mathrm{pi}}^2} = \begin{cases} \dfrac{1}{2}\dfrac{(\boldsymbol{k} \cdot \boldsymbol{v}_{\mathrm{E}})^2}{\omega_0^2 - \omega_{\mathrm{pe}}^2} - \dfrac{1}{2}\dfrac{[\boldsymbol{k}, \boldsymbol{v}_{\mathrm{E}}]^2}{k^2 c^2} & \text{for} \quad \omega \gg k v_{\mathrm{Te}} \,, \\[2ex] k^2 r_{\mathrm{De}}^2 + \dfrac{1}{2}\dfrac{(\boldsymbol{k} \cdot \boldsymbol{v}_{\mathrm{E}})^2}{\omega_0^2 - \omega_{\mathrm{pe}}^2} - \dfrac{1}{2}\dfrac{[\boldsymbol{k}, \boldsymbol{v}_{\mathrm{E}}]^2}{k^2 c^2} & \text{for} \quad k v_{\mathrm{Ti}} \ll \omega \ll k v_{\mathrm{Te}} \,, \end{cases} \tag{7.6.4}$$

$$\frac{\omega}{\omega_{\mathrm{pi}}} = -\mathrm{i}\sqrt{\frac{2}{\pi}}\frac{1}{k r_{\mathrm{Di}}}\left[k^2 (r_{\mathrm{De}}^2 + r_{\mathrm{Di}}^2) + \frac{1}{2}\frac{(\boldsymbol{k} \cdot \boldsymbol{v}_{\mathrm{E}})^2}{\omega_0^2 - \omega_{\mathrm{pe}}^2} - \frac{1}{2}\frac{[\boldsymbol{k}, \boldsymbol{v}_{\mathrm{E}}]^2}{k^2 c^2}\right] \quad \text{for} \quad \omega \ll k v_{\mathrm{Ti}} \,.$$

These expressions also show that the last terms describing the transverse components of the perturbation are essential only for $\boldsymbol{k} \cdot \boldsymbol{v}_{\mathrm{E}} = 0$, when the potential corrections of the plasma oscillation spectrum are absent. Moreover, longitudinal contributions lead to a buildup of oscillations in the range of plasma opacity, $\omega_0 < \omega_{\mathrm{pe}}$, whereas the transverse contributions destabilize the oscillations also in the range of transparency, $\omega_0 > \omega_{\mathrm{pe}}$. It is appropriate to mention here that our analysis of the plasma parametric instability with respect to longitudinal oscillations was restricted to the nonisothermal plasma with $T_{\mathrm{e}} \gg T_{\mathrm{i}}$ in the range of transparency $\omega_0 > \omega_{\mathrm{pe}}$, however, in a rather wide frequency range. The analysis of the transverse oscillations is not subject to this restriction, though the increments are small, of the order of $\mathrm{Im}\{\omega\} \sim \omega_{\mathrm{pi}} v_{\mathrm{E}}/c \ll \omega_{\mathrm{pi}}$. Physically, the transverse parametric instability of the plasma in a high-frequency field is analogous to the interchange instability in a constant field (the plasma carrying a current or penetrated by a beam, see Exercises 7.7.5 and 6.7.6).

Concluding this section it should be noted that (7.6.4) is also valid for the degenerate solid-state plasma if v_{Te} and v_{Ti} are replaced by v_{Fe} and v_{Fi}, respectively. An analogous substitution should be made for r_{De} and r_{Di}.

7.7 Exercises

7.7.1. Estimate the number of the electrons which can run away in the completely ionized plasma for $E_0 < E_{\mathrm{cr}}$.

Solution. For the posed problem the distribution function (7.1.13) is valid on the average for the bulk electrons. However, a small group of run-away electrons is continuously accelerated and involved in the process of escape when $E_0 < E_{cr}$. Apparently the electrons with initial velocities parallel to the electric field greater than the thermal velocity can become run-aways. The motion of these electrons in the electric field is described by

$$\frac{du}{dt} = \frac{eE_0}{m} - \nu_e u = \frac{eE_0}{m} - \frac{\nu_{ei}(T_e)\, v_{Te}^3}{u^3} u \; . \qquad (7.7.1)$$

Introducing the critical velocity

$$u_{cr}^2 = \frac{\nu_{ei}(T_e)\, m}{eE_0} v_{Te}^3 = v_{Te}^2 \frac{E_{cr}}{E_0} > v_{Te}^2 \, , \qquad (7.7.2)$$

(7.7.1) can be written as

$$u^2 \frac{du}{dt} = \frac{eE_0}{m} (u^2 - u_{cr}^2) \; . \qquad (7.7.3)$$

We see that electrons with an initial velocity $u_0 > u_{cr}$ are continuously accelerated.

Thus, under the condition $E_0 < E_{cr}$, the number of the escaping electrons N_b follows from

$$\frac{N_b}{N_e} = \frac{1}{(2\pi T_e m)^{3/2}} \int_{v_z \geq u_{cr}} dp \exp\left(-\frac{mv^2}{2T_e}\right)$$

$$= \sqrt{\frac{m}{2\pi T_e}} \int_{u_{cr}}^{\infty} dv \exp\left(-\frac{v^2}{2v_{Te}^2}\right) = \frac{1}{2}\left[1 - \Phi\left(\frac{u_{cr}}{\sqrt{2}\, v_{Te}}\right)\right] , \qquad (7.7.4)$$

where

$$\Phi(x) = (2/\sqrt{\pi}) \int_0^x dx\, e^{-x^2}$$

is the error function.

In the limit of weak fields $E_{cr} \gg E_0$, i.e., $u_{cr} \gg v_{Te}$, we have with a high degree of accuracy

$$\frac{N_b}{N_e} \approx \frac{1}{\sqrt{2\pi}} \exp\left(-\frac{u_{cr}^2}{2v_{Te}^2}\right) = \frac{1}{\sqrt{2\pi}} \exp\left(-\frac{E_{cr}}{E_0}\right) . \qquad (7.7.5)$$

With increasing field strengths E_0 the number of the escaping electrons grows exponentially.

7.7.2. Show that a high-frequency instability of the nonmagnetized plasma in a strong electric field still exists when the conditions (7.2.3) are violated. Use the model equation (7.2.5) and discuss whether the instability is hydrodynamic or kinetic.

Solution. In strong electric fields, when the drift velocity of the electrons obeys $u \gg v_{\mathrm{Te}}$ we obtain from (7.2.5) in the nonrelativistic case

$$1-\frac{\omega_{\mathrm{pe}}^2}{(\omega-\boldsymbol{k}\cdot\boldsymbol{u})^2}\left[1-\mathrm{i}\sqrt{\frac{\pi}{2}}\frac{(\omega-\boldsymbol{k}\cdot\boldsymbol{u})^3}{k^3v_{\mathrm{Te}}^3}\exp\left(-\frac{(\omega-\boldsymbol{k}\cdot\boldsymbol{u})^2}{2k^2v_{\mathrm{Te}}^2}\right)\right] -\frac{\omega_{\mathrm{pi}}^2}{\omega^2}=0\,. \tag{7.7.6}$$

Under the condition $(\boldsymbol{k}\cdot\boldsymbol{u})^2 \gg \omega_{\mathrm{pe}}^2$, which is inverse to the condition (7.2.3), we find ($\omega \rightarrow \omega + \mathrm{i}\delta$)

$$\omega^2=\frac{\omega_{\mathrm{pi}}^2}{1-\omega_{\mathrm{pe}}^2/(\boldsymbol{k}\cdot\boldsymbol{u})^2}\approx\omega_{\mathrm{pi}}^2\,,\qquad \delta=\sqrt{\frac{\pi}{8}}\frac{M\omega^3}{mk^3v_{\mathrm{Te}}^3}(\boldsymbol{k}\cdot\boldsymbol{u})\exp\left(-\frac{(\boldsymbol{k}\cdot\boldsymbol{u})^2}{2k^2v_{\mathrm{Te}}^2}\right)\,. \tag{7.7.7}$$

We see that the increment of this instability is exponentially small. To treat the problem properly a kinetic description should be used. Consequently the instability is a kinetic one.

7.7.3. Show that ion-cyclotron oscillations can be excited in the isothermal magneto-active plasma with $T_{\mathrm{e}} \sim T_{\mathrm{i}}$. The drift velocity of the electrons is smaller than the thermal velocity in this case.

Solution. It can easily be shown that the excitation of short-wavelength ($k_{\perp}\varrho_{\lambda\mathrm{i}} \gg 1$) ion-cyclotron oscillations is possible under the given conditions. The oscillations propagate almost across the magnetic field ($|\omega-\Omega_{\mathrm{i}}| \gg k_z v_{\mathrm{Ti}}$) and obey the dispersion equation

$$1+\mathrm{i}\sqrt{\frac{\pi}{2}}\frac{\omega-\boldsymbol{k}\cdot\boldsymbol{u}}{|k_z|\,v_{\mathrm{Te}}}+\frac{T_{\mathrm{e}}}{T_{\mathrm{i}}}\left(1-\frac{\omega}{\omega-\Omega_{\mathrm{i}}}A_1(k_{\perp}^2\varrho_{\lambda\mathrm{i}}^2)\right) +\mathrm{i}\sqrt{\frac{\pi}{2}}\frac{A_1(k_{\perp}^2\varrho_{\lambda\mathrm{i}}^2)\,\omega}{|k_z|\,v_{\mathrm{Ti}}}\exp\left(-\frac{(\omega-\Omega_{\mathrm{i}})^2}{2k_z^2v_{\mathrm{Ti}}^2}\right)=0\,, \tag{7.7.8}$$

where $A_1(k_\perp^2\varrho_{\lambda i}^2) \approx (\sqrt{2\pi}\,k_\perp\varrho_{\lambda i})^{-1}$. We obtain the oscillation spectrum

$$\omega = \Omega_i\left(1 + \frac{A_1(k_\perp^2\varrho_{\lambda i}^2)}{1 + T_i/T_e}\right),$$

$$\delta = -\sqrt{\frac{\pi}{2}}\,\frac{\Omega_i A_1(k_\perp^2\varrho_{\lambda i}^2)}{2 + T_i/T_e + T_e/T_i} \tag{7.7.9}$$

$$\times\left[\frac{\omega - \boldsymbol{k}\cdot\boldsymbol{u}}{|k_z|\,v_{Te}} + \frac{T_e\omega A_1(k_\perp^2\varrho_{\lambda i}^2)}{T_i|k_z|\,v_{Ti}}\exp\left(-\frac{(\omega-\Omega_i)^2}{2k_z^2 v_{Ti}^2}\right)\right].$$

The oscillations become unstable ($\mathrm{Im}\{\delta\} > 0$) if

$$u > \frac{\Omega_i}{|k_z|}\left[1 + \frac{T_e v_{Te}}{T_i v_{Ti}} A_1(k_\perp^2\varrho_{\lambda i}^2)\exp\left(-\frac{\Omega_i^2 A_1^2(k_\perp^2\varrho_{\lambda i}^2)}{2k_z^2 v_{Ti}^2(1 + T_i/T_e)^2}\right)\right]. \tag{7.7.10}$$

Minimizing this expression by varying $k_\perp$ and k_z we can estimate the threshold of the electron drift velocity for $k_\perp\varrho_{\lambda i} \sim 1$:

$$u_{th} \simeq 4\sqrt{\pi}\,v_{Ti}(1 + T_i/T_e)\sqrt{\ln\left[\frac{M}{m}\left(\frac{T_e}{T_i}\right)^{3/2}\right]}. \tag{7.7.11}$$

It should be noted that this instability also occurs in the degenerate solid-state plasma. Thus, for example, in a plasma with degenerate electrons and nondegenerate ions the dispersion equation is of the form

$$1 + \mathrm{i}\frac{\pi}{2}\frac{\omega - \boldsymbol{k}\cdot\boldsymbol{u}}{|k_z|\,v_{Fe}} + \frac{1}{3}\frac{\mathscr{E}_{Fe}}{T_i}\left[1 - \frac{\omega}{\omega - \Omega_i}A_1(k_\perp^2\varrho_{\lambda i}^2)\right.$$

$$\left. + \mathrm{i}\sqrt{\frac{\pi}{2}}\frac{\omega A_1(k_\perp^2\varrho_{\lambda i}^2)}{|k_z|\,v_{Ti}}\exp\left(-\frac{(\omega-\Omega_i)^2}{2k_z^2 v_{Ti}^2}\right)\right] = 0, \tag{7.7.12}$$

which is similar to (7.7.8). The corollaries (7.7.9–11) remain valid after the substitution $T_e \to \mathscr{E}_{Fe}$.

7.7.4. Show that the ion-acoustic instability of the weakly ionized nonisothermal plasma with a current can develop in the limit of frequent collisions, too. The mean free path of the electrons and the ions is shorter than the wavelength of the ion-acoustic oscillations in this case.

Solution. For simplicity we only treat the low-frequency ($\omega \ll \Omega_i$) and long-wavelength ($k_\perp v_{Ti} \ll \Omega_i$) oscillations. The general dispersion equation

$$1+\sum_{\alpha=\mathrm{e,i}}\frac{\omega_{\mathrm{p}\alpha}^2}{k^2v_{\mathrm{T}\alpha}^2}\left[1-\sum_n\frac{\omega-\boldsymbol{k}\cdot\mathrm{u}_\alpha+\mathrm{i}\nu_\alpha}{\omega-\boldsymbol{k}\cdot\boldsymbol{u}_\alpha+\mathrm{i}\nu_\alpha-n\Omega_\alpha}A_n(z_\alpha)I_+(\beta_{n\alpha})\right]$$

$$\times\left[1-\sum_n\frac{\mathrm{i}\nu_\alpha}{\omega-\boldsymbol{k}\cdot\boldsymbol{u}_\alpha+\mathrm{i}\nu_\alpha-n\Omega_\alpha}A_n(z_\alpha)I_+(\beta_{n\alpha})\right]=0\,, \qquad (7.7.13)$$

$$z_\alpha=\frac{k_\perp^2v_{\mathrm{T}\alpha}^2}{\Omega_\alpha^2}\,,\qquad \beta_{n\alpha}=\frac{\omega-\boldsymbol{k}\cdot\boldsymbol{u}_\alpha+\mathrm{i}\nu_\alpha-n\Omega_\alpha}{|k_z|\,v_{\mathrm{T}\alpha}}$$

is simplified and takes the form

$$1+\sum_{\alpha=\mathrm{e,i}}\frac{\omega_{\mathrm{p}\alpha}^2}{k^2v_{\mathrm{T}\alpha}^2}\cdot\frac{(\omega-\boldsymbol{k}\cdot\boldsymbol{u}_\alpha+\mathrm{i}\nu_\alpha)\left(\dfrac{k_\perp^2v_{\mathrm{T}\alpha}^2}{\Omega_\alpha^2}-\dfrac{k_z^2v_{\mathrm{T}\alpha}^2}{(\omega+\mathrm{i}\nu_\alpha)^2}\right)}{\omega-\boldsymbol{k}\cdot\boldsymbol{u}_\alpha+\mathrm{i}\nu_\alpha\left(\dfrac{k_\perp^2v_{\mathrm{T}\alpha}^2}{\Omega_\alpha^2}-\dfrac{k_z^2v_{\mathrm{T}\alpha}^2}{(\omega+\mathrm{i}\nu_\alpha)^2}\right)}\,. \qquad (7.7.14)$$

For small electron drift velocities $u \ll v_{\mathrm{Te}}$ and frequent collisions ($\nu_\alpha \gg k_z v_{\mathrm{T}\alpha}$) the unstable frequency range is $\nu_\mathrm{e} \gg \omega \gg \nu_\mathrm{i}$. We get for the current-carrying plasma

$$1+\frac{\omega_{\mathrm{pe}}^2}{k^2v_{\mathrm{Te}}^2}\left(1+\mathrm{i}\frac{(\omega-\boldsymbol{k}\cdot\boldsymbol{u})\,\nu_\mathrm{e}}{k_z^2v_{\mathrm{Te}}^2}\right)-\frac{k_z^2\,\omega_{\mathrm{pi}}^2}{k^2\omega^2}\left(1+\mathrm{i}\frac{\nu_\mathrm{i}}{\omega}\right)=0\,. \qquad (7.7.15)$$

Deriving this equation it was tacitly assumed that $\omega\nu_\mathrm{e} \ll k_z^2v_{\mathrm{Te}}^2$ and $\omega\nu_\mathrm{i} \gg k_z^2v_{\mathrm{Ti}}^2$ since only under these conditions, corresponding to a high electron thermal conductivity and a low ion thermal conductivity, can ion-acoustic waves occur in the weakly ionized plasma (Chap. 4).

We obtain from (7.7.15)

$$\omega^2=\omega_{\mathrm{pi}}^2\frac{k_z^2}{k^2}\left(1+\frac{\omega_{\mathrm{pe}}^2}{k^2v_{\mathrm{Te}}^2}\right)^{-1},\qquad \delta=-\frac{\nu_\mathrm{i}}{2}-\frac{m}{2M}\,\frac{\nu_\mathrm{e}\omega^3(\omega-\boldsymbol{k}\cdot\boldsymbol{u})}{k_z^4v_{\mathrm{Te}}^4}\,. \qquad (7.7.16)$$

Consequently the system becomes unstable for $u > \omega/k_z \simeq v_\mathrm{s}$ when the sign of the diffusion absorption changes.

7.7.5. Show that even for $u < v_{\mathrm{Ti}}$ the collisionless nonmagnetized plasma with a current is unstable with respect to transverse perturbations.

Solution. The dispersion equation (7.2.1) for arbitrary oscillations of the collisionless nonmagnetized plasma with a current is written as ($u \ll c$):

$$D(\omega, \boldsymbol{k}) \equiv \left\{k^2 - \frac{\omega^2}{c_2}\left[\delta\varepsilon_{\mathrm{i}}^{\mathrm{tr}}(\omega, k) + 1 + \frac{\omega'^2}{\omega^2}\delta\varepsilon_{\mathrm{e}}^{\mathrm{tr}}(\omega', k)\right]\right\}[\delta\varepsilon_{\mathrm{e}}^{\mathrm{lo}}(\omega', k)$$

$$+ \delta\varepsilon_{\mathrm{i}}^{\mathrm{lo}}(\omega, k) + 1] - \frac{k^2u^2 - (\boldsymbol{k}\cdot\boldsymbol{u})^2}{c^2}\left\{\delta\varepsilon_{\mathrm{e}}^{\mathrm{lo}}(\omega', k) + \frac{\omega'^2}{k^2c^2}[\delta\varepsilon_{\mathrm{e}}^{\mathrm{tr}}(\omega', k)\right.$$

$$\left.- \delta\varepsilon_{\mathrm{e}}^{\mathrm{lo}}(\omega', k)]\right\}\left\{\delta\varepsilon_{\mathrm{i}}^{\mathrm{lo}}(, k) + \frac{\omega^2}{k^2c^2}[\delta\varepsilon_{\mathrm{i}}^{\mathrm{tr}}(\omega, k) - \delta\varepsilon_{\mathrm{i}}^{\mathrm{lo}}(\omega, k)]\right\} = 0\,. \tag{7.7.17}$$

Here $\delta\varepsilon_{\alpha}^{\mathrm{lo}}(\omega, k)$ and $\delta\varepsilon_{\alpha}^{\mathrm{tr}}(\omega, k)$ for $\alpha = \mathrm{e, i}$ are the contributions of the electrons and the ions to the longitudinal and the transverse permittivities defined by (7.4.16).

In the limit of very low frequencies, $\omega \ll k_z v_{\mathrm{Ti}}$, and for wave propagation transverse to the current, $\boldsymbol{u}\cdot\boldsymbol{k} = 0$, this equation is of the form

$$D(0, k) + \omega\frac{\partial D(0, k)}{\partial\omega} = 0\,, \quad \text{where} \tag{7.7.18}$$

$$D(0, k) = k^2\left(1 + \sum_{\alpha=\mathrm{e,i}}\frac{\omega_{\mathrm{p}\alpha}^2}{k^2v_{\mathrm{T}\alpha}^2}\right) - \frac{u^2}{c^2}\frac{\omega_{\mathrm{pe}}^2\omega_{\mathrm{pi}}^2}{k^2v_{\mathrm{Te}}^2v_{\mathrm{Ti}}^2}\,,$$

$$\frac{\partial D(0, k)}{\partial\omega} = \mathrm{i}\sqrt{\frac{\pi}{2}}\left(\frac{\omega_{\mathrm{pi}}^2}{v_{\mathrm{Ti}}^2} - \frac{u^2}{c^2}\frac{\omega_{\mathrm{pe}}^2\omega_{\mathrm{pi}}^2}{k^2v_{\mathrm{Te}}^2v_{\mathrm{Ti}}^2}\right)\frac{1}{kv_{\mathrm{Ti}}}\,. \tag{7.7.19}$$

Perturbations with a wave vector k obeying

$$k \lesssim k_0 \approx \frac{u}{c}\frac{\omega_{\mathrm{pe}}\omega_{\mathrm{pi}}}{\sqrt{\omega_{\mathrm{pe}}^2v_{\mathrm{Ti}}^2 + \omega_{\mathrm{pi}}^2v_{\mathrm{Te}}^2}} \tag{7.7.20}$$

are unstable for arbitrarily small velocities $u < v_{\mathrm{Ti}}$. They increase aperiodically with the increment

$$\mathrm{Im}\{\omega\} = (k_0 - k)\,v_{\mathrm{Ti}} = k_0v_{\mathrm{Ti}}\left(1 - \frac{k}{k_0}\right)\,. \tag{7.7.21}$$

Note that a strong external magnetic field aligned with the current stabilizes this instability.

7.7.6. Analyze the stability of the magnetized collisionless plasma in a strong electric field with respect to electrostatic oscillations in the adiabatic approximation.

Solution. Using the dielectric tensor (6.3.1) it is easy to obtain the dispersion equation for the electrostatic oscillations:

$$\varepsilon(\omega,\boldsymbol{k}) = \frac{k_i k_j}{k^2}\varepsilon_{ij}(\omega,\boldsymbol{k}) = \frac{k_\perp^2}{k^2}\left(1-\frac{\omega_{\mathrm{pe}}^2\gamma^{-1}}{(\omega-\boldsymbol{k}\cdot\boldsymbol{u})^2-\Omega_{\mathrm{e}}^2/\gamma^2}-\frac{\omega_{\mathrm{pi}}^2}{\omega^2-\Omega_{\mathrm{i}}^2}\right)$$

$$+\frac{k_z^2}{k^2}\left(1-\frac{\omega_{\mathrm{pi}}^2}{\omega^2}-\frac{\omega_{\mathrm{pe}}^2\gamma^{-3}}{(\omega-\boldsymbol{k}\cdot\boldsymbol{u})^2}\right)=0\,. \tag{7.7.22}$$

Considering nonmagnetized ions only ($\omega \gg \Omega_{\mathrm{i}}$), we get

$$\frac{\omega_{\mathrm{pi}}^2}{\omega^2} = 1+\delta\varepsilon_{\mathrm{e}}(\omega-\boldsymbol{k}\cdot\boldsymbol{u})\,, \qquad \text{where} \tag{7.7.23}$$

$$\delta\varepsilon_{\mathrm{e}}(\omega-\boldsymbol{k}\cdot\boldsymbol{u}) = -\frac{k_\perp^2}{k^2}\frac{\omega_{\mathrm{pe}}^2\gamma^{-1}}{(\omega-\boldsymbol{k}\cdot\boldsymbol{u})^2-\Omega_{\mathrm{e}}^2/\gamma^2}-\frac{k_z^2\omega_{\mathrm{pe}}^2\gamma^{-3}}{k^2(\omega-\boldsymbol{k}\cdot\boldsymbol{u})^2} \tag{7.7.24}$$

is the longitudinal electron dielectric permittivity. Due to (7.7.23) unstable modes exist in the frequency range $\omega \ll \boldsymbol{k}\cdot\boldsymbol{u}$ under the condition

$$1+\delta\varepsilon_{\mathrm{e}}(\boldsymbol{k}\cdot\boldsymbol{u}) \leqslant 0\,. \tag{7.7.25}$$

The increment is maximal when (7.7.26) holds with the equality sign, i.e., when the Doppler frequency coincides with a natural frequency of the longitudinal electron oscillation spectrum (Sect. 5.2):

$$\omega = \frac{-1+\mathrm{i}\sqrt{3}}{2}\left[\omega_{\mathrm{Li}}^2\Big/\frac{\partial\delta\varepsilon_{\mathrm{e}}(\boldsymbol{k}\cdot\boldsymbol{u})}{\partial(\boldsymbol{k}\cdot\boldsymbol{u})}\right]^{1/3}. \tag{7.7.26}$$

In the absence of a magnetic field ($\Omega_{\mathrm{e}} \to 0$), (7.7.25, 26) take the form

$$\omega_{\mathrm{pe}}^2\left(\frac{k_z^2+k_\perp^2\gamma^2}{k^2}\right) \geqslant (\boldsymbol{k}\cdot\boldsymbol{u})^2\gamma^3\,, \tag{7.7.27}$$

$$\omega = \frac{-1+\mathrm{i}\sqrt{3}}{2}\left(\frac{m}{2M}\right)^{1/3}(\boldsymbol{k}\cdot\boldsymbol{u})\gamma\left(\frac{k^2}{\mathrm{k}_\perp^2\gamma^2+k_z^2}\right)^{1/3}. \tag{7.7.28}$$

For $k_\perp = 0$ they are identical with (7.2.3, 4), i.e., for $k_\perp = 0$ the electrostatic approximation is valid.

In the opposite limit of an infinitely strong magnetic field (7.7.25, 26) read

$$\omega_{\mathrm{pe}}^2 \geqslant k^2u^2\gamma^3\,, \tag{7.7.29}$$

$$\omega = \frac{-1+\mathrm{i}\sqrt{3}}{2}k_z u\gamma\left(\frac{mk^2}{2Mk_z^2}\right)^{1/3}. \tag{7.7.30}$$

In this case they are identical with (7.3.2, 3), implying that oscillations of the strongly magnetized plasma are longitudinal with high accuracy under the condition $\omega < k_z u$.

7.7.7. Use the model of independent particles to study the stability of a plasma in a constant strong electric field with respect to electrostatic perturbations in the nonadiabatic approximation.

Solution. We linearize the basic equations of the model of independent particles

$$\frac{\partial N_\alpha}{\partial t} + \operatorname{div} N_\alpha \boldsymbol{V}_\alpha = 0\,,$$
$$\left[\frac{\partial}{\partial t} + (\boldsymbol{V}_\alpha \cdot \nabla)\right]\frac{\boldsymbol{V}_\alpha}{\sqrt{1 - V_\alpha^2/c^2}} = \frac{e_\alpha}{m_\alpha}\left\{\boldsymbol{E} + \frac{1}{c}\,[\boldsymbol{V}_\alpha, \boldsymbol{B}]\right\} \tag{7.7.31}$$

for small perturbations $\delta\boldsymbol{V}_\alpha$ and δN_α, assuming the field to be longitudinal, $\boldsymbol{E} = -\nabla\Phi$, $\boldsymbol{B} = 0$, $\boldsymbol{U}_\mathrm{e} = \boldsymbol{U}(t) = e\boldsymbol{E}_0 t/m$ and $\boldsymbol{u}(0) = 0$. The perturbed quantities are written in the form of $f(t)\exp(\mathrm{i}\boldsymbol{k}\cdot\boldsymbol{r})$ and, for simplicity, only oscillations parallel to the electric field are considered, i.e., $k_\perp = 0$. For both the isotropic and the magnetized plasma we obtain

$$\frac{\partial^2 \delta N_\mathrm{i}}{\partial t^2} + \omega_\mathrm{pi}^2\left(1 - \frac{\omega_\mathrm{pe}^2}{\gamma^3[\boldsymbol{k}\cdot\boldsymbol{u}(t)]^2}\right)\delta N_\mathrm{i} = 0\,. \tag{7.7.32}$$

The validity of (7.7.32) is given when the growth time of the instability is much greater than ω_pe^{-1} [more exactly $\partial/\partial t \ll \boldsymbol{k}\cdot\boldsymbol{u}(t)$]. This follows from the analysis of the problem in the adiabatic approximation (Sects. 7.2 and 3).

Under the condition

$$[\boldsymbol{k}\cdot\boldsymbol{u}(t)]^2 \gg \omega_\mathrm{pe}^2\gamma^{-3} \tag{7.7.33}$$

(when the value of k is given the time after switching on the field E_0 must be long) (7.7.32) has oscillating solutions

$$\delta N_\mathrm{i} = c\sin(\omega_\mathrm{pi} t + \phi)\,. \tag{7.7.34}$$

Under the condition, complementary to (7.7.33), we have

$$\frac{\partial^2 \delta N_\mathrm{i}}{\partial t^2} - \frac{m}{M}\,\frac{k^2 e^2 E_0^2 t^2 \delta N_\mathrm{i}}{m^2} = 0 \quad \text{for} \quad u \ll c\,, \quad \gamma \approx 1\,,$$
$$\frac{\partial^2 \delta N_\mathrm{i}}{\partial t^2} - \frac{m}{M}\,\frac{k^2 e^3 E_0^3 t^3 \delta N_\mathrm{i}}{c m^3} = 0 \quad \text{for} \quad u \approx c\,, \quad \gamma \gg 1 \tag{7.7.35}$$

with solutions of the form

$$\delta N_i = \begin{cases} \sqrt{t}\left[c_1 I_{1/4}\left(\sqrt{\frac{m}{M}}\frac{keE_0 t^2}{2m}\right) + c_2 I_{-1/4}\left(\sqrt{\frac{m}{M}}\frac{keE_0 t^2}{2m}\right)\right] & \text{for } \gamma \approx 1 , \\ \sqrt{t}\left[c_1 I_{1/5}\left(\sqrt{\frac{k^2e^3E_0^3}{Mm^2c}}\frac{2}{5}t^{5/2}\right) + c_2 I_{-1/5}\left(\sqrt{\frac{k^2e^3E_0^3}{Mm^2c}}\frac{2}{5}t^{5/2}\right)\right] & \text{for } \gamma \gg 1 . \end{cases} \tag{7.7.36}$$

Hence, for small times (small arguments of the Bessel functions $ku\gamma^{3/2}t \ll 1$) the perturbation of the plasma density grows linearly with time

$$\delta N_i \approx c_1 + c_2 t . \tag{7.7.37}$$

For long times ($ku\gamma^{3/2}t \gg 1$), however, the perturbations grow exponentially

$$\delta N_i = \begin{cases} \frac{1}{\sqrt{t}} \exp\left(\sqrt{\frac{m}{M}}\frac{keE_0}{m}\frac{t^2}{2}\right) & \text{for } \gamma \approx 1 , \\ \frac{1}{\sqrt[4]{t^3}} \exp\left(\sqrt{\frac{k^2e^3E_0^3}{Mm^2c}}\frac{2}{5}t^{5/2}\right) & \text{for } \gamma \gg 1 . \end{cases} \tag{7.7.38}$$

7.7.8. Show that an electron plasma in an external SHF field is unstable with respect to the parametric buildup of Langmuir oscillations. Take account of relativistic effects in oscillating velocity of the electrons.

Solution. We use the linearized relativistic equations for the electron plasma in a SHF electric field $E_0 \sin \omega_0 t$ in the model of independent particles

$$\begin{aligned} &\frac{\partial \delta N}{\partial t} + \mathrm{i}\boldsymbol{k}(N_0 \delta \boldsymbol{V} + \delta N \boldsymbol{u}) = 0 , \\ &\left(\frac{\partial}{\partial t} + \mathrm{i}\boldsymbol{k}\cdot\boldsymbol{u}\right)\left[\frac{\delta \boldsymbol{V}}{\sqrt{1-u^2/c^2}} + \frac{\boldsymbol{u}(\boldsymbol{u}\cdot\delta\boldsymbol{V})}{c^2(1-u^2/c^2)^{3/2}}\right] = -\mathrm{i}\frac{e}{m}\boldsymbol{k}\Phi , \\ &k^2\Phi = 4\pi e \delta N . \end{aligned} \tag{7.7.39}$$

Here N_0 is the homogeneous density of the ion background compensating the equilibrium electron charge. The electrons are oscillating in the SHF field with a velocity given by the relation

$$\frac{\boldsymbol{u}(t)}{\sqrt{1-u^2/c^2}} = -\frac{e\boldsymbol{E}_0}{m\omega_0}\cos\omega_0 t = -\boldsymbol{u}_0\cos\omega_0 t\,. \tag{7.7.40}$$

Considering oscillations parallel to the SHF field, $\boldsymbol{k}\,\|\,\boldsymbol{u}$, only, and eliminating $\delta\boldsymbol{V}$ and Φ from (7.7.39), we obtain ($\delta\boldsymbol{V}\,\|\,\boldsymbol{u}\,\|\,\boldsymbol{k}$)

$$\left(\frac{\partial}{\partial t} + \mathrm{i}\boldsymbol{k}\cdot\boldsymbol{u}\right)\frac{1}{(1-u^2/c^2)^{3/2}}\left(\frac{\partial}{\partial t} + \mathrm{i}\boldsymbol{k}\cdot\boldsymbol{u}\right)\delta N = -\,\omega_{\mathrm{pe}}^2\delta N\,. \tag{7.7.41}$$

Substituting

$$y = \frac{1}{(1-u^2/c^2)^{3/2}}\frac{\partial}{\partial t}\delta N\exp\left(\mathrm{i}\int dt\,ku\right), \quad \text{we get} \tag{7.7.42}$$

$$\frac{\partial^2 y}{\partial t^2} + \omega_{\mathrm{pe}}^2\left(1-\frac{u^2(t)}{c^2}\right)^{3/2} y = 0\,. \tag{7.7.43}$$

When the relativistic effects are weak we obtain

$$\frac{\partial^2 y}{\partial t^2} + \omega_{\mathrm{pe}}^2\left(1-\frac{3}{4}\frac{u_0^2}{c^2}\right)\left(1\;-\frac{3}{4}\frac{u_0^2}{c^2}\cos 2\,\omega_0 t\right)y = 0\,. \tag{7.7.44}$$

Introducing the time variable $\tau = \omega_0 t$ this equation is the well-known Mathicu cquation

$$\frac{d^2y}{d\tau^2} + (a-2q\cos 2\tau)\,y = 0\,, \quad \text{where} \tag{7.7.45}$$

$$a = \frac{\omega_{\mathrm{pe}}^2}{\omega_0^2}\left(1-\frac{3}{4}\frac{u_0^2}{c^2}\right), \qquad q = \frac{3}{8}a\frac{u_0^2}{c^2}\,.$$

Equation (7.7.45) depends on the parameters a and q and has both stable and unstable solutions. The instability range is hatched in Fig. 7.5. For very small values of q the instability appears when $a = n^2$, or

$$n^2\omega_0^2 \simeq \omega_{\mathrm{pe}}^2\left(1-\frac{3}{4}\frac{u_0^2}{c^2}\right). \tag{7.7.46}$$

The increment of y [i.e., of the density $\delta N \sim \exp(\delta t)$] is equal to

$$\delta = \frac{3}{16}\omega_{\mathrm{pe}}\frac{u_0^2}{c^2}\,. \tag{7.7.47}$$

The quantity δ determines the width of the parametric resonance for small values of q, i.e., in the nonrelativistic limit.

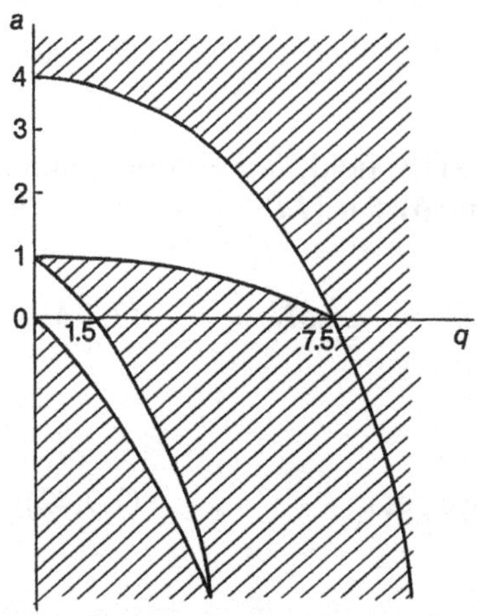

Fig. 7.5. Variation ranges of the parameters q and a corresponding to the stable and unstable solutions

7.7.9. Use (7.5.1) to study the parametric kinetic instability of the nonmagnetized nonisothermal plasma in a strong SHF electric field. Assume $v_E = \omega_0 r_E \gg v_{Te}$, and that the wavelength of the excited oscillation is small against the electron Debye length $kr_{De} \gg 1$ but large against the ion Debye length $kr_{Di} \ll 1$.

Solution. Under these conditions (7.5.1) can be written as

$$1 + \frac{1}{\delta\varepsilon_i(\omega, \boldsymbol{k})} = \frac{1}{k^2 r_{De}^2} \sum_n J_n^2(\boldsymbol{k}\cdot\boldsymbol{r}_E)\left[1 - I_+\left(\frac{n\omega_0 + \omega}{kv_{Te}}\right)\right]. \tag{7.7.48}$$

In the frequency range $kv_{Ti} \ll \omega \ll kv_{Te}$ we have

$$1 - \frac{\omega^2}{\omega_{pi}^2}\left[1 + i\sqrt{\frac{\pi}{2}}\frac{\omega^3}{k^3 v_{Ti}^3}\exp\left(-\frac{\omega^2}{2k^2 v_{Ti}^2}\right)\right]$$
$$= i\sqrt{\frac{\pi}{2}}\frac{1}{k^2 r_{De}^2}\sum_n J_n^2(\boldsymbol{k}\cdot\boldsymbol{r}_E)\frac{\omega + n\omega_0}{kv_{Te}}\exp\left[-\frac{(\omega + n\omega_0)^2}{2k^2 v_{Te}^2}\right]. \tag{7.7.49}$$

Taking the sum over n, which is possible in the limit $\omega \ll kv_{Te}$, $\boldsymbol{k}\cdot\boldsymbol{r}_E \gg 1$ we get

$$1 - i\sqrt{\frac{\pi}{2}}\frac{\omega^5}{k^3 v_{Ti}^3 \omega_{pi}^2}\exp\left(-\frac{\omega^2}{2k^2 v_{Ti}^2}\right) - \frac{\omega^2}{\omega_{pi}^2} + i\frac{2\omega\omega_{pe}^2}{(\boldsymbol{k}\cdot\boldsymbol{v}_E)^3} = 0\,, \tag{7.7.50}$$

where $\boldsymbol{v}_E = \omega_0 \boldsymbol{r}_E$.

The complex zero of this equation is

$$\omega^2 = \omega_{pi}^2\,, \qquad \delta = -\sqrt{\frac{\pi}{8}}\frac{\omega_{pi}}{(kr_{Di})^3}\exp\left(-\frac{\omega_{pi}^2}{2k^2 v_{Ti}^2}\right) + \frac{\omega_{pi}^2\omega_{pe}^2}{(\boldsymbol{k}\cdot\boldsymbol{v}_E)^3}. \tag{7.7.51}$$

The first term in the expression for δ describes the wave absorption by the ions and the second one the inverse absorption (amplification) by the electrons. We get the result that a wave can be amplified ($\delta > 0$), meaning that the system is unstable.

Note that this instability occurs in the range of plasma transparency, i.e., for $\omega_0 > \omega_{pe}$. However, here the following condition should be satisfied:

$$\omega_0 < \sqrt{\frac{T_e}{T_i}}\,\omega_{pe} \, .$$

7.7.10. Using (7.5.22) and taking account of electron-electron collisions determine the threshold of the aperiodic parametric instability (7.5.26).

Solution. Taking account of electron-electron collisions we have for $\omega_0 \approx \omega_{pe} \gg \nu_e$:

$$\varepsilon(\omega \pm \omega_0) \approx 1 - \frac{\omega_{pe}^2}{\omega_0^2} \pm \left(\frac{i\nu_e}{\omega_0} + 2\frac{\omega}{\omega_0}\right) . \tag{7.7.52}$$

Substituting this expession into (7.5.22) we obtain, cf. (7.5.24),

$$\Delta^2 - \left(\frac{i\nu_e}{\omega} + 2\frac{\omega}{\omega_0}\right)^2 - \eta\Delta\left(1 + i\sqrt{\frac{\pi}{2}}\frac{\omega}{kv_{Ti}}\frac{T_i}{T_e + T_i}\right) = 0 \, . \tag{7.7.53}$$

In the range of very low frequencies we find the spectrum

$$\omega = i\frac{\Delta^2 - \eta\Delta + \nu_e^2/\omega_0^2}{\eta\Delta} kv_{Ti}\sqrt{\frac{2}{\pi}}\frac{T_e + T_i}{T_i} \, , \tag{7.7.54}$$

which goes over to the spectrum (7.5.25) in the limit $\nu_e \to 0$. The instability occurs under the condition

$$\eta \geqslant \frac{\Delta^2 + \nu_e^2/\omega_0^2}{\Delta} \, . \tag{7.7.55}$$

Minimizing (7.7.55) over Δ, we obtain the threshold of the instability (7.5.26).

7.7.11. Analyze the parametric instability of the cold magneto-active plasma in a SHF electric field when the frequency ω_0 of the field is close to the sum of the frequencies of two longitudinal electron plasma oscillations $\omega_0 \approx \omega_1 + \omega_2$.

Solution. The longitudinal dielectric permittivity of the cold electron plasma can be written as

$$1 + \delta\varepsilon_e(\omega, k) = \frac{(\omega^2 - \omega_1^2)(\omega^2 - \omega_2^2)}{\omega^2(\omega^2 - \Omega_e^2)} , \tag{7.7.56}$$

where ω_α^2 (for $\alpha = 1, 2$) is defined by (7.5.14).

We introduce the frequency difference Δ of the external field frequency with respect to the resonance frequency

$$\omega_0 = \omega_1 + \omega_2 + \Delta \tag{7.7.57}$$

and try to get a solution of (7.5.22) in the form

$$\omega = \omega_1 + \delta . \tag{7.7.58}$$

We assume $\omega_1 > \omega_2$, but, due to the symmetry, we may also seek for a solution of the form $\omega = \omega_0 + \delta$. As a result we have

$$-\frac{\omega_1^2}{\omega_{pi}^2} + \frac{\omega_1(\omega_1^2 - \Omega_e^2)}{2\delta(\omega_1^2 - \omega_2^2)} + \frac{(\boldsymbol{k}\cdot\boldsymbol{r}_E)^2}{8}\frac{\omega_2(\omega_2^2 - \Omega_e^2)}{(\delta + \Delta)(\omega_1^2 - \omega_2^2)} = 0 . \tag{7.7.59}$$

When the frequency shift $\Delta \gg \delta$ satisfies the condition

$$\frac{\omega_1^2}{\omega_{pi}^2}\Delta = \frac{(\boldsymbol{k}\cdot\boldsymbol{r}_E)^2}{8}\frac{\omega_2(\omega_2^2 - \Omega_e^2)}{\omega_1^2 - \omega_2^2} , \tag{7.7.60}$$

we obtain from (7.7.59) the growth rate of the parametric instability:

$$\frac{\delta^2}{\Delta^2} = \frac{4}{(\boldsymbol{k}\cdot\boldsymbol{r}_E)^2}\frac{\omega_1}{\omega_2}\frac{\omega_1^2 - \Omega_e^2}{\omega_2^2 - \Omega_e^2} . \tag{7.7.61}$$

Because of $\omega_2^2 < \Omega_e^2$ and $\omega_1^2 > \Omega_e^2$, we have $\delta^2 < 0$, i.e., the plasma is always unstable. However, the account of collisions leads to the existence of an instability threshold. For $\delta > \Delta\delta_1, \Delta\delta_2$ [the damping decrements of the upper electron branch are determined by (5.7.3)] we obtain the condition

$$\frac{(\boldsymbol{k}\cdot\boldsymbol{r}_E)^2}{4}\cdot\frac{\omega_2\omega_{pi}^4(\omega_1^2 - \Omega_e^2)(\omega_2^2 - \Omega_e^2)}{(\omega_1^2 - \omega_2^2)^2\omega_1^3} > \nu_e^2\left(\frac{(\omega_1^2 - \Omega_e^2)\cos^2\theta + (\omega_1^2 + \Omega_e^2\sin^2\theta)^2}{(\omega_1^2 - \Omega_e^2)^2\cos^2\theta + \omega_1^4\sin^2\theta}\right) , \tag{7.7.62}$$

where ν_e is the electron-electron collision frequency.

In the case of a dense plasma ($\omega_{pe}^2 > \Omega_e^2$), (7.7.59, 61) are significantly simplified. Here, for $\theta \neq 0$, we have

$$\omega_1^2 \approx \omega_{pe}^2 , \quad \omega_2^2 = \Omega_e \cos^2\theta ,$$

$$\Delta = -\frac{m}{M}\frac{(\boldsymbol{k}\cdot\boldsymbol{r}_E)^2}{8}\frac{\Omega_e^3\cos\theta\sin^2\theta}{\omega_{pe}^2} ,$$

$$\delta^2 = -\left(\frac{m}{M}\right)^2\frac{(\boldsymbol{k}\cdot\boldsymbol{r}_E)^2}{16}\frac{\Omega_e^3\cos\theta\sin^2\theta}{\omega_{pe}} , \tag{7.7.63}$$

$$(\boldsymbol{k}\cdot\boldsymbol{r}_E)^2 > \frac{4\nu_e^2\omega_{pe}}{\sin^2\theta\cos\theta\,\Omega_e^3}\left(\frac{M}{m}\right)^2 ,$$

Note that the growth of the amplitude of the upper and the lower hybrid mode occurs simultaneously during the development of the instability.

7.7.12. Study the parametric instability of the degenerate magnetized solid-state plasma at the frequency ω_0 close to the lower hybrid frequency of the electrons, $\omega_0 \approx \omega_2$. The ions should be considered nondegenerate.

Solution. Starting with (7.7.56) of the last problem we try to get a solution in the range of the ion-acoustic frequencies $\omega_0 \sim \omega_2 \gg kv_{Fe} \gg \omega \gg kv_{Ti}$. Then, (7.5.22) takes the form

$$-\frac{\omega^2}{\omega_{pi}^2} + \frac{k^2v_{Fe}^2}{3\omega_{pe}^2}\left(1 - i\frac{\pi}{2}\frac{\omega}{|k_z|\,v_{Fe}}\right) + \frac{(\boldsymbol{k}\cdot\boldsymbol{r}_E)^2}{4}\frac{\omega_2(\omega_2^2-\Omega_e^2)\,\Delta}{(\omega_2^2-\omega_1^2)(\Delta^2-\omega^2)} = 0 , \tag{7.7.64}$$

where $\Delta = \omega_0 - \omega_2$ is the frequency shift of the SHF field. Hence, under the condition of the parametric resonance (the decay of the pump wave into a lower hybrid and an ion-acoustic wave) we get

$$\omega = kv_s = \Delta , \tag{7.7.65}$$

where $v_s^2 = 3v_{Fe}^2 m/M$. Ion-acoustic waves are excited in the plasma and the increment of the instability is

$$\delta = \frac{(\omega_0-\omega_2)(\Omega_e^2-\omega_2^2)\,\omega_2}{4\pi\omega^2(\omega_1^2-\omega_2^2)}\frac{\omega_{pi}^2}{k^2v_s^2}(\boldsymbol{k}\cdot\boldsymbol{r}_E)^2|k_z|\,v_{Fe} . \tag{7.7.66}$$

We see that the plasma is unstable for $\omega_0 > \omega_2$ (i.e., $\omega_0 = \omega_2 + \omega$) and the threshold is determined by $|\delta| > \nu_e$, which yields

$$\eta = \frac{E_0^2\cos^3\theta}{8\pi N\mathscr{E}_{Fe}} > 12\pi\frac{\nu_e}{\omega_{pe}}\frac{\omega_2^3}{\omega_{pe}^3}\sqrt{\frac{3M}{m}}\frac{\omega_1^2-\omega_2^2}{\Omega_e^2-\omega_2^2} . \tag{7.7.67}$$

In the dense plasma, where $\omega_{pe}^2 \gg \Omega_e^2$, (7.7.66, 67) are significantly simplified (for $\theta \neq 0$):

$$\omega_1^2 \approx \omega_{pe}^2 , \quad \omega_2^2 \approx \Omega_e^2 \cos^2\theta ,$$

$$\delta = \frac{m}{M} \frac{(\omega_0 - \Omega_e \cos\theta)\, \Omega_e^3 \sin^2\theta \cos\theta}{4\pi\omega^2 k^2 v_s^2} (\boldsymbol{k} \cdot \boldsymbol{r}_E)^2 |k_z| v_{Fe} , \tag{7.7.68}$$

$$\eta > 12\pi \frac{\nu_e}{\omega_{pe}} \sqrt{\frac{3M}{m} \frac{\Omega_e \cos^3\theta}{\omega_{pe} \sin\theta}} .$$

8. Electromagnetic Properties of Inhomogeneous Plasmas

The method of geometrical optics is described for spatially inhomogeneous media, with the spatial dispersion, the eikonal equation and the quasi-classical quantization rules being derived. In the geometrical optics approximation the expression for the dielectric tensor of a weakly inhomogeneous plasma is also obtained. The spectra of drift and convective instabilities of the inhomogeneous plasma are studied in detail.

8.1 Inhomogeneous Media Without Spatial Dispersion. Approximation of Geometrical Optics

In the preceding chapters we considered only homogeneous plasmas with parameters which do not depend on the coordinates. Real plasmas, being bounded, are however inhomogeneous. In this chapter, we adopt the model of a spatially unbounded inhomogeneous plasma. The method to describe spatially bounded plasmas will be studied in the next chapter. The characteristic length scale of the inhomogeneity of laboratory plasmas usually is the dimension of the experimental set-up. For example, in controlled thermonuclear fusion devices or in gas discharges the charged particle density varies significantly at distances of the order of the dimension of a plasma filament (the radius of the discharge tube). The characteristic length of the density inhomogeneity L_N is of the order of 1 to 10 cm. The charged particle temperature can be independent of the coordinates. In the ionospheric plasma the characteristic lengths of the regular inhomogeneity are $L_N \approx 10^7$ cm for the charged particle density, $L_T \approx 5 \cdot 10^7$ cm for the temperature and $L_B \approx 10^8$ to 10^9 cm for the inhomogeneity of the earth's magnetic field. In solid-state plasmas the characteristic length for the inhomogeneity is often determined by the method of creating the charge carriers. It is of the order of 10^{-1} to 1 cm.

We begin our study of the electromagnetic properties of the inhomogeneous plasma with the formulation of its dielectric permittivity $\varepsilon_{ij}(\omega, \boldsymbol{k})$. Since in the case of an inhomogeneous plasma the kernels of the material equations (2.2.1, 2) are not functions of the difference of the coordinates $\boldsymbol{r}$ and $\boldsymbol{r}'$, see

(2.2.4, 5), they depend both on $\boldsymbol{r}-\boldsymbol{r}'$ and on $\boldsymbol{r}$ and $\boldsymbol{r}'$ separately. Thus, one cannot describe the general electromagnetic properties within the concept of the tensor of the dielectric permittivity (or conductivity), as it has been defined in (2.2.6–9).

8.1.1 Field Equations for an Inhomogeneous Medium Without Spatial Dispersion

We begin the analysis of the electromagnetic properties of the inhomogeneous plasma with the simplest model where the spatial dispersion can be ignored. In this case for the homogeneous plasma the operator $\varepsilon_{ij}(t-t')$ does not depend on the difference $\boldsymbol{r}-\boldsymbol{r}'$. For the inhomogeneous plasma the operator $\varepsilon_{ij}(t-t',\boldsymbol{r})$ can depend on $\boldsymbol{r}$ only. We can apply the decomposition

$$\varepsilon_{ij}(\omega,\boldsymbol{r}) = \int_0^\infty dt_1\,\varepsilon_{ij}(t_1,\boldsymbol{r})\,\mathrm{e}^{\mathrm{i}\omega t_1}, \tag{8.1.1}$$

i.e., actually the expressions for the dielectric tensor $\varepsilon_{ij}(\omega,0)$ of the plasma without spatial dispersion in the limit $k/\omega \to 0$. However, the corresponding parameters (density, temperature, etc.) should be regarded as functions of the coordinates. Even in this simplest model the formulation of the propagation theory of electromagnetic waves presents a complex problem since it is necessary to solve the field equation

$$\Delta\boldsymbol{E} - \operatorname{grad}\operatorname{div}\boldsymbol{E} + \frac{\omega^2}{c^2}\boldsymbol{D} = 0\,, \quad \text{where} \tag{8.1.2}$$

$$D_i = \varepsilon_{ij}(\omega,\boldsymbol{r})\,E_j\,.$$

Equation (8.1.2) is the basic equation of the theory of electromagnetic wave propagation in inhomogeneous media. For the most common case of normal incidence of a wave on a plane stratified medium it has the same form as the stationary Schrödinger equation. When the medium is isotropic, $\varepsilon_{ij}(\omega,\boldsymbol{r}) = \varepsilon(\omega,\boldsymbol{r})\,\delta_{ij}$, and inhomogeneous along the x-axis only, we write $\varepsilon(\omega,\boldsymbol{r}) = \varepsilon(\omega,x)$ and $\boldsymbol{E}(\omega,\boldsymbol{r}) = \boldsymbol{E}(\omega,x)$. Then, for the transverse waves ($\boldsymbol{E} \perp 0x$) Eq. (8.1.2) simplifies to

$$\frac{d^2\boldsymbol{E}}{dx^2} + \frac{\omega^2}{c^2}\varepsilon(\omega,x)\,\boldsymbol{E} = 0\,. \tag{8.1.3}$$

Comparing with the one-dimensional Schrödinger equation

$$\frac{d^2\Psi}{dx^2} + \frac{2m}{\hbar^2}[W - V(x)]\,\Psi = 0\,, \tag{8.1.4}$$

where Ψ is the wave function, W the total energy of a particle, and $V(x)$ its potential energy, we see that (8.1.3, 4) are closely related.

8.1.2 The Method of Geometrical Optics and the WKB Method

A thorough investigation of possible ways to solve equations of the type of (8.1.3, 4) and, particularly, to derive exact solutions for specific functions $\varepsilon(\omega, x)$ shows that they are fundamental in physics. Exact solutions are known for linear, parabolic and some other dependences of $\varepsilon(\omega, x)$ on x. Approximate methods of solving the wave equation for an arbitrary dependence of $\varepsilon(\omega, x)$ on x are elaborated in the theory of electromagnetic wave propagation and quantum mechanics. The main methods are the method of *geometrical optics* in electrodynamics and the WKB *(Wentzel-Kramers-Brillouin) method* in quantum mechanics. We summarize the essentials of this method, which is of great importance for our further analysis.

The method of geometrical optics can be applied when the medium is weakly inhomogeneous on the scale of the wavelength of the electromagnetic oscillations, i.e., when the wavelength λ is smaller than the characteristic length L_0 of the inhomogeneity

$$\lambda/L_0 \ll 1 .$$

In this case the wave propagation is similar to that in a homogeneous medium with the corresponding parameters. For example, plane waves are eigensolutions of the wave equation in a homogeneous unbounded medium. In an inhomogeneous medium this is not true, but, if the properties of the medium do not vary much on the scale λ, the wave behaves as nearly plane.

In the approximation of geometrical optics any quantity characterizing the wave is of the form

$$\boldsymbol{E} = \boldsymbol{E}_0 \exp[-\mathrm{i}\omega t + \mathrm{i}\Psi(\boldsymbol{r})] , \tag{8.1.5}$$

$\Psi(\boldsymbol{r})$ being called the *eikonal*. Physically it is the phase of the wave dependent on the coordinates. The numerical value of the eikonal is large since it must take the value 2π at $r = \lambda$, and since the approximation of geometrical optics corresponds to the limit $\lambda \to 0$.

In homogeneous media we have

$$\Psi(\boldsymbol{r}) = \boldsymbol{k} \cdot \boldsymbol{r} = \frac{\omega}{c} \boldsymbol{n} \cdot \boldsymbol{r} . \tag{8.1.6}$$

For an inhomogeneous medium we assume quite analogously

$$\nabla\Psi(\boldsymbol{r}) = \boldsymbol{k}(\boldsymbol{r}) = \frac{\omega}{c} \boldsymbol{n}(\boldsymbol{r}) . \tag{8.1.7}$$

In weakly inhomogeneous media $\boldsymbol{k}(\boldsymbol{r})$ is a slowly varying function of $\boldsymbol{r}$ which is determined by the variations of the properties of the medium in space. We can suppose that the scale of the inhomogeneity of $\boldsymbol{k}(\boldsymbol{r})$ coincides with L_0, i.e.,

$$\frac{\partial}{\partial r}k(r) \sim \frac{k(r)}{L_0} .$$

We call $\boldsymbol{k}(\boldsymbol{r})$ a wave vector in a weakly inhomogeneous plasma, $\lambda = 2\pi/k$ a wavelength and $\boldsymbol{n}(\boldsymbol{r})$ a refractive index. Since the wave vector depends weakly on the coordinates, we can still construct the solutions of electrodynamic problems for weakly inhomogeneous media in the form of an expansion in the parameter λ/L_0. In the zero-order approximation the wave is considered plane, i.e., all terms of the order λ/L_0 and of higher order are fully neglected, in the first-order approximation only the terms of the first order of λ/L_0 are accounted for, etc. In other words, we neglect all derivatives of $\boldsymbol{k}(\boldsymbol{r})$ in the zero-order approximation, the first-order derivatives are accounted for in the first-order approximation, etc. When calculating the terms of higher order one can simultaneously solve the field equations with any desired degree of accuracy.

We apply the described method to solve (8.1.3), assuming the field E in the form of

$$E(x) = E_0 \exp[\mathrm{i}\Psi(x)] . \tag{8.1.8}$$

Substituting this expression into (8.1.3) we obtain an equation for $\Psi(x)$ which is called the *eikonal equation*:

$$\Psi'^2 - \mathrm{i}\Psi'' = \varepsilon(\omega, x)\frac{\omega^2}{c^2} . \tag{8.1.9}$$

The prime denotes differentiation with respect to the coordinate. Since the condition $\lambda/L_0 \ll 1$ was assumed, one can expand the solutions $\Psi(x)$ in powers of this small parameter

$$\Psi = \Psi_0 + \Psi_1 + \Psi_2 + \dots . \tag{8.1.10}$$

Here Ψ_0 is the value of Ψ in the zero-order approximation of geometrical optics, following from (8.1.9) when the second term on the left-hand side is ignored:

$$\Psi_0'^2 = \frac{\omega^2}{c^2}\varepsilon(\omega, x) , \quad \text{or} \tag{8.1.11}$$

$$\Psi_0(x) = \pm\frac{\omega}{c}\int_0^x \sqrt{\varepsilon(\omega, x')}\, dx' . \tag{8.1.12}$$

In the theory of wave propagation the domain where $\varepsilon(\omega, x) > 0$ holds is of special interest, since $\Psi_0(x)$ is a real function in this domain. The field $E(x)$ has a wave (oscillatory) character with the wavelength

$$\lambda \sim 1/\Psi_0' \sim c/[\omega\sqrt{\varepsilon(\omega, x)}] .$$

At first sight one could think that the dependence of $\varepsilon(\omega, x)$ on the coordinate x should be completely neglected in the zero-order approximation of geometrical optics. However, this is not true, since [in spite of the weak inhomogeneity of $\varepsilon(\omega, x)$] the integration range in (8.1.12) can be very large and even significantly exceed the characteristic scale L_0 of the inhomogeneity. Therefore, the function $\varepsilon(\omega, x)$ cannot be regarded constant.

From the condition which allows to neglect the second term of (8.1.9) we obtain an inequality which defines the range of applicability of the zero-order approximation (8.1.12):

$$\frac{\Psi_0''}{(\Psi_0')^2} \sim \frac{c}{\omega}\frac{d}{dx}\frac{1}{\sqrt{\varepsilon(\omega, x)}} \sim \frac{\lambda}{L_0} \ll 1 . \tag{8.1.13}$$

Expectedly, this equation gives us the condition of applicability for the approximation of geometrical optics. Equation (8.1.13) is violated when $\lambda/L_0 \sim 1$. This is possible for finite values of the dielectric $\varepsilon(\omega, x) \sim 1$ and not too small values of $c/(\omega L_0) \equiv \lambda_0/L_0 \gtrsim 1$ or for any value of $\lambda_0/L_0 \ll 1$ near the points where $\varepsilon(\omega, x) \approx 0$.

Thus, according to (8.1.12), the field can be written in the zero-order approximation as

$$E = C_+ \exp\left[\mathrm{i}\frac{\omega}{c}\int^x \sqrt{\varepsilon(\omega, x')}\,dx'\right] + C_- \exp\left[-\mathrm{i}\frac{\omega}{c}\int^x \sqrt{\varepsilon(\omega, x')}\,dx'\right] , \tag{8.1.14}$$

or, using the wave vector which is defined in the weakly inhomogeneous plasma by (8.1.7),

$$E = C_+ \exp\left[\mathrm{i}\int^x dx' k(x')\right] + C_- \exp\left[-\mathrm{i}\int^x dx' k(x')\right] . \tag{8.1.15}$$

In order to calculate the next term of the expansion (8.1.10) we substitute this expansion into (8.1.9), take account of (8.1.11), and keep all the first-order terms. As a result we obtain

$$2\Psi_0'\Psi_1' - \mathrm{i}\Psi_0'' = 0 , \tag{8.1.16}$$

hence

$$\Psi_1 = \frac{\mathrm{i}}{2}\ln\Psi_0' . \tag{8.1.17}$$

Accounting for this correction to (8.1.8) we obtain, by a simple calculation, the expression for the field which is correct up to the terms of the first-order approximation of geometrical optics:

$$E(x) = \frac{C_1}{\sqrt[4]{\varepsilon(\omega, x)}} \exp\left[\mathrm{i}\frac{\omega}{c} \int^{x} \sqrt{\varepsilon(\omega, x')}\, dx'\right] + \frac{C_2}{\sqrt[4]{\varepsilon(\omega, x)}} \exp\left[-\mathrm{i}\frac{\omega}{c} \int^{x} \sqrt{\varepsilon(\omega, x')}\, dx'\right]. \tag{8.1.18}$$

As stated above, the ranges where the function $\varepsilon(\omega, x)$ is real and positive are of special importance in the theory of wave propagation in inhomogeneous media. The electromagnetic field (8.1.18) has an oscillatory character. In other words, propagation of waves is possible. These ranges are called the *ranges of transparency* in geometrical optics. In contrast to them, the ranges where the function $\varepsilon(\omega, x)$ is negative are called the *ranges of opacity*. In these domains the field E varies exponentially with the coordinate x, either increasingly or decreasingly. In quantum mechanics the range of transparency corresponds to the range where the function $U(x) = W - V(x)$ is positive, i.e., where motion of the particles can be described classically, see (8.1.4). The ranges where $U(x) < 0$ holds are inaccessible for a classical particle, however. Therefore the point $U(x) = 0$ is called a *turning point*. By analogy, the points which separate the ranges of transparency and opacity, i.e., the points where $\varepsilon(\omega, x) = 0$, are called turning points in geometrical optics, too. An electromagnetic wave propagating in the transparency range is reflected at these points.

Since the approximation of geometrical optics is inapplicable near the turning points, the solution (8.1.18) loses its sense. However, near the turning point the function $\varepsilon(\omega, x)$ can be expanded in a series which represents the exact solution of (8.1.3) when it tends asymptotically to the term (8.1.18) far from this point. We assume that the range of classically accessible solutions, or the range of transparency, lies between the turning points a and b, which are the solutions of the equation $\varepsilon(\omega, x) = 0$, i.e., $\varepsilon(\omega, x) \geq 0$ for $a \leq x \leq b$. Then the asymptotic solution of (8.1.3) at the left of the point b which goes over into the damped solution (8.1.18) for $x > b$ is of the form $(x < b)$

$$E = \frac{C}{\sqrt[4]{\varepsilon(\omega, x)}} \sin\left[\frac{\omega}{c} \int_x^b dx' \sqrt{\varepsilon(\omega, x')} + \frac{\pi}{4}\right]. \tag{8.1.19}$$

The asymptotic solution of (8.1.3) at the right of the point b, passing over into the damped solution (8.1.18) for $x < a$, is written analogously in the form

$$E = \frac{C'}{\sqrt[4]{\varepsilon(\omega, x)}} \sin\left[\frac{\omega}{c} \int_a^x dx' \sqrt{\varepsilon(\omega, x')} + \frac{\pi}{4}\right]. \tag{8.1.20}$$

8.1.3 The Bohr-Sommerfeld Quasiclassical Quantization Rules

Naturally, (8.1.19, 20) must coincide in the entire range $a \leq x \leq b$. To ensure this, the sum of their phases must be an integer multiple of π. This is in fact given if

$$\frac{\omega}{c}\int_a^b dx \sqrt{\varepsilon(\omega, x)} = \pi\left(n + \frac{1}{2}\right), \tag{8.1.21}$$

n being an arbitrary integer: $n = 0, \pm 1, \pm 2, \ldots$ The integration constants C and C' are connected by the relation $C = (-1)^n C'$.

In the approximation of geometrical optics, due to (8.1.13), the eikonal has a large value, i.e., $|n| \gg 1$. Taking account of this relation (8.1.21) can be written approximately as

$$\frac{\omega}{c}\int_a^b dx \sqrt{\varepsilon(\omega, x)} = \int_a^b dx\, k(\omega, x) = \pi n\,, \tag{8.1.22}$$

with $n \gg 1$.

Note that (8.1.22) can also be obtained when an additional circular integral along the closed contour around the turning points in the complex k-plane is considered in the solution (8.1.15). The phase of the solution (8.1.15) grows, the additional phase being $\mathrm{i}\oint k(\omega, x)\, dx$, and since the field $E(x)$ must remain unchanged the following condition (the condition of a single-valued solution)

$$\oint k(\omega, x)\, dx = 2\pi n$$

should be satisfied. In the case of two turning points (8.1.22) follows.

In geometrical optics (8.1.22) is the dispersion equation defining the frequency spectrum ω, or the wave vector spectrum $\boldsymbol{k}$ of electromagnetic waves which are trapped in the transparency range of the medium. Thus, in the approximation of geometrical optics, when there are two turning points present, the eigenvalue spectrum of the wave equation is discrete in the inhomogeneous medium. This is the important qualitative difference of the electromagnetic properties of inhomogeneous and homogeneous media.

Equation (8.1.22) has a simple physical meaning: there must be place for an integer number of halfwaves between the turning points in the transparency range of the medium (Fig. 8.1). Oscillations of a string with fixed end points are analogous to this situation, since (8.1.3) is identical with the equation of string oscillations of a medium with a weakly inhomogeneous modulus of elasticity. The fixed end points of the string may be identified with the turning points, since the solutions of the wave equation decrease exponentially in the complementary domain outside these points. An analogous situation occurs in quantum mechanics in the quasiclassical limit. Therefore

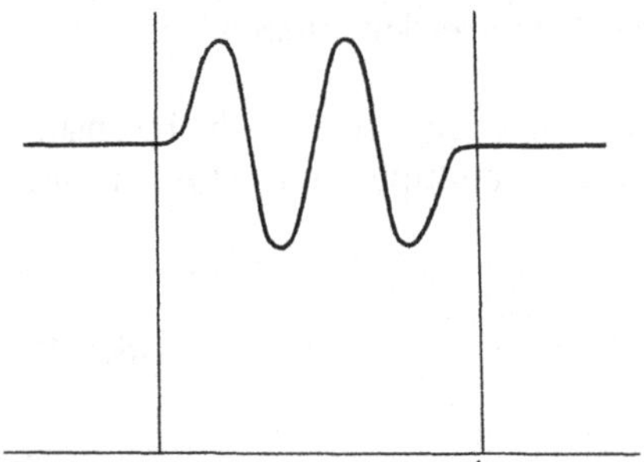

Fig. 8.1

(8.1.21, 22) are called the *quasiclassical quantization rules* and the integrals on the left-hand sides of these relations are known as *Bohr-Sommerfeld phase integrals*.

In the above conclusions we considered the function $\varepsilon(\omega, x)$ real. However, it becomes complex when the medium is dissipative, e.g., due to particle collisions. As stated in the foregoing chapters, weakly damped electromagnetic waves can exist in the medium only when the dissipation is weak. Then $\varepsilon(\omega, x)$ is an almost real function and it is not difficult to generalize (8.1.22) for this case of weakly dissipative inhomogeneous media. Due to $\mathrm{Re}\{k(\omega, x)\} \gg \mathrm{Im}\{k(\omega, x)\}$ and because the electromagnetic oscillations are thus weakly damped ($\omega \to \omega + \mathrm{i}\delta$, $\omega \gg \delta$), the integrand takes the form

$$k(\omega + \mathrm{i}\delta, x) \approx \mathrm{Re}\{k(\omega, x)\} + \mathrm{i}\,\mathrm{Im}\{k(\omega, x)\} + \mathrm{i}\delta \frac{\partial \mathrm{Re}\{k(\omega, x)\}}{\partial \omega} .$$

Substituting this expansion into (8.1.22) and isolating its real and imaginary parts yields

$$\int_a^b \mathrm{Re}\{k(\omega, x)\}\, dx = \pi n , \qquad \delta = \frac{\int_a^b \mathrm{Im}\{k(\omega, x)\}\, dx}{\int_a^b \frac{\partial}{\partial \omega} \mathrm{Re}\{k(\omega, x)\}\, dx} . \tag{8.1.23}$$

Here, the turning points a and b are the points where $\mathrm{Re}\{k^2(\omega, x)\} = 0$.

In homogeneous media without spatial dispersion (8.1.23) determines the frequency spectrum and the damping decrement of electromagnetic waves in the approximation of geometrical optics. The second equation (8.1.23) should be used only when the frequency ω, defined by the first equation, is real.

To summarize the above, we have considered the simplest case of an inhomogeneous medium with two turning points, between which the range of transparency lies. Electromagnetic waves can propagate in the medium between these turning points. Therefore they are often referred to as *trapped oscillations* and the corresponding solutions of the field equation as *finite solutions*. Under real conditions other situations are possible in inhomogeneous media, for example, when there exists only one turning point in the medium or, conversely, when there are several ranges of transparency, separated from each other and included between pairs of corresponding turning points, or when there is no turning point at all and the medium is transparent in the entire space. From this variety of cases we shall treat only those which are most frequently realized in a plasma.

8.2 Approximation of Geometrical Optics for Inhomogeneous Media with Spatial Dispersion

Using the method of geometrical optics presented in Sect. 8.1, we can formulate the material equation for an arbitrary inhomogeneous medium and, in particular, introduce the concept of the dielectric tensor for inhomogeneous plasmas. We assume the plasma to be weakly inhomogeneous, i.e., the wavelength to be significantly smaller than the characteristic dimensions of inhomogeneity of the medium. Here the material equation relating the dielectric displacement to the electric field strength, cf. (2.2.4), is of the form

$$D_i(t,\boldsymbol{r}) = \int_{-\infty}^{t} dt' \int d\boldsymbol{r}' \varepsilon_{ij}(t-t',\boldsymbol{r}-\boldsymbol{r}',\boldsymbol{r})\, E_j(t',\boldsymbol{r}') \,. \tag{8.2.1}$$

With respect to a weak inhomogeneity of the medium, it is allowable to keep the dependence of the kernel of (8.2.1) on $\boldsymbol{r}-\boldsymbol{r}'$, only, and to neglect that on $\boldsymbol{r}$. The dependence on $\boldsymbol{r}-\boldsymbol{r}'$ is related to the wavelength of the oscillation and in a homogeneous medium thus the dependence of $\varepsilon_{ij}(\omega,\boldsymbol{k})$ on $\boldsymbol{k}$ arises after the Fourier transformation. The dependence on $\boldsymbol{r}$ is determined by the inhomogeneity of the medium and related to the characteristic dimension L_0 of the inhomogeneity. Since the parameter λ/L_0 is small, we can again apply the approximation of geometrical optics, considering the wave vector $\boldsymbol{k}$ as a weak function of the coordinates in the zero-order approximation. Then (8.2.1) can be written as

$$D_i(\omega,\boldsymbol{k}) = \varepsilon_{ij}(\omega,\boldsymbol{k},\boldsymbol{r})\, E_j(\omega,\boldsymbol{k}) \,, \quad \text{where} \tag{8.2.2}$$

$$\varepsilon_{ij}(\omega,\boldsymbol{k},\boldsymbol{r}) = \int_{0}^{\infty} dt_1 \int d\boldsymbol{R}\; \varepsilon_{ij}(t_1,\boldsymbol{R},\boldsymbol{r}) \exp(\mathrm{i}\omega t_1 - \mathrm{i}\boldsymbol{R}\cdot\boldsymbol{k}) \tag{8.2.3}$$

is the dielectric tensor of the weakly inhomogeneous medium taking account of spatial dispersion in the zero-order approximation of geometrical optics.

8.2.1 Eikonal Equation for an Inhomogeneous Medium with Spatial Dispersion

Writing the Maxwell equations in the zero-order approximation of geometrical optics:

$$\begin{aligned} [\boldsymbol{k}, \boldsymbol{B}] &= -\frac{\omega}{c}\boldsymbol{D}\,, \qquad & \boldsymbol{k}\cdot\boldsymbol{B} &= 0\,, \\ [\boldsymbol{k}, \boldsymbol{E}] &= \frac{\omega}{c}\boldsymbol{B}\,, \qquad & \boldsymbol{k}\cdot\boldsymbol{D} &= 0\,, \end{aligned} \tag{8.2.4}$$

yields together with (8.2.2) the condition for the existence of solutions of this system of equations

$$\left| k^2\delta_{ij} - k_ik_j - \frac{\omega^2}{c^2}\varepsilon_{ij}(\omega, \boldsymbol{k}, \boldsymbol{r})\right| = 0\,. \tag{8.2.5}$$

This relation can be obtained from (8.1.2) for media without spatial dispersion, too, when the method of geometrical optics in the zero-order approximation is applied.

Note that in the homogeneous medium an equation of the type (8.2.5) is a dispersion equation, see (2.4.3), defining the frequency spectrum $\omega(\boldsymbol{k})$ of the natural electromagnetic oscillations. If (8.2.5) is solved for ω, the "frequency of the natural oscillations of the medium", the eigenmodes depend on the coordinates, i.e., $\omega = \omega(\boldsymbol{k}, \boldsymbol{r})$. This is not possible physically. The principal difference between (8.2.5) and (2.4.3) is the following. The former does not constitute a dispersion relation but simply defines $\boldsymbol{k}(\boldsymbol{r})$ or the eikonal $\Psi(\boldsymbol{r})$. Therefore it is called the *eikonal equation*. It generalizes (8.1.11) in the zero-order approximation of geometrical optics for weakly inhomogeneous media with spatial dispersion.

When the electric field of a wave in the inhomogeneous medium is longitudinal, (8.2.5) simplifies to

$$\varepsilon(\omega, \boldsymbol{k}, \boldsymbol{r}) = \frac{k_ik_j}{k^2}\varepsilon_{ij}(\omega, \boldsymbol{k}, \boldsymbol{r}) = 0\,. \tag{8.2.6}$$

Here the quantity $\varepsilon(\omega, \boldsymbol{k}, \boldsymbol{r})$ is known as the longitudinal dielectric permittivity of a weakly inhomogeneous medium and (8.2.6) constitutes the eikonal equation for longitudinal waves in the zero-order approximation of geometrical optics.

It is convenient to calculate the electric field as a function of the coordinates with the aid of the eikonal equation, i.e., to solve the boundary value problem. In order to solve the initial value problem, i.e., to calculate the frequency spectra of the electromagnetic oscillations of the weakly inhomogeneous medium, the eikonal equation is not sufficient. Dispersion equations of the type (8.1.23), derived for media without spatial dispersion, must be derived in this case.

8.2.2 Quantization Rules

In general, when the inhomogeneous medium is treated three-dimensionally, one must know the spectrum of eigenvalues of partial integro-differential equations of higher order in order to derive the dispersion equation. This problem has not been solved yet. At present, it is more or less solved for media with a one-dimensional inhomogeneity. In this case we can determine the projection of the wave vector in the direction of the inhomogeneity $k_x(\omega, k_y, k_z, x)$. In the most interesting frequency ranges (8.2.5, 6) have the pair of nondegenerate roots $\pm\, k_{x,s}(\omega, x)$. In the zero-order approximation of geometrical optics the differential equation

$$\frac{d^2y}{dx^2} + k_{xs}^2(\omega, x)\, y = 0 \tag{8.2.7}$$

can describe each pair. The general theory, given in Sect. 8.1, can be applied here. To determine the frequency spectrum we can write relations defining the frequency and the damping decrement in analogy with (8.1.23) $(\omega \to \omega + \mathrm{i}\delta)$:

$$\int\limits_{x_\mu}^{x_\nu} \mathrm{Re}\,\{k_{xs}(\omega, x)\}\, dx = \pi n\,, \qquad \delta = -\frac{\int\limits_{x_\mu}^{x_\nu} \mathrm{Im}\,\{k_{xs}(\omega, x)\}\, dx}{\int\limits_{x_\mu}^{x_\nu} \frac{\partial}{\partial\omega} \mathrm{Re}\,\{k_{xs}(\omega, x)\}\, dx}\,. \tag{8.2.8}$$

In these relations the domain of integration is the range where the medium is transparent with respect to these oscillations and the turning points x_μ and x_ν are determined from $\mathrm{Re}\,\{k_{xs}^2(\omega, x)\} = 0$.

Thus, the basic idea for calculating the electromagnetic wave propagation in a weakly inhomogeneous dispersive medium is similar to that applied to obtain the dispersion of the medium. The method of solution is to calculate from the eikonal equation (8.2.5), or (8.2.6), the wave vectors $k_{xs}^2(\omega, x)$ corresponding to the S-branches of oscillations, their frequency spectra and damping decrements being determined by the quantization rules (8.2.8).

Finally, note that the transition to unbounded homogeneous media is included formally. In this case there are no turning points but we can introduce the arbitrary points a and b. With the aid of them (8.1.22) is written as

$$\int\limits_a^b k_{xs}(\omega, x)\, dx = k_{xs}(\omega)(b - a) = \pi n\,. \tag{8.2.9}$$

Assuming $\pi n/(b-a) = \text{const}$, we obtain the equation determining the mode spectrum of the homogeneous plasma:

$$k_{xs}(\omega) = k_{xs} = \text{const}. \tag{8.2.10}$$

Naturally, the roots of this equation $\omega(\boldsymbol{k})$ are identical with those of (8.2.5), which is the dispersion equation for homogeneous media.

8.3 Dielectric Tensor of Weakly Inhomogeneous Plasmas in the Approximation of Geometrical Optics

We now come to the explicit calculation of the dielectric tensor of weakly inhomogeneous plasmas $\varepsilon_{ij}(\omega, \boldsymbol{k}, x)$, and start the analysis with the collisionless magneto-active plasma. As commonly done, we apply the kinetic equation with a self-consistent field (Vlasov's equation) for particles of the type α:

$$\frac{\partial f_\alpha}{\partial t} + \boldsymbol{v}\frac{\partial f_\alpha}{\partial \boldsymbol{r}} + e_\alpha\left\{\boldsymbol{E} + \frac{1}{c}[\boldsymbol{v}, \boldsymbol{B}]\right\}\frac{\partial f_\alpha}{\partial \boldsymbol{p}_\alpha} = 0\,. \tag{8.3.1}$$

8.3.1 Distribution Function for the Equilibrium Inhomogeneous Plasma

The external magnetic field $\boldsymbol{B}_0$ is assumed to be oriented along the z-axis and the gradient of the plasma inhomogeneity is taken along the x-axis, i.e., across the field $\boldsymbol{B}_0$. The distribution function $f_{0\alpha}(\boldsymbol{v}, x)$ for the stationary state, where $\boldsymbol{E}_0 = 0$, $\boldsymbol{B}_0 \parallel 0z$, should be determined first. In the nonrelativistic plasma where $\boldsymbol{p}_\alpha = m_\alpha \boldsymbol{v}$ we obtain from (8.3.1) for $f_{0\alpha}(\boldsymbol{v}, x)$

$$v_\perp \cos\phi \frac{\partial f_{0\alpha}}{\partial x} - \Omega_\alpha(x)\frac{\partial f_{0\alpha}}{\partial \phi} = 0\,. \tag{8.3.2}$$

We used cylindrical coordinates in the velocity space ($v_x = v_\perp \cos\phi$, $v_y = v_\perp \sin\phi$, v_z) and introduced $\Omega_\alpha(x) = e_\alpha B_0(x)/(m_\alpha c)$, the inhomogeneous cyclotron frequency of particles of the type α.

Any function of the characteristics $\mathscr{E}_\alpha$ and C_α

$$f_{0\alpha}(\boldsymbol{v}, x) = f_{0\alpha}(\mathscr{E}_\alpha, C_\alpha)$$

is a general solution of (8.3.2). Here $\mathscr{E}_\alpha = m_\alpha v^2/2$ is the energy and C_α follows from the characteristic equation

$$\frac{dx}{v_\perp \cos\phi} = -\frac{d\phi}{\Omega_\alpha(x)}\,, \tag{8.3.3}$$

which has solutions of the form

$$C_\alpha = v_\perp \sin\phi + \int^x \Omega_\alpha(x')\,dx' \ . \tag{8.3.4}$$

Hence,

$$f_{0\alpha} = f_{0\alpha}\left[\mathscr{E}_\alpha, v_y + \int^x \Omega_\alpha(x')\,dx'\right] . \tag{8.3.5}$$

In real plasmas the characteristic length of the inhomogeneity significantly exceeds the Larmor radius of the particles. This allows us to introduce the small parameter

$$v_{T\alpha}/(\Omega_\alpha L_0) \ll 1 \tag{8.3.6}$$

and to expand the solution (8.3.5) in powers of it. We can write

$$f_{0\alpha}(\mathscr{E}_\alpha, C_\alpha) = \left(1 + \frac{v_\perp \sin\phi}{\Omega_\alpha}\frac{\partial}{\partial x}\right) F_\alpha(\mathscr{E}_\alpha, x) \ , \tag{8.3.7}$$

where $F_\alpha(\mathscr{E}_\alpha, x)$ is an arbitrary function of $\mathscr{E}_\alpha$ depending, in addition, parametrically on x. For the nondegenerate plasma it is natural to choose the local Maxwellian distribution function with inhomogeneous density and temperature and to write

$$F_\alpha(\mathscr{E}_\alpha, x) = \frac{N_\alpha(x)}{[2\pi m_\alpha T_\alpha(x)]^{3/2}} \exp\left(-\frac{\mathscr{E}_\alpha}{T_\alpha(x)}\right) . \tag{8.3.8}$$

Due to the inhomogeneity the plasma in local equilibrium obtains principally new properties which we study in more detail now. We calculate the density of the electric charge ϱ_0 and of the current $\boldsymbol{j}_0$ in local equilibrium. Assuming $\boldsymbol{E}_0 = 0$ we have

$$\varrho_0 = \sum_\alpha e_\alpha \int f_{0\alpha}\,d\boldsymbol{p} = \sum_\alpha e_\alpha \int F_\alpha\,d\boldsymbol{p} = \sum_\alpha e_\alpha N_\alpha(x) = 0 \ , \tag{8.3.9}$$

which is the condition for plasma quasineutrality.

The current density in local equilibrium is

$$\boldsymbol{j}_0 = \sum_\alpha e_\alpha \int d\boldsymbol{p}\,\boldsymbol{v} f_{0\alpha} \ . \tag{8.3.10}$$

From (8.3.7) it follows for $f_{0\alpha}$ that only the second term, proportional to $v_\perp \sin\phi = v_y$, contributes to the current. The current flows parallel to the $0y$-axis; explicitly we have

$$j_{0x} = j_{0z} = 0 \, ,$$

$$j_{0y} = \sum_{\alpha} e_{\alpha} \int v_y f_{0\alpha} \, d\boldsymbol{p} = \sum_{\alpha} e_{\alpha} \int d\boldsymbol{p} \frac{v_y^2}{\Omega_{\alpha}} \frac{\partial F_{\alpha}}{\partial x} \tag{8.3.11}$$

$$= \sum_{\alpha} \frac{e_{\alpha}}{\Omega_{\alpha} m_{\alpha}} \frac{\partial N_{\alpha} T_{\alpha}}{\partial x} = \frac{c}{B_0} \sum_{\alpha} \frac{\partial N_{\alpha} T_{\alpha}}{\partial x} = \frac{c}{B_0} \frac{\partial p_0}{\partial x} \, ,$$

where $p_0 = \sum_{\alpha} N_{\alpha} T_{\alpha}$ is the total plasma pressure.

8.3.2 Magnetic Confinement of the Inhomogeneous Plasma

Substituting (8.3.11) into the Maxwell equation

$$\operatorname{curl} B_0 = \frac{4\pi}{c} j_0$$

gives the following condition for the plasma equilibrium:

$$\frac{\partial}{\partial x} \left(\frac{B_0^2}{8\pi} + p_0 \right) = 0 \, . \tag{8.3.12}$$

This is the well-known MHD condition for the ideal plasma equilibrium which has a clear physical meaning: the inhomogeneous plasma is confined by the magnetic field pressure, and as a result the gradients of the magnetic and hydrodynamic pressure compensate each other. There is no need to analyze the corollaries, following from this condition, in detail. Note, however, that one of these corollaries is a conclusion about the possibility to confine hot plasmas in various thermonuclear devices. This conclusion is of great practical importance. In the following we need only the equilibrium condition (8.3.12), which implies that for low-pressure plasmas with

$$\beta = \frac{8\pi p_0}{B_0^2} \ll 1$$

the characteristic length of the inhomogeneity of the magnetic field L_B must greatly exceed the characteristic inhomogeneity scale of the kinetic pressure L_p (or L_N, L_T). Actually, (8.3.12) can be written as

$$\frac{d}{dx} \ln \frac{B_0^2}{8\pi} + \beta \frac{d}{dx} \ln p_0 = 0 \, . \tag{8.3.13}$$

Hence, it follows that $(L_p/L_B) \sim \beta$ since $d[\ln(B_0^2/8\pi)]/dx \sim 1/L_B$ and $d(\ln p_0)/dx \sim L_p^{-1}$.

Low β plasmas $\beta \ll 1$ occur in a large number of practical cases, for example in many devices for thermonuclear fusion, in many gas-discharges and ionospheric plasmas, and in degenerate solid-state plasmas when the magnetic field is strong. Below we restrict ourselves to these low β plasmas, neglecting the magnetic field inhomogeneity as compared with the particle density and temperature scales.

8.3.3 The Dielectric Tensor of Weakly Inhomogeneous Plasma

Admitting a small perturbation δf_α of the local equilibrium in the form of an oscillation

$$\delta f_\alpha = \delta f_\alpha(x) \exp(-\mathrm{i}\omega t + \mathrm{i}k_y y + \mathrm{i}k_z z) . \tag{8.3.14}$$

we obtain from (8.3.1)

$$(\omega - k_y v_y - k_z v_z)\,\delta f_\alpha + \mathrm{i}v_x \frac{\partial \delta f_\alpha}{\partial x} - \mathrm{i}\Omega_\alpha \frac{\partial \delta f_\alpha}{\partial \phi}$$

$$= -\mathrm{i}e_\alpha\left\{\boldsymbol{E} + \frac{1}{c}[\boldsymbol{v}, \boldsymbol{B}]\right\} \frac{\partial f_{0\alpha}}{\partial \boldsymbol{p}_\alpha} . \tag{8.3.15}$$

Note that the characteristic of this inhomogeneous partial differential equation is given by (8.3.4). Neglecting the magnetic field inhomogeneity ($\beta \ll 1$) thus yields

$$v_\perp \sin\phi + \Omega_\alpha x = v_\perp \sin\phi' + \Omega_\alpha x' = C_\alpha . \tag{8.3.16}$$

With account of this relation the general solution of (8.3.15) can be written as, cf. (5.1.7):

$$\delta f_\alpha = \frac{e_\alpha}{m_\alpha \Omega_\alpha} \int_\infty^\phi d\phi' \left\{\boldsymbol{E}(x') + \frac{1}{c}[\boldsymbol{v}, \boldsymbol{B}(x')]\right\}$$

$$\times \frac{\partial f_{0\alpha}(x', \phi')}{\partial \boldsymbol{v}(\phi')} \exp\left[\frac{\mathrm{i}}{\Omega_\alpha}\int_\phi^{\phi'} d\phi''(\omega - k_y v_\perp \sin\phi'' - k_z v_z)\right] . \tag{8.3.17}$$

Here the coordinate x' depends on x and ϕ through the characteristic equation (8.3.16).

The function $\delta f_\alpha(x)$ and the fields $\boldsymbol{E}(x)$, $\boldsymbol{B}(x)$ can be presented in the form of

$$\exp\left[\mathrm{i}\int^x k_x(x')\,dx'\right],$$

too. Confining our consideration to the zero-order approximation of geometrical optics, i.e., when differentiating, taking account of the terms proportional to $k_x(x)$ only, and ignoring the terms proportional to the spatial derivatives $k_x'(x)$, we can proceed further. Taking the derivative $\partial/\partial x$ of (8.3.17) and of the field equations and multiplying these expressions by k_x, we can eliminate due to the field equation $\partial \boldsymbol{B}/\partial t = -c \operatorname{curl} \boldsymbol{E}$ the magnetic induction $\boldsymbol{B}(x')$ from (8.3.17) which gives for δf_a:

$$\begin{aligned}\delta f_\alpha(\boldsymbol{k}, x) &= \frac{e_\alpha}{\Omega_\alpha} \int_\infty^\phi d\phi' \left[\left(1 - \frac{\boldsymbol{k}\cdot\boldsymbol{v}}{\omega}\right)\delta_{ij} + \frac{v_i k_j}{\omega}\right]_{\phi'} \\ &\times \frac{\partial f_{0\alpha}(x', \phi')}{\partial p_{\alpha j}} E_i(\boldsymbol{k}, \omega) \exp\left[\frac{\mathrm{i}}{\Omega_\alpha}\int_\phi^{\phi'} d\phi''(\omega - \boldsymbol{k}\cdot\boldsymbol{v})_{\phi''}\right] . \end{aligned} \tag{8.3.18}$$

Here the vector $\boldsymbol{k}$ is three-dimensional $\boldsymbol{k} = (k_x, k_y, k_z)$.

Taking into account that

$$\frac{\partial f_{0\alpha}}{\partial p_{\alpha j}} = \frac{\partial f_{0\alpha}}{\partial \mathscr{E}_\alpha} v_j + \frac{\partial f_{0\alpha}}{\partial C_\alpha}\frac{\partial C_\alpha}{\partial p_{\alpha j}} = \frac{\partial f_{0\alpha}}{\partial \mathscr{E}_\alpha} v_j + \frac{\delta_{yj}}{m_\alpha \Omega_\alpha}\frac{\partial f_{0\alpha}}{\partial x} , \tag{8.3.19}$$

and substituting (8.3.19) into (8.3.18) we obtain by simple transformations completely analogous to those made for homogeneous plasmas in Sect. 5.1

$$\delta f(\boldsymbol{k}, x) = \delta f_1(\boldsymbol{k}, x) + \delta f_2(\boldsymbol{k}, x) . \tag{8.3.20}$$

For the inhomogeneous Maxwellian plasma distributed according to (8.3.7, 8) we have

$$\begin{aligned}\delta f_1(\boldsymbol{k}, x) &= \frac{1}{T}\left(1 - \frac{k_y v_T^2}{\Omega\omega}\frac{\partial}{\partial x}\right) T\delta f^{(0)}(\boldsymbol{k}, x) , \\ \delta f_2(\boldsymbol{k}, x) &= -\frac{\mathrm{i}eE_y}{m\Omega\omega}\frac{\partial F(x, \boldsymbol{v})}{\partial x}\sum_{s,n} \mathrm{J}_s\left(\frac{k_\perp v_\perp}{\Omega}\right)\mathrm{J}_n\left(\frac{k_\perp v_\perp}{\Omega}\right) \mathrm{e}^{\mathrm{i}(n-s)(\xi-\phi)} .\end{aligned} \tag{8.3.21}$$

The species index α is omitted for simplicity here, and the following notations are introduced: ξ is the polar angle of the vector $\boldsymbol{k}$, i.e., $\boldsymbol{k} = (k_x = k_\perp \cos\xi, k_y = k_\perp \sin\xi, k_z)$; δf_0 is a function coinciding in form with the correction of the equilibrium distribution function of the homogeneous Maxwellian plasma (5.1.4), the only difference being that N and T are now assumed to depend on the coordinate x.

It is easy to show that δf_2 does not contribute to the current density induced in the plasma. Due to the plasma quasineutrality after summation over the species index of the charged particles the contribution of δf_2 to the density of the induced charge density vanishes, too. Thus, the densities of the space charge and the current and therefore the dielectric tensor are determined by the correction $\delta f_1(\boldsymbol{k}, x)$ alone. From the explicit form (8.3.21) of $\delta f_1(\boldsymbol{k}, x)$ the dielectric tensor of a weakly inhomogeneous nondegenerate plasma follows

$$\varepsilon_{ij}(\omega, \boldsymbol{k}, x) = \delta_{ij} + \sum_\alpha \frac{1}{T_\alpha}\left(1 - \frac{k_y v_{T\alpha}^2}{\omega \Omega_\alpha}\frac{\partial}{\partial x}\right) T_\alpha \left[\varepsilon_{ij}^\alpha(\omega, \boldsymbol{k}, x) - \delta_{ij}\right]. \qquad (8.3.22)$$

As in the preceding chapters $\varepsilon_{ij}^\alpha(\omega, \boldsymbol{k}, x)$ is the partial contribution of the particles of type α to the dielectric tensor. It coincides normally with the dielectric tensor of the homogeneous plasma (5.1.10). However, N_α and T_α are functions of the coordinate x, here.

8.3.4 The Larmor Drift Frequency

It should be noted that the tensor (5.1.10) is used in a coordinate system where the wave vector has the components $\boldsymbol{k} = (k_\perp, 0, k_z)$. In (8.3.22) the orientation of $\boldsymbol{k}$ is arbitrary, however: $\boldsymbol{k} = (k_x, k_y, k_z)$. Therefore, the tensor (5.1.10) must be transformed according to the general transformation rule applicable when the frame of coordinates is rotated. In general, however, this is not necessary as we will see in the next section. Note that the longitudinal dielectric permittivity is invariant with respect to the transformation of the coordinate system. Thus, in order to calculate the longitudinal dielectric permittivity we can use (5.1.13), keeping in mind that we have to apply the operator

$$\sum_\alpha \frac{1}{T_\alpha}\left(1 - \frac{k_y v_{T\alpha}^2}{\omega \Omega_\alpha}\frac{\partial}{\partial x}\right) T_\alpha \qquad (8.3.23)$$

to the components of this tensor in the case of an inhomogeneous plasma. As a result, we obtain

$$\varepsilon(\omega, \boldsymbol{k}, x) = 1 + \sum_\alpha \frac{\omega_{p\alpha}^2}{k^2 v_{T\alpha}^2}\left\{1 - \sum_n \frac{\omega}{\omega - n\Omega_\alpha}\left[1 - \frac{k_y v_{T\alpha}^2}{\omega \Omega_\alpha}\right.\right.$$
$$\left.\left.\times\left(\frac{\partial \ln N_\alpha}{\partial x} + \frac{\partial T_\alpha}{\partial x}\frac{\partial}{\partial T_\alpha}\right)\right] A_n\left(\frac{k_\perp^2 v_{T\alpha}^2}{\Omega_\alpha^2}\right) I_+\left(\frac{\omega - n\Omega_\alpha}{k_z v_{T\alpha}}\right)\right\}. \qquad (8.3.24)$$

Here $k_\perp$ is given by $k_\perp = \sqrt{k_x^2 + k_y^2}$.

As it is seen from (8.3.22, 24), there appears a new characteristic frequency

$$\omega_{dr\alpha} = k_y v_{dr\alpha} \sim \frac{k_y v_{T\alpha}^2}{\Omega_\alpha L_0} \qquad (8.3.25)$$

in the inhomogeneous Maxwellian plasma with the characteristic inhomogeneity scale L_0, which is called the *Larmor drift frequency* (we shall analyze the meaning and the physical nature of $\omega_{\mathrm{dr}\alpha}$ later). For high frequencies $\omega \gg \omega_{\mathrm{dr}\alpha}$ the terms of the tensor components $\varepsilon_{ij}(\omega, \boldsymbol{k}, x)$ which contain space derivatives can be neglected. Then the remaining components exactly coincide with the corresponding expressions for the components of the dielectric permittivity of the homogeneous plasma with space dependent N_α and T_α, however. Moreover, from the derivation of (8.3.22) it follows that in the limit $\omega \gg \omega_{\mathrm{dr}\alpha}$ the relations for the dielectric tensor components, taking account of particle collisions, are also valid (Sects. 5.5, 7). This holds true as well for the longitudinal and for the transverse dielectric permittivity of the isotropic plasma without external fields (Sects. 4.1, 5, 6).

8.3.5 The Case of the Degenerate Plasma

The relation (8.3.22) between the dielectric tensors of the weakly inhomogeneous and of the homogeneous plasma can be generalized to include the degenerate plasma case where the function $F_\alpha(\mathscr{E}_\alpha, x)$ has the form of the Fermi distribution function:

$$F_\alpha(\mathscr{E}_\alpha, x) = \begin{cases} \dfrac{2}{(2\pi\hbar)^3} & \text{for } \mathscr{E}_\alpha < \mathscr{E}_{\mathrm{F}\alpha}(x)\,, \\[2ex] 0 & \text{for } \mathscr{E}_\alpha > \mathscr{E}_{\mathrm{F}\alpha}(x)\,, \end{cases} \tag{8.3.26}$$

$\mathscr{E}_{\mathrm{F}\alpha}(x) = (3\pi^2)^{2/3}\, \hbar^2/2m_\alpha N_\alpha^{2/3}(x)/(2m_\alpha)$ being the local Fermi energy of the particles of type α with an inhomogeneous density $N_\alpha(x)$. The equilibrium condition (8.3.12) remains valid if p_0 is replaced by the gas-kinetic pressure of the degenerate plasma

$$p_0 = \sum_\alpha \frac{(3\pi^2)^{2/3}\hbar^2 N_\alpha^{5/3}}{5m_\alpha}\,. \tag{8.3.27}$$

The derivation of the dielectric tensor of the low-pressure degenerate plasma with $\beta = 8\pi p_0/B_0^2 \ll 1$ is completely analogous to the corresponding calculation for the nondegenerate plasma. Note that the relation

$$\frac{\partial f_{0\alpha}}{\partial \mathscr{E}_\alpha} + \frac{k_y}{m\Omega_\alpha\omega}\frac{\partial f_{0\alpha}}{\partial x} = \left(1 - \frac{k_y}{m\Omega_\alpha\omega}\frac{\partial \mathscr{E}_{\mathrm{F}\alpha}}{\partial x}\right)\frac{\partial f_{0\alpha}}{\partial \mathscr{E}_\alpha} \tag{8.3.28}$$

is valid for Fermian particles of the type α, distributed according to (8.3.26). Substituting (8.3.28) into (8.3.18) results in, cf. (8.3.22):

$$\varepsilon_{ij}(\omega, \boldsymbol{k}, x) = \delta_{ij} + \sum_\alpha \left(1 - \frac{2}{3}\frac{k_y v_{\mathrm{F}\alpha}^2}{\omega\Omega_\alpha}\frac{\partial \ln N_\alpha}{\partial x}\right)[\varepsilon_{ij}^{(\alpha)}(\omega, \boldsymbol{k}, x) - \delta_{ij}]\,, \tag{8.3.29}$$

where $\varepsilon_{ij}^{\alpha}(\omega, \boldsymbol{k}, x)$ is the partial dielectric tensor of the particles of the type α, coinciding in form with the dielectric tensor of the homogeneous degenerate plasma (5.1.14) with $N_\alpha = N_\alpha(x)$.

Finally, we present the longitudinal dielectric permittivity describing longitudinal oscillations of the degenerate plasma:

$$\varepsilon(\omega, \boldsymbol{k}, x) = 1 + \sum_\alpha \frac{3\,\omega_{p\alpha}^2}{k^2 v_{F\alpha}^2}\left[1 - \frac{\omega}{2}\sum_n \left(1 - \frac{2}{3}\frac{k_y v_{F\alpha}^2}{\omega\Omega_\alpha}\frac{\partial \ln N_\alpha}{\partial x}\right)\right.$$
$$\left.\times \int_0^\pi \frac{d\theta \sin\theta\, J_n^2(k_\perp v_{F\alpha}\sin\theta/\Omega_\alpha)}{\omega - k_z v_{F\alpha}\cos\theta - n\Omega_\alpha}\right] . \qquad (8.3.30)$$

As in the nondegenerate plasma there appears a new characteristic frequency, the drift frequency, which is here equal to

$$\omega_{dr\alpha} = k_y v_{dr\alpha} = \frac{2}{3}\frac{k_y v_{F\alpha}^2}{\Omega_\alpha L_0} , \qquad (8.3.31)$$

where L_0 is the characteristic length of the inhomogeneity of the particle density. For frequencies $\omega \gg \omega_{dr\alpha}$ the drift terms in (8.3.29, 30) can be neglected. In this limit the dielectric tensor has the same form as for the homogeneous plasma case. However, the particle density still depends on the parameter of the space coordinate.

8.4 Spectra of HF-Oscillations in Weakly Inhomogeneous Plasmas

We apply the general results, obtained in the preceding sections, to the analysis of high-frequency ($\omega \gg \omega_{dr\alpha}$) oscillations of weakly inhomogeneous plasma. We are especially interested in the limiting cases where analytical relations for the frequency spectra can be obtained.

8.4.1 Transverse Oscillations of Weakly Inhomogeneous Isotropic Plasma

To begin with, the transverse oscillations in the isotropic plasma are analyzed. In the limit of sufficiently infrequent particle collisions $\omega \gg \nu_e$, using (4.5.12) and (4.6.9), which define $\varepsilon^{tr}(\omega, \boldsymbol{k}, x)$, we get the eikonal equation for these oscillations in the form of

$$k^2 - \frac{\omega^2}{c^2}\left[1 - \frac{\omega_{pe}^2(x)}{\omega^2}\left(1 - i\frac{\nu_e(x)}{\omega}\right)\right] = 0 . \qquad (8.4.1)$$

The collision frequency is $\nu_e = \nu_{en}$ or $\nu_e = \nu_{eff}$ for the weakly or completely ionized plasma, respectively. Hence, we find

$$k_x^2(\omega, x) = -k_y^2 - k_z^2 + \frac{\omega^2}{c^2}\left(1 - \frac{\omega_{pe}^2}{\omega^2} + i\frac{\omega_{pe}^2 \nu_e}{\omega^3}\right). \tag{8.4.2}$$

To calculate the frequency spectrum and the damping decrement of these oscillations according to the technique described above, one has to determine $\mathrm{Re}\{k_x\}$ and $\mathrm{Im}\{k_x\}$ from (8.4.2) and to substitute the result into (8.2.8). Then, taking account of $\mathrm{Re}\{k_x\} \gg \mathrm{Im}\{k_x\}$, we obtain the dispersion equation of the transverse electromagnetic waves in an isotropic inhomogeneous plasma in the approximation of geometrical optics:

$$\int dx\, \mathrm{Re}\{k_x\} = \int dx \left(-k_y^2 - k_z^2 + \frac{\omega^2}{c^2} - \frac{\omega_{pe}^2}{c^2}\right)^{1/2} = \pi n\,,$$

$$\delta = -\frac{1}{2}\int \frac{dx}{\mathrm{Re}\{k_x\}} \frac{\omega_{pe}^2 \nu_e}{\omega^2} \left(\int \frac{dx}{\mathrm{Re}\{k_x\}}\right)^{-1}. \tag{8.4.3}$$

It can be seen from (8.4.3) that the frequency of the transverse waves always exceeds the local value of the electron plasma frequency $\omega^2 > \omega_{pe}^2(x)$ in the weakly inhomogeneous plasma. The range of transparency is determined by the condition that $\omega_{pe}^2(x)$ has to be smaller than the local value $\omega_{pe}^2(b)$ at the turning point $x = b$:

$$(k_y^2 + k_z^2)\, c^2 + \omega_{pe}^2(b) = \omega^2\,. \tag{8.4.4}$$

In the majority of real experiments the distribution of the charged particle density achieves a maximum at some point along the direction of the inhomogeneity and then falls off smoothly. Examples are the radial distribution of the charged particle density in the plasma of a gas-discharge, in the thermonuclear fusion plasma (Fig. 8.2) and also in the ionospheric plasma. It

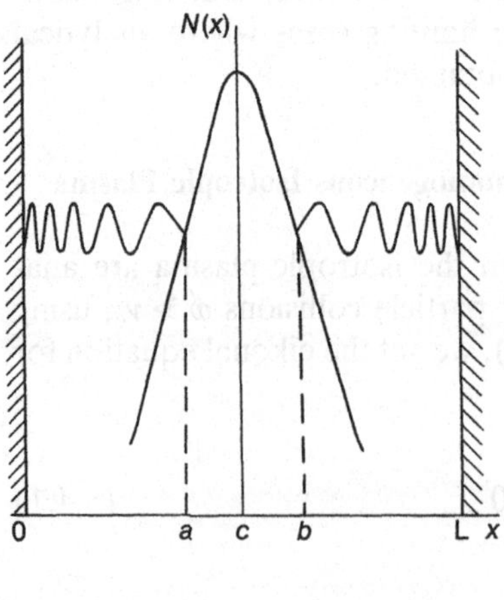

Fig. 8.2. Transparence range for transverse waves in an isotropic plasma with density decreasing towards the periphery

is easily seen that there exist two turning points a and b for frequencies $\omega^2 < \omega_{\rm pe}^2(c) \approx \omega_{\rm pe,\,max}$, the range of plasma opacity lying between them. For frequencies $\omega^2 > \omega_{\rm pe,\,max}^2$ the plasma is totally transparent, i.e., the *electromagnetic waves propagate freely* (the oscillations are untrapped) and their frequency spectrum is not discrete. The number of oscillations is unbounded, too, in the ranges of transparency lying to the left and to the right from the turning points a and b, respectively. Note that those assumptions are valid for the spatially infinite plasma with a density which smoothly goes to zero at infinity. The points 0 and L (metallic surfaces bounding the plasma) where the electromagnetic waves are reflected cannot be described within the model. Due to the presence of such points, the oscillations become trapped between the points 0 and a and between b and L. Therefore, the quantized spectra apply to these regions. Electromagnetic oscillations of longer wavelengths can be bound strongly due to the spectrum quantization.

8.4.2 The Langmuir Oscillations. The Tonks-Dattner Resonances

Next we consider the case of longitudinal waves, in particular high-frequency ($\omega \gg \nu_{\rm e}$, $kv_{\rm Te}$) longitudinal plasma oscillations. Using (4.5.12) and (4.6.8) in the high-frequency limit we obtain from the general eikonal equation (8.2.6)

$$1 - \frac{\omega_{\rm pe}^2}{\omega^2}\left(1 + 3\,\frac{k^2 v_{\rm Te}^2}{\omega^2} - {\rm i}\frac{\nu_{\rm e}}{\omega}\right) + {\rm i}\sqrt{\frac{\pi}{2}}\,\frac{\omega\omega_{\rm pe}^2}{k^3 v_{\rm Te}^3}\exp\left(-\frac{\omega^2}{2k^2 v_{\rm Te}^2}\right) = 0\,. \tag{8.4.5}$$

Solving this equation for $\mathrm{Re}\,\{k_x\}$ and $\mathrm{Im}\,\{k_x\}$ and substituting the result into (8.4.3) gives the dispersion equation for high-frequency longitudinal waves in the inhomogeneous plasma:

$$\int dx\,\mathrm{Re}\,\{k_x\} = \int dx\left(-k_y^2 - k_z^2 + \frac{\omega^2 - \omega_{\rm pe}^2}{3\,v_{\rm Te}^2}\right)^{1/2} = \pi n\,,$$

$$\delta = -\int \frac{dx}{\mathrm{Re}\,\{k_x\}}\,\frac{1}{v_{\rm Te}^2}\left[\sqrt{\frac{\pi}{2}}\,\nu_{\rm e} + \frac{\omega_{\rm pe}^4}{k^3 v_{\rm Te}^3}\exp\left(-\frac{3}{2} - \frac{1}{2k^2 r_{\rm De}^2}\right)\right]$$
$$\times\left(\int \frac{dx}{\mathrm{Re}\,\{k_x\}}\,\frac{1}{v_{\rm Te}^2}\right)^{-1}. \tag{8.4.6}$$

In the expression for δ the quantity k^2 means $k^2 = \mathrm{Re}\,\{k_x^2\} + k_y^2 + k_z^2$. The integration is performed over the range of transparency, i.e., over the range with $\mathrm{Re}\,\{k_x^2\} > 0$ and, consequently, $\omega^2 > \omega_{\rm pe}^2$. As for the transverse modes there exist two turning points, a and b, determined by $\omega^2 = 3\,(k_y^2 + k_z^2)\,v_{\rm Te}^2 + \omega_{\rm pe}^2(x)$ when the spatial density distribution is of the type shown in Fig. 8.2. Of course the range of plasma opacity lies between these points.

Thus, longitudinal plasma waves can exist only in the peripheral range of the plasma. However, in contrast to the transverse waves, the range of transparency is rather narrow, since a strong collisionless damping occurs when ω^2 differs slightly from ω_{pe}^2. As a result, the range of transparency is limited by the turning point on the one hand, and by the range of strong absorption, on the other hand. Thus, the oscillation spectrum is not quantized. The only exception is a particular plasma configuration where the plasma density changes smoothly first and then falls steeply near the boundaries. The plasma waves can reach the boundaries before they are damped in this case, and they are reflected from the wall. Then, we have to integrate in (8.4.6) over the domain lying between the plasma boundaries and the turning points a and b. As a result the oscillation spectrum becomes quantized. In gas-discharge devices these oscillations are called the *Tonks-Dattner resonances* when they are resonantly excited by external electric fields.

8.4.3 Ion-Acoustic Oscillations of the Inhomogeneous Isotropic Plasma

Another longitudinal mode is the low-frequency ion-acoustic mode existing in the inhomogeneous isotropic plasma with $T_e \gg T_i$. When the space charge depends on the coordinates as shown in Fig. 8.2, they are trapped in the plasma. The eikonal equation (8.2.6) for these waves (in the frequency range $kv_{Ti} \ll \omega \ll kv_{Te}$, see Sects. 4.5, 6) takes the form

$$1-\frac{\omega_{pi}^2}{\omega^2}\left(1-\mathrm{i}\alpha\frac{\nu_i}{\omega}\right)+\frac{\omega_{pe}^2}{k^2v_{Te}^2}\left(1+\mathrm{i}\sqrt{\frac{\pi}{2}}\frac{\omega}{kv_{Te}}\right)=0\,, \qquad (8.4.7)$$

where we have $\alpha = 1$, $\nu_i = \nu_{in}$ or $\alpha = 8k^2v_{Ti}^2/(5\omega^2)$, $\nu_i = \nu_{ii}$ for the weakly or completely ionized plasma, respectively.

We obtain from (8.4.7) for the ion-acoustic waves of the inhomogeneous plasma the following dispersion equation:

$$\int dx\,\mathrm{Re}\,\{k_x\} = \int dx\left(-k_y^2-k_z^2+\frac{\omega^2}{v_s^2}\frac{\omega_{pi}^2}{\omega_{pi}^2-\omega^2}\right)^{1/2} = \pi n\,,$$

$$\delta = -\frac{\omega^2}{2}\left[\int dx\frac{\omega_{pi}^2}{\mathrm{Re}\,\{k_x\}(\omega_{pi}^2-\omega^2)}\left(\sqrt{\frac{\pi}{2}}\frac{1}{kv_{Te}v_s^2}+\alpha\frac{k^2\nu_i}{\omega^4}\right)\right] \qquad (8.4.8)$$

$$\times\left[\int\frac{dx}{\mathrm{Re}\,\{k_x\}}\frac{\omega_{Li}^4}{v_s^2(\omega_{pi}^2-\omega^2)^2}\right]^{-1}.$$

According to the condition of weak damping, $k^2 = \mathrm{Re}\,\{k_x^2\} + k_y^2 + k_z^2$ has to be taken in the expression for δ. From (8.4.8) it follows that these oscillations exist in the frequency range $\omega^2 \ll \omega_{pi}^2(x)$. In the case of a bell-shaped density distribution (Fig. 8.3) they are trapped in the plasma between the points a

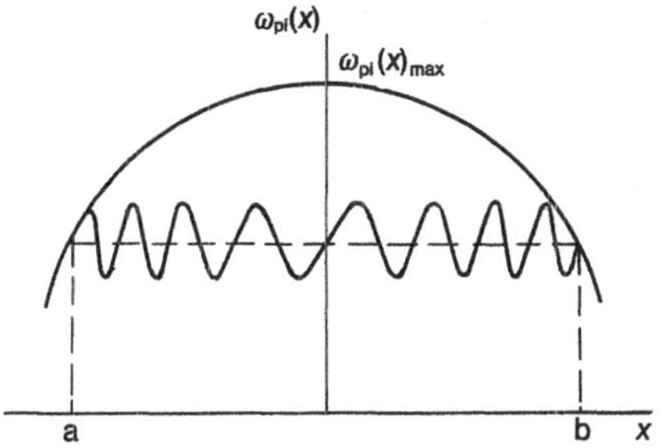

Fig. 8.3. Range of transparency of low-frequency oscillations of an isotropic plasma with bell-shaped density distribution

and b given by $\omega_{pi}^2(a) = \omega_{pi}^2(b) = \omega^2$. However, these points are not the turning points of the system. Near them $\mathrm{Re}\{k_x^2\} \to \infty$ holds and the applicability conditions of geometrical optics are not violated. On the contrary, they are fulfilled even better since the wavelength sharply decreases when the wave approaches these points. Consequently they are called the *clustering points*. The ion-acoustic waves are strongly damped near these points and, therefore, they are nonquantized. Only if there are walls of the plasma-limiting container between a and b reflecting the waves do they become quantized.

8.4.4 The Case of the Degenerate Isotropic Plasma

Note that the dispersion equations for transverse waves (8.4.4) and for longitudinal electron plasma oscillations (8.4.6) are also valid for the degenerate plasma if v_{Fe}^2 is substituted for v_{Te}^2 in (8.4.6). The exponentially small term corresponding to the Cherenkov absorption of the longitudinal waves by the plasma electrons has to be omitted in the expression for the damping decrement δ, however. In the degenerate plasma such an absorption mechanism is absent.

As for the ion-acoustic waves, we have in a plasma with degenerate electrons and nondegenerate ions the eikonal equation

$$1 - \frac{\omega_{pi}^2}{\omega^2}\left(1 + \mathrm{i}\alpha\frac{\nu_i}{\omega}\right) + \frac{3\,\omega_{pe}^2}{k^2 v_{Fe}^2}\left(1 + \mathrm{i}\frac{\pi}{2}\frac{\omega}{k v_{Fe}}\right) = 0 \tag{8.4.9}$$

instead of (8.4.7). The resulting dispersion equation changes from (8.4.8) to

$$\begin{aligned} &\int dx\,\mathrm{Re}\{k_x\} = \int dx\left(-k_y^2 - k_z^2 + \frac{\omega^2}{v_s^2}\frac{\omega_{pi}^2}{\omega_{pi}^2 - \omega^2}\right)^{1/2} = \pi n\,, \\ &\delta = -\frac{\omega^2}{2}\left[\int\frac{dx}{\mathrm{Re}\{k_x\}}\frac{\omega_{pi}^2}{\omega_{pi}^2 - \omega^2}\left(\frac{\pi}{2}\frac{1}{k v_{Fe} v_s^2} + \alpha\frac{k^2\nu_i}{\omega^4}\right)\right] \\ &\quad\times\left[\int\frac{dx}{\mathrm{Re}\{k_x\}}\frac{\omega_{pi}^2}{(\omega_{pi}^2 - \omega^2)\,v_s^2}\right]^{-1}, \end{aligned} \tag{8.4.10}$$

$v_s^2 = 3v_{Fe}^2 m/M$ being the ion-acoustic velocity of sound in the plasma with degenerate electrons. Here, as in the case of the nondegenerate plasma, oscillations can be excited in the range $\omega^2 < \omega_{pi}^2(x)$. In a plasma layer with a density distribution going to zero at the surface, these oscillations are trapped in the plasma between the clustering points a and b given by $\omega^2 = \omega_{pi}^2(a) = \omega_{pi}^2(b)$. Finally, as in the nondegenerate plasma, the spectrum of the ion-acoustic waves is not discrete in the degenerate plasma due to strong absorption near the clustering points.

The high-frequency $\omega \gg \omega_{dr\alpha}$ oscillation spectra of the weakly inhomogeneous magneto-active plasma are of great interest, too. As stated above, the terms with the space derivations can be neglected in the dielectric tensor of the weakly inhomogeneous plasma. Consequently, the eikonal equation takes the same form as the dispersion equation for the oscillations of the homogeneous plasma, the only difference being that due to the inhomogeneity of the plasma density and the temperature, the components of the tensor $\varepsilon_{ij}(\omega, \boldsymbol{k}, x)$ depend on the coordinate x.

8.4.5 Oscillations of the Weakly Inhomogeneous Magneto-Active Plasma

We confine our investigation to the cold magneto-active plasma when (5.2.1) is satisfied. Since the oscillation spectra do not depend on the thermal velocity in this limit, the results obtained above remain valid both for the non-degenerate and degenerate plasmas. The eikonal equation (8.2.5) is conveniently written as

$$k_\perp^4 \varepsilon_{xx} + k_\perp^2 \left[\left(k_z^2 - \frac{\omega^2}{c^2} \varepsilon_{xx} \right) (\varepsilon_{xx} + \varepsilon_{zz}) - \frac{\omega^2}{c^2} \varepsilon_{xy}^2 - \frac{\omega^2}{c^2} (\varepsilon_{xx} \varepsilon_{yy} - \varepsilon_{xx}^2) \right]$$

$$+ \varepsilon_{zz} \left[\left(k_z^2 - \frac{\omega^2}{c^2} \varepsilon_{xx} \right) \left(k_z^2 - \frac{\omega^2}{c^2} \varepsilon_{yy} \right) + \frac{\omega^4}{c^4} \varepsilon_{xy}^2 \right] = 0 , \qquad (8.4.11)$$

where $k_\perp^2 = k_x^2 + k_y^2$, and the components of the tensor $\varepsilon_{ij}(\omega, \boldsymbol{k}, x)$ are given by (5.2.3, 3.7, 5.6) and (5.6.7). It is important that these components do not depend on the perpendicular projection of the wave vector $k_\perp$. Thus, from (8.4.11) we find the two solutions $k_{x1,2}^2(\omega, x)$, which correspond to the ordinary and extraordinary waves in the cold magneto-active plasma. Neglecting the small dissipation associated with the anti-Hermitian part of the tensor $\varepsilon_{ij}(\omega, \boldsymbol{k}, x)$ and determining $\mathrm{Re}\{k_x(\omega, x)\}$ from (8.4.11), we obtain with use of the quantization rule (8.2.8) the dispersion equation:

$$\int dx \, \mathrm{Re}\{k_x\} = \int dx \, (-k_y^2 - p \pm \sqrt{p^2 - q})^{1/2} = \pi n . \qquad (8.4.12)$$

Here the following notations are introduced:

$$p = \frac{1}{2\varepsilon_\perp^{\mathrm{H}}}\left[\left(k_z^2 - \frac{\omega^2}{c^2}\varepsilon_\perp^{\mathrm{H}}\right)(\varepsilon_\perp^{\mathrm{H}} + \varepsilon_\parallel^{\mathrm{H}}) + \frac{\omega^2}{c^2}g^{\mathrm{H}2}\right],$$

$$q = \frac{\varepsilon_\parallel^{\mathrm{H}}}{\varepsilon_\perp^{\mathrm{H}}}\left[\left(k_z^2 - \frac{\omega^2}{c^2}\varepsilon_\perp^{\mathrm{H}}\right)^2 - \frac{\omega^4}{c^4}g^{\mathrm{H}2}\right]. \tag{8.4.13}$$

The frequency dependence of these oscillation branches can be determined from (8.4.12) (Sects. 5.2–7). We consider in detail the spectra of the Alfvén, the fast magneto-sonic, and the helical waves only.

To analyze the Alfvén and the fast magneto-sonic waves, we have to study the frequency range $\omega \ll \Omega_\mathrm{i}$. It is easily seen that (8.4.12) goes over to

$$\int dx\,\mathrm{Re}\,\{k_x\} = \int dx\left[-k_y^2 - k_z^2 + \frac{\omega^2}{c^2}\left(1 + \frac{c^2}{v_\mathrm{A}^2}\right)\right]^{1/2} = \pi n\,, \tag{8.4.14}$$

$$\int dx\,\mathrm{Re}\,\{k_x\} = \int dx\left[-k_y^2 - \frac{\omega_{\mathrm{pe}}^2}{c^2}\left(1 - \frac{k_z^2 v_\mathrm{A}^2}{\omega^2(1 + v_\mathrm{A}^2/c^2)}\right)\right]^{1/2} = \pi n \tag{8.4.15}$$

in this limit. In the homogeneous plasma the first oscillation branch corresponds to the fast magneto-sonic wave and the second one to the Alfvén wave. The frequency spectra are given by (5.2.16) in this case. Further, it follows from (8.4.14, 15) that the turning points of fast magneto-sonic waves are determined by $\omega^2 = (k_y^2 + k_z^2)\,v_\mathrm{A}^2(x)$ and that the range of transparency is given by the inequality $\omega^2 > (k_y^2 + k_z^2)\,v_\mathrm{A}^2(x)$. For a bell-shaped spatial distribution of the plasma density (Fig. 8.2) this implies that the fast magneto-sonic waves are trapped inside the plasma between the turning points, their spectra thus being quantized. The Alfvén waves of the inhomogeneous plasma can exist in the range $\omega^2 < k_z^2 v_\mathrm{A}^2(x)$, i.e., there are two ranges of transparency. These ranges are situated in the plasma periphery at values of the space coordinate x, smaller or larger than the respective coordinate of the turning points given by $\omega^2 = k_z^2 v_\mathrm{A}^2(x)$. In the plasma with a free surface these ranges are of infinite extension and the oscillations are not trapped. Therefore, their spectra are not quantized. In fact, under laboratory conditions, there always exist walls confining the plasma and limiting the peripheral range of transparency with respect to the propagation of the Alfvén wave. Incidentally, with increasing distance from the centre of the plasma layer the condition $\omega^2 < k_z^2 v_\mathrm{A}^2(x)$ can be violated due to the growth of the Alfvén velocity $v_\mathrm{A}(x)$ and (8.4.15) may lose its sense.

Finally, we consider the helical waves, also called the helicons, for the weakly inhomogeneous cold plasma case. As shown in Sects. 5.2–7, these waves can exist only in the intermediate frequency range $\Omega_\mathrm{i} \ll \omega \ll \Omega_\mathrm{e}$ and if the plasma is sufficiently dense. Under the condition $\omega_{\mathrm{pe}}^2 \gg \omega\Omega_\mathrm{e}$ it is easy to obtain from (8.4.11) an approximation for $\mathrm{Re}\{k_x(\omega, x)\}$ and to determine the frequency spectrum of the helical waves from the dispersion equation:

$$\int dx\, \mathrm{Re}\{k_x\} = \int dx \left(-k_y^2 - k_z^2 + \frac{\omega_{\mathrm{pe}}^4 \omega^2}{c^4 k_z^2 \Omega_{\mathrm{e}}^2}\right)^{1/2} = \pi n. \tag{8.4.16}$$

This relation follows directly from (8.4.12) if we account for the explicit form of the dielectric tensor in the given frequency range. Equation (8.4.16) shows that the helical waves can propagate in the inhomogeneous plasma only in the frequency ranges where

$$\omega_{\mathrm{pe}}^2(x) \geqslant \sqrt{c^4 k_z^2 (k_y^2 + k_z^2) \Omega_{\mathrm{e}}^2 / \omega^2}\,.$$

When the spatial distribution of the plasma density is bell-shaped, the range of transparency of these waves lies inside the plasma between the turning points a and b given by $\omega_{\mathrm{pe}}^2(a) = \omega_{\mathrm{pe}}^2(b) = \sqrt{c^4 k_z^2 (k_y^2 + k_z^2) \Omega_{\mathrm{e}}^2 / \omega^2}$. Therefore, the helical waves are trapped in the plasma and their spectra are quantized.

We completely neglected dissipative effects in our analysis of the spectra of the Alfvén, magneto-sonic and helical waves. The account of dissipation leads to the appearance of an imaginary term in (8.4.11) and thus to an imaginary part $\mathrm{Im}\{k_x(\omega, k)\}$ of the wave vector. Also a damping decrement δ of the complex wave spectrum appears. Since the explicit expressions for δ are rather complicated, we do not give them here. Moreover, there exist no principally new effects differing from those studied in Chap. 5 for the case of a spatially homogeneous plasma.

8.5 Drift Oscillations of a Weakly Inhomogeneous Collisionless Plasma

As shown above, the inhomogeneity of the plasma does not give rise to the appearance of new oscillation spectra in the high-frequency range satisfying the condition $\omega \gg \omega_{\mathrm{dr}\alpha}$, where $\omega_{\mathrm{dr}\alpha}$ is defined by (8.3.25). The reverse is true for the low-frequency range

$$\omega \lesssim \omega_{\mathrm{dr}\alpha} \approx \frac{k_y v_{\mathrm{T}\alpha}^2}{\Omega_\alpha L_0} \tag{8.5.1}$$

which we investigate in this section. The new characteristic frequency appearing in (8.5.1) is called the Larmor drift frequency $\omega_{\mathrm{dr}\alpha}$. We will show that a new mode of oscillation can be excited here, in particular when the frequencies are close to $\omega_{\mathrm{dr}\alpha}$.

8.5.1 Larmor Drift in the Inhomogeneous Plasma

To begin with, we discuss in detail the physical meaning of this drift frequency. Considering a steady state inhomogeneous plasma immersed into a magnetic field, it follows from (8.3.11) that the plasma current differs from zero and that it flows perpendicularly to the magnetic field $\boldsymbol{B}_0 \parallel 0z$ and to the direction of the plasma inhomogeneity (to the $0z$-axis) as well. In general the current density can be written as

$$j_{\mathrm{i}} = \sum_{\alpha} j_{\mathrm{i}\alpha} = \frac{c\,[\boldsymbol{B}_0, \nabla p_0]_{\mathrm{i}}}{B_0^2} . \tag{8.5.2}$$

This current can be compared with the effective drift velocity of the particles of type α oriented parallel to the $0y$-axis:

$$\begin{aligned} \boldsymbol{v}_{\mathrm{dr}\alpha} &= \frac{\boldsymbol{j}_\alpha}{e_\alpha N_\alpha} = \frac{c}{e_\alpha N_\alpha} \frac{[\boldsymbol{B}_0, \nabla p_{0\alpha}]}{B_0^2} , \\ v_{\mathrm{dr}\alpha} &\sim \frac{cT_\alpha}{e_\alpha B_0 L_0} \sim \frac{v_{\mathrm{T}\alpha}^2}{\Omega_\alpha} \cdot \frac{1}{L_0} . \end{aligned} \tag{8.5.3}$$

This drift motion is the well-known *Larmor drift* of the particles. We see that charged particles with different sign of the charge are drifting in opposite directions. Note that this drift does not correspond to a real motion of the guiding centres of the charged particles (the centres of the Larmor circles which are the particle orbits in the magnetic field). From Fig. 8.4 the nature of such a drift can be seen clearly. Here we show two circular Larmor orbits and two elementary currents in the inhomogeneous plasma. There results a current perpendicular to the magnetic field (the $0z$-axis perpendicular to the plane of the drift) and perpendicular to the direction of the plasma inhomogeneity (the $0x$-axis) in the plasma. The density of this current

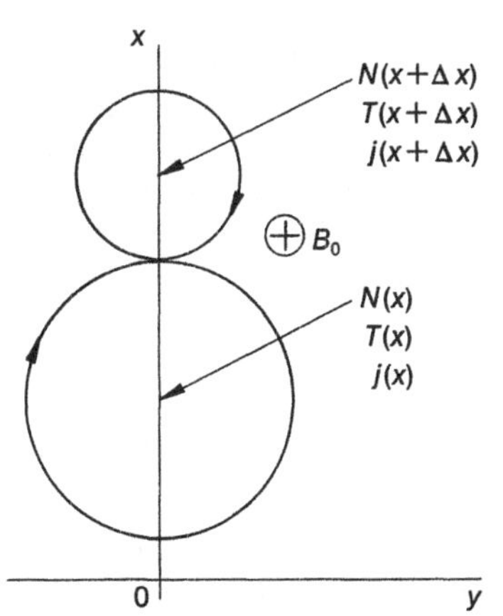

Fig. 8.4. Larmor drift of particles in a magnetized inhomogeneous plasma

$$j_{y\alpha} \sim [e_\alpha N_\alpha(x+\Delta x)\,v_{T\alpha}(x+\Delta x) - e_\alpha N_\alpha(x)\,v_{T\alpha}(x)]$$
$$\sim e_\alpha \Delta x \,\frac{\partial N_\alpha v_{T\alpha}}{\partial x} \sim e_\alpha \varrho_{\lambda\alpha}\frac{\partial N_\alpha v_{T\alpha}}{\partial x} \sim e_\alpha N_\alpha \frac{v_{T\alpha}^2}{\Omega_\alpha L_0} \tag{8.5.4}$$

coincides with the current density of the form of (8.5.2), $\varrho_{\lambda\alpha} = v_{T\alpha}/\Omega_\alpha$ being the Larmor radius of the particles of the type α.

The description given shows that this current exists in the steady state due to the diamagnetic effect and not due to a real motion of the charges. Then $j_{y\alpha}$ must be interpreted as a differential diamagnetic current in the inhomogeneous plasma. Nevertheless, the Larmor drift can lead to the development of specific instabilities, as for example the beam instability. The instabilities associated with drift motions in a plasma are called *drift instabilities*. They possess qualitatively new characteristics. In particular, they can develop even in a plasma with Maxwellian velocity distributions of the particles. The density or temperature must be inhomogeneous, however.

The physical nature of the quantity $\omega_{\mathrm{dr}\alpha}$ becomes clear from (8.5.1–4). It can be regarded as the Doppler shift of the frequency due to the Larmor drift of the particles. In the thermonuclear plasma ($N \sim 10^{14}$–10^{15} cm^{-3}, $T \sim 10^8$ K, $L_0 \sim 10$ cm, $B_0 \sim 10^5$ Gauss) we have $v_{\mathrm{dr}} \sim 10^5$ cm/s and $\omega_{\mathrm{dr}} \sim 10^4$ s^{-1} for $k_y \sim L_0^{-1}$. In the plasma of a gas-discharge ($N \sim 10^{10}$–10^{12} cm^{-3}, $T \sim 10^4$–10^5 K, $L_0 \sim 1$ cm and $B_0 \sim 10^3$–10^4 Gauss) the order of magnitude is $v_{\mathrm{dr}} \sim 10^5$–$10^6$ cm/s and $\omega_{\mathrm{dr}} \sim 10^5$–$10^6$ s^{-1}. In the ionospheric plasma ($N \sim 10^7$ cm^{-3}, $T \sim 10^4$ K, $L_0 \sim 10$–30 km and $B_0 \sim 1$ Gauss) we obtain $v_{\mathrm{dr}} \sim 10^2$ cm/s and $\omega_{\mathrm{dr}} \sim 10^4$ s^{-1}. Finally, in the degenerate solid state plasma with a Fermi energy $\mathscr{E}_{\mathrm{F}} \sim 0.1$–1 eV (typical for semiconductors and metals) situated in a magnetic field of $B_0 \sim 10^4$ Gauss and with an inhomogeneity scale $L_0 \sim 1$ cm we have $v_{\mathrm{dr}} \sim 10^4$–$10^5$ cm/s and $\omega_{\mathrm{dr}} \sim 10^4$–$10^5$ s^{-1}. It follows from these estimates that the drift frequencies are much smaller than the electron and ion Larmor frequencies, which significantly simplifies the analysis of the drift oscillations of the inhomogeneous magnetized plasma.

To simplify as much as possible we make the following assumptions.

Firstly, we confine our interest to the low-pressure plasma $\beta \ll 1$. In this limit the low-frequency drift oscillations are longitudinal, with a high degree of accuracy. Our analysis of the oscillation spectra of the homogeneous magneto-active plasma confirms this fact (Chap. 5). Further, for $\beta \ll 1$ the plasma oscillations cannot significantly perturb a strong external magnetic field, which consequently remains constant. For longitudinal plasma oscillations the eikonal equation is of the form (8.2.6), i.e.,

$$\varepsilon(\omega, \boldsymbol{k}, x) = 0\,, \tag{8.5.5}$$

where the longitudinal dielectric permittivity $\varepsilon(\omega, \boldsymbol{k}, x)$ is given by (8.3.24) for the nondegenerate plasma and by (8.3.29) for the degenerate plasma.

Secondly, we are interested in the analogs of the local spectra only, which are directly defined by (8.5.5). It is important to obtain the qualitatively new results first, i.e., the spectra and the stability properties of the inhomogeneous plasma. Exact quantitative results can be easily obtained later by means of the quantization rules. In the approximation of geometrical optics the use of the eikonal equation as a local dispersion equation is justified for very short wavelengths compared to the scale of the plasma inhomogeneity. This approximate analysis corresponds to the evaluation of exact integral relations with the aid of the mean-value theorem.

8.5.2 The Dispersion Equation for Drift Oscillations

Consequently, the local dispersion equation for the drift oscillations of the inhomogeneous nondegenerate plasma is

$$\varepsilon(\omega, \boldsymbol{k}, x) = 1 + \sum_{a} \frac{\omega_{\mathrm{p}a}^{2}}{k^{2} v_{\mathrm{T}a}^{2}} \left\{ 1 - \sum_{n} \frac{\omega}{\omega - n\Omega_{a}} \left[1 - \frac{k_{y} v_{\mathrm{T}a}^{2}}{\omega \Omega_{a}} \right.\right.$$

$$\left.\left. \times \left(\frac{\partial \lambda v N_{a}}{\partial x} + \frac{\partial T_{a}}{\partial x} \frac{\partial}{\partial T_{a}} \right) \right] A_{n} \left(\frac{k_{\perp}^{2} v_{\mathrm{T}a}^{2}}{\Omega_{a}^{2}} \right) I_{+} \left(\frac{\omega - n\Omega_{a}}{k_{z} v_{\mathrm{T}a}} \right) \right\} = 0 \,. \quad (8.5.6)$$

The evaluation of this equation is rather complex (Chap. 5). Therefore, the complete analysis is impossible. We assume that the frequency of the drift oscillations satisfies the inequality $\omega \ll \Omega_{\mathrm{i}}$ and that the longitudinal wavelength is much larger than the Larmor radius $k_{z} v_{\mathrm{T}a} \ll \Omega_{a}$. Under these conditions only the term $n = 0$ appears to be significant in (8.5.6). The contributions of the higher harmonics can be neglected and the eikonal equation for longitudinal oscillations becomes

$$\varepsilon(\omega, \boldsymbol{k}, x) = 1 + \sum_{a} \frac{\omega_{\mathrm{p}a}^{2}}{k^{2} v_{\mathrm{T}a}^{2}} \left\{ 1 - \left[1 - \frac{k_{y} v_{\mathrm{T}a}^{2}}{\omega \Omega_{a}} \left(\frac{\partial \ln \mathrm{N}_{a}}{\partial x} + \frac{\partial T_{a}}{\partial x} \frac{\partial}{\partial T_{a}} \right) \right] \right.$$

$$\left. \times A_{0} \left(\frac{k_{\perp}^{2} v_{\mathrm{T}a}^{2}}{\Omega_{a}^{2}} \right) I_{+} \left(\frac{\omega}{k_{z} v_{\mathrm{T}a}} \right) \right\} = 0 \,. \quad (8.5.7)$$

Note that (8.5.7) has no zero for $\omega \ll k_{z} v_{\mathrm{Ti}}$. Oscillations of the plasma are impossible in this range. Instead, screening of a potential field by the plasma occurs. In the collisionless magnetized plasma the quantity $1/(k_{z} v_{\mathrm{Ti}}) \sim \lambda_{\parallel}/v_{\mathrm{Ti}}$ characterizes the time interval during which the density and the temperature of the electrons as well as the ions relax towards equilibrium at the distance of the wavelength due to the free flight of the particles. Thus, the condition $\omega \ll k_{z} v_{\mathrm{Ti}}$ implies that the relaxation time of the density and temperature perturbations is short compared to the oscillation period. Obviously, longitudinal waves cannot exist under these conditions. Thus, it is sufficient to analyze the case $\omega \gg k_z v_{\mathrm{Ti}}$ and to neglect the ion Landau damping which is exponentially small here.

8.5.3 Spectra of the Fast Long-Wavelength Drift Oscillations

Further, we consider drift oscillations of long wavelengths only which satisfy the condition $\lambda_\perp \gg \varrho_{\lambda\mathrm{i}}$ $(\lambda_\perp \sim 1/k_\perp)$. From the point of view of creating a stable plasma state, the long-wavelength oscillations are most dangerous since they can perturb rather large plasma regions in radial direction. Oscillations of short wavelengths are less dangerous since they perturb comparatively small plasma regions.

Another important parameter in the analysis of drift oscillations is the ratio of their phase velocity to the thermal velocity of the particles. One distinguishes the fast, $\omega/k_z \gg v_{\mathrm{Te}}$, and the slow $v_{\mathrm{Ti}} \ll \omega/k_z \ll v_{\mathrm{Te}}$ drift oscillations. We begin our analysis with the fast long-wavelength oscillations $\lambda_\perp \gg \varrho_{\lambda\mathrm{i}}$, $\omega/k_z \gg v_{\mathrm{Te}}$. In this case the eikonal equation (8.5.7) is of the form

$$1-\frac{\omega_{\mathrm{pe}}^2}{\omega^2}\frac{k_z^2}{k^2}\left[1-\frac{k_y v_{\mathrm{Te}}^2}{\omega\Omega_{\mathrm{e}}}\frac{\partial\ln(NT_{\mathrm{e}})}{\partial x}\right]+\frac{k_\perp^2\omega_{\mathrm{pi}}^2}{k^2\Omega_{\mathrm{i}}^2}\left[1-\frac{k_y v_{\mathrm{Ti}}^2}{\omega\Omega_{\mathrm{i}}}\frac{\partial\ln(NT_{\mathrm{i}})}{\partial x}\right]=0\,. \tag{8.5.8}$$

Under the condition $c^2 \gg v_{\mathrm{A}}^2$ $(\omega_{\mathrm{pi}}^2 \gg \Omega_{\mathrm{i}}^2)$ we have in the low-frequency range $\omega \ll \omega_{\mathrm{dra}}$

$$\frac{k_\perp^2 c^2}{v_{\mathrm{A}}^2}\frac{k_y v_{\mathrm{Ti}}^2}{\Omega_{\mathrm{i}}}\frac{\partial\ln(NT_{\mathrm{i}})}{\partial x}-k_z^2\frac{\omega_{\mathrm{pe}}^2}{\omega^2}\frac{k_y v_{\mathrm{Te}}^2}{\Omega_{\mathrm{e}}}\frac{\partial\ln(NT_{\mathrm{e}})}{\partial x}=0\,. \tag{8.5.9}$$

Using this relation as local dispersion equation we obtain the following local spectrum[1]:

$$\omega^2=-\frac{k_z^2}{k_\perp^2}\frac{T_{\mathrm{e}}}{T_{\mathrm{i}}}\frac{M}{m}\Omega_{\mathrm{i}}^2\frac{\partial\ln(NT_{\mathrm{e}})}{\partial\ln(NT_{\mathrm{i}})}\,. \tag{8.5.10}$$

In (8.5.9, 10) $k_\perp^2$, as usual, denotes $k_\perp^2 = k_y^2 + \mathrm{Re}\,\{k_x^2\} \approx k_y^2 + \pi^2 n^2/L_\perp^2$, where $L_\perp$ is the extension of the device in the direction of the $0x$-axis since this is at the same time the order of the characteristic length of the plasma inhomogeneity L_0.

1 Here and in the following we use the abbreviation

$$\frac{\partial\ln A}{\partial\ln B}=\frac{\partial\ln A}{\partial x}\bigg|\frac{\partial\ln B}{\partial x}=\frac{B}{A}\frac{\partial A}{\partial x}\bigg|\frac{\partial B}{\partial x}\,.$$

8.5.4 Universal Instability of the Inhomogeneous Plasma

We have obtained a qualitatively new oscillation branch which does not occur in the spatially homogeneous plasma. This branch is aperiodically unstable when the inequality

$$\frac{\partial \ln (NT_e)}{\partial \ln (NT_i)} > 0 \tag{8.5.11}$$

is satisfied. Almost all kinds of real plasmas are subject to this instability. Since the electron and ion pressure generally fall from the centre to the plasma boundaries, and since the particle temperature T decreases on a longer scale than the density N, the derived instability is called *universal*. However, for the excitation of this instability the condition $\omega^2 \ll \Omega_i^2$ leads to

$$\frac{k_z^2}{k_\perp^2} \sim \frac{L_\perp^2}{L_\parallel^2} \ll \frac{m}{M}\frac{T_i}{T_e}\frac{\partial \ln (NT_i)}{\partial \ln (NT_e)} \sim \frac{m}{M} \, . \tag{8.5.12}$$

The instability can develop in sufficiently long devices only, with a longitudinal plasma extension at least $\sqrt{M/m} \gtrsim 40$ times larger than the transverse one.

The universal instability is purely hydrodynamic and not related to the Cherenkov energy dissipation. Due to the fact that this instability is caused by the guiding centre drift in the plasma it is one special case of the *drift instabilities*.

8.5.5 Spectra of the Slow Long-Wavelength Drift Oscillations

The requirement regarding the longitudinal plasma dimension appears less restrictive for the domain of existence of long-wavelength drift oscillations in the range of phase velocities $v_{Ti} \ll \omega/k_z \ll v_{Te}$. In this frequency range the eikonal equation (8.5.7) takes the form

$$\begin{aligned} 1 &+ \frac{\omega_{pe}^2}{k^2 v_{Te}^2}\left[1 + \frac{k_y v_s^2}{\omega \Omega_i}\frac{\partial \ln N}{\partial x} - \frac{k_z^2 v_s^2}{\omega^2}\left(1 - \frac{k_y v_{Ti}^2}{\omega \Omega_i}\frac{\partial \ln (NT_i)}{\partial x}\right)\right] \\ &+ \mathrm{i}\frac{\omega_{pe}^2}{k^2 v_{Te}^2}\sqrt{\frac{\pi}{2}}\frac{\omega}{|k_z| v_{Te}}\left(1 - \frac{k_y v_{Te}^2}{\omega \Omega_e}\frac{\partial}{\partial x}\ln \frac{N}{\sqrt{T_e}}\right) = 0 \, . \end{aligned} \tag{8.5.13}$$

For the isothermal plasma ($T_e \sim T_i$) the last term in the square brackets can be neglected. Then, it is easy to obtain the local spectrum for $\omega \gg k_z v_s$:

$$\begin{aligned} \omega_1 &= -\frac{k_y v_s^2}{\Omega_i}\frac{\partial \ln N}{\partial x} \, , \\ \delta_1 &= \sqrt{\frac{\pi}{2}}\frac{\omega_1^2}{|k_z| v_{Te}}\left(k^2 r_{De}^2 + \frac{k_\perp^2 v_s^2}{\Omega_i^2} - \frac{1}{2}\frac{\partial \ln T_e}{\partial \ln N}\right) . \end{aligned} \tag{8.5.14}$$

8.5.6 The Drift-Dissipative and Drift-Temperature Instabilities

We see that a kinetically unstable slow oscillation mode of long wavelength can be excited if

$$\frac{\partial \ln T_e}{\partial \ln N} < 2\left(k^2 r_{De}^2 + \frac{k_\perp^2 v_s^2}{\Omega_i}\right) . \tag{8.5.15}$$

These oscillations of the inhomogeneous plasma are called the *drift dissipative oscillations*. Here, the Cherenkov mechanism of dissipation by the electrons is responsible for the buildup of the oscillations. In the range of the drift frequencies the Cherenkov term can have the opposite sign and produce a buildup of the oscillations. Note that these oscillations have a frequency larger than the ion-acoustic frequency: $\omega_1 \gg k_z v_s$. In the homogeneous isothermal plasma ($T_e \sim T_i$) no oscillations can arise in this frequency range.

The last term in the square brackets of (8.5.13), which is proportional to $k_z v_s/\omega$, becomes significant in the nonisothermal plasma with $T_e \gg T_i$. This term can drive oscillations in the frequency range $\omega^2 \ll k_z v_s^2$, i.e., at the frequencies smaller than the ion-acoustic frequency. Under these conditions it is easy to obtain from (8.5.4) the local spectrum ($\omega \to \omega + i\delta$)

$$\omega_2 = \frac{k_z^2 \Omega_i}{k_y\, \partial \ln N/\partial x} , \qquad \delta_2 = -\sqrt{\frac{\pi}{2}}\frac{\omega_2^2}{|k_z|\, v_{Te}}\left(1 - \frac{1}{2}\frac{\partial \ln T_e}{\partial \ln N}\right) . \tag{8.5.16}$$

It can be interpreted as the continuation of the acoustic branch into the low-frequency range $\omega \ll k_z v_s$. From (8.5.16) we see that the electron Cherenkov dissipation drives these oscillations unstable if

$$\frac{\partial \ln T_e}{\partial \ln N} > 2 . \tag{8.5.17}$$

As in the special case before, the instability is kinetic and can be called a drift-dissipative instability, too.

Finally, in the range of very low frequencies $\omega \ll \omega_{dra}$ we obtain from (8.5.13) two more hydrodynamically unstable oscillation modes:

$$\omega_3^2 = -k_z^2 v_{Ti}^2 \frac{\partial \ln T_i}{\partial \ln N} , \tag{8.5.18}$$

$$\omega_4^2 = -k_z^2 v_s^2 \frac{k_y v_{Ti}^2}{\Omega_i} \frac{\partial \ln T_i}{\partial x} . \tag{8.5.19}$$

They can exist when $(\partial \ln T_i/\partial \ln N) \gg 1$ and therefore are called *drift-temperature instabilities*.

We already mentioned that in the range of intermediate phase velocities $v_{Ti} \ll \omega/k_z \ll v_{Te}$ the long-wavelength drift oscillations can be excited in relatively short plasma devices. From $\omega \sim \omega_{dra} \gg k_z v_{Ti}$ it follows that the condition $L_\parallel/L_\perp \sim L_\perp/\varrho_{\lambda i} > 1$ is sufficient for their occurrance. On the other hand, the neglect of particle collisions ($\nu_\alpha \ll k_z v_{T\alpha}$) implies that the longitudinal extension of the system should be smaller than the mean free path of the particles, which gives $L_\parallel < v_{T\alpha}/\nu_\alpha \equiv l_\alpha$. Thus, the condition of validity for the collisionless description of long-wavelength drift oscillations can be written as

$$1 < \frac{L_\perp}{\varrho_{\lambda i}} < \frac{L_\parallel}{L_\perp} < \frac{v_{T\alpha}}{\nu_\alpha}\frac{1}{L_\perp} \sim \frac{l_\alpha}{L_\perp}\,. \tag{8.5.20}$$

Strictly speaking, the following conditions must be satisfied to ensure the validity of the picture of this section. Either the frequencies and increments of the drift oscillations must greatly exceed the frequencies of the particle collisions, $|\omega| \gg \nu_\alpha$, or the mean free path of the particles must exceed the longitudinal wavelength of the drift oscillations, $\nu_\alpha \ll k_z v_{T\alpha}$. Only under these conditions can the Vlasov kinetic equation, which completely ignores particle collisions, be applied. Since in real devices the drift frequencies are of the order of 10^4 to $10^6\ \mathrm{s}^{-1}$, this limitation can be satisfied in high-temperature nondegenerate plasmas at relatively small charged particle densities, only. In the degenerate cold solid-state plasma this condition is not fulfilled, therefore we do not discuss such a plasma here. However, it will be shown in the next section that the drift instabilities can develop in a dense plasma with a large number of particle collisions, too. Even more, in the inhomogeneous plasma the collisional friction, especially due to the electron collisions, can become the driving force which excites drift instabilities.

8.6 Influence of Charged Particle Collisions on the Spectra of Drift Oscillations in Weakly Inhomogeneous Plasmas

In order to investigate the effect of particle collisions on the spectra of drift oscillations in the weakly inhomogeneous plasma, we first formulate the dielectric permittivity taking account of particle collisions. As in Sect. 8.3, we assume the plasma kinetic pressure to be low as compared to the magnetic pressure, i.e., $\beta \ll 1$. This allows us to neglect the inhomogeneity of the external magnetic field and to confine our consideration to the derivation of the longitudinal dielectric permittivity $\varepsilon(\omega, \boldsymbol{k}, x)$ which describes the spectra of the longitudinal oscillations.

8.6.1 Weakly Ionized Plasma

To begin with, we deduce the dielectric permittivity $\varepsilon(\omega, \boldsymbol{k}, x)$ for the case of a weakly ionized nondegenerate plasma when the collisions between the charged particles and the neutrals prevail. Under these circumstances the basic equation is the kinetic equation with the model BGK integral (Sect. 3.5). Since we are concerned with drift oscillations we may assume $\Omega_\alpha \gg \nu_\alpha$, here. Only in this limit can one speak of the Larmor rotation and, consequently, of the guiding centre drift of the particles.

The calculation of the equilibrium distribution function $f_{0\alpha}$ is similar to the analogous calculation in the collisionless approximation. Neglecting the collision integral, the distribution function $f_{0\alpha}$ can be written as (see Sect. 8.3)

$$f_{0\alpha}(C_\alpha, \mathscr{E}_\alpha) = \frac{N_\alpha(C_\alpha)}{[2\pi m_\alpha T_\alpha(C_\alpha)]^{3/2}} \exp\left(-\frac{m_\alpha v^2}{2T_\alpha(C_\alpha)}\right), \tag{8.6.1}$$

where $C_\alpha = x + v_y/\Omega_\alpha$ is the characteristic of the collisionless kinetic equation.

To investigate a small deviation of the distribution function from equilibrium

$$\delta f_\alpha = \delta f_\alpha(x) \exp(-\mathrm{i}\omega t + \mathrm{i}k_y y + \mathrm{i}k_z z) \tag{8.6.2}$$

in the zero-order approximation of geometrical optics, we study the linearized kinetic equation

$$-\mathrm{i}(\omega - \boldsymbol{k}\cdot\boldsymbol{v})\,\delta f_\alpha - \Omega_\alpha \frac{\partial \delta f_\alpha}{\partial \phi} = -\mathrm{i}\frac{e_\alpha}{m_\alpha}\Phi \frac{\partial f_{0\alpha}}{\partial \boldsymbol{v}}\cdot \boldsymbol{k} - \nu_{\alpha n}(\delta f_\alpha - \eta_\alpha f_{0\alpha}). \tag{8.6.3}$$

Here the perturbed density

$$\eta_\alpha = \frac{1}{N_\alpha}\int d\boldsymbol{p}\, \delta f_\alpha \tag{8.6.4}$$

and the potential Φ ($\boldsymbol{E} = -\mathrm{i}\boldsymbol{k}\Phi$) are introduced. We have applied the cylindrical coordinate system in velocity space.

Further, taking account of [see (8.3.19)]

$$\frac{\partial f_{0\alpha}}{\partial v_\mathrm{i}} = \left(-\frac{v_\mathrm{i}}{v_{T\alpha}^2} + \frac{\delta_{y\mathrm{i}}}{\Omega_\alpha}\frac{\partial}{\partial x}\right) f_{0\alpha}(x), \tag{8.6.5}$$

we find from (8.6.3)

$$\delta f_\alpha = -\frac{1}{\Omega_\alpha}\int_\infty^\phi d\phi' \left[\frac{\mathrm{i}e_\alpha}{m_\alpha}\Phi\left(-\frac{\boldsymbol{k}\cdot\boldsymbol{v}}{v_{T\alpha}^2} + \frac{k_y}{\Omega_\alpha}\frac{\partial}{\partial x}\right) f_{0\alpha} + \nu_{\alpha n}\eta_\alpha f_{0\alpha}\right] \exp\left[\frac{\mathrm{i}}{\Omega_\alpha}\int_\phi^{\phi'} d\phi''(\omega - \boldsymbol{k}\cdot\boldsymbol{v} + \mathrm{i}\nu_{\alpha n})\right]. \tag{8.6.6}$$

The evaluation of this integral is analogous to the evaluation of the homogeneous plasma case (Sect. 5.2).

It is easy to find from δf_α the charge densities ϱ_α

$$\varrho_\alpha = e_\alpha \int dp\, \delta f_\alpha \tag{8.6.7}$$

and to determine the longitudinal dielectric permittivity according to the relation

$$\varepsilon(\omega, \boldsymbol{k}, x) = 1 - \sum_\alpha \frac{4\pi\varrho_\alpha}{k^2\Phi}\,. \tag{8.6.8}$$

Doing so, we finally obtain in the zero-order approximation of geometrical optics

$$\begin{aligned}\varepsilon(\omega, \boldsymbol{k}, x) = 1 + \sum_\alpha \frac{\omega_{\mathrm{p}\alpha}^2}{k^2 v_{\mathrm{T}\alpha}^2} & \left[1 - \sum_n \frac{\mathrm{i}\nu_{\alpha n}}{\omega - n\Omega_\alpha + \mathrm{i}\nu_{\alpha n}} A_n\left(\frac{k_\perp^2 v_{\mathrm{T}\alpha}^2}{\Omega_\alpha^2}\right)\right.\\ & \left.\times I_+\left(\frac{\omega + \mathrm{i}\nu_{\alpha n} - n\Omega_\alpha}{k_z v_{\mathrm{T}\alpha}}\right)\right]^{-1} \left\{1 - \sum_n \frac{\omega + \mathrm{i}\nu_{\alpha n}}{\omega + \mathrm{i}\nu_{\alpha n} - n\Omega_\alpha}\right.\\ & \times \left[1 - \frac{k_y v_{\mathrm{T}\alpha}^2}{(\omega + \mathrm{i}\nu_{\alpha n})\Omega_\alpha}\left(\frac{\partial \ln N_\alpha}{\partial x} + \frac{\partial T_\alpha}{\partial x}\frac{\partial}{\partial T_\alpha}\right)\right]\\ & \left.\times A_n\left(\frac{k_\perp^2 v_{\mathrm{T}\alpha}^2}{\Omega_\alpha^2}\right) I_+\left(\frac{\omega + \mathrm{i}\nu_{\alpha n} - n\Omega_\alpha}{|k_z|\, v_{\mathrm{T}\alpha}}\right)\right\}.\end{aligned} \tag{8.6.9}$$

One can analogously calculate the dielectric permittivity of the weakly ionized degenerate plasma. The only difference is that we have instead of (8.6.5)

$$\frac{\partial f_{0\alpha}}{\partial v_{\mathrm{i}}} = m v_{\mathrm{i}} \frac{\partial f_{0\alpha}}{\partial \mathscr{E}_\alpha} + \frac{\delta_{y\mathrm{i}}}{\Omega_\alpha}\frac{\partial f_{0\alpha}}{\partial x} = \left(m v_{\mathrm{i}} - \frac{\delta_{y\mathrm{i}}}{\Omega_\alpha}\frac{\partial \mathscr{E}_{\mathrm{F}\alpha}}{\partial x}\right)\frac{\partial f_{0\alpha}}{\partial \mathscr{E}_\alpha}\,. \tag{8.6.10}$$

As a result we obtain in the zero-order approximation of geometrical optics

$$\begin{aligned}\varepsilon(\omega, \boldsymbol{k}, x) = 1 + \sum_\alpha & \frac{3\,\omega_{\mathrm{p}\alpha}}{k^2 v_{\mathrm{F}\alpha}^2}\\ \times & \left[1 - \sum_n \frac{\mathrm{i}\nu_{\alpha n}}{2}\int_0^\pi \frac{\sin\theta\, d\theta \mathrm{J}_n^2\left(\frac{k_\perp v_{\mathrm{F}\alpha}}{\Omega_\alpha}\sin\theta\right)}{\omega + \mathrm{i}\nu_{\alpha n} - k_z v_{\mathrm{F}\alpha}\cos\theta - n\Omega_\alpha}\right]^{-1}\left[1 - \sum_n \frac{\omega + \mathrm{i}\nu_{\alpha n}}{2}\right.\\ \times & \left.\left(1 - \frac{2}{3}\frac{k_y v_{\mathrm{F}\alpha}^2}{\omega\Omega_\alpha}\frac{\partial \ln N_\alpha}{\partial x}\right)\int_0^\pi \frac{\sin\theta\, d\theta \mathrm{J}_n^2\left(\frac{k_\perp v_{\mathrm{F}\alpha}}{\Omega_\alpha}\sin\theta\right)}{\omega + \mathrm{i}\nu_{\alpha n} - k_z v_{\mathrm{F}\alpha}\cos\theta - n\Omega_\alpha}\right].\end{aligned} \tag{8.6.11}$$

The general expressions (8.6.9, 11) allow us to investigate the drift oscillations of both the nondegenerate and the degenerate weakly ionized inhomogeneous plasma for any special case. In the following, however, we analyze the low-frequency $\omega < \Omega_i$ and long-wavelength $\lambda_\perp \gg \varrho_{\lambda i}$ drift oscillations only.

8.6.2 Completely Ionized Plasma

As shown in Sect. 6.5, it is impossible to obtain a general dielectric tensor of the type (8.6.9 or 11) in the case of a completely ionized plasma. This is due to the fact that then the collision integral of the charged particles is a complex integral expression. Consequently, we study the various limiting cases only.

If the Larmor frequency of the particles is large against the characteristic collision frequencies the equilibrium distribution function of the inhomogeneous plasma can be obviously written in the form (8.6.1). When the perturbation of the equilibrium distribution function has the form (8.6.2), one obtains a kinetic equation for δf_α which is similar to (5.6.2):

$$-\mathrm{i}(\omega - k_y v_y - k_z v_z)\,\delta f_\alpha + v_x \frac{\partial \delta f_\alpha}{\partial x} - \Omega_\alpha \frac{\partial \delta f_\alpha}{\partial \phi}$$

$$= \frac{e_\alpha}{m_\alpha} \nabla \Phi \frac{\partial f_{0\alpha}}{\partial \boldsymbol{v}} + \sum_\beta \left(\frac{\partial f_\alpha}{\partial t}\right)^{\alpha\beta}_{\mathrm{col}} . \tag{8.6.12}$$

Here $(\partial f_\alpha/\partial t)^{\alpha\beta}_{\mathrm{col}}$ is the linearized collision integral for collisions of charged particles of the type α with particles of the type β

$$\left(\frac{\partial f_\alpha}{\partial t}\right)^{\alpha\beta}_{\mathrm{col}} = 2\pi e_\alpha e_\beta L \frac{\partial}{\partial p_i} \int d\boldsymbol{p}' \frac{u^2 \delta_{ij} - u_i u_j}{u^3}$$

$$\times \left(f_{0\beta} \frac{\partial \delta f_\alpha}{\partial p_j} + \delta f_\beta \frac{\partial f_{0\alpha}}{\partial p_j} - f_{0\alpha} \frac{\partial \delta f_\beta}{\partial p_j'} - \delta f_\alpha \frac{\partial f_{0\beta}}{\partial p_j'}\right) . \tag{8.6.13}$$

The notations are $\boldsymbol{u} = \boldsymbol{v}_\alpha - \boldsymbol{v}_\beta$ for the relative velocity and L for the Coulomb logarithm.

In the following we analyze (8.6.12) separately for the electron distribution function δf_e and the ion distribution function δf_i.

First, we consider the electron contribution to the dielectric permittivity $\delta\varepsilon_e(\omega, \boldsymbol{k}, x)$. The integral equation (8.6.12) for the electrons, taking account of (8.6.5), is easily solved in the approximation of geometrical optics. It is possible to integrate the kinetic equation over ϕ since we have assumed $\omega \ll \Omega_e$, $kv_{Te} \ll \Omega_e$ (the drift oscillations can occur under these conditions). Actually, as mentioned in the foregoing section, the zero-order harmonic contributes dominantly to δf_e. Therefore the perturbation δf_e may be considered independent of ϕ. We obtain from (8.6.12)

$$(\omega - k_z v_z)\,\delta f_e = \frac{e\Phi}{T_e}\left(k_z v_z - \frac{k_y v_{Te}^2}{\Omega_e}a_e\right) \times f_{0e} + \mathrm{i}\left[\left(\frac{\partial f_e}{\partial t}\right)_{col}^{ee} + \left(\frac{\partial f_e}{\partial t}\right)_{col}^{ei}\right], \quad \text{where} \tag{8.6.14}$$

$$a_e = \frac{\partial \ln N_e}{\partial x} + \frac{\partial \ln T_e}{\partial x}\left(-\frac{3}{2} + \frac{v^2}{2v_{Te}^2}\right).$$

Incidentally, since the ion thermal velocity is very small in comparison with the electron thermal velocity, the terms depending on the perturbation of the ion distribution function δf_i are negligible in the electron-ion collision integral $(\partial f_e/\partial t)_{col}^{ei}$. Then

$$\left(\frac{\partial f_e}{\partial t}\right)_{col}^{ei} = 2\pi e^2 e_i^2 L N_i \frac{\partial}{\partial p_i}\frac{v^2\delta_{ij} - v_i v_j}{v^3}\frac{\partial \delta f_e}{\partial p_j}. \tag{8.6.15}$$

The integral $(\partial f_e/\partial t)_{col}^{ee}$ can be taken in the form of (8.6.13).

In order to solve (8.6.14), we restrict ourselves to the case of frequent electron-electron collisions, $\nu_e \gg \omega, k_z v_{Te}$. Then the collision term is the main term in (8.6.14) and the Chapman-Enskog method can be applied. To simplify the notation we introduce the function F_e by

$$\delta f_e = -\frac{k_y v_{Te}^2}{\omega\Omega_e}\frac{e}{T_e}a_e \Phi f_{0e} + F_e. \tag{8.6.16}$$

Substituting this equation into (8.6.14) yields

$$\frac{ie}{T_e}k_z v_z \Phi f_{0e}\left\{1 - \frac{k_y v_{Te}^2}{\omega\Omega_e}\left[\frac{\partial \ln(NT_e)}{\partial x} + \frac{\partial \ln T_e}{\partial x}\left(\frac{5}{2} - \frac{v^2}{2v_{Te}^2}\right)\right]\right\} = \left(\frac{\partial F_e}{\partial t}\right)_{col}^{ee} + \left(\frac{\partial F_e}{\partial t}\right)_{col}^{ei}. \tag{8.6.17}$$

Expanding F_e in the Sonin-Laguerre polynomials and using only the first two terms of the expansion:

$$F_e = v_z f_{0e}\left[a_0 + a_1\left(\frac{5}{2} - \frac{v^2}{2v_{Te}^2}\right)\right], \tag{8.6.18}$$

we obtain from (8.6.17) a system of algebraic equations for the expansion coefficients a_0 and a_1:

$$\frac{ek_z\Phi}{T_e}\left[1-\frac{k_y v_{Te}^2}{\omega\Omega_e}\frac{\partial\ln(NT_e)}{\partial x}\right]=i\nu_{eff}\left(a_0+\frac{3}{2}a_1\right),$$
$$\frac{ek_z\Phi}{T_e}\frac{5}{2}\frac{k_y v_{Te}^2}{\omega\Omega_e}\frac{\partial\ln(NT_e)}{\partial x}=i\nu_{eff}\left(\frac{3}{2}a_0+\frac{13+4\sqrt{2}}{4}a_1\right). \qquad (8.6.19)$$

The electron contribution $\delta\varepsilon_e(\omega,\boldsymbol{k},x)$ to the dielectric function is calculated as follows. Firstly, we use the derived expression for F_e to calculate the current δj_z in terms of F_e. The charge associated with F_e can be obtained then by means of the continuity equation $\delta\varrho_e=k_z\delta j_z/\omega$. Finally, the expression for $\delta\varepsilon_e(\omega,\boldsymbol{k},x)$ is obtained as usual:

$$\delta\varepsilon_e(\omega,\boldsymbol{k},x)=\frac{1}{k^2r_{De}^2}\left[\frac{k_y v_{Te}^2}{\omega\Omega_e}\frac{\partial\ln N}{\partial x}+i\,1.96\,\frac{k_z v_{Te}^2}{\omega\nu_{eff}}\left(1-\frac{k_y v_{Te}^2}{\omega\Omega_e}\frac{\partial\ln NT_e^{1.7}}{\partial x}\right)\right]$$

for $\nu_{eff}\omega\gg k_z^2v_{Te}^2$ and (8.6.20)

$$\delta\varepsilon_e(\omega,\boldsymbol{k},x)=\frac{1}{k^2r_{De}^2}\left[1+i\,1.44\,\frac{\omega\nu_{eff}}{k_z^2v_{Te}^2}\times\left(1-\frac{k_y v_{Te}^2}{\omega\Omega_e}\frac{\partial\ln NT_e^{-0.56}}{\partial x}\right)\right] \qquad (8.6.21)$$

for $\nu_{eff}\omega\ll k_z^2v_{Te}^2$.

Next, we calculate the ion contribution to the dielectric permittivity. The equation defining δf_i is analogous to (8.6.12) in principle. However, the ion-electron collision integral $(\partial f_i/\partial t)_{col}^{ie}$ can be neglected here, since $\nu_{ie}\sim\nu_{ii}\sqrt{m/M}\,(T_i/T_e)^{3/2}$. Its contribution to the dissipative processes is negligibly small under the condition $(T_e/T_i)>(m/M)^{1/3}$ which is always satisfied in real plasmas. This fact greatly simplifies the calculation of δf_i.

Introducing the function F_i by

$$\delta f_i=\frac{e_i}{T_i}\Phi f_{0i}+F_i\,, \qquad (8.6.22)$$

we obtain for F_i in the zero-order approximation of geometrical optics the following equation

$$(\omega-\boldsymbol{k}\cdot\boldsymbol{v})F_i-\Omega_i\frac{\partial F_i}{\partial\phi}=-\frac{e_i}{T_i}\left(\omega-\frac{k_y v_{Ti}^2}{\Omega_i}a_i\right)f_{0i}+i\left(\frac{\partial f_i}{\partial t}\right)_{st}^{ii}, \qquad (8.6.23)$$

where

$$a_i=\frac{\partial\ln N_i}{\partial x}+\frac{\partial\ln T_i}{\partial x}\left(-\frac{3}{2}+\frac{v^2}{2v_{Ti}^2}\right).$$

In order to solve this equation, we consider the long-wavelength limit $k_\perp \varrho_{\lambda i} \approx \varrho_{\lambda i}/\lambda_\perp \ll 1$ only, and assume $\Omega_i \gg \omega \gg \nu_i$, $k_z v_{Ti}$. Though it is a small term one must account for the collision integral in (8.6.23). The solution can be easily found by the method of successive approximations, the ratio ν_i/ω being the expansion parameter. A lengthy calculation leads to

$$\begin{aligned}\delta\varepsilon_i(\omega, k, x) &= \frac{1}{k^2 r_{Di}^2}\left\{\frac{k_y v_{Ti}^2}{\omega\Omega_i}\frac{\partial \ln N}{\partial x} + \frac{k_\perp^2 v_{Ti}^2}{\Omega_i^2}\left(1 - \frac{k_y v_{Ti}^2}{\omega\Omega_i}\frac{\partial \ln NT_i}{\partial x}\right)\right. \\ &\quad + \frac{i\nu_{ii} v_{Ti}^4}{10\,\omega}\left[\left(\frac{16\,k_z^4}{\omega^4} + 28\,\frac{k_z^2 k_\perp^2}{\omega^2\Omega_i^2} + 7\,\frac{k_\perp^4}{\Omega_i^4}\right)\left(1 - \frac{k_y v_{Ti}^2}{\omega\Omega_i}\frac{\partial \ln N}{\partial x}\right)\right. \\ &\quad \left.\left. - \left(24\,\frac{k_z^4}{\omega^4} + \frac{33}{2}\frac{k_z^2 k_\perp^2}{\omega^2\Omega_i^2} - \frac{3}{4}\frac{k_\perp^4}{\Omega_i^4}\right)\frac{k_y v_{Ti}^2}{\omega\Omega_i}\frac{\partial \ln T_i}{\partial x}\right]\right\}. \qquad (8.6.24)\end{aligned}$$

8.6.3 Spectra of the Hydrodynamic Drift-Dissipative Oscillations

In the completely ionized collisional plasma, (8.6.20, 21) and (8.6.24) describe the drift oscillations under the conditions $\nu_e \gg \omega$, $k_z v_{Te}$ and $\omega \gg \nu_i$, $k_z v_{Ti}$. Taking these oscillations as an example, we want to show that particle collisions do not stabilize the drift instabilities of the inhomogeneous plasma. On the contrary, they may even cause their development. Under the conditions given above the expression for $\varepsilon(\omega, k, x)$ simplifies both for the weakly and the completely ionized plasma. If we assume additionally $\omega\nu_e \gg k_z^2 v_{Te}^2$, the eikonal equation can be written as

$$\begin{aligned}\varepsilon(\omega, k, x) &= 1 + \frac{\omega_{pi}^2 k_\perp^2}{\Omega_i^2 k^2}\left(1 - \frac{k_y v_{Ti}^2}{\omega\Omega_i}\frac{\partial \ln NT_i}{\partial x}\right) \\ &\quad + i\alpha\frac{k_z^2\omega_{pe}^2}{k^2\omega\nu_e}\left(1 - \frac{k_y v_{Te}^2}{\omega\Omega_e}\frac{\partial \ln NT_e^\beta}{\partial x}\right) = 0\,. \qquad (8.6.25)\end{aligned}$$

Here $\nu_e = \nu_{en}$ and $\alpha = \beta = 1$ for the weakly ionized plasma and $\nu_e = \nu_{eff}$ and $\alpha = 1.96$, $\beta = 1.71$ for the completely ionized plasma. For simplicity, we have taken account only of dissipation due to the electron-electron collisions and neglected the ion-ion collisions.

Equation (8.6.25) becomes simplest for oscillations with wavelengths larger than the Debye length

$$1 - \frac{k_y v_{Ti}^2}{\omega\Omega_i}\frac{\partial \ln NT_i}{\partial x} + i\frac{\omega_s}{\omega}\left(1 - \frac{k_y v_{Te}^2}{\omega\Omega_e}\frac{\partial \ln NT_e^\beta}{\partial x}\right) = 0\,. \qquad (8.6.26)$$

Here, the following notation is introduced

$$\omega_s = \alpha \frac{k_z^2}{k^2} \frac{M}{m} \frac{\Omega_i^2}{\nu_e} . \tag{8.6.27}$$

It is not difficult to obtain the general solution of (8.6.26) which is quadratic in the frequency ω. We write down the solutions only in the limiting cases where the oscillations are unstable, however. For example, in the limit $\omega_{dr,e} \ll \omega_s$ we obtain from (8.6.26) the local spectrum

$$\omega_1 = \frac{k_y v_{Te}^2}{\Omega_e} \frac{\partial \ln NT_e^\beta}{\partial x} , \qquad \delta_1 = \frac{\omega_1^2}{\omega_s} \left(\frac{\partial \ln NT_i}{\partial \ln NT_e^\beta} \frac{T_i}{T_e} + 1 \right) \tag{8.6.28}$$

and in the limit $\omega_{dr,i} \gg \omega_s$ we have

$$\omega_2 = i\omega_s \frac{T_e}{T_i} \frac{\partial \ln NT_e^\beta}{\partial \ln NT_i} . \tag{8.6.29}$$

Obviously, these oscillations are always unstable since

$$\frac{\partial \ln NT_e^\beta}{\partial \ln NT_i} > 0 \tag{8.6.30}$$

is always satisfied and thus $\operatorname{Im}\{\omega\} > 0$. These instabilities exist in the frequency range $\omega \ll \nu_e$ and their increments increase proportionally to the electron collision frequency. It is seen from (8.6.25, 26) that the change in sign of the collisional wave absorption by the plasma electrons is responsible for the buildup of oscillations in the inhomogeneous plasma.

The spectra (8.6.28, 29) constitute two new branches of drift-dissipative oscillations, which do not exist in the collisionless plasma. In contrast to the drift-dissipative instabilities discussed before, these oscillations are sometimes called *hydrodynamic drift-dissipative instabilities*, which stresses that their buildup is associated with collisional dissipation.

The oscillations existing in the frequency range $\nu_i \ll \omega \ll \nu_e$ are also referred to as hydrodynamic drift-dissipative oscillations, but under the condition $\omega \nu_e \ll k_z^2 v_{Te}^2$, only. It is natural to assume $|\omega + i\nu_\alpha| \gg k_z v_{T\alpha}$. The eikonal equation then becomes

$$1 + \frac{\omega_{pe}^2}{k^2 v_{Te}^2} \left[1 + \frac{k_y v_s^2}{\omega \Omega_i} \frac{\partial \ln N}{\partial x} - \frac{k_z^2 v_s^2}{\omega^2} \left(1 - \frac{k_y v_{Ti}^2}{\omega \Omega_i} \frac{\partial \ln NT_i}{\partial x} \right) \right]$$
$$+ i\alpha_1 \frac{\omega_{pe}^2 \omega \nu_e}{k^2 k_z^2 v_{Te}^4} \left(1 - \frac{k_y v_{Te}^2}{\omega \Omega_e} \frac{\partial}{\partial x} \ln \frac{N}{T_e^{\beta_1}} \right) = 0 . \tag{8.6.31}$$

Here $\nu_e = \nu_{en}$ and $\alpha_1 = 1, \beta_1 = 0$ for the weakly ionized plasma and $\nu_e = \nu_{eff}$, $\alpha_1 = 1.44, \beta_1 = 0.56$ for the completely ionized plasma. As before, we have completely neglected the ion-ion collisions in this equation which is allowable if we have $\omega \gg \nu_i$.

Equation (8.6.31) differs from (8.5.14) only by the small dissipative term. In (8.5.14) the dissipative term is associated with the collisionless Cherenkov absorption of waves by the electrons, whereas in (8.6.31) it is determined by the electron diffusion and the thermal conductivity, i.e., by electron-electron collisions. Naturally, the real parts of the local oscillation frequencies following from (8.6.31) coincide with the frequency spectra of the collisionless plasma, i.e., the frequencies ω_1, ω_2 given by (8.5.15, 17) respectively, remain unchanged. The expressions for the increments are different, however, and take the form

$$\delta_1 = \alpha_1 \frac{\omega_1^2 \nu_e}{k_z^2 v_{Te}^2} \left(k^2 r_{De}^2 + \frac{k_\perp^2 v_s^2}{\Omega_i^2} - \frac{\partial \ln T_e^{\beta_1}}{\partial \ln N} \right), \tag{8.6.32}$$

$$\delta_2 = -\alpha_1 \frac{\omega_2^2 \nu_e}{k_z^2 v_{Te}^2} \left(1 - \frac{\partial \ln T_e^{\beta_1}}{\partial \ln N} \right) \tag{8.6.33}$$

instead of (8.5.15, 17). Hence the oscillations belonging to the first branch are unstable if

$$\frac{\partial \ln T_e^{\beta_1}}{\partial \ln N} < \left(k^2 r_{De}^2 + \frac{k_\perp^2 v_s^2}{\Omega_i^2} \right). \tag{8.6.34}$$

The second branch which exists in the nonisothermal plasma with $T_e \gg T_i$, only, becomes unstable under the condition

$$\frac{\partial \ln T_e^{\beta_1}}{\partial \ln N} > 1\,. \tag{8.6.35}$$

On both branches the electron-electron collisions are responsible for the excitation of drift oscillations.

8.6.4 The Effect of Ion Collisions on the Development of Drift Oscillations

Concluding this section, we discuss the role of ion-ion collisions in the development of the drift instabilities. We have assumed above $\omega \gg \nu_i$ that the ion-ion collisions have been completely neglected. It is not difficult to take account of them, however. For $\omega \gg \nu_i$ these collisions produce small corrections only, and their influence can be both stabilizing and nonstabilizing. The question if drift oscillations can exist in the frequency range $\omega \ll \nu_i$ is of greater interest. We study this question using the weakly ionized plasma as an example. Assuming $\nu_i \ll \Omega_i$ for the nondegenerate plasma, we obtain from (8.6.9) the following eikonal equation

$$1 + \sum_\alpha \frac{\omega_{p\alpha}^2}{k^2 v_{T\alpha}^2} \left[1 - \frac{i\nu_\alpha}{\omega + i\nu_\alpha} A_0\left(\frac{k_\perp^2 v_{T\alpha}^2}{\Omega_\alpha^2}\right) I_+\left(\frac{\omega + i\nu_\alpha}{k_z v_{T\alpha}}\right)\right]^{-1}$$
$$\times \left\{1 - \left[1 - \frac{k_y v_{T\alpha}^2}{(\omega + i\nu_\alpha)\Omega_\alpha}\left(\frac{\partial \ln N_\alpha}{\partial x} + \frac{\partial T_\alpha}{\partial x}\frac{\partial}{\partial T_\alpha}\right)\right]\right.$$
$$\left.\times A_0\left(\frac{k_\perp^2 v_{T\alpha}^2}{\Omega_\alpha^2}\right) I_+\left(\frac{\omega + i\nu_\alpha}{k_z v_{T\alpha}}\right)\right\} = 0\,. \tag{8.6.36}$$

As before, we restrict ourselves to the limit of long wavelengths $k_\perp^2 v_{T\alpha}^2 \ll \Omega_\alpha^2$ and low frequencies $\omega \ll \nu_i$. In the limit of frequent collisions, when the mean free paths of the particles are smaller than the wavelengths of the longitudinal waves, i.e., $\nu_\alpha \gg k_z v_{T\alpha}$, (8.6.36) has a solution only when the ion diffusion is weak, $\omega\nu_i \gg k_z^2 v_{Ti}^2$. The electron diffusion can be either weak or strong. For a sufficiently dense plasma and for $\omega\nu_e \gg k_z^2 v_{Te}^2$ we obtain from (8.6.36)

$$1 + \frac{\omega_{pe}^2 k_\perp^2}{\Omega_i^2 k^2}\left(1 - \frac{k_y v_{Ti}^2}{\omega\Omega_i}\frac{\partial \ln NT_i}{\partial x}\right)$$
$$+ i\frac{\omega_{pe}^2 k_z^2}{\omega\nu_e k^2}\left(1 - \frac{k_y v_{Te}^2}{\omega\Omega_e}\frac{\partial \ln NTe}{\partial x}\right) = 0\,. \tag{8.6.37}$$

This equation is identical with (8.6.25). Therefore, the spectra of the drift oscillations (8.6.28, 29) and the condition for their buildup (8.6.30) remain unchanged in the limit of frequent collisions, $\omega \sim \omega_{dr} \ll \nu_i$.

Equation (8.6.37) describes the drift instabilities of the degenerate plasma, too. This can be found from the eikonal equation for the drift oscillations of the plasma with nondegenerate ions and degenerate electrons. In the frequency range $\omega \ll \nu_i$, $\omega\nu_i \gg k_z^2 v_{Ti}^2$, $\omega\nu_e \gg k_z^2 v_{Fe}^2$ it is written as

$$1 + \frac{\omega_{pi}^2 k_\perp^2}{\Omega_i^2 k^2}\left(1 - \frac{k_y v_{Ti}^2}{\omega\Omega_i}\frac{\partial \ln NT_i}{\partial x}\right)$$
$$+ i\frac{\omega_{pe}^2 k_z^2}{\omega\nu_e k^2}\left(1 - \frac{2}{3}\frac{k_y v_{Fe}^2}{\omega\Omega_e}\frac{\partial \ln N}{\partial x}\right) = 0\,. \tag{8.6.38}$$

This equation has the same structure as (8.6.25) and (8.6.37). Therefore, the spectra of the drift oscillations of the degenerate plasma in the assumed frequency range can be written analogously to (8.6.25) and (8.6.29):

$$\omega_1 = \frac{2}{3}\frac{k_y v_{\mathrm{Fe}}^2}{\Omega_\mathrm{e}}\frac{\partial \ln N}{\partial x}\,, \qquad \delta_1 = \frac{\omega_1^2}{\omega_\mathrm{s}}\left(1 + \frac{3}{2}\frac{M}{m}\frac{v_{\mathrm{Ti}}^2}{v_{\mathrm{Fe}}^2}\frac{\partial \ln NT_\mathrm{i}}{\partial \ln N}\right) \tag{8.6.39}$$

for $\omega_1 \sim \omega_{\mathrm{dr,e}} \ll \omega_\mathrm{s} = k_z^2 \Omega_\mathrm{i}^2 M/(k_\perp^2 \nu_\mathrm{e} m)$, and

$$\omega_2 = \mathrm{i}\frac{2}{3}\,\omega_\mathrm{s}\,\frac{m}{M}\frac{v_{\mathrm{Fe}}^2}{v_{\mathrm{Ti}}^2}\frac{\partial \ln N}{\partial \ln NT_\mathrm{i}} \tag{8.6.40}$$

for $\omega_2 \sim \omega_\mathrm{s} \ll \omega_{\mathrm{dr},\alpha}$. These formulas show that the degenerate plasma, like the nondegenerate plasma, is always unstable since the instability condition, cf. (8.6.30),

$$\frac{\partial \ln N}{\partial \ln NT_\mathrm{i}} > 0 \tag{8.6.41}$$

is always satisfied in real plasmas.

Our analysis of the effect of particle collisions on the spectra of drift oscillations in the spatially inhomogeneous plasma has shown the following. Under the condition of frequent collisions, i.e., that the mean free path of the particles is smaller than the longitudinal dimension of the plasma ($\nu_\alpha \gg k_z v_{\mathrm{T}\alpha} \gtrsim v_{\mathrm{T}\alpha}/L_\parallel$), and under the condition that the drift frequencies are smaller than the collision frequencies ($\nu_\alpha \gg \omega_{\mathrm{dr},\alpha} \gtrsim \omega$), long-wavelength unstable drift oscillations can develop in the plasma if $\omega_{\mathrm{dr,i}}\nu_\mathrm{i} \gg k_z^2 v_{\mathrm{Ti}}^2 \gtrsim v_{\mathrm{Ti}}^2/L_\parallel^2$ or, cf. (8.5.20),

$$L_\parallel/L_\perp > \sqrt{\Omega_\mathrm{i}/\nu_\mathrm{i}} > 1\,. \tag{8.6.42}$$

In the opposite limit, the inhomogeneities of the particle density of the perturbations will be equalized due to diffusion, which is the faster process than the development of the drift instability then.

8.7 Convective Instabilities of the Inhomogeneous Plasma

In the preceding we have considered the oscillations of the inhomogeneous plasma in an external homogeneous magnetic field $B_0 = B_{0z}$ with Maxwellian distributed particle velocities. The field lines have been assumed to be straight and aligned with the $0z$-axis which is perpendicular to the direction of the plasma inhomogeneity (the $0x$-axis). As we have shown, such a plasma can be unstable and support drift oscillations with frequencies and increments smaller than the frequency associated with the Larmor drift of the particles. The driving force of these instabilities is the spatial inhomogeneity of the plasma, which, though being in force equilibrium with the magnetic field (the gas-kinetic pressure is compensated by the magnetic pressure), still tends to reach the thermodynamic equilibrium by the Larmor drift of the particles.

We want to show here that quickly growing instabilities can arise due to a nonequilibrium of another nature, for example, the gravitational drift caused by the curvilinearity of the magnetic field lines or by the current driven by the longitudinal electric field. Here, the increments significantly exceed the drift frequencies even for very small deviations from equilibrium.

8.7.1 Inhomogeneous Plasma in a Curvilinear Magnetic Field

We begin our analysis of this type of plasma instabilities with the case when the magnetic field lines have a weak curvature, i.e., there is a small inhomogeneous component B_{0x} of the magnetic field, parallel to the direction of the plasma inhomogeneity, in addition to the main longitudinal field B_{0z}. This geometry models the configuration in experimental thermonuclear devices. Figure 8.5, for example, shows the form of the magnetic field in systems with magnetic mirrors, or mirror machines. In the collisionless low-pressure plasma charged particles are tied to the magnetic field lines, i.e., they can move freely only along the field lines and their transverse motion is a bound rotation. However, an average transverse force acts on a particle of the type α (see Exercise 3.8)

$$\boldsymbol{F}_\alpha = \boldsymbol{n}\frac{m_\alpha}{R}\left(v_{\|\alpha}^2 + \frac{v_{\perp\alpha}^2}{2}\right), \tag{8.7.1}$$

where $\boldsymbol{n}$ is the unit vector directed from the centre of curvature to the particle and R is the radius of curvature of the magnetic field lines. Since most of the plasma particles move with thermal velocities, we can further substitute T_α for $m_\alpha v_{\|\alpha}^2$ and $m_\alpha v_{\perp\alpha}^2/2$. Then, the transverse force (8.7.1) approximately takes the form

$$F_\alpha \simeq \frac{m_\alpha v_{T\alpha}^2}{R} = \frac{T_\alpha}{R}. \tag{8.7.2}$$

Moreover, for simplicity this force will be referred to as constant, i.e., independent of the coordinate x. Strictly speaking, then the temperature of the plasma is also assumed to be independent of x.

Thus, the effect of the curvature of the magnetic field lines can be studied by introducing an effective gravity field $g_\alpha = v_{T\alpha}^2/R$ which is different for different types of particles and oriented along the outward normal to magnetic field lines. Note that for a magnetic field of the form shown in Fig. 8.5 and for decreasing plasma density in the outward direction, the gravity force is oriented oppositely to the density gradient and independent of the sign of the charge of a particle. Figure 8.6 shows the layer of an inhomogeneous plasma in the curved magnetic field.

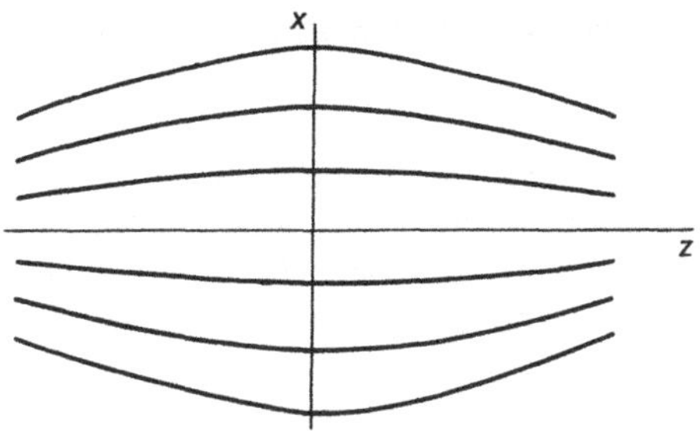

Fig. 8.5. Configuration of the magnetic field in mirror-machine-type systems

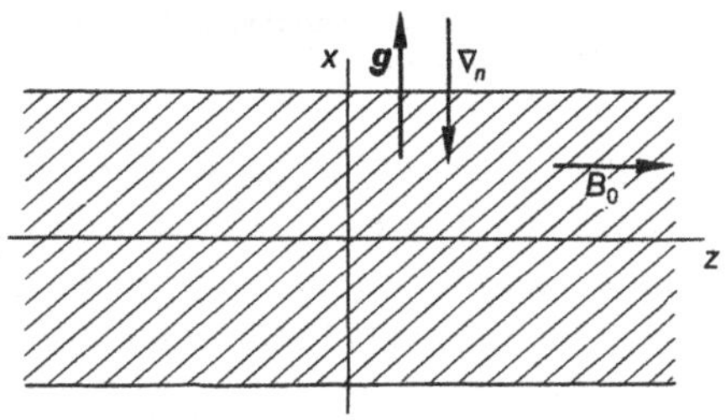

Fig. 8.6. Forces affecting a particle in an inhomogeneous plasma in the magnetic field of a mirror machine

8.7.2 The Gravitational Drift of Plasma Particles

In the presence of a gravity field $\boldsymbol{g}_\alpha$ parallel to the $0x$-axis the equilibrium distribution function of the particles in an inhomogeneous plasma follows from

$$v_x \frac{\partial f_{0\alpha}}{\partial x} + g_\alpha \frac{\partial f_{0\alpha}}{\partial v_x} + \frac{e_\alpha}{m_\alpha c} [\boldsymbol{v}, \boldsymbol{B}_0] \frac{\partial f_{0\alpha}}{\partial \boldsymbol{v}} = 0 . \quad (8.7.3)$$

Only the case when the fields $\boldsymbol{B}_0$ and $\boldsymbol{g}_\alpha$ are homogeneous will be analyzed (this condition is satisfied for plasmas with $\beta \ll 1$). Then, the solution of (8.7.3) can be easily written as:

$$f_{0\alpha}(\boldsymbol{v} - \boldsymbol{u}_\alpha) = \left[1 + \frac{(\boldsymbol{v} - \boldsymbol{u}_\alpha)_y}{\Omega_\alpha} \left(\frac{\partial N_\alpha}{\partial x} \frac{\partial}{\partial N_\alpha} + \frac{\partial T_\alpha}{\partial x} \frac{\partial}{\partial T_\alpha}\right)\right]$$

$$\times \frac{N_\alpha}{(2\pi m_\alpha T_\alpha)^{3/2}} \exp\left[-\frac{m_\alpha (\boldsymbol{v} - \boldsymbol{u}_\alpha)^2}{2 T_\alpha}\right] , \quad (8.7.4)$$

where $\boldsymbol{u}_\alpha = [\boldsymbol{g}_\alpha, \boldsymbol{h}]/\Omega_\alpha = (0, -g_\alpha/\Omega_\alpha, 0)$ is the velocity of the gravitational drift arising due to the crossed fields $\boldsymbol{B}_0$ and $\boldsymbol{g}_\alpha$, and $\boldsymbol{h}$ is the unit vector in the direction of $\boldsymbol{B}_0$. In contrast to the Larmor drift studied in Sect. 8.3, this drift is a real drift of the particles. Therefore, it appears explicitly in the distribution function (8.7.4).

We see that charged particles perform an ordered motion in equilibrium. Under these conditions instabilities of the type of the beam and the current instabilities can develop in the plasma. Before we come to the analysis of these instabilities, we want to make a remark. Fig. 8.6 shows that an analogous situation can also occur in an ordinary compressible fluid immersed in a gravity field (Exercise 8.8.6). In some cases this state of the fluid is unstable and the development of the instability is accompanied by convection. Because of this, the instabilities described below are generally called *convective*.

8.7.3 Dielectric Permittivity of the Inhomogeneous Plasma in a Curvilinear Magnetic Field

In order to determine the oscillation spectra, we must calculate the dielectric permittivity. The form (8.7.4) of the distribution function $f_{0\alpha}$ suggests that it is convenient to use the frame moving together with the particles of the given type at the velocity of their gravitational drift. The calculation is analogous to that presented in Sect. 8.3. As a result, we obtain in the approximation of geometrical optics for the longitudinal dielectric permittivity

$$\varepsilon(\omega, \boldsymbol{k}, x) = 1 + \sum_\alpha \frac{\omega_{p\alpha}^2}{k^2 v_{T\alpha}^2} \left\{1 - \sum_n \frac{\omega'}{\omega' - n\Omega_\alpha} \left[1 - \frac{k_y v_{T\alpha}^2}{\omega' \Omega_\alpha} \right.\right.$$

$$\left.\left. \times \left(\frac{\partial N_\alpha}{\partial x} \frac{\partial}{\partial N_\alpha} + \frac{\partial T_\alpha}{\partial x} \frac{\partial}{\partial T_\alpha}\right)\right] A_n\left(\frac{k_\perp^2 v_{T\alpha}^2}{\Omega_\alpha^2}\right) I_+\left(\frac{\omega' - n\Omega_\alpha}{k_z v_{T\alpha}}\right)\right\} . \quad (8.7.5)$$

Compared with (8.3.24) only the substitution $\omega \to \omega' = \omega - \boldsymbol{k} \cdot \boldsymbol{u}_\alpha = \omega + k_y g_\alpha / \Omega_\alpha$ is made, where ω' is the frequency taking account of the Doppler shift due to the gravitational drift. Naturally, (8.7.5) coincides with (8.3.24) for $g_\alpha = 0$.

We shall not analyze the complete oscillation spectrum of the inhomogeneous plasma since the variety of modes is greater than in the case of the drift oscillation spectrum. Only some new qualitative effects associated with the presence of the gravity field $\boldsymbol{g}_\alpha$ will be considered.

Restricting ourselves to the low-frequency range $\omega' \ll \Omega_\alpha$, only the terms with $n = 0$ must be kept in (8.7.6). The wavelengths are assumed to be long $(k_\perp^2 v_{T\alpha}^2 \ll \Omega_\alpha^2)$, the phase velocities to be large $(\omega' \gg k_z v_{T\alpha})$ and the frequencies under consideration to be much greater than the drift frequencies

$$\omega' \gg \omega_{dr,\alpha} \sim \frac{k_y v_{T\alpha}^2}{\Omega_\alpha L_0} , \quad (8.7.6)$$

where L_0 is the characteristic length of the inhomogeneity of the plasma density. Then, we obtain from (8.7.5) the following eikonal equation:

$$\varepsilon(\omega, \boldsymbol{k}, x) = 1 - \sum_\alpha \frac{\omega_{p\alpha}^2}{k^2} \left(\frac{k_z^2}{\omega'^2} - \frac{k_\perp^2}{\Omega_\alpha^2} - \frac{k_y}{\omega' \Omega_\alpha} \frac{\partial \ln N_\alpha}{\partial x}\right) = 0 . \quad (8.7.7)$$

Particles with thermal velocities are excluded from this equation. Therefore, it can be obtained within the model of independent particles, or, what is the same, on the basis of the two-fluid magnetohydrodynamic equations of the cold plasma.

Notice that the gravitational drift velocities of the electrons and the ions have the same order of magnitude if the temperatures are approximately equal. This allows us to simplify (8.7.7):

$$1-\frac{k_z^2\omega_{pe}^2}{k^2\omega'^2}+\frac{k_\perp^2\omega_{pi}^2}{k^2\Omega_i^2}+\frac{k_y\boldsymbol{k}\cdot(\boldsymbol{u}_e-\boldsymbol{u}_i)\,\omega_{pi}^2}{k^2(\omega-\boldsymbol{k}\cdot\boldsymbol{u}_e)(\omega-\boldsymbol{k}\cdot\boldsymbol{u}_i)\,\Omega_i}\,\frac{\partial\ln N}{\partial x}=0\,. \quad (8.7.8)$$

In most plasmas the condition $R \gg L_0$ is satisfied. Then, it follows from (8.7.5–8) that $\omega \gg \boldsymbol{k}\cdot\boldsymbol{u}_\alpha$. Hence, we have

$$1-\frac{k_z^2\omega_{pe}^2}{k^2\omega^2}+\frac{k_\perp^2\omega_{pi}^2}{k^2\Omega_i^2}-\frac{k_y^2}{k^2}\,\frac{g_{\rm eff}\,\omega_{pi}^2}{\omega^2\Omega_i^2}\,\frac{\partial\ln N}{\partial x}=0\,, \quad (8.7.9)$$

where the following notation is introduced:

$$g_{\rm eff}=\frac{v_{Ti}^2+v_s^2}{R}\,. \quad (8.7.10)$$

8.7.4 The Flute (Interchange) Instability

The solution of (8.7.9), correct up to terms of the order of m/M, leads to the following local spectrum:

$$\omega^2=\frac{k_y^2\,g_{\rm eff}(\partial\ln N/\partial x+k_z^2\omega_{pe}^2v_A^2/c^2)}{k^2(1+v_A^2/c^2)}\,. \quad (8.7.11)$$

The first term in the numerator of (8.7.11) is always negative (since $\partial\ln N/\partial x<0$) and the second one is positive. For modes with $k_z=0$ the second term vanishes and the resulting spectrum is

$$\omega^2=\frac{k_y^2}{k^2}\,\frac{g_{\rm eff}\,\partial\ln N/\partial x}{1+v_A^2/c^2}<0\,. \quad (8.7.12)$$

The corresponding hydrodynamic instability of the plasma is called the *flute instability* since the $k_z=0$ perturbations are homogeneous in the direction of the magnetic field lines and have the form of a flute. It is also called the *interchange instability* because the plasma is transposed across the magnetic field B_0 along the whole length of a field line during the development of this instability: the plasma and the magnetic field interchange their positions.

The time for the development of the interchange instability in devices for thermonuclear fusion is rather short. It is equal to $(\mathrm{Im}\{\omega\})^{-1}\approx\sqrt{L_0/g_{\rm eff}}\sim 10^{-5}$–$10^{-6}$ s. However, this instability can develop within a plasma with a sufficiently large longitudinal extension, only:

$$\frac{L_\parallel}{L_\perp}>\sqrt{\frac{ML_0R}{m\varrho_{\lambda i}^2}}\,. \quad (8.7.13)$$

The dielectric permittivity (8.7.5) is appropriate to describe the inhomogeneous plasma not only when the drift is associated with a gravitational field but also in all other cases with a real particle drift. It allows one to describe correctly the behaviour of the plasma with beams of charged particles and in particular of the plasma with a current.

8.7.5 The Current-Convective Instability

We consider the inhomogeneous plasma with a current ($\boldsymbol{u}_e = \boldsymbol{u} \parallel 0z$, $\boldsymbol{u}_i = 0$) aligned with a strong external magnetic field. For simplicity the gravitation associated with the curvature of the magnetic field lines will be ignored. We restrict ourselves to the low-frequency oscillations of the cold plasma (hydrodynamic limit), assuming

$$k_\perp^2 v_{T\alpha}^2 \ll \Omega_\alpha^2; \qquad \Omega_e \gg \omega' \gg k_z v_{Te}, \omega_{dr,e}; \qquad \Omega_i \gg k_z v_{Ti}\ , \tag{8.7.14}$$

where $\omega' = \omega - \boldsymbol{k} \cdot \boldsymbol{u}$ and $\boldsymbol{u}$ is the electron drift velocity. Then, the eikonal equation which is similar to (8.7.10) can be easily obtained:

$$k_\perp^2\left(1 + \frac{c^2}{v_A^2}\right) + k_z^2\left(1 - \frac{\omega_{pe}^2}{(\omega - \boldsymbol{k}\cdot\boldsymbol{u})^2}\right) - \frac{k_y \omega_{pi}^2}{\omega \Omega_i}\,\frac{\boldsymbol{k}\cdot\boldsymbol{u}}{\omega - \boldsymbol{k}\cdot\boldsymbol{u}}\,\frac{\partial \ln N}{\partial x} = 0\ . \tag{8.7.15}$$

Therefore, in the limit $\omega \gg \boldsymbol{k} \cdot \boldsymbol{u}$, we obtain a local spectrum similar to (8.7.1):

$$\omega^2 = \frac{- k_y \boldsymbol{k}\cdot\boldsymbol{u}\,\Omega_i \dfrac{\partial \ln N}{\partial x} + k_z^2 \omega_{pe}^2 \dfrac{v_A^2}{c^2}}{k^2(1 + v_A^2/c^2)}\ . \tag{8.7.16}$$

The first term in the numerator can be negative and lead to an instability, the second one is always positive and thus plays a stabilizing part. It is easy to show that this instability sets in when the inequality

$$\frac{u}{\Omega_e L_0} > \left|\frac{k_z}{k_y}\right| \sim \frac{L_\perp}{L_\parallel} \tag{8.7.17}$$

is satisfied, i.e., when the electron drift velocities are sufficiently high or when the longitudinal plasma dimension is large.

Equation (8.7.17) shows that the decrease of the magnetic field facilitates the development of the instability. However, this holds only up to some limit. From the condition $\omega^2 \ll \Omega_i^2$ it follows that this instability can arise only for field strengths exceeding some critical value which is defined by the inequality

$$\frac{L_\perp}{L_\parallel} \sim \left|\frac{k_z}{k_y}\right| < \frac{\Omega_i L_0}{u}\ . \tag{8.7.18}$$

With increasing $L_{\parallel}$ the critical field decreases. The inequalities (8.7.17, 18) show that the longitudinal plasma dimension must be sufficiently large for the existence of the instability: $L_{\parallel}/L_{\perp} > \sqrt{M/m}$. However, the characteristic time of the development is $(\mathrm{Im}\{\omega\})^{-1} \sim \sqrt{L_{\parallel}/u\Omega_{\mathrm{i}}}$ and for shorter systems the increment decreases.

In literature the discussed instability is known as the *current-convective instability*.

8.7.6 The Effect of Particle Collisions on Convective Instabilities of the Plasma

Closing this section we briefly consider the effect of particle collisions on the plasma convective instabilities. To begin with, we write down the dielectric tensor of the collisional plasma, taking account of the gravitational or current-convective particle drift. For simplicity, we confine ourselves to the nondegenerate weakly ionized plasma. Then, we obtain from (8.7.7) by the substitution $\omega \to \omega' = \omega - k_y u_{\mathrm{g}\alpha}$, where $\boldsymbol{u}_{\mathrm{g}\alpha}$ is the gravitational drift velocity of the particles of the type α, or by the substitution $\omega \to \omega' = \omega - k_z u_\alpha$, where $\boldsymbol{u}_\alpha$ is the current drift velocity, the eikonal equation

$$
\begin{aligned}
\varepsilon(\omega, \boldsymbol{k}, x) &= 1 + \sum_\alpha \frac{\omega_{\mathrm{p}\alpha}^2}{k^2 v_{\mathrm{T}\alpha}^2} \\
&\times \left[1 - \sum_n \frac{\mathrm{i}\nu_{\alpha n}}{\omega' - n\Omega_\alpha + \mathrm{i}\nu_{\alpha n}} A_n\left(\frac{k_\perp^2 v_{\mathrm{T}\alpha}^2}{\Omega_\alpha^2}\right) I_+\left(\frac{\omega' + \mathrm{i}\nu_{\alpha n} - n\Omega_\alpha}{k_z v_{\mathrm{T}\alpha}}\right)\right]^{-1} \\
&\times \left\{1 - \sum_n \frac{\omega' + \mathrm{i}\nu_{\alpha n}}{\omega' + \mathrm{i}\nu_{\alpha n} - n\Omega_\alpha}\left[1 - \frac{k_y v_{\mathrm{T}\alpha}^2}{(\omega' + \mathrm{i}\nu_{\alpha n})\Omega_\alpha}\right.\right. \\
&\times \left.\left.\left(\frac{\partial \ln N}{\partial x} + \frac{\partial T_\alpha}{\partial x}\frac{\partial}{\partial T_\alpha}\right)\right] A_n\left(\frac{k_\perp^2 v_{\mathrm{T}\alpha}^2}{\Omega_\alpha^2}\right) I_+\left(\frac{\omega' + \mathrm{i}\nu_{\alpha n} - n\Omega_\alpha}{k_z v_{\mathrm{T}\alpha}}\right)\right\} = 0 \,.
\end{aligned}
\tag{8.7.19}
$$

As in the foregoing section, we consider only the case when the inequalities $\omega \gg \omega_{\mathrm{dr},\alpha}$, $|\omega' + \mathrm{i}\nu_\alpha| \gg k_z v_{\mathrm{T}\alpha}$, $\Omega_\alpha^2 \gg k_\perp^2 v_{\mathrm{T}\alpha}^2$, $\nu_{\mathrm{e}} \gg \omega' \gg \boldsymbol{k}\cdot\boldsymbol{u}$, ν_{i} are satisfied which is nearest to the hydrodynamic case. After straightforward calculations (8.7.19) can be reduced to

$$
\left(1 + \frac{v_{\mathrm{A}}^2}{c^2}\right) k_\perp^2 \frac{\omega^2}{4\pi\sigma} + \mathrm{i}\omega\frac{k_z^2 v_{\mathrm{A}}^2}{c^2} + \frac{\boldsymbol{k}\cdot\boldsymbol{u} k_y \Omega_{\mathrm{i}}}{4\pi\sigma}\frac{\partial \ln N}{\partial x} = 0 \,, \tag{8.7.20}
$$

where $\sigma = \omega_{\mathrm{pe}}^2/4\pi\nu_{\mathrm{e}}$ is the low-frequency (static) plasma conductivity.

It follows from (8.7.20) that its second term can be neglected for the interchange mode, i.e., for k_z so small that $k_z^2/k_\perp^2 \ll \omega/\sigma$. The resulting spectra are independent of the plasma conductivity

$$\omega^2 = -\frac{\boldsymbol{k}\cdot\boldsymbol{u}}{k_\perp^2}\frac{k_y\Omega_\mathrm{i}}{1+v_\mathrm{A}^2/c^2}\frac{\partial \ln N}{\partial x} \simeq \begin{cases} g_\mathrm{eff}\dfrac{\partial \ln N}{\partial x}, \\ -\dfrac{k_y k_z u}{k_\perp^2}\Omega_\mathrm{i}\dfrac{\partial \ln N}{\partial x}. \end{cases} \tag{8.7.21}$$

The upper case corresponds to unstable interchange oscillations of the plasma in the gravitational field with $u_\mathrm{i} = -\delta_{y\mathrm{i}}\, g_\mathrm{eff}/\Omega_\mathrm{i}$. The lower case describes the current-convective oscillations with $u_\mathrm{i} = u\delta_{z\mathrm{i}}$. The spectra (8.7.21) are very similar to the spectra of the interchange and the current-convective instabilities of the collisionless plasma [(8.7.12 and 16), respectively]. They can be regarded as a continuation of these spectra into the frequency range $\omega < \nu_\mathrm{e}$.

Equation (8.7.20) has unstable solutions not only in case of interchange perturbations which have a long wavelength parallel to the magnetic field, but also in the limit of short-wavelength perturbations with $k_z^2/k_\perp^2 \gg \omega/\sigma$. Then, we obtain from (8.7.20) the local spectra:

$$\omega = \frac{\mathrm{i}\Omega_\mathrm{i} c^2 \boldsymbol{k}\cdot\boldsymbol{u} k_y}{4\pi\sigma v_\mathrm{A}^2 k_z^2}\frac{\partial \ln N}{\partial x} = \begin{cases} -\mathrm{i}\dfrac{c^2 k_y^2 g_\mathrm{eff}}{v_\mathrm{A}^2 4\pi\sigma k_z^2}\dfrac{\partial \ln N}{\partial x}, \\ \mathrm{i}\dfrac{c^2\Omega_\mathrm{i} k_y u}{4\pi\sigma v_\mathrm{A}^2 k_z}\dfrac{\partial \ln N}{\partial x}. \end{cases} \tag{8.7.22}$$

Here, the upper case also corresponds to unstable oscillations of the plasma in the effective gravitational field and the lower one to the short-wavelength branch of the current-convective instability of the collisional plasma. In literature this instability is also called the *helical instability* of the magnetized plasma with a current.

8.8 Exercises

8.8.1. Using the approximation of geometrical optics derive the *quasiclassical quantization rules for the fourth-order differential equations* in the presence of branch points.

Solution. In the first-order approximation of geometrical optics the fourth-order equation describing small oscillations of the weakly inhomogeneous plasma without account of dissipative effects can be written in the general form

$$y'''' + 2p(\omega,x)y'' + 2\varepsilon(\omega,x)y' + q(\omega,x)y = 0\,. \tag{8.8.1}$$

Here $p(\omega, x)$ and $q(\omega, x)$ are slowly varying real functions of x, satisfying the inequality

$$\eta \sim \frac{p'}{p^{3/2}} \sim \frac{q'}{q^{5/4}} \ll 1 \tag{8.8.2}$$

and the real function $\varepsilon(\omega, x)$ is of the first order in the small quantity η.

Assuming the WKB-Ansatz

$$y = C\exp\left(\mathrm{i}\int^x k(\omega, x)\,dx\right), \tag{8.8.3}$$

for the solution of (8.8.1) we obtain in the zero-order approximation of geometrical optics (i.e., in the parameter η) for $k(\omega, x)$

$$k_{1,2}^2(x) = p \pm \sqrt{p^2 - q}\,. \tag{8.8.4}$$

To the first order in η the correction is

$$\delta k_{1,2} = \frac{\mathrm{i}}{2}\left\{[\ln k_{1,2}(p^2 - q)]' + \frac{p' - \varepsilon}{k_{1,2}^2 - p}\right\}. \tag{8.8.5}$$

The approximation of geometrical optics breaks down near the points

$$k_{1,2}(\omega, x) = 0\,, \quad p^2(\omega, x) = q(\omega, x)\,, \tag{8.8.6}$$

since the correction δk goes to infinity there. The first point is a turning point and the second one a branch point. We have already studied the behaviour of solutions near turning points in Sect. 8.1. Here, we investigate the behaviour near branch points where k_1^2 and k_2^2 become equal and the corresponding eigensolutions coalesce. Therefore, the roots k_1 and k_2 are involved in the zero-order approximation of geometrical optics already, leading to different quantization rules.

To investigate the behaviour near the branch points we consider the case when two branch points a and b occur and no other singularity exists between them. Further we assume $p(x) > 0$ (Fig. 8.7). In the range II (the range of transparency), far from the branch points, the general solution of (8.8.1) can be written to the first-order in the smallness parameter η as

$$\begin{aligned} y^{\mathrm{II}} &= \frac{1}{\sqrt[4]{k_1^2(p^2 - q)}} \exp\left(-\frac{1}{2}\int^x dx \frac{p' - \varepsilon}{\sqrt{p^2 - q}}\right) \sum_{\pm} C_{1\pm} \exp\left(\pm\mathrm{i}\int^x k_1 dx\right) \\ &+ \frac{1}{\sqrt[4]{k_2^2(p^2 - q)}} \exp\left(\frac{1}{2}\int^x dx \frac{p' - \varepsilon}{\sqrt{p^2 - q}}\right) \sum_{\pm} C_{2\pm} \exp\left(\pm\mathrm{i}\int^x k_z dx\right). \end{aligned} \tag{8.8.7}$$

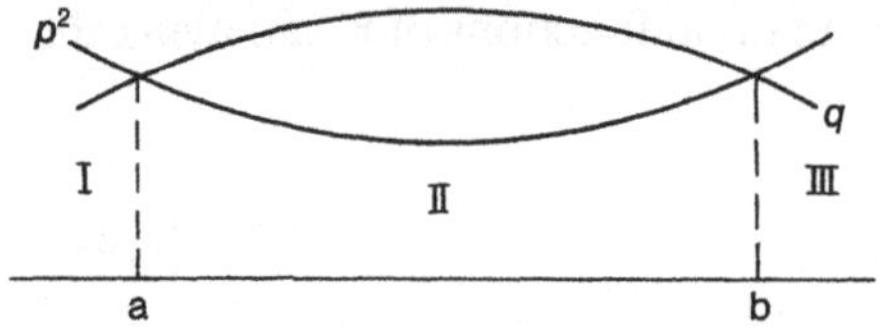

Fig. 8.7.

Since the solutions must be finite for $x \to \pm\infty$ we can omit the increasing solutions in the ranges I and III (the ranges of opacity). Then

$$y^{\mathrm{I}} = \frac{C'_{1+}}{\sqrt{\bar{k}_1^2\,(q-p)^2}} \exp\left(\mathrm{i}\int_a^x \bar{k}_1\,dx + \frac{\mathrm{i}}{2}\int_a^x \frac{p'-\varepsilon}{\sqrt{q-p^2}}\,dx\right) + \frac{C'_{2-}}{\sqrt{\bar{k}_2^2\,(q-p)^2}} \times \exp\left(-\mathrm{i}\int_a^x \bar{k}_2\,dx - \frac{\mathrm{i}}{2}\int_a^x \frac{p'-\varepsilon}{\sqrt{q-p^2}}\,dx\right); \quad (8.8.8)$$

$$y^{\mathrm{III}} = \frac{C'_{1-}}{\sqrt{\bar{k}_1^2\,(q-p)^2}} \exp\left(-\mathrm{i}\int_b^x \bar{k}_1\,dx - \frac{\mathrm{i}}{2}\int_b^x \frac{p'-\varepsilon}{\sqrt{q-p^2}}\,dx\right) + \frac{C'_{2+}}{\sqrt{\bar{k}_2^2\,(q-p)^2}} \times \exp\left(\mathrm{i}\int_b^x \bar{k}_2\,dx + \frac{\mathrm{i}}{2}\int_b^x \frac{p'-\varepsilon}{\sqrt{q-p^2}}\,dx\right), \quad \text{where} \quad (8.8.9)$$

$$\bar{k}_{1,2}^2 = p \pm \mathrm{i}\sqrt{q-p^2}\,.$$

We can find the relation between the integration constants $C_{i\pm}$ and $C'_{i\pm}$ ($i = 1,2$) by connecting these solutions by means of an integral passing around the points a and b in the complex x-plane. Each of the two linearly independent solutions in the ranges of opacity can be expanded into two independent solutions in the range of transparency. Since these solutions must be single-valued, we finally obtain quantization rules defining the spectrum of the eigenvalue ω:

$$\int_a^b (k_1 - k_2)\,dx \pm \int_a^b \frac{p'-\varepsilon}{\sqrt{p^2-q}}\,dx = 2\pi\left(n + \frac{1}{2}\right). \quad (8.8.10)$$

Here n is an integer. The presence of imaginary terms in these quantization rules indicates that energy can be transferred from one oscillation branch to the other at the branch points where k_1^2 and k_2^2 coincide.

If only one branch point a exists and if nondissipative boundary conditions $y(b) = y'(b) = 0$ are given at the other boundary b of the range of transparency, then the quantization rule can be written as

$$\int_a^b (k_2 - k_1)\,dx = \pi n\,. \quad (8.8.11)$$

Here we have omitted small real terms of the order of η.

8.8.2. Using (8.4.16) calculate the frequency spectrum of the helical waves for $N = N_0\sqrt{1 - x^2/x_0^2}$. Compare it with the local spectrum.

Solution. The local spectrum of the helical waves can be written as, see (5.2.17):

$$\omega = \frac{k^2c^2\Omega_e}{\omega_{pe}^2}\frac{|k_z|}{k}, \tag{8.8.12}$$

where $k^2 = k_x^2 + k_y^2 + k_z^2$ and ω_{pe} is some averaged electron Langmuir frequency.

From (8.4.16) we obtain

$$\int_0^{x_1} \sqrt{1 - \frac{(k_y^2 + k_z^2)\,c^4k_z^2\Omega_e^2}{\omega_{p0}^4\,\omega^2} - \frac{x^2}{x_0^2}}\,dx = \frac{\pi n c^3|k_z|\,\Omega_e}{2\,\omega\omega_{p0}^2}, \tag{8.8.13}$$

where ω_{p0} is the electron Langmuir frequency at the point $x = 0$, and $x_1 < x_0$ is a turning point:

$$\frac{x_1^2}{x_0^2} = 1 - \frac{(k_y^2 + k_z^2)\,c^4k_z^2\Omega_e^2}{\omega_{p0}^4\,\omega^2}. \tag{8.8.14}$$

The integration of (8.8.13) results in

$$1 - \frac{(k_y^2 + k_z^2)\,c^4k_z^2\Omega_e^2}{\omega^2\omega_{p0}^4} = \frac{2n}{x_0}\frac{c^2\Omega_e|k_z|}{\omega\omega_{p0}^2}. \tag{8.8.15}$$

Hence we find the frequency spectrum of the oscillations

$$\omega = \frac{c^2|k_z|\,\Omega_e}{\omega_{p0}^2}\left(\frac{n}{x_0} + \sqrt{k_y^2 + k_z^2 + \frac{n^2}{x_0^2}}\right). \tag{8.8.16}$$

The comparison of (8.8.12) with (8.8.16) shows that the local spectrum is close to the exact one for

$$k_y^2 + k_z^2 \gg n^2/x_0^2 \approx k_x^2\,.$$

8.8.3. Study the low-frequency short-wavelength drift oscillations of the collisionless inhomogeneous plasma on the basis of (8.5.7).

Solution. Oscillations of a wavelength short against the ion Larmor radius, i.e., $k_\perp\varrho_{\lambda i} \gg 1$, are considered here. We further assume $k_\perp\varrho_{\lambda e} \ll 1$ and $v_{Ti} \ll \omega/k_z \ll v_{Te}$. Under these conditions the homogeneous plasma cannot oscillate. In the inhomogeneous plasma we obtain from (8.5.7)

$$\varepsilon(\omega, \boldsymbol{k}, x) = 1 + \frac{\omega_{\mathrm{pi}}^2}{k^2 v_{\mathrm{Ti}}^2}\left(1 + \frac{k_y v_{\mathrm{Ti}}}{\sqrt{2\pi}\, k_\perp \omega}\frac{\partial}{\partial x}\ln\frac{N}{\sqrt{T_{\mathrm{i}}}}\right)$$

$$+ \frac{\omega_{\mathrm{pe}}^2}{k^2 v_{\mathrm{Te}}^2}\left(1 + \mathrm{i}\sqrt{\frac{\pi}{2}}\frac{k_y v_{\mathrm{s}}^2}{|k_z|\, v_{\mathrm{Te}} \Omega_{\mathrm{i}}}\frac{\partial}{\partial x}\ln\frac{N}{\sqrt{T_{\mathrm{i}}}}\right) = 0\,. \tag{8.8.17}$$

Taking account of the small term associated with the Cherenkov dissipation by the electrons, we obtain the spectrum

$(\omega \to \omega + \mathrm{i}\delta)$

$$\omega = -\frac{T_{\mathrm{e}} k_y v_{\mathrm{Ti}}}{T_{\mathrm{e}} + T_{\mathrm{i}}(1 + k^2 r_{\mathrm{De}}^2)}\frac{1}{\sqrt{2\pi} k_\perp}\frac{\partial}{\partial x}\ln\frac{N}{\sqrt{T_{\mathrm{e}}}}\,, \tag{8.8.18}$$

$$\delta = \frac{T_{\mathrm{e}}^2}{[T_{\mathrm{e}} + T_{\mathrm{i}}(1 + k^2 r_{\mathrm{De}}^2)]^2}\frac{k_y^2 v_{\mathrm{Ti}}^3}{2|k_z|\, k_\perp v_{\mathrm{Te}} \Omega_{\mathrm{i}}}\left(\frac{\partial}{\partial x}\ln\frac{N}{\sqrt{T_{\mathrm{e}}}}\right)\left(\frac{\partial}{\partial x}\ln\frac{N}{\sqrt{T_{\mathrm{i}}}}\right)$$

Thus, the frequency of the short-wavelength drift oscillations is smaller than the drift frequency $\omega \sim \omega_{\mathrm{dr}} \lambda_\perp / \varrho_{\lambda\mathrm{i}} \ll \omega_{\mathrm{dr}}$. The increment is positive, i.e., an instability occurs under the condition

$$1 + \frac{1}{2}\frac{\partial \ln(N/\sqrt{T_{\mathrm{e}}})}{\partial \ln(N/\sqrt{T_{\mathrm{i}}})} > 0\,, \tag{8.8.19}$$

which is satisfied practically always. Therefore, the short-wavelength drift instability has a universal character. Moreover, it also develops in short systems. The longitudinal dimensions must be larger than the ion Larmor radius, only.

8.8.4. Show that the short-wavelength drift oscillations extend up to the harmonics of the ion-cyclotron frequency. Study their spectrum and the buildup condition for the collisionless inhomogeneous plasma.

Solution. In the range of the ion-cyclotron frequencies, $\omega \approx \omega_{\mathrm{dr}} \approx n\Omega_{\mathrm{i}}$, the drift oscillations always have short wavelengths, $k_\perp \varrho_{\lambda\mathrm{i}} \gg 1$. Assuming $k_\perp \varrho_{\lambda\mathrm{e}} \ll 1$ and $v_{\mathrm{Ti}} \ll \omega/k_z \ll v_{\mathrm{Te}}$ we obtain from the general equation (8.5.6)

$$\varepsilon(\omega, \boldsymbol{k}, x) = 1 + \frac{\omega_{\mathrm{pi}}^2}{k^2 v_{\mathrm{Ti}}^2}\left[1 - \frac{1}{\sqrt{2\pi}\, k_\perp \varrho_{\lambda\mathrm{i}}}\frac{\omega}{\omega - n\Omega_{\mathrm{i}}}\right.$$

$$\left.\times\left(1 - \frac{k_y v_{\mathrm{Ti}}^2}{\omega \Omega_{\mathrm{i}}}\frac{\partial}{\partial x}\ln\frac{N}{\sqrt{T_{\mathrm{i}}}}\right)\right]$$

$$+ \frac{\omega_{\mathrm{pe}}^2}{k^2 v_{\mathrm{Te}}^2}\left[1 + \mathrm{i}\sqrt{\frac{\pi}{2}}\frac{k_y v_{\mathrm{s}}^2}{\Omega_{\mathrm{i}} |k_z|\, v_{\mathrm{Te}}}\frac{\partial}{\partial x}\ln\frac{N}{\sqrt{T_{\mathrm{e}}}}\right] = 0\,. \tag{8.8.20}$$

Hence, the following spectrum of the drift ion-cyclotron oscillations is found ($\omega \to \omega + \mathrm{i}\delta$):

$$\omega = n\Omega_\mathrm{i} + \Delta \; ,$$

$$\begin{aligned} \mathrm{Re}\,\{\Delta\} &= \frac{T_\mathrm{e}}{T_\mathrm{e} + T_\mathrm{i}(1 + k^2 r_\mathrm{De}^2)} \frac{1}{\sqrt{2\pi}\, k_\perp \varrho_{\lambda\mathrm{i}}} \\ &\quad \times \left(n\Omega_\mathrm{i} - \frac{k_y v_\mathrm{Ti}^2}{\Omega_\mathrm{i}} \frac{\partial}{\partial x} \ln \frac{N}{\sqrt{T_\mathrm{i}}} \right) , \end{aligned} \tag{8.8.21}$$

$$\begin{aligned} \mathrm{Im}\,\{\Delta\} &= \delta = -\frac{T_\mathrm{i}}{T_\mathrm{e}} \frac{\pi k_\perp \varrho_{\lambda\mathrm{i}}}{n\Omega_\mathrm{i} - \dfrac{k_y v_\mathrm{Ti}^2}{\Omega_\mathrm{i}} \dfrac{\partial}{\partial x} \ln \dfrac{N}{\sqrt{T_\mathrm{i}}}} \\ &\quad \times \frac{\mathrm{Re}\,\{\Delta^2\}}{|k_z| v_\mathrm{Te}} \left(n\Omega_\mathrm{i} + \frac{k_y v_\mathrm{s}^2}{\Omega_\mathrm{i}} \frac{\partial}{\partial x} \ln \frac{N}{\sqrt{T_\mathrm{e}}} \right) . \end{aligned}$$

These oscillations are stable for $\omega \approx n\Omega_\mathrm{i} \gg \omega_\mathrm{dr}$ and unstable for $\omega_\mathrm{dr} \gtrsim n\Omega_\mathrm{i}$. The instability has a universal character.

8.8.5. Show that the drift instabilities of the inhomogeneous plasma can be stabilized by the shear of the magnetic field lines, i.e., by a small strongly inhomogeneous transverse component of the magnetic field, $B_{0y}(x) \ll B_{0z}$.

Solution. In the presence of the fields B_{0y} and B_{0z} it is suitable to introduce the total field $B_0 = (0, B_{0y}, B_{0z})$ and

$$k_\parallel(x) = k_z \frac{B_{0z}}{B_0} + k_y \frac{B_{0y}}{B_0} \approx k_z + k_y \frac{B_{0y}}{B_0} = k_z + k_y \Theta(x) \; . \tag{8.8.22}$$

In the zero-order approximation of geometrical optics, accounting for a shear of the magnetic field, the dielectric tensor of the inhomogeneous plasma is of the same form as when there is no shear. One has to substitute $k_\parallel(x)$ for k_z, only. Taking account of the strong inhomogeneity of $\Theta(x)$ we have $k_{\parallel\mathrm{min}} \approx k_y \Theta(x)$. The condition for stabilizing drift oscillations by the shear follows immediately. The drift instability occurs in the collisionless plasma if

$$\omega \approx \omega_\mathrm{dr} \approx \frac{k_y}{\Omega_\mathrm{i}} \frac{T_\mathrm{e} + T_\mathrm{i}}{M L_0} > k_\parallel v_\mathrm{Ti} \gtrsim k_y \Theta v_\mathrm{Ti} \; . \tag{8.8.23}$$

When these conditions are violated the plasma is stable. Therefore the condition for stabilizing the drift instability by the shear can be written as

$$\Theta > \left(1 + \frac{T_\mathrm{e}}{T_\mathrm{i}}\right) \frac{\varrho_{\lambda\mathrm{i}}}{L_0} \; . \tag{8.8.24}$$

In the limit of frequent collisions the criterion is modified. Then drift instabilities are possible if

$$\omega\nu_i \approx \omega_{dr}\nu_i \approx \frac{k_y}{\Omega_i}\frac{T_e + T_i}{ML_0}\nu_i > k_\parallel^2 v_{Ti}^2 \gtrsim k_y^2 v_{Ti}^2 \Theta^2 . \tag{8.8.25}$$

Consequently, the drift instability is stabilized in the collisional plasma under the condition

$$\Theta > \sqrt{\left(1 + \frac{T_e}{T_i}\right)\frac{\nu_i}{\Omega_i}} . \tag{8.8.26}$$

8.8.6. Investigate the convective instability of the inhomogeneous plasma in the limit of frequent collisions. Take account of the curvature and the shear of the field lines of the confining magnetic field.

Solution. As in the problem treated before, we can introduce $k_\parallel(x)$ in the form of (8.8.22) and write the eikonal equation for the convective oscillations as (Sect. 8.7):

$$\left(1 + \frac{v_A^2}{c^2}\right)\frac{\eta}{4\pi}k_\perp^2\omega^2 + i\omega k_\parallel^2 v_A^2 - \frac{\eta}{4\pi}k_y^2 g_{eff}\frac{\partial \ln N}{\partial x} = 0 , \tag{8.8.27}$$

where $\eta = c^2/\sigma$, $\sigma = \alpha\omega_{pe}^2/4\pi\nu_e$. We have $\alpha = 1$, $\nu_e = \nu_{en}$ for the weakly ionized plasma and $\alpha = 1.96$, $\nu_e = \nu_{eff}$ for the completely ionized plasma. We introduce the notation $\Theta(x) = sx$ and the new space variable $\xi = x + k_z s/k_y$. Then, due to the quantization rule, we obtain the dispersion equation

$$\int d\xi\left[-k_y^2 + \frac{4\pi}{\eta\omega^2}\frac{k_y^2}{1 + \dfrac{v_A^2}{c^2}}\left(i\omega\xi^2 s^2 v_A^2 - \frac{\eta}{4\pi}g_{eff}\frac{\partial \ln N}{\partial x}\right)\right]^{1/2} = \pi n . \tag{8.8.28}$$

Here, we have to integrate over the range of transparency where the integrand is positive.

Neglecting the inhomogeneity of the plasma density as compared to the inhomogeneity of the magnetic shear $\Theta(x)$ and assuming s = const, we obtain from (8.8.28)

$$k_y^2\left(1 - \frac{g_{eff}(\partial \ln N/\partial x)}{\omega^2(1 + v_A^2/c^2)}\right) + 4n\left(\frac{k_y^2 s^2 \pi v_A^2}{i\omega\eta(1 + v_A^2/c^2)}\right)^{1/2} = 0 . \tag{8.8.29}$$

Hence, in the case of a weak shear ($s \to 0$) we find the already known spectrum of the unstable interchange (convective) oscillations (Sect. 8.7). In the opposite limit of a strong shear there appears a new branch of unstable convective oscillations in the frequency range $\omega^2 \ll |g_{\rm eff} \cdot \partial \ln N/\partial x|$ with the spectrum

$$\omega = \mathrm{i}\left[\frac{\eta k_y^2 g_{\rm eff}}{(4\pi)^2 n^2 s^2 v_{\rm A}^2 (1 + v_{\rm A}^2/c^2)}\left(\frac{\partial \ln N}{\partial x}\right)^2\right]^{1/3}. \tag{8.8.30}$$

Thus, although the increment decreases with increasing s, the convective instability of the inhomogeneous plasma in the curved magnetic field is not stabilized by the shear.

8.8.7. Study the instability of the magnetized straight electron beam with an inhomogeneous profile of the ordered velocity *(slipping instability)* for the relativistic case in the approximation of geometrical optics.

Solution. We assume that the Langmuir frequency of the beam $\omega_{\rm pe}(x)$ is small compared to the Larmor frequency $\Omega_{\rm e}$. Confining our analysis to longitudinal oscillations we can write the eikonal equation in the zero-order approximation of geometrical optics as

$$k_\perp^2 + k_z^2\left(1 - \frac{\omega_{\rm pe}^2}{\gamma^3(\omega - k_z u)^2}\right) + k_y \frac{\partial}{\partial x}\frac{\omega_{\rm pe}^2}{\Omega_{\rm e}(\omega - k_z u)} = 0\,, \tag{8.8.31}$$

where $\gamma = (1 - u^2/c^2)^{-1/2}$. Hence, the local spectrum is determined by

$$k^2\omega'^2 + k_y \frac{\omega_{\rm pe}^2}{\Omega_{\rm e}}\frac{\partial \ln N}{\partial x}\,\omega' - \frac{k_z^2\omega_{\rm pe}^2}{\gamma^3}\left(1 - \frac{k_y\gamma^3 u}{k_z\Omega_{\rm e}}\frac{\partial \ln u}{\partial x}\right) = 0\,, \tag{8.8.32}$$

where $\omega' = \omega - k_z u(x)$.

Under the condition

$$\frac{k_y}{k_z}\frac{u\gamma^3}{\Omega_{\rm e}}\frac{\partial \ln u}{\partial x} > 1 \tag{8.8.33}$$

the electron beam may become unstable. But the density inhomogeneity plays a stabilizing part and the beam is stable under the condition

$$\frac{\omega_{\rm pe}^2\gamma^3}{\Omega_{\rm e}^2}\left(\frac{\partial}{\partial x}\ln\frac{N}{B}\right)^2 > 4k_z^2\left|1 - \frac{k_y}{k_z}\frac{\gamma^3 u}{\Omega_{\rm e}}\frac{\partial \ln u}{\partial x}\right|. \tag{8.8.34}$$

The instability occurring in the reverse case is called the slipping instability due to the slipping of different layers of the beam, one along the other. The growth rate of this instability is of the order of

$$\delta = \mathrm{Im}\,\{\omega\} \approx \frac{\omega_{\rm pe}^2}{\gamma^{3/2}}\sqrt{\left|\frac{k_z k_y \gamma^3 u}{k^2\Omega_{\rm e}}\frac{\partial \ln u}{\partial x}\right|} \gtrsim \frac{\omega_{\rm pe}}{\gamma^{3/2}}\frac{l_\parallel}{L_\perp}\,. \tag{8.8.35}$$

8.8.8. Show that the strongly magnetized cold nonrelativistic electron beam with an inhomogeneous profile of the ordered velocity can sustain symmetric oscillation modes with $k_y = 0$. How large is the damping decrement due to the resonant interaction of these oscillations with the electron beam (this is an analog of the Landau damping)?

Solution. The eikonal equation for these longitudinal modes can be written as

$$k_x^2 + k_z^2\left(1 - \frac{\omega_{pe}^2(x)}{[\omega - k_z u(x)]^2}\right) = 0 . \tag{8.8.36}$$

Hence we obtain the following dispersion equation in the frequency range $(\omega - k_z u)^2 \ll \omega_{pe}^2$

$$\int dx \frac{\omega_{pe}(x)}{\omega - k_z u(x)} = \frac{\pi n}{|k_z|} , \tag{8.8.37}$$

where n is an integer and where the integration is performed over the range of transparency defined by $\omega_{pe}^2 > (\omega - k_z u)^2$. We assume that this condition is satisfied in the whole region covered by the plasma, $|x| \leq a$, and that nondissipative boundary conditions are given at the plasma boundaries. Then the integral in (8.8.37) extends over the range $-a \leq x \leq a$. One should take account of the pole at $\omega = k_z u(x)$ which corresponds to the resonant interaction of the wave with the inhomogeneous electron beam at the points where the beam velocity coincides with the phase velocity of the wave. The result of this interaction is the resonant absorption of the wave at the resonance points of the beam.

For the special case of a beam with a homogeneous density $N = \text{const}$ and an inhomogeneous velocity $u(x) = u_0 x/a$ we obtain from (8.8.37)

$$\frac{\omega_{pe} a}{u_0}\left[\ln\left|\frac{\omega + k_z u_0}{\omega - k_z u_0}\right| - i\frac{\pi}{2}\left(1 + \frac{k_z u_0 - \omega}{|k_z u_0 - \omega|}\right)\right] = \pi n . \tag{8.8.38}$$

Hence, the applicability condition of geometrical optics reads

$$\omega_{pe} a \gg u_0 . \tag{8.8.39}$$

Equation (8.8.38) shows that under the given conditions oscillations can be excited in the frequency range $\omega > k_z u_0$, where the imaginary term vanishes and where there is no wave absorption. Then the oscillation spectrum is

$$\omega = k_z u_0 \coth\frac{\pi n u_0}{2\omega_{pe} a} . \tag{8.8.40}$$

In the frequency range $\omega \leq k_z u_0$ oscillations are impossible due to the strong absorption at the resonance points.

Another special case is the case when the beam density is also inhomogeneous. We assume $N = N_0 \exp(-x^2/L_0^2)$ and $L_0^2 \ll a^2$. Then (8.8.37) yields

$$1 - \sqrt{\frac{2}{\pi}} \frac{\omega_{\mathrm{pe}}(0)|k_z| L_0}{\omega n} I_+ \left(\frac{\omega a}{k_z u_0 L_0}\right) = 0 . \tag{8.8.41}$$

It is easy to see that oscillations of the beam are impossible in the limit $\omega < k_z u_0 L_0/a$. The reason is the strong absorption at the resonance points. In the frequency range $\omega \gg k_z u_0 L_0/a$ (8.8.41) leads to the following spectrum of oscillations with a small damping decrement:

$$\omega = \sqrt{\frac{2}{\pi}} \frac{\omega_{\mathrm{pe}}(0)|k_z|}{n} L_0 , \quad \frac{\delta}{\omega} = -\frac{\omega_{\mathrm{pe}}(0)\, a}{n u_0} \exp\left(-\frac{\omega_{\mathrm{pe}}^2(0)\, a^2}{\pi n^2 u_0^2}\right) . \tag{8.8.42}$$

The absorption mechanism is analogous to the Cherenkov mechanism in the case of the thermal motion of the particles in the plasma. One can speak of Landau damping here, too. However, in the present case the absorption significantly depends on the form of the function $u(x)$.

9. Linear Electromagnetic Phenomena in Bounded Plasmas

The analysis pertains to the problem of electromagnetic waves in a semi-bounded plasma with a mirror reflection of charged particles from the plasma boundary. A surface impedance of such a plasma is calculated and on its basis a dispersion equation for surface waves is obtained and their spectra. A special study is undertaken to solve the problem of the stability of a magnetically confined plasma boundary with respect to the development of surface drift and convective oscillations. Illustrated by an electron plasma, electromagnetic waves are analyzed in a plasma waveguide.

9.1 Surface Electromagnetic Waves in Semi-Bounded Plasmas

In the previous chapter we have dealt with short-wavelength electromagnetic oscillations in a spatially inhomogeneous plasma, their wavelengths being much smaller than the characteristic dimension of the plasma inhomogeneity. To describe these oscillations, the plasma can be treated as effectively infinite and the approximation of geometrical optics can be applied. Let us consider the opposite limit, when the scales of the plasma inhomogeneity are sufficiently sharp compared to the wavelength. We shall study only the simplest problems of the electrodynamics of plasmas with sharp inhomogeneities, i.e., surface electromagnetic waves in a semi-bounded plasma conterminous with vacuum. Surface waves are qualitatively a new type of electromagnetic oscillations of a bounded medium. They are waves travelling along the medium surface and damping in the perpendicular direction.

It is obvious that the character of surface waves essentially depends on the properties of the plasma surface, or, more exactly, on the boundary conditions, considered in Sect. 2.1., which must supplement the field equations. From a variety of models of plasma surfaces only two cases of different physical nature will be studied. This section deals with the first model in which the existence of a sufficiently sharp plasma surface is presupposed. Here, all plasma quantities with the dimension of length (wavelengths, Debye and Larmor radii of particles, mean free paths, etc.) greatly exceed the size of the density variation near the plasma surface. A similar situation occurs, for example, in a gaseous plasma confined in a glass vessel with electromagnetic properties insignificantly differing from the vacuum properties, or in a solid-state plasma whose surface structure is defined by a crystal lattice. The second model assumes the plasma to be confined by a strong external magnetic field, its surface being the boundary layer with thickness of the order of the Larmor radius of particles. This surface can be treated as sharp for surface waves much longer than the Larmor radius of particles. This situation is characteristic of a high-temperature plasma in controlled thermonuclear fusion devices where the plasma is isolated from the metal walls by the magnetic field and appears to be spatially confined and contiguous to vacuum. This model of a plasma surface will be studied in the next section.

9.1.1 Solution of the Vlasov Equation for the Semi-Bounded Isotropic Plasma

We shall limit our analysis of surface waves in a plasma with a sharp boundary to the case of the isotropic collisionless semi-bounded plasma in the absence of external electric and magnetic fields. As the unperturbed distribution function of particles of type $\alpha\,(\alpha = \mathrm{e}, \mathrm{i})$, the nonrelativistic Maxwellian distribution function

$$f_{0\alpha} = \frac{N_{0\alpha}}{(2\pi m_\alpha T_\alpha)^{3/2}} \exp\left(-\frac{m_\alpha v^2}{2T_\alpha}\right) \tag{9.1.1}$$

will be taken for a nondegenerate plasma, or the Fermi distribution function

$$f_{0\alpha} = \begin{cases} \dfrac{2}{(2\pi\hbar)^3} & \text{for} \quad p < p_{\mathrm{F}\alpha} = (3\pi^2)^{1/3}\hbar N_{0\alpha}^{1/3}\,, \\[2ex] 0 & \text{for} \quad p > p_{\mathrm{F}\alpha} \end{cases} \tag{9.1.2}$$

for a degenerate one. Here $N_{0\alpha} = \text{const}$ for $x > 0$ (in the plasma, see Fig. 9.1) and $N_{0\alpha} = 0$ for $x < 0$ (in vacuum). Then the solution of the kinetic equation for the perturbation of the distribution function

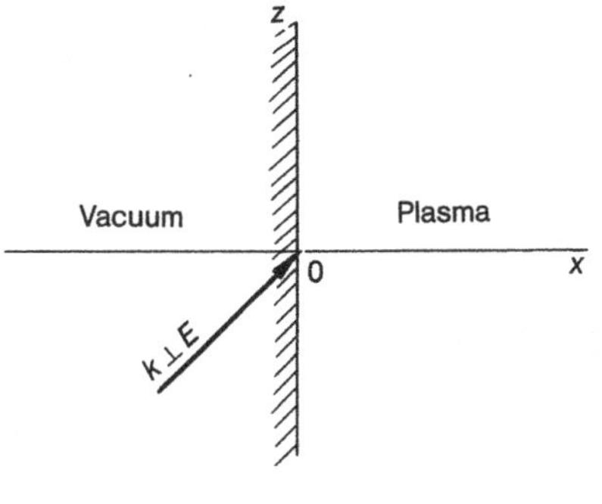

Fig. 9.1.

$$\frac{\partial \delta f_\alpha}{\partial t} + (\boldsymbol{v} \cdot \nabla)\, \delta f_\alpha + \frac{e_\alpha}{m_\alpha} \boldsymbol{E} \frac{\partial f_{0\alpha}}{\partial \boldsymbol{v}} = 0 \tag{9.1.3}$$

without any restrictions on generality can be sought in the form

$$\delta f_\alpha = \delta f_\alpha (x) \exp(-\mathrm{i}\omega t + \mathrm{i}k_z z)\ , \tag{9.1.4}$$

assuming the wave vector to be parallel to the plasma surface. As a result, from (9.1.3) we obtain

$$-\mathrm{i}(\omega - \boldsymbol{k} \cdot \boldsymbol{v})\, \delta f_\alpha + v_x \frac{\partial \delta f_\alpha}{\partial x} + \frac{e_\alpha}{m_\alpha} \boldsymbol{E} \frac{\partial f_{0\alpha}}{\partial \boldsymbol{v}} = 0\ . \tag{9.1.5}$$

To solve (9.1.5) it is necessary to set the boundary condition for $\delta f_\alpha(x)$ for $x = 0$, i.e., on the plasma surface. This very condition must contain all information on the character of interaction of charged particles with the surface, confining the plasma. Henceforth, we shall assume charged particles to undergo a mirror reflection from this surface. This implies that for $x = 0$ the condition

$$\delta f_\alpha (0, v_x > 0) = \delta f_\alpha (0, v_x < 0) \tag{9.1.6}$$

is satisfied. Besides, when studying plasma surface waves (or the field penetration into the plasma), we assume the electromagnetic field $\boldsymbol{E}$ and $\boldsymbol{B}$ to vanish far from the plasma surface.

To solve (9.1.5), it is convenient to write

$$\delta f_\alpha (x, v_x) = \delta f_\alpha^+ (x, v_x) + \delta f_\alpha^- (x, v_x),$$

$$\delta f_\alpha^\pm (x, v_x) = \begin{cases} \delta f_\alpha (x, v_x > 0), \\ \delta f_\alpha (x, v_x < 0)\ . \end{cases} \tag{9.1.7}$$

Each of the quantities $\delta f_\alpha^\pm$ also satisfies (9.1.5). Therefore, using $\delta f_\alpha^- (x, v_x) \to 0$ for $x \to \infty$ we obtain

$$\delta f_\alpha^- (x, v_x) = \frac{e_\alpha}{m_\alpha} \frac{\partial f_{0\alpha}}{\partial \boldsymbol{v}} \frac{1}{v_x} \int_x^\infty dx' \boldsymbol{E}(x') \times \exp\left[\mathrm{i}\frac{(x - x')}{v_x} (\omega - k_z v_z)\right]. \tag{9.1.8}$$

To find $\delta f_\alpha^+ (x, v_x)$ let us rewrite the condition (9.1.6) in the form of

$$\delta f_\alpha^+ (0, v_x) = \delta f_\alpha^- (0, -v_x)\ . \tag{9.1.6a}$$

Then from (9.1.5) we have

$$\delta f_a^+(x, v_x) = -\frac{e_a}{m_a}\frac{\partial f_0}{\partial \boldsymbol{v}}\frac{1}{v_x}\left\{\int_0^x dx' \boldsymbol{E}(x') \exp\left[\mathrm{i}\frac{(x-x')}{v_x}(\omega - k_z v_z)\right]\right.$$

$$\left. + \int_0^\infty dx' \boldsymbol{E}(x') \exp\left[\mathrm{i}\frac{(x-x')}{v_x}(\omega + k_z v_z)\right]\right\} . \tag{9.1.9}$$

Substituting (9.1.8) and (9.1.9) into the formula for the current density

$$\boldsymbol{j}(x) = \sum_a e_a \int d\boldsymbol{p}\, \boldsymbol{v}\delta f_a(x, \boldsymbol{v}) = \sum_a e_a \int_{v_x>0} d\boldsymbol{p}\, \boldsymbol{v}\delta f_a^+ + \int_{v_x<0} d\boldsymbol{p}\, \boldsymbol{v}\delta f_a^- \tag{9.1.10}$$

we have after simple calculations

$$j_i^{(x)} = \int_0^\infty dx' \left[K_{ij}(|x-x'|) + K_{ij}(|x+x'|)\right] E_j(x') . \tag{9.1.11}$$

Here

$$K_{ij}(|x|) = -\sum_a \frac{e_a^2}{m_a} \int_{v_x \geqslant 0} dp \frac{v_i}{v_x}\frac{\partial f_{0a}}{\partial v_j} \exp\left[\mathrm{i}\frac{|x|}{v_x}(\omega - k_z v_z)\right] . \tag{9.1.12}$$

9.1.2 Solution of Field Equations

We can proceed to solve the field equations which in the present geometry can be written as

$$\left.\begin{aligned} \frac{i\omega}{c}\boldsymbol{B} &= \left\{\begin{matrix} -\mathrm{i}k_z E_y \\ \mathrm{i}k_z E_x - \dfrac{\partial E_z}{\partial x} \\ \dfrac{\partial E_y}{\partial x} \end{matrix}\right\} , \\ -\frac{i\omega}{c}\boldsymbol{E} + \frac{4\pi}{c}\boldsymbol{j} &= \left\{\begin{matrix} -\mathrm{i}k_z B_y \\ \mathrm{i}k_z B_x - \dfrac{\partial B_z}{\partial x} \\ \dfrac{\partial B_y}{\partial x} \end{matrix}\right\} \end{aligned}\right\} . \tag{9.1.13}$$

This system of equations is valid for $x \geq 0$ (plasma) as well as for $x < 0$ (vacuum) since $\boldsymbol{j}(x<0) = 0$. They follow from (9.1.11) and (9.1.12) accounting for the plasma dependence on x.

Noting the symmetry of the tensor $K_{ij}(|x|)$, it is easy to show that the system of equations (9.1.13) splits into two independent subsystems for the field components E_x, E_z, B_y and B_x, B_z, E_y, respectively. The latter admits no solutions in the form of surface waves. Therefore we confine our analysis to the equations for the field components E_x, E_z and B_y:

$$\begin{aligned}
&\frac{\partial E_z}{\partial x} - ik_z E_x + i\frac{\omega}{c}B_y = 0\,,\\
&ik_z B_y - i\frac{\omega}{c}E_x + \frac{4\pi}{c}j_x = 0\,,\\
&\frac{\partial B_y}{\partial x} + i\frac{\omega}{c}E_z - \frac{4\pi}{c}j_z = 0\,.
\end{aligned} \tag{9.1.14}$$

By integrating these equations over an infinitely narrow intermediate layer near the plasma-vacuum interface, we obtain the boundary conditions which link the fields $\boldsymbol{E}$ and $\boldsymbol{B}$ in vacuum and plasma. As a corollary to the finiteness of the fields $\boldsymbol{E}, \boldsymbol{B}$ and the current density $\boldsymbol{j}$, we obtain the general electrodynamic boundary condition, i.e., the continuity of the tangential components of the fields $\boldsymbol{E}$ and $\boldsymbol{B}$ for $x = 0$:

$$\{E_z\}_{x=0} = \{B_y\}_{x=0} = 0\,, \tag{9.1.15}$$

where the notation $\{A\}_{x=0} = A\,(x \to +0) - A\,(x \to -0)$ is introduced.

In order to solve this boundary problem, we shall apply the following method. We extend (9.1.14) into the range $x < 0$ assuming

$$\begin{aligned}
&E_x(x) = -E_x(-x)\,, \qquad N_{0\alpha}(x) = N_{0\alpha}(-x)\,,\\
&E_z(x) = E_z(-x)\,, \qquad B_y(x) = -B_y(-x)\,.
\end{aligned} \tag{9.1.16}$$

Then in the whole range $-\infty \leq x \leq +\infty$ we have

$$j_i(x) = \int_{-\infty}^{\infty} dx'\sigma_{ij}(|x-x'|)\,E_j(x')\,, \quad \text{where} \tag{9.1.17}$$

$$\sigma_{ij}(|x|) = -\sum_\alpha \frac{e_\alpha^2}{m_\alpha}\int d\boldsymbol{p}\,\frac{v_i}{v_x}\frac{\partial f_{0\alpha}}{\partial v_j}\exp\left[i\frac{|x|}{v_x}(\omega - k_z v_z)\right] \tag{9.1.18}$$

is the tensor connecting $\boldsymbol{j}(x)$ and $\boldsymbol{E}(x)$ in a spatially infinite isotropic plasma. Its Fourier transform

$$\sigma_{ij}(\omega, \boldsymbol{k}) = \int_{-\infty}^{\infty} dx\,\sigma_{ij}(|x|)\,e^{-ik_x x} = -i\sum_\alpha \frac{e_\alpha^2}{m_\alpha}\int\frac{d\boldsymbol{p}\,v_i}{\omega - \boldsymbol{k}\cdot\boldsymbol{v}}\frac{\partial f_{0\alpha}}{\partial v_j} \tag{9.1.19}$$

coincides with the conductivity tensor of the isotropic plasma (4.1.8).

Thus, the material equation

$$j_i(\boldsymbol{k}) = \sigma_{ij}(\omega, \boldsymbol{k})\, E_j(\boldsymbol{k}) \tag{9.1.20}$$

connecting the Fourier components of the current and electric field density, in the case of the mirror reflection of particles from the surface for the semi-bounded plasma, is of the same form as for the spatially infinite isotropic case. This is clear, since for the mirror reflection of particles from the plasma surface, their motion is the same as in the infinite plasma and thus the perturbation, developed in the system due to the electromagnetic field, is independent of the surface presence.

It should be noted that when applying the continuation (9.1.16), the components E_x and B_y (oddly continued) for $x = 0$ undergo an abrupt change. One can easily see this when substituting (9.1.17) into the system (9.1.14) and integrating it over a thin layer near $x = 0$. At the same time if one proceeds from (9.1.11) and assumes $j(x) = 0$ for $x < 0$, then the integration of (9.1.17) over an intermediate layer near the plasma surface results in the continuity condition for the tangential components of the fields $\boldsymbol{E}$ and $\boldsymbol{B}$ for $x = 0$, shown by the boundary conditions (9.1.15). This is natural since the systems (9.1.14) and (9.1.17) are valid only in the region of the plasma with $x \geq 0$, the applied continuation being only a mathematical method for solving this problem. On the other hand, this fact should be accounted for when solving (9.1.14) which is a system of integro-differential equations with differential kernels after the substitution of the expressions for the current density (9.1.17). Such a system should be solved with the aid of the Fourier transformation

$$A(x) = \int_{-\infty}^{\infty} dk_x e^{ik_x x} A(k_x)\,, \qquad A(k_x) = \frac{1}{2\pi} \int_{-\infty}^{\infty} dx\, e^{-ik_x x} A(x) \tag{9.1.21}$$

accounting for possible jumps of the quantities E_x and B_y at $x = 0$.

Fourier transforming (9.1.14) and taking into account the continuity of the function $E_z(x)$ and the discontinuity of the function $B_y(x)$ at $x = 0$, we obtain the algebraic equations

$$\left.\begin{aligned} &ik_x E_z(k_x) - ik_z E_x(k_x) + i\frac{\omega}{c} B_y(k_x) = 0\,, \\ &ik_z B_y(k_x) - i\frac{\omega}{c} E_x(k_x) + \frac{4\pi}{c} j_x(k_x) = 0\,, \\ &ik_x B_y(k_x) - \frac{1}{\pi} B_y(x=0) + i\frac{\omega}{c} E_z(k_x) \\ &\qquad\qquad - \frac{4\pi}{c} j_z(k_x) = 0\,. \end{aligned}\right\}. \tag{9.1.22}$$

Hence

$$E_z(k_x) = \frac{-ic}{\pi\omega k^2} B_y(x=0) \left(\frac{k_z^2}{\varepsilon^{\mathrm{lo}}(\omega, k)} - \frac{k_x^2\omega^2}{c^2k^2 - \omega^2\varepsilon^{\mathrm{tr}}(\omega, k)} \right) . \qquad (9.1.23)$$

Here $k^2 = k_x^2 + k_z^2$ and $\varepsilon^{\mathrm{lo}}(\omega, k)$ and $\varepsilon^{\mathrm{tr}}(\omega, k)$ are respectively the longitudinal and transverse dielectric permittivities of an isotropic plasma, determined by (4.1.14).

9.1.3 Surface Impedance

After substituting (9.1.23) into the expression for the Fourier transformation (9.1.21), we shall determine the so-called *surface impedance* of the semi-bounded isotropic plasma:

$$\begin{aligned} Z_s &= \frac{4\pi}{c} \frac{E_z(x=0)}{B_y(x=0)} \\ &= -\frac{8i\omega}{c^2} \int_0^\infty \frac{dk_x}{k^2} \left(\frac{k_z^2c^2}{\omega^2\varepsilon^{\mathrm{lo}}(\omega, k)} - \frac{k_x^2c^2}{c^2k^2 - \omega^2\varepsilon^{\mathrm{tr}}(\omega, k)} \right) . \end{aligned} \qquad (9.1.24)$$

The system of equations (9.1.14) may analogously be solved for the vacuum, i.e., for $\boldsymbol{j}(x) = 0$. One must only continue the fields $\boldsymbol{E}$ and $\boldsymbol{B}$ into the region $x > 0$ according to (9.1.16). Thus, the surface impedance of the vacuum half-space is

$$Z_v = -\frac{4\pi}{c} \frac{E_z(x=0)}{B_y(x=0)} = \frac{4\pi i}{c} \sqrt{\frac{k_z^2c^2}{\omega^2} - 1} . \qquad (9.1.25)$$

For $\varepsilon^{\mathrm{lo}} = \varepsilon^{\mathrm{tr}} = 1$ this formula can also be obtained from (9.1.24).

9.1.4 Dispersion Equation for Surface Waves

Now we can apply the boundary conditions (9.1.15) and equate Z_s and Z_v. This equality yields the dispersion equation for surface waves in the semi-bounded isotropic plasma:

$$\sqrt{\frac{k_z^2c^2}{\omega^2} - 1} + \frac{2\omega}{\pi c} \int_0^\infty \frac{dk_x}{k^2} \left(\frac{k_z^2c^2}{\omega^2\varepsilon^{\mathrm{lo}}(\omega, k)} - \frac{k_x^2c^2}{k^2c^2 - \omega^2\varepsilon^{\mathrm{tr}}(\omega, k)} \right) = 0 . \qquad (9.1.26)$$

In the limit $c \to \infty$, from this equation we obtain a dispersion equation for longitudinal (potential) surface waves in the semi-bounded plasma:

$$1 + \frac{2}{\pi} \int_0^\infty \frac{dk_x |k_z|}{k^2\varepsilon^{\mathrm{lo}}(\omega, k)} = 0 . \qquad (9.1.27)$$

9.1.5 Surface Waves in Cold Semi-Bounded Plasma

Let us analyze the frequency spectra of the surface electromagnetic waves in the semi-bounded plasma. We start with the cold plasma, neglecting the thermal motion of particles. Accounting for

$$\varepsilon^{\mathrm{tr}}(\omega) = \varepsilon^{\mathrm{lo}}(\omega) = \varepsilon(\omega) = 1 - \frac{\omega_{\mathrm{pe}}^2}{\omega^2}\,, \tag{9.1.28}$$

from the general dispersion equation (9.1.26), we obtain

$$\sqrt{k_z^2c^2 + \omega_{\mathrm{pe}}^2 - \omega^2} + \left(1 - \frac{\omega_{\mathrm{pe}}^2}{\omega^2}\right)\sqrt{k_z^2c^2 - \omega^2} = 0\,. \tag{9.1.29}$$

It shows that in the semi-bounded, cold, isotropic plasma, surface waves exist only in the frequency range $\omega < \omega_{\mathrm{pe}}$, their phase velocity being always smaller than that of light, $\omega/k_z < c$. The general solution of (9.1.29) is graphically shown in Fig. 9.2. Here the dashed continuation of the dispersion curve corresponds to the frequency range where the electron thermal motion ($\omega \sim k_z v_{\mathrm{Te}}$) becomes significant and oscillations are strongly damped. The analytical solution can easily be written in limiting cases of long and short wavelengths:

$$\omega^2 = \begin{cases} k_z^2c^2 & \text{for} \quad k_z^2c^2 \ll \omega_{\mathrm{pe}}^2\,, \\ \omega_{\mathrm{pe}}^2/2 & \text{for} \quad k_z^2c^2 \gg \omega_{\mathrm{pe}}^2\,. \end{cases} \tag{9.1.30}$$

In the long-wavelength limit, the phase velocity of the surface waves is near the velocity of light and oscillations are closely transverse, while in the short-wavelength limit their phase velocity is small compared to that of light, oscillations being almost longitudinal (potential). Potentiality of short (slow) surface waves in a cold plasma is also verified by the solution of (9.1.27) which takes the form

$$\varepsilon(\omega) = 1 - \frac{\omega_{\mathrm{pe}}^2}{\omega^2} = -1 \tag{9.1.31}$$

and has a solution identical to the second solution (9.1.30).

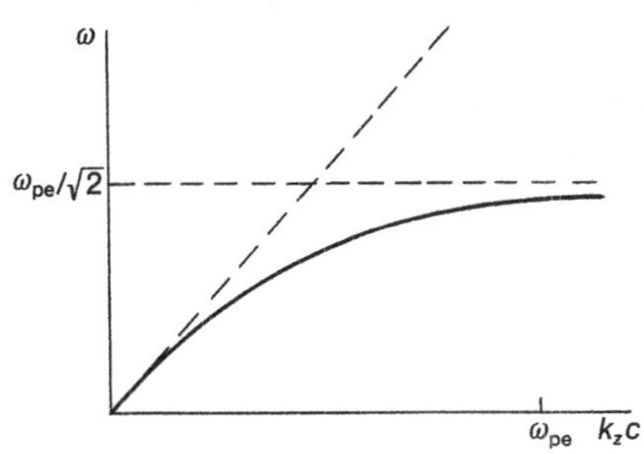

Fig. 9.2. Spectrum of surface electromagnetic waves in a cold semi-bounded plasma

9.1.6 Cherenkov Damping of Surface Waves

In the above, the thermal motion of particles and the collisionless Cherenkov damping of surface waves have been ignored. Slow waves, being closely longitudinal, are obviously most strongly damped. Thus, to analyze the collisionless damping of surface waves, we shall confine our analysis to (9.1.27). It can be analyzed only by means of numerical integration. However, in the range $\omega \gg k_z v_{Te}$, we can approximate (9.1.27) as

$$2 - \frac{\omega_{pe}^2}{\omega^2} + \mathrm{i}\sqrt{\frac{2}{\pi}}\,|k_z| \int_0^\infty \frac{dk_x \omega_{pe}^2 \omega}{k^5 v_{Te}^3} \exp\left(-\frac{\omega^2}{2k^2 v_{Te}^2}\right) = 0\,. \tag{9.1.32}$$

Small values of the imaginary parts specified by the Cherenkov absorption of surface waves by plasma electrons have been accounted for when obtaining (9.1.32). Since for $\omega \gg k_z v_{Te}$ the main contribution in the integral (9.1.32) gives the range of large values of k_x, then k can be replaced by k_x in the integrand with a high degree of accuracy. As a result, we have from (9.1.32)

$$2 - \frac{\omega_{pe}^2}{\omega^2}\left(1 - 2\mathrm{i}\sqrt{\frac{2}{\pi}}\frac{|k_z| v_{Te}}{\omega}\right) = 0\,. \tag{9.1.33}$$

Hence, (9.1.30) again gives the frequency spectrum of longitudinal surface waves. For the damping decrement ($\omega \to \omega + \mathrm{i}\delta$) we obtain[1]:

$$\delta = -\sqrt{\frac{2}{\pi}}\,|k_z| v_{Te}\,. \tag{9.1.34}$$

Damping of high-frequency surface waves, in contrast to that of volume Langmuir waves (Sect. 5.2), is not exponentially small, though their phase velocity greatly exceeds the thermal velocity of electrons. This is a corollary to the fact that the resonance interaction of particles with a wave occurs for $\omega \gg k_z v_{Te}$ in the frequency range $\omega = k_x v_{Te}$. Since, according to (9.1.27), k_x has arbitrarily large values (in particular, $k_x > \omega/v_{Te}$ for surface waves), the main bulk of electrons and not just an exponentially small portion of the particles (as in the case of volume Langmuir waves in an unbounded plasma) takes part in their absorption.

It is not difficult to account for the collisionless Cherenkov dissipation of high-frequency longitudinal surface waves in a degenerate electron plasma. This can be done by substituting the expression for the longitudinal dielectric

1 In the damping decrement of surface waves an exact numerical solution of (9.1.27) gives the factor 0.125 instead of $\sqrt{2/\pi} \approx 0.8$.

permittivity of the degenerate plasma (4.3.1) into (9.1.27). Then in the frequency range $\omega \gg k_z v_{Fe}$ we obtain, cf. (9.1.33),

$$2 - \frac{\omega_{pe}^2}{\omega^2}\left(1 - i\frac{3}{4}\frac{|k_z| v_{Fe}}{\omega}\right) = 0 \,. \tag{9.1.35}$$

In this range, the frequency spectrum of longitudinal surface waves in the degenerate plasma is the same as that in the nondegenerate plasma. From (9.1.35) for the damping decrement ($\omega \to \omega + i\delta$) we obtain

$$\delta = -\frac{3}{8}|k_z| v_{Fe} \,. \tag{9.1.36}$$

A numerical calculation of the damping decrement gives the factor 0.021 instead of 3/8. Moreover, a numerical integration of (9.1.27) for the degenerate plasma enables us to take account of the thermal correction to the frequency spectrum (9.1.30). As a result, we have

$$\begin{aligned} \omega &= \frac{\omega_{pe}}{\sqrt{2}} + 0.4\, k_z v_{Fe} \,, \\ \delta &= -0.021\, k_z v_{Fe} \,. \end{aligned} \tag{9.1.37}$$

This spectrum is given in Fig. 9.3. The character of collisionless absorption of surface waves significantly differs from that of volume waves. As shown in Sect. 4.3, high-frequency volume waves do not damp in the collisionless limit in a degenerate plasma since their phase velocity exceeds the Fermi velocity and the electrons cannot interact with these waves.

9.1.7 Surface Ion-Acoustic Waves

Let us now proceed to the analysis of surface ion-acoustic waves in a semi-bounded isotropic plasma. Volume ion-acoustic waves exist in an unbounded

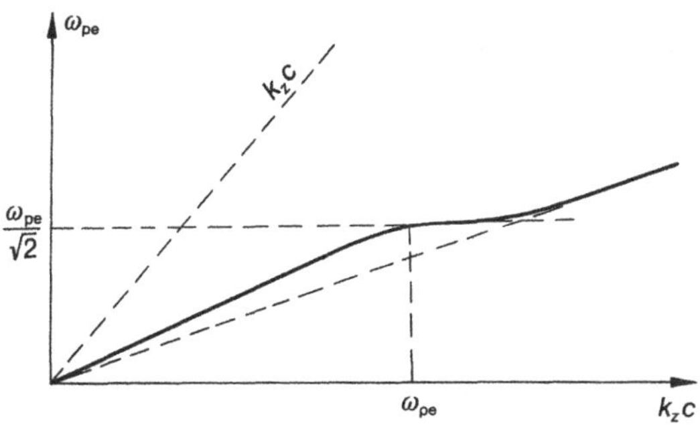

Fig. 9.3. Spectrum of surface longitudinal waves in a degenerate isotropic plasma

plasma only when $T_e \gg T_i$ (Sect. 4.2). This is also necessary for surface waves, since similar to space waves, surface ion-acoustic waves are longitudinal, described by (9.1.27). For the degenerate plasma we have

$$\varepsilon^{lo}(\omega, k) = 1 - \frac{\omega_{pi}^2}{\omega^2} + \frac{\omega_{pe}^2}{k^2 v_{Te}^2}\left(1 + i\sqrt{\frac{\pi}{2}}\frac{\omega}{k v_{Te}}\right). \tag{9.1.38}$$

Since $v_{Te} \gg \omega/k \gg v_{Ti}$, we can account only for the Cherenkov dissipation on electrons. The ion thermal motion is completely ignored.

After substituting (9.1.38) and integrating over (9.1.27), we obtain

$$\varepsilon_i + \left(1 + \frac{\omega_{pi}^2}{k_z^2 v_s^2 \varepsilon_i}\right)^{-1/2} - i\sqrt{\frac{2m}{\pi M}}\frac{\omega_{pi}^2 \omega}{|k_z|^3 v_s^3}$$
$$\times \frac{1}{\varepsilon_i}\int_0^\infty \frac{dx}{\sqrt{x^2+1}\,(x^2 + 1 + \omega_{pi}^2/k_z^2 v_s^2 \varepsilon_i)^2} = 0\,, \tag{9.1.39}$$

where $v_s = \sqrt{T_e/m}$ and $\varepsilon_i = 1 - \omega_{pi}^2/\omega^2$. Though the integral here may be calculated in terms of elementary functions to analyze the spectra of ion-acoustic waves, it is more convenient to leave it in this form.

In (9.1.39), the imaginary part specified by the Cherenkov wave dissipation on plasma electrons is small in comparison to the real part. Thus, in the first-order approximation it can be neglected. Surface ion-acoustic waves are possible only in the frequency range where

$$\varepsilon_i(\omega) = 1 - \frac{\omega_{pi}^2}{\omega^2} < -\frac{\omega_{pi}^2}{k_z^2 v_s^2}\,,$$

i.e., $\omega^2 < \omega_{pi}^2$. The frequency spectrum of oscillations and their damping decrement (accounting for the imaginary part) can be obtained from the real part of (9.1.39). In general, the formulas for ω and δ ($\omega \to \omega + i\delta$) are rather unwieldy, but they take a simple form in the limiting cases of long ($k_z^2 v_s^2 \ll \omega_{pi}^2$) and short ($k_z^2 v_s^2 \gg \omega_{pi}^2$) surface waves:

$$\omega^2 = \begin{cases} k_z^2 v_s^2 & \text{for } k_z^2 v_s^2 \ll \omega_{pi}^2\,, \\ \omega_{pi}^2/2 & \text{for } k_z^2 v_s^2 \gg \omega_{pi}^2\,, \end{cases}$$
$$\frac{\delta}{\omega} = \begin{cases} -\sqrt{\dfrac{\pi}{8}\dfrac{m}{M}} & \text{for } k_z^2 v_s^2 \ll \omega_{pi}^2\,, \\ -\dfrac{1}{6}\sqrt{\dfrac{m}{\pi M}}\dfrac{\omega_{pi}^3}{k_z^3 v_s^3} & \text{for } k_z^2 v_s^2 \gg \omega_{pi}^2\,. \end{cases} \tag{9.1.40}$$

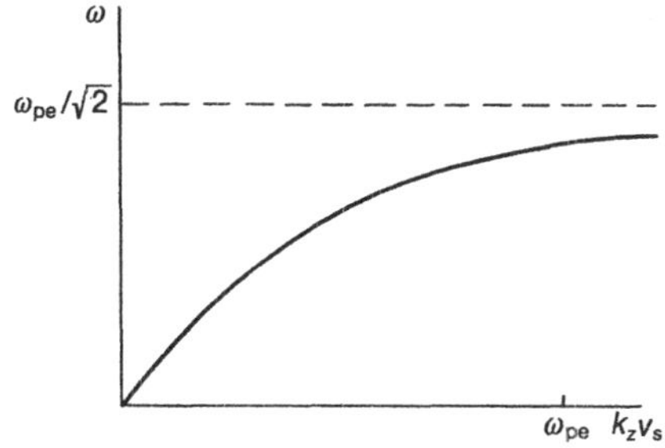

Fig. 9.4. Spectrum of surface ion-acoustic wave in a semi-bounded isotropic plasma

Figure 9.4 shows ω versus $k_z v_s$ for *surface ion-acoustic waves* in the semi-bounded isotropic plasma.

Finally, note that the picture of the propagation of surface ion-acoustic waves in the nonisothermal gaseous plasma remains qualitatively the same for the degenerate plasma in the frequency range $k_z v_{\mathrm{Ti}} \ll \omega \ll k_z v_{\mathrm{Fe}}$. Under these conditions, the expression for the longitudinal dielectric permittivity (Sect. 4.3) is

$$\varepsilon^{\mathrm{lo}}(\omega, k) = 1 - \frac{\omega_{\mathrm{pi}}^2}{\omega^2} + \frac{3\,\omega_{\mathrm{pe}}^2}{k^2 v_{\mathrm{Fe}}^2}\left(1 + \mathrm{i}\frac{\pi}{2}\frac{\omega}{k v_{\mathrm{Fe}}}\right). \tag{9.1.41}$$

This relation is similar to (9.1.38). Therefore, (9.1.39, 40) conserve their form if v_s^2 is replaced by $v_s^2 = 3\,v_{\mathrm{Fe}}^2 m/M$, and their imaginary parts (in particular, δ) are multiplied by $\sqrt{\pi/2}$.

We now study surface waves in a magnetized plasma with a sharp boundary and with the characteristic dimension of the boundary inhomogeneity shorter than the Larmor radius. Under the condition of mirror reflection from the plasma surface the general dispersion equation for surface waves in a magnetized plasma is analogous to that in the absence of the magnetic field. But the derivation is rather unwieldy and requires the calculation of the inverse dielectric tensor. Thus let us confine our consideration to quasi-longitudinal waves since their electric field is derived from a potential field to a good approximation. We study them in the following section where the stability of surface waves in a plasma confined by a strong magnetic field will be analyzed.

9.2 Instability of the Boundary of Magnetically Confined Plasma

Let us analyze a semi-bounded plasma confined by a strong magnetic field parallel to the plasma surface and oriented along the z-axis. The boundary inhomogeneity of such a plasma has a characteristic dimension greatly exceeding the Larmor radii of particles. This boundary is assumed to be set

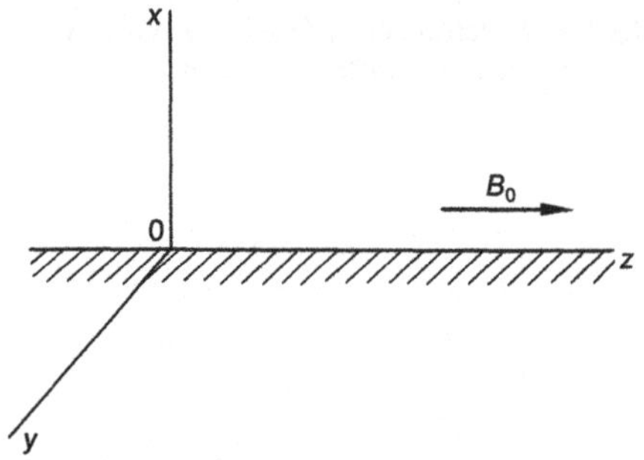

Fig. 9.5.

near the plane $x = 0$ (Fig. 9.5). For simplicity, collisions of charged particles in the plasma can be neglected, assuming $\Omega_\alpha \gg \nu_\alpha$, where $\alpha = \mathrm{e, i}$. The distribution function of particles α that is unperturbed by an electromagnetic field of oscillations can be obtained analogously to that in Sect. 8.3:

$$f_{0\alpha} = \left(1 + \frac{v_\perp \sin\phi}{\Omega_\alpha} \frac{\partial}{\partial x}\right) F_{0\alpha}(\mathscr{E}, x) , \tag{9.2.1}$$

where $F_{0\alpha}(\mathscr{E}, x)$ is either Maxwellian (9.1.1) or Fermian (9.1.2) with x-dependent temperature and density.

In contrast to Chap. 8, where the plasma has been considered to be smoothly inhomogeneous, here the plasma is inhomogeneous within a thin layer near the plane $x = 0$ (plasma surface). The difference of the distribution function (9.2.1) from that in thermodynamic equilibrium is manifested in this layer where the diamagnetic currents, due to the Larmor rotation of particles in the spatially inhomogeneous plasma, are localized. As shown in Sect. 8.5, in the region of plasma inhomogeneity, diamagnetic currents may cause the excitation of short-wavelength (compared to the inhomogeneity dimension) drift oscillations described in the framework of geometrical optics approximation. Furthermore, it will be shown that diamagnetic currents can also excite oscillations with a wavelength being significantly greater than the dimension of the plasma boundary inhomogeneity, i.e., surface waves damping deep into the plasma.

9.2.1 Poisson's Equation for the Magnetically Confined Inhomogeneous Plasma

For surface waves, the plasma boundary can be regarded as infinitely thin, diamagnetic currents being the boundary conditions for the electromagnetic field equations. Thus, the aim of our investigations is the derivation of the field equations and their boundary conditions accounting for the inhomogeneity of the plasma surface layer and diamagnetic currents. We solve this problem using longitudinal (potential) waves as an example, since in the case of magnetically confined plasmas the magnetic pressure greatly exceeds the gas pressure and the plasma oscillations can be regarded as longitudinal with a high degree of accuracy. For this reason the inhomogeneity of the magnetic field, compared to the plasma inhomogeneity, will also be ignored (as in Chap. 8).

Under the given restrictions, the kinetic equation for the nonequilibrium addition to the distribution function (9.2.1), dependent on time and coordinates as

$$\delta f_\alpha = \delta f_\alpha(x)\exp(-\mathrm{i}\omega t + \mathrm{i}k_y y + \mathrm{i}k_z z)\,, \tag{9.2.2}$$

is of the form

$$(\omega - k_y v_y - k_z v_z)\,\delta f_\alpha + \mathrm{i}v_x\frac{\partial \delta f_\alpha}{\partial x} - \mathrm{i}\Omega_\alpha\frac{\partial \delta f_\alpha}{\partial \phi} = -\,e_\alpha \boldsymbol{E}\cdot\frac{\partial f_{0\alpha}}{\partial \boldsymbol{p}}\,, \tag{9.2.3}$$

$\boldsymbol{E}$ being the potential of the perturbations $\boldsymbol{E} = -\nabla\Phi$.

Equation (9.2.3) is solved by integrating it over the characteristic (Sect. 8.3)

$$v_\perp \sin\phi + \Omega_\alpha x = \text{const.} \tag{9.2.4}$$

Then

$$\delta f_\alpha(x) = -\frac{e_\alpha}{m_\alpha \Omega_\alpha}\int_\infty^\phi d\phi'\,\nabla\Phi(x')\frac{\partial f_{0\alpha}(x')}{\partial v} \times \exp\left[\frac{\mathrm{i}}{\Omega_\alpha}\int_\phi^{\phi'} d\phi''\,(\omega - k_\perp v_\perp \sin\phi'' - k_z v_z)\right]. \tag{9.2.5}$$

Here x', x and ϕ' are mutually related by (9.2.4).

Substituting (9.2.5) into the formula for the charge density

$$\varrho(x) = \sum_\alpha e_\alpha \int d\boldsymbol{p}\,\delta f_\alpha \tag{9.2.6}$$

and using Poisson's equation

$$\Delta\Phi = -4\pi\varrho(x)\,, \tag{9.2.7}$$

after rather unwieldy calculations (analogous to those given in Sect. 8.3), in the case of the nondegenerate plasma, we finally obtain the equation for the potential of the oscillation field:

$$\Delta\Phi = \sum_\alpha \int dk_x \Phi(k_x)\,e^{\mathrm{i}k_x x}\frac{\omega_{\mathrm{p}\alpha}^2}{v_{\mathrm{T}\alpha}^2}\left\{1 - \sum_s \frac{\omega}{\omega - s\Omega_\alpha}\left[1 - \frac{k_y v_{\mathrm{T}\alpha}^2}{\omega\Omega_\alpha}\frac{\partial^0}{\partial x}\left(1 - \frac{s\omega}{z_\alpha \Omega_\alpha}\right) - \mathrm{i}\frac{k_x v_{\mathrm{T}\alpha}^2}{\Omega_\alpha^2}\frac{\partial^0}{\partial x}\frac{A_s'(z_\alpha)}{A_s(z_\alpha)}\right] A_s(z_\alpha)\,I_+\left(\frac{\omega - s\Omega_\alpha}{k_z v_{\mathrm{T}\alpha}}\right)\right\}, \tag{9.2.8}$$

where

$$z_\alpha = \frac{k_\perp^2 v_{T\alpha}^2}{\Omega_\alpha^2}, \qquad \frac{\partial^0}{\partial x} = \frac{\partial \ln N_\alpha}{\partial x} + \frac{\partial T_\alpha}{\partial x}\frac{\partial}{\partial T_\alpha}.$$

The operator $\partial^0/\partial x$ acts on all the values to its right.

Equation (9.2.8) is valid for the whole space both in the plasma ($x \geq 0$) and in vacuum ($x < 0$). Thus, there is no need to set special boundary conditions which can be obtained by means of integrating (9.2.8) over a physically infinitely thin intermediate layer near the plasma surface, this layer being small compared to the wavelength of surface waves.

9.2.2 Surface Oscillations of the Cold Magneto-Active Plasma with a Sharp Boundary

We start the analysis of (9.2.8) with the cold plasma when diamagnetic currents in the surface layer are totally neglected. In other words, we study oscillations with the phase velocity much higher than thermal velocities of particles and with the wavelength exceeding their Larmor radii. The limit $T \to 0$ should be taken in (9.2.8). Then the equation

$$\left(1 - \sum_\alpha \frac{\omega_{p\alpha}^2}{\omega^2 - \Omega_\alpha^2}\right)\left(\frac{\partial^2}{\partial x^2} - k_y^2\right)\Phi - \left(1 - \sum_\alpha \frac{\omega_{p\alpha}^2}{\omega^2}\right)k_z^2\Phi$$
$$+ \frac{\partial \Phi}{\partial x}\frac{\partial}{\partial x}\left(1 - \sum_\alpha \frac{\omega_{p\alpha}^2}{\omega^2 - \Omega_\alpha^2}\right) + k_y \Phi \frac{\partial}{\partial x}\sum_\alpha \frac{\omega_{p\alpha}^2 \Omega_\alpha}{\omega(\omega^2 - \Omega_\alpha^2)} = 0, \qquad (9.2.9)$$

being independent of the form of the distribution function of particles, is valid both for nondegenerate and degenerate plasmas. Besides, in the approximation of the cold plasma, the diamagnetic currents are totally neglected in the surface layer, to say nothing of the fact that here the surface layer together with the Larmor radius of particles vanishes. Therefore, (9.2.9) is valid for describing oscillations of plasmas with an arbitrarily sharp boundary, in particular, of a plasma confined by the walls of a real dielectric vessel (glass).

In the plasma volume (for $x \geq 0$), where the density may be regarded as homogeneous from (9.2.9), we have

$$\left(1 - \sum_\alpha \frac{\omega_{p\alpha}^2}{\omega^2 - \omega_\alpha^2}\right)\left(\frac{\partial^2}{\partial x^2} - k_y^2\right)\Phi_1 - \left(1 - \sum_\alpha \frac{\omega_{p\alpha}^2}{\omega^2}\right)k_z^2\Phi_1 = 0. \qquad (9.2.10)$$

In the vacuum ($x < 0$), Eq. (9.2.9) is reduced to the Laplace equation

$$\Delta\Phi_2 = 0. \qquad (9.2.11)$$

Finally, by integrating (9.2.9) we obtain the boundary conditions relating Φ_1 to Φ_2 on the surface of the plasma-vacuum partition ($x = 0$)

$$\{\Phi\}_{x=0} = 0 ,$$

$$\left\{\left(1 - \sum_\alpha \frac{\omega_{p\alpha}^2}{\omega^2 - \Omega_\alpha^2}\right)\frac{\partial \Phi}{\partial x} + k_y \Phi \sum_\alpha \frac{\omega_{p\alpha}^2 \Omega_\alpha}{\omega(\omega^2 - \Omega_\alpha^2)}\right\}_{x=0} = 0 . \tag{9.2.12}$$

In the ranges $x \geq 0$ and $x < 0$ the field equations and their boundary conditions are known, they can be solved and the solutions jointed. Then

$$\begin{aligned} \Phi_1(x) &= C_1 \exp\left(-\sqrt{k_y^2 + k_z^2 \frac{\varepsilon_\parallel}{\varepsilon_\perp}}\, x\right) \quad &&\text{for} \quad x \geq 0 , \\ \Phi_2(x) &= C_2 \exp\left(\sqrt{k_y^2 + k_z^2}\, x\right) &&\text{for} \quad x < 0 , \end{aligned} \tag{9.2.13}$$

where

$$\varepsilon_\parallel = 1 - \sum_\alpha \frac{\omega_{p\alpha}^2}{\omega^2} , \qquad \varepsilon_\perp = 1 - \sum_\alpha \frac{\omega_{p\alpha}^2}{\omega^2 - \Omega_\alpha^2} . \tag{9.2.14}$$

On substituting these solutions into the boundary conditions (9.2.12), we obtain the system of homogeneous algebraic equations for the constants C_1 and C_2, its solvability condition yields the *dispersion equation for surface waves* in the semi-bounded plasma with a sharp boundary:

$$\varepsilon_\perp \sqrt{k_y^2 + k_z^2 \frac{\varepsilon_\parallel}{\varepsilon_\perp}} + k_y g + \sqrt{k_y^2 + k_z^2} = 0 . \tag{9.2.15}$$

Here

$$g = \sum_\alpha \frac{\omega_{p\alpha}^2 \Omega_\alpha}{\omega(\omega^2 - \Omega_\alpha^2)} . \tag{9.2.16}$$

In the absence of magnetic field, when $\Omega_\alpha \to 0$, from (9.2.15) we obtain the dispersion equation for longitudinal surface waves in the semi-bounded isotropic plasma (9.1.31). The frequency spectrum of these waves is given by the second expression (9.1.30). The external magnetic field, parallel to the plasma surface, essentially modifies the derived frequency spectrum if $\Omega_e \gtrsim \omega = \omega_{pe}/\sqrt{2}$. For the modes with $k_y = 0$ according to (9.2.15) surface waves exist in a purely electron plasma only for $\omega_{pe}^2 > \Omega_e^2$ and in the frequency range $\omega_{pe}^2 > \omega^2 > \Omega_e^2$. Here the frequency spectrum of these waves is determined by the relation

$$\omega^2 = \tfrac{1}{2}(\omega_{pe}^2 + \Omega_e^2) . \tag{9.2.17}$$

For $k_y = 0$ surface waves can also exist in the frequency range $\omega < \Omega_e$. For example, for the flute modes with $k_z = 0$ in the case of a purely electron plasma from (9.2.15) we obtain

$$\omega = \frac{k_y}{2|k_y|}\,\Omega_e \pm \frac{1}{2}\sqrt{\Omega_e^2 + 2\,\omega_{pe}^2}\,. \tag{9.2.18}$$

Hence low-frequency surface waves are possible in strong magnetic fields when $\Omega_e^2 \gg \omega_{pe}^2$ and $\omega \approx \omega_{pe}^2/\Omega_e$.

It is straightforward to analyze (9.2.15) also in the low-frequency range $\omega \ll \omega_{pi}$, when the ion motion becomes significant. In the range of the lowest frequencies $\omega \ll \Omega_i$, both plasma electrons and ions are strongly magnetized, and $g \to 0$, $\varepsilon_\perp \to 0$. As a result, (9.2.5) has no solutions, i.e., in the cold strongly magnetized plasma with a sharp boundary in the frequency range $\omega \ll \Omega_i$ surface waves are nonexistent. They are possible only in the frequency range $\omega \gtrsim \Omega_i$. Actually, in the low-frequency range $\varepsilon_\parallel \gg \varepsilon_\perp$ and from (9.2.15) we obtain

$$\omega^2 = \omega_{pi}^2 + \Omega_i^2 \tag{9.2.19}$$

for the waves, propagating not strictly perpendicular to the magnetic field ($k_z \neq 0$). For $k_z = 0$ (flute modes), the spectrum of low-frequency waves is determined by

$$\omega = \frac{k_y}{|k_y|}\left(\Omega_i + \frac{\omega_{pi}^2}{2\,\Omega_i}\right). \tag{9.2.20}$$

The above analysis shows that the surface waves of a plasma with sharp boundary are always stable in collisionless magnetized plasmas when the thermal motion of particles is ignored. Moreover, accounting for particle collisions results in their damping (Exercise 9.6.3). This is not surprising since in the approximation discussed the diamagnetic currents, which lead to oscillation buildup, are totally ignored in the inhomogeneous surface layer of the plasma.

9.2.3 Instability of the Surface of the Magnetically Confined Plasma

Accounting for the finite plasma temperature and the finite Larmor radius of the particles, the situation is qualitatively altered. Besides dissipative effects, specified by the collisionless Cherenkov absorption and wave radiation by plasma particles, the diamagnetic currents which can cause the instability of plasma surface waves become considerable in the inhomogeneous surface layer of the plasma. Let us consider (9.2.8) in the range of low-frequency ($\omega^2 \ll \Omega_\alpha^2$) and short-wavelength ($k_\perp^2 v_{T\alpha}^2 \ll \Omega_\alpha^2$) oscillations. On expanding the function $A_s(z_\alpha)$ in a power series of z_α and keeping only the term with $s=0$, in the sum over cyclotron harmonics, from the integro-differential equation (9.2.8) we obtain the second-order differential equation[2]

$$\Delta\Phi = \sum_\alpha \frac{\omega_{p\alpha}^2}{v_{T\alpha}^2}\left\{\left(1 - I_+\left(\frac{\omega}{k_z v_{T\alpha}}\right) + \frac{k_y v_{T\alpha}^2}{\omega\Omega_\alpha}\frac{\partial^0}{\partial x} I_+\left(\frac{\omega}{k_z v_{T\alpha}}\right)\right]\Phi\right.$$
$$\left. + \frac{v_{T\alpha}^2}{\Omega_\alpha^2} I_+\left(\frac{\omega}{k_z v_{T\alpha}}\right)\left(k_y^2 - \frac{\partial^2}{\partial x^2}\right)\Phi - \frac{\partial\Phi}{\partial x}\frac{1}{\Omega_\alpha^2}\frac{\partial^0}{\partial x} v_{T\alpha}^2 I_+\left(\frac{\omega}{k_z v_{T\alpha}}\right)\right\}. \tag{9.2.21}$$

When deriving this equation, we also assumed $\omega^2 k_\perp^2 \ll k_z^2\Omega_i^2$.

Integrating (9.2.21) over the intermediate layer near the plasma boundary gives the boundary conditions for the potential Φ. As a result we have

$$\{\Phi\}_{x=0} = 0\,,$$
$$\left\{\frac{\partial\Phi}{\partial x} + \sum_\alpha \frac{\omega_{p\alpha}^2}{\Omega_\alpha^2} I_+\left(\frac{\omega}{k_z v_{T\alpha}}\right)\left(\frac{\partial\Phi}{\partial x} - \frac{\Omega_\alpha}{\omega}k_y\Phi\right)\right\}_{x=0} = 0\,. \tag{9.2.22}$$

With these boundary conditions, (9.2.21) can now be written as:

for the plasma $(x \geq 0, \Phi = \Phi_1)$

$$\Delta\Phi_1 = \sum_\alpha \frac{\omega_{p\alpha}^2}{v_{T\alpha}^2}\left[1 - I_+\left(\frac{\omega}{k_z v_{T\alpha}}\right) + \frac{v_{T\alpha}^2}{\Omega_\alpha^2} I_+\left(\frac{\omega}{k_z v_{T\alpha}}\right)\left(k_y^2 - \frac{\partial^2}{\partial x^2}\right)\right]\Phi_1\,, \tag{9.2.23}$$

and for the vacuum $(x < 0, \Phi = \Phi_2)$:

$$\Delta\Phi_2 = 0\,. \tag{9.2.24}$$

Their solutions, damped for $x \to \pm\infty$, are

$$\Phi_1 = C_1 e^{-\kappa x}\,, \quad \Phi_2 = C_2 \exp\left(\sqrt{k_y^2 + k_z^2}\,x\right)\,, \quad \text{where} \tag{9.2.25}$$

$$\kappa^2 = k_y^2 + \frac{k_z^2 + \sum_\alpha \frac{\omega_{p\alpha}^2}{v_{T\alpha}^2}\left[1 - I_+\left(\frac{\omega}{k_z v_{T\alpha}}\right)\right]}{1 + \sum_\alpha \frac{\omega_{p\alpha}^2}{\Omega_\alpha^2} I_+\left(\frac{\omega}{k_z v_{T\alpha}}\right)}\,. \tag{9.2.26}$$

---

2 In the approximation of geometrical optics from (9.2.8) we obtain the eikonal equation (9.5.6) which is reduced to (8.5.8) in the low-frequency limit.

Substitution of the solutions (9.2.25) into the boundary conditions (9.2.22) yields a system of homogeneous algebraic equations for the constants C_1 and C_2. The solvability condition of this system is the dispersion equation for surface waves in a semi-bounded plasma accounting for the diamagnetic current on its surface:

$$\left[1+\sum_{\alpha}\frac{\omega_{p\alpha}^2}{\Omega_\alpha^2}I_+\left(\frac{\omega}{k_z v_{T\alpha}}\right)\right]\kappa+\sum_{\alpha}\frac{\omega_{p\alpha}^2}{\omega\Omega_\alpha}k_y\,I_+\left(\frac{\omega}{k_z v_{T\alpha}}\right)+\sqrt{k_y^2+k_z^2}=0\,. \tag{9.2.27}$$

We note that surface waves are possible only in the frequency range $\kappa^2>0$. In the frequency range $\omega \gg k_z v_{Te}$, the thermal motion of the particles can be neglected, and (9.2.27) reduces to (9.2.15). In this limit, the diamagnetic currents on the plasma surface are also neglected, and the oscillations do not grow with time. As mentioned above, accounting for diamagnetic currents can result in the buildup of surface oscillations. To verify this, we study the solutions of (9.2.27) in the frequency range $k_z v_{Ti} \ll \omega \ll k_z v_{Te}$, where the effects of the thermal motion of the electrons are significant. Here we obtain from (9.2.27)

$$\left(1+\frac{\omega_{pi}^2}{\Omega_i^2}\right)\kappa+\frac{\omega_{pi}^2 k_y}{\omega\Omega_i}\left(1+i\sqrt{\frac{\pi}{2}}\frac{\omega}{|k_z|v_{Te}}\right)+\sqrt{k_y^2+k_z^2}=0\,, \tag{9.2.28}$$

and

$$\kappa^2=k_y^2+\frac{k_z^2\left(1-\dfrac{\omega_{pi}^2}{\omega^2}\right)+\dfrac{\omega_{pi}^2}{v_s^2}\left(1+i\sqrt{\dfrac{\pi}{2}}\dfrac{\omega}{|k_z|v_{Te}}\right)}{1+\omega_{pi}^2/\Omega_i^2}\,. \tag{9.2.29}$$

For the modes with $k_y^2 \gg k_z^2$ under the condition $\omega_{pi}^2 \gg \Omega_i^2$ (usually satisfied with ample reserve in a real plasma confined by a magnetic field) from (9.2.28) we obtain the spectrum of slowly increasing oscillations ($\omega \to \omega + i\delta$):

$$\omega = k_y v_s\,,\qquad \delta=\sqrt{\frac{\pi}{8}}\frac{\omega^2}{|k_z|v_{Te}}\,. \tag{9.2.30}$$

Hence it follows that under the influence of diamagnetic currents in the inhomogeneous surface layer the buildup of surface ion-acoustic waves occurs in a semi-bounded plasma confined by a magnetic field. These waves propagate along the plasma boundary at a large angle to the magnetic field and subside deep into the plasma with the characteristic space scale of the order v_s/Ω_i. In strongly nonisothermal plasmas with $T_e \gg T_i$, this scale is many times greater than the ion Larmor radius and the dimension of the plasma inhomogeneity.

Finally, we note that the instabilities of surface waves may be more dangerous for the problem of magnetic confinement than drift instabilities, since the latter lead to the excitation of short-wavelength oscillations localized in the inhomogeneous layer near the plasma surface while unstable surface waves can extend much deeper into the plasma. This is particularly manifested when either a current or gravitational drift, caused by curvature of field lines of the confining magnetic field, is present in the plasma (Exercise 9.6.7).

9.3 Plasma Waveguide

After the analysis of the semi-bounded plasma, we now study a plasma bounded at all sides, for example, a plasma cylinder with the radius r_0 bounded by a metallic vessel with the radius $R \geq r_0$ (Fig. 9.6). Let us place the whole system in a longitudinal magnetic field along the z-axis. The plasma is treated as homogeneous with a sharp boundary, i.e., the velocity distribution of particles will be regarded as the Maxwellian (9.1.1) or Fermian (9.1.2) distribution function where

$$N_\alpha(r) = \begin{cases} N_{0\alpha} = \text{const} & \text{if} \quad r \leq r_0 \, , \\ 0 & \text{if} \quad r > r_0 \, . \end{cases} \tag{9.3.1}$$

However, such a distribution function can be applied only when the plasma is confined by a hard wall, impermeable for particles. When it is confined by the longitudinal magnetic field, in order to determine the equilibrium distribution function of particles, one should solve the kinetic equation

$$\cos(\phi - \psi) v_\perp \frac{\partial f_{0\alpha}}{\partial r} - \Omega_\alpha \frac{\partial f_{0\alpha}}{\partial \phi} + \sin(\phi - \psi) \frac{v_\perp}{r} \frac{\partial f_{0\alpha}}{\partial \psi} = 0 \, , \tag{9.3.2}$$

considering the plasma inhomogeneous along the radius and strongly magnetized ($\Omega_\alpha \gg \nu_\alpha$). Here the cylindrical system of coordinates is applied both to the velocities $v = (v_\perp, \phi, v_z)$ and to the space coordinates $r = (r, \psi, z)$.

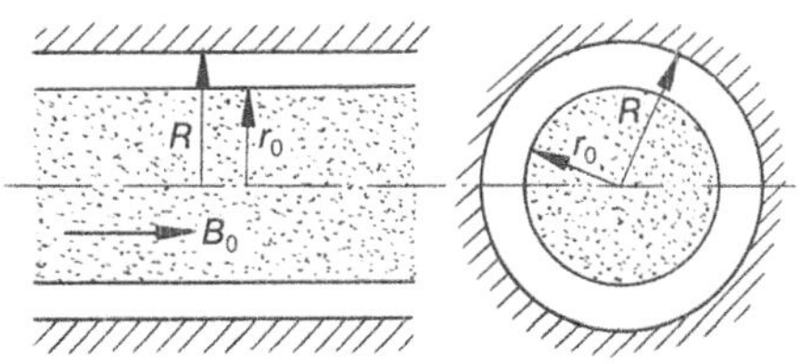

Fig. 9.6. Plasma waveguide

The general solution of (9.3.2) is an arbitrary function of its characteristics, in particular

$$C_\alpha = rv_\perp \sin(\phi - \psi) + \int^r r\, dr \Omega_\alpha(r) . \tag{9.3.3}$$

On writing the solution in the form of $f_{0\alpha}(\mathscr{E}_\alpha, C_\alpha)$ and assuming the characteristic dimension of plasma inhomogeneity to be larger than the Larmor radius of particles, we obtain, cf. (9.2.1),

$$f_{0\alpha}(\mathscr{E}_\alpha, C_\alpha) = \left(1 + \frac{v_\perp \sin(\phi - \psi)}{\Omega_\alpha} \frac{\partial}{\partial r}\right) F_{0\alpha}(\mathscr{E}_\alpha, r) . \tag{9.3.4}$$

Here the Maxwell or Fermi distribution functions with inhomogeneous temperature and density can be used for $F_{0\alpha}(\mathscr{E}_\alpha, r)$.

The equilibrium distribution function (9.3.4) allows us to study small oscillations of the plasma cylinder, filling incompletely the metal waveguide. To analyze such oscillations, we assume all nonequilibrium values to depend on time and space in the form

$$A(r) = A(r) \exp(-\mathrm{i}\omega t + \mathrm{i}l\phi + \mathrm{i}k_z z) . \tag{9.3.5}$$

Then in the collisionless plasma, for a small perturbation of the distribution function (9.3.4), we have

$$\begin{aligned} &\mathrm{i}\left(\omega - k_z v_z - \frac{l}{r} v_\perp \sin\chi\right) \delta f_\alpha - v_\perp \cos\chi \frac{\partial \delta f_\alpha}{\partial r} \\ &+ \left(\Omega_\alpha + \frac{v_\perp \sin\chi}{r}\right) \frac{\partial \delta f_\alpha}{\partial \chi} = \frac{e_\alpha}{m_\alpha} \left\{\boldsymbol{E} + \frac{1}{c}[\boldsymbol{v}, \boldsymbol{B}]\right\} \frac{\partial f_{0\alpha}}{\partial \boldsymbol{v}} , \end{aligned} \tag{9.3.6}$$

where $\boldsymbol{E}(\boldsymbol{r})$ and $\boldsymbol{B}(\boldsymbol{r})$ are the electric and magnetic fields of oscillations and $\chi = \phi - \psi$.

The characteristic of (9.3.6), as for (9.3.2), is given by (9.3.3). Therefore, regarding for simplicity the magnetic field as homogeneous all over the plasma section, the general solution can be written as

$$\begin{aligned} \delta f_\alpha(r) &= \frac{e_\alpha}{m_\alpha \Omega_\alpha} \int_\infty^\chi d\chi' \frac{\left\{\boldsymbol{E}(r') + \frac{1}{c}[\boldsymbol{v}, \boldsymbol{B}(r')]\right\} \dfrac{\partial f_{0\alpha}}{\partial \boldsymbol{v}(\chi')}}{1 + v_\perp \sin\chi'/(r'\Omega_\alpha)} \\ &\times \exp\left(\frac{\mathrm{i}}{\omega_\alpha} \int_\chi^{\chi'} d\chi'' \frac{\omega - k_z v_z - \dfrac{l v_\perp}{r''} \sin\chi''}{1 + v_\perp \sin\chi''/(r''\Omega_\alpha)}\right) . \end{aligned} \tag{9.3.7}$$

Here the values r', r'' and χ' are connected with r, χ and χ'' by the equation of characteristic (9.3.3).

Substituting (9.3.7) into the formula for the current density

$$j = \sum_{\alpha} e_{\alpha} \int dp\, \boldsymbol{v} \delta f_{\alpha} \tag{9.3.8}$$

results in the material equation, connecting $\boldsymbol{j}(\boldsymbol{r})$ with the electric field $\boldsymbol{E}(\boldsymbol{r})$ through the integral relation. In the zero-order approximation of geometrical optics, this material equation determines the conductivity tensor and thus the dielectric tensor analogous to (8.3.22) (for a nondegenerate plasma) and to (8.3.29) (for a degenerate plasma) by the substitution

$$k_y \to \frac{l}{r} , \tag{9.3.9}$$

due to the cylindrical geometry. Naturally, because of this substitution, all the conclusions about drift oscillations of the inhomogeneous collisionless plasma derived in the previous chapter are also valid in the case of a plasma cylinder.

9.3.1 Field Equation for the Cylindrical Plasma Waveguide

We should expect the appearance of qualitatively new effects distinguishing the cylindrical geometry from the plane one only in the limit of long wavelengths, not described in the approximation of geometrical optics. Therefore we consider this limit. For simplicity, our analysis will be confined to high-frequency waves with phase velocities greatly exceeding the thermal velocities of the particles in the plasma. For such waves the thermal motion of particles in the first-order approximation may be neglected. Here, irrespective of the form of the equilibrium distribution function (i.e., for nondegenerate as well as degenerate plasmas), the material equation appears local and

$$D_i(\omega, k_z, l, r) = \varepsilon_{ij}(r)\, E_j(\omega, k_z, l, r) , \tag{9.3.10}$$

$\varepsilon_{ij}(r)$ being the dielectric tensor of the plasma:

$$\varepsilon_{ij}(r) = \begin{pmatrix} \varepsilon_{rr} & \varepsilon_{r\phi} & 0 \\ \varepsilon_{\phi r} & \varepsilon_{\phi\phi} & 0 \\ 0 & 0 & \varepsilon_{zz} \end{pmatrix} . \tag{9.3.11}$$

Here

$$\begin{aligned} \varepsilon_{rr} &= \varepsilon_{\phi\phi} = \varepsilon_{\perp} = 1 - \sum_{\alpha} \frac{\omega_{p\alpha}^2(r)}{\omega^2 - \Omega_{\alpha}^2} , \\ \varepsilon_{r\phi} &= \varepsilon_{\phi r} = ig = -\mathrm{i} \sum_{\alpha} \frac{\omega_{p\alpha}^2(r)\, \Omega_{\alpha}}{\omega(\omega^2 - \Omega_{\alpha}^2)} , \\ \varepsilon_{zz} &= \varepsilon_{\parallel} - \sum_{\alpha} \frac{\omega_{p\alpha}^2(r)}{\omega^2} . \end{aligned} \tag{9.3.12}$$

It is easily seen that this tensor can be obtained for the cold inhomogeneous collisionless plasma with the model of independent particles (Exercise 9.6.8).

The field equations are

$$\frac{l}{r}E_z - k_z E_\phi = \frac{\omega}{c}B_r\,, \qquad \frac{l}{r}B_z - k_z B_\phi = -\frac{\omega}{c}(\varepsilon_\perp E_r + \mathrm{i}gE_\phi)\,,$$

$$k_z E_r + \mathrm{i}\frac{\partial E_z}{\partial r} = \frac{\omega}{c}B_\phi\,, \qquad k_z B_r + \mathrm{i}\frac{\partial B_z}{\partial r} = -\frac{\omega}{c}(\varepsilon_\perp E_\phi - \mathrm{i}gE_r)\,,$$

$$\frac{\mathrm{i}}{r}\frac{\partial}{\partial r}rE_\phi + \frac{l}{r}E_r = -\frac{\omega}{c}B_z\,, \qquad \frac{\mathrm{i}}{r}\frac{\partial}{\partial r}rB_\phi + \frac{l}{r}B_r = \frac{\omega}{c}\varepsilon_\parallel E_z\,. \tag{9.3.13}$$

This system of equations is valid both in the plasma ($r \leq r_0$) and in the vacuum ($r_0 \leq r \leq R$). Therefore the boundary conditions can be obtained directly from these equations by integrating them over an infinitely thin intermediate layer of plasma-vacuum. As a result we have

$$\{E_z\}_{r=r_0} = \{E_\phi\}_{r=r_0} = \{B_z\}_{r=r_0} = \{B_\phi\}_{r=r_0} = 0\,. \tag{9.3.14}$$

Finally, on the surface of the metal waveguide the following condition should be satisfied

$$E_z|_{r=R} = E_\phi|_{r=R} = 0\,. \tag{9.3.15}$$

Equations (9.3.13) together with the boundary conditions (9.3.14, 15) determine the spectrum of natural electromagnetic oscillations in the plasma waveguide in the limit of a cold plasma, when phase velocities of waves greatly exceed the thermal velocities of the particles such that the latter can be neglected. Then the values $\varepsilon_\perp$, g and $\varepsilon_\parallel$ are defined by (9.3.12).

We analyze the formulated boundary problem for a pure electron plasma, the ion motion being neglected. Besides, let us confine our consideration to two opposite limiting cases:

a) $r_0 = R$ and the plasma fills in the waveguide completely;
b) $R \to \infty$ and the plasma waveguide with a free surface is available.

9.3.2 Spectrum of Oscillations of the Isotropic Plasma Waveguide

Let us start our analyis of an isotropic electron plasma assuming $\Omega_\alpha \to 0$ (i.e., $B_0 \to 0$) in (9.3.12). Excluding the components E_r, B_r, E_ϕ and B_ϕ from the system (9.3.13), and using the relations

$$E_r = \kappa^{-2}\left(-\mathrm{i}k_z\frac{\partial E_z}{\partial r} + \frac{\omega}{c}\frac{l}{r}B_z\right),$$

$$E_\phi = \kappa^{-2}\left(k_z\frac{l}{r}E_z + \mathrm{i}\frac{\omega}{c}\frac{\partial B_z}{\partial r}\right),$$

$$B_r = -\kappa^{-2}\left(\frac{\omega}{c}\frac{l}{r}\varepsilon E_z + \mathrm{i}k_z\frac{\partial B_z}{\partial r}\right), \tag{9.3.16}$$

$$B_\phi = -\kappa^{-2}\left(\mathrm{i}\frac{\omega}{c}\varepsilon\frac{\partial E_z}{\partial r} - k_z\frac{l}{r}B_z\right), \quad \text{where}$$

$$\kappa^2 = k_z^2 - \frac{\omega^2}{c^2}\varepsilon, \qquad \varepsilon(\omega) = 1 - \frac{\omega_{\mathrm{pe}}^2}{\omega^2}, \tag{9.3.17}$$

we obtain two independent equations for E_z and B_z:

$$\varepsilon\left(\Delta E_z + \frac{\omega^2}{c^2}\varepsilon E_z\right) = 0, \qquad \Delta B_z + \frac{\omega^2}{c^2}\varepsilon B_z = 0. \tag{9.3.18}$$

For $B_z = 0$, $E_z \neq 0$ oscillations are called E-type waves and for $B_z \neq 0$ and $E_z = 0$ B-type waves.

According to the first equation (9.3.18), in the absence of the external magnetic field, irrespective of the degree of plasma filling in the waveguide, there always exist purely longitudinal (potential) Langmuir oscillations for which

$$\varepsilon(\omega) = 1 - \frac{\omega_{\mathrm{pe}}^2}{\omega^2} = 0, \qquad \omega^2 = \omega_{\mathrm{pe}}^2. \tag{9.3.19}$$

These oscillations are of space type and localized inside the plasma cylinder.

Along with longitudinal oscillations, transverse (nonpotential) waves of E- and B-types can exist in the isotropic plasma in the frequency range where $\varepsilon(\omega) \neq 0$. According to (9.3.18) for such waves we have

$$E_z(r) = E_{z0}\mathrm{J}_l(\mathrm{i}\kappa r), \qquad B_z = B_{z0}\mathrm{J}_l(\mathrm{i}\kappa r), \tag{9.3.20}$$

for $r \lesssim r_0$. If the plasma fills in the waveguide completely ($r_0 = R$), substitution of the solutions (9.3.20) into the boundary conditions (9.3.15) and accounting for (9.3.16) leads to the following oscillation spectra for E- and B-type waves, respectively:

$$\omega^2 = k_z^2c^2 + \frac{\mu_{ls}^2c^2}{R^2} + \omega_{\mathrm{pe}}^2, \qquad \omega^2 = k_z^2c^2 + \frac{\mu_{ls}'^2c^2}{R^2} + \omega_{\mathrm{pe}}^2, \tag{9.3.21}$$

where μ_{ls} and μ_{ls}' are the roots of the Bessel function and its derivative, respectively, i.e., $\mathrm{J}_l(\mu_{ls}) = 0$ and $\mathrm{J}_l'(\mu_{ls}') = 0$.

From (9.3.19, 21) it follows that the phase velocities of transverse electromagnetic waves in the waveguide, completely filled with isotropic plasma, always exceed the velocity of light, while the phase velocities of longitudinal waves can be both larger or smaller than that of light.

We now consider electromagnetic waves in the plasma cylinder with a free surface. The solutions of (9.3.18), limited at the waveguide axis, $r \to 0$ and for $r \to \infty$, are written as

$$\begin{aligned} E_{z\mathrm{I}}(r) &= E_{z\mathrm{I}0}\,\mathrm{I}_l(\kappa_\mathrm{I} r)\,, \quad B_{z\mathrm{I}}(r) = B_{z\mathrm{I}0}\,\mathrm{I}_l(\kappa_\mathrm{I} r) \qquad r \le r_0\,, \\ E_{z\mathrm{II}}(r) &= E_{z\mathrm{II}0}\,K_l(\kappa_\mathrm{II} r)\,, \quad B_{z\mathrm{II}}(r) = B_{z\mathrm{II}0}\,K_l(\kappa_\mathrm{II} r) \qquad r > r_0\,, \end{aligned} \tag{9.3.22}$$

where

$$\kappa_\mathrm{I}^2 = k_z^2 - \frac{\omega^2}{c^2}\varepsilon(\omega)\,, \qquad \kappa_\mathrm{II}^2 = k_z^2 - \frac{\omega^2}{c^2} \ge 0\,. \tag{9.3.23}$$

Substituting these solutions into the boundary conditions (9.3.14) yields the following dispersion equation for E-type surface electromagnetic waves in the plasma cylinder with a free surface:

$$\begin{aligned} &\left[\frac{1}{\kappa_\mathrm{II}}\mathrm{I}_l(\kappa_\mathrm{I} r_0)\,K_l'(\kappa_\mathrm{II} r_0) - \frac{1}{\kappa_\mathrm{I}}\mathrm{I}_l'(\kappa_\mathrm{I} r_0)\,K_l(\kappa_\mathrm{II} r_0)\right] \\ &\quad + \left[\frac{1}{\kappa_\mathrm{II}}\mathrm{I}_l(\kappa_\mathrm{I} r_0)\,K_l'(\kappa_\mathrm{II} r_0) - \frac{\varepsilon}{\kappa_\mathrm{I}}\mathrm{I}_l'(\kappa_\mathrm{I} r_0)\,K_l(\kappa_\mathrm{II} r_0)\right] \\ &\quad - \frac{c^2 l^2 k_z^2}{\omega^2 r_0^2}\mathrm{I}_l^2(\kappa_\mathrm{I} r_0)\,K_l^2(\kappa_\mathrm{II} r_0)\left(\frac{1}{\kappa_\mathrm{I}^2} - \frac{1}{\kappa_\mathrm{II}^2}\right)^2 = 0\,. \end{aligned} \tag{9.3.24}$$

We analyze (9.3.24) only for symmetric modes with $l = 0$. In the short-wavelength limit, i.e., when $\kappa_{\mathrm{I,II}}^2 r_0^2 \gg 1$, we obtain from (9.3.24)

$$1/\kappa_\mathrm{II} + \varepsilon/\kappa_\mathrm{I} = 0\,, \tag{9.3.25}$$

or in the explicit form

$$\sqrt{k_z^2 c^2 + \omega_\mathrm{pe}^2 - \omega^2} + \left(1 - \frac{\omega_\mathrm{pe}^2}{\omega^2}\right)\sqrt{k_z^2 c^2 - \omega^2} = 0\,. \tag{9.3.26}$$

It exactly coincides with (9.1.29) derived for the semi-bounded plasma. The geometry of the plasma surface is unimportant in the short-wavelength limit. From the conditions $\kappa_{\mathrm{I,II}}^2 r_0^2 \gg 1$ it follows that in the plasma cylinder such short-wavelength surface oscillations exist only in case of a sufficiently dense plasma when $\omega_\mathrm{pe}^2 r_0^2 \gg c^2$.

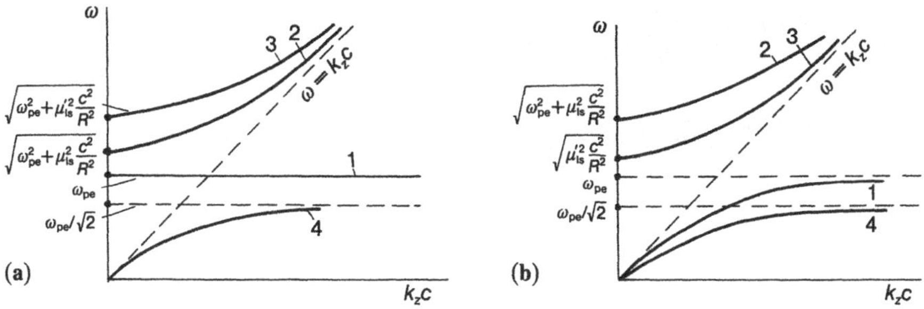

Fig. 9.7 a, b. Spectra of electromagnetic waves in a plasma waveguide: (**a**) isotropic plasma; (**b**) magnetoactive plasma

In the rarefied plasma where $\omega_{pe}^2 r_0^2 \ll 1$, only long-wavelength surface oscillations with $\kappa_{I,II}^2 r_0^2 \ll 1$ are possible. For axially symmetric modes ($l = 0$) of long-wavelength oscillations (9.3.24) is of the form

$$\left(1 - \frac{\omega_{pe}^2}{\omega^2}\right) \ln \left(r_0 \sqrt{k_z^2 - \frac{\omega^2}{c^2}}\right) - \frac{2}{r_0^2} \left(k_z^2 - \frac{\omega^2}{c^2}\right)^{-1} = 0 \,. \tag{9.3.27}$$

It is seen that long-wavelength oscillations, as well as short-wavelength ones, exist only in the frequency range $\omega < k_z c,\ \omega_{pe}$. If the strong inequalities are satisfied, the oscillations are potential with a high degree of accuracy, and their spectrum is determined by

$$\omega^2 \approx k_z^2 c^2 \frac{\omega_{pe}^2 r_0^2}{2} \ln \left(\frac{1}{|k_z| r_0}\right) . \tag{9.3.28}$$

The solution of (9.3.27) in the general case is graphically shown in Fig. 9.7a (curve 4). Here the spectra of space waves described by (9.3.19, 21) (curves 1–3) are shown, too.

9.3.3 Spectrum of Oscillations of the Magnetized Plasma Waveguide

Let us study electromagnetic waves in the plasma cylinder in the presence of the external longitudinal magnetic field, and for simplicity confine our analysis to the infinitely strong field when $B_0 \to \infty$ when the tensor (9.3.11) takes the form

$$\varepsilon_{ij}(\omega) = \begin{pmatrix} 1 & 0 & 0 \\ 0 & 1 & 0 \\ 0 & 0 & \varepsilon \end{pmatrix} . \tag{9.3.29}$$

Then from (9.3.13) we obtain

$$
\begin{aligned}
E_r &= \kappa^{-2}\left(-\mathrm{i}k_z\frac{\partial E_z}{\partial r} + \frac{\omega}{c}\frac{l}{r}B_z\right),\\
E_\phi &= \kappa^{-2}\left(k_z\frac{l}{r}E_z + \mathrm{i}\frac{\omega}{c}\frac{\partial B_z}{\partial r}\right),\\
B_r &= -\kappa^{-2}\left(\frac{\omega}{c}\frac{l}{r}E_z + \mathrm{i}k_z\frac{\partial B_z}{\partial r}\right),\\
B_\phi &= -\kappa^{-2}\left(\mathrm{i}\frac{\omega}{c}\frac{\partial E_z}{\partial r} - k_z\frac{l}{r}B_z\right).
\end{aligned}
\tag{9.3.30}
$$

Here $\kappa^2 = k_z^2 - \omega^2/c^2$, and the components E_z and B_z satisfy the independent equations

$$
\Delta_\perp E_z + \kappa^2\varepsilon E_z = 0\,, \quad \Delta B_z + \frac{\omega^2}{c^2}B_z = 0\,, \quad \text{where} \tag{9.3.31}
$$

$$
\Delta_\perp \equiv \frac{1}{r}\frac{\partial}{\partial r}r\frac{\partial}{\partial r} - \frac{l^2}{r^2}\,, \qquad \varepsilon \equiv \varepsilon_\parallel = 1 - \frac{\omega_{\mathrm{pe}}^2}{\omega^2}\,. \tag{9.3.32}
$$

Here we also distinguish between E-type $(B_z = 0, E_z \neq 0)$ and B-type $(B_z \neq 0, E_z = 0)$ waves.

Comparing (9.3.32) with (9.3.18), we find that strictly longitudinal oscillations do not exist in the limit of a strong magnetic field as is the case for isotropic plasma [see the spectrum (9.3.19)]. Moreover, the B-type waves are unimportant, and their dispersion law is the same as that for the vacuum waveguide. Therefore, we deal only with the E-type waves.

The solution of the first equation (9.3.32) in the region $r \le r_0$ is written as

$$
E_z(r) = E_{z0}\mathrm{J}_l(\mathrm{i}\kappa\sqrt{\varepsilon}\,r)\,. \tag{9.3.33}
$$

For a plasma completely filling the waveguide ($r_0 = R$), the substitution of this solution into the boundary conditions (9.3.15) gives the following dispersion equation for determining the spectrum of space E-type waves:

$$
\left(k_z^2 - \frac{\omega^2}{c^2}\right)\varepsilon(\omega) + \frac{\mu_{ls}}{R^2} = 0\,, \tag{9.3.34}
$$

where μ_{ls} are the roots of the Bessel function, i.e., $\mathrm{J}_l(\mu_{ls}) = 0$. Hence we obtain the frequency spectrum of oscillations

$$
\omega_{1,2}^2 = \frac{1}{2}\left[\omega_{\mathrm{Le}}^2 + \left(k_z^2 + \frac{\mu_{ls}^2}{R^2}\right)c^2 \pm \sqrt{\left(\omega_{\mathrm{pe}}^2 + k_z^2c^2 + \frac{\mu_{ls}^2c^2}{R^2}\right)^2 + 4k_z^2c^2\omega_{\mathrm{pe}}^2}\;\right]. \tag{9.3.35}
$$

In the magneto-active plasma, in constrast to the isotropic one, there exist two branches of space E-type waves: the high-frequency and the low-frequency branches. At the upper branch the phase velocity of waves is higher than the velocity of light and the oscillations are transverse with a high degree of accuracy; at the lower branch $v_{\text{ph}} = \omega/k_z < c$ and the wave field has an essential longitudinal component.

In the plasma cylinder with a free surface ($R \to \infty$) in a strong magnetic field, surface waves subsiding outside the plasma cylinder are possible along with the space waves. Actually for symmetric modes (with $l = 0$) of a surface E-type wave, the solutions of (9.3.32) have the form

$$\begin{aligned} E_{z\mathrm{I}}(r) &= E_{z\mathrm{I}0}\,\mathrm{I}_0(\kappa\sqrt{\varepsilon}\,r) \quad \text{if} \quad r \leq r_0\,, \\ E_{z\mathrm{II}}(r) &= E_{z\mathrm{II}0}\,K_0(\kappa r) \quad\quad \text{if} \quad r > r_0\,, \end{aligned} \tag{9.3.36}$$

where $\kappa^2 = k_z^2 - \omega^2/c^2 > 0$ $(\omega < k_z c)$. Substituting these solutions into the boundary conditions (9.3.14) we obtain

$$\mathrm{I}_0(\sqrt{\varepsilon}\,\kappa r_0)\,K_0'(\kappa r_0) - \sqrt{\varepsilon}\;\mathrm{I}_0'(\sqrt{\varepsilon}\,\kappa r_0)\,K_0(\kappa r_0) = 0\,. \tag{9.3.37}$$

It is easily shown that this equation has no solutions in the short-wavelength limit $\kappa^2 r_0^2 \gg 1$, i.e., short-wavelength surface oscillations in the plasma cylinder are impossible in a strong magnetic field. In the long-wavelength limit, when $\kappa^2 r_0^2 \ll 1$, $|\varepsilon|\kappa^2 r_0^2 \ll 1$ from (9.3.37) we obtain

$$\varepsilon(\omega)\ln(\kappa r_0) - 2(\kappa r_0)^{-2} = 0\,. \tag{9.3.38}$$

It exactly coincides with (9.3.27). Thus the analysis for the isotropic plasma also remains valid in case of the plasma cylinder with a free surface in a strong magnetic field. The only difference is that in the isotropic plasma the field E_z weakly damps radially from the surface of the cylinder, while in the magneto-active plasma it grows slowly and reaches maximum on the waveguide axis.

Figure 9.7 b shows the spectra of space and surface waves of the plasma cylinder in a strong magnetic field.

In conclusion, it should be stressed that the obtained relations are valid for both nondegenerate and degenerate plasmas, since under the given conditions of large phase velocities (compared to thermal velocities) the form of the equilibrium distribution function is insignificant.

9.4 Stability of Spatially Bounded Nonequilibrium Plasma

9.4.1 Buneman Instability in the Plasma Waveguide

In the previous section we have studied electromagnetic properties of a plasma cylinder, the distribution (Maxwellian or Fermian) of charged particle velocities (9.3.4) being assumed to be hydrodynamically in equilibrium. Now let us analyze the properties of a nonequilibrium spatially bounded electron-ion plasma with a current, regarding the electron distribution function $F_{0\alpha}$ as Maxwellian with a nonzero longitudinal velocity $\boldsymbol{u}$, and ions as stationary. The calculation of the perturbed distribution function δf_α presents no difficulty; this function is derived from (9.3.7) by substituting $\omega \to \omega - \boldsymbol{k} \cdot \boldsymbol{u}_\alpha$, where $\boldsymbol{u}_e = \boldsymbol{u}$ and $\boldsymbol{u}_i = 0$. We can also easily calculate the perturbed current and charge densities, and write a system of field equations. Below we analyze these equations by considering only potential field oscillations since in Chap. 7 it has been shown that the characteristic instabilities (Buneman and ion acoustic) of the current-carrying plasma are realized in the frequency range $\omega \ll k_z u$, and are longitudinal with a high degree of accuracy. Besides, for simplicity, the system is assumed to be imbedded in a sufficiently strong longitudinal magnetic field such that the electrons are magnetized ($\Omega_e \gg \omega_{pe}$), but the ions are not ($\omega_{pi} \gg \Omega_i$).

For a sufficiently large $\boldsymbol{u}$ exceeding the electron and ion thermal velocities, Poisson's equation

$$\Delta\Phi = -\sum_\alpha 4\pi\varrho_\alpha = -4\pi\sum_\alpha e_\alpha \int d\boldsymbol{p}\, \delta f_\alpha \tag{9.4.1}$$

together with (9.3.7) and the substitution $\omega \to \omega - \boldsymbol{k} \cdot \boldsymbol{u}_\alpha$ as well as accounting for the field potentiality ($\boldsymbol{E} = -\nabla\Phi$), takes the form

$$\left(1 - \frac{\omega_{pi}^2}{\omega^2}\right)\Delta\Phi - \frac{\omega_{pe}^2 k_z^2}{(\omega - k_z u)^2}\,\Phi = 0\,. \tag{9.4.2}$$

When deriving this equation, the waveguide has been assumed to be completely filled with a homogeneous plasma (Fig. 9.6), and all perturbed quantities to be time- and coordinate-dependent in the form of (9.3.5). Note that this equation can also be obtained from the model of independent particles by applying the dielectric tensor operator derived in Exercise 9.6.8. Moreover, in this model (9.4.2) can easily be generalized for the case of the relativistic electron drift velocity $\boldsymbol{u}$. Thus, we can study the problem of the stability of a relativistic electron beam, its charge being compensated by ions. The generalization is effected by the simple substitution

$$\omega_{pe}^2 \to \omega_{pe}^2\gamma^{-3}, \tag{9.4.3}$$

where $\gamma = (1 - u^2/c^2)^{-1/2}$. Furthermore, we shall take into account this substitution in the analysis of (9.4.2).

Now we write the solution of (9.4.2) in the form

$$\Phi(r) = \Phi_0 \mathrm{J}_l\left(\mu_{ls}\frac{r}{R}\right), \tag{9.4.4}$$

when the boundary condition

$$\Phi(R) = 0 \tag{9.4.5}$$

is satisfied. Here μ_{ls} are the roots of the Bessel function, $\mathrm{J}_\lambda(\mu_{ls}) = 0$. As a result, we obtain the following dispersion equation for the oscillation spectrum

$$\left(1 - \frac{\omega_{\mathrm{pi}}^2}{\omega^2}\right)\left(\frac{\mu_{ls}}{R^2} + k_z^2\right) - \frac{\omega_{\mathrm{pe}}^2 k_z^2 \gamma^{-3}}{(\omega - k_z u)^2} = 0 . \tag{9.4.6}$$

Its solution corresponds to unstable oscillations existing in the frequency range $\omega \ll k_z u$ (because of this condition we can confine ourselves to the electrostatic approximation) as well as the condition

$$\omega_{\mathrm{pe}}^2 > u^2\gamma^3\left(k_z^2 + \frac{\mu_{ls}^2}{R^2}\right) . \tag{9.4.7}$$

The increment of the resulting aperiodic instability, known as the Buneman instability (Sect. 7.2), is equal to

$$\delta = \mathrm{Im}\{\omega\} = \frac{\sqrt{3}}{2} k_z u \left(\frac{m}{M}\frac{\mu_{ls}}{R^2 k_z^2}\gamma^3\right)^{1/3} > \omega_{\mathrm{pi}} . \tag{9.4.8}$$

It follows from (9.4.7) that the instability under discussion can develop if the electron current exceeds a threshold value given by

$$\begin{aligned} I_{\mathrm{th}} &= eN_e u\pi R^2 = \min\left[\frac{mc^3}{4e}(\mu_{ls}^2 + k_z^2R^2)(\gamma^2 - 1)^{3/2}\right] \\ &\approx 1{,}4\frac{mc^3}{e}(\gamma^2 - 1)^{3/2} . \end{aligned} \tag{9.4.9}$$

This current is more than $(\gamma^2 - 1)^{3/2}/(\gamma^{2/3} - 1)^{3/2}$ times greater than the *limiting vacuum current* of the electron beam in the waveguide (Exercise 9.6.9).

9.4.2 Pierce Instability of the Plasma with a Current in Longitudinally Bounded Systems

Note that the increment of the Buneman instability given by (9.4.8) depends on the ion mass and vanishes in the limit $m/M \to 0$. It does not imply, however, that the electron beam with its charge, compensated by a background of infinitely heavy ions ($\omega_{\rm pi} \to 0$), is stable. The development of the instability specified by finite longitudinal dimensions of the system appears possible in such a beam. In a longitudinally bounded system the solution of the field equation cannot be represented by (9.3.5); it should be sought in the form of the sum

$$\Phi(r) = \mathrm{J}_l\left(\mu_{ls}\frac{r}{R}\right)\sum_{n=1}^{4}\Phi_n \mathrm{e}^{\mathrm{i}k_{zn}z}\,, \tag{9.4.10}$$

where k_{zn} $(n = 1, 2, 3, 4)$ are the four roots of (9.4.6) treated as a characteristic equation.

The solution (9.4.10) satisfies the radial boundary condition (9.4.5). But this is insufficient for the derivation of a dispersion equation: the assignment of longitudinal boundary conditions (their number must be equal to 4) is necessary. We choose for these conditions

$$\Phi|_{z=0,\mathrm{L}} = 0, \qquad \varrho_{\rm e}|_{z=0} = j_{z\rm e}|_{z=0} = 0\,, \tag{9.4.11}$$

where $z = 0$ and $z = L$ are the coordinates of the metallic end planes of a finite waveguide. The second condition (9.4.11) means that at the front $(z = 0)$ of the system the beam is unperturbed.

Now we can derive the unknown dispersion equation by determining k_{zn} from (9.4.6), and substituting the resulting solution (9.4.10) into the boundary conditions (9.4.11). For the threshold of the instability, we consider the frequency range $\omega \ll \omega_{\rm pe}$, where the roots of (9.4.6) are given by

$$\begin{aligned} k_{z1,2} &\approx \pm\frac{1}{u}\left(\frac{\omega_{\rm pe}^2}{\gamma^3} - k_\perp^2 u^2\right)^{1/2} + \frac{\omega\omega_{\rm pe}^2}{\gamma^3 u}\left(\frac{\omega_{\rm pe}^2}{\gamma^3} - k_\perp^2 u^2\right)^{-1}, \\ k_{z3,4} &\approx \omega\left(u \mp \frac{\omega_{\rm pe}}{k_\perp \gamma^{3/2}}\right)^{-1}, \end{aligned} \tag{9.4.12}$$

where $k_\perp \equiv \mu_{ls}/R$. As a result the dispersion equation defining the frequency spectrum can be written as

$$\begin{aligned} &\left(\frac{\omega_{\rm pe}^2}{\gamma^3} - k_\perp^2 u^2\right)^{3/2}(\mathrm{e}^{\mathrm{i}k_{z1}L} - \mathrm{e}^{\mathrm{i}k_{z2}L}) - \frac{2\,\omega\omega_{\rm pe}^2}{\gamma^3}(\mathrm{e}^{\mathrm{i}k_{z1}L} + \mathrm{e}^{\mathrm{i}k_{z2}L} - \mathrm{e}^{\mathrm{i}k_{z3}L} - \mathrm{e}^{\mathrm{i}k_{z4}L}) \\ &\quad + \frac{\omega\omega_{\rm pe}^2}{\gamma^{3/2}k_\perp u}\left(\frac{\omega_{\rm pe}^2}{\gamma^3} - k_\perp^2 u^3\right)(\mathrm{e}^{\mathrm{i}k_{z3}L} - \mathrm{e}^{\mathrm{i}k_{z4}L}) = 0\,. \end{aligned} \tag{9.4.13}$$

Equation (9.4.13) has solutions with $\mathrm{Im}\{\omega\} > 0$, corresponding to growing oscillations at the interval

$$(2n-1)\frac{\pi u}{L} < \sqrt{\frac{\omega_{pe}^2}{\gamma^3} - k_\perp^2 u^2} < 2n\frac{\pi u}{L}\,, \tag{9.4.14}$$

where $n = 1, 2, 3\ldots$ By assuming $n = 1$ and $k_{\perp\min} = 2.4/R$, we may obtain the threshold electron current above which an instability develops in the system. For $L \gg R$, it coincides with the current (9.4.9) and the instability increment is

$$\delta = \mathrm{Im}\{\omega\} \approx u/L\,. \tag{9.4.15}$$

The discussed current-driven instability of plasmas in longitudinally bounded systems is known as the *Pierce instability*. The threshold current for the development of the Pierce instability is the same as for the Buneman instability. Nevertheless, a comparison of their increments (9.4.8 and 15) reveals that the physical nature of these instabilities is different. In particular, when the threshold current (9.4.9) is surpassed in systems with $L < u/\omega_{pi}$, the Pierce instability develops, and for $L > u/\omega_{pi}$ the Buneman instability occurs.

9.4.3 Ion-Acoustic Instability of the Bounded Plasma with a Current

To conclude, let us analyze an instability of bounded plasmas with a current when the electron drift velocity is small, $v_{Te} \gg u \gg v_{Ti}$. As shown in Chap. 7, under these conditions the development of the ion-acoustic instability is possible in a nonisothermal ($T_e \gg T_i$) spatially unbounded plasma with a current at the drift velocity $\boldsymbol{u}$ exceeding the phase velocity of the ion-acoustic waves. Let us consider the possibility of this instability in a plasma waveguide.

As before, we proceed from the solution of the kinetic equation (9.3.7) for δf_α, with the substitution $\omega \to \omega - \boldsymbol{k}\cdot\boldsymbol{u}_\alpha$. The ions are assumed to be cold, $\omega \gg k v_{Ti}$, and electrons hot $|\omega - \boldsymbol{k}\cdot\boldsymbol{u}| \ll k v_{Te}$. We consider the frequency range $\omega \gg \Omega_e$. Poisson's equation (9.4.1) may be reduced to the following integro-differential equation

$$\left(1 - \frac{\omega_{pi}^2}{\omega^2}\right)\Delta\Phi - \frac{\omega_{pi}^2}{v_s^2}\Phi = \mathrm{i}\frac{\omega_{pi}^2}{\omega^2}\int d\boldsymbol{r}'\,Q(\boldsymbol{r}-\boldsymbol{r}')(\omega - \boldsymbol{k}\cdot\boldsymbol{u})\,\Phi(\boldsymbol{r}')\,, \tag{9.4.16}$$

where $v_s = \sqrt{T_e/M}$ and the kernel $Q(r)$ is of the form

$$Q(r) = \sqrt{\frac{\pi}{2}}\frac{1}{(2\pi)^3}\frac{1}{v_{Te}}\int\frac{d\boldsymbol{k}}{k}\,e^{i\boldsymbol{k}\cdot\boldsymbol{r}}\,. \tag{9.4.17}$$

Assuming the waveguide to be longitudinally unbounded and filled with plasma, we can solve (9.4.16) accounting for only the boundary condition

(9.4.5). Besides, the integral term specified by the Cherenkov energy dissipation of the oscillations on the electrons is small, and therefore (9.4.16) can be solved by means of successive approximations. In the first-order approximation, i.e., neglecting the electron dissipation term, we obtain (9.4.4) for the eigenfunction satisfying the boundary condition (9.4.5). By substituting this function into the right-hand side of (9.4.16), we obtain the dispersion equation of the oscillations:

$$1 + \frac{\omega_{\mathrm{pi}}^2}{k^2 v_{\mathrm{s}}^2}\left(1 + \mathrm{i}\sqrt{\frac{\pi}{2}}\frac{\omega - \boldsymbol{k}\cdot\boldsymbol{u}}{k v_{\mathrm{Te}}}\right) - \frac{\omega_{\mathrm{pi}}^2}{\omega^2} = 0\,, \tag{9.4.18}$$

where $k = \sqrt{k_\perp^2 + k_z^2} = \sqrt{\mu_{ls}/R^2 + k_z^2}$. Equation (9.4.18) coincides with (7.2.6) if the Cherenkov dissipation on ions is neglected and $k_\perp$ is taken to be μ_{ls}/R. Therefore, the analysis given in Sect. 7.2 is also valid in the case of bounded plasmas if such a substitution is made. Moreover, accounting for ion Cherenkov dissipation and collisions does not alter this conclusion.

9.5 Excitation of a Plasma Resonator by Relativistic Electron Beam

In Sect. 9.3 we have dealt with natural electromagnetic oscillations of a plasma waveguide. Now let us study the possibility of exciting such oscillations by means of relativistic electron beams. We confine our consideration to the case when a plasma is penetrated by a dense electron beam completely filling a metallic waveguide. The magnetic field is assumed to be sufficiently strong, magnetizing both the plasma and the beam, i.e., $\Omega_{\mathrm{e}} \gg \omega_{\mathrm{pe}}, \omega_{\mathrm{b}}\sqrt{\gamma}$. Such excitation of electromagnetic waves in the plasma is possible because of Cherenkov and cyclotron interactions between the beam and the wave field (Chap. 6). We assume that the distribution of the monoenergetic electron beam is of the form (6.1.11):

$$f_{0\mathrm{b}} = \frac{N_{\mathrm{b}}}{2\pi p_{\perp 0}}\,\delta(p_\perp - p_{\perp 0})\,\delta(p_\parallel - p_{\parallel 0}) \tag{9.5.1}$$

and $u_\perp = p_{\perp 0}/m\gamma \ll c$, where $\gamma = (1 - u^2/c^2)^{-1/2}$. The plasma is assumed to be cold and purely electronic.

9.5.1 Cherenkov Wave Excitation

We start the analysis of electromagnetic wavegeneration in a plasma waveguide by the Cherenkov mechanism of interaction between the electron beam and the wave field. We limit our consideration to the case of a strong magnetic field when the inequalities

$$\Omega_e \gg \omega_{pe}\,, \frac{c\gamma}{R}\,, \qquad \omega \gg \frac{\omega_b}{\gamma^{3/2}} \tag{9.5.2}$$

are satisfied. Under these conditions, the dielectric tensor of the plasma-beam system given in Sect. 6.4 takes a very simple form

$$\varepsilon_{ij}(\omega) = \begin{pmatrix} 1 & 0 & 0 \\ 0 & 1 & 0 \\ 0 & 0 & \varepsilon_{\parallel} \end{pmatrix}, \qquad \text{where} \tag{9.5.3}$$

$$\varepsilon_{\parallel} = 1 - \frac{\omega_{pe}^2}{\omega^2} - \frac{\omega_b^2 \gamma^{-1} \gamma_{\parallel}^{-2}}{(\omega - k_z u_{\parallel})^2}\,. \tag{9.5.4}$$

The tensor (9.5.3) is analogous to the tensor (9.3.29), and its substitution into the system of the Maxwell equations naturally yields two equations for the E- and B-type waves, respectively. The beam interacts only with an E-wave which satisfies the equation

$$\left(\frac{1}{r}\frac{\partial}{\partial r} r \frac{\partial}{\partial r} - \frac{l^2}{r^2}\right) E_z - \left(k_z^2 - \frac{\omega^2}{c^2}\right) \varepsilon_{\parallel}(\omega) E_z = 0\,. \tag{9.5.5}$$

The only boundary condition to this equation is

$$E_z|_{r=R} = 0\,. \tag{9.5.6}$$

By substituting the general solution of (9.5.5) limited to the waveguide axis

$$E_z(r) = E_{z0} \mathrm{J}_l (\mathrm{i}\sqrt{\kappa^2 \varepsilon_{\parallel}}\, r),\ \kappa^2 = k_z^2 - \omega^2/c^2 \tag{9.5.7}$$

into the boundary condition (9.5.6), we obtain the dispersion equation for the E-wave interacting with an electron beam in a plasma waveguide

$$\left(k_z^2 - \frac{\omega^2}{c^2}\right)\left(1 - \frac{\omega_{pe}^2}{\omega^2} - \frac{\omega_b^2 \gamma^{-1} \gamma_{\parallel}^{-2}}{(\omega - k_z u_{\parallel})^2}\right) + \frac{\mu_{ls}}{R^2} = 0\,, \tag{9.5.8}$$

where μ_{ls} are the zeros of the Bessel function $\mathrm{J}_l(\mu_{ls})$.

As to be expected, (9.5.8) coincides with (6.3.10) if in the latter $k_\perp$ is replaced by μ_{ls}/R. Therefore, the analysis of the Cherenkov beam instability in Sect. 6.3 is also valid in the case of the spatially bounded plasma. In particular, in the first-order approximation of (9.5.8), we obtain two branches of E-waves in a plasma waveguide which have been studied in detail

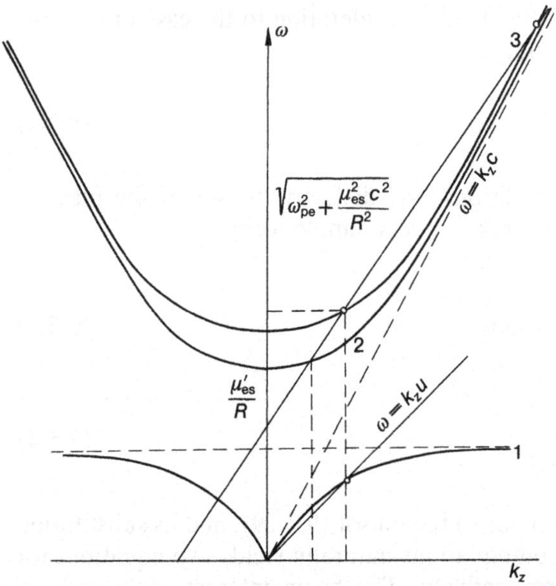

Fig. 9.8. Condition for the resonance interaction of an electron beam with electromagnetic waves in a plasma waveguide

in Sect. 9.3. Here a small beam summand is ignored. The beam resonantly interacts only with the low-frequency oscillation branch when the Cherenkov resonance condition $\omega_1 = k_z u_{\|}$ is satisfied (Fig. 9.8). Then we obtain the frequency of the wave excited by the beam

$$\omega_0 = \sqrt{\omega_{pe}^2 - \mu_{ls}^2 u_{\|}^2 \gamma_{\|}^2 / R^2} \; . \tag{9.5.9}$$

Hence the excitation of the μ_{ls}-mode is possible only under the condition

$$\omega_{pe}^2 > \mu_{ls}^2 u_{\|} \gamma_{\|}^2 / R^2 \; . \tag{9.5.10}$$

If the plasma density satisfies the inequalities

$$3.8 > \frac{\omega_{pe} R}{u_{\|} \gamma_{\|}} > 2.4 \; , \tag{9.5.11}$$

then only one axially symmetric mode with $\mu_{0I} = 2.4$ can be excited.

The temporal oscillation increment is derived accounting for the beam summand in (9.5.8). One obtains ($\omega = \omega_0 + \delta$)

$$\frac{\delta}{\omega_0} = \frac{-1 + \mathrm{i}\sqrt{3}}{2} \left(\frac{N_b}{2\gamma\gamma_{\|}^2 N_e} \right)^{1/3} \left(1 + \frac{\mu_{ls}^2 u_{\|}^2 \gamma_{\|}^2 (\gamma_{\|}^2 - 1)}{R^2 \omega_{pe}^2} \right)^{-1/3} . \tag{9.5.12}$$

Naturally, this relation also coincides with (6.3.13) under the substitution $k_\perp = \mu_{ls}/R$.

As shown in Chap. 6, the Cherenkov beam instability is of a washing-away (drift) character and therefore in systems of finite length it may not have time to develop. To make sure, let us consider a metallic waveguide of a finite but rather large length, $L \gg R$. The waveguide is filled with a strongly magnetized plasma penetrated by a relativistic electron beam. Let the electromagnetic radiation be entirely reflected (e.g., due to a limiting concentration of the waveguide) by the left end plane of the waveguide $z = 0$ from which the beam is injected. A partial reflection of electromagnetic waves with the reflection index $\kappa(\omega)$ and a partial radiation out of the system occur at the right end plane $z = L$, transparent for an electron beam. In the latter's absence such a system is an electromagnetic wave resonator with finite quality (finite time of radiation energy outflow from the system). A sufficiently intense electron beam can excite such a resonator and then the system turns into a generator of electromagnetic radiation.

This problem of the excitation of a plasma resonator by an electron beam is a boundary value problem, and while solving it, (9.5.8) is to be treated as a characteristic equation for the longitudinal wave numbers k_{zn} of the normal waves

$$E_z(r) = \mathrm{J}_l\left(\mu_{ls}\frac{r}{R}\right)\sum_{n=1}^{4} E_{z0n}\exp(\mathrm{i}k_{zn}z)\,. \tag{9.5.13}$$

Equation (9.5.8) shows that in the case considered there are four such waves, and for all these waves $k_{zn}(\omega)$ $(n = 1, 2, 3, 4)$ can easily be obtained if the small beam term in (9.5.8) is taken into account:

$$k_{z1,2,3}(\omega) = k_{z0}(\omega) + \delta k_{z1,2,3}, \qquad k_{z4}(\omega) = -k_{z0}(\omega)\,. \tag{9.5.14}$$

Here $k_{z0}(z)$ is a solution of (9.5.8) satisfying the Cherenkov resonance condition in the absence of the beam:

$$k_{z0}(\omega_0) = \frac{\omega_0}{u_\parallel} = \sqrt{\frac{\omega_0^2}{c^2} + \frac{\mu_{ls}^2}{R^2}\left(\frac{\omega_{pe}^2}{\omega_0^2} - 1\right)^{-1}} \tag{9.5.15}$$

and the quantities $\delta k_{z1,2,3}$ are given by the expressions

$$\begin{aligned}\delta k_{z1} &= -k_{z0}\left(\frac{\omega_b^2R^2}{2\gamma_\parallel^6\gamma\mu_{ls}^2u_\parallel^2}\right)^{1/3},\\ \delta k_{z2,3} &= \frac{1\pm\mathrm{i}\sqrt{3}}{2}k_{z0}\left(\frac{\omega_b^2R^2}{2\gamma_\parallel^6\gamma\mu_{ls}^2u_\parallel^2}\right)^{1/3}.\end{aligned} \tag{9.5.16}$$

Equations (9.5.14, 16) show that in the system there exist four waves, three of them $k_{z1,2,3}(\omega)$ travelling along the beam and so are passing waves (therefore interacting with the beam), the fourth one $k_{z4}(\omega)$ being counter and not interacting with it. The wave $k_{z3}(\omega)$ is amplified in the beam's direction, the wave $k_{z2}(\omega)$ damps, and the waves $k_{z1}(\omega)$ and $k_{z4}(\omega)$ remain neutral (i.e., neither amplified nor damped).

To obtain the dispersion equation and to determine the excitation condition for the resonator, one must substitute (9.5.13) into the longitudinal boundary conditions at $z=0$. Since we intend to take account of electromagntic wave radiation from the end plane $z=L$, we must know the solution of the field equation in the range $z>L$. This fact significantly complicates the derivation of the dispersion equation. However, if the resonator length is large, so that $|\delta k_{z3}L|>1$, the problem is simplified and reduced to the inner one, i.e., knowledge of (9.5.13) appears significant. Actually, by this assumption a complete reflection of the electromagnetic wave occurs at the left end plane of the resonator $z=0$, i.e., a counterwave E_{z04} is totally converted into each of the passing waves $E_{z01,2,3}$ with the same probability. At the right end plane $z=L$ an electromagnetic wave is radiated out of the system and only a part of it is reflected with the reflection index $\kappa(\omega)$ and is transformed into a counterwave. Since $|\delta k_{z3}L|>1$ from three passing waves at the end plane $z=L$ only a wave E_{z03}, reinforced by the beam, is sufficient to be accounted for. Then

$$E_{z03}=\frac{1}{3}E_{z04}, \qquad \kappa E_{z03}\,\mathrm{e}^{\mathrm{i}k_{z3}L}=E_{z04}\,\mathrm{e}^{\mathrm{i}k_{z4}L}\,. \tag{9.5.17}$$

Hence the dispersion equation immediately follows

$$\exp[\mathrm{i}(k_{z3}-k_{z4})L]=3/\kappa\,. \tag{9.5.18}$$

From the latter, accounting for the solution (9.5.14), we obtain the frequency spectrum of oscillations of a plasma resonator ($\omega\to\omega_0+\mathrm{i}\delta$)

$$\begin{aligned}\omega &= \omega_0-\min_n v_{g0}\left[k_{z0}(\omega_0)-\frac{\pi n}{L}+\frac{1}{4}k_{z0}(\omega_0)\right.\\ &\quad\times\left.\left(\frac{\omega_b^2R^2}{2\gamma_\parallel^6\gamma\mu_{ls}^2u_\parallel^2}\right)^{1/3}+\frac{\arg\{\kappa(\omega_0)\}}{2L}\right],\\ \delta &= \left[\frac{\sqrt{3}}{2}k_{z0}(\omega_0)\left(\frac{\omega_b^2R^2}{2\gamma_\parallel^6\gamma\mu_{ls}^2u_\parallel^2}\right)^{1/3}-\frac{1}{L}\ln\frac{3}{\kappa}\right]\left(\frac{1}{W}+\frac{1}{v_{g0}}\right)^{-1},\end{aligned} \tag{9.5.19}$$

where v_{g0} is a group velocity of the beam-excited wave

$$\begin{aligned}v_{g0} &= \left(\frac{\partial k_0(\omega)}{\partial\omega}\right)^{-1}=\frac{u_\parallel}{u_\parallel^2/c^2+\omega_{pe}^2R^2/(\gamma_\parallel^4u_\parallel^2\mu_{ls}^2)}\,,\\ W &= \frac{3u_\parallel v_{g0}}{u_\parallel+2v_{g0}}\end{aligned} \tag{9.5.20}$$

and $n = 1, 2, 3\ldots$ determines the number of halfwaves set on the resonator length, i.e., $k_{z0}(\omega_0) \approx \pi n/L$.

For $\delta > 0$ the electromagnetic wave with the frequency $\omega \approx \omega_0$ in the plasma resonator will exponentially grow with time since the field amplification in the resonator due to the electron beam exceeds the losses specified by electromagnetic wave radiation out of the resonator. The excitation of the resonator is possible when the beam current is larger than a threshold value known as a *starting current*

$$I_{st} = \frac{4}{3\sqrt{3}} \frac{mc^3}{e} \frac{\gamma\gamma_\parallel^6 \mu_{ls}^2}{k_{z0}^3 L^3} \left(\ln \frac{3}{|\kappa|}\right)^3 . \tag{9.5.21}$$

When (9.5.11) is satisfied, and the excitation only of the main axially symmetric mode with $\mu_{01} = 2.4$ is possible, the starting current (9.5.21) appears to be minimal.

9.5.2 Cyclotron Wave Excitation

Now let us consider cyclotron wave excitation by a rotating electron beam in a spatially bounded system. As shown in Chap. 6 such a beam may excite electromagnetic waves in the absence of a plasma, i.e., in vacuum. Therefore we confine our analysis to the interaction of a rotating electron beam with a vacuum metallic waveguide. Moreover, the beam will be regarded as sufficiently rarefied so that the conditions

$$\Omega_e \gtrsim \frac{c\gamma}{R} \sim \omega \gg \omega_{pe}, \omega_b\sqrt{\gamma} \tag{9.5.22}$$

are satisfied. Under these conditions, the equation for E- and B-type waves which are in cyclotron resonance with the electron beam [accounting for the beam terms containing only the second-order cyclotron poles $k_z u_\parallel + \Omega_e/\gamma = \omega$ in the dielectric tensor (6.4.3)] are respectively

$$\begin{aligned} \Delta E_z + \frac{\omega^2}{c^2} E_z &= \frac{u_\perp^2 u_\parallel^2}{4c^4}\left(\frac{k_z^2 c^2}{\omega^2} - 1\right)^2 \frac{\omega_b^2 \gamma^{-1}}{(\omega - k_z u_\parallel - \Omega_e/\gamma)^2} \Delta_\perp E_z , \\ \Delta B_z + \frac{\omega^2}{c^2} B_z &= \frac{u_\perp^2}{4c^2} \frac{(k_z^2 - \omega^2/c^2)\,\omega_b^2\gamma^{-1}}{(\omega - k_z u_\parallel - \Omega_e/\gamma)^2} B_z . \end{aligned} \tag{9.5.23}$$

In the case of a longitudinally infinite waveguide the boundary conditions for these equations are

$$E_z\Big|_{r=R} = \frac{\partial B_z}{\partial r}\Big|_{r=R} = 0 . \tag{9.5.24}$$

The solutions of (9.5.23) satisfying the boundary conditions (9.5.24) are

$$E_z(r) = E_{z0}\,\mathrm{J}_l\left(\mu_{ls}\frac{r}{R}\right), \qquad B_z = B_{z0}\,\mathrm{J}_l\left(\mu'_{ls}\frac{r}{R}\right), \tag{9.5.25}$$

where μ_{ls} and μ'_{ls} are the roots of the Bessel function and its derivative, $\mathrm{J}_l(\mu_{ls}) = 0$ and $\mathrm{J}'_l(\mu'_{ls}) = 0$, respectively. After substituting (9.5.25) into (9.5.23) the following dispersion equations for E- and B-waves result:

$$\begin{aligned} \frac{\mu_{ls}^2}{R^2} + k_z^2 - \frac{\omega^2}{c^2} &= \frac{u_\perp^2 u_\parallel^2}{4c^4}\left(\frac{k_z^2 c^2}{\omega^2} - 1\right)^2 \frac{\omega_b^2 \mu_{ls}^2/R^2}{\gamma(\omega - k_z u_\parallel - \Omega_e/\gamma)^2}\,, \\ \frac{\mu_{ls}'^2}{R^2} + k_z^2 - \frac{\omega^2}{c^2} &= \frac{u_\perp^2}{4c^2}\left(\frac{\omega^2}{c^2} - k_z^2\right) \frac{\omega_b^2}{\gamma(\omega - k_z u_\parallel - \Omega_e/\gamma)^2}\,. \end{aligned} \tag{9.5.26}$$

These equations differ from the corresponding dispersion equations from Sect. 6.4 by the replacement of $k_\perp$ for μ_{ls}/R for E-waves and of μ'_{ls}/R for B-waves.

The structure of both equations (9.5.26) is the same, therefore we analyze only one of them, i.e., the second equation which describes the interaction of the electron beam with a B-wave of a vacuum waveguide. Note that the interaction of the lines $\omega = k_z u_\parallel + \Omega_e/\gamma$ and $\omega(k_z) = \sqrt{k_z^2c^2 + \mu_{ls}'^2 c^2/R^2}$ determines two frequencies of B-waves excited by the beam (see Fig. 9.8)

$$\omega_{01,2} = \frac{\gamma_\parallel^2 \Omega_e}{\gamma}\left(1 \pm \frac{u_\parallel}{c}\sqrt{1 - \frac{\mu_{ls}'^2 c^2 \gamma^2}{R^2 \Omega_e^2 \gamma_\parallel^2}}\right), \tag{9.5.27}$$

their group velocity and longitudinal wave number being respectively

$$v_{g01,2} = c\frac{k_{z01,2}\,c}{\omega_{01,2}}\,, \qquad k_{z01,2} = \frac{1}{u_\parallel}\left(\omega_{01,2} - \frac{\Omega_e}{\gamma}\right). \tag{9.5.28}$$

It is easy to see that under the condition $\gamma_\parallel > \mu'_{ls} c\gamma/(\Omega_e R) > 1$ both roots $k_{01,2} > 0$, i.e., both high- and low-frequency beam-excited B-waves are passing waves. As shown in Sect. 6.4, the cyclotron instability here has a washing-away character. If $\mu'_{ls} c\gamma/(R\Omega_e) < 1$, then $k_{z01} > 0$ and $k_{z02} < 0$. Hence the high-frequency B-wave is passing and the low-frequency one is counter. In the latter case, the cyclotron instability is of absolute character. Equation (9.5.27) also shows that under the conditions

$$3.8 > \frac{R\gamma_\parallel \Omega_e}{c\gamma} > 1.8 \tag{9.5.29}$$

only one radial mode with a minimal value $\mu_{11} = 1.8$ (the first asymmetric mode of the B_{11}-wave) will be excited in the system.

The analysis of the cyclotron excitation of the passing B-wave in a longitudinally bounded vacuum resonator is carried out in the same way as has been done above in analyzing the Cherenkov excitation of a plasma resonator (excitation of the counterwave, see Exercise 9.6.12). Thus, it is unnecessary to render again the above discussion. Note that the values v_{g0} and k_{z0} in (9.5.14, 18–20) should be replaced by the corresponding expressions (9.5.28), and the amplification and damping coefficients of waves interacting with a beam are

$$\begin{aligned} \delta k_{z1} &= -k_{z0}\left(\frac{\omega_b^2 \mu_{ls}'^2 u_\perp^2}{8R^2c^2\gamma k_{z0}^4 u_\parallel^2}\right)^{1/3}, \\ \delta k_{z2,3} &= \frac{1 \pm \mathrm{i}\sqrt{3}}{2} k_{z0}\left(\frac{\omega_b^2 \mu_{ls}'^2 u_\perp^2}{8R^2c^2\gamma k_{z0}^4 u_\parallel^2}\right)^{1/3}. \end{aligned} \tag{9.5.30}$$

Finally, for the *starting current of the resonator excitation* on the passing B-wave by the rotation electron beam we obtain

$$I_{st} = \frac{16}{3\sqrt{3}} \frac{mc^3}{e} \frac{u_\parallel^2}{u_\perp^2} \frac{R^4|k_{z0}|}{L^3\mu_{ls}'^2} \frac{u_\parallel}{c}\left(\ln\frac{3}{|\kappa|}\right)^3. \tag{9.5.31}$$

Hence it is more difficult to excite the resonator at the high frequency of the B-wave than at the low frequency since always $|k_{z01}| > |k_{z02}|$.

9.6 Exercises

9.6.1 Find the frequency of electrostatic (potential) natural oscillations of the cold electron plasma layer $2d$ thick, placed between the plane capacitor plates $2L$. Assume the electric field of oscillations to be normal to the surface of the layer.

Solution. In all three ranges the field equation is of the same form:

$$\Delta\Phi = \frac{\partial^2\Phi}{\partial x^2} = 0\,. \tag{9.6.1}$$

Therefore its solution may be written as

$$\Phi(x) = \begin{cases} a_1x + a_2 & \text{if } -L \le x < -d\,, \\ b_1x + b_2 & \text{if } |x| \le d\,, \\ c_1x + c_2 & \text{if } d \le x \le L\,. \end{cases} \tag{9.6.2}$$

Substituting these solutions into the boundary conditions

$$\Phi(-L) = \Phi(L) = 0 ,$$
$$\{\Phi\}_{x=-d} = \{\Phi\}_{x=d} = 0 , \tag{9.6.3}$$
$$\left\{\varepsilon \frac{\partial \Phi}{\partial x}\right\}_{x=-d} = \left\{\varepsilon \frac{\partial \Phi}{\partial x}\right\}_{x=d} = 0 ,$$

where $\varepsilon(\omega) = 1 - \omega_{pe}^2(x)/\omega^2$ and $\omega_{pe}(x) \neq 0$ only for $|x| \leq d$, yields the dispersion relation

$$\varepsilon(\omega) - \frac{d}{L-d} = 0 . \tag{9.6.4}$$

Hence we obtain the frequency spectrum of oscillations

$$\omega^2 = \frac{L-d}{L}\omega_{pe}^2 \leq \omega_{pe}^2 . \tag{9.6.5}$$

9.6.2 Find the penetration of the quasistatic monochromatic field with the frequency ω into the semi-bounded isotropic plasma, assuming mirror reflection of particles from the plasma surface.

Solution. Regarding the field as normal to the plasma surface we can write for the region $x \geq 0$ (in the plasma)

$$\frac{\partial D(x)}{\partial x} = 0 , \quad \text{or} \tag{9.6.6}$$

$$D(x) = E_0 , \tag{9.6.7}$$

E_0 being the electric field density on the plasma surface.

On the other hand, we can determine $D(x)$ by continuing it evenly into the region $x < 0$ and carrying out the Fourier transformation:

$$D(k) = \frac{1}{2\pi}\int_{-\infty}^{\infty} D(x)\,e^{-ikx}dx , \qquad D(x) = \int_{-\infty}^{\infty} D(k)\,e^{ikx}dk . \tag{9.6.8}$$

As a result we obtain from (9.6.7)

$$\int_{-\infty}^{\infty} D(k)\,e^{ikx}dx = \frac{E_0}{\pi i}\int_{-\infty}^{\infty}\frac{e^{ikx}}{k}dk . \tag{9.6.9}$$

Then taking into account that for mirror reflection of particles from the plasma surface, for the longitudinal field

$$D(k) = \varepsilon^{lo}(\omega, k)\, E(k) \;, \tag{9.6.10}$$

we obtain

$$E(k) = \frac{1}{\pi i} \frac{E_0}{k\varepsilon^{lo}(\omega, k)} \;, \tag{9.6.11}$$

or substituting it into (9.6.8)

$$E(x) = \frac{E_0}{\pi i} \int_{-\infty}^{\infty} \frac{e^{ikx}}{k\varepsilon^{lo}(\omega, k)}\, dk \;. \tag{9.6.12}$$

This expression determines the penetration of the quasistatic field into the semi-bounded isotropic plasma normal to its surface.

The integral (9.6.12) is essentially determined by the poles of the integrand

$$k = 0 \;, \quad \varepsilon^{lo}(\omega, k) = 0 \;. \tag{9.6.13}$$

At large distances from the plasma surface, $x \gg v_{0e}/|\omega + i\nu_e|$, where v_{0e} is the velocity of the electron chaotic motion, only the first of these poles gives a contribution to the integral (9.6.12). Then

$$E(x) = \frac{E_0}{\varepsilon^{lo}(\omega, 0)} = \frac{E_0}{1 - \omega_{pe}^2 [\omega(\omega + i\nu_e)]^{-1}} \;. \tag{9.6.14}$$

At small distances from the plasma surface, $x \ll v_{0e}/|\omega + i\nu_e|$, the poles determined by the zeros of the dielectric permittivity $\varepsilon^{lo}(\omega, k)$ become significant. Assuming $|\omega + i\nu_e| \ll \omega_{pe}$ for the purely electron plasma ($\omega^2 \gg \omega_{pi}^2$), these poles are obtained from the equation

$$1 + \frac{1}{k^2 r_{De}^2} \left(1 + i\alpha \frac{\omega}{|k| v_{0e}}\right) = 0 \;, \tag{9.6.15}$$

where $\alpha = \sqrt{\pi/2}$, $r_{De} = v_{Te}/\omega_{pe}$, $v_{0e} = v_{Te}$ for the nondegenerate electron plasma and $\alpha = \pi/2$, $r_{De} = v_{Fe}/(\sqrt{3}\,\omega_{pe})$, $v_{0e} = v_{Fe}$ for the degenerate one. As a result

$$E(x) = E_0 e^{-x/r_{De}} \left(1 + i\frac{\alpha}{2} \frac{\omega x}{v_{0e}}\right) . \tag{9.6.16}$$

The notion of an impedance of the plasma layer d is frequently introduced:

$$Z = \frac{\int_0^d E(x)\,dx}{\omega E_0/4\pi\mathrm{i}} \,. \tag{9.6.17}$$

The voltage drop on the layer d appears in the numerator and the current penetrating this layer is present in the denominator. The value $\mathrm{Re}\{Z\}$ characterizes the field absorption in the layer d.

At large distances from the plasma surface, the impedance Z characterizes the volume absorption, and according to (9.6.14)

$$Z_\mathrm{v} = \frac{4\pi\mathrm{i}(\omega + \mathrm{i}\nu_\mathrm{e})}{\omega(\omega + \mathrm{i}\nu_\mathrm{e}) - \omega_\mathrm{pe}^2}\, d \,. \tag{9.6.18}$$

Using (9.6.16), we can introduce the notion of a surface impedance at small distances from the plasma surface:

$$\begin{aligned} Z_\mathrm{s} &= \frac{4\pi\mathrm{i}}{\omega}\int_0^\infty dx\, \mathrm{e}^{-x/r_\mathrm{De}}\left(1 + \mathrm{i}\frac{\alpha}{2}\frac{\omega x}{v_\mathrm{0e}}\right) \\ &= \frac{4\pi\mathrm{i}}{\omega} r_\mathrm{De}\left(1 + \mathrm{i}\frac{\alpha}{2}\frac{\omega r_\mathrm{De}}{v_\mathrm{0e}}\right) . \end{aligned} \tag{9.6.19}$$

9.6.3 In an electron plasma, find the collisional correction to the damping decrement of high-frequency surface waves.

Solution. Accounting for electron collisions, irrespective of the degree of plasma degeneracy, we must use in the dispersion equation (9.1.26)

$$\varepsilon(\omega) = 1 - \frac{\omega_\mathrm{pe}^2}{\omega(\omega + \mathrm{i}\nu_\mathrm{e})} \,, \tag{9.6.20}$$

where $\nu_\mathrm{e} = \nu_\mathrm{ei}$ and $\nu_\mathrm{e} = \nu_\mathrm{en}$ for the completely and weakly ionized plasmas, respectively. Finally, from (9.1.26) we obtain ($\omega \to \omega + \mathrm{i}\delta, \omega \gg \nu_\mathrm{e}$)

$$\omega^2 = \begin{cases} k_z^2c^2 & \text{if } k_z^2c^2 \ll \omega_\mathrm{pe}^2 \,, \\ \omega_\mathrm{pe}^2/2 & \text{if } k_z^2c^2 \gg \omega_\mathrm{pe}^2 \,, \end{cases} \tag{9.6.21}$$

$$\delta \approx -\frac{\nu_\mathrm{e}}{2}\begin{cases} k_z^2c^2/\omega_\mathrm{pe}^2 & \text{if } k_z^2c^2 \ll \omega_\mathrm{pe}^2 \,, \\ 1 & \text{if } k_z^2c^2 \gg \omega_\mathrm{pe}^2 \,. \end{cases} \tag{9.6.22}$$

The limit $k_z^2c^2 \gg \omega_{pe}^2$ corresponds to longitudinal oscillations, therefore here δ can also be obtained from (9.1.31) when account is taken of (9.6.20). Note that the same collisional correction appears in (9.1.34) and (9.1.36).

9.6.4 On the basis of the general expression (9.1.24) calculate the surface impedance for the normal incidence of an electromagnetic wave on the surface of a semi-bounded plasma when $\nu_e, \omega \ll \omega_{pe}$. Find the relationship between the impedance and the reflection and damping indices of the wave.

Solution. For normal wave incidence $k_z = 0$ and

$$Z_p = 8\,\mathrm{i}\omega \int_0^\infty \frac{dk}{k^2c^2 - \omega^2\varepsilon^{tr}(\omega, k)} \,. \tag{9.6.23}$$

Then the field penetration into the plasma is determined by

$$E_z(x) = \frac{\mathrm{i}\omega c}{\pi} B_y(0) \int_{-\infty}^{\infty} \frac{e^{\mathrm{i}kx}dx}{k^2c^2 - \omega^2\varepsilon^{tr}(\omega, k)} \,. \tag{9.6.24}$$

At large distances from the plasma surface, $x \gg v_{0e}(|\omega + \mathrm{i}\nu_e|)^{-1}$, the spatial dispersion in $\varepsilon^{tr}(\omega, k)$ can be neglected, then

$$\varepsilon^{tr}(\omega) \to \varepsilon(\omega) = 1 - \frac{\omega_{pe}^2}{\omega(\omega + \mathrm{i}\nu_e)} \,. \tag{9.6.25}$$

Thus,

$$E_z(x) = -\frac{B_y(0)}{4\sqrt{\varepsilon(\omega)}} e^{\mathrm{i}\frac{\omega}{c}\sqrt{\varepsilon(\omega)}x} \,. \tag{9.6.26}$$

Hence, for $\omega_{pe} \gg \omega \gg \nu_e$ the field damps in the plasma by the exponential law and the damping length (the penetration depth) is

$$\lambda_{sk} = \mathrm{i}\frac{c}{\omega}\sqrt{\frac{1}{\varepsilon(\omega)}} \approx \frac{c}{\omega_{pe}} \,. \tag{9.6.27}$$

If $\omega \ll \nu_e$, one must introduce the complex damping length:

$$\lambda_{sk} = \mathrm{i}\frac{c}{\omega}\sqrt{\frac{1}{\varepsilon(\omega)}} = \frac{(\mathrm{i}+1)c}{\sqrt{2\omega\omega_{pe}^2/\nu_e}} = \frac{(\mathrm{i}+1)c}{\sqrt{8\pi\sigma\omega}} \,, \tag{9.6.28}$$

where $\sigma = \omega_{pe}^2/(4\pi\nu_e)$ is the static plasma conductivity.

Note that for a cold plasma substituting (9.6.25) into (9.6.24) results in

$$Z_p = \frac{4\pi i\omega}{c^2}\lambda_{sk} = -\frac{4\pi}{c\sqrt{\varepsilon(\omega)}} . \tag{9.6.29}$$

Therefore for the normal incidence of an electromagnetic wave on the surface of a semi-bounded plasma with dielectric permittivity $\varepsilon(\omega)$, the *reflection index* is determined by the modulus of a complex value

$$r = \frac{1-\sqrt{\varepsilon(\omega)}}{1+\sqrt{\varepsilon(\omega)}} = -\frac{1+cZ_p/(4\pi)}{1-cZ_p/(4\pi)} . \tag{9.6.30}$$

At small distances from the plasma surface, $x \ll v_{0e}\,|\omega + i\nu_e|^{-1}$, when calculating the integral (9.6.24), strong spatial dispersion of $\varepsilon^{tr}(\omega, k)$ becomes significant. Here

$$\varepsilon^{tr}(\omega, k) = 1 + i\alpha\frac{\omega_{pe}^2}{\omega|k|v_{0e}} , \tag{9.6.31}$$

where $\alpha = \sqrt{\pi/2}$ and $\alpha = 3\pi/4$ for the nondegenerate and degenerate plasmas, respectively. Substituting (9.6.31) into (9.6.24) and (9.6.23) gives an exponential decay of the field deep in the plasma and

$$\begin{gathered} E_z(x) = e^{-x/\lambda_{sk}}, \qquad \lambda_{sk} = \frac{2}{3}\left(1+\frac{i}{\sqrt{3}}\right)\left(\frac{c^2 v_{0e}}{\alpha\omega\omega_{pe}^2}\right)^{1/3}, \\ Z_p = \frac{4\pi i\omega}{c^2}\lambda_{sk} , \qquad r = -\frac{1+cZ_p/(4\pi)}{1-cZ_p/(4\pi)} . \end{gathered} \tag{9.6.32}$$

9.6.5 Study potential surface waves on the boundary of two isotropic media.

Solution. According to the general derivation of the dispersion equation for surface waves (Sect. 9.1) and on equalizing the surface impedance between two isotropic plasmas, for potential surface waves we obtain

$$\int_0^\infty \frac{dk_x}{k^2\varepsilon_1^{lo}(\omega, k)} + \int_0^\infty \frac{dk_x}{k^2\varepsilon_2^{lo}(\omega, k)} = 0 , \tag{9.6.33}$$

where $\varepsilon_1^{lo}(\omega, k)$ and $\varepsilon_2^{lo}(\omega, k)$ are the longitudinal dielectric permittivities of the first and the second plasma medium, respectively.

When the spatial dispersion is neglected, i.e., $\varepsilon_{1,2}^{lo}(\omega, k) = \varepsilon_{1,2}(\omega)$, (9.6.33) yields the known equation for high-frequency surface waves on the boundary between two media:

$$\varepsilon_1(\omega) + \varepsilon_2(\omega) = 0 . \tag{9.6.34}$$

Let one plasma medium be the nondegenerate cold plasma, i.e., $\omega \gg kv_{\mathrm{Te}}, \nu_{\mathrm{e1}}$, and the other be the dense degenerate plasma where $\nu_{\mathrm{e2}}\omega \ll kv_{\mathrm{Fe2}}$, then

$$\begin{aligned} \varepsilon_1^{\mathrm{lo}}(\omega, k) &= 1 - \frac{\omega_{\mathrm{pe1}}^2}{\omega^2}\left(1 - \mathrm{i}\frac{\nu_{\mathrm{e1}}}{\omega}\right), \\ \varepsilon_2^{\mathrm{lo}}(\omega, k) &= 1 + 3\frac{\omega_{\mathrm{pe2}}^2}{kv_{\mathrm{Fe2}}}\left(1 + \mathrm{i}\frac{\pi}{2}\frac{\omega}{kv_{\mathrm{Fe2}}}\right). \end{aligned} \tag{9.6.35}$$

By substituting (9.6.35) into (9.6.33) we obtain

$$\begin{aligned} &1 - \frac{\omega_{\mathrm{pe1}}^2}{\omega^2} + \sqrt{1 + 3\frac{\omega_{\mathrm{pe2}}^2}{k_z^2 v_{\mathrm{Fe2}}^2}} + \mathrm{i}\frac{\omega_{\mathrm{pe1}}^2 \nu_{\mathrm{e1}}}{\omega^2} \\ &\quad + 3\frac{\omega_{\mathrm{pe2}}^2\omega}{v_{\mathrm{Fe2}}^3}\left(1 + 3\frac{\omega_{\mathrm{pe2}}^2}{k_z^2 v_{\mathrm{Fe2}}^2}\right)\int_0^\infty \frac{|k_z|dk}{k\left(k^2 + 3\dfrac{\omega_{\mathrm{pe2}}^2}{v_{\mathrm{Fe2}}^2}\right)^2} = 0\,. \end{aligned} \tag{9.6.36}$$

Finally, accounting for the small imaginary terms, we arrive at ($\omega \to \omega + \mathrm{i}\delta$):

$$\begin{aligned} \omega^2 &= \frac{\omega_{\mathrm{pe1}}^2}{\sqrt{1 + 3\,\omega_{\mathrm{pe2}}^2/(k_z^2 v_{\mathrm{Fe2}}^2)} + 1}, \\ \delta &= -\frac{\nu_{\mathrm{e1}}}{2} - \frac{3}{2}\frac{\omega_{\mathrm{pe2}}^2}{\omega_{\mathrm{pe1}}^2}\frac{\omega^4}{v_{\mathrm{Fe2}}^3}\left(1 + 3\frac{\omega_{\mathrm{pe2}}^2}{k_z^2 v_{\mathrm{Fe2}}^2}\right)\int_0^\infty \frac{|k_z|dk_x}{k\,(k^2 + 3\,\omega_{\mathrm{pe2}}^2/v_{\mathrm{Fe2}}^2)^2}\,. \end{aligned} \tag{9.6.37}$$

In (9.6.36, 37), the integration may be carried out in a general form. The result is rather unwieldy, therefore we analyze only the long- and short-wavelength limits:

$$\omega = \begin{cases} \omega_{\mathrm{pe1}}\sqrt{\dfrac{|k_z|v_{\mathrm{Fe2}}}{\sqrt{3}\,\omega_{\mathrm{pe2}}}} & \text{for} \quad |k_z|v_{\mathrm{Fe}} \ll \omega_{\mathrm{pe2}}\,, \\ \dfrac{\omega_{\mathrm{pe1}}}{\sqrt{2}} & \text{for} \quad |k_z|v_{\mathrm{Fe2}} \gg \omega_{\mathrm{pe2}}\,, \end{cases} \tag{9.6.38}$$

$$\delta = -\frac{\nu_{\mathrm{e1}}}{2} - \begin{cases} \dfrac{1}{6}\dfrac{\omega_{\mathrm{pe1}}^2|k_z|v_{\mathrm{Fe2}}}{\omega_{\mathrm{pe2}}^2}\ln\dfrac{12\,\omega_{\mathrm{pe2}}^2}{k_z^2 v_{\mathrm{Fe2}}^2} & \text{for} \quad |k_z|v_{\mathrm{Fe2}} \ll \omega_{\mathrm{pe2}}\,, \\ \omega_{\mathrm{pe2}}\dfrac{\omega_{\mathrm{pe1}}^3}{4\,|k_z^3|v_{\mathrm{Fe2}}^3} & \text{for} \quad |k_z|v_{\mathrm{Fe2}} \gg \omega_{\mathrm{pe2}}\,. \end{cases} \tag{9.6.39}$$

In the long-wavelength limit, the frequency spectrum of surface waves is $\omega \sim \sqrt{k_z}$. Consequently, phase and group velocities of these waves sharply increase when the wavelength grows. Therefore, the damping decrement δ, specified by the Cherenkov wave absorption by electrons of the degenerate plasma, sharply decreases.

9.6.6 In the semi-bounded isotropic plasma consider the excitation of high-frequency waves by a monoenergetic nonrelativistic electron beam travelling above the plasma surface.

Solution. To solve the problem, it is sufficient to confine ourselves to longitudinal (potential) waves in the model of independent particles since the waves excited by the nonrelativistic beam possess nonrelativistic phase velocities. From the linearized system of equations of this model, i.e.,

$$\begin{aligned} &\frac{\partial N_1}{\partial t} + \operatorname{div}(N_1 V_0 + N_0 V_1) = 0\,, \\ &\frac{\partial V_1}{\partial t} + (V_1 \cdot \nabla)\, V_0 + (V_0 \cdot \nabla)\, V_1 = -\frac{e}{m} \nabla \Phi - \nu_e V_1\,, \\ &\Delta \Phi = -4\pi e N_1\,, \end{aligned} \tag{9.6.40}$$

where N_0, V_0 are the unperturbed electron densities and velocities and N_1, V_1 their small perturbations, we obtain

$$\left(\frac{\partial}{\partial x}\varepsilon\frac{\partial}{\partial x} - k_z^2\varepsilon\right)\Phi = 0\,. \tag{9.6.41}$$

Here

$$\varepsilon = 1 - \frac{\omega_{pe}^2}{(\omega - k_z V_0)(\omega - k_z V_0 + i\nu_e)}\,. \tag{9.6.42}$$

Equation (9.6.41) is valid both in the plasma region ($x > 0$) and in the beam region ($x \le 0$). In the plasma ω_{pe} is the Langmuir frequency of the plasma electrons, ν_e their collision frequency and $V_0 = 0$; in the beam $\omega_{pe} = \omega_b$ is the Langmuir frequency of the beam electrons moving with the velocity V_0 and considered collisionless.

The boundary conditions for (9.6.41) are

$$\{\Phi\}_{x=0} = \left\{\varepsilon\frac{\partial \Phi}{\partial x}\right\}_{x=0} = 0\,. \tag{9.6.43}$$

Substituting the solution of (9.6.41)

$$\Phi = \begin{cases} C_1 e^{-|k_z|x} & \text{if } x > 0\,, \\ C_2 e^{|k_z|x} & \text{if } x \le 0 \end{cases} \tag{9.6.44}$$

into the boundary conditions (9.6.43) results in the dispersion equation

$$2 - \frac{\omega_b^2}{(\omega - k_z V_0)^2} - \frac{\omega_{pe}^2}{\omega(\omega + i\nu_e)} = 0 \,. \tag{9.6.45}$$

Hence we obtain the frequency spectrum and the increment ($\omega \to \omega + i\delta = k_z V_0 + i\delta$) of surface waves excited in the plasma for $\omega_{pe} \gg \nu_e$:

$$\omega = \frac{\omega_{pe}}{\sqrt{2}} k_z V_0$$

$$\frac{\delta}{\omega} = \begin{cases} \dfrac{i + \sqrt{3}}{2} \left(\dfrac{N_b}{2N_p}\right)^{1/3} & \text{if} \quad |\delta| > \nu_e \,, \\[2ex] (1 + i) \sqrt{\dfrac{\omega}{\nu_e}} \left(\dfrac{N_b}{2N_p}\right)^{1/2} & \text{if} \quad |\delta| < \nu_e \,. \end{cases} \tag{9.6.46}$$

The upper expression for δ corresponds to a nondissipative instability and the lower to a dissipative one. The latter is possible in the plasma with frequent particle collisions, particularly in the solid-state plasma.

9.6.7 Show that the surface of the plasma, confined by a magnetic field with a positive curvature of the field lines (i.e., with a positive normal oriented away from the plasma surface), is unstable with respect to low-frequency flute perturbations.

Solution. The curvature of the magnetic field lines can be accounted for by introducing a centrifugal acceleration of the particles of type α. The acceleration is oriented along the positive normal to the plasma surface and is equal to $v_{T\alpha}^2/R$, R being the radius of the curvature of the magnetic field lines. This results in a particle drift along the plasma surface (0y-axis) with the velocity $u_\alpha = -v_{T\alpha}^2/(\Omega_\alpha R)$. Finally, in (9.2.15) $\omega - k_y u_\alpha$ must be substituted for the value ω, and for flute oscillation modes in the low-frequency $|\omega - k_y u_\alpha| \ll \Omega_\alpha$ we obtain

$$2 + \frac{\omega_{pi}^2}{\omega_i^2} + |k_y| \frac{\omega_{pi}^2}{\Omega_i^2} \frac{v_s^2 + v_{Ti}^2}{R\omega^2} = 0 \,, \tag{9.6.47}$$

where $v_s^2 = T_e/M$ is the ion-acoustic velocity. Hence

$$\omega^2 = -\frac{|k_y| \, g_{eff}}{1 + 2v_A^2/c^2} \,. \tag{9.6.48}$$

Here $g_{eff} = (v_s^2 + v_{Ti}^2)/R$ is an effective gravitational field accounting for the curvature of field lines of the magnetic field, confining the plasma.

Since $\omega^2 < 0$, the surface of the plasma, confined by the magnetic field with the positive curvature of field lines, is always unstable. This instability is analogous to the convective (flute) instability of volume waves in the inhomogeneous magnetized plasma (Sect. 8.7).

9.6.8 Obtain the operator of the dielectric tensor of a cylindrically inhomogeneous multicomponent plasma by proceeding from the model of independent particles.

Solution. Let us linearize the system of equations

$$\begin{gathered} \frac{\partial N_\alpha}{\partial t} + \operatorname{div} N_\alpha \boldsymbol{V}_\alpha = 0 \,, \\ \left(\frac{\partial}{\partial t} + \boldsymbol{V}_\alpha \cdot \nabla\right) \frac{\boldsymbol{V}_\alpha}{\sqrt{1 - V_\alpha^2/c^2}} = \frac{e_\alpha}{m_\alpha} \left\{\boldsymbol{E} + \frac{1}{c} [\boldsymbol{V}_\alpha, \boldsymbol{B}]\right\} \end{gathered} \tag{9.6.49}$$

with respect to the equilibrium state with the ordered particle velocity $\boldsymbol{u}_\alpha \| 0z$, $\boldsymbol{B}_0 \| 0z$ and $N_{0\alpha}(r)$ and take into account the Maxwell equations

$$\begin{aligned} \operatorname{curl} \boldsymbol{B} &= \frac{1}{c}\frac{\partial \boldsymbol{E}}{\partial t} + 4\pi \sum_\alpha (e_\alpha N_\alpha \boldsymbol{V}_\alpha - e_\alpha N_{0\alpha} \boldsymbol{u}_\alpha) \,, \quad \operatorname{div} \boldsymbol{B} = 0 \,, \\ \operatorname{curl} \boldsymbol{E} &= -\frac{1}{c}\frac{\partial \boldsymbol{B}}{\partial t} \,, \qquad \operatorname{div} \boldsymbol{E} = 4\pi \sum_\alpha (e_\alpha N_\alpha - e_\alpha N_{0\alpha}) \,. \end{aligned} \tag{9.6.50}$$

For perturbations, dependent on time and coordinates in the form of

$$A(\boldsymbol{r}) = A(r) \exp(-\,i\omega t + il_\phi + ik_z z) \,, \tag{9.6.51}$$

we obtain

$$\operatorname{curl}\, \operatorname{curl} \boldsymbol{E} - \frac{\omega^2}{c^2} \boldsymbol{D} = 0, \quad D_i = \varepsilon_{ij} E_j \,,$$

where ε_{ij} is an operator of the dielectric tensor with the components:

$$\varepsilon_{rr} = \varepsilon_{\phi\phi} = 1 - \sum_\alpha \frac{\omega_{p\alpha}^2 \omega_\alpha'^2}{\gamma_\alpha \omega^2 (\omega_\alpha'^2 - \Omega_\alpha^2)} \,,$$

$$\varepsilon_{r\phi} = -\,\varepsilon_{\phi r} = -\,i \sum_\alpha \frac{\omega_{p\alpha}^2 \omega_\alpha' \Omega_\alpha}{\gamma_\alpha \omega^2 (\omega_\alpha'^2 - \Omega_\alpha^2)} \,,$$

$$\varepsilon_{rz} = \sum_\alpha \alpha_{rz} \left(-\frac{l}{r}\frac{\Omega_\alpha}{\omega_\alpha'} + \frac{\partial}{\partial r}\right) ,$$

$$\varepsilon_{zr} = \sum_\alpha \left(\frac{l}{r}\frac{\Omega_\alpha}{\omega'_\alpha} + \frac{1}{r}\frac{\partial}{\partial r}r\right)\alpha_{rz}\,,$$

$$\varepsilon_{\phi z} = \mathrm{i}\sum_\alpha \alpha_{rz}\left(\frac{l}{r} - \frac{\Omega_\alpha}{\omega'_\alpha}\frac{\partial}{\partial r}\right),\tag{9.6.52}$$

$$\varepsilon_{z\phi} = \mathrm{i}\sum_\alpha \left(\frac{l}{r} + \frac{\Omega_\alpha}{\omega'_\alpha}\frac{1}{r}\frac{\partial}{\partial r}r\right)\alpha_{rz}\,,$$

$$\varepsilon_{zz} = 1 - \sum_\alpha\left[\frac{\omega_{\mathrm{p}\alpha}^2}{\gamma_\alpha\omega_\alpha'^2} + \frac{1}{r}\frac{\partial}{\partial r}r\alpha_{zz}\frac{\partial}{\partial r} - \frac{l^2}{r^2}\alpha_{zz} + \frac{l}{r}\frac{\Omega_\alpha}{\omega'_\alpha}\left(\frac{\partial}{\partial r}\alpha_{zz} - \alpha_{zz}\frac{\partial}{\partial r}\right)\right].$$

Here

$$\alpha_{rz} = \frac{\mathrm{i}u_\alpha\omega_{\mathrm{p}\alpha}^2\omega'_\alpha}{\omega^2(\omega_\alpha'^2-\Omega_\alpha^2)^2}\,,\qquad \alpha_{zz} = \frac{u_\alpha^2\omega_{\mathrm{p}\alpha}^2\gamma_\alpha}{\omega^2(\omega_\alpha'^2-\Omega_\alpha^2)}\,,$$

$$\omega_{\mathrm{p}\alpha}^2 = \frac{4\pi e^2 N_{0\alpha}}{m_\alpha}\,,\qquad \Omega_\alpha = \frac{e_\alpha B_0}{m_\alpha c}\,,\tag{9.6.53}$$

$$\omega'_\alpha = (\omega - \boldsymbol{k}\cdot\boldsymbol{u}_\alpha)\gamma_\alpha\,,\qquad \gamma_\alpha = \left(1-\frac{u_\alpha^2}{c^2}\right)^{-1/2}.$$

The operator $\partial/\partial r$ acts on all the values to its right.

9.6.9 Find the *limiting current* of the monoenergetic relativistic electron beam travelling through the equipotential vacuum drift space in the strong external longitudinal magnetic field. Analyze cases of plane and cylindrical geometry.

Solution. When travelling in the equipotential drift space, beam electrons produce a spatially distributed charge which causes a spatial modification of the potential and so therefore brake the beam. As a result the distribution of the potential with a maximum in the centre will be established in the plane drift space (Fig. 9.9 a). Poisson's equation for the potential distribution $\Phi(x)$ can be written as

$$\frac{d^2\Phi}{dx^2} = -\frac{4\pi j}{v} = -\frac{4\pi j}{c}\left[1-\left(\gamma-\frac{e\Phi}{mc^2}\right)^{-2}\right].\tag{9.6.54}$$

Here j is the current density of the electron beam, $\gamma = (1-v^2/c^2)^{-1/2}$ the relativistic factor, and v the velocity of the injected electrons. When deriving (9.6.54) the integral of motion

$$mc^2(1-v^2/c^2)^{-1/2} + e\Phi = mc^2\gamma\tag{9.6.55}$$

is taken into account.

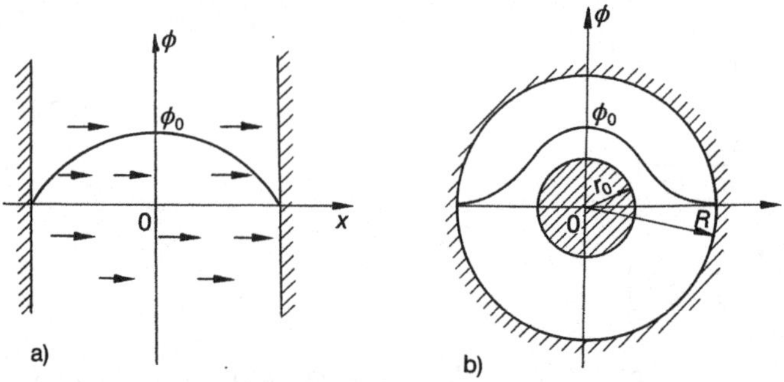

Fig. 9.9. Distribution of the beam potential: (**a**) in plane drift space; (**b**) in cylindrical drift space

Equation (9.6.54) must be supplemented with the boundary conditions (Fig. 9.9 a)

$$\Phi|_{x=\pm d} = 0\,, \quad \Phi|_{x=0} = \Phi_0\,. \tag{9.6.56}$$

The given problem is overdetermined: the potential on the axis is determined uniquely for the assigned current density j. Therefore the solution permits us to relate j to Φ_0:

$$j = F(\Phi_0)\,. \tag{9.6.57}$$

In order to determine the limiting current density one must maximize this relation over Φ_0 and thus finds $j_0 = j_{\max}$.

The analytical solution of (9.6.54) is found in the limiting cases of non-relativistic ($e\Phi \ll mc^2, \gamma \approx 1$) and ultrarelativistic ($e\Phi \gg mc^2$, $\gamma \gg 1$) beam, where for the limiting current density j_0, we obtain

$$j_0 = \frac{mc^3}{e}\frac{1}{2\pi d^2}\begin{cases} \dfrac{2\sqrt{2}}{9}(\gamma-1)^{3/2} & \text{for} \quad \gamma = 1 + \dfrac{u^2}{2c^2}\,, \\ \gamma & \text{for} \quad \gamma \gg 1\,. \end{cases} \tag{9.6.58}$$

Using interpolation, we can write these two equations by a single formula

$$j_0 = \frac{mc^3}{e}\frac{(\gamma^{2/3}-1)^{3/2}}{2\pi d^2}\,. \tag{9.6.59}$$

For $\gamma \gg 1$, this expression coincides with (9.6.58), and for $\gamma = 1 + u^2/(2c^2)$, it differs from the latter by the factor $\sqrt{3}$.

In case of the beam travelling along the axis of the cylindrical drift space (Fig. 9.9 b), the problem is formulated and solved analogously:

$$\frac{1}{r}\frac{d}{dr}r\frac{d\Phi}{dr} = -\frac{4\pi j(r)}{v} = -\frac{4\pi j(r)}{c}\left[1 - \left(\gamma - \frac{e\Phi}{mc^2}\right)^{-2}\right], \tag{9.6.60}$$

$$\Phi|_{r=R} = 0, \quad \Phi|_{r=0} = \Phi_0 . \tag{9.6.61}$$

Then the current distribution over a radius is

$$j(r) = \begin{cases} 0 & \text{for} \quad r > r_0 , \\ \text{const} & \text{for} \quad r \leq r_0 . \end{cases} \tag{9.6.62}$$

When solving this problem, we must take account of the continuity of the potential and its derivative at the beam boundary for $r = r_0$. Hence

$$I_0 = \pi r_0^2 j_0 = \frac{mc^3}{e}\frac{(\gamma^{2/3} - 1)^{3/2}}{1 + 2\ln(R/r_0)} . \tag{9.6.63}$$

Finally, note that such an estimate for a thin hollow beam with an average radius r_0 and thickness $a \ll r_0$ leads to the following expression for the limiting current:

$$I_0 = \frac{mc^3}{e}\frac{(\gamma^{2/3} - 1)^{3/2}}{\dfrac{a}{r_0} + 2\ln\dfrac{R}{r_0}} . \tag{9.6.64}$$

Here $mc^3/e \approx 17$ kA.

9.6.10 Show that the plane layer of the nonrelativistic electron beam confined by conductive planes $-a < x < a$ with a linear transverse velocity profile $u(x) = u_0 \cdot x/a_0$, $a_0 > a$ may be unstable in the external longitudinal magnetic field. Calculate the marginal condition for the instability (a *slipping instability*).

Solution. According to the method described in Sect. 9.2 we obtain instead of (9.2.8) when the distribution function $F_{0\alpha}(\boldsymbol{v} - \boldsymbol{u})$ is a Maxwellian with the inhomogeneous ordered velocity $u(x)$

$$\Delta\Phi = \frac{\omega_{\rm pe}^2}{v_{\rm Te}^2}\int dk_x \Phi(k_x)\,{\rm e}^{ik_x x}\left\{1 - \sum_n\left[\omega - k_z u - \frac{k_y v_{\rm Te}^2}{\Omega_{\rm e}}\frac{\partial u}{\partial x}\right.\right.$$
$$\times \frac{\partial}{\partial u}\left(1 - \frac{n}{z}\frac{\omega - k_z u}{\Omega_{\rm e}}\right) - \frac{ik_x v_{\rm Te}^2}{\Omega_{\rm e}^2}\frac{\partial u}{\partial x}\frac{\partial}{\partial u}(\omega - k_z u)\left.\frac{A_n'(z)}{A_n(z)}\right]$$
$$\left.\times \frac{A_n(z)}{\omega - k_z u - n\Omega_{\rm e}} I_+\left(\frac{\omega - k_z u - n\Omega_{\rm e}}{k_z v_{\rm Te}}\right)\right\}, \tag{9.6.65}$$

where $z=(k_x^2+k_y^2)\,v_{\mathrm{Te}}^2/\Omega_{\mathrm{e}}^2$. The operator $\partial/\partial u$ acts on all the quantities to its right.

Here high-frequency unstable oscillations with higher phase velocities than the electron thermal velocity are of interest. Thus, in (9.6.65) we deal with the limit $T\to 0$. As a result we obtain

$$\Delta\Phi+\frac{\omega_{\mathrm{pe}}^2}{\Omega_{\mathrm{e}}^2-\omega'^2}\left(\frac{\partial^2}{\partial x^2}-k_y^2\right)\Phi+\frac{\omega_{\mathrm{pe}}^2k_z^2}{\omega'^2}\Phi$$
$$+\frac{\partial\Phi}{\partial x}\frac{\partial}{\partial x}\left(\frac{\omega_{\mathrm{pe}}^2}{\Omega_{\mathrm{e}}^2-\omega'^2}\right)-k_y\Phi\frac{\partial}{\partial x}\left(\frac{\omega_{\mathrm{pe}}^2\Omega_{\mathrm{e}}}{\omega'(\Omega_{\mathrm{e}}^2-\omega'^2)}\right)=0\,, \qquad (9.6.66)$$

where $\omega'=\omega-k_zu(x)$. When deriving this equation from (9.6.65), we essentially assume

$$\beta=\frac{\omega-k_zu}{k_zv_{\mathrm{Te}}}\gg 1\,,$$

$\arg\{\beta\}$ lying outside the sector

$$-\frac{3\pi}{4}\le\arg\{\beta\}\le-\frac{\pi}{4}\,.$$

The model of independent particles leading to (9.6.66) is valid for this problem only under given conditions.

Let us consider the limit of a strong field $\Omega_{\mathrm{e}}\gg\omega_{\mathrm{pe}}$. Assuming $\omega'\ll\Omega_{\mathrm{e}}$, we obtain from (9.6.66)

$$\frac{\partial^2\Phi}{\partial x^2}+U(x,\omega)\,\Phi=0\,,$$
$$U(x,\omega)=-k_y^2-k_z^2+\left(1-\frac{k_yu_0}{k_z\Omega_{\mathrm{e}}a_0}\right)\frac{\omega_{\mathrm{pe}}^2a_0^2}{u_0^2\left(x-\dfrac{\omega a_0}{k_zu_0}\right)^2}\,. \qquad (9.6.67)$$

It should be supplemented with boundary conditions on the conductive planes

$$\Phi|_{x=\pm a}=0\,. \qquad (9.6.68)$$

The general solution of (9.6.67) is of the form

$$\Phi(x)=\sqrt{s}\,[c_1\mathrm{J}_\nu(iks)+c_2\mathrm{J}_{-\nu}(iks)]\,, \qquad (9.6.69)$$

where

$$k = \sqrt{k_y^2 + k_z^2}\,, \qquad s = x - \frac{\omega a_0}{k_z u_0}\,,$$
$$\nu^2 = \frac{1}{4} - \frac{\omega_{\rm pe}^2 a_0^2}{u_0^2}\left(1 - \frac{k_y u_0}{k_z \Omega_{\rm e} a_0}\right). \tag{9.6.70}$$

The restriction on $\arg\{\beta\}$ results in the fact that the obtained solution has meaning only outside the sector

$$\pi/4 \le \arg\{s\} \le 3\pi/4\,.$$

Substituting (9.6.69) into (9.6.68) gives

$$\mathrm{J}_\nu(\mathrm{i}ks_-)J_{-\nu}(\mathrm{i}ks_+) - \mathrm{J}_{-\nu}(\mathrm{i}ks_-)\mathrm{J}_\nu(\mathrm{i}ks_+) = 0\,,$$
$$s_\pm = \pm\, a - \frac{\omega a_0}{k_z u_0}\,. \tag{9.6.71}$$

Unstable oscillations should first emerge in the long-wavelength limit $ks_\pm \to 0$. Then (9.6.71) is reduced to

$$s_+^{2\nu} - s_-^{2\nu} = 0\,. \tag{9.6.72}$$

Hence $\mathrm{Re}\{\omega\}$ and

$$\arg\{s_+\} - \arg\{s_-\} = \frac{\pi n}{\nu}\,, \qquad n = \pm 1, \pm 2, \ldots, \tag{9.6.73}$$

where

$$\arg\{s_\pm\} = \mp \arctan\frac{a_0 \mathrm{Im}\{\omega\}}{a k_z u_0} - \pi x \begin{cases} 0\,, \\ 1\,. \end{cases} \tag{9.6.74}$$

The angles $\arg\{s_\pm\}$ lying in the third and fourth quarters of the complex variable s, i.e., $\arg\{s_+\} - \arg\{s_-\} < \pi$, correspond to unstable oscillations. According to (9.6.73) this is possible only under the condition $\nu^2 > n^2$, i.e., the marginal condition of the slipping instability in the plane-parallel electron beam in the longitudinal magnetic field is of the form

$$\nu^2 = \frac{1}{4} - \frac{\omega_{\rm pe}^2 a_0^2}{u_0^2}\left(1 - \frac{k_y u_0}{k_z \Omega_{\rm e} a_0}\right) > n^2\,. \tag{9.6.75}$$

For $n \gg 1$ this condition corresponds to the excitation of high oscillation modes described in the approximation of geometrical optics and takes the form

$$\frac{k_y u_0}{k_z \Omega_{\rm e} a} = \frac{k_y}{k_z}\frac{1}{\Omega_{\rm e}}\frac{\partial u}{\partial x} > 1\,. \tag{9.6.76}$$

Then the increment of the slipping instability is equal to

$$\delta = \operatorname{Im}\{\omega\} = k_z u_0 \frac{a}{a_0}\left(\operatorname{ctan}\frac{\pi n}{\nu} + \frac{1}{\left|\sin\frac{\pi n}{\nu}\right|}\right) \approx k_z a \frac{\partial u}{\partial x} . \tag{9.6.77}$$

Finally note that in the absence of the external magnetic field under the condition $\omega' \ll \omega_{pe}$ (9.6.66) is reduced to

$$\left(\frac{\partial^2}{\partial x^2} - k_y^2\right)\Phi + \frac{2}{\frac{\omega a_0}{k_z u_0} - x}\frac{\partial \Phi}{\partial x} = 0 . \tag{9.6.78}$$

Its general solution

$$\Phi(x) = s^{3/2}\left[c_1 J_{3/2}(iks) + c_2 J_{-3/2}(iks)\right] \tag{9.6.79}$$

always satisfies the condition (9.6.75) for $n = 1$.

Thus, in the absence of the external magnetic field, the electron beam with an inhomogeneous velocity profile is always unstable and there the main oscillation mode is excited, the approximation of geometrical optics being invalid for this description.

9.6.11 Show that in a drift space a magnetized hollow electron beam rotating as a whole is unstable with respect to electrostatic oscillations *(diocotron instability)*.

Solution. On the basis of the model of independent particles let us assume an equilibrium electron beam to possess an inhomogeneous density $n_b(r)$, homogeneous longitudinal velocitiy $u_\parallel$ and azimuthal rotation velocity u_ϕ. For flute electrostatic perturbations along the magnetic field and dependent on time and coordinates in the form

$$f(r) = \exp(-i\omega t + il_\phi) ,$$

we obtain

$$\delta v_r = -\frac{ilc}{rB_0}\Phi , \qquad \delta v_\phi = \frac{c}{B_0}\frac{\partial \Phi}{\partial r} ,$$

$$\left(\omega - \frac{l}{r}u_\phi\right)\delta n = -\frac{i}{r}\frac{\partial}{\partial r}(r n_b \delta v_r) + \frac{n_b l}{r}\delta v_\phi . \tag{9.6.80}$$

Taking account of these relations Poisson's equation

$$\Delta\Phi = -4\pi e\delta n \tag{9.6.81}$$

can be written as

$$\frac{1}{r}\frac{\partial}{\partial r}r\frac{\partial\Phi}{\partial r} - \frac{l^2}{r^2}\Phi - \frac{l}{r}\frac{\Phi}{\Omega_e(\omega - lu_\phi/r)}\frac{\partial\omega_b^2}{\partial r} = 0\,, \tag{9.6.82}$$

its boundary condition being

$$\Phi|_{r=R} = 0\,. \tag{9.6.83}$$

The stated boundary problem admits the existence of eigenvalues with $\mathrm{Im}\{\omega\}$ corresponding to unstable oscillations only when the value $\partial\omega_b^2/\partial r$ reverses its sign in the range $0 < r < R$. This occurs in hollow electron beams separated from the walls of the drift tube

$$n_b(R) = \begin{cases} \text{const} & \text{for} \quad r_0 \leq r \leq r_1\,, \\ 0 & \text{for} \quad r < r_0 \text{ and } r_1 < r \leq R. \end{cases} \tag{9.6.84}$$

Then (9.6.82) can be solved in the separate beam homogeneity regions and the derived solutions can be connected with account taken of the conditions at the beam's boundaries

$$\begin{aligned} &\{\Phi\}_{r=r_0} = \{\Phi\}_{r=r_1} = 0\,, \\ &\left\{r\frac{\partial\Phi}{\partial r} - \frac{l\Phi\omega_b^2}{\Omega_e(\omega - lu_\phi/r)}\right\}_{r=r_0} = 0\,, \\ &\left\{r\frac{\partial\Phi}{\partial r} + \frac{l\Phi\omega_b^2}{\Omega_e(\omega - lu_\phi/r)}\right\}_{r=r_1} = 0\,. \end{aligned} \tag{9.6.85}$$

By substituting the general solution

$$\Phi(r) = \begin{cases} c_1 r^l & \text{for} \quad r < r_0\,, \\ c_2 r^l + \dfrac{c_3}{r^l} & \text{for} \quad r_0 \leq r \leq r_1\,, \\ c_4 r^l + \dfrac{c_5}{r^l} & \text{for} \quad r_1 < r \leq R \end{cases} \tag{9.6.86}$$

into (9.6.85) and taking account of (9.6.83), we obtain after simple calculations

$$\left(\frac{\omega}{\omega_{\mathrm{d}}}\right)^2 - b\left(\frac{\omega}{\omega_{\mathrm{d}}}\right) + C = 0\,, \quad \text{where} \tag{9.6.87}$$

$$\omega_{\mathrm{d}} = \omega_{\mathrm{b}}^2/(2\,\Omega_{\mathrm{e}})\,, \quad b \equiv l\left(1-\frac{r_0^2}{r_1^2}\right) + \left(\frac{r_1^{2l}}{R^{2l}} - \frac{r_0^{2l}}{R^{2l}}\right),$$
$$C \equiv l\left(1-\frac{r_0^2}{r_1^2}\right)\left(1-\frac{r_0^{2l}}{R^{2l}}\right) - \left(1-\frac{r_0^{2l}}{r_1^{2l}}\right)\left(1-\frac{r_1^{2l}}{R^{2l}}\right). \tag{9.6.88}$$

Equation (9.6.87) has solutions with $\mathrm{Im}\,\{\omega\} > 0$ when $b^2 - 4c = 0$. It can be shown that for $r_0 = 0$ (a solid beam) or $r_1 = R$ (a hollow beam pressed to a wall) this inequality is violated, i.e., such beams are stable with respect to the discussed oscillations. The mode with $l = 1$ always appears stable. Thus, oscillations with $l = 2$ are unstable only in the hollow electron beam separated from the conductive wall of the drift tube. The maximum increment of oscillations achieves the value $\mathrm{Im}\,\{\omega\} \simeq \omega_{\mathrm{pe}}^2/(2\,\Omega_{\mathrm{e}})$.

9.6.12 Analyze the possibility of the excitation of a *B*-counterwave by a relativistic beam of rotating electrons in a vacuum resonator.

Solution. The beam-excited low-frequency *B*-wave appears counter when $\mu_{ls}' c\gamma/(\Omega_{\mathrm{e}} R) < 1$ since k_{z02} and $v_{\mathrm{g}02} < 0$, see (9.5.27, 28). Such a wave may be excited by a beam even in the complete absence of reflection from the right end plane of the resonator or at full consistence of the emitter with the resonator. Excitation is possible at sufficiently large frequency differences when

$$|\Delta_\omega| = \left|\omega - k_{z02} u_\parallel - \frac{\Omega_{\mathrm{e}}}{\gamma}\right| \gg \left(\frac{u_\perp^2}{8c^2}\frac{\mu_{ls}^2\omega_{\mathrm{b}}^2}{\gamma|k_{z02}|R^2}\,u_\parallel\right)^{1/3}. \tag{9.6.89}$$

The solutions of (9.5.26) for a *B*-wave yield

$$k_{z1,2,3} = k_{z02}(\omega) + \delta k_{z1,2,3}\,, \quad k_{z4} = -k_{z02}(\omega)\,, \quad \text{where} \tag{9.6.90}$$

$$\delta k_{z1} = \frac{1}{\Delta_\omega^2}\frac{u_\perp^2}{8c^2}\frac{\mu_{ls}^2\omega_{\mathrm{b}}^2}{\gamma|k_{z02}|R^2}\,,$$
$$\delta k_{z2,3} = \frac{\Delta_\omega}{u_\parallel}\left[1 \mp \mathrm{i}\left(\frac{u_\perp^2}{8c^2}\frac{\mu_{ls}^2\omega_{\mathrm{b}}^2 u_\parallel}{\gamma|k_{z02}|R^2\Delta_\omega^2}\right)^{1/2}\right]. \tag{9.6.91}$$

Now we can substitute the solutions of the field equations as a sum of normal waves

$$B_z(r) = \sum_{n=1}^{4} B_{z0n}\, e^{ik_{zn}(\omega)\, z} \tag{9.6.92}$$

into the boundary conditions at the resonator end planes $z = 0, L$. A total wave reflection is assumed to occur at the end plane $z = 0$ and total radiation at the end plane $z = L$. As a result we have

$$\sum_{n=1}^{3} \alpha_i\, e^{ik_{zn}(\omega)\, L} = 0\,, \quad \text{where} \tag{9.6.93}$$

$$\alpha_{1,2} = \pm \frac{\delta k_{z2,1}\delta k_{z1}}{(\delta k_{z2} - \delta k_{z1})(\delta k_{z3} - \delta k_{z1,2})}\,, \qquad \alpha_3 = \frac{\delta k_{z1}\delta k_{z2}}{(\delta k_{z3} - \delta k_{z1})(\delta k_{z3} - \delta k_{z2})}\,. \tag{9.6.94}$$

Hence we obtain the oscillation spectrum ($\omega \to \omega + i\delta$):

$$\begin{aligned} \omega &= \omega_{02} - |v_{g02}| \frac{\pi}{2L} (4n - 1)\left(1 + \frac{u_\|}{v_{g02}}\right)^{-1}, \\ \delta &= \frac{u_\|}{L} \left(\frac{L^3}{u_\|^2} \frac{u_\perp^2}{8c^2} \frac{\mu_{ls}^2 \omega_b^2}{\gamma |k_{z02}| R^2} - 1 \right) \left(1 + \frac{u_\|}{|v_{g02}|}\right)^{-1}. \end{aligned} \tag{9.6.95}$$

Finally, from the condition $\delta > 0$ we obtain the starting current for the resonator excitation on the passing B-wave:

$$I_{st} = 16 \frac{mc^3}{e} \frac{R^3}{L^3} \frac{\gamma |k_{z02}| R u_\|^3}{\mu_{ls}^2 u_\perp^2 c}\,. \tag{9.6.96}$$

9.6.13 Solve the Fresnel problem, i.e., the problem of reflection and refraction, for the electromagnetic wave on the plane boundary of the homogeneous collisionless isotropic cold plasma with vacuum.

Solution. Since the tangential component of the electric field is equal to the normal component of the electric induction on the plasma boundary with vacuum, we can find the reflection index: a) for the S-polarized wave, i.e., the H-wave with the electric vector parallel to the plasma surface,

$$r_s = \frac{\cos\theta - \sqrt{\varepsilon(\omega) - \sin^2\theta}}{\cos\theta + \sqrt{\varepsilon(\omega) - \sin^2\theta}}\,, \tag{9.6.97}$$

and b) for the P-polarized wave, i.e., the E-wave with the electric vector lying in the incidence plane,

$$r_p = \frac{\varepsilon(\omega)\cos\theta - \sqrt{\varepsilon(\omega) - \sin^2\theta}}{\varepsilon(\omega)\cos\theta + \sqrt{\varepsilon(\omega) - \sin^2\theta}} . \tag{9.6.98}$$

Here θ is the angle of incidence, $\varepsilon(\omega) = 1 - \omega_p^2/\omega^2$, $|r_{s,p}|^2$ is the ratio of the electromagnetic energy flux reflected from the surface to the incident flux, and $1 - |r_{s,p}|^2$ is the electromagnetic energy flux passing into the plasma.

For $\varepsilon(\omega) < \sin^2\theta \leqslant 1$ the waves of both polarizations are totally reflected from the plasma surface. This occurs at any frequency ω of the field when $\theta \to \pi/2$. Note also that the ratio of normal components of the electric field in the plasma and in vacuum is

$$\frac{E_{ni}}{E_{ne}} = \frac{1}{\sqrt{\varepsilon(\omega)}} . \tag{9.6.99}$$

When $\omega \to \omega_p$ the normal component of the electric field in the plasma is significantly amplified.

Part III

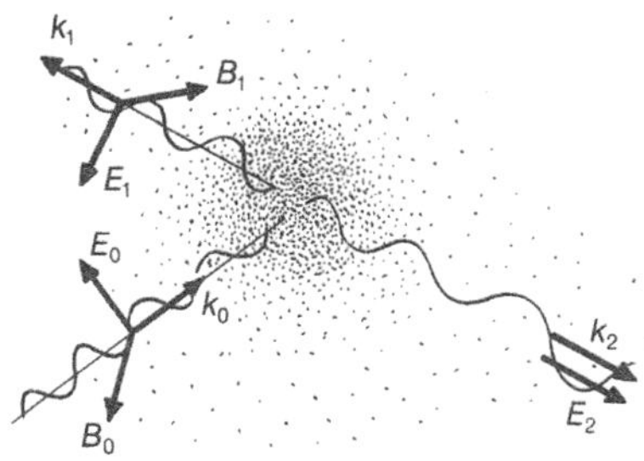

Principles of Nonlinear Electrodynamics of Plasma

Part III gives the basic principles of the nonlinear electrodynamics of a plasma. The theories of electrodynamic plasma fluctuations, the nonlinear wave interaction in the plasma and the quasilinear theory of plasma oscillations are presented. Along with the consideration of the main provisions of the general theory, a number of concrete nonlinear phenomena in the plasma are discussed.

10. Electromagnetic Fluctuations in Plasma and Wave Scattering

Using microscopic considerations, the expression for a spectral distribution correlator of current density fluctuations in the plasma is derived for the general case of a variable distribution function of charged particles. The analysis concerns fluctuations of the basic parameters of the plasma and the electromagnetic field in equilibrium plasmas and quasi-equilibrium nonisothermal plasmas, and in a plasma with a beam. The chapter contains the theory of electromagnetic wave scattering and transformation, along with the theory of charged-particle scattering on fluctuations of the electromagnetic field in the plasma.

10.1 Correlation Functions of the System of Charged Particles. General Analysis

So far, a plasma was referred to as a system of particles characterized by a set of macroscopic parameters which are the mean values of the corresponding microscopic quantities. At the same time, the parameters of any system can experience a deviation from their mean values which are called *fluctuations* of physical quantities.

Here we consider fluctuation phenomena as well as some processes associated with them. Besides the self-dependent interest (since, for example, thermal fluctuations of the electromagnetic field determine the "noise" level in plasma), it is also of importance to study fluctuations from other points of view. In particular, such processes as the *scattering* and *transformation* of waves in plasma can occur due to thermal fluctuations. As mentioned in Chap. 3, fluctuations determine the collision integral in the plasma, and thus allow one to study transfer processes. Here, we discuss the plasma fluctuation theory on the basis of macroscopic characteristics of plasma electromagnetic properties (dielectric permittivity and conductivity). As discussed in the previous chapters, we proceed making use of the most general plasma model, i.e., the kinetic equation with a self-consistent field. However, an alternative approach to electromagnetic phenomena is also possible, i.e., to determine the plasma dielectric permittivity by means of the microscopic fluctuation theory.

The correlation functions will be introduced for a quantitative description of fluctuations. Let us consider fluctuations of some quantity, for example, assume the current density $\boldsymbol{j}(t, \boldsymbol{r})$ to be real, but its mean value equals zero

$$\langle \boldsymbol{j}(t, \boldsymbol{r}) \rangle = 0 \, . \tag{10.1.1}$$

In a general case, averaging must be done over all possible quantum mechanical states of the system and also the probability of their statistical distribution. In other words, one takes a statistical ensemble which is equivalent to time averaging.

A *space-time correlation function* (or simply a *correlator*) is defined as a mean value of the product of fluctuations of a quantity $\boldsymbol{j}(t, \boldsymbol{t})$ at different space points and at different time moments. When the medium is homogeneous both in space and time[1], then the quadratic space-time correlation function takes the form

$$\langle j_i j_j \rangle_{t,r} = \langle j_i(t_1, \boldsymbol{r}_1) j_j(t_2, \boldsymbol{r}_2) \rangle \, , \tag{10.1.2}$$

where $t = t_2 - t_1$ and $\boldsymbol{r} = \boldsymbol{r}_2 - \boldsymbol{r}_1$.

Since we consider a spatially homogeneous and stationary plasma, the Fourier transformation may be applied[2]

$$\begin{aligned} \boldsymbol{j}(t, \boldsymbol{r}) &= \frac{1}{(2\pi)^4} \int d\omega \, d\boldsymbol{k} \, \mathrm{e}^{-\mathrm{i}\omega t + \mathrm{i}\boldsymbol{k} \cdot \boldsymbol{r}} \boldsymbol{j}(\omega, \boldsymbol{k}) \, , \\ \boldsymbol{j}(\omega, \boldsymbol{k}) &= \int dt \, d\boldsymbol{r} \, \mathrm{e}^{\mathrm{i}\omega t - \mathrm{i}\boldsymbol{k} \cdot \boldsymbol{r}} \boldsymbol{j}(t, \boldsymbol{r}) \, , \end{aligned} \tag{10.1.3}$$

and a *spectral distribution* of the space-time correlation function can be defined in the form

$$\langle j_i j_j \rangle_{\omega,k} = \int dt \, d\boldsymbol{r} \, \mathrm{e}^{\mathrm{i}\omega t - \mathrm{i}\boldsymbol{k} \cdot \boldsymbol{r}} \langle j_i j_j \rangle_{t,r} \, . \tag{10.1.4}$$

The above value is called a *spectral density* of the correlation function. Using (10.1.3, 4) it is easy to obtain

$$\langle j_i^*(\omega, \boldsymbol{k}) j_j(\omega', \boldsymbol{k}') \rangle = (2\pi)^4 \delta(\omega - \omega') \delta(\boldsymbol{k} - \boldsymbol{k}') \langle j_i j_j \rangle_{\omega,k} \, , \tag{10.1.5}$$

---

1 Here, we focus our attention on a spatially infinite homogeneous plasma, since plasma finiteness or inhomogeneity do not bring new results.

2 In contrast to other chapters, here we apply the Fourier expansion in which a factor $(2\pi)^{-4}$ appears in the direct [the first relation (10.1.3)] expansion. This is accepted in the presentation of the fluctuation theory.

which relates the mean value of the product of the Fourier components of fluctuating quantities to the spectral distribution of the correlation function.

Fluctuations can be conveniently characterized by *space*

$$\langle j_i(t, \boldsymbol{r}_1) j_j(t, \boldsymbol{r}_2) \rangle \equiv \langle j_i j_j \rangle_{\boldsymbol{r}} , \tag{10.1.6}$$

or *time* (autocorrelation)

$$\langle j_i(t_1, \boldsymbol{r}) j_j(t_2, \boldsymbol{r}) \rangle \equiv \langle j_i j_j \rangle_t \tag{10.1.7}$$

correlation functions being the mean values of fluctuations either at a given time moment but at a different space point or at a given space point but at a different time moment.

For the spectral components of the correlation functions, we take

$$\langle j_i j_j \rangle_{\boldsymbol{k}} = \frac{1}{2\pi} \int d\omega \, \langle j_i j_j \rangle_{\omega,\boldsymbol{k}} , \qquad \langle j_i j_j \rangle_{\omega} = \frac{1}{(2\pi)^3} \int d\boldsymbol{k} \, \langle j_i j_j \rangle_{\omega,\boldsymbol{k}} , \tag{10.1.8}$$

which imply that the spectral density of the autocorrelation function is the integral for the spectral density of the space-time fluctuation distribution over all wave vectors.

By using the autocorrelation function, we can define the characteristic fluctuation frequency averaged over the whole spectrum

$$\langle \omega^2 \rangle = \frac{\int \omega^2 \langle j_i j_j \rangle_{\omega} d\omega}{\int \langle j_i j_j \rangle_{\omega} d\omega} . \tag{10.1.9}$$

Because of the space correlation function, we can analogously obtain the characteristic length at which the correlation among fluctuations of a given quantity takes place.

An important corollary follows from (10.1.1). Let us rewrite this equation so that the mean value of amplitudes of the Fourier components of fluctuating quantities equals zero:

$$\langle \boldsymbol{j}(\omega, \boldsymbol{k}) \rangle = 0 . \tag{10.1.10}$$

The complex amplitude $\boldsymbol{j}(\omega, \boldsymbol{k})$ can be represented as

$$j_i(\omega, \boldsymbol{k}) = |j_i(\omega, \boldsymbol{k})| e^{i\phi} , \tag{10.1.11}$$

where ϕ is a phase of the corresponding Fourier component of a plane wave. Then, (10.1.10) shows that phases of the Fourier components of fluctuating quantities are random and the averaging in (10.1.10) has been performed over random phases. This is the main difference between fluctuation perturbations of a medium in an equilibrium state and those perturbations with a regular phase corresponding to natural oscillations of the medium, as studied before.

10.1.1 Fluctuations of Charge and Current Densities in the System of Noninteracting Particles

The chief aim of the general fluctuation theory in material media consists in determining the relationship between correlation functions for various physical quantities and macroscopic medium characteristics, i.e., dielectric permittivity and conductivity. For a plasma as a system of charged particles, this problem is accurately solved because of a small parameter η which is the ratio of the interaction energy of particles to the mean value of their thermal energy. In the first-order approximation, charged plasma particles may be considered non-interacting.

Before we calculate fluctuations of various specific plasma parameters, it should be noted that the distribution function (3.2.1) is the mean statistical value of the microscopic distribution function

$$f(\boldsymbol{r},\boldsymbol{p},t) = \sum_{s=1}^{N} \delta[\boldsymbol{r}-\boldsymbol{r}_s(t)]\,\delta[\boldsymbol{v}-\boldsymbol{v}_s(t)] \tag{10.1.12}$$

which characterizes the microscopic distribution of electrons over a phase space, i.e.,

$$f(\boldsymbol{r},\boldsymbol{p},t) = \langle f_M(\boldsymbol{r},\boldsymbol{p},t)\rangle \;. \tag{10.1.13}$$

Here averaging is carried out over the distribution of various possible states of the statistical ensemble consisting of N particles. In other words, over the so-called Liouville distribution function which depends on coordinates and velocities of all N particles constituting the ensemble, $F_N(\boldsymbol{r}_1, \ldots, \boldsymbol{r}_N, \boldsymbol{p}_1, \ldots, \boldsymbol{p}_N, t)$. It is natural, therefore, to study first the fluctuation of the distribution function itself. The latter is defined as a difference between the microscopic distribution in the phase space (10.1.12) and its mean value

$$\delta f(\boldsymbol{r},\boldsymbol{p},t) \equiv f_M(\boldsymbol{r},\boldsymbol{p},t) - f(\boldsymbol{r},\boldsymbol{p},t) \;. \tag{10.1.14}$$

Evidently, in the absence of external fields in a system of noninteracting particles their trajectories are straight lines. Hence, in (10.1.12) one assumes $\boldsymbol{v}_s^0(t) = \boldsymbol{v}_s^0 = \text{const}$ and $\boldsymbol{r}_s^0(t) = \boldsymbol{r}_s^0 + \boldsymbol{v}_s^0(t-t_0)$, where the superscript 0 means that the given quantity refers to a system without interaction, $\boldsymbol{r}_s^0$, $\boldsymbol{v}_s^0$ and $\boldsymbol{p}_s^0$ being the initial position, velocity, and momentum of a particle, respectively. Then, the Liouville distribution function can be written as

$$F_N^0(\boldsymbol{r}_1^0, \ldots, \boldsymbol{r}_N^0, \boldsymbol{p}_1, \ldots, \boldsymbol{p}_N^0) = \prod_{s=1}^{N} f(\boldsymbol{p}_s^0) \;. \tag{10.1.15}$$

Here, the single-particle distribution functions $f(\boldsymbol{p}_s^0)$ depend only on the momentum $\boldsymbol{p}_s^0 = \boldsymbol{p}$ owing to the spatial homogeneity and the stationarity of the system.

Forming the correlator for fluctuations of the distribution of particles of the same type and averaging it over the function (10.1.15), we may show that the space-time correlation function for fluctuations of the distribution function for noninteracting particles is[3]

$$\begin{aligned}\langle \delta f^0(\boldsymbol{r},\boldsymbol{p},t)\,\delta f^0(\boldsymbol{r}',\boldsymbol{p}',t)\rangle &\equiv \langle \delta f(\boldsymbol{p})\,\delta f(\boldsymbol{p}')\rangle_{\boldsymbol{r}-\boldsymbol{r}',\,t-t'}\\ &= \delta(\boldsymbol{v}-\boldsymbol{v}')\,\delta[\boldsymbol{r}-\boldsymbol{r}'-\boldsymbol{v}(t-t')]f(\boldsymbol{p})\ .\end{aligned} \tag{10.1.16}$$

Proceeding to the Fourier transforms in the absence of interparticle interaction for the spectral distribution of fluctuations of the distribution functions from (10.1.16), we obtain

$$\langle \delta f(\boldsymbol{p})\,\delta f(\boldsymbol{p}')\rangle^0_{\omega,\boldsymbol{k}} = 2\pi\delta(\boldsymbol{v}-\boldsymbol{v}')\,\delta(\omega-\boldsymbol{k}\cdot\boldsymbol{v})f(\boldsymbol{p})\ , \tag{10.1.17}$$

where $f(\boldsymbol{p})$ is an arbitrary nonequilibrium distribution function.

By means of the correlator (10.1.17) it is easy to find spectral distributions of fluctuations for all parameters characterizing a system of noninteracting charged particles. For example, multiplying (10.1.17) by e^2, then by $e^2 v_i v_j$ and integrating it twice over momenta, we obtain spectral distributions of the space-time correlation functions for the charge, particle, and current densities, in the absence of particle interaction, respectively:

$$\langle \varrho^2\rangle^0_{\omega,\boldsymbol{k}} = e^2\langle \delta N^2\rangle^0_{\omega,\boldsymbol{k}} = 2\pi e^2\int d\boldsymbol{p}\, f(\boldsymbol{p})\,\delta(\omega-\boldsymbol{k}\cdot\boldsymbol{v})\ , \tag{10.1.18}$$

$$\langle j_i j_j\rangle^0_{\omega,\boldsymbol{k}} = 2\pi e^2\int d\boldsymbol{p}\ v_i v_j f(\boldsymbol{p})\,\delta(\omega-\boldsymbol{k}\cdot\boldsymbol{v})\ . \tag{10.1.19}$$

To calculate correlation functions in the presence of external field, one must know the law of particle motion. In particular, in the presence of an external magnetic field in a plasma we find

$$\boldsymbol{r}(t) = \left\{-\frac{v_\perp\gamma}{\Omega}\cos\left(\frac{\Omega t}{\gamma}+\phi\right),\ \frac{v_\perp}{\Omega}\gamma\sin\left(\frac{\Omega t}{\gamma}+\phi\right),\ v_\parallel t\right\}\ , \tag{10.1.20}$$

where γ is the relativistic factor; $\Omega = eB_0/(mc)$ is the cyclotron frequency; $v_\perp$ and $v_\parallel$ are the transverse and longitudinal (with respect to the external magnetic field $\boldsymbol{B}_0$) components of the velocity of charged particles, respectively; ϕ is the phase at the initial time $t = 0$; external magnetic field $\boldsymbol{B}_0$ is considered to be along the $0z$-axis.

---

3 Note that averaging the microscopic function $f_M(\boldsymbol{r},\boldsymbol{p},t)$ over the distribution (10.1.15) leads to the relation $\langle f^0_M(\boldsymbol{r},\boldsymbol{p},t)\rangle = f(\boldsymbol{p}^0_\alpha)$.

Forming the correlation fluctuation function of the current density according to (10.1.14) as in the isotropic plasma, after performing calculations analogous to those of Sect. 5.1, for the dielectric tensor of the magnetoactive plasma we obtain

$$\langle j_i j_j \rangle^0_{\omega,k} = 2\pi e^2 \sum_{n=-\infty}^{\infty} \int d\boldsymbol{p}\, f(\boldsymbol{p})\, \Pi^n_{ij}\, \delta\left(\omega - \frac{n\Omega}{\gamma} - k_\parallel v_\parallel\right), \tag{10.1.21}$$

where the tensor Π^n_{ij} is defined by means of (5.1.8). An expression for the spectral distribution of charge density fluctuations appears to be similar:

$$\langle \varrho^2 \rangle^0_{\omega,k} = 2\pi e^2 \sum_{n=-\infty}^{\infty} \int d\boldsymbol{p}\, f(\boldsymbol{p})\, \mathrm{J}_n\left(\frac{k_\perp v_\perp \gamma}{\Omega}\right) \delta\left(\omega - \frac{n\Omega}{\gamma} - k_\parallel v_\parallel\right). \tag{10.1.22}$$

The correlation function of particle density fluctuations can be obtained from (10.1.22) if we divide it by e^2.

10.1.2 Plasma Fluctuations in the First-Order Approximation of Interparticle Interactions

In a system of noninteracting particles, (10.1.18–22) relate the charge and current density fluctuations to the distribution function $f(\boldsymbol{p})$. In this approximation (interparticle interaction being totally ignored), fluctuations in electron and ion components are independent and the derived relations can be considered separately. Then, the interparticle interaction in a plasma should be taken into account by means of a self-consistent field. The density of a fluctuation current $\boldsymbol{j}^0(\omega, \boldsymbol{k})$ must be substituted for an external source in the system of Maxwell equations and the field produced by this current be determined. As a result we find

$$\left[k^2\delta_{ij} - k_i k_j - \frac{\omega^2}{c^2}\varepsilon_{ij}(\omega, \boldsymbol{k})\right] E_j(\omega, \boldsymbol{k})$$

$$= \Lambda_{ij}(\omega, \boldsymbol{k})\, E_j(\omega, \boldsymbol{k}) = \frac{4\pi i\omega}{c^2} \sum_\alpha j_i^{\alpha 0}(\omega, \boldsymbol{k}). \tag{10.1.23}$$

Here, the summation extends over all types of charged particles $\alpha = e, i$, and $\varepsilon_{ij}(\omega, \boldsymbol{k})$ is the dielectric tensor.

The solution of (10.1.23) may be formally written as

$$E_i(\omega, \boldsymbol{k}) = \frac{4\pi i\omega}{c^2} \sum_\alpha \Lambda^{-1}_{ij}(\omega, \boldsymbol{k})\, j_i^{\alpha 0}(\omega, \boldsymbol{k}). \tag{10.1.24}$$

Hence, on using (10.1.5), we can obtain the spectral distribution of fluctuations of the self-consistent field in a plasma:

$$\langle E_i E_j \rangle_{\omega,k} = \frac{16\pi^2\omega^2}{c^2} \Lambda_{ik}^{*-1}(\omega, \boldsymbol{k}) \Lambda_{jl}^{-1}(\omega, \boldsymbol{k}) \sum_\alpha \langle j_k^\alpha j_l^\alpha \rangle_{\omega,k} \, . \tag{10.1.25}$$

In obtaining this relation within the approximation of noninteracting particles, we have accounted for the relation

$$\langle j_i^\alpha j_j^\beta \rangle_{\omega,k}^0 = \delta_{\alpha\beta} \langle j_i^\alpha j_j^\alpha \rangle_{\omega,k}^0 \, . \tag{10.1.26}$$

Equations (10.1.21, 25) relate the correlator $\langle E_i E_j \rangle_{\omega,k}$ to the distribution function of particles in a plasma.

When the correlation function of fluctuations of the self-consistent field is known, we can then obtain the corrections to current and charge fluctuations due to the self-consistent interparticle interaction in a plasma. Thus, under the influence of the self-consistent field there appears a correction to current density fluctuations of particles of the type α:

$$\delta j_i^\alpha(\omega, \boldsymbol{k}) = -\frac{i\omega}{4\pi} \delta\varepsilon_{ij}^\alpha(\omega, \boldsymbol{k}) E_j(\omega, \boldsymbol{k}) \, . \tag{10.1.27}$$

Hence, the total density of the fluctuation current of particles of type α is given by

$$j_i^\alpha(\omega, \boldsymbol{k}) = j_i^{\alpha 0}(\omega, \boldsymbol{k}) + \frac{\omega^2}{c^2} \delta\varepsilon_{i\mu}^\alpha(\omega, \boldsymbol{k}) \, \Lambda_{\mu\nu}^{-1}(\omega, \boldsymbol{k}) \sum_\beta j_\nu^{\beta 0}(\omega, \boldsymbol{k}) \, . \tag{10.1.28}$$

Next, we obtain the relationship between the spectral distribution of density fluctuations of the current and the correlator of density fluctuations of the current of noninteracting particles:

$$\langle j_i^\alpha j_j^\beta \rangle_{\omega,k} = \sum_{\alpha'} \left[\delta_{i\mu}\delta_{\alpha\alpha'} - \frac{\omega^2}{c^2} \delta\varepsilon_{is}^\alpha(\omega, \boldsymbol{k}) \Lambda_{s\mu}^{-1}(\omega, \boldsymbol{k}) \right]$$

$$\times \left[\delta_{j\nu}\delta_{\alpha'\beta} - \frac{\omega^2}{c^2} \delta\varepsilon_{js}^\beta(\omega, \boldsymbol{k}) \Lambda_{s\nu}(\omega, \boldsymbol{k}) \right] \langle j_\mu^{\alpha'} j_\nu^{\alpha'} \rangle_{\omega,k}^0 \, . \tag{10.1.29}$$

Accounting for the self-consistent interparticle interaction, the spectral distribution of the charge density correlations can be easily obtained by using the continuity equation. Thus

$$\langle \varrho^\alpha \varrho^\beta \rangle_{\omega,k} = \frac{k_i k_j}{\omega^2} \langle j_i^\alpha j_j^\beta \rangle_{\omega,k} \, . \tag{10.1.30}$$

Dividing it by $e_\alpha e_\beta$ we obtain the correlator of particle density fluctuations $\langle \delta N_\alpha \delta N_\beta \rangle_{\omega,k}$.

Analogously, one can obtain the correlation functions of fluctuations of the magnetic field, the mean value of energy, and other physical quantities.

All of them are expressed by the correlator of current density fluctuations of noninteracting particles. The correlator, in turn, is determined by the integral of the distribution of particles over momenta according to (10.1.19, 21).

Note that the relations derived above are of general character and are applicable both to the nonequilibrium as well as to the equilibrium plasmas. The only restriction is the requirement of plasma stability which includes the assumption of its spatial homogeneity and stationarity of its equilibrium state.

10.2 Fluctuations in Equilibrium Plasma. Fluctuation-Dissipation Theorem

Let us apply the general results obtained so far to a plasma which is in thermodynamic equilibrium. We start with the nondegenerate isotropic collisionless plasma with the Maxwellian distribution function $f_{0\alpha}(p)$ and temperature T for all types of particles. The contribution of the particles of type α to the dielectric permittivity is given by (4.1.11), from which it follows that

$$\delta\varepsilon_{ij}^{aa}(\omega,\boldsymbol{k}) = \mathrm{i}\pi\frac{4\pi e_\alpha^2}{\omega T}\int d\boldsymbol{p}\; v_i v_j f_{0\alpha}(\boldsymbol{p})\,\delta(\omega-\boldsymbol{k}\cdot\boldsymbol{v})\,. \tag{10.2.1}$$

On comparing this expression with (10.1.19) we obtain the relationship between the current correlator of noninteracting particles and the anti-Hermitian part of the partial dielectric tensor

$$\langle j_i^a j_j^a\rangle_{\omega,\boldsymbol{k}}^0 = -\frac{\mathrm{i}\omega T}{2\pi}\,\delta\varepsilon_{ij}^{aa}(\omega,\boldsymbol{k})\,. \tag{10.2.2}$$

This equation shows that the fluctuations in the plasma are proportional to its temperature and to the anti-Hermitian part of the dielectric tensor which is responsible for the electromagnetic field dissipation in the plasma.

10.2.1 Fluctuation-Dissipation Theorem for the Thermodynamically Equilibrium Isotropic Plasma

Then, for an isotropic plasma, we have

$$\begin{aligned}\Lambda_{ij}(\omega,\boldsymbol{k}) = &-\frac{\omega^2}{c^2}\frac{k_ik_j}{k^2}\varepsilon^{\mathrm{lo}}(\omega,k)\\ &+\left(\delta_{ij}-\frac{k_ik_j}{k^2}\right)\left(k^2-\frac{\omega^2}{c^2}\varepsilon^{\mathrm{tr}}(\omega,k)\right).\end{aligned} \tag{10.2.3}$$

On the basis of (10.1.29) and (10.2.2), the correlator of fluctuations of the total current density in the isotropic plasma can be written as

$$\langle j_i j_j \rangle_{\omega,k} = \frac{T_\omega}{2\pi}\left[\frac{k_i k_j}{k^2}\frac{\operatorname{Im}\{\varepsilon^{\mathrm{lo}}(\omega,k)\}}{|\varepsilon^{\mathrm{lo}}(\omega,k)|^2} + \left(\delta_{ij}-\frac{k_i k_j}{k^2}\right)\left(1-\frac{k^2c^2}{\omega^2}\right)\frac{\operatorname{Im}\{\varepsilon^{\mathrm{tr}}(\omega,k)\}}{|\varepsilon^{\mathrm{tr}}(\omega,k)-k^2c^2/\omega^2|^2}\right]. \quad (10.2.4)$$

10.2.2 The Case of the Anisotropic Plasma

Equation (10.2.4) represents the so-called *fluctuation-dissipation theorem* for the nondegenerate thermodynamically equilibrium isotropic plasma which is a particular case of a more general fluctuation-dissipation theorem for the thermodynamically equilibrium anisotropic plasma

$$\begin{aligned}\langle j_i j_j \rangle_{\omega,k} &= -\frac{\mathrm{i}c^2T}{4\pi\omega}\Lambda^0_{is}(\Lambda^{-1}_{ls}-\Lambda^{-1*}_{sl})\Lambda^0_{lj}\\ &= -\frac{\mathrm{i}c^2T}{4\pi\omega}\frac{1}{\Lambda}\Lambda^0_{is}(\Delta_{ls}-\Delta^*_{sl})\Lambda^0_{lj}\,,\end{aligned} \quad (10.2.5)$$

where $\Lambda = |\Lambda_{ij}|$ is the determinant formed with the elements of the matrix Λ_{ij}; Δ_{ij} is the algebraic complement $\Delta_{is}\Lambda_{sj} = \Lambda\delta_{ij}$ and

$$\Lambda^0_{ij}(\omega,\boldsymbol{k}) = \lim_{\varepsilon_{ij}\to\delta_{ij}}\Lambda_{ij}(\omega,\boldsymbol{k}) = \left[\left(k^2-\frac{\omega^2}{c^2}\right)\delta_{ij}-k_ik_j\right]. \quad (10.2.6)$$

The fluctuation-dissipation theorem in the form of (10.2.5) is applicable to both nondegenerate and degenerate anisotropic plasmas. The only requirement is that the plasma should be in a thermodynamically equilibrium state with the same temperature T for all types of charged particles. Besides, while deriving this theorem, the inequality $\hbar\omega \ll T$ is assumed to be satisfied, thereby taking the classical consideration to be valid. In quantum statistical physics, the fluctuation-dissipation theorem is proved in a more general form without assuming this inequality:

$$\langle j_i j_j \rangle_{\omega,k} = -\frac{\mathrm{i}}{4\pi}\frac{\hbar c^2}{\exp(\hbar\omega/T)-1}\frac{1}{\Lambda}\Lambda^0_{is}(\Delta_{ls}-\Delta^*_{sl})\Lambda^0_{lj}\,. \quad (10.2.7)$$

In the limit $\hbar\omega/T\to 0$, it goes over to (10.2.5).

While using the fluctuation-dissipation theorem, one must substitute $\Lambda(\omega,\boldsymbol{k})$, $\Lambda_{ij}(\omega,\boldsymbol{k})$ and $\Delta_{ij}(\omega,\boldsymbol{k})$ into (10.2.5, 7). To describe the charged particle motion, we have applied the classical kinetic equation with the self-consistent interaction. This equation is valid when $\hbar\omega \ll \mathscr{E}_{av}$, where $\mathscr{E}_{av}$ is the mean value of energy of the particle chaotic motion[4]. This implies that $\hbar\omega \ll T$ for a nondegenerate plasma and $\hbar\omega \ll \mathscr{E}_F$ for a degenerate one. The ratio $\hbar\omega/T$ may be arbitrary, however. Therefore, we apply the fluctuation-dissipation theorem (10.2.5) for a nondegenerate plasma and (10.2.7) for a degenerate one.

A very important general conclusion follows from (10.2.5) and (10.2.7): The spectral density of fluctuations sharply grows when the equality

$$\Lambda = \left| k^2\delta_{ij} - k_i k_j - \frac{\omega^2}{c^2}\varepsilon_{ij}(\omega, \boldsymbol{k}) \right| = 0 \tag{10.2.8}$$

is realized. It is simply the dispersion equation leading to the spectra of natural oscillations in a plasma, (2.4.4). Thus, fluctuations have sharp maxima near the frequencies of natural plasma oscillations.

In the equilibrium plasma, due to (10.2.4–7) fluctuations have been expressed by a macroscopic characteristic of its electromagnetic properties, i.e., dielectric permittivity [more exactly its anti-Hermitian part $\varepsilon_{ij}^{a}(\omega, \boldsymbol{k})$]. But we can act in the opposite way, i.e., according to (10.2.2) express the anti-Hermitian part of the dielectric permittivity by the correlator of the current density for the system of noninteracting particles. Then due to the Kramers-Kronig formulas (2.2.17) we can also find the Hermitian part $\varepsilon_{ij}^{H}(\omega, \boldsymbol{k})$, i.e., we obtain the complete expression for the tensor of complex dielectric permittivity. The corresponding relations have the form

$$\begin{aligned} \varepsilon_{ij}^{a}(\omega, \boldsymbol{k}) &= \frac{2\pi^2 i}{\omega T} \langle j_i j_i \rangle_{\omega,\boldsymbol{k}}^{0} , \\ \varepsilon_{ij}^{H}(\omega, \boldsymbol{k}) - \delta_{ij} &= \frac{\mathscr{P}}{\pi} \int_{-\infty}^{\infty} \frac{\varepsilon_{ij}^{a}(\omega', \boldsymbol{k})}{\omega' - \omega} d\omega' . \end{aligned} \tag{10.2.9}$$

The relations constitute the mathematical content of the so-called *inversion of the fluctuation-dissipation theorem* due to which one can obtain the plasma dielectric tensor if the current density fluctuations in the plasma are predetermined on the basis of the microscopic theory.

4 The condition $\hbar\omega \ll \mathscr{E}_{av}$ follows from the approximation $p \gg \hbar k$ used when deriving the kinetic equation, taking into account the interparticle collisions (Chap. 3). Multiplying this inequality by $v_{ph} = \omega/k$, which is of the order of the particle thermal velocities for plasma waves, we arrive at the initial condition.

10.3 Spectra Distribution of Fluctuations in Equilibrium Collisionless Plasma

Let us consider fluctuations in the equilibrium collisionless electron plasma starting with the simplest case of the isotropic nondegenerate plasma. We proceed from (10.2.4) for the correlator of current density fluctuations. Using the relationship between the self-consistent field and the current density

$$
\begin{aligned}
\boldsymbol{j}^{\mathrm{lo}}(\omega,\boldsymbol{k}) &= -\frac{\mathrm{i}\omega}{4\pi}\delta\varepsilon^{\mathrm{lo}}(\omega,k)\boldsymbol{E}^{\mathrm{lo}}(\omega,\boldsymbol{k})\ ,\\
\boldsymbol{j}^{\mathrm{tr}}(\omega,\boldsymbol{k}) &= -\frac{\mathrm{i}\omega}{4\pi}\delta\varepsilon^{\mathrm{tr}}(\omega,k)\boldsymbol{E}^{\mathrm{tr}}(\omega,\boldsymbol{k})\ ,
\end{aligned}
\tag{10.3.1}
$$

the correlator of fluctuations of the self-consistent field can be obtained due to (10.2.4):

$$
\langle E_iE_j\rangle_{\omega,\boldsymbol{k}} = \frac{8\pi T}{\omega}\left[\frac{k_ik_j}{k^2}\frac{\mathrm{Im}\{\varepsilon^{\mathrm{lo}}(\omega,k)\}}{|\varepsilon^{\mathrm{lo}}(\omega,k)|^2} + \left(\delta_{ij}-\frac{k_ik_j}{k^2}\right)\frac{\mathrm{Im}\{\varepsilon^{\mathrm{tr}}(\omega,k)\}}{\left|\varepsilon^{\mathrm{tr}}(\omega,k)-\dfrac{k^2c^2}{\omega^2}\right|^2}\right] . \tag{10.3.2}
$$

Equations (10.2.4) and (10.3.2) show the correlators of fluctuations of the current density and electric field to split up into two summands characterizing longitudinal and transverse fluctuations. Hence, longitudinal and transverse fluctuations in thc isotropic plasma appear to be independent.

10.3.1 Fluctuations of Charge Densities and Longitudinal Electric Field

From the continuity equation

$$
\omega\varrho = \boldsymbol{k}\cdot\boldsymbol{j}(\omega,\boldsymbol{k}) \tag{10.3.3}
$$

with account taken of (10.2.4), the correlator of charge density fluctuations can be easily obtained

$$
\langle\varrho^2\rangle_{\omega,\boldsymbol{k}} = \frac{k^2}{16\pi^2}\langle(E^{\mathrm{lo}})^2\rangle_{\omega,\boldsymbol{k}} = \frac{k^2T}{2\pi\omega}\frac{\mathrm{Im}\{\varepsilon^{\mathrm{lo}}(\omega,k)\}}{|\varepsilon^{\mathrm{lo}}(\omega,k)|^2} . \tag{10.3.4}
$$

Naturally charge density fluctuations depend only on the longitudinal dielectric permittivity.

Using the explicit form of the dielectric permittivity, let us express correlation functions in terms of plasma parameters. We initially study only the case of a purely electron nondegenerate plasma using (4.1.4) for $\varepsilon^{\mathrm{lo}}(\omega, k)$ and $\varepsilon^{\mathrm{tr}}(\omega, k)$.

First of all, we consider charge density fluctuations in the range of high-frequencies (high phase velocities) $\omega \gg k v_{\mathrm{Te}}$ where weakly damped longitudinal Langmuir waves with the spectrum determined by the dispersion equation $\varepsilon^{\mathrm{lo}}(\omega, k) = 0$ exist. In this frequency range, (10.3.4) yields

$$\langle \varrho^2 \rangle_{\omega,k} = \frac{k^2}{16\pi^2} \langle (E^{\mathrm{lo}})^2 \rangle_{\omega,k} \approx \frac{k^2 T}{2\omega} \delta [\mathrm{Re}\{\varepsilon^{\mathrm{lo}}(\omega, k)\}] . \tag{10.3.5}$$

Here, we have applied a possible formal substitution

$$\frac{\mathrm{Im}\{\varepsilon^{\mathrm{lo}}(\omega, k)\}}{|\varepsilon^{\mathrm{lo}}(\omega, k)|^2} = -\pi\delta [\mathrm{Re}\{\varepsilon^{\mathrm{lo}}(\omega, k)\}] , \tag{10.3.6}$$

valid in the range of transparency where $\mathrm{Re}\{\varepsilon^{\mathrm{lo}}(\omega, k)\} \gg \mathrm{Im}\{\varepsilon^{\mathrm{lo}}(\omega, k)\}$.

On substituting

$$\mathrm{Re}\{\varepsilon^{\mathrm{lo}}(\omega, k)\} \approx 1 - \frac{\omega_{\mathrm{pe}}^2}{\omega^2}\left(1 + 3\frac{k^2 v_{\mathrm{Te}}^2}{\omega^2}\right) , \tag{10.3.7}$$

into (10.3.5) we finally obtain

$$\begin{aligned} \langle \varrho^2 \rangle_{\omega,k} &= \frac{k^2 T}{2\omega} \omega_{\mathrm{pe}}^2 \delta (\omega^2 - \omega_{\mathrm{pe}}^2 - 3k^2 v_{\mathrm{Te}}^2) , \\ \langle (E^{\mathrm{lo}})^2 \rangle_{\omega,k} &= \frac{8\pi^2 T}{\omega} \omega_{\mathrm{pe}}^2 \delta (\omega^2 - \omega_{\mathrm{pe}}^2 - 3k^2 v_{\mathrm{Te}}^2) . \end{aligned} \tag{10.3.8}$$

Here, the spectrum of longitudinal waves, lying in the narrow frequency range $\omega^2 \gtrsim \omega_{\mathrm{pe}}^2$, has been taken into account.

In Fig. 10.1, we present spectral distributions of the quantity $2\langle \varrho^2 \rangle_{\omega,k}/e^2 N_{\mathrm{e}}$ versus the dimensionless frequency $\omega/(k v_{\mathrm{Te}})$. The maxima are seen to be sharp only for small values of $k^2 r_{\mathrm{De}}^2$ and they disappear with the decrease of the wavelength (increase of a wave vector $|\boldsymbol{k}|$) due to the electron Landau damping.

Using the asymptotic form of the function $I_+[\omega/(k v_{\mathrm{Te}})]$ for small values of $\omega/(k v_{\mathrm{Te}})$, we may also study the low-frequency branch. However, the corresponding calculations do not give simple expressions and it is not of much interest to present them here.

To characterize fluctuations further, it is convenient to consider the space correlation of fluctuations with the given wavelength which is obtained by

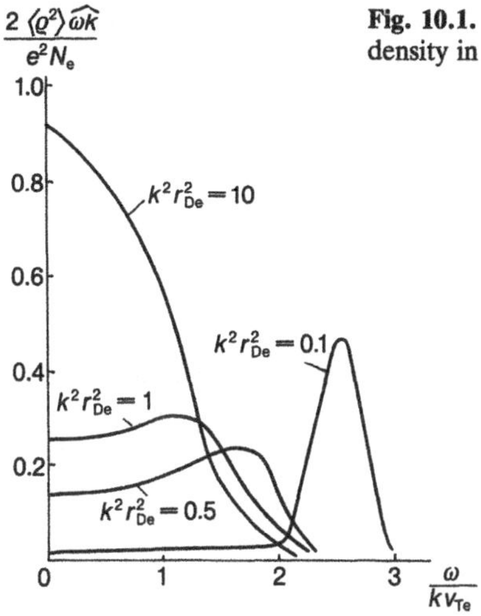

Fig. 10.1. Spectral distribution of fluctuations of the charge density in an equilibrium plasma

integrating the spectral distribution of the correlation function over frequencies:

$$\langle \varrho^2 \rangle_k = \frac{1}{2\pi} \int_{-\infty}^{\infty} \langle \varrho^2 \rangle_{\omega,k} d\omega, \quad \langle (E^{lo})^2 \rangle_k = \frac{1}{2\pi} \int_{-\infty}^{\infty} \langle (E^{lo})^2 \rangle_{\omega,k} d\omega . \tag{10.3.9}$$

Here, the integration can easily be accomplished by means of the Kramers-Kronig formula to be applied to the function $1/\varepsilon^{lo}(\omega, k)$ (Exercise 2.8.5):

$$\frac{\mathrm{Re}\,\{\varepsilon^{lo}(\omega, k)\}}{|\varepsilon^{lo}(\omega, k)|^2} - 1 = \frac{\mathscr{P}}{\pi} \int_{-\infty}^{\infty} d\omega' \frac{\mathrm{Im}\,\{\varepsilon^{lo}(\omega', k)\}}{|\varepsilon^{lo}(\omega', k)|^2 (\omega' - \omega)} \tag{10.3.10}$$

When in (10.3.10) $\omega = 0$ and the condition $\mathrm{Im}\,\{\varepsilon^{lo}(\omega, k)\} = 0$ holds, then we find from (10.3.9)

$$\langle \varrho^2 \rangle_k = \frac{k^2 T}{4\pi} \left(1 - \frac{1}{\varepsilon^{lo}(0, k)}\right), \quad \langle (E^{lo})^2 \rangle_k = 4\pi T \left(1 - \frac{1}{\varepsilon^{lo}(0, k)}\right) . \tag{10.3.11}$$

Thus, the spectral density of the space correlation function of the charge density and longitudinal electric field fluctuations is completely determined by the longitudinal static dielectric permittivity. Substituting the value of the latter for an isotropic plasma into (10.3.11), we obtain

$$\langle \varrho^2 \rangle_k = \frac{e^2 N_e k^2}{k^2 + 1/r_D^2}, \quad \langle (E^{lo})^2 \rangle_k = \frac{16\pi^2 e^2 N_e}{k^2 + 1/r_D^2}, \tag{10.3.12}$$

where (4.2.13) has been used.

Applying the inverse Fourier transformation, space correlation functions of the charge density and longitudinal electric field fluctuations[5] may be easily found:

$$\langle \varrho^2 \rangle_r = 2e^2 N_e \left[\delta(r) - \frac{1}{4\pi} \frac{\exp(-r/r_D)}{r r_D^2} \right],$$
$$\langle (E^{lo})^2 \rangle_r = 8\pi e^2 N_e \frac{\exp(-r/r_D)}{r}. \tag{10.3.13}$$

Note that the Debye length is the characteristic distance at which the correlation between fluctuations of the charge density or the longitudinal electric field occurs. Therefore, for $\eta \ll 1$, the energy of longitudinal field fluctuations is small, as compared to the electron thermal energy:

$$\frac{\langle (E^{lo})^2 \rangle_r}{4\pi N_e T} \approx \left(\frac{e^2 N_e^{1/3}}{T} \right)^{3/2} = \eta^{3/2} \ll 1. \tag{10.3.14}$$

The mean square fluctuation frequency is determined by (10.1.9):

$$\langle \omega^2 \rangle = \frac{\int_{-\infty}^{\infty} \omega^2 \langle \varrho^2 \rangle_{\omega,k} d\omega}{\int_{-\infty}^{\infty} \langle \varrho^2 \rangle_{\omega,k} d\omega}. \tag{10.3.15}$$

Here the integration can be conveniently carried out also by means of the Kramers-Kronig formula which provides

$$\langle \omega^2 \rangle = \lim_{\omega \to \infty} \omega^2 [1 - \varepsilon^{lo}(\omega, k)] \left(1 - \frac{1}{\varepsilon^{lo}(0, k)} \right)^{-1}. \tag{10.3.16}$$

It has been repeatedly shown and it also follows from the general physical assumptions that the spatial dispersion is insignificant in the high-frequency domain. Accordingly,

$$\varepsilon^{lo}(\omega, k) \approx 1 - \frac{\omega_{pe}^2}{\omega^2}$$

is valid for $\omega \gg k v_{Te}$. Making use of this, the mean square fluctuation frequency is given by

$$\langle \omega^2 \rangle \approx \frac{\omega_{pe}^2}{1 - 1/\varepsilon^{lo}(\omega, k)} = \omega_{pe}^2 + \frac{k^2 v_{Te}^2}{2}. \tag{10.3.17}$$

5 Henceforth, for simplicity, ions are regarded as singly charged.

Hence it is smaller than the frequency of natural plasma oscillations $\omega^2 = \omega_{pe}^2 + 3k^2v_{Te}^2$. From (10.3.17) it also follows that all the derived expressions based on (10.3.4) are valid under the condition $\hbar\omega_{pe} \ll T$. The latter is usually satisfied in a nondegenerate plasma.

10.3.2 Fluctuations in the Degenerate Electron Plasma

Let us now analyze peculiarities of fluctuations in the degenerate electron plasma. We proceed from (10.2.7) for the correlator $\langle j_i j_j \rangle_{\omega,k}$. By using the continuity equation (10.3.3), we may obtain an expression for the correlator of electron density fluctuations in the isotropic plasma:

$$\langle \delta N_e^2 \rangle_{\omega,k} = \frac{k^2}{2\pi e^2} \frac{\hbar}{\exp\left(\frac{\hbar\omega}{T} - 1\right)} \frac{\mathrm{Im}\{\varepsilon^{lo}(\omega,k)\}}{|\varepsilon^{lo}(\omega,k)|^2} . \tag{10.3.18}$$

Note that according to the fluctuation-dissipation theorem in the classical limit, i.e., when $\hbar\omega \ll T$ and (10.2.5) is valid, fluctuations are impossible for $T = 0$. This is not true for the degenerate plasma. Actually, (10.3.18) shows fluctuations to be possible also for $T = 0$ (complete degeneracy) but only with negative frequencies $\omega < 0$. It physically follows from the fact that for $T = 0$ the particles in the system occupy all the possible lowest energy levels, i.e., the system of Fermi particles is in the lowest energy state and allows transfers into the state with higher energy. This corresponds to the absorption of an energy quantum, i.e., to the negative frequency. For $T = 0$ fluctuations are purely quantum and proportional to the Planck constant $\hbar$.

Let us analyze (10.3.18) for an arbitrary ratio $\hbar\omega/T$ using (4.1.18) for $\varepsilon^{lo}(\omega,k)$. In the high-frequency range ($\omega \gg kv_{Fe}$) the imaginary part of the dielectric permittivity in the collisionless limit [according to (4.3.3)] is strictly equal to zero, fluctuations being impossible in the plasma. But taking account of rare interparticle collisions leads to the appearance of a nonzero imaginary part of $\varepsilon^{lo}(\omega,k)$ and also of fluctuations at the frequencies of natural plasma oscillations with the spectrum (4.3.4):

$$\begin{aligned}\langle \delta N_e^2 \rangle_{\omega,k} &= \frac{k^2}{2e^2} \frac{\hbar}{\exp(\hbar\omega/T) - 1} \delta[\varepsilon^{lo}(\omega,k)] \\ &\approx \frac{k^2\omega_{pe}^2}{2e^2} \frac{\hbar}{\exp(\hbar\omega/T) - 1} \delta\left[\omega^2 - \omega_{pe}^2\left(1 + \frac{3}{5}\frac{k^2v_{Fe}^2}{\omega_{pe}^2}\right)\right] .\end{aligned} \tag{10.3.19}$$

In the low-frequency domain ($\omega \ll kv_{Fe}$) we have the asymptotic representation (4.3.6) for $\varepsilon^{lo}(\omega,k)$. On substituting it into (10.3.18), we obtain

$$\langle \delta N_e^2 \rangle_{\omega,k} = \frac{k\omega}{4e^2v_{Fe}} \frac{\hbar}{\exp(\hbar\omega/T) - 1} \frac{1}{1 + k^2v_{Fe}^2/3\omega_{pe}^2} \tag{10.3.20}$$

This expression shows that in the low-frequency limit (of low phase velocities) the long-wavelength fluctuation excitations with $kv_{\mathrm{Fe}} \ll \omega_{\mathrm{pe}}$ are predominant, whereas the short-wavelength fluctuations $kv_{\mathrm{Fe}} \gg \omega_{\mathrm{pe}}$ are relatively weaker in the system of noninteracting Fermi particles.

Finally, note that it is easy to find the fluctuations of the charge density and the longitudinal electric field in the plasma when the electron density fluctuations are known:

$$\langle (E^{\mathrm{lo}})^2 \rangle_{\omega,k} = \frac{16\pi^2}{k^2} \langle \varrho^2 \rangle_{\omega,k} = \frac{16\pi^2}{k^2} e^2 \langle \delta N_{\mathrm{e}}^2 \rangle_{\omega,k} . \tag{10.3.21}$$

According to this equation, the energy associated with longitudinal field fluctuations is small as compared to that of the electron chaotic motion, as in the case of the nondegenerate plasma. Then, in the limit $\hbar\omega \ll T$, we find

$$\frac{\langle (E^{\mathrm{lo}})^2 \rangle_r}{4\pi N_{\mathrm{e}} \mathscr{E}_{\mathrm{Fe}}} \approx \frac{T_{\mathrm{e}}}{\mathscr{E}_{\mathrm{Fe}}} \left(\frac{e^2 N_{\mathrm{e}}^{1/3}}{\mathscr{E}_{\mathrm{Fe}}} \right)^{3/2} \ll 1 . \tag{10.3.22}$$

Thus, high-frequency density fluctuations of the charge and longitudinal electric field at the frequencies near the Langmuir electron ones are possible in both the degenerate as well as the nondegenerate isotropic plasma. In contrast to the nondegenerate plasma, these fluctuations occur in the degenerate one both for $\hbar\omega_{\mathrm{pe}} < T$, when they are determined by temperature (similar to the nondegenerate plasma), and for $\hbar\omega_{\mathrm{pe}} > T \to 0$, when they are of quantum character.

10.3.3 Fluctuations in the Equilibrium Magneto-Active Plasma

In conclusion, we briefly analyze fluctuations in the nondegenerate thermodynamically equilibrium magneto-active plasma. Here, the correlator of density fluctuations of the current is determined by means of the fluctuation-dissipation theorem (10.2.5). The constituting quantities are expressed by the dielectric tensor of the magneto-active plasma. Equation (5.1.10) with $T_{\mathrm{e}} = T_{\mathrm{i}} = T$ must be applied to this tensor. In a general case the analysis of fluctuations in the magneto-active plasma is associated with unwieldy calculations. Nevertheless, the form of the spectral distribution of correlation functions remains qualitatively the same: it is characterized by the presence of sharp maxima near natural frequencies of oscillations of the magneto-active plasma studied in detail in Chap. 5. Thus we discuss a simple example. Let us calculate the correlation function of charge density fluctuations in the limit of the cold plasma. Due to (10.2.5) and the continuity equation (10.3.3), we obtain

$$\langle \varrho^2 \rangle_{\omega,k} = -\frac{k^2 T}{2\pi\omega} \,\mathrm{Im}\left\{\frac{P(\omega, k)}{\Lambda(\omega, k)}\right\}, \quad \text{where} \tag{10.3.23}$$

$$\begin{aligned} P(\omega, k) &= \frac{k^4 c^4}{\omega^4} - (\varepsilon_\perp + \varepsilon_\perp \cos^2\theta + \varepsilon_\| \sin^2\theta)\frac{k^2 c^2}{\omega^2} \\ &\quad + (\varepsilon_\perp^2 - g^2)\cos^2\theta + \varepsilon_\| \varepsilon_\perp \sin^2\theta , \\ \Lambda(\omega, k) &= \frac{k^4 c^4}{\omega^4}(\varepsilon_\perp \sin^2\theta + \varepsilon_\| \cos^2\theta) \\ &\quad - \frac{k^2 c^2}{\omega^2}[2\varepsilon_\perp \varepsilon_\| + (\varepsilon_\perp^2 - g^2 - \varepsilon_\perp \varepsilon_\|)\sin^2\theta] + \varepsilon_\| (\varepsilon_\perp^2 - g^2) \end{aligned} \tag{10.3.24}$$

(θ being the angle included between the vectors $\boldsymbol{k}$ and $\boldsymbol{B}_0$), and for the purely electron plasma the dielectric tensor components $\varepsilon_\perp$, $\varepsilon_\|$ and g are given by, see (5.2.5):

$$\varepsilon_\perp = 1 - \frac{\omega_{\mathrm{pe}}^2}{\omega^2 - \Omega_{\mathrm{e}}^2}, \quad \varepsilon_\| = 1 - \frac{\omega_{\mathrm{pe}}^2}{\omega^2}, \quad g = -\frac{\omega_{\mathrm{pe}}^2 \Omega_{\mathrm{e}}}{\omega(\omega^2 - \Omega_{\mathrm{e}}^2)} . \tag{10.3.25}$$

In the limit $k^2c^2 \gg \omega^2$, when the waves in the magneto-active plasma may be regarded as potential, (10.3.23) can be simplified:

$$\langle \varrho^2 \rangle_{\omega,k} = -\frac{k^2 T}{2\pi\omega} \,\mathrm{Im}\left\{\frac{1}{\varepsilon_\perp \sin^2\theta + \varepsilon_\| \cos^2\theta}\right\}. \tag{10.3.26}$$

Accounting for the vanishing parts $\varepsilon_\perp$ and $\varepsilon_\|$, caused for example by interparticle collisions, we obtain from (10.3.26)

$$\begin{aligned} \langle \varrho^2 \rangle_{\omega,k} &= \frac{k^2 T \omega^2}{4\omega_{\mathrm{pe}}^2} \frac{(\omega^2 - \Omega_{\mathrm{e}}^2)^2}{\omega^4 \sin^2\theta + (\omega^2 - \Omega_{\mathrm{e}}^2)^2 \cos^2\theta} \\ &\quad \times [\delta(\omega - \omega_1) + \delta(\omega + \omega_1) + \delta(\omega - \omega_2) + \delta(\omega + \omega_2)] , \end{aligned} \tag{10.3.27}$$

where the frequencies ω_1 and ω_2 are determined by the roots of the dispersion equation for the longitudinal Langmuir waves in the magneto-active plasma, see (5.2.20):

$$\omega_{1,2}^2 = \frac{\omega_{\mathrm{pe}}^2 + \Omega_{\mathrm{e}}^2}{2} \pm \frac{1}{2}\sqrt{(\omega_{\mathrm{pe}}^2 + \Omega_{\mathrm{e}}^2)^2 - 4\omega_{\mathrm{pe}}^2 \Omega_{\mathrm{e}}^2 \cos^2\theta} . \tag{10.3.28}$$

From the applicability condition of the theorem (10.2.5) it follows that (10.3.23, 26, 27) are valid for $\hbar\omega_{1,2} \ll T$. Since they are concerned with the cold magneto-active plasma, they are applicable both to the degenerate and nondegenerate plasmas. Moreover, these formulas may be easily generalized for the case of the degenerate plasma with an arbitrary ratio $\hbar\omega_{1,2}/T$. Here $\hbar\omega\,[\exp(\hbar\omega/T)-1]^{-1}$ should be substituted for T. Analogously to the isotropic plasma, purely quantum fluctuations are possible in the limit $T\to 0$ of the magneto-active degenerate plasma, their frequencies also being negative.

10.4 Fluctuations in Nonequilibrium Plasmas. Nonisothermal Plasma and Plasma with a Beam

The general expressions for fluctuations in the nonequilibrium plasma obtained in Sect. 10.1 are too complex for our analysis. For the isotropic plasma they can be slightly simplified. Considering only fluctuations of longitudinal fields and currents (and density fluctuations of charges and particles of the given type, expressed in their terms), the spectral distribution of the fluctuation correlator of the longitudinal electric field can be easily obtained from (10.1.25):

$$\langle (E^{\mathrm{lo}})^2\rangle_{\omega,k} = \sum_\alpha \frac{32\pi^3 e_\alpha^2}{k^2|\varepsilon^{\mathrm{lo}}(\omega,k)|^2}\int d\boldsymbol{p}\, f_{0\alpha}(\boldsymbol{p})\,\delta(\omega-\boldsymbol{k}\cdot\boldsymbol{v})\,. \tag{10.4.1}$$

Equation (10.1.29) provides

$$\begin{aligned}\langle j_e^2\rangle_{\omega,k} = \frac{2\pi e^2}{|\varepsilon^{\mathrm{lo}}(\omega,k)|^2}\,&[|1+\delta\varepsilon_{\mathrm{i}}^{\mathrm{lo}}(\omega,k)|^2\int d\boldsymbol{p}\, f_{0\mathrm{e}}(\boldsymbol{p})\,v^2\delta(\omega-\boldsymbol{k}\cdot\boldsymbol{v})\\ &+|\delta\varepsilon_{\mathrm{e}}^{\mathrm{lo}}(\omega,k)|^2\int d\boldsymbol{p}\, f_{0\mathrm{i}}(\boldsymbol{p})\,v^2\delta(\omega-\boldsymbol{k}\cdot\boldsymbol{v})]\end{aligned} \tag{10.4.2}$$

for the spectral distribution of the correlator of current density fluctuations of particles of type α (for definiteness of the electron current density). Hence, for the continuity equation, we obtain the spectral distribution of the correlator of electron density fluctuations:

$$\begin{aligned}\langle \delta N_{\mathrm{e}}^2\rangle_{\omega,k} = \frac{2\pi}{|\varepsilon^{\mathrm{lo}}(\omega,k)|^2}\,&[|1+\delta\varepsilon_{\mathrm{i}}^{\mathrm{lo}}(\omega,k)|^2\int d\boldsymbol{p}_{\mathrm{e}} f_{0\mathrm{e}}(\boldsymbol{p}_{\mathrm{e}})\,\delta(\omega-\boldsymbol{k}\cdot\boldsymbol{v})\\ &+|\delta\varepsilon_{\mathrm{e}}^{\mathrm{lo}}(\omega,k)|^2\int d\boldsymbol{p}_{\mathrm{i}} f_{0\mathrm{i}}(\boldsymbol{p}_{\mathrm{i}})\,\delta(\omega-\boldsymbol{k}\cdot\boldsymbol{v})]\,.\end{aligned} \tag{10.4.3}$$

Ion fluctuations are determined by the same formulas as those of electrons under the substitution of the index e for i and vice versa.

From the last two expressions it follows that the electron current and the density fluctuations depend on the distribution function of both the electrons and the ions; this dependence is determined not only by the contribution from the corresponding particles to the dielectric permittivity but also from the electron and ion distribution functions over velocities in the direction of the wave vector $\boldsymbol{k}$.

To characterize fluctuations in the nonequilibrium plasma, we can conveniently introduce the concept of the *effective fluctuation temperature* T_{eff}. Let us illustrate this by the example of longitudinal field fluctuations by writing (10.4.1) as

$$\langle (E^{\text{lo}})^2 \rangle_{\omega,\boldsymbol{k}} = \frac{F(\omega,\boldsymbol{k})}{|\varepsilon^{\text{lo}}(\omega,\boldsymbol{k})|^2}, \tag{10.4.4}$$

where the notation

$$F(\omega,\boldsymbol{k}) = \sum_\alpha \frac{32\pi^3 e_\alpha^2}{k^2} \int d\boldsymbol{p}\, f_{0\alpha}(\boldsymbol{p})\, \delta(\omega - \boldsymbol{k}\cdot\boldsymbol{v}) \tag{10.4.5}$$

is introduced. For the equilibrium plasma (10.4.4) reduces to (10.3.4). Comparing these expressions for the fluctuation correlators in the nonequilibrium and equilibrium plasmas, respectively, we determine the *effective temperature of nonequilibrium plasma*:

$$T_{\text{eff}} = \frac{\omega F(\omega,\boldsymbol{k})}{8\pi \operatorname{Im}\{\varepsilon^{\text{lo}}(\omega,\boldsymbol{k})\}} \tag{10.4.6}$$

The effective temperature T_{eff} may clearly characterize not only fluctuations, but also longitudinal oscillations in the nonequilibrium plasma.

Let us apply the results thus obtained to two specific cases of the nonequilibrium plasma: a) quasi-equilibrium nonisothermal (two-temperature) electron-ion plasma; b) plasma with a beam.

10.4.1 Fluctuations in the Quasi-Equilibrium Nonisothermal Plasma

We consider the quasi-equilibrium plasma where electrons and ions possess the Maxwellian velocity distributions but with different temperatures. Then the Maxwellian distribution functions with the temperature T_α, $\alpha = \text{e, i}$ should be substituted for $f_{0\alpha}(\boldsymbol{p})$. Using (10.4.1) for the spectral distribution of the longitudinal field, we obtain

$$\langle (E^{\text{lo}})^2 \rangle_{\omega,\boldsymbol{k}} = \frac{8\pi}{\omega} \frac{1}{|\varepsilon^{\text{lo}}(\omega,\boldsymbol{k})|^2} \sum_\alpha T_\alpha \operatorname{Im}\{\delta\varepsilon_\alpha^{\text{lo}}(\omega,\boldsymbol{k})\}, \tag{10.4.7}$$

i.e., longitudinal field fluctuations in the nonisothermal plasma under the substitution of

$$\sum_\alpha T_\alpha \operatorname{Im}\{\delta\varepsilon_\alpha^{\text{lo}}(\omega,\boldsymbol{k})\} \quad \text{for} \quad T \operatorname{Im}\{\varepsilon^{\text{lo}}(\omega,\boldsymbol{k})\}.$$

Performing similar calculations, it is easy to obtain the expression for the fluctuations of other quantities in the nonisothermal plasma. In particular, the spectral distribution of electron density fluctuations is found to be

$$\langle \delta N_{\mathrm{e}}^2 \rangle_{\omega,k} = \frac{2N_{\mathrm{e}}}{\omega} \frac{k^2 v_{\mathrm{Te}}^2}{\omega_{\mathrm{pe}}^2} \frac{1}{[\varepsilon^{\mathrm{lo}}(\omega,k)]^2} \left[|1 + \delta\varepsilon_{\mathrm{i}}^{\mathrm{lo}}(\omega,k)|^2 \operatorname{Im}\{\delta\varepsilon_{\mathrm{e}}^{\mathrm{lo}}(\omega,k)\} + \frac{T_{\mathrm{i}}}{T_{\mathrm{e}}} |\delta\varepsilon_{\mathrm{e}}^{\mathrm{lo}}(\omega,k)|^2 \operatorname{Im}\{\delta\varepsilon_{\mathrm{i}}^{\mathrm{lo}}(\omega,k)\} \right]. \tag{10.4.8}$$

In (10.4.7, 8) $\varepsilon^{\mathrm{lo}}(\omega, k)$ and $\delta\varepsilon_{\alpha}^{\mathrm{lo}}(\omega, k)$ are determined by means of (4.1.14).

Let us consider the transparency ranges of the nonisothermal plasma in order to analyze more thoroughly the spectral distribution of electron density fluctuations. Two such ranges are known to exist. The first one corresponds to the high-frequency range (high phase velocities), where the spectrum of weakly damped plasma oscillations exists. In this range, the plasma behaves as an electron gas and one can apply the results obtained in the previous section for its description.

The second range of transparency corresponds to the low-frequency branch of the weakly damped longitudinal oscillations with phase velocities satisfying the condition $v_{\mathrm{Ti}} \ll \omega/k < v_{\mathrm{Te}}$ in a highly nonisothermal ($T_{\mathrm{e}} \gg T_{\mathrm{i}}$) plasma. For $k^2 r_{\mathrm{De}}^2 \ll 1$, the spectrum of these oscillations is governed by, see (4.2.10):

$$\omega^2 = k^2 Z \frac{T_{\mathrm{e}}}{M} \left(1 + \frac{3T_{\mathrm{i}}}{ZT_{\mathrm{e}}}\right) = k^2 v_{\mathrm{s}}^2. \tag{10.4.9}$$

In this frequency range (10.4.8) yields

$$\langle \delta N_{\mathrm{e}}^2 \rangle_{\omega,k} = \frac{\pi}{2} N_{\mathrm{e}} k^4 r_{\mathrm{De}}^4 [\delta(\omega + k v_{\mathrm{s}}) + \delta(\omega - k v_{\mathrm{s}})]. \tag{10.4.10}$$

It follows that fluctuations have a sharp maximum near the resonance ion-acoustic frequencies $\omega = \pm k v_{\mathrm{s}}$. According to an exponential law the quantity $\langle \delta N_{\mathrm{e}}^2 \rangle_{\omega,k}$ vanishes when the frequency grows since in this frequency range, see (4.1.16):

$$\operatorname{Im}\{\delta\varepsilon_{\mathrm{e}}^{\mathrm{lo}}(\omega,k)\} \sim \exp\left(-\frac{\omega^2}{2k^2 v_{\mathrm{s}}^2}\right).$$

In the low-frequency range, the quantity $\langle \delta N_{\mathrm{e}}^2 \rangle_{\omega,k}$ also decreases. For $\omega \ll k v_{\mathrm{Ti}}$, the last summand in (10.4.8) gives the main contribution, and we have approximately

$$\langle \delta N_{\mathrm{e}}^2 \rangle_{\omega,k} \approx \sqrt{8\pi}\, N_{\mathrm{e}} \frac{k^3 r_{\mathrm{Di}}^3}{\omega_{\mathrm{pi}}}. \tag{10.4.11}$$

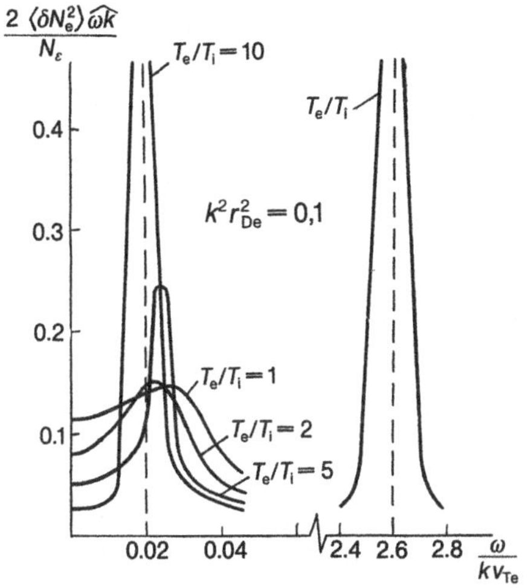

Fig. 10.2. Spectral distribution of fluctuations of the electron density in a nonisothermal plasma

Returning to the isothermal plasma, we note that the ion-acoustic maximum practically vanishes, but the absolute value of the correlator of electron density fluctuations in the low-frequency range for $T_e = T_i$ increases, as compared to a nonisothermal plasma. Figure 10.2 shows the spectral distribution of the quantity $2\langle\delta N_e^2\rangle_{\omega,k}/N_e$ versus the dimensionless frequency ω/kv_{Te} for $k^2r_{De}^2 \ll 1$ and for different values of T_e/T_i.

For a nonisothermal plasma we obtain from (10.4.6)

$$T_{eff} = \frac{\sum\limits_\alpha N_\alpha e_\alpha^2 m_\alpha^{1/2} T_\alpha^{-1/2} \exp\left(-\dfrac{\omega^2}{2k^2v_{T\alpha}^2}\right)}{\sum\limits_\alpha N_\alpha e_\alpha^2 m_\alpha^{1/2} T_\alpha^{-3/2} \exp\left(-\dfrac{\omega^2}{2k^2v_{T\alpha}^2}\right)} . \qquad (10.4.12)$$

Naturally, for the ion-acoustic range of transparency of the highly nonisothermal plasma with $T_e \gg T_i$, (10.4.12) shows that $T_{eff} \simeq T_e$. Thus, the effective fluctuation temperature is completely determined by the electron temperature both in high- and low-frequency ranges of transparency.

10.4.2 Fluctuations in the Plasma with an Electron Beam

Let us now consider the plasma penetrated by the nonrelativistic beam of charged particles (electrons) with the ordered velocity $\boldsymbol{u}$. We assume beam particles to possess Maxwellian velocity distribution in the moving frame. Plasma electrons have Maxwellian distribution in the laboratory frame. Then for the longitudinal dielectric permittivity of the plasma-beam system $\varepsilon(\omega, \boldsymbol{k})$ we can apply (6.3.14):

$$\varepsilon(\omega, \boldsymbol{k}) = \frac{k_i k_j}{k^2} \varepsilon_{ij}(\omega, \boldsymbol{k}) = 1 + \sum_\alpha \frac{\omega_{p\alpha}^2}{k^2 v_{T\alpha}^2}\left[1 - I_+\left(\frac{\omega - \boldsymbol{k}\cdot\boldsymbol{u}_\alpha}{k v_{T\alpha}}\right)\right]. \tag{10.4.13}$$

The summation extends over plasma and beam electrons. The expression for the function $F(\omega, k)$, which according to (10.4.6) determines the effective fluctuation temperature, can also be easily written as

$$F(\omega, k) = \sum_\alpha \sqrt{32\pi^3}\frac{\omega_{p\alpha}^2}{k^3 v_{T\alpha}^3} \exp\left(-\frac{(\omega - \boldsymbol{k}\cdot\boldsymbol{u}_\alpha)^2}{2k^2 v_{T\alpha}^2}\right). \tag{10.4.14}$$

As shown in Chap. 6, the plasma-beam system is stable for sufficiently small beam velocities. Therefore, the above-discussed general fluctuation theory of the nonequilibrium plasma is applicable to it[6]. Various types of instabilities begin to develop for the beam velocity exceeding some critical value. First, with the growth of the beam velocity there appear kinetic instabilities associated with a change in sign in the imaginary part of the dielectric permittivity Im $\{\varepsilon(\omega, \boldsymbol{k})\}$ and then hydrodynamic instabilities, their increment being independent of the thermal motion of particles (consequently, of Im $\{\varepsilon(\omega, \boldsymbol{k})\}$). Therefore, we consider fluctuations of a plasma which is in a subcritical state. Here we confine ourselves to the range of transparency where (10.4.4) can be applied to the fluctuations of the longitudinal field. Equation (10.4.4) shows that fluctuations infinitely increase in such a plasma which approaches a kinetically unstable state. The effective temperature must also infinitely increase and according to (10.4.6, 12, 13), we have

$$T_{\text{eff}} \approx \frac{\dfrac{N_e}{\sqrt{T_b}} \exp\left(-\dfrac{(\omega - \boldsymbol{k}\cdot\boldsymbol{u})^2}{2k^2 v_{Tb}^2}\right) + \dfrac{N_e}{\sqrt{T_e}} \exp\left(-\dfrac{\omega^2}{2k^2 v_{Te}^2}\right)}{\dfrac{N_e}{\sqrt{T_b^3}} \dfrac{(\omega - \boldsymbol{k}\cdot\boldsymbol{u})}{\omega} \exp\left(-\dfrac{(\omega - \boldsymbol{k}\cdot\boldsymbol{u})^2}{2k^2 v_{Tb}^2}\right) + \dfrac{N_e}{\sqrt{T_b^3}} \exp\left(-\dfrac{\omega^2}{2k^2 v_{Te}^2}\right)}, \tag{10.4.15}$$

where the frequency and the wave vector are related by the dispersion equation.

Further, the beam density is regarded as low, $N_b \ll N_e$. The beam contribution to the real part of the dielectric permittivity can be neglected. But the beam contribution to its imaginary part as well as to the function $F(\omega, k)$

6 Strictly speaking, this is valid only for electrostatic perturbations. In other words, the plasma-beam system is stable for small beam velocities with respect to longitudinal (electrostatic) fluctuations of the field.

becomes decisive when the plasma approaches a kinetically unstable state. Let us consider the high-frequency range of transparency $\omega \gg kv_{\mathrm{Te}}$ with the plasma oscillation spectrum

$$\omega^2 = \omega_{\mathrm{pe}}^2 + 3k^2v_{\mathrm{Te}}^2 \,. \tag{10.4.16}$$

Then for the beam with a broad thermal spread of particles over velocities when $|\omega - \boldsymbol{k}\cdot\boldsymbol{u}| < kv_{\mathrm{Tb}}$ (10.4.15) is simplified:

$$T_{\mathrm{eff}} = T_{\mathrm{e}} \frac{1 + \dfrac{N_{\mathrm{b}}}{N_{\mathrm{e}}}\sqrt{\dfrac{T_{\mathrm{e}}}{T_{\mathrm{b}}}}\exp\left(-\dfrac{\omega^2}{2k^2v_{\mathrm{Te}}^2}\right)}{\left[1 + \dfrac{N_{\mathrm{b}}}{N_{\mathrm{e}}}\left(\dfrac{T_{\mathrm{e}}}{T_{\mathrm{b}}}\right)^{3/2}\exp\left(-\dfrac{\omega^2}{2k^2v_{\mathrm{Te}}^2}\right)\right]\left(1 - \dfrac{u}{u_{\mathrm{cr}}}\cos\theta\right)} \,. \tag{10.4.17}$$

Here θ is the angle included between the vectors $\boldsymbol{u}$ and $\boldsymbol{k}$, and

$$u_{\mathrm{cr}} = \frac{\omega}{k}\left[1 + \frac{N_{\mathrm{e}}}{N_{\mathrm{b}}}\left(\frac{T_{\mathrm{b}}}{T_{\mathrm{e}}}\right)^{3/2}\exp\left(-\frac{\omega^2}{2k^2v_{\mathrm{Te}}^2}\right)\right] . \tag{10.4.18}$$

Equation (10.4.18) shows an infinite increase of the effective fluctuation temperature for the beam velocity u approaching the critical velocity u_{cr}. The latter corresponds to the beginning of the kinetic instability development in the system.

An increase of fluctuations in the plasma, approaching an unstable state, is called the *critical opalescence* and specifies a sharp increase of scattering of electromagnetic waves in such a plasma (Sect. 10.6).

10.5 Fluctuations and Interparticle Collisions in Plasma

In the previous sections plasma fluctuations have been considered when interparticle collisions are neglected. Let us analyze the relationship between collisional and fluctuation processes, which is vividly manifested in a plasma composed of charged particles with long-range interaction forces. This is a two-aspect problem. The first aspect lies in the fact that collisions must essentially influence the form of fluctuation correlators of electromagnetic fields and plasma parameters. The reason is that fluctuations are determined by the electromagnetic energy dissipation in plasma, in particular, due to the interparticle collisions. Then collisions can be taken into account by means of general methods. Indeed, since the anti-Hermitian part of the dielectric permittivity completely determines fluctuations (at least, in the thermodynamically equilibrium plasma), their calculation can be carried out on the basis of the relations given in the previous sections, e.g., (10.2.2 and 4) for an equilibrium plasma where the expressions for the tensor $\varepsilon_{ij}(\omega, \boldsymbol{k})$ including the interparticle collisions should be substituted. Thus, one should apply (4.5.8), (5.5.6, 7) for a weakly ionized nondegenerate plasma, and (4.5.19), (5.5.6, 11) for a degenerate one. For a fully ionized plasma with interparticle collisions, the tensor $\varepsilon_{ij}(\omega, \boldsymbol{k})$ is determined in various limits by the formulas from Sects. 4.6 and 5.6.

Principally, a further consideration of fluctuations of electromagnetic fields and plasma parameters is not difficult but it is associated with cumbersome calculations. Physically, the influence of collisions on correlators of various quantities is quite clear. Actually, the collisional dissipation must lead to effects of two kinds. Far from the ranges of transparency fluctuations must grow due to the emergence of the additional dissipation mechanism. On the other hand, the spectral distribution of fluctuations near resonance frequencies must also vary. In the above, the fluctuations have been assumed to be proportional to δ-functions of corresponding frequencies, i.e., increase infinitely near natural frequencies of weakly damped oscillations of the plasma. However, due to the Landau damping, fluctuations are naturally finite even for resonance frequencies, and resonance peaks possess not only finite amplitudes but also finite width. The account of collisions must amplify these effects, namely decrease resonance fluctuation amplitudes and widen the resonance.

Let us take as an example fluctuations of the charge in the isotropic plasma accounting for the interparticle collisions near the frequencies of natural plasma oscillations. Assuming the collision frequency to be small, $\omega \gg \nu_e$, kv_{Te}, the collisional correction to the longitudinal dielectric permittivity both for the weakly and fully ionized plasma is of the form, see (4.5.10), (4.6.5):

$$\operatorname{Im}\{\delta\varepsilon^{\mathrm{lo}}_{\mathrm{col}}\} = \frac{\omega_{\mathrm{pe}}^2 \nu_e}{\omega^3}, \tag{10.5.1}$$

where $\nu_e = \nu_{en}$ for the weakly ionized plasma and $\nu_e = \nu_{\mathrm{eff}}$ for the fully ionized one. Substitution of this expression into (10.3.4) results in the following relation for charge density fluctuations near the resonance frequency $\omega^2 \approx \omega_{\mathrm{res}}^2 = \omega_{\mathrm{pe}}^2 + 3k^2 v_{\mathrm{Te}}^2$

$$\langle \varrho^2 \rangle_{\omega,k} \approx \frac{k^2 T}{2\pi} \frac{\omega_{\mathrm{pe}}^2 \nu_e}{(\omega^2 - \omega_{\mathrm{pe}}^2 - 3k^2 v_{\mathrm{Te}}^2)^2 + (\omega_{\mathrm{pe}}^2 \nu_e/\omega)^2}. \tag{10.5.2}$$

This relation shows that the quantity $\langle \varrho^2 \rangle_{\omega,k}$ decreases with increasing ν_e and the width of the resonance grows. For $\nu_e \to 0$ (10.5.2) reduces to (10.3.8).

10.5.1 Fokker-Planck Equation

The other aspect of the discussed problem lies in the close relationship between collisions and fluctuations. It is obvious for collisions with neutrals in the weakly ionized plasma since interaction forces strongly subside with distance and a collision directly means a close convergence of particles. But with respect to a spatially homogeneous distribution of particles this is a fluctuation. More complex is the case of the completely ionized plasma where interparticle interaction forces slowly subside with distance. This has been briefly discussed when we analyzed the possibility of the plasma description by means of the collisionless Vlasov kinetic equation. Actually, on defining the field in terms of the distribution function of particles averaged over fluctuations, fluctuation effects also associated with the Coulomb interparticle interaction might be placed on the right-hand side of the kinetic equation and can be identified as collisions. In the kinetic gas theory this identification is quite valid and applicable to an ordinary gas where particles are neutral on the whole; as a consequence of this, the interaction forces between them quickly subside with distance. In the completely ionized plasma, when interaction forces of particles slowly subside with distance, the contribution to the fluctuation interaction associated with simultaneous interaction of a lot of rather distant particles appears more significant than single convergence of particles as binary collisions. Therefore, in the completely ionized plasma a collision motion has a conventional meaning. To describe fluctuation interaction, it is more convenient to apply the continuity model which treats the interaction not as a discrete process of binary collisions but as a continuously acting force of the Coulomb friction. Hence, the calculation of the collision integral for the completely ionized plasma (see Sect. 3.3) gives the Fokker-Planck equation (3.3.12), which is the basic equation for the description of various random processes in the continuity model. To illustrate this fact, let us obtain the Fokker-Planck equation by means of a general method.

Let us consider the behaviour of the distribution function of charged particles in the plasma under the action of the fluctuation interaction, or, in other words, under the influence of the Coulomb collisions. At the time moment t the distribution function is $f(\boldsymbol{v}-\Delta\boldsymbol{v},t)$ and at $t+\Delta t$ it is modified due to the fluctuation interaction and becomes $f(\boldsymbol{v},t+\Delta t)$. If the probability of such a transfer from one state into the other (from one phase trajectory to the other) is equal to $W(\Delta\boldsymbol{v},\Delta t)$, then we can write

$$f(\boldsymbol{v},t+\Delta t)=\int f(\boldsymbol{v}-\Delta v,t)\,W(\Delta\boldsymbol{v},\Delta t)\,d(\Delta\boldsymbol{v})\ . \tag{10.5.3}$$

By expanding the left-hand side of this equation in the Taylor power series in Δt and the right-hand side in $\Delta\boldsymbol{v}$, and confining ourselves to the first expansion terms, we can transform it into the differential equation of the form

$$\frac{\partial f(\boldsymbol{v},t)}{\partial t}=\frac{\partial f(\boldsymbol{v},t)}{\partial v_i}\frac{\langle\Delta v_i\rangle}{\Delta t}+\frac{1}{2}\frac{\partial^2 f(\boldsymbol{v},t)}{\partial v_i\partial v_j}\frac{\langle\Delta v_i\Delta v_j\rangle}{\Delta t}\ , \tag{10.5.4}$$

which is called the Fokker-Planck equation. Here

$$\begin{aligned} &\int W(\Delta \boldsymbol{v}, \Delta t)\, d(\Delta \boldsymbol{v}) = 1\,, \\ &\int W(\Delta \boldsymbol{v}, \Delta t)\, \Delta v_i\, d(\Delta \boldsymbol{v}) = \langle \Delta v_i \rangle\,, \\ &\int W(\Delta \boldsymbol{v}, \Delta t)\, \Delta v_i \Delta v_j\, d(\Delta \boldsymbol{v}) = \langle \Delta v_i \Delta v_j \rangle\,. \end{aligned} \tag{10.5.5}$$

The first relation means the normalization condition and the others are common definitions of the mean values.

Equation (10.5.4) can also be written as

$$\frac{\partial f(\boldsymbol{v}, t)}{\partial t} = \frac{\partial}{\partial v_i}\left[D_{ij}\frac{\partial f(\boldsymbol{v}, t)}{\partial v_j} - A_i f(\boldsymbol{v}, t)\right], \tag{10.5.6}$$

where the quantities

$$A_i = -\frac{\langle \Delta v_i \rangle}{\Delta t}, \quad D_{ij} = \frac{1}{2}\frac{\langle \Delta v_i \Delta v_j \rangle}{\Delta t} \tag{10.5.7}$$

are called the *Fokker-Planck coefficients*. Equation (10.5.6) coincides in form with (3.3.12) but the Fokker-Planck coefficients *(coefficients of dynamic friction and diffusion in a velocity space)* are directly expressed in terms of velocity fluctuations[7].

Note that in the kinetic gas theory the terms which we neglected in (10.5.4) contain higher orders of Δv_i. These describe binary collisions during which the velocity undergoes sharp jumps and their summation leads to the Boltzmann collision integral. Physical reasons show that the coefficient of dynamic friction in the velocity space A_i characterizes an average variation in the absolute value of the particle velocity per unit time. Consequently, it can be expressed in terms of energy losses of the ordered velocity of charged plasma particles (Exercise 2.8.4). The coefficient D_{ij} is of purely fluctuative origin and characterizes an ordered variation of particle velocity due to scattering on fluctuation fields in the plasma. Therefore, it is called the diffusion coefficient in a velocity space.

10.5.2 Correlation Coefficients of the Dynamic Friction and Diffusion with Plasma Fluctuation Fields

The coefficients A_i and D_{ij} can be calculated from (10.5.7) and from the equations of motion of charged particles in fluctuation fields. However, we act otherwise. Let us express the coefficients A_i and D_{ij} in terms of longitudi-

7 In (10.5.6) the coefficients D_{ij} and A_i differ from those in (3.3.12) by the factors $1/m^2$ and a/m, respectively.

nal field fluctuations and energy losses of a test particle on excitation of this field in a plasma by comparing (3.3.12, 13), (3.4.5), (10.4.1) and (2.8.29). For simplicity, we focus our attention on the nondegenerate isotropic electron plasma taking account only of electron-electron collisions. According to (3.3.13) and (3.4.5), the coefficients D_{ij} and A_i have the form

$$\begin{aligned} D_{ij}(\boldsymbol{v}) &= 2e^4 \int d\boldsymbol{p}' f(\boldsymbol{p}') \int d\boldsymbol{k} \frac{k_i k_j \delta(\boldsymbol{k}\cdot\boldsymbol{v} - \boldsymbol{k}\cdot\boldsymbol{v}')}{k^2 |\varepsilon^{\mathrm{lo}}(\boldsymbol{k}\cdot\boldsymbol{v}, \boldsymbol{k})|^2}, \\ A_i(\boldsymbol{v}) &= e^4 \int d\boldsymbol{p}' \frac{\partial f(\boldsymbol{p}')}{\partial p_j'} \int d\boldsymbol{k} \frac{k_i k_j \delta(\boldsymbol{k}\cdot\boldsymbol{v} - \boldsymbol{k}\cdot\boldsymbol{v}')}{k^2 |\varepsilon^{\mathrm{lo}}(\boldsymbol{k}\cdot\boldsymbol{v}, \boldsymbol{k})|^2}. \end{aligned} \tag{10.5.8}$$

On the other hand, for the electron plasma (10.4.1) leads to

$$\langle (E^{\mathrm{lo}})^2 \rangle_{\boldsymbol{k}\cdot\boldsymbol{v},\boldsymbol{k}} = \frac{32\pi^3 e^2}{k^2 |\varepsilon^{\mathrm{lo}}(\boldsymbol{k}\cdot\boldsymbol{v}, \boldsymbol{k})|^2} \int d\boldsymbol{p}' f(\boldsymbol{p}') \delta(\boldsymbol{k}\cdot\boldsymbol{v} - \boldsymbol{k}\cdot\boldsymbol{v}'). \tag{10.5.9}$$

Comparing (10.5.8 and 9), we find

$$D_{ij}(\boldsymbol{v}) = \frac{e^2}{16\pi^3} \int d\boldsymbol{k} \frac{k_i k_j}{k^2} \langle (E^{\mathrm{lo}})^2 \rangle_{\boldsymbol{k}\cdot\boldsymbol{v},\boldsymbol{k}}. \tag{10.5.10}$$

Clearly the diffusion coefficient is expressed in terms of longitudinal field fluctuations in the plasma.

The dynamic friction coefficient A_i can be expressed here in terms of polarization losses of the test particle energy in the plasma. In order to accomplish this goal, we write the expression for the longitudinal field radiation due to a moving electron in an isotropic plasma, see (2.7.29):

$$\boldsymbol{E}^{\mathrm{lo}}(\boldsymbol{k}\cdot\boldsymbol{v}, \boldsymbol{k}) = -\frac{4\pi \mathrm{i} e}{k^2} \frac{\boldsymbol{k}}{(2\mathrm{p})^3 \varepsilon^{\mathrm{lo}}(\boldsymbol{k}\cdot\boldsymbol{v}, \boldsymbol{k})}. \tag{10.5.11}$$

Since (4.1.11) gives

$$\mathrm{Im}\{\varepsilon^{\mathrm{lo}}(\omega, \boldsymbol{k})\} = -\mathrm{i}\pi \frac{4\pi e^2}{\omega} \int d\boldsymbol{p} \frac{(\boldsymbol{k}\cdot\boldsymbol{v})\boldsymbol{k}}{k^2} \frac{\partial f_0}{\partial \boldsymbol{p}} \delta(\omega - \boldsymbol{k}\cdot\boldsymbol{v}) \tag{10.5.12}$$

we finally obtain

$$A_i(\boldsymbol{v}) = \frac{1}{32\pi^3} \int d\boldsymbol{k}\, k_i E^{\mathrm{lo}}(\boldsymbol{k}\cdot\boldsymbol{v}, \boldsymbol{k}) E^{*\mathrm{lo}}(\boldsymbol{k}\cdot\boldsymbol{v}, \boldsymbol{k}) \mathrm{Im}\{\varepsilon^{\mathrm{lo}}(\boldsymbol{k}\cdot\boldsymbol{v}, \boldsymbol{k})\}. \tag{10.5.13}$$

Hence the dynamic friction coefficient is expressed in terms of the longitudinal field produced in a plasma by a moving electron. This field is responsible for the polarization energy losses of a moving electron.

Note that A_i and D_{ij} are expressed in terms of the longitudinal field in a plasma. But actually transverse fields also make a contribution, i.e., losses on wave radiation and fluctuations of a transverse field, though their contribution is negligible in a nonrelativistic plasma. In an isotropic plasma, transverse waves are not radiated at all and the transverse field fluctuations are c^3/v_{Te}^3 times smaller than those of the longitudinal field (Exercise 10.8.2). Therefore, scattering of particles on the transverse field fluctuations may also be neglected.

Thus, the collision integral in the kinetic equation for charged particles in a fully ionized nonrelativistic plasma is shown to be expressed in terms of longitudinal field fluctuations. This implies that variations of momentum of charged particles due to the Coulomb collisions can be treated as energy losses and diffusion in the velocity space due to particle scattering in random electrostatic fields specified by thermal fluctuations in the plasma. Such a close relationship between thermal fluctuations and collisions of charged particles in the thermodynamically equilibrium fully ionized plasma can be seen from (10.3.14) which is rewritten in the form

$$\frac{\langle (E^{\mathrm{lo}})^2\rangle_r}{4\pi N_{\mathrm{e}} T_{\mathrm{e}}} \approx \left(\frac{e^2 N_{\mathrm{e}}^{1/3}}{T_{\mathrm{e}}}\right)^{3/2} \approx \frac{\nu_{\mathrm{eff}}}{\omega_{\mathrm{pe}}} \ll 1 \,. \tag{10.5.14}$$

It follows that the relative energy of the longitudinal field fluctuation (with respect to the thermal energy) is equal to the relative effective frequency of electron-electron collisions in the plasma (with respect to the natural frequency of plasma oscillations).

10.6 Scattering of Electromagnetic Waves in a Plasma

When electromagnetic waves propagate in media with random inhomogeneities, there appear waves with frequencies and wave vectors which are different from the frequency and wave vector of the fundamental wave. Here, the so-called scattering process occurs. If the medium is spatially homogeneous but parameters defining its electromagnetic properties experience fluctuations, then scattering must occur on these fluctuations, the latter being random inhomogeneities.

Induced charges and currents leading to radiating new scattered waves emerge in a medium under the influence of the fundamental wave, thereby initiating the appearance of scattered waves. However, as it has been mentioned, within the linear approximation induced charges and currents in the homogeneous medium represent only the modification of wave propagation characteristics in a medium, as compared to vacuum, i.e., modification of the complex refractive index. Here, the frequency and the dispersion law of a propagating wave are strictly constant. In the presence of fluctuations in a medium the situation is significantly altered. Thus, since the density of charged particles fluctuates, the induced current will also fluctuate, and a fluctuating addition may cause radiation of waves with new properties, i.e., frequency and propagation orientation, *(scattering)*, or even the occurrence of waves of another kind *(transformation)*. In turn, new waves alter the plasma state giving rise to induced currents associated with them and they are also able to influence the propagation of the fundamental wave. The process of complex nonlinear interaction between fields and currents, earlier completely ignored, occurs in the plasma. Some properties of the nonlinear processes will be treated in this and the next sections. Then we can deal separately with incident and scattered waves and assume the field of the incident wave and plasma parameters to be assigned. Note that fluctuations obey quadratic forms (Sect. 10.1), therefore, nonlinear equations must be applied to the interaction processes of fluctuation fields with those of regular waves.

Here, we consider the scattering process of transverse electromagnetic waves in a plasma; this process is of self-important value from the point of view of studying wave propagation and absorption. For simplicity, we deal only with the isotropic plasma. Then we can assert that for the region of transparency of propagating waves $\omega > \omega_{pe}$ the plasma can be referred to as a purely electron gas, and effects associated with spatial dispersion of electric permittivity can be neglected, i.e., a simple expression

$$\varepsilon(\omega) = 1 - \frac{\omega_{pe}^2}{\omega^2} \tag{10.6.1}$$

can be used for dielectric permittivity.

Let a transverse electromagnetic wave propagate in the plasma

$$\begin{pmatrix} \boldsymbol{E}_0(t, \boldsymbol{r}) \\ \boldsymbol{B}_0(t, \boldsymbol{r}) \end{pmatrix} = \begin{pmatrix} \boldsymbol{E}_0 \\ \boldsymbol{B}_0 \end{pmatrix} \exp(-i\omega_0 t + i\boldsymbol{k}_0 \cdot \boldsymbol{r}) , \tag{10.6.2}$$

where $k_0^2 = \omega_0^2 \varepsilon(\omega_0)/c^2$. Then we consider its scattering on density fluctuations of electrons, i.e., calculate the field of a scattered wave in the form

$$\boldsymbol{E}(t, \boldsymbol{r}) = \boldsymbol{E} \exp(-i\omega t + i\boldsymbol{k} \cdot \boldsymbol{r}) . \tag{10.6.3}$$

This field can be determined from the wave equation

$$\operatorname{curl}\operatorname{curl} \boldsymbol{E} + \frac{\varepsilon(\omega)}{c^2} \frac{\partial^2 \boldsymbol{E}}{\partial t^2} = -\frac{4\pi}{c} \frac{\partial \boldsymbol{j}}{\partial t} , \tag{10.6.4}$$

where $\boldsymbol{j}$ is a part of the induced current stimulating the emergence of a scattered wave.

Generally, the density and velocity of charged particles in the plasma can be presented as

$$\begin{aligned} N(t, \boldsymbol{r}) &= N_{0e} + \delta N_{in} + \delta N_f + \delta N_{sc} \, , \\ \boldsymbol{u}(t, \boldsymbol{r}) &= \delta \boldsymbol{u}_{in} + \delta \boldsymbol{u}_f + \delta \boldsymbol{u}_{sc} \, . \end{aligned} \tag{10.6.5}$$

Here N_{0e} is a mean (over fluctuations) equilibrium value of the electron density in the plasma; δN_{in} and $\delta \boldsymbol{u}_{in}$ are electron density and velocity perturbations produced by the field of an incident wave; δn_f and $\delta \boldsymbol{u}_f$ are density and velocity fluctuations of particles; δN_{sc} and $\delta \boldsymbol{u}_{sc}$ are electron density and velocity perturbations associated with a scattered wave.

If thermal effects are ignored, i.e., the phase velocities of incident and scattered waves are considered larger than the thermal velocities of particles, then the hydrodynamic description can be applied to the analysis of plasma processes. Thus, in (10.6.5) $\boldsymbol{u}$ means a hydrodynamic electron velocity in the plasma. The quantities δN_{in} and $\delta \boldsymbol{u}_{in}$ are determined from the linearized equations of motion and continuity accounting for an incident wave

$$\delta \boldsymbol{u}_{in} = \frac{ie\boldsymbol{E}_0(t, \boldsymbol{r})}{m\omega_0} \, , \qquad \delta N_{in} = \frac{ie\boldsymbol{k}_0 \boldsymbol{E}_0(t, \boldsymbol{r})}{m\omega_0^2} \, . \tag{10.6.6}$$

The quantities $\delta \boldsymbol{u}_{sc}$ and δN_{sc} satisfy the equations of motion and continuity:

$$\begin{aligned} \frac{\partial \delta \boldsymbol{u}_{sc}}{\partial t} &= \frac{e}{m}\boldsymbol{E} + \frac{e}{mc}[\delta \boldsymbol{u}_f, \boldsymbol{B}_0] - (\delta \boldsymbol{u}_{in} \cdot \nabla)\, \delta \boldsymbol{u}_f - (\delta \boldsymbol{u}_f \cdot \nabla)\, \delta \boldsymbol{u}_{in} \, , \\ \frac{\partial \delta N_{sc}}{\partial t} &+ N_{0e} \operatorname{div} \delta \boldsymbol{u}_{sc} + \delta N_{in} \operatorname{div} \delta \boldsymbol{u}_f + \delta N_f \operatorname{div} \delta \boldsymbol{u}_{in} = 0 \, . \end{aligned} \tag{10.6.7}$$

Velocity and density fluctuations are also associated with the continuity equation. In (10.6.7) it has been taken into account that the quantities δN_f, $\delta \boldsymbol{u}_f$, δN_{sc} and $\delta \boldsymbol{u}_{sc}$ are small against the quantities δN_{in} and $\delta \boldsymbol{u}_{in}$. Assuming that δN_{in} and $\delta \boldsymbol{u}_{in}$ are dependent on time and coordinates in the same way as the incident wave (10.6.2), i.e., $\sim \exp(i\omega_0 t + i\boldsymbol{k}_0 \cdot \boldsymbol{r})$ and δN_{sc} and $\delta \boldsymbol{u}_{sc}$ as the scattered wave (10.6.3), i.e., $\sim \exp(-i\omega t + i\boldsymbol{k} \cdot \boldsymbol{r})$, then δN_f and $\delta \boldsymbol{u}_f$ will be $\sim \exp(-i\Delta\omega t + i\boldsymbol{q} \cdot \boldsymbol{r})$, where $\boldsymbol{q} = \boldsymbol{k} - \boldsymbol{k}_0$, $\Delta\omega = \omega - \omega_0$. As a result we obtain from (10.6.7)

$$\delta \boldsymbol{u}_{sc} = \frac{ie}{m\omega}\left\{\boldsymbol{E} + \frac{1}{c}[\delta \boldsymbol{u}_f, \boldsymbol{B}_0]\right\} + \frac{1}{\omega}[(\boldsymbol{q} \cdot \delta \boldsymbol{u}_{in})\, \delta \boldsymbol{u}_f + (\boldsymbol{k} \cdot \delta \boldsymbol{u}_f)\, \delta \boldsymbol{u}_{sc}] \, . \tag{10.6.8}$$

Let us consider further only longitudinal fluctuations for which $\delta \boldsymbol{u}_f \parallel \boldsymbol{q}$. Therefore, from the continuity equation it follows that

$$\delta \boldsymbol{u}_{\mathrm{f}} = \frac{\Delta\omega \boldsymbol{q}}{N_{0e} q^2}\, \delta N_{\mathrm{f}} . \tag{10.6.9}$$

The density of the current associated with the scattered wave is

$$\boldsymbol{j}_{\mathrm{sc}} = e\,(N_{0e}\delta \boldsymbol{u}_{\mathrm{sc}} + \delta N_{\mathrm{in}}\delta \boldsymbol{u}_{\mathrm{f}} + \delta N_{\mathrm{f}}\,\delta \boldsymbol{u}_{\mathrm{in}}) . \tag{10.6.10}$$

Then we must subtract from this expression a value

$$\boldsymbol{j}_{\mathrm{sc}}|_{\delta N_{\mathrm{f}}=0} = eN_0 \delta \boldsymbol{u}_{\mathrm{sc}}|_{\delta N_{\mathrm{f}}=0} = \frac{\mathrm{i}e^2 N_{0e}}{m\omega} \boldsymbol{E} , \tag{10.6.11}$$

which corresponds to the density of the current associated with the scattered wave in the absence of fluctuations. Thus, we can obtain the current density of the scattering wave, which participates in (10.6.4)

$$\boldsymbol{j} = \boldsymbol{j}_{\mathrm{sc}} - \boldsymbol{j}_{\mathrm{sc}}|_{\delta N_{\mathrm{f}}=0} . \tag{10.6.12}$$

Substituting (10.6.10, 11) into (10.6.12) we find

$$\boldsymbol{j}(\omega, \boldsymbol{k}) = \frac{\mathrm{i}e^2}{m\omega_0} \left[\boldsymbol{E}_0 + \frac{\Delta\omega}{\omega}\frac{\boldsymbol{k}}{q^2}(\boldsymbol{q}\cdot\boldsymbol{E}_0) + \frac{\Delta\omega}{\omega_0}\frac{\boldsymbol{q}}{q^2}(\boldsymbol{k}_0\cdot\boldsymbol{E}_0) \right] \delta N_{\mathrm{f}}(\Delta\omega, \boldsymbol{q}) . \tag{10.6.13}$$

This expression shows that the current density possesses both transverse (the first term) and longitudinal components (the second and third term), the latter (for transverse $\boldsymbol{E}_0$ and thus $\delta \boldsymbol{u}_{\mathrm{in}}$) emerging due to the nonlinearity of the equation of motion. Here, the transverse component is the origin of the scattered transverse wave and the longitudinal component is associated with a longitudinal wave, i.e., the transformation of the transverse wave into a longitudinal one.

Let us factor out the transverse (with respect to the vector $\boldsymbol{k}$) component of the current density from (10.6.13) and substitute it into (10.6.4) as an external current. Then the solution of this equation gives the field of the scattering wave

$$\boldsymbol{E}(\omega, \boldsymbol{k}) = \frac{4\pi\mathrm{i}\omega}{c^2} \frac{\boldsymbol{j}_{\perp}(\omega, \boldsymbol{k})}{k^2 - \omega^2\varepsilon(\omega)/c^2} , \tag{10.6.14}$$

where we obtain from (10.6.13) for $\boldsymbol{j}_{\perp}(\omega, \boldsymbol{k})$

$$\boldsymbol{j}_{\perp}(\omega, k) = \frac{\mathrm{i}e^2}{\mathrm{m}\omega_0} \delta N_{\mathrm{f}}(\Delta\omega, \boldsymbol{q})\, \boldsymbol{E}_0 . \tag{10.6.15}$$

Let us determine the energy increment of scattered waves per unit time:

$$\delta Q = -\tfrac{1}{2}\mathrm{Re}\{\int d\mathbf{r}\,\mathbf{E}(t,\mathbf{r})\mathbf{j}^*(t,\mathbf{r})\}\ . \tag{10.6.16}$$

On passing over to the Fourier components and averaging taking account of (10.6.14, 15) and (10.1.5), we obtain the expression for a mean energy increment of scattered waves:

$$\delta Q = \frac{\mathscr{V}}{(2\pi)^3}\frac{e^4}{m^2c^2\omega_0^2}\mathrm{Im}\left\{\int d\omega\, d\mathbf{k}\,\omega\,\frac{\langle\delta N_e^2\rangle_{\Delta\omega,\mathbf{q}}\,E_{0\perp}^2}{k^2-\varepsilon(\omega)\dfrac{\omega^2}{c^2}}\right\}\ . \tag{10.6.17}$$

Here $\mathscr{V}$ is a volume,

$$\mathbf{E}_{0\perp} = \frac{1}{k^2}[\mathbf{k},[\mathbf{k},\mathbf{E}_0]]$$

and $\langle\delta N_e^2\rangle_{\Delta\omega,\mathbf{q}}$ obeys (10.4.8).

Since the integrand is real, only its poles contribute to the imaginary part of (10.6.17). Formally substituting[8]

$$[k^2-\omega^2\varepsilon(\omega)/c^2]^{-1} \rightarrow \mathrm{i}\pi\delta\,[k^2-\omega^2\varepsilon(\omega)/c^2] \tag{10.6.18}$$

and integrating over the modulus $\mathbf{k}$ by means of the δ-function in (10.6.17), we obtain

$$\delta Q = \frac{c\mathscr{V}}{16\pi^2}\left(\frac{e^2}{mc^2}\right)^2\frac{\omega^2}{\omega_0^2}\sqrt{\varepsilon(\omega)}\,E_{0\perp}^2\,\langle\delta N_e^2\rangle_{\Delta\omega,\mathbf{q}}\,d\omega\, d\Omega\ , \tag{10.6.19}$$

where $d\Omega$ is the element of the solid angle. Note that in this equation the wave vector $\mathbf{k}$ and frequency ω are related by a general dispersion equation for a transverse wave $k^2 = \omega^2\varepsilon(\omega)/c^2$ and they can be treated as the wave vector and frequency of the scattered wave. Then the quantities $\mathbf{q}$ and $\Delta\omega$ become variations of the wave vector and frequency due to the scattering.

Let us choose the coordinate system so that the vector $\mathbf{k}$ oriented along the z-axis and the vector $\mathbf{k}_0$ lie in the yz-plane. Here, the angle included between these vectors will be denoted by θ (Fig. 10.3). Then the electric vector component of the incident wave $\mathbf{E}_{0\perp}$ perpendicular to the vector $\mathbf{k}$ is expressed by

$$E_{0\perp}^2 = E_0^2(\cos^2\phi + \cos^2\theta\sin^2\phi)\ , \tag{10.6.20}$$

---

8 The existence of the small positive imaginary part of $\varepsilon(\omega)$ specified by interparticle collisions enables this substitution.

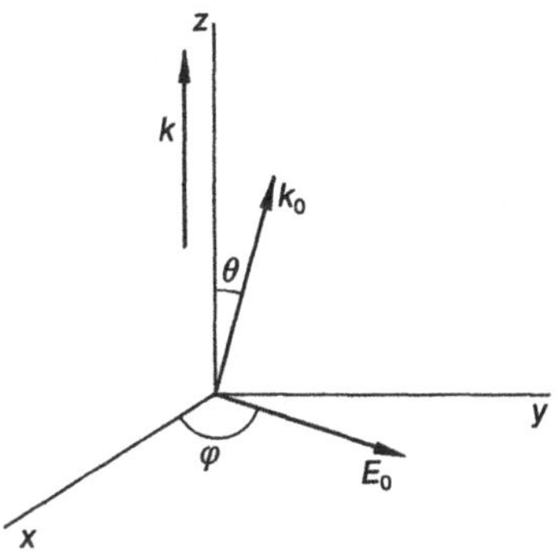

Fig. 10.3.

where ϕ is the angle included between the x-axis and the vector $\boldsymbol{E}_0$. For a nonpolarized incident wave this expression must be averaged over ϕ and thus

$$\overline{E}_{0\perp} = \tfrac{1}{2}(1 + \cos^2\theta)\, E_0^2\,. \tag{10.6.21}$$

10.6.1 Differential Cross Sections of Scattering of Transverse Electromagnetic Waves in the Nonisothermal Plasma

To characterize the dispersive properties of the medium, one usually introduces the notion of a *differential cross section of scattering*, defined as a ratio of the intensity of scattered waves (in the frequency range $d\omega$ and at the element $d\Omega$ of the solid angle) to the energy flux falling on the scattering volume V. The value of this current can be obtained by multiplying the Poynting vector by the volume V; therefore, the differential cross section of scattering for a nonpolarized wave, taking account of (10.6.20, 21), is[9]

$$d\Sigma = \frac{8\pi \overline{dQ}}{c\mathcal{V}\sqrt{\varepsilon(\omega_0)}\,E_0^2} = \frac{1}{4\pi}\left(\frac{e^2}{mc^2}\right)^2 \frac{\omega^2}{\omega_0^2}\sqrt{\frac{\varepsilon(\omega)}{\varepsilon(\omega_0}} \times (1+\cos^2\theta)\,\langle\delta N_e^2\rangle_{\Delta\omega,q}\, d\omega\, d\Omega\,. \tag{10.6.22}$$

It is valid for any values of $\Delta\omega$, i.e., for an arbitrary frequency variation due to the scattering. This formula can be simplified in the most interesting case of the Rayleigh scattering, i.e., scattering with small frequency variation, $\Delta\omega \ll \omega_0$. Finally, we obtain the well-known expression for the differential cross section of scattering on density fluctuations of electrons:

9 Note that the dimension of the scattering cross section so obtained differs from that of the generally accepted differential scattering cross section by the factor cm^{-3}. In the theory of scattering on fluctuations the scattering of radiation is considered per unit volume of the medium on N scattering centres.

$$d\Sigma = \frac{3}{32\pi^2}\sigma_T(1+\cos^2\theta)\langle\delta N_e^2\rangle_{\Delta\omega,q}\,d\omega\,d\Omega\,, \quad \text{where} \tag{10.6.23}$$

$$\sigma_T = \frac{8\pi}{3}(e^2/mc^2)^2$$

is the *Thomson cross section* of scattering of electromagnetic waves on free electrons which is proportional to the square of the classical electron radius. Then $|\boldsymbol{k}| \approx |\boldsymbol{k}_0|$ and

$$|\boldsymbol{q}| = \frac{2\omega_0}{c^2}\sin\frac{\theta}{2}\sqrt{\varepsilon(\omega_0)} = 2|\boldsymbol{k}_0|\sin\frac{\theta}{2}\,. \tag{10.6.24}$$

In order to analyze scattering of electromagnetic waves on density fluctuations of electrons in a nonisothermal plasma, (10.6.22 or 23) must be substituted into (10.4.8) for the spectral distribution of these fluctuations. Being extremely cumbersome, the final expression for $d\Sigma$ cannot be analytically studied in a general situation. Thus, we carry out the analysis only for the most interesting limiting cases, noting before that though scattering occurs on plasma electrons, by virtue of the dependence of $\langle\delta N_e^2\rangle_{\Delta\omega,q}$ on ion characteristics the character of scattering also appears to be dependent on them.

First, let us consider short wavelengths ($q^2 r_{De}^2 \gg 1$), i.e., $\omega_0 \sim \omega \gg \omega_{pe}c/v_{Te}$. Under these limits, we have $\varepsilon^{lo}(\Delta\omega, \boldsymbol{q}) \simeq 1$. Using (10.4.8) and (4.2.3) we obtain from (10.6.23)

$$d\Sigma = \frac{3N_{0e}}{(8\pi)^{3/2}}\sigma_T\frac{1+\cos^2\theta}{qv_{Te}}\exp\left(-\frac{(\Delta\omega)^2}{2q^2v_{Te}^2}\right)d\omega\,d\Omega\,, \tag{10.6.25}$$

i.e., here the scattering occurs in isolated electrons with the thermal spread over velocities. Line broadening due to the scattering is caused by the Doppler effect and is governed by the thermal electron velocity. The spectral distribution of scattering (or else a contour of the scattering line) is of Gaussian character.

In the opposite limit, when $q^2 r_{De}^2 \ll 1$, i.e.,

$$\omega_{pe} \ll \omega \sim \omega_0 \ll \omega_{pe}c/v_{Te}\,,$$

there appear collective effects associated with the self-consistent interaction between electron and ion plasma components. Let us analyze the spectral distribution of scattering for various frequency intervals.

For very small values of $\Delta\omega$ ($\Delta\omega \ll qv_{Ti}$), i.e., near ω_0 (the centre of the scattering line), charges with small velocities must mainly contribute to the intensity. Since for comparable electron and ion temperatures the number of ions with small velocities is many times ($\approx\sqrt{M/m}$) larger than the number of electrons, the form of the scattering line will be determined by the ion contribution to $\langle \delta N_e^2 \rangle_{\Delta\omega,q}$ (ion polarization clouds). Then,

$$d\Sigma = \frac{3N_{0e}}{(8\pi)^{3/2}}\sigma_T \frac{1}{qv_{Ti}}\left\{1 + \left[\frac{m}{M}\left(\frac{T_e}{T_i}\right)^3\right]^{1/2}\right\} + \left(1 + \frac{T_e}{T_i}\right)^{-2}(1 + \cos^2\theta)\, d\omega\, d\Omega\,. \qquad (10.6.26)$$

Hence, for a strongly nonisothermal plasma with $T_e \gg T_i$ the cross section of scattering with a small frequency variation can be much smaller than that of the isothermal plasma.

Slightly farther from the centre of the scattering line, where the condition $\Delta\omega \gg qv_{Ti}$ appears satisfied, the number of resonance ions exponentially falls and consequently the scattering cross section sharply decreases. Naturally, under these conditions the width of the scattering line will be determined by the thermal velocity. In the frequency range $qv_{Ti} \ll \Delta\omega \lesssim qv_{Te}$ of the strongly nonisothermal plasma, the scattering cross section has sharp maxima near the natural plasma ion-acoustic frequencies $|\Delta\omega| \approx qv_s$. Here, $\langle \delta N_e^2 \rangle_{\Delta\omega,q}$ should be applied according to (10.4.10), hence

$$d\Sigma = \frac{3N_0}{16\pi}\sigma_T(1 + \cos^2\theta)\,[\delta(\Delta\omega - qv_s) + \delta(\Delta\omega + qv_s)]\, d\omega\, d\Omega\,. \qquad (10.6.27)$$

The presence of maxima for $\Delta\omega = \pm qv_s$ (ion-acoustic satellites) corresponds to resonance scattering of transverse electromagnetic waves accompanied by the creation of the ion-acoustic waves. Alternatively, one also encounters a *decay* of an incident transverse wave with a frequency ω_0 into a daughter wave with a frequency ω plus an ion-acoustic wave with a frequency $\pm qv_s$, i.e., $\omega_0 = \omega \pm qv_s$, $\boldsymbol{k}_0 = \boldsymbol{k} + \boldsymbol{q}$.

Finally, for $\Delta\omega \gg qv_{Te}$ the scattering cross section has sharp maxima near the plasma oscillation frequencies $(\Delta\omega)^2 \approx \omega_{pe}^2$. Here the ion contribution may be ignored. Then

$$d\Sigma = \frac{3N_{0e}}{(8\pi)^{3/2}}\sigma_T \frac{(1 + \cos^2\theta)}{qv_{Te}} \cdot \frac{r_{De}^4 q^2 \exp(-\beta^2/2)}{(r_{De}^2 q^2 - 1/\beta^2)^2 + \frac{\pi}{2}\beta^2 \exp(-\beta^2)}\,, \qquad (10.6.28)$$

where $\beta = \Delta\omega/qv_{Te}$, i.e., the so-called plasma peak (the Langmuir or plasma satellite) with the contour shape of the dispersion character[10] is available near the frequencies $(\Delta\omega)^2 \approx \omega_{pe}^2$. The presence of such a peak corresponds to the

10 In optics the shape of a contour of a scattering or absorption line, being of the form $f(\omega) \sim [(\omega - \omega_0)^2 + \alpha^2]^{-1}$, is commonly called a dispersion curve.

resonance decay of an incident transverse wave when scattering into transverse and plasma waves with the conservation laws $\omega_0 = \omega \pm \omega_{pe}$, $\boldsymbol{k}_0 = \boldsymbol{k} \pm \boldsymbol{q}$ is realized.

The general character of the spectral distribution of Rayleigh scattering ($\Delta\omega \ll \omega_0$) follows the dependence of the spectral distribution of electron density fluctuations on the dimensionless frequency $\Delta\omega/(qv_{Te})$, see (10.6.23), and is described by diagrams, analogous to those in Fig. 10.2. The difference lies in the fact that since $\Delta\omega$ can be either larger or smaller than zero, the diagrams in Fig. 10.2 must be mirror reflected around the axis of ordinates to obtain a full shape of the scattering contour. Thus, Fig. 10.2 shows that part of the contour of scattering corresponds to the appearance of the scattered wave with a frequency $\omega > \omega_0$. Thus, the scattering character for $q^2 r_{De}^2 \ll 1$ (in Fig. 10.2. it corresponds to $k^2 r_{De}^2 \approx 0.1$) greatly differs for an isothermal ($T_e = T_i$) and a strongly nonisothermal ($T_e \gg T_i$) plasma. In an isothermal plasma, in the scattering spectrum there is a central maximum and a high-frequency lateral peak [for $\Delta\omega/(qv_{Te}) \approx 2.6$] caused by scattering on the Langmuir oscillations, see (10.3.8). The central maximum is associated with incoherent scattering on electrons, but its Doppler width is determined by the ion thermal velocity due to the self-consistent interaction between electron and ion components (the presence of ion polarization clouds around electrons). For a strongly nonisothermal plasma, the central maximum vanishes and there appear two low-frequency peaks placed symmetrically around the centre of the scattering line at the frequencies of natural ion-acoustic oscillations $\Delta\omega \approx \pm qv_s$ (naturally, in Fig. 10.2 we see only the peak corresponding to $\Delta\omega \approx qv_s$). The form of plasma peaks is conserved here.

Finally, we write down the cross section, integrated over the whole spectrum of scattered frequencies, at an angle θ. It is easily obtained by means of integrating the spectral distribution (10.6.23) over ω:

$$\Delta\Sigma = \frac{3N_{0e}}{16\pi}\sigma_T(1+\cos^2\theta)\,d\Omega \begin{cases} \dfrac{1+q^2 r_{De}^2}{2+q^2 r_{De}^2} & \text{if} \quad T_e \lesssim T_i\,, \\[2ex] 1 & \text{if} \quad T_e \gg T_i\,. \end{cases} \tag{10.6.29}$$

Note that the frequency range $\Delta\omega \lesssim qv_{Te}$ (in particular, ion-acoustic satellites in case of the nonisothermal plasma) gives the main contribution to this cross section; the contribution of plasma (Langmuir) satellites to the integral cross section is negligibly small.

10.6.2 Effect of Collisions on Scattering

Let us now analyze the influence of collisions on the spectral distribution of the scattering line for a weakly ionized plasma. The scattering cross section is defined, as usual, by (10.6.22) in which the correlator $\langle \delta N_e^2 \rangle_{\Delta\omega,\boldsymbol{q}}$ calculated from (4.5.8) must be substituted (Sect. 10.5). Here, the analytical calculation of the scattering contour is associated with lengthy computations, therefore, it is usually done by means of numerical methods. The results of such a calculation are shown in Fig. 10.4. The effect of collisions is to narrow the main scattering contour; here the Gauss distribution over velocities goes over to a dispersion-like shape. In the high-frequency range of the scattering line contour, the effect of collisions is manifested in widening the plasma peak according to (10.5.3) from which one can immediately obtain the spectral distribution of the correlator $\langle \delta N_e^2 \rangle_{\Delta\omega, q}$.

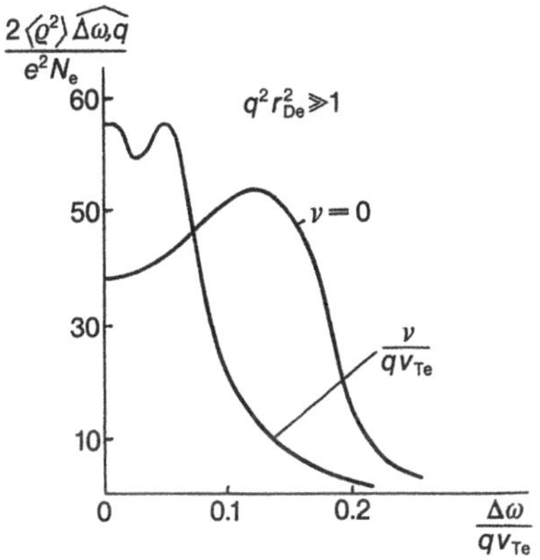

Fig. 10.4. Spectral distribution of fluctuations of the charge density in a weakly ionized plasma

In conclusion, we briefly discuss the problem of wave scattering in the nonequilibrium plasma with a beam where the instability development is possible. When the beam velocity is smaller than the critical one and the plasma is unstable, then the scattering cross section is determined by (10.6.22, 23) where (10.6.8) should be substituted for the electron density fluctuations in view of (10.6.13) for the longitudinal dielectric permittivity of the plasma with a beam. The contribution of ion summands can be neglected. Without a detailed analysis of the scattering cross section let us make only one general remark. As shown in Sect. 10.5, when the beam velocity approaches a critical value, plasma fluctuations sharply increase, thus the scattering cross section must also sharply grow. This phenomenon is associated with the plasma transition from one state (stable) into the other (with a developed instability). This can be interpreted as an analog of a phase transition in an ordinary medium. A sharp increase of scattering near the point of a phase transition is known in optics and is called the *phenomenon of critical opalescence*. Thus, it also exists in the nonequilibrium plasma near its stability limit.

10.7 Wave Transformation in Plasmas

Let us consider the case when a new type of wave, i.e., longitudinal, appears due to nonlinear interaction of the transverse electromagnetic wave $\boldsymbol{E}_0(t, \boldsymbol{r}) = \boldsymbol{E}_0 \exp(-\mathrm{i}\omega_0 t + \mathrm{i}\boldsymbol{k}_0 \cdot \boldsymbol{r})$ propagating in a plasma with fluctuations. We denote the field of the longitudinal wave as $\boldsymbol{E}^{\mathrm{lo}}(t, \boldsymbol{r}) = \boldsymbol{E}^{\mathrm{lo}} \exp(-\mathrm{i}\omega t + \mathrm{i}\boldsymbol{k} \cdot \boldsymbol{r})$. The field amplitude $\boldsymbol{E}^{\mathrm{lo}}$ should be determined from

$$\frac{\varepsilon^{\mathrm{lo}}(\omega, k)}{c^2} \frac{\partial^2 \boldsymbol{E}^{\mathrm{lo}}}{\partial t^2} = -\frac{4\pi}{c^2} \frac{\partial \boldsymbol{j}^{\mathrm{lo}}}{\partial t}, \tag{10.7.1}$$

where $\boldsymbol{j}^{\mathrm{lo}}$ is the longitudinal component of the current density (10.6.13):

$$\boldsymbol{j}^{\mathrm{lo}} = \frac{\mathrm{i}e^2}{m\omega_0}\left(1 + \frac{\Delta\omega}{\omega}\frac{k^2}{q^2}\right) \boldsymbol{E}_0^{\mathrm{lo}}\, \delta N_{\mathrm{e}}(\Delta\omega, \boldsymbol{q}) \,. \tag{10.7.2}$$

Here $\boldsymbol{E}_0^{\mathrm{lo}} = (\boldsymbol{k} \cdot \boldsymbol{E}_0)/k$ is the component of the vector $\boldsymbol{E}_0$, longitudinal to the vector $\boldsymbol{k}$, i.e., in the direction of propagation of the longitudinal wave $\boldsymbol{E}^{\mathrm{lo}}$.

The expression for $\boldsymbol{E}^{\mathrm{lo}}$ can be obtained by means of these formulas. Substituting it into (10.6.16) and averaging over time yields the increment of the longitudinal wave energy:

$$dQ = \frac{\mathscr{V}}{(2\pi)^3} \frac{3}{8\pi} \sigma_{\mathrm{T}} \mathrm{Im}\left\{\int \frac{E^{\mathrm{lo}2} d\omega\, d\boldsymbol{k}}{\varepsilon^{\mathrm{lo}}(\omega, k)\,\omega}\left(1 + \frac{\Delta\omega}{\omega}\frac{k^2}{q^2}\right)\right\} \langle \delta N_{\mathrm{e}}^2 \rangle_{\Delta\omega, \boldsymbol{q}} \,. \tag{10.7.3}$$

It follows that only frequencies satisfying the condition $\varepsilon^{\mathrm{lo}}(\omega, k) = 0$, i.e., frequencies of longitudinal natural oscillations of the plasma, contribute to the integral. Thus, the transformation effect consists in exciting natural longitudinal oscillations of the plasma as a result of absorption of the incident transverse electromagnetic wave. The integration over $d\boldsymbol{k}$ is easily carried out in a routine way by means of the δ-function, analogously to that done when deriving (10.6.20). Then, the ratio of dQ to the plasma volume and to the energy flux density of the incident transverse wave gives the expression for the *transformation coefficient* of the transverse wave into the longitudinal one.

10.7.1 Transformation of Transverse into Longitudinal Wave in an Isotropic Plasma

For the transformation of a transverse wave into a plasma wave, we can easily obtain

$$d\Sigma^{\mathrm{tr,lo}} = \frac{1}{32\sqrt{3}\pi^2}\sigma_{\mathrm{T}}\left(\frac{c}{v_{\mathrm{Te}}}\right)^3\frac{\omega^2}{\omega_0^2}\sqrt{\frac{\varepsilon(\omega)}{\varepsilon(\omega_0)}}\left(1+\frac{\Delta\omega}{\omega}\cdot\frac{k^2}{q^2}\right)$$

$$\times\sin^2\theta\,\langle\delta N_{\mathrm{e}}^2\rangle_{\Delta\omega,\boldsymbol{q}}\,d\omega\,d\Omega\,. \tag{10.7.4}$$

Here the incident wave is assumed to be nonpolarized, i.e., $(E^{\mathrm{lo}})^2 = 0.5\,E_0^2\sin^2\theta$, damping of plasma waves is neglected, and $k^2 \approx \varepsilon(\omega)\,\omega^2/(3\,v_{\mathrm{Te}}^2)$, or, $\omega^2 \approx \omega_{\mathrm{pe}}^2 + 3\,k^2v_{\mathrm{Te}}^2$ is used. The process under discussion may be treated as a decay of the incident transverse wave with the frequency ω_0 into two longitudinal waves with the frequencies $\omega \approx \omega_{\mathrm{pe}}$. Therefore, $\omega_0 \approx 2\,\omega_{\mathrm{pe}}$, i.e., $\Delta\omega \approx \omega_{\mathrm{pe}}$, $\boldsymbol{k}_0 = \boldsymbol{k} + \boldsymbol{q}$ and $k_0^2c^2 \approx 3\,\omega_{\mathrm{pe}}^2$.

One can analogously determine the transformation coefficient of the transverse wave into the longitudinal ion-acoustic wave.

Note that in fact both the scattering and transformation processes occur simultaneously during the wave propagation. Therefore, it is important to know a relative contribution of these processes causing attenuation of an incident wave. The ratio of the transformation coefficient (10.7.4) to the scattering cross section is given by

$$\frac{d\Sigma^{\mathrm{tr,lo}}}{d\Sigma} = \left(\frac{c^2}{3\,v_{\mathrm{Te}}^2}\right)^{3/2}\frac{\sin^2\theta}{1+\cos^2\theta}\left(1+\frac{\Delta\omega}{\omega}\frac{k^2}{q^2}\right)^2. \tag{10.7.5}$$

Clearly, the ratio can exceed unity because of the large factor c^2/v_{Te}^2. In other words, the wave absorption associated with its transformation can exceed the wave attenuation due to the scattering.

10.7.2 Transformation of Longitudinal into Transverse Wave

Finally let us consider the transformation of an incident Langmuir wave $\boldsymbol{E}_0$ into a transverse electromagnetic wave $\boldsymbol{E}^{\mathrm{tr}}$. This process is important in view of its responsibility for the electromagnetic radiation going out from a plasma. The radiation is a result of plasma waves excited by means of some mechanism, e.g., the current buildup. The field of the transverse wave satisfies

$$\left[k^2 - \frac{\omega^2}{c^2}\varepsilon^{\mathrm{tr}}(\omega,k)\right]\boldsymbol{E}^{\mathrm{tr}} = -\frac{4\pi}{c}\boldsymbol{j}^{\mathrm{tr}}\,, \tag{10.7.6}$$

where $\boldsymbol{j}^{\mathrm{tr}}$ is a component of the current density (10.6.13) transverse to the wave vector $\boldsymbol{k}$. Assuming $\boldsymbol{k}_0\boldsymbol{E}_0 = k_0E_0$ and $[\boldsymbol{k}_0, \boldsymbol{E}_0] = 0$, we find

$$\boldsymbol{j}^{\mathrm{tr}} = \frac{\mathrm{i}e^2}{m\omega_0}\left(1-\frac{\Delta\omega}{\omega}\frac{k^2}{q^2}\right)\boldsymbol{E}_0^{\mathrm{tr}}\,\delta N_{\mathrm{e}}(\Delta\omega,\boldsymbol{q})\,. \tag{10.7.7}$$

The intensity of the transverse wave emerging due to the transformation of longitudinal plasma waves on fluctuations can be obtained similarly to the previous case. The cross section of this process is defined as the ratio of the intensity of the transverse wave to the volume and the energy flux density of the incident longitudinal wave which is expressed in terms of the group velocity

$$\frac{d\omega_0}{dk_0} = \frac{3\,k_0 v_{\mathrm{Te}}^2}{\omega_{\mathrm{pe}}} = \frac{v_{\mathrm{Te}}}{\sqrt{3}}\sqrt{\varepsilon(\omega_0)} \tag{10.7.8}$$

of the plasma wave as

$$S_0 = \frac{v_{\mathrm{Te}}}{8\sqrt{3}\pi}\sqrt{\varepsilon(\omega_0)}\,E_0^2\,. \tag{10.7.9}$$

Finally, for the cross section of the process we have

$$d\Sigma^{\mathrm{lo,\,tr}} = \frac{3\sqrt{3}}{16\pi^2}\sigma_{\mathrm{T}}\frac{c}{v_{\mathrm{Te}}}\frac{\omega^2}{\omega_0^2}\sqrt{\frac{\varepsilon(\omega)}{\varepsilon(\omega_0)}}\left(1-\frac{\Delta\omega}{\omega_0}\frac{k^2}{q^2}\right)^2$$
$$\times\sin^2\theta\,\langle\delta N_{\mathrm{e}}^2\rangle_{\Delta\omega,q}\,d\omega\,d\Omega\,. \tag{10.7.10}$$

Since density fluctuations have maxima for $\Delta\omega \approx \omega_{\mathrm{pe}}$ and $\omega_0 \approx \omega_{\mathrm{pe}}$, then $\omega = \omega_0 + \Delta\omega \approx 2\,\omega_{\mathrm{pe}}$ and transverse waves mainly have the frequencies $\omega \approx 2\,\omega_{\mathrm{pe}}$. Thus, the existence of maxima near the doubled plasma frequencies should be a characteristic of the electromagnetic self-radiation of a plasma. This process of a plasma wave transformation into a transverse electromagnetic wave is a *coalescence* of two longitudinal waves into a transverse wave with a doubled plasma frequency.

10.8 Exercises

10.8.1 Study the initial value problem for fluctuations of the distribution function in an isotropic collisionless plasma. The field in the plasma is referred to as potential.

Solution. The exact microscopic fluctuation distribution function, see (10.1.12), of particles of the type α

$$f_{\mathrm{m}}^{\alpha}(t,\boldsymbol{r},\boldsymbol{p}) = \sum_{a=1}^{N}\delta[\boldsymbol{r}-\boldsymbol{r}_a(t)]\,\delta[\boldsymbol{p}-\boldsymbol{p}_a(t)] \tag{10.8.1}$$

satisfies the kinetic equation

$$\frac{\partial f_{\mathrm{m}}^{a}}{\partial t} + \boldsymbol{v}\frac{\partial f_{\mathrm{m}}^{a}}{\partial \boldsymbol{r}} - e_a \frac{\partial \phi_{\mathrm{m}}}{\partial \boldsymbol{r}} \frac{\partial f_{\mathrm{m}}}{\partial \boldsymbol{p}} = 0 \,, \tag{10.8.2}$$

where the potential of the microscopic electric field ϕ_{m} is determined from Poisson's equation

$$\Delta \phi_{\mathrm{m}} = -4\pi \sum_a e_a \int d\boldsymbol{p}\, f_{\mathrm{m}}^{a} \,. \tag{10.8.3}$$

By (10.1.14) a fluctuation of the distribution function is defined as

$$\delta f_{\mathrm{m}}^{a}(t, \boldsymbol{r}, \boldsymbol{p}_a) = f_{\mathrm{m}}^{a}(t, \boldsymbol{r}, \boldsymbol{p}_a) - f^{a}(t, \boldsymbol{r}, \boldsymbol{p}_a) \,, \tag{10.8.4}$$

where $f^a(t, \boldsymbol{r}, \boldsymbol{p})$ is a mean statistical value of the distribution function. Analogously

$$\delta\phi = \phi_{\mathrm{m}} - \phi \,, \tag{10.8.5}$$

where ϕ is a mean statistical value of ϕ_{m}, i.e., the potential of an average macroscopic field.

By substituting f_{m}^{a} and ϕ_{m} from (10.8.4, 5) into (10.8.2, 3) and averaging we obtain the following system of equations

$$\frac{\partial f^{a}}{\partial t} + \boldsymbol{v}\frac{\partial f^{a}}{\partial \boldsymbol{r}} - e_a \frac{\partial \phi}{\partial \boldsymbol{r}} \frac{\partial f^{a}}{\partial \boldsymbol{p}} = e_a \left\langle \frac{\partial \delta\phi}{\partial \boldsymbol{r}} \frac{\partial \delta f^{a}}{\partial \boldsymbol{p}} \right\rangle , \tag{10.8.6}$$

$$\Delta \phi = -4\pi \sum_a e_a \int d\boldsymbol{p}\, f^{a} \,. \tag{10.8.7}$$

The left-hand side of (10.8.6), taking account of (10.8.7), is nothing else but Vlasov's equation and the right-hand side is a collision integral for a fully ionized plasma.

By subtracting (10.8.6, 7) from the exact equations (10.8.2 and 3), respectively, neglecting collisions and assuming the plasma to be spatially homogeneous [$f^a = f^a(\boldsymbol{p})$, $\phi = 0$], we obtain

$$\frac{\partial \delta f^{a}}{\partial t} + \boldsymbol{v}\frac{\partial \delta f^{a}}{\partial \boldsymbol{r}} - e_a \frac{\partial \delta\phi}{\partial \boldsymbol{r}} \frac{\partial f^{a}}{\partial \boldsymbol{p}} = 0 \,, \tag{10.8.8}$$

$$\Delta \delta\phi = -4\pi \sum_a e_a \int \delta f^{a}\, d\boldsymbol{p} \,. \tag{10.8.9}$$

Owing to (10.8.8, 9), the function $\delta f^a(t, \boldsymbol{r}, \boldsymbol{p})$ can be derived at an arbitrary time if its value is known at the initial time $t = 0$.

Let us solve the formulated initial value problem by means of the one-sided Fourier transformation:

$$\delta f^a_{\omega,k}(p) = \int dr \int_0^\infty dt\, e^{i\omega t - ik\cdot r}\, \delta f^a(t, r, p) \,, \tag{10.8.10}$$

$$\delta f^a(t, r, p) = \frac{1}{(2\pi)^4} \int_{-\infty+i\sigma}^{\infty+i\sigma} d\omega e^{-i\omega t} \int dk\, e^{ik\cdot r}\, \delta f^a_{\omega,k}(p) \,. \tag{10.8.11}$$

As a result, we find from (10.8.8, 9)

$$i(k\cdot v - \omega)\,\delta f^a_{\omega,k} - ie_a k \frac{\partial f^a}{\partial p}\,\delta\phi_{\omega,k} = \delta f^a_k(0, p) \,, \tag{10.8.12}$$

$$k^2 \delta\phi_{\omega,k} = 4\pi \sum_a e_a \int dp\, \delta f^a_{\omega,k} \,, \tag{10.8.13}$$

where $\delta f^a_k(0, p)$ is a Fourier transform of the initial function

$$\delta f^a(0, r, p) = f^a_{\mathrm{m}}(t, r, p)|_{t=0} - f^a(p) \,. \tag{10.8.14}$$

From (10.8.12, 13) $\delta f^a_{\omega,k}$ and $\delta\phi_{\omega,k}$ may be expressed in terms of $\delta f^a_k(0, p)$. By forming them into quadratic forms we can find the spectral distribution of the corresponding correlation functions. Thus, for $\delta\phi_{\omega,k}$ we have

$$\delta\phi_{\omega,k} = \frac{4\pi}{k^2 \varepsilon^{\mathrm{lo}}(\omega, k)} \sum_a \int \frac{e_a \delta f^a_k(0, p)\, dp}{i(\omega - k\cdot v)} \,, \tag{10.8.15}$$

where $\varepsilon^{\mathrm{lo}}(\omega, k)$ is the longitudinal dielectric permittivity of the plasma.

Now we can easily write the unknown correlation function

$$\begin{aligned} \langle \delta\phi_{\omega,k}\delta\phi_{\omega',k'} \rangle &= -\frac{16\pi^2}{k^4 \varepsilon^{\mathrm{lo}}(\omega, k)\, \varepsilon^{\mathrm{lo}}(\omega', k')} \sum_a e_a e_\beta \\ &\times \int \frac{\langle \delta f^a_k(0, p)\, \delta f^a_{k'}(0, p') \rangle}{(k\cdot v - \omega)(k'\cdot v' - \omega')}\, dp\, dp' \,. \end{aligned} \tag{10.8.16}$$

By analogy with (10.1.5) the initial correlator of noninteracting particles from the integrand (10.8.16) can be represented in the form

$$\begin{aligned} \langle \delta f^a_k(0, p)\, \delta f^\beta_{k'}(0, p') \rangle &= (2\pi)^3 \delta(k + k') \langle \delta f^a \delta f^\beta \rangle_k \\ &= (2\pi)^3 \delta(k + k')\, \delta_{a\beta} f^a(p)\, \delta(p - p') \,. \end{aligned} \tag{10.8.17}$$

Substituting this expression into (10.8.16) yields

$$\langle \delta\phi_{\omega,k}\delta\phi_{\omega' k'}\rangle = (2\pi)^4\,\delta(\omega+\omega')\,\delta(\boldsymbol{k}+\boldsymbol{k}')\,\langle\delta\phi^2\rangle_{\omega,k} \tag{10.8.18}$$

where

$$\langle\delta\phi^2\rangle_{\omega,k} = \frac{32\pi^2}{k^4|\varepsilon^{\mathrm{lo}}(\omega,k)|^2}\sum_\alpha e_\alpha^2\int f^\alpha(\boldsymbol{p})\,\delta(\omega-\boldsymbol{k}\cdot\boldsymbol{v})\,d\boldsymbol{p}\,. \tag{10.8.19}$$

Fluctuations of other quantities are calculated in the same way, in particular, the correlator

$$\langle\delta\phi\cdot\delta f^\alpha\rangle_{\omega,k} = \frac{e_\alpha\boldsymbol{k}}{\boldsymbol{k}\cdot\boldsymbol{v}-\omega+\mathrm{i}0}\,\frac{\partial f^\alpha}{\partial\boldsymbol{p}}\,\langle\delta\phi^2\rangle_{\omega,k} + \frac{8\pi^2 e_\alpha}{k^2\varepsilon^{\mathrm{lo}}(\omega,k)}f^\alpha(\boldsymbol{p})\,\delta(\omega-\boldsymbol{k}\cdot\boldsymbol{v}) \tag{10.8.20}$$

and the correlator

$$\begin{aligned}\langle\delta f^\alpha\delta f^\beta\rangle_{\omega,k} &= 2\pi\delta_{\alpha\beta}\,\delta(\boldsymbol{p}_1-\boldsymbol{p}_2)f^\alpha(\boldsymbol{p}_1)(\omega-\boldsymbol{k}\cdot\boldsymbol{v}_1)\\ &+\frac{e_\alpha e_\beta\langle\delta\phi^2\rangle_{\omega,k}}{(\omega-\boldsymbol{k}\cdot\boldsymbol{v}_1+\mathrm{i}0)(\omega-\boldsymbol{k}\cdot\boldsymbol{v}_2-\mathrm{i}0)}\left(\boldsymbol{k}\cdot\frac{\partial f^\alpha}{\partial\boldsymbol{p}_1}\right)\left(\boldsymbol{k}\cdot\frac{\partial f^\alpha}{\partial\boldsymbol{p}_2}\right)-\frac{8\pi^2 e_\alpha e_\beta}{k^2}\\ &\times\left[\left(\boldsymbol{k}\cdot\frac{\partial f^\alpha}{\partial\boldsymbol{p}_1}\right)f^\beta\frac{\delta(\omega-\boldsymbol{k}\cdot\boldsymbol{v}_2)}{\varepsilon^{\mathrm{lo}}(\omega,k)(\omega-\boldsymbol{k}\cdot\boldsymbol{v}_1+\mathrm{i}0)}+\left(\boldsymbol{k}\cdot\frac{\partial f^\beta}{\partial\boldsymbol{p}_2}\right)\right.\\ &\left.\times f^\alpha\frac{\delta(\omega-\boldsymbol{k}\cdot\boldsymbol{v}_1)}{\varepsilon^{\mathrm{lo}}(\omega,k)(\omega-\boldsymbol{k}\cdot\boldsymbol{v}_2-\mathrm{i}0)}\right].\end{aligned} \tag{10.8.21}$$

10.8.2 Derive the expression for magnetic fluctuations in an equilibrium isotropic plasma.

Solution. Let us determine from (10.3.2) the fluctuation correlator of the transverse electric field in an isotropic plasma:

$$\langle E^{\mathrm{tr}2}\rangle_{\omega,k} = \frac{8\pi T}{\omega}\,\frac{\mathrm{Im}\{\varepsilon^{\mathrm{tr}}(\omega,k)\}}{|\varepsilon^{\mathrm{tr}}(\omega,k)-k^2c^2/\omega^2|^2}\,. \tag{10.8.22}$$

On using the relationship $\boldsymbol{B} = c/\omega\,[\boldsymbol{k},\boldsymbol{E}]$ we obtain

$$\langle B^2\rangle_{\omega,k} = \frac{k^2c^2}{\omega^2}\langle E^{\mathrm{tr}2}\rangle_{\omega,k} = \frac{8\pi Tk^2c^2\,\mathrm{Im}\{\varepsilon^{\mathrm{tr}}(\omega,k)\}}{\omega^3\left[\varepsilon^{\mathrm{tr}}(\omega,k)-\dfrac{k^2c^2}{\omega^2}\right]^2}\,. \tag{10.8.23}$$

In the plasma transparency range, i.e., in the limit $\omega \gg kv_{\mathrm{Te}}$, when $\varepsilon^{\mathrm{tr}}(\omega, k) \approx 1 - \omega_{\mathrm{pe}}^2/\omega^2$ we have

$$\langle B^2 \rangle_{\omega,k} = \frac{16\pi^2 T k^2 c^2}{|\omega|} \delta(\omega^2 - \omega_{\mathrm{pe}}^2 - k^2c^2) . \tag{10.8.24}$$

Integrating this expression over frequencies yields

$$\langle B^2 \rangle_k = \frac{8\pi^2 T k^2 c^2}{\omega_{\mathrm{pe}}^2 + k^2c^2} . \tag{10.8.25}$$

Hence it is easy to derive a space correlation function

$$\langle B^2 \rangle_r = 8\pi^2 T \left[\delta(\boldsymbol{r}) - \frac{1}{4\pi} \frac{\exp(-r/r_{\mathrm{cor}})}{r r_{\mathrm{cor}}^2} \right] , \tag{10.8.26}$$

where $r_{\mathrm{cor}} = c/\omega_{\mathrm{pe}}$ is the correlation length of magnetic fluctuations in an isotropic plasma.

For $r \approx r_{\mathrm{cor}}$ we have from (10.8.26)

$$\frac{\langle B^2 \rangle_r}{4\pi N_{\mathrm{e}} T} \sim \frac{\langle E^{\mathrm{tr}2} \rangle_r}{4\pi N_{\mathrm{e}} T} \sim \frac{1}{r_{\mathrm{cor}}^3 N_{\mathrm{e}}} \sim \frac{v_{\mathrm{Te}}^3}{c^3} \frac{1}{r_{\mathrm{De}}^3 N_{\mathrm{e}}} \simeq \frac{v_{\mathrm{Te}}^3}{c^3} \frac{\langle (E^{\mathrm{lo}})^2 \rangle_r}{4\pi N_{\mathrm{e}} T} . \tag{10.8.27}$$

Thus, the fluctuation energy of the transverse field is c^3/v_{Te}^3 times smaller than that of the longitudinal field.

10.8.3 In the equilibrium electron plasma, determine the relative contribution of the high-frequency fluctuation intensity to the total intensity of charge density fluctuations.

Solution. The integral contribution of high-frequency fluctuations is determined by integrating (10.3.8) over the frequencies:

$$\langle \tilde{\varrho}^2 \rangle_k = \int_{-\infty}^{\infty} d\omega \frac{e^2 N_{\mathrm{e}} k^2 v_{\mathrm{Te}}^2}{|\omega|} \delta(\omega^2 - \omega_{\mathrm{pe}}^2 - 3k^2 v_{\mathrm{Te}}^2) \approx \frac{e^2 N_{\mathrm{e}} k^2}{2(3k^2 + 1/r_{\mathrm{De}}^2)} . \tag{10.8.28}$$

The unknown ratio of the high-frequency intensity to the total one is obtained by applying (10.3.12) for the total intensity:

$$\frac{\langle \tilde{\varrho}^2 \rangle_k}{\langle \varrho^2 \rangle_k} = \frac{1 + k^2 r_{\mathrm{De}}^2}{1 + 3k^2 r_{\mathrm{De}}^2} . \tag{10.8.29}$$

Hence, for $k^2 r_{\mathrm{De}}^2 \ll 1$, only high-frequency fluctuations at the plasma frequency remain in the fluctuation spectrum. When the wavelength decreases, low-frequency fluctuations become more essential and for $k^2 r_{De}^2 \gg 1$ they give the main contribution to the total intensity.

10.8.4 Derive the expression for the scattering cross section of longitudinal waves on density fluctuations of the equilibrium plasma.

Solution. Let us proceed from the expression for the density of the scattering longitudinal current

$$j = \frac{ie^2 \delta N_{\Delta\omega,q}}{m\omega_0} \left(\cos\theta + \frac{\Delta\omega}{\omega} \frac{k^2\cos\theta - k_0 k}{q^2} + \frac{\Delta\omega}{\omega_0} \frac{k_0 k - k_0^2 \cos\theta}{q^2} \right) \boldsymbol{E}_0 \tag{10.8.30}$$

where $[\boldsymbol{k}, \boldsymbol{j}] = 0$, $\boldsymbol{k}_0 \cdot \boldsymbol{k} = k_0 k \cos\theta$ and $\boldsymbol{E}_0$ – is the incident longitudinal wave $[\boldsymbol{k}_0, \boldsymbol{E}_0] = 0$. Substituting it into (10.7.1) gives the following expression for the scattering cross section of longitudinal waves on plasma density fluctuations:

$$d\Sigma = \frac{3}{16\pi^2} \sigma_T \left(\frac{c^2}{3v_{Te}^2} \right)^2 \frac{\omega^2}{\omega_0^2} \sqrt{\frac{\varepsilon(\omega)}{\varepsilon(\omega_0)}} \left(\cos\theta + \frac{\Delta\omega}{\omega} \frac{k^2\cos^2\theta - kk_0}{q^2} \right.$$
$$\left. + \frac{\Delta\omega}{\omega_0} \frac{kk_0 - k^2\cos\theta}{q^2} \right)^2 \langle \delta N_e^2 \rangle_{\Delta\omega,q}\, d\Omega\, d\omega \,. \tag{10.8.31}$$

Here, we have used (10.7.8, 9) for the group velocity and the current density of the incident Langmuir wave energy.

Comparing (10.8.31) to (10.7.10) for the transformation coefficient of the longitudinal wave into the transverse one results in the fact that its ratio to the scattering cross section for longitudinal waves is equal to $\cot^2\theta c^3/v_{Te}^3$, i.e., it can be rather large.

11. Principles of the Quasilinear Theory of Plasma Oscillations

A closed system of quasilinear equations for an isotropic and magnetized plasma is derived. Proceeding from their solutions, the problem of quasilinear relaxation of plasma oscillations and beam instability in the plasma is considered. A detailed analysis concerns the deformation of the distribution function in the process of quasilinear relaxation, plateau creation and beam heating.

11.1 Basic Equations

A plasma is often called a system with an infinite number of degrees of freedom. In fact, the number of the oscillation modes or wave branches sustained by the plasma is infinitely large. If the plasma is in thermodynamic equilibrium, homogeneous and stationary and not influenced by external fields, these oscillations are damped, i.e., small perturbations of the equilibrium state relax with time. If a stationary plasma state is not in thermodynamic equilibrium, then some of the small perturbations may increase with time, i.e., the plasma may be unstable. In Chaps. 6–9 we considered examples of a nonequilibrium plasma and it was shown that most nonequilibrium states can cause plasma instabilities. In particular, the rarefied high-temperature plasma with a small number of collisions is subject to instabilities since the particle collisions mostly inhibit the development of an instability and cause the relaxation of the plasma to the equilibrium state. Instabilities develop only when their time increment exceeds the inverse relaxation time of the plasma into the thermodynamic equilibrium state which is spatially homogeneous.

The linear theory of oscillations and plasma stability can answer the question of the development of small perturbations only for the initial stage, as long as the amplitudes are infinitesimal. In the unstable plasma it is of great interest to know how long an initial perturbation can increase, how it influences the equilibrium plasma, i.e., how the latter is modified due to the response to the increasing perturbations. This problem is answered by the nonlinear theory of plasma oscillations.

The formulation of the nonlinear theory of plasma oscillations or the theory of the *turbulent plasma* with a high level of instabilities presents a rather complex problem. At present, there is no adequate theoretical method to describe in general the strongly turbulent plasma. However, such a theory can be given for the weakly turbulent plasma when the oscillations resulting from an instability possess a small amount of energy compared to the thermal energy of the plasma particles, only. A smallness parameter of the theory of the *weak turbulent plasma* thus is

$$W/(N\mathscr{E}) \ll 1 , \qquad (11.1.1)$$

where $N\mathscr{E}$ is the thermal energy of the particles ($\mathscr{E} = T$ for the nondegenerate plasma, and $\mathscr{E} = \mathscr{E}_F$ for the degenerate plasma);

$$W = \sum_k W_k = \sum_k E_k^2/8\pi$$

is the energy of the oscillations of the plasma.

We assume, however, that the energy of the plasma oscillations is sufficiently large and greatly exceeds the energy of thermal electromagnetic fluctuations in the plasma, cf. (10.3.13, 14):

$$\frac{W}{N\mathscr{E}} \gg \frac{1}{N_D} \approx \left(\frac{e^2 N^{1/3}}{\mathscr{E}}\right)^{3/2} \approx \frac{\nu_{ei}}{\omega_{pe}} . \qquad (11.1.2)$$

Here, N_D is the number of particles in the Debye interaction sphere which is large according to the gas approximation condition, i.e., $N_D \gg 1$. If we have $N_D \approx (10^6$ to $10^7)$, e.g., for $N \approx (10^{10}$ to $10^{12})$ cm^{-3} and $T \approx (1$ to $10)$ eV, then (11.1.2) is satisfied for weak oscillation fields with a field strength $E_k = (10^{-1}$ to $1)$ V/cm, already. Though these fields are small, they greatly exceed the thermal fluctuation level. We may call them the fields of *overthermal oscillations* (fluctuations) of the weakly turbulent plasma.

The condition (11.1.2) allows to neglect the Coulomb particle collisions as compared to the particle interaction with the overthermal oscillation fields. Thus, when studying nonlinear effects, we can consider the plasma as collisionless. On the other hand, (11.1.1) implies that the overthermal oscillation fields are sufficiently small and cannot significantly modify such parameters of the equilibrium plasma as the particle density, the temperature and the thermal energy. In this sense, (11.1.1) ensures the weakness of the plasma turbulence.

11.1.1 Quasilinear Equations for the Isotropic Plasma

Having formulated the main restrictions of the modern nonlinear theory of plasma oscillations, we may come to the analysis of the first and most explicit nonlinear effect, namely the influence of the overthermal fluctuation fields on the velocity distribution function of the particles. We shall see that the influence of the oscillations on the zero-order plasma state modifies the temporal development of the modes, in turn, and in some cases leads to a stabilization of the instability. Confining our consideration to longitudinal oscillations in the absence of a magnetic field, the plasma is described by the nonlinear Vlasov equations for the electrons and the ions

$$\frac{\partial f_a}{\partial t} + \boldsymbol{v}\frac{\partial f_a}{\partial \boldsymbol{r}} + e_a \boldsymbol{E}\frac{\partial f_a}{\partial \boldsymbol{p}} = 0 \tag{11.1.3}$$

together with the Poisson equation

$$\operatorname{div}\boldsymbol{E} = 4\pi \sum_a e_a \int d\boldsymbol{p}\, f_a(\boldsymbol{p}) \,. \tag{11.1.4}$$

Instead of the Poisson equation it is often convenient to apply the equivalent Maxwell equation

$$\frac{\partial \boldsymbol{E}}{\partial t} + 4\pi \sum_a e_a \int d\boldsymbol{p}\, \boldsymbol{v} f_a(\boldsymbol{p}) \,. \tag{11.1.5}$$

For small deviations of the particle distribution functions from some zero-order distribution $f_{0a}(\boldsymbol{p})$ the longitudinal plasma oscillations obey the system of equations (11.1.3–5) in the linear approximation. These oscillations can be either damped when $f_{0a}(\boldsymbol{p})$ is the Maxwellian distribution describing the thermodynamic equilibrium, or growing, e.g., when $f_{0a}(\boldsymbol{p})$ is the distribution function with a directed velocity describing a beam plasma system. We analyze the conditions when the longitudinal waves are slowly damped or weakly unstable, i.e., when $\omega \gg \delta$, where ω is the oscillation frequency and δ the time decrement or increment. Only when this inequality is given can one speak of periodic plasma oscillations and a complete system of equations describing also the response of the zero-order state is obtained. Note that this property is characteristic of the Langmuir and ion-acoustic oscillations of the isotropic plasma with Maxwellian distributed particle velocities for the unstable oscillations excited by a low-density beam penetrating a denser plasma, for the ion-acoustic oscillations of the nonisothermal plasma, when the electrons are drifting with respect to the ions with $u < v_{\mathrm{Te}}$, and also for the kinetically unstable oscillations which are determined by the explicit form of the function $f_{0a}(\boldsymbol{p})$.

In order to derive the equations in the quasilinear approximation we split up the distribution functions $f_{0a}(\boldsymbol{p})$ into a large slowly and a small fast developing part:

$$\begin{aligned} f_a(\boldsymbol{p}) &= f_{0a}(\boldsymbol{p},\mu t) + f_{1a}(\boldsymbol{p},t) = f_{0a}(\boldsymbol{p},\mu t) \\ &\quad + \sum_{\boldsymbol{k}} \operatorname{Re}\{f_{1a\boldsymbol{k}}(\boldsymbol{p},\mu t)\exp(-\mathrm{i}\omega t + \mathrm{i}\boldsymbol{k}\cdot\boldsymbol{r})\}\,, \end{aligned} \tag{11.1.6}$$

where $f_{0\alpha}(\boldsymbol{p}, \mu t) \gg f_{1\alpha}(\boldsymbol{p}, t)$. The electric field in the plasma is of the form

$$\boldsymbol{E}(t, \boldsymbol{r}) = \sum_{\boldsymbol{k}} \operatorname{Re}\{\boldsymbol{E}_{\boldsymbol{k}}(\mu t, \omega, \boldsymbol{k}) \exp(-\mathrm{i}\omega t + \mathrm{i}\boldsymbol{k} \cdot \boldsymbol{r})\} . \tag{11.1.7}$$

These expressions take account of the slow time dependence of the distribution function $f_{0\alpha}(\boldsymbol{p}, \mu t)$ of the amplitudes of the electric field oscillations if we assume $\mu \ll 1$. We further assume that there is no average slowly varying electric field present in the plasma.

We substitute the decompositions (11.1.6, 7) into (11.1.3) and take a time average over the fast oscillations. The time interval τ must satisfy the condition

$$\frac{2\pi}{\omega} \ll \tau \sim \tau_{\mathrm{r}} . \tag{11.1.8}$$

Here τ_{r} is the characteristic time of the *quasilinear relaxation,* i.e., the time during which the oscillations influence the equilibrium plasma state. Then we have

$$f_{0\alpha} = \langle f_\alpha \rangle = \frac{1}{\tau} \int_0^\tau f_\alpha \, dt . \tag{11.1.9}$$

Due to averaging, (11.1.3) separates into two equations for the slow and fast parts of the distribution function:

$$\frac{\partial f_{0\alpha}}{\partial \mu t} + e_\alpha \left\langle \boldsymbol{E} \frac{\partial f_{1\alpha}}{\partial \boldsymbol{p}} \right\rangle = 0 , \qquad \frac{\partial f_{1\alpha}}{\partial t} + \boldsymbol{v} \frac{\partial f_{1\alpha}}{\partial \boldsymbol{r}} + e_\alpha \boldsymbol{E} \frac{\partial f_{0\alpha}}{\partial \boldsymbol{p}} = 0 . \tag{11.1.10}$$

Of course, the inequality $f_{0\alpha}(\boldsymbol{p}, \mu t) \gg f_{1\alpha}(\boldsymbol{p}, t)$ was needed and small nonlinear terms were neglected in the derivation of (11.1.10).

Applying the Fourier "ansatz" (11.1.6, 7) neglecting the slow time variation of $f_{1\alpha k}$ and E_k, and taking account of the condition (11.1.8) we obtain from the second equation of (11.1.10):

$$f_{1\alpha k} = - \frac{\mathrm{i} e_\alpha \boldsymbol{E}_k}{\omega - \boldsymbol{k} \cdot \boldsymbol{v}} \frac{\partial f_{0\alpha}}{\partial \boldsymbol{p}} . \tag{11.1.11}$$

Substituting this equation into the first equation of (11.1.10) gives the *quasilinear equation for the slow* (quasistationary) *part* of the distribution function

$$\frac{\partial f_{0\alpha}}{\partial t} = \frac{\partial}{\partial p_i} D_{\alpha ij} \frac{\partial f_{0\alpha}}{\partial p_j} , \quad \text{where} \tag{11.1.12}$$

$$\begin{aligned} D_{\alpha ij} &= - \frac{e_\alpha^2}{2} \sum_{\boldsymbol{k}} \frac{k_i k_j}{k^2} |\boldsymbol{E}_k|^2 \operatorname{Im}\left\{ \frac{1}{\omega - \boldsymbol{k} \cdot \boldsymbol{v}} \right\} \\ &= \frac{e_\alpha^2}{2} \sum_{\boldsymbol{k}} \frac{k_i k_j}{k^2} |\boldsymbol{E}_k|^2 \frac{\delta_k}{(\omega - \boldsymbol{k} \cdot \boldsymbol{v})^2 + \delta_k^2} . \end{aligned} \tag{11.1.13}$$

Here the complex frequency $\omega \rightarrow \omega + i\delta_k$ has been introduced and the relation $E_k = E_k \cdot k/k$ has been invoked.

Note that

$$\frac{\delta_k}{(\omega - k \cdot v)^2 + \delta_k^2} = \begin{cases} \dfrac{1}{\delta_k} & \text{for} \quad (\omega - k \cdot v)^2 \ll \delta_k^2 \,, \\ \pi\delta(\omega - k \cdot v) & \text{for} \quad (\omega - k \cdot v)^2 \gg \delta_k^2 \rightarrow 0 \,. \end{cases} \tag{11.1.14}$$

The first relation will be applied when studying the development of hydrodynamic instabilities of the plasma which are free of dissipative processes; the second one is convenient for the analysis of the kinetically unstable oscillations which are dominated by dissipative processes.

Due to (11.1.11) the fast part f_{1ak} of the distribution function is expressed in terms of the slow one f_{0a}; the latter in turn, is related to the electric field of the oscillations E_k by (11.1.12). Thus, in order to close the system, one must engage an equation for E_k. For systems with electrostatic interaction (11.1.5) closes the system of equations. Taking the scalar product with E_k^* and averaging over the fast oscillations with account taken of (11.1.11), we obtain after simple calculations

$$\frac{\partial |E_k|^2}{\partial \mu t} = 2\delta_k |E_k|^2 \,, \quad \text{where} \tag{11.1.15}$$

$$\delta_k = -\frac{1}{2}\sum_a \frac{4\pi e_a^2 \omega}{k^2} \int k \frac{\partial f_{0a}}{\partial p} \operatorname{Im}\left\{\frac{1}{\omega - k \cdot v}\right\} dp \,. \tag{11.1.16}$$

In the hydrodynamically unstable plasma the quantity δ_k in (11.1.15) means a time increment of the oscillations. If the buildup (or damping) of waves in the plasma is given by dissipative effects, then (11.1.15) describes the variation of the field energy in the medium, and δ_k is the imaginary part of the longitudinal dielectric permittivity of the isotropic collisionless plasma.

Equations (11.1.12, 15) form the *closed system of quasilinear equations* for the longitudinal plasma oscillations in the absence of an external magnetic field. The dispersion law is given by the well-known equation

$$\varepsilon(\mu t, \omega, k) = 1 + \sum_a \frac{4\pi e_a^2}{k^2} \int \frac{k \cdot \partial f_{0a}/\partial p}{\omega - k \cdot v} dp = 0 \,. \tag{11.1.17}$$

11.1.2 Conservation Laws of the Quasilinear Theory

It is easy to show that the system of quasilinear equations satisfies the conservation laws of energy, momentum and number of particles. Actually, multiplying (11.1.12) by $\mathscr{E}_a = m_a c^2 \gamma_a$, integrating over the momentum and summing up over the particle species yields

$$\frac{\partial}{\partial \mu t} \sum_{a} \int f_{0a} \mathscr{E}_a d\boldsymbol{p} = \frac{1}{2} \sum_{a} \sum_{\boldsymbol{k}} \int \frac{e_a^2 |\boldsymbol{E}_{\boldsymbol{k}}|^2 \omega}{k^2} \left(\boldsymbol{k} \cdot \frac{\partial f_{0a}}{\partial \boldsymbol{p}} \right) \operatorname{Im} \left\{ \frac{1}{\omega - \boldsymbol{k} \cdot \boldsymbol{v}} \right\} . \tag{11.1.18}$$

Using (11.1.15) we finally obtain the conservation law of energy:

$$\frac{d}{d\mu t} \left(\sum_{a} \int \mathscr{E}_a f_{0a} d\boldsymbol{p} + \sum_{\boldsymbol{k}} \frac{|\boldsymbol{E}_{\boldsymbol{k}}|^2}{8\pi} \right) = 0 . \tag{11.1.19}$$

The conservation laws of momentum and number of particles are derived analogously:

$$\frac{d}{d\mu t} \left(\sum_{a} \int \boldsymbol{p} f_{0a} d\boldsymbol{p} + \sum_{\boldsymbol{k}} \frac{\boldsymbol{k} |\boldsymbol{E}_{\boldsymbol{k}}|^2}{8\pi\omega} \right) = 0 , \quad \frac{d}{d\mu t} \int f_{0a} d\boldsymbol{p} = 0 . \tag{11.1.20}$$

11.1.3 Quasilinear Equations for Magnetized Plasma

It is straightforward to generalize the quasilinear equations to include magneto-active plasmas, too. As before, we confine our interest to longitudinal plasma oscillations and separate the distribution functions into slowly and fast varying parts according to (11.1.6, 7). Using the Vlasov equation of the magneto-active plasma

$$\frac{\partial f_a}{\partial t} + \boldsymbol{v} \frac{\partial f_a}{\partial \boldsymbol{r}} + e_a \left\{ \boldsymbol{E} + \frac{1}{c} [\boldsymbol{v}, \boldsymbol{B}_0] \right\} \frac{\partial f_a}{\partial \boldsymbol{p}} = 0 , \tag{11.1.21}$$

we obtain for f_{0a} and f_{1a}

$$\frac{\partial f_{0a}}{\partial \mu t} + e_a \left\langle \boldsymbol{E} \cdot \frac{\partial f_{1a}}{\partial \boldsymbol{p}} \right\rangle + \frac{e_a}{c} [\boldsymbol{v}, \boldsymbol{B}_0] \frac{\partial f_{0a}}{\partial \boldsymbol{p}} = 0 , \tag{11.1.22}$$

$$\frac{\partial f_{1a}}{\partial t} + \boldsymbol{v} \frac{\partial f_{1a}}{\partial \boldsymbol{r}} + e_a \boldsymbol{E} \frac{\partial f_{0a}}{\partial \boldsymbol{p}} + \frac{e_a}{c} [\boldsymbol{v}, \boldsymbol{B}_0] \frac{\partial f_{1a}}{\partial \boldsymbol{p}} = 0 . \tag{11.1.23}$$

In the following we restrict ourselves to axially symmetric functions $f_{0a} = f_{0a}(p_z, p_\perp)$. Such a situation occurs, for example, during the interaction of a beam of particles with the plasma, or when the plasma is imbedded in an electric field[1]. Then, using (11.1.7) with account taken of the potential field $\boldsymbol{E} = -\nabla\Phi$, we obtain the solution of (11.1.3)

1 The function $f_{0a}(\boldsymbol{p})$ may also be a function of ϕ, for instance, in the case of an inhomogeneous magneto-active plasma. Then the quasilinear theory is significantly complicated (Exercise 11.4.2).

$$f_{1ak} = -\frac{ie_a\gamma_a}{\Omega_a}\Phi_k \int_{\infty}^{\phi} d\phi' \left(k\frac{\partial f_{0a}}{\partial p}\right) \exp\left[i\frac{\gamma_a}{\Omega_a}\int_{\phi}^{\phi'} d\phi''(\omega - k\cdot v)_{\phi''}\right]$$

$$= e_a \sum_{n,m} \frac{J_n\left(\frac{k_\perp v_\perp \gamma_a}{\Omega_a}\right) J_m\left(\frac{k_\perp v_\perp \gamma_a}{\Omega_a}\right)}{\omega - k_z v_z - n\Omega_a/\gamma_a}\Phi_k$$

$$\times \left(\frac{n\Omega_a}{v_\perp \gamma_a}\frac{\partial f_{0a}}{\partial p_\perp} + k_z\frac{\partial f_{0a}}{\partial p_z}\right)\exp\left[i(m-n)\phi\right]. \tag{11.1.24}$$

We substitute this solution into (11.1.22) and average the latter over time. Taking into account that the terms $m \neq n$ vanish for axially symmetric functions f_{0a}, after averaging over ϕ we finally obtain the quasilinear equation for the slow part of the distribution function:

$$\frac{\partial f_{0a}}{\partial \mu t} = -\frac{e_a^2}{2}\sum_k\sum_n |\Phi_k|^2 \left(k_z\frac{\partial}{\partial p_z} + \frac{n\Omega_a}{v_\perp\gamma_a}\frac{\partial}{\partial p_\perp}\right) J_n^2\left(\frac{k_\perp v_\perp \gamma_a}{\Omega_a}\right)$$

$$\times \operatorname{Im}\left\{\frac{1}{\omega - k_z v_z - n\Omega_a/\gamma_a}\right\}\left(k_z\frac{\partial}{\partial p_z} + \frac{n\Omega_a}{v_\perp\gamma_a}\frac{\partial}{\partial p_\perp}\right) f_{0a}. \tag{11.1.25}$$

In the limit $\Omega_a \to 0$ it goes over into (11.1.12), but only for a one-dimensional motion (along the 0z-axis).

The quasilinear equation for the field $E = -\nabla\Phi$ follows from (11.1.5). Taking the scalar product with E and averaging over the fast oscillations gives (11.1.15), where

$$\delta_k = \frac{1}{2}\sum_a\sum_n \frac{4\pi e_a^2\omega}{k^2}\int dp\, \operatorname{Im}\left\{\frac{1}{\omega - k_z v_z - n\Omega_a/\gamma_a}\right\}$$

$$\times J_n^2\left(\frac{k_\perp v_\perp\gamma_a}{\Omega_a}\right)\left(\frac{n\Omega_a}{v_\perp\gamma_a}\frac{\partial f_{0a}}{\partial p_\perp} + k_z\frac{\partial f_{0a}}{\partial p_z}\right) \tag{11.1.26}$$

is the time increment (damping decrement) of the oscillations.

Equations (11.1.15, 25) form the closed system of quasilinear equations for longitudinal oscillations of the magneto-active plasma. Their spectrum is given by

$$\varepsilon(\mu t, \omega, k) = 1 + \sum_a\sum_n \frac{4\pi e_a^2}{k^2}\int dp \frac{J_n^2\left(\frac{k_\perp v_\perp\gamma_a}{\Omega_a}\right)}{\omega - k_z v_z - n\Omega_a/\gamma_a}$$

$$\times\left(k_z\frac{\partial f_{0a}}{\partial p_z} + \frac{n\Omega_a}{v_\perp\gamma_a}\frac{\partial f_{0a}}{\partial p_\perp}\right) = 0. \tag{11.1.27}$$

11.2 Quasilinear Relaxation of Plasma Oscillations

As a first example of the application of the quasilinear equations we investigate the relaxation of high-frequency plasma oscillations in the nonrelativistic isotropic plasma. Within the framework of the linear theory of the oscillations of an isotropic plasma (Chap. 4) it has been shown that small longitudinal oscillations damp with time in the collisionless plasma due to the Cherenkov absorption by the plasma electrons. The characteristic time of damping is given by the Landau damping decrement. In the linear theory, however, the effect of the oscillations on the equilibrium distribution function of the electrons is not accounted for. Thus the linear damping decrement determines only the time of damping in the limit of infinitely weak oscillations. In the case of finite oscillation amplitudes the distribution function of the bulk plasma varies with time during the development of the oscillation and the expression for the linear damping decrement is no longer valid. The damping process can be described in the framework of the quasilinear theory, however, since here the change of the equilibrium distribution function under the influence of oscillations is taken into account. Evidently, the oscillations must influence the distribution function of the main plasma in such a way as to hinder the wave absorption (i.e., to decrease the Cherenkov effect). Therefore, along with the growth of the oscillation amplitude the damping must decrease, and in the collisionless plasma (when no "Maxwellization" of the distribution function occurs) there can be established a stationary distribution function so that the wave absorption is completely vanishing and an undamped wave remains. According to the linear theory the wave damping in the plasma is given by the quantity $\partial f_0/\partial v|_{v=\omega/k}$. Thus, a stationary state is possible if a "plateau", i.e., a region with $\partial f_0/\partial v|_{v=\omega/k} = 0$, develops in the equilibrium distribution function due to the effect of the wave and if the latter is not entirely absorbed up to this moment of time. In order to consider the relaxation of the plasma oscillations, we restrict ourselves to the one-dimensional case where the vectors $\boldsymbol{k}$ of all waves are parallel.

11.2.1 Relaxation of the Distribution Function

We assume that in a nondegenerate equilibrium plasma (with a Maxwellian distribution function) there are excited plasma oscillations with the spectral energy density $W_k(0) = |E_k(0)|^2/8\pi$ uniformly in space at some narrow interval of wave vectors $\boldsymbol{k}$ initially. The level of the spectral energy density is supposed to exceed the thermal noise level significantly (consequently, particle collisions are of no importance). In the one-dimensional case the quasilinear equations (11.1.12, 15) take on the simple form:

$$\frac{\partial F_0}{\partial t} = \frac{\partial}{\partial p} D \frac{\partial F_0}{\partial p}\,, \quad D = -\frac{e^2}{2}\sum_k |E_k|^2 \operatorname{Im}\left\{\frac{1}{\omega - kv}\right\} = \frac{\pi}{2}\frac{e^2|E_k|^2}{kv}\,,$$

$$\frac{\partial |E_k|^2}{\partial t} = 2\delta_k |E_k|^2\,, \tag{11.2.1}$$

$$\delta_k = -\frac{\omega_{p0}^3}{2k^2}\int dv\, k\, \frac{\partial F_0}{\partial v} \operatorname{Im}\left\{\frac{1}{\omega - kv}\right\} = \frac{\pi}{2}\frac{\omega_{p0}^3}{k^2}\frac{\partial F_0}{\partial v}\bigg|_{v=\omega/k}\,.$$

Here the function $F_0(v, t)$ is introduced for convenience. It results by integrating the function $f_0(\boldsymbol{v}, t)$ over the components of the momentum perpendicular to the vector $\boldsymbol{k}$. We normalize this function by unity: $\int F_0 dv = 1$. Incidentally, electrons with $kv = \omega_{p0} = (4\pi e^2 N_0/m)^{1/2}$, N_0 being the equilibrium plasma density, contribute to the absorption of the plasma waves.

The initial conditions for (11.2.1) can be written as

$$F_0(v, 0) = \sqrt{\frac{m}{2\pi T}}\exp\left(-\frac{mv^2}{2T}\right)\,,$$

$$W_k(0) = \frac{|E_k(0)|^2}{8\pi} = \begin{cases} 0 & \text{for } \dfrac{\omega}{k} = v < v_1\,, \\ W_0 & \text{for } v_1 \leqslant v = \dfrac{\omega}{k} \leqslant v_2\,, \\ 0 & \text{for } v = \dfrac{\omega}{k} > v_2\,. \end{cases} \tag{11.2.2}$$

Only in this narrow range of velocities is the diffusion coefficient $D(v, t = 0)$ nonzero.

11.2.2 Plateau Creation

As a result of the quasilinear relaxation the system passes over into a new stationary state with $\partial F_0/\partial v \to 0$ in the range $v_1 \leq v \leq v_2$, i.e., a "plateau" is established (Fig. 11.1). This is immediately seen from the second equation of the system (11.2.1). The spectral density of the oscillation energy is lower than initially $W_k(\infty) < W_k(0)$. The quasilinear equations (11.2.1) allow to calculate $W_k(\infty)$ and $F_0(v, \infty)$ in terms of $W_k(0)$ and $F_0(v, 0)$. We obtain the relation

$$\frac{\partial}{\partial t}\left[F_0(v, t) - \frac{\partial}{\partial v} W_k(t)\frac{4\pi e^2}{m^2k^2v^4}\right] = 0\,, \tag{11.2.3}$$

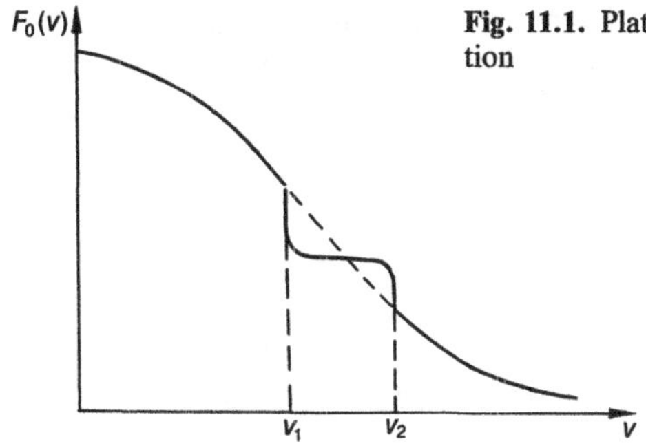

Fig. 11.1. Plateau creation in the particle distribution function

which shows that the quantity

$$F_0(v,t) - \frac{\partial}{\partial v} W_k(t) \frac{4\pi e^2}{m^2 k^2 v^4}$$

is conserved during the quasilinear relaxation process. Therefore

$$F_0(v,0) - F_0(v,\infty) = \frac{\partial}{\partial v} W_k(0) \frac{4\pi e^2}{m^2 k^2 v^4} - \frac{\partial}{\partial v} W_k(\infty) \frac{4\pi e^2}{m^2 k^2 v^4}, \quad (11.2.4)$$

or, after integration over the velocities

$$W_k(\infty) - W_k(0) = \frac{m^2 k^2 v^4}{4\pi e^2} \int_{v_1}^{v} dv [F_0(v,\infty) - F_0(v,0)] . \quad (11.2.5)$$

Evidently,

$$\int_{v_1}^{v_2} [F_0(v,\infty) - F_0(v,0)]\, dv = 0 , \quad (11.2.6)$$

since the particle interaction occurs in the range $v_1 \le v \le v_2$ and the distribution function is distorted in this range only. Finally, since $F_0(v,\infty) = \text{const}$, we have

$$F_0(v,\infty) = (v_2 - v_1)^{-1} \int_{v_1}^{v_2} dv\, F_0(v,0) . \quad (11.2.7)$$

Substituting (11.2.7) into (11.2.5) we obtain the spectral energy density of the oscillations in the final state of the system:

$$W_k(\infty) - W_k(0)$$
$$= \frac{m^2 k^2 v^4}{4\pi e^2} \int_{v_1}^{v_2} dv \left[\frac{1}{v_2 - v_1} \int_{v_1}^{v_2} dv\, F_0(v,0) - F_0(v,0) \right] . \quad (11.2.8)$$

The initial energy may be insufficient to establish a "plateau" in the distribution function of the electrons. Then the stationary state is not achieved and the field energy is completely absorbed by the plasma.

11.2.3 Time of Quasilinear Relaxation

We now evaluate the time of the quasilinear relaxation of the oscillations at the initial stage when the wave amplitude still remains large, as compared with the thermal noise. Since the function f_0 becomes flat due to the quasilinear relaxation process, i.e., since the derivative $\partial f_0/\partial v$ decreases, the characteristic time of wave absorption must increase and its inverse, the damping decrement, must decrease. Therefore, the time of quasilinear relaxation greatly exceeds the inverse damping decrement of the waves at the initial stage. The *time of quasilinear relaxation of the oscillations,* i.e., the time during which the "plateau" is created, follows from the first equation of the system (11.2.1). In the case of plasma waves with $v \simeq \omega_{p0}/k \gg v_{Te}$ and $E_k \sim k\Phi_k$ it is of the order of

$$t_q \sim \frac{m^2 v^2}{D} \sim \frac{1}{\omega_{p0}} \frac{v^2}{v_{Te}^2} \frac{NT}{W_k} \sim \frac{1}{\omega_{p0}} \left(\frac{T}{e\Phi_k}\right)^2 \frac{v^4}{v_{Te}^4} . \tag{11.2.9}$$

The inequalities (11.1.1, 2) can be written as

$$\left(\frac{T}{e^2 N^{1/3}}\right)^{3/2} \approx N_D \gg \frac{NT}{W_k} \sim \left(\frac{T}{e\Phi_k}\right)^2 \frac{v^2}{v_{Te}^2} \gg 1 . \tag{11.2.10}$$

Then we obtain from (11.2.9)

$$t_q \omega_{p0} \sim \left(\frac{T}{e\Phi_k}\right)^2 \frac{v^4}{v_{Te}^4} \gg 1 . \tag{11.2.11}$$

The requirement that the inverse damping decrement of the waves must be small against the time of quasilinear relaxation is even more restrictive:

$$\begin{aligned} t_q \delta_k &\sim \tau_q \frac{\omega_{p0} \exp(-1/2\, k^2 r_{De}^2)}{k^3 r_{De}^3} \\ &\sim \left(\frac{T}{e\Phi_k}\right)^2 \frac{\exp(-1/2\, k^2 r_{De}^2)}{k^7 r_{De}^7} \gtrsim 1 . \end{aligned} \tag{11.2.12}$$

Hence, a restriction for the potential of the plasma oscillations, for which the quasilinear theory is valid, is obtained:

$$\frac{e\Phi_k}{T} \lesssim \frac{1}{(k r_{De})^7} \exp\left(-\frac{1}{4 k^2 r_{De}^2}\right) . \tag{11.2.13}$$

Note that a similar restriction follows from the requirement that there is no electron capture in the field of the plasma oscillations $e\Phi_k \ll m\delta_k^2/k^2$, too (Exercise 11.4.3):

$$\frac{e\Phi_k}{T} \ll \frac{\pi}{8\,(kr_{\mathrm{De}})^8} \exp\left(-\frac{1}{k^2 r_{\mathrm{De}}^2} - 3\right). \tag{11.2.14}$$

Under this condition only the linear approximation for the high-frequency plasma oscillations of the collisionless isotropic equilibrium plasma is applicable.

Finally, we give the condition for neglecting the particle collisions during the quasilinear relaxation process of the high-frequency plasma oscillations of the completely ionized plasma. This condition also determines the range of applicability of the derived formulas and specifies (11.2.1) for the case of plasma oscillations

$$t_q \nu_{\mathrm{ei}} \sim \left(\frac{T}{e\Phi_k}\right)^2 \frac{1}{k^4 r_{\mathrm{De}}^4} \left(\frac{e^2 N_0^{1/3}}{T}\right)^{3/2} \ll 1\,. \tag{11.2.15}$$

11.3 Quasilinear Relaxation of the Beam Instability

We apply the quasilinear equations to study the development of the beam instability. In our analysis of the interaction of a low-density electron beam with the plasma it was shown that in the absence of an external magnetic field two types of instabilities can develop: the hydrodynamic and the kinetic instabilities (Chap. 6). If the beam of nonrelativistic electrons is sufficiently monoenergetic, so that $v_{\mathrm{Tb}} \ll u\,(N_{\mathrm{b}}/2\,N_0)^{1/3}$, where v_{Tb} is the thermal velocity spread of beam electrons, $\boldsymbol{u}$ their ordered velocity and N_{b}/N_0 the ratio of the beam density to the plasma density, then the instability has a hydrodynamic character and plasma waves with the frequency

$$\omega(\boldsymbol{k}) = \boldsymbol{k} \cdot \boldsymbol{u} \approx \omega_{\mathrm{p0}} \tag{11.3.1}$$

are excited. Their increment is

$$\delta = \delta_k \approx \frac{\sqrt{3}}{2} \left(\frac{N_{\mathrm{b}}}{2\,N_0}\right)^{1/3} \omega_{\mathrm{p0}} \ll \omega\,. \tag{11.3.2}$$

The determining factors of the hydrodynamic beam instability are macroscopic parameters like the density, velocity, etc., of the system. On the other hand, the beam instability is kinetic for electrons with a large thermal velocity spread, when $v_{\mathrm{Tb}} > u\,(N_{\mathrm{b}}/2\,N_0)^{1/3}$. Then the instability is determined by the explicit form of the velocity distribution function of the electrons.

11.3.1 Quasilinear Dynamics of the Hydrodynamical Beam Instability

We begin our analysis of the quasilinear problem of beam instability with the hydrodynamic case. Confining ourselves to the one-dimensional situation the quasilinear equations can be written as

$$\frac{\partial F_0}{\partial t} = \frac{\partial}{\partial p} D \frac{\partial F_0}{\partial p}, \qquad \frac{\partial |E_k|^2}{\partial t} = 2\delta_k |E_k|^2, \quad \text{where} \tag{11.3.3}$$

$$D = -\frac{e^2}{2} \sum_k |E_k|^2 \frac{\delta_k}{(\omega - kv)^2 + \delta_k^2} . \tag{11.3.4}$$

At the hydrodynamic stage of the growth of the oscillations we have $(\omega - kv) \approx \delta_k/\sqrt{3} \gg kv_{\text{Tb}}$, the increment δ_k being independent of the wave vector. Hence,

$$D = -\frac{1}{2}\frac{e^2}{\delta} \sum_k |E_k|^2 , \qquad \frac{\partial D}{\partial t} = 2\delta D , \qquad \frac{\partial F_0}{\partial t} = D \frac{\partial^2 F_0}{\partial p^2} . \tag{11.3.5}$$

In order to solve time development of the beam instability, the system (11.3.5) must be complemented by initial conditions. In the case of a mono-energetic electron beam we have

$$F_0(v, 0) = \delta(v - u) . \tag{11.3.6}$$

We introduce the substitution

$$\frac{d\tau}{dt} = D(t) , \qquad \frac{d}{dt} = \frac{d\tau}{dt}\frac{d}{d\tau} .$$

Then (11.3.5) can be written as

$$\frac{\partial F_0}{\partial \tau} - \frac{\partial^2 F_0}{\partial p^2} = 0 , \qquad \frac{\partial D}{\partial \tau} = 2\delta , \quad D = -4\pi e^2 W , \tag{11.3.7}$$

where $W = \sum_k |E_k|^2/8\pi$ is the energy of the oscillation field.

The solution of the system (11.3.7) with the initial condition (11.3.6) is

$$F_0(v, \tau) = \frac{A}{\sqrt{\tau}} \exp\left(-\frac{m(v - u)^2}{4\tau}\right) . \tag{11.3.8}$$

From the normalization condition $\int F_0 dv = 1$ we have $A = \sqrt{m/(4\pi)}$.

Thus, during the development of the beam instability the initially mono-energetic distribution (11.3.6) spreads and the effective temperature grows with time: $T = 2\tau \sim t$. This velocity diffusion, however, stops when the applicability condition of the hydrodynamic approach $\delta > kv_{\mathrm{Tb}}$ or $v_{\mathrm{Tb}} = \sqrt{T/m} = \sqrt{2\tau/m} < u(N_{\mathrm{b}}/2N_0)^{1/3}$ is violated. As a result we have

$$T_{\max} = 2\tau_{\max} = mu^2\left(\frac{N_{\mathrm{b}}}{2N_0}\right)^{2/3}. \tag{11.3.9}$$

Hence, the part $2(N_{\mathrm{b}}/2N_0)^{1/3}$ of the kinetic energy of the beam is spent on heating the beam, i.e.,

$$\frac{T_{\max}}{mu^2/2} = 2\left(\frac{N_{\mathrm{b}}}{2N_0}\right)^{2/3}. \tag{11.3.10}$$

The energy of the oscillation field can be evaluated easily, too. From (11.3.7) we have $D \sim 2\delta\tau \sim 3\pi e^2 W/\delta$ and therefore $W_{\max} \sim 2\delta^2\tau_{\max}/(3\pi e^2)$ or

$$\frac{W_{\max}}{N_{\mathrm{b}}mu^2/2} = \frac{1}{4\pi}\frac{\delta^2}{me^2N_{\mathrm{b}}}\left(\frac{N_{\mathrm{b}}}{2N_0}\right)^{2/3} \approx \left(\frac{N_{\mathrm{b}}}{2N_0}\right)^{1/3}. \tag{11.3.11}$$

The part $(N_{\mathrm{b}}/2N_0)^{1/3}$ of the beam energy is transferred into oscillation energy.

The equations (11.3.10, 11) show that the energy loss of the beam is small at the hydrodynamic stage. The energy is used mainly to excite waves and, to a less degree, to heat the beam.

Finally, we evaluate the time required for the quasilinear relaxation of the hydrodynamic beam instability. On account of $D \sim 2\delta\tau \sim 2\delta Dt$ it is easy to show that

$$t_q \sim \frac{1}{\delta} \approx \frac{2}{\sqrt{3}}\left(\frac{2N_0}{N_{\mathrm{b}}}\right)^{1/3}\frac{1}{\omega_{\mathrm{p}0}}. \tag{11.3.12}$$

The nonstationarity of the hydrodynamic beam instability is its most characteristic peculiarity since the increment δ is independent of the form of the distribution function of the beam electrons and does not vary with time. Therefore, the instability cannot be stabilized during its development. Since, due to the spread of the beam electron velocities, the inequality $v_{\mathrm{Tb}} < u(N_{\mathrm{b}}/2N_0)^{1/3}$ is violated at the time $1/\delta$, the instability finally acquires a kinetic character.

11.3.2 Relaxation of the Beam in the Kinetic Stage

In order to investigate the beam behaviour at the kinetic stage of the instability, we analyze the quasilinear relaxation of the spread beam. We assume that the total electron distribution function is initially of the form shown in

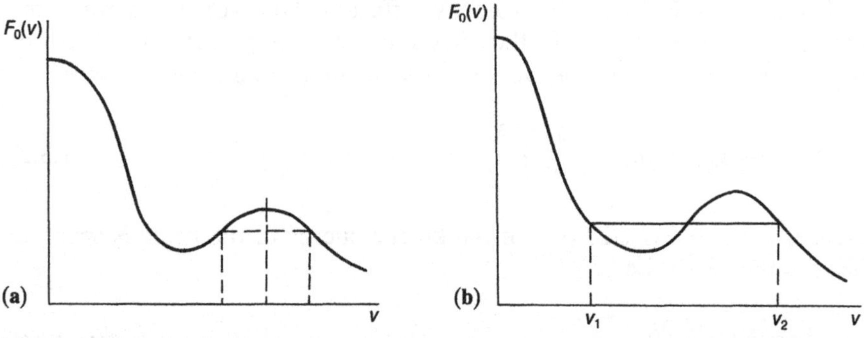

Fig. 11.2. (a) Initial distribution function in the plasma beam system; (b) Distortion of the distribution function due to the quasi-linear relaxation of the plasma beam instability

Fig. 11.2a. Then, in the one-dimensional case we can write (usually assuming $T_b \gg T_0$)

$$F_0(v,0) = \left(\frac{m}{2\pi T_0}\right)^{1/2} \exp\left(-\frac{mv^2}{2T_0}\right) - \frac{N_b}{N_0}\left(\frac{m}{2\pi T_b}\right)^{1/2} \exp\left(-\frac{m(v-u)^2}{2T_b}\right). \tag{11.3.13}$$

In contrast to the hydrodynamic instability, the increment of the kinetic instability is completely determined by the velocity distribution function of the electrons (Chap. 6)

$$\delta_k = \frac{\pi}{2}\frac{\omega_{p0}^3}{k^2}\frac{\partial F_0}{\partial v}\bigg|_{v=\omega/k}. \tag{11.3.14}$$

For $N_b \ll N_0$ the frequency spectrum of the oscillations is identical with the spectrum of the Langmuir plasma oscillations

$$\omega \approx \omega_{p0}\left(1 + \frac{3}{2}\frac{k^2 v_{Te}^2}{\omega_{p0}^2}\right). \tag{11.3.15}$$

The system of quasilinear equations for the kinetically unstable oscillations can be written as

$$\frac{\partial F_0}{\partial t} = \frac{\partial}{\partial p} D \frac{\partial F_0}{\partial p}, \quad \frac{\partial |E_k|^2}{\partial t} = 2\delta_k |E_k|^2,$$

$$D = -\frac{e^2}{2}\sum_k |E_k|^2 \,\mathrm{Im}\left\{\frac{1}{\omega - kv}\right\} = \frac{\pi e^2}{2}\sum_k |E_k|^2 \delta(\omega - kv). \tag{11.3.16}$$

In the derivation of this system from (11.3.3, 4) we have adopted the limit $|\omega - kv| \gg \delta_k \to +0$, which is always satisfied for the kinetic beam instability, and used

$$\frac{\delta_k}{(\omega - kv) + \delta_k^2} \to \pi\delta(\omega - kv) . \tag{11.3.17}$$

The set of equations (11.3.16) admits a stationary solution for $\partial F_0/\partial v|_{\omega/k=v} \to 0$, i.e., where a plateau in the velocity distribution function of the electrons is formed in the velocity range $v \approx \omega/k$. More exactly, the plateau is formed in the velocity range $v_1 \leq v \leq v_2$, shown in Fig. 11.2b. The problem of the quasilinear theory consists in determining the values of v_1 and v_2 and the height $F_0(v, \infty)$ in this range. We have three relations to solve this problem: the conservation law of the number of particles

$$\int_{v_1}^{v_2} F_0(v,0)\,dv = \int_{v_1}^{v_2} F_0(v,\infty)\,dv = F_0(v,\infty)\,(v_2 - v_1) \tag{11.3.18}$$

and the two obvious equalities

$$F_0(v_1, 0) = F_0(v_2, 0) = F_0(v, \infty) . \tag{11.3.19}$$

We want to evaluate now which part of the beam energy passes over into oscillation energy at the quasilinear stage of the kinetic beam instability. Similar to (11.2.3) and (11.3.16) we have

$$\frac{\partial}{\partial t}\left(F_0 - \frac{\partial}{\partial v}\frac{e^2|E_k|^2}{2m^2k^2v^4}\right) = 0 . \tag{10.3.20}$$

Hence

$$F_0(v,t) = F_0(v,0) + \frac{\partial}{\partial v}\frac{e^2|E_k|^2}{2m^2k^2v^4} . \tag{11.3.21}$$

Integrating this relation over the velocity we finally obtain

$$|E_k(\infty)|^2 = \frac{2m^2k^2v^4}{e^2}\int_{v_1}^{v_2} dv[F_0(v,\infty) - F_0(v,0)] , \tag{11.3.22}$$

where $v = \omega/k \approx \omega_{pe}/k \approx u$. The quantity $|E_k(\infty)|^2$ is nonzero only in the range of phase velocities $v_1 < \omega/k = v < v_2$, since the electron beam can excite plasma oscillations in this range, only.

Equation (11.3.22) shows that the energy density of the oscillations is distributed nonuniformly over the phase velocities in the final state of the quasilinear relaxation. Oscillations with large velocities prevail in the energy distribution. Besides, the ratio

$$\frac{|E_k(\infty)|^2}{N_b m u^2/2} \approx \frac{v_2 - v_1}{u} \sim \left(\frac{N_b}{N_0}\right)^{1/3} \tag{11.3.23}$$

can easily be estimated from (11.3.22). Thus, by the quasilinear relaxation of the kinetic beam instability a small part of the beam energy, of the order of $(N_b/N_0)^{1/3}$, is transferred into plasma oscillations. The quasilinear relaxation time of the kinetic beam instability is of the order of

$$t_q \sim \frac{m^2 v^2}{D} \sim \frac{m^2 v^2 \omega_{p0}}{e^2 |E_k|^2} \sim \frac{1}{\omega_{p0}} \left(\frac{N_0}{N_b}\right)^{4/3} . \tag{11.3.24}$$

It exceeds $N_0/N_b \gg 1$ times the relaxation time of the beam instability at the hydrodynamic stage.

In contrast to the hydrodynamic stage, in the kinetic stage there is a slowing down of the electrons since $\partial F_0/\partial v$ decreases due to the quasilinear relaxation. This process becomes very slow at the final stage. Hence, phenomena which restore the slope of the distribution function, e.g., particle collisions, become essential (Exercise 11.4.1).

11.4 Exercises

11.4.1 With the example of the Langmuir oscillations of the isotropic plasma show that a velocity distribution with a finite slope in the resonance range is established due to quasilinear relaxation when electron collisions are accounted for.

Solution. Particle collisions tend to convert the distribution function into the equilibrium Maxwellian distribution. Since the Langmuir oscillations distort the particle distribution in the range $v \gg v_{Te}$ in the case of a completely ionized plasma the complete quasilinear equation for the electron distribution function, taking account of electron-electron collisions, is of the form

$$\frac{\partial f_0}{\partial t} = \frac{\partial}{\partial p_\parallel} D \frac{\partial f_0}{\partial p_\parallel} + \frac{\partial}{\partial p_\parallel} D_e \frac{\partial}{\partial p_\parallel} (f_0 - f_M) , \tag{11.4.1}$$

with $p_\parallel = m v_\parallel$ and $p_\perp = m v_\perp$, where $v_\parallel$ and $v_\perp$ are the velocity projections along and across the direction of the wave propagation, respectively. f_M is the Maxwellian distribution function. Further we have

$$D = -\frac{e^2}{2} \sum_k |E_k|^2 \, \mathrm{Im}\left\{\frac{1}{\omega - k v_\parallel}\right\} , \qquad D_e = \nu_e \frac{2 m^2 v_{Te}^3}{v_\parallel^3} (v_\perp^2 + v_{Te}^2) ,$$
$$\nu_e = \frac{2\pi e^4 N_0 L}{m^2 v_{Te}^3} . \tag{11.4.2}$$

Equation (11.4.1) shows that due to electron-electron collisions the time of relaxation of the distribution function towards the Maxwellian equals $\tau_e \approx (\Delta v_\parallel / v_{Te})^2 / \nu_e \gg \nu_e^{-1}$, $\Delta v_\parallel$ being the width of the resonance range where the electron distribution function is distorted under the influence of the oscillations. Thus, in (11.4.1) both the quasilinear distortion of the electron distribution function (creation of the plateau) and its collisional relaxation to the Maxwellian are taken into account. The competition of these two processes results in establishing a stationary distribution (the oscillation field is assumed to be maintained by a stationary external source):

$$\frac{\partial f_0}{\partial v_\parallel} = \frac{\partial f_M}{\partial v_\parallel} \frac{1}{1 + D/D_e} . \tag{11.4.3}$$

In the absence of collisions $D_e \to 0$ and thus $\partial f_0 / \partial v_\parallel \to 0$.

Using the kinetic equation with the BGK collision integral instead of (11.4.1) for the case of the weakly ionized plasma we obtain

$$\frac{\partial f_0}{\partial t} = \frac{\partial}{\partial p_\parallel} D \frac{\partial f_0}{\partial p_\parallel} - \nu_e (f_0 - f_M) , \tag{11.4.4}$$

where $\nu_e = \text{const}$. Here also $\partial f_0 / \partial v_\parallel = 0$ for $\nu_e \neq 0$.

11.4.2 Derive the quasilinear equations for the low-frequency drift oscillations of the collisionless magneto-active inhomogeneous plasma in the zero-order approximation of geometrical optics.

Solution. We can write the quasilinear equations only for the slowly increasing kinetically unstable drift oscillations, since $\omega \sim \omega_{dr} \gg \delta$ holds for them. The Cherenkov wave dissipation by the plasma electrons is responsible for the buildup of these oscillations in the collisionless plasma. The electron distribution function is distorted as a result of the development of these instabilities. According to Sect. 11.1 and with account taken of the explicit form of the electron distribution function in the inhomogeneous plasma (Sect. 8.3), we obtain instead of (11.1.25) for the low-frequency drift oscillations $\omega \ll \Omega_e$

$$\begin{aligned} \frac{\partial f_0}{\partial t} = &-\frac{e^2}{2m^2} \sum_k |\Phi_k|^2 \left(k_\parallel \frac{\partial}{\partial v_\parallel} + \frac{k_y}{\Omega_e} \frac{\partial}{\partial x} \right) \mathrm{J}_0^2 \left(\frac{k_\perp v_\perp}{\Omega_e} \right) \\ &\times \operatorname{Im} \left\{ \frac{1}{\omega - k_\parallel v_\parallel} \right\} \left(k_\parallel \frac{\partial}{\partial v_\parallel} + \frac{k_y}{\Omega_e} \frac{\partial}{\partial x} \right) f_0 . \end{aligned} \tag{11.4.5}$$

For kinetically unstable oscillations we have

$$\operatorname{Im} \left\{ \frac{1}{\omega - k_\parallel v_\parallel} \right\} = - \pi \delta (\omega - k_\parallel v_\parallel) .$$

Hence, from (11.4.5) the equation for the stationary distribution function follows:

$$\left(\frac{\omega}{v_\parallel}\frac{\partial}{\partial v_\parallel}+\frac{k_y}{\Omega_e}\frac{\partial}{\partial x}\right)f_0=0\,. \tag{11.4.6}$$

Then, when the drift instabilities are developed, a state without a plateau, i.e., with $\partial f_0/\partial v_\parallel \neq 0$, is established at the quasilinear stage. Assuming that f_0 does not deviate much from the Maxwellian distribution function, we may write approximately

$$\frac{\partial f_0}{\partial v_\parallel}\approx-\frac{k_y v_\parallel}{\omega\Omega_e}\frac{\partial f_M}{\partial x}\,. \tag{11.4.7}$$

11.4.3 Using the model of independent particles investigate the one-dimensional stationary monochromatic waves with finite amplitude of a magnetized plasma penetrated by a straight monoenergetic relativistic low-density electron beam.

Solution. As shown in Chap. 6, the straight low-density beam excites longitudinal waves with a phase velocity close to the beam velocity in the magnetized plasma (Cherenkov instability). Thus, the finite amplitude waves of the plasma-beam system are of some interest. Naturally, the nonlinearity of these waves first of all affects the beam electron motion. The interaction of the waves with the plasma electrons can be regarded as linear and the plasma can be regarded as cold.

In the model of independent particles the motion of the beam electrons obeys the system of the continuity (3.6.4) and the Euler (3.6.8) equation. In the field of a monochromatic wave all quantities will be treated as functions of the coordinate $\xi = t - zk/\omega$; here the wave propagation is along the z-axis. Further, we shall treat only the one-dimensional problem, where the velocity of the beam electrons is assumed to be nonrelativistic and parallel to the Oz-axis in the intrinsic frame of the wave. The field $\boldsymbol{E}$ is assumed to be longitudinal, i.e.,

$$E=\frac{k}{\omega}\frac{d\Phi}{d\xi}\,, \tag{11.4.8}$$

$\Phi(\xi)$ being the potential of the wave field. Under the given restrictions the integrals of the equations of motion are

$$N_0\left(\frac{\omega}{k}-v\right)=N_b\left(\frac{\omega}{k}-u\right),$$

$$\frac{kc}{\omega}\sqrt{p^2+m^2c^2}-p+\frac{ek}{\omega}\Phi=\frac{kc}{\omega}\,mc\gamma\left(1-\frac{u}{c}\frac{\omega}{kc}\right), \tag{11.4.9}$$

where $p = mv(1 - v^2/c^2)^{-1/2}$ is the momentum and $\gamma = (1 - u^2/c^2)^{-1/2}$; u and N_b are the velocity and density of the beam electrons at the points where $\Phi = 0$.

Deriving the electron density from the system (11.4.9) and substituting the result into the Poisson equation yields

$$\left(1 - \frac{\omega_{p0}^2}{\omega^2}\right)\frac{d^2\Phi}{d\xi^2} = \omega_b^2\frac{\omega^2 m}{k^2 e}\left(\frac{(u - \omega/k)}{\sqrt{(u - \omega/k)^2 - \dfrac{2e\Phi}{m\gamma^3}}} - 1\right) . \tag{11.4.10}$$

In the linear approximation, i.e., in the limit of $m\gamma^3(u - \omega/k)^2 \gg 2e\Phi$, the solution of this equation takes the form of $\Phi(\xi) = \Phi_0 \cos(\omega\xi)$ with the known dispersion equation which describes time-increasing (consequently, nonstationary) potential waves with an infinitesimal amplitude (Chap. 6). The waves with a finite amplitude remain nonstationary if the kinetic energy of the beam electrons exceeds the potential of the wave field in the intrinsic frame of the wave, meaning that the electrons are not trapped by the wave, and can overcome the humps of the potential.

A nonlinear wave with the amplitude

$$2e\Phi_0 = m\gamma^3(u - \omega/k)^2 \tag{11.4.11}$$

however, becomes stationary. Under this condition the wave field captures the beam electrons, and there is no energy exchange. Substituting the solution $\Phi(\xi) = \Phi_0\cos(\omega\xi)$ into (11.4.10) and averaging over ξ results in

$$e\Phi_0 = \frac{mu^2\gamma}{\sqrt[3]{2}}\left(\frac{N_b}{N_0}\right)^{2/3} . \tag{11.4.12}$$

The energy of the oscillation field is also easily determined:

$$\frac{E_0^2}{8\pi N_b mc^2(\gamma - 1)} = \frac{1}{2}\left(\frac{N_b}{N_0}\right)^{2/3}(\gamma + 1) . \tag{11.4.13}$$

Note that in the nonrelativistic limit $e\Phi_0 \ll m\delta_k^2/k^2$, where δ_k is the Landau damping decrement for a longitudinal wave in the isotropic plasma, we obtain from the condition (11.2.14) that there is no electron capture

$$\frac{e\Phi_0}{T} \ll \frac{\pi}{8(kr_{De})^8}\exp\left(-\frac{1}{k^2r_{De}^2} - 3\right) . \tag{11.4.14}$$

11.4.4 Analyze the one-dimensional relaxation of a straight monoenergetic low-density electron beam injected into a semi-bounded ($Z > 0$) strongly magnetized plasma.

Solution. For $\omega_{p0} \gg \nu_e$ the electron beam excites plasma oscillations and loses its energy. As a result, the beam velocity decreases, $u = u_0 - \delta u$, and the electron density grows, $N_b = N_{b0}(1 + \delta u/u_0)$. According to the conservation law of energy we have

$$v_g W_E = N_{b0} m u_0^2 \gamma_0^3 \delta u \,. \tag{11.4.15}$$

Here $\gamma_0 = (1 - u_0^2/c^2)^{-1/2}$, $W_E = \omega\,(\partial\,\mathrm{Re}\,\{\varepsilon\}/\partial\omega)\,E^2/8\pi$ is the energy density of the beam-excited wave with the group velocity $v_g = \partial\,\mathrm{Re}\,\{\varepsilon\}/\partial k$ and

$$\varepsilon(\omega, k) = 1 - \frac{\omega_{p0}^2}{\omega^2}\left(1 + \frac{3k^2 v_{Te}^2}{\omega^2} - \mathrm{i}\frac{\nu_e}{\omega}\right) - \frac{\omega_b^2}{\gamma^3(\omega - ku_0)^2} \tag{11.4.16}$$

is the dielectric permittivity of the system. The beam will relax due to the growth of W_E and δu until the beam electrons are captured by the wave field. During the Cherenkov beam instability with $\omega \approx \omega_{p0} \approx ku_0$, the trapping occurs when the field reaches the value

$$E_{\mathrm{capt}} = \frac{m(\delta u)^2}{2eu_0}\,\omega_{p0}\gamma_0^3 \,. \tag{11.4.17}$$

Hence we obtain from (11.4.15)

$$\begin{aligned} \frac{\delta u_{\mathrm{capt}}}{u_0} &= \frac{2}{\gamma_0}\left(\frac{u_0}{v_g}\frac{N_{b0}}{N_0}\,\omega\frac{\partial\mathrm{Re}\,\{\varepsilon\}}{\partial\omega}\right)^{1/3}, \\ W_{\mathrm{capt}} &= 2N_{b0}mc^2(\gamma_0^2 - 1)\left(\frac{v_g^4}{u_0}\frac{N_{b0}}{N_b}\,\omega\frac{\partial\mathrm{Re}\,\{\varepsilon\}}{\partial\omega}\right)^{-1/3}. \end{aligned} \tag{11.4.18}$$

In order to complete this solution we must determine the group velocity v_g of the wave in the case of the Cherenkov resonance. Taking account of the dispersion equation $\varepsilon(\omega, k) = 0$ we obtain

$$\begin{aligned} \omega\frac{\partial\mathrm{Re}\,\{\varepsilon\}}{\partial\omega} &= 2\left(1 + \frac{\omega_b^2\omega_{p0}}{\gamma^3(\omega - ku_0)}\right), \\ v_g &= \left(3\frac{v_{Te}^2}{u_0} + \frac{\omega_b^2\omega_{p0}u_0}{\gamma^3(\omega - ku_0)^3}\right)\left(1 + \frac{\omega_b^2\omega_{p0}}{\gamma_0^3(\omega - ku_0)^3}\right)^{-1}. \end{aligned} \tag{11.4.19}$$

In the collisionless limit ($\nu_e \to 0$), assuming $k = k_0 + \kappa$ in the dispersion equation $\varepsilon(\omega, k) = 0$, and $k_0 u_0 = \omega_{p0}$ we obtain

$$\kappa = \left(\frac{\omega_b^2 u_0^2}{6\gamma_0^3\omega_{p0}^2 v_{Te}^2}\right)^{1/3} k_0 \frac{1 - \mathrm{i}\sqrt{3}}{2} \,. \tag{11.4.20}$$

The imaginary part is an inverse length of the beam relaxation and

$$v_{\mathrm{g}} \approx 3\,\frac{v_{\mathrm{Te}}^2}{u_0^2}\,u_0 \ll u_0\,, \qquad \omega\,\frac{\partial \operatorname{Re}\{\varepsilon\}}{\partial \omega} = 2\,,$$

$$\frac{W_{\mathrm{capt}}}{N_{\mathrm{b0}} m c^2 (\gamma - 1)} = 2(\gamma_0 + 1)\left(\frac{u_0^2}{9 v_{\mathrm{Te}}^2}\right)^{4/3}\left(\frac{N_{\mathrm{b0}}}{2 N_0}\right)^{1/3}. \tag{11.4.21}$$

Note that the field which captures the electrons when the beam is injected into the semi-bounded plasma and which is determined by the last relation of (11.4.21) greatly exceeds the corresponding field in the spatially unbounded plasma-beam system which has been estimated in Exercise 11.4.3. This fact is explained by the accumulation effect of beam-excited oscillations owing to their small group velocity $v_{\mathrm{g}} \ll u_0$, which is much smaller than the group velocity in the case of the unbounded plasma-beam system (Sect. 6.3).

The collisionless limit is valid when the inequality

$$\eta = \frac{\nu_{\mathrm{e}}^{3/2} \gamma_0^{3/2}}{6\,\omega_{\mathrm{b}} \sqrt{\omega_{\mathrm{p0}}}}\,\frac{u_0^2}{v_{\mathrm{Te}}^2} \ll 1 \tag{11.4.22}$$

is satisfied. For frequent collisions, $\eta \gg 1$, the thermal motion of the plasma electrons can be neglected, assuming $v_{\mathrm{T}} = 0$. Hence the dispersion equation $\varepsilon(\omega, k) = 0$ results in $(k = k_0 + \kappa)$

$$\kappa = \frac{1 - \mathrm{i}}{\sqrt{2}}\,\frac{\omega_{\mathrm{b}}}{u_0 \gamma_0^{3/2}}\left(\frac{\omega_{\mathrm{p0}}}{\nu_{\mathrm{e}}}\right)^{1/2},$$

$$v_{\mathrm{g}} = u_0\,\frac{\nu_{\mathrm{e}}^{3/2} \gamma_0^{3/2}}{\omega_{\mathrm{b}} \sqrt{2\,\omega_{\mathrm{p0}}}} \ll u_0\,, \qquad \omega\,\frac{\partial \operatorname{Re}\{\varepsilon\}}{\partial \omega} = 2\,. \tag{11.4.23}$$

Thus, though the group velocity fulfils $v_{\mathrm{g}} \ll u_0$ it greatly exceeds the value (11.4.21) derived in the collisionless limit. Therefore, the collision frequency grows and the accumulation effect decreases. This, in turn, results in a growth of the electron beam relaxation length and a decrease of the trapping field

$$\frac{W_{\mathrm{capt}}}{N_{\mathrm{b0}} m c^2 (\gamma - 1)} = 4(\gamma + 1)\left(\frac{2 N_{\mathrm{b0}}}{N_0}\right)^{1/3}\left(\frac{\omega_{\mathrm{b}}^2 \omega_{\mathrm{p0}}}{\gamma_0^3 \nu_{\mathrm{e}}^3}\right)^{2/3}. \tag{11.4.24}$$

11.4.5 Derive the velocity and the profile of a nonlinear one-dimensional solitary ion-acoustic wave *(ion-acoustic soliton)*.

Solution. We write the system of equations of motion of the plasma in the range of the ion-acoustic waves in the form of (Sect. 3.6):

$$\frac{\partial v}{\partial t} + v\frac{\partial v}{\partial x} = -\frac{e}{m}\frac{\partial \Phi}{\partial x}\,, \qquad \frac{\partial N_{\mathrm{i}}}{\partial t} + \frac{\partial (N_{\mathrm{i}} v)}{\partial x} = 0\,, \tag{11.4.25}$$

$$\frac{\partial^2 \Phi}{\partial x} = -4\pi e\,(N_{\mathrm{i}} - N_0 \mathrm{e}^{e\Phi/T_{\mathrm{e}}})\,.$$

Here N_0 is the density of the plasma electrons and ions for $\Phi = 0$ (for simplicity we assume $Z = 1$).

We look for a solution of the system (11.4.25) in the form of a function of $\xi = x - vt$ and assume u to be constant. Then, the first two equations of this system are reduced to

$$(v-u)\,v' = -\frac{e}{m}\Phi'\,, \qquad (N_{\mathrm{i}} v)' - uN_{\mathrm{i}}' = 0\,. \tag{11.4.26}$$

For a solitary wave (soliton) we have the boundary conditions $v \to 0$, $N_{\mathrm{i}} \to N_0$ for $\xi \to \infty$. Hence

$$e\Phi = \frac{Mu^2}{2} - \frac{M(u-v)^2}{2}\,, \qquad N_{\mathrm{i}} = \frac{N_0 u}{\sqrt{u^2 - 2e\Phi/M}}\,. \tag{11.4.27}$$

The last equation of the system (11.4.25) can be solved now and assumes the form

$$\Phi'^2 = -8\pi e N_0 \int\limits_0^{\Phi} d\Phi \left(\frac{u}{\sqrt{u^2 - 2e\Phi/M}} - \mathrm{e}^{e\Phi/T_{\mathrm{e}}}\right). \tag{11.4.28}$$

Consequently, the velocity of the solitary wave u is determined by the wave amplitude, i.e., the maximum value of $\Phi(\xi)$. Expressing the wave amplitude in terms of Φ_{m} ($\Phi' = 0$ for $\Phi = \Phi_{\mathrm{m}}$) we obtain from (11.4.28) the relationship between u and Φ_{m}

$$\frac{Mu^2}{T_{\mathrm{e}}}\left(1 - \sqrt{1 - \frac{2e\Phi_{\mathrm{m}}}{Mu^2}}\right) = \mathrm{e}^{e\Phi_{\mathrm{m}}/T_{\mathrm{e}}} - 1\,. \tag{11.4.29}$$

Evidently, $e\Phi_{\mathrm{m}} < Mu^2/2$ must be given.

Equation (11.4.29) is significantly simplified for a weakly nonlinear solitary ion-acoustic wave when $e\Phi_{\mathrm{m}} \ll T_{\mathrm{e}}$. Then we get

$$u \approx v_{\mathrm{s}}\left(1 + \frac{8}{15}\sqrt{\frac{e\Phi_{\mathrm{m}}}{\pi T_{\mathrm{e}}}}\right). \tag{11.4.30}$$

In this case it is easy to find the wave profile from (11.4.28)

$$\Phi(\xi) = \frac{\Phi_{\mathrm{m}}}{\mathrm{ch}^2\left(\dfrac{\xi}{\sqrt{15}\, r_{\mathrm{De}}}\sqrt{\dfrac{e\Phi_{\mathrm{m}}}{\pi T_{\mathrm{e}}}}\right)}\,. \tag{11.4.31}$$

12. Nonlinear Interaction of Waves in a Plasma

Basic principles of nonlinear electrodynamics of media with spatial dispersion are presented; the multi-index dielectric tensors and the method of reduced equations are introduced. Having obtained an expression for the multi-index tensors of a homogeneous isotropic plasma, the problem of the nonlinear interaction of longitudinal waves is solved (Coulomb and delayed scattering). The process of induced scattering of a transverse wave in an isotropic plasma is discussed.

12.1 Principles of Nonlinear Electrodynamics of Material Media

So far we have treated nonlinear phenomena involving the effect of electromagnetic waves on the plasma and its distribution function. There, wave interaction has been neglected. Such a nonlinear phenomenon is commonly known as quasilinear. As shown before, the quasilinear modification of the plasma is characterized by a relaxation time t_q. The quasilinear approximation is valid if nonlinear wave interaction has no time to grow noticeably during t_q.

To evaluate the characteristic time of nonlinear wave interaction in the plasma, the opposite limit shall be considered. Assuming the plasma state to be unstable, let us study the nonlinear effects associated with wave interaction, and evaluate their interaction time t_n. The condition $t_n < t_q$ justifies such a consideration. The quasilinear approximation is valid in the opposite limit.

Before turning to a systematic exposition of the nonlinear effects of wave interaction in the plasma, let us briefly review the general principles of nonlinear electrodynamics of material media. As mentioned in Chap. 1, field equations in the medium may contain nonlinearities associated with the material equations (2.1.6 or 7), establishing the relationship between $\boldsymbol{j}$ and $\boldsymbol{E}$ or $\boldsymbol{D}$ and $\boldsymbol{E}$. In linear electrodynamics, we restrict ourselves to linear material equations. For sufficiently weak fields, the expansion in a power series in the field accounting for both linear (linear electrodynamics) and nonlinear terms to any order can be applied here. The last terms describe the wave interaction. One can introduce the concept of eigenmodes in the medium and describe the wave interaction by means of small nonlinear terms only in the case of weak fields (the concrete plasma model explains its meaning) when the expansion in a power series in the field is valid and linear terms are predominant.

The material equation in a general form for a spatially homogeneous and stationary medium expanded in a power series, and accounting for the causality principle, is written in the form:

$$D_i(t,\boldsymbol{r}) = \sum_{n=1}^{\infty} \int_{-\infty}^{t} dt_1 \int d\boldsymbol{r}_1 \int_{-\infty}^{t_1} dt_2 \int d\boldsymbol{r}_2 \ldots \int_{-\infty}^{t_{n-1}} dt_n \int d\boldsymbol{r}_n$$
$$\times\, \varepsilon_{ij_1,\ldots,j_n}(t-t_1, \boldsymbol{r}-\boldsymbol{r}_1, \ldots, t_{n-1}-t_n, \boldsymbol{r}_{n-1}-\boldsymbol{r}_n)\, E_{j1}(t_1,\boldsymbol{r}_1)\ldots E_{jn}(t_n,\boldsymbol{r}_n)\,. \tag{12.1.1}$$

We can obtain the material equation of the linear theory (2.2.2) by confining ourselves to the first term of this series. Then we take into account terms of higher orders, notably those which are quadratic and cubic in the field. Furthermore, we deal only with the nonlinear effects in a spatially homogeneous and stationary medium where the time- and coordinate-dependence of the tensors $\varepsilon_{ij_1,\ldots,j_n}$ is purely differential, i.e., there is no explicit dependence on t_n and $\boldsymbol{r}_n$. Most characteristic nonlinear effects of wave interaction in a weakly nonlinear plasma are manifested in this approximation.

12.1.1 Multi-Index Dielectric Tensors

Together with the material equation (2.1.5), the Maxwell equation (12.1.1) forms a complete system of electrodynamic equations for a weakly nonlinear medium. In the case of a spatially homogeneous and stationary medium the fields can be conveniently expanded in the Fourier integral:

$$\boldsymbol{E}(t,\boldsymbol{r}) = \int d\omega \int d\boldsymbol{k}\, \boldsymbol{E}(\omega,\boldsymbol{k})\, \mathrm{e}^{-\mathrm{i}\omega t + \mathrm{i}\boldsymbol{k}\cdot\boldsymbol{r}}\,. \tag{12.1.2}$$

Here the concept of a *multi-index dielectric tensor* can be introduced:

$$\varepsilon_{ij_1\ldots j_n}(\omega,\boldsymbol{k};\omega_1,\boldsymbol{k}_1;\ldots;\omega_{n-1},\boldsymbol{k}_{n-1}) = \int_0^{\infty} dt\, \mathrm{e}^{\mathrm{i}\omega t} \int d\boldsymbol{r}\, \mathrm{e}^{-\mathrm{i}\boldsymbol{k}\cdot\boldsymbol{r}} \int dt_1\, \mathrm{e}^{\mathrm{i}\omega_1 t_1}$$
$$\times \int d\boldsymbol{r}_1\, \mathrm{e}^{-\mathrm{i}\boldsymbol{k}_1\cdot\boldsymbol{r}_1} \ldots \int_0^{\infty} dt_{n-1}\, \mathrm{e}^{\mathrm{i}\omega_{n-1}t_{n-1}}$$
$$\times \int d\boldsymbol{r}_{n-1}\, \mathrm{e}^{-\mathrm{i}\boldsymbol{k}_{n-1}\cdot\boldsymbol{r}_{n-1}}\, \varepsilon_{ij_1\ldots j_n}(t,\boldsymbol{r};t_1,\boldsymbol{r}_1;\ldots;t_{n-1}\boldsymbol{r}_{n-1})\,. \tag{12.1.3}$$

In particular, a double-index tensor $\varepsilon_{ij}(\omega,\boldsymbol{k})$ is an ordinary dielectric tensor used in linear electrodynamics.

In terms of multi-index tensors, (12.1.1) can be written as

$$D_i(\omega, \boldsymbol{k}) = \varepsilon_{ij}(\omega, \boldsymbol{k}) E_j(\omega, \boldsymbol{k}) + \sum_{n=2}^{\infty} \int d\omega_1 \, d\boldsymbol{k}_1 \ldots d\omega_{n-1} \, d\boldsymbol{k}_{n-1}$$
$$\times \varepsilon_{ij_1 \ldots j_n}(\omega, \boldsymbol{k}; \ldots; \omega_{n-1}, \boldsymbol{k}_{n-1}) E_{j_1}(\omega - \omega_1, \boldsymbol{k} - \boldsymbol{k}_1) \ldots$$
$$\ldots E_{j_{n-1}}(\omega_{n-2} - \omega_{n-1}, \boldsymbol{k}_{n-2} - \boldsymbol{k}_{n-1}) E_{j_n}(\omega_{n-1}, \boldsymbol{k}_{n-1}) . \tag{12.1.4}$$

Then the Maxwell equation

$$\operatorname{curl}\operatorname{curl}\boldsymbol{E} + \frac{1}{c^2}\frac{\partial^2 \boldsymbol{D}}{\partial t^2} = 0 \tag{12.1.5}$$

in the $(\omega, \boldsymbol{k})$ space takes the form

$$\left[\frac{k^2c^2}{\omega^2}\left(\delta_{ij} - \frac{k_ik_j}{k^2}\right) - \varepsilon_{ij}(\omega, \boldsymbol{k})\right] E_j(\omega, \boldsymbol{k}) = \sum_{n=2}^{\infty} \int d\omega_1 \, d\boldsymbol{k}_1 \ldots d\omega_{n-1} \, d\boldsymbol{k}_{n-1}$$
$$\times \varepsilon_{ij_1 \ldots j_n}(\omega, \boldsymbol{k}, \ldots, \omega_{n-1}, \boldsymbol{k}_{n-1}) E_{j_1}(\omega - \omega_1, \boldsymbol{k} - \boldsymbol{k}_1) \ldots E_{j_{n-1}}$$
$$\times (\omega_{n-2} - \omega_{n-1}, \boldsymbol{k}_{n-2} - \boldsymbol{k}_{n-1}) E_{j_n}(\omega_{n-1}, \boldsymbol{k}_{n-1}) . \tag{12.1.6}$$

In the linear approximation, (12.1.6) reduces to the linear electrodynamic equation (2.4.3).

12.1.2 Averaged Equation of Nonlinear Electrodynamics

Equation (12.1.6) can provide the basis for nonlinear electrodynamics of a medium only for weak nonlinearity when the field representation in the form of a combination of plane waves (12.1.2) is close to the exact solution of the problem. Here, the terms on the right-hand side are small, and they describe a small deviation of the linear waves due to their nonlinear interaction. In the first-order approximation, when nonlinear terms are neglected, plane waves represent exact solutions of (2.4.3), provided that ω is related to $\boldsymbol{k}$ by the general dispersion equation

$$\left| k^2\delta_{ij} - k_ik_j - \frac{\omega^2}{c^2}\varepsilon_{ij}(\omega, \boldsymbol{k}) \right| = 0 . \tag{12.1.7}$$

Thus, in the linear approximation, we obtain information only about the type of waves which can exist in the medium. Taking account of small nonlinear terms results in the wave amplitudes being weak functions of time and coordinate. To provide for slow amplitude modification, as in the preceding chapter, a small parameter μ is introduced:

$$\boldsymbol{E}(t, \boldsymbol{r}) = \boldsymbol{E}(\mu t, \mu \boldsymbol{r}, \omega, \boldsymbol{k}) \, \mathrm{e}^{-\mathrm{i}\omega t + \mathrm{i}\boldsymbol{k}\cdot\boldsymbol{r}} + \boldsymbol{E}^*(\mu t, \mu \boldsymbol{r}, \omega, \boldsymbol{k}) \, \mathrm{e}^{\mathrm{i}\omega t - \mathrm{i}\boldsymbol{k}\cdot\boldsymbol{r}} . \tag{12.1.8}$$

In order to describe this weak time- and coordinate-dependence (compared to $1/\omega$ and $1/k$) we apply averaging, as has been done in Sect. 2.3 when deriving the expression for the field energy in the medium. Thus

$$\begin{aligned}\frac{\partial D_i(t,\boldsymbol{r})}{\partial t} &\approx -\,i\omega \mathrm{e}^{-i\omega t+i\boldsymbol{k}\cdot\boldsymbol{r}}\varepsilon_{ij}(\omega,\boldsymbol{k})\,E_j(\mu t,\mu\boldsymbol{r},\omega,\boldsymbol{k})\\ &+\mathrm{i}\omega e^{i\omega t-i\boldsymbol{k}\cdot\boldsymbol{r}}\varepsilon_{ij}(\omega,\boldsymbol{k})\,E_j^*(\mu t,\mu\boldsymbol{r},\omega,\boldsymbol{k})\\ &+\frac{\partial}{\partial\omega}\left[\omega\varepsilon_{ij}^{\mathrm{H}}(\omega,\boldsymbol{k})\right]\frac{\partial E_j(\mu t,\mu\boldsymbol{r},\omega,\boldsymbol{k})}{\partial\mu t}\,\mathrm{e}^{-i\omega t+i\boldsymbol{k}\cdot\boldsymbol{r}}\\ &+\frac{\partial}{\partial\omega}\left[\omega\varepsilon_{ij}^{\mathrm{H}}(\omega,\boldsymbol{k})\right]\frac{\partial E_j^*(\mu t,\mu\boldsymbol{r},\omega,\boldsymbol{k})}{\partial\mu t}\,\mathrm{e}^{i\omega t-i\boldsymbol{k}\cdot\boldsymbol{r}}\\ &-\frac{\partial}{\partial\boldsymbol{k}}\left[\omega\varepsilon_{ij}^{\mathrm{H}}(\omega,\boldsymbol{k})\right]\frac{\partial E_j(\mu t,\mu\boldsymbol{r},\omega,\boldsymbol{k})}{\partial\mu\boldsymbol{r}}\,\mathrm{e}^{-i\omega t+i\boldsymbol{k}\cdot\boldsymbol{r}}\\ &-\frac{\partial}{\partial\boldsymbol{k}}\left[\omega\varepsilon_{ij}^{\mathrm{H}}(\omega,\boldsymbol{k})\right]\frac{\partial E_j^*(\mu t,\mu\boldsymbol{r},\omega,\boldsymbol{k})}{\partial\mu\boldsymbol{r}}\,\mathrm{e}^{i\omega t-i\boldsymbol{k}\cdot\boldsymbol{r}}\\ &-\mathrm{i}\omega\delta D_i(\omega,\boldsymbol{k})+\mathrm{i}\omega\delta D_i^*(\omega,\boldsymbol{k})\,.\end{aligned}\tag{12.1.9}$$

Here $\delta D_i(\omega,\boldsymbol{k})$ is a nonlinear part of the induction vector [i.e., a nonlinear term in (12.1.4)] where weak t and $\boldsymbol{r}$ dependences are ignored.

Substituting (12.1.9) into (2.3.1) for the field energy (for $\boldsymbol{j}_0=0$)

$$\frac{1}{4\pi}\left(\boldsymbol{E}\frac{\partial\boldsymbol{D}}{\partial t}+\boldsymbol{B}\frac{\partial\boldsymbol{B}}{\partial t}\right)+\frac{c}{4\pi}\operatorname{div}[\boldsymbol{E},\boldsymbol{B}]=0\tag{12.1.10}$$

and averaging the results in

$$\begin{aligned}&\frac{\partial}{\partial t}\left\{\left[\frac{\partial\omega\varepsilon_{ij}^{\mathrm{H}}(\omega,\boldsymbol{k})}{\partial\omega}+\frac{k^2c^2}{\omega^2}\left(\delta_{ij}-\frac{k_ik_j}{k^2}\right)\right]E_j(\omega,\boldsymbol{k})\,E_i^*(\omega,\boldsymbol{k})\right\}\\ &-\frac{\partial}{\partial r_s}\left\{\left[\frac{\partial\omega\varepsilon_{ij}^{\mathrm{H}}(\omega,\boldsymbol{k})}{\partial k_s}+(k_i\delta_{sj}+k_j\delta_{is}-2k_s\delta_{ij})\frac{c^2}{\omega}\right]E_j(\omega,\boldsymbol{k})\,E_i^*(\omega,\boldsymbol{k})\right\}\\ &=\mathrm{i}\omega\left[\varepsilon_{ij}(\omega,\boldsymbol{k})\,E_j(\omega,\boldsymbol{k})\,E_i^*(\omega,\boldsymbol{k})-\varepsilon_{ij}^*(\omega,\boldsymbol{k})\,E_j^*(\omega,\boldsymbol{k})\,E_i(\omega,\boldsymbol{k})\right]\\ &+\mathrm{i}\omega\sum_{n=2}^{\infty}\int d\omega_1\,d\boldsymbol{k}_1\ldots d\omega_{n-1}\,d\boldsymbol{k}_{n-1}\,\varepsilon_{ij_1\ldots j_n}(\omega,\boldsymbol{k},\ldots,\omega_{n-1},\boldsymbol{k}_{n-1})\\ &\times E_i^*(\omega,\boldsymbol{k})\,E_{j_1}(\omega-\omega_1,\boldsymbol{k}-\boldsymbol{k}_1)\ldots E_{j_n}(\omega_{n-1},\boldsymbol{k}_{n-1})-\mathrm{i}\omega\sum_{n=2}^{\infty}\int d\omega_1\,d\boldsymbol{k}_1\\ &\ldots,d\omega_{n-1}\,d\boldsymbol{k}_{n-1}\,\varepsilon_{ij_1,\ldots,j_n}^*(\omega,\boldsymbol{k},\ldots,\omega_{n-1},\boldsymbol{k}_{n-1})\,E_i(\omega,\boldsymbol{k})\\ &\times E_{j_1}^*(\omega-\omega_1,\boldsymbol{k}-\boldsymbol{k}_1)\ldots E_{j_n}^*(\omega_{n-1},\boldsymbol{k}_{n-1})\,.\end{aligned}\tag{12.1.11}$$

In fact, this equation means the conservation of energy in the system. Its left-hand side describes a wave amplitude variation due to the wave radiation and absorption, and the right-hand side due to their nonlinear interaction. In the right-hand side, only several terms usually remain (in nonlinear optics we restrict ourselves to three expansion terms). In case of such a cut-off of the series, the so-derived equation is called *shortened*.

12.1.3 Shortened Equation for Waves with Chaotic Phases

In deriving (12.1.11), there were no restrictions on wave phases. Therefore, it can be applied in the analysis of nonlinear waves both with fixed and random phases. For random (chaotic) wave phases, (12.1.11) should be averaged over phases, and thus a transfer to statistic electrodynamics is done. Then we confine ourselves to terms up to the fourth order in the field, and take into account the relations (Chap. 10):

$$\begin{aligned} &\langle \boldsymbol{E}(\omega, \boldsymbol{k})\rangle = 0 \,, \\ &\langle E_i(\omega, \boldsymbol{k})\, E_j^*(\omega', \boldsymbol{k}')\rangle = \delta(\omega-\omega')\,\delta(\boldsymbol{k}-\boldsymbol{k}')\langle E_i E_j\rangle_{\omega,\boldsymbol{k}} \,, \end{aligned} \tag{12.1.12}$$

where $\langle E_i E_j\rangle_{\omega,k}$ is a spectral density of electric field fluctuations.

Strictly speaking, (12.1.12) is valid only for a stationary and spatially homogeneous medium. In nonlinear electrodynamics the field amplitudes of the medium slowly vary in space and time according to (12.1.11), therefore the relations summarized in (12.1.12) are approximate. But in linear electrodynamics they are exact, and by omitting terms of the third and higher orders from (12.1.11), we obtain

$$\frac{\partial W(\omega, \boldsymbol{k})}{\partial t} + \operatorname{div} \boldsymbol{S}(\omega, \boldsymbol{k}) = -2\,\delta(\omega, \boldsymbol{k})\, W(\omega, \boldsymbol{k}) \,, \tag{12.1.13}$$

where we have introduced the notations

$$\begin{aligned} &\frac{1}{4\pi}\frac{\partial}{\partial\omega}[\omega M_{ij}^{\mathrm{H}}(\omega, \boldsymbol{k})]\langle E_i E_j\rangle_{\omega,\boldsymbol{k}} = W(\omega, \boldsymbol{k}) \,, \\ &M_{ij}(\omega, \boldsymbol{k}) = \varepsilon_{ij}(\omega, \boldsymbol{k}) - \frac{k^2c^2}{\omega^2}\left(\delta_{ij} - \frac{k_i k_j}{k^2}\right) = \frac{c^2}{\omega^2}\Lambda_{ij}(\omega, \boldsymbol{k}) \,, \\ &\boldsymbol{S}(\omega, \boldsymbol{k}) = -\frac{1}{4\pi}\frac{\partial}{\partial \boldsymbol{k}}[\omega M_{ij}^{\mathrm{H}}(\omega, \boldsymbol{k})]\langle E_i E_j\rangle_{\omega,\boldsymbol{k}} \,, \\ &\delta(\omega, \boldsymbol{k})\, W(\omega, \boldsymbol{k}) = -\frac{\mathrm{i}\omega}{4\pi}\frac{\varepsilon_{ij}^{\mathrm{a}}(\omega, \boldsymbol{k})\langle E_i E_j\rangle_{\omega,\boldsymbol{k}}}{W(\omega, \boldsymbol{k})} \,. \end{aligned} \tag{12.1.14}$$

When deriving an analogous equation of nonlinear electrodynamics, we must know ternary and quadruple correlators along with binary ones (2.1.12). In linear electrodynamics, the ternary correlators are identically zero, and the quadruple ones split up into products of binary correlators. In the nonlinear approximation (with accuracy up to the fourth-order terms in the field) quadruple correlators can be represented in the form of products of binary correlators:

$$\begin{aligned}&\langle E_i(\omega,\boldsymbol{k})\,E_s(\omega',\boldsymbol{k}')\,E_j(\omega'',\boldsymbol{k}'')\,E_r(\omega''',\boldsymbol{k}''')\rangle\\&\quad=\langle E_iE_s\rangle_{\omega,\boldsymbol{k}}\langle E_jE_r\rangle_{\omega'',\boldsymbol{k}''}\,\delta(\omega+\omega')\,\delta(\boldsymbol{k}+\boldsymbol{k}')\,\delta(\omega''+\omega''')\,\delta(\boldsymbol{k}''+\boldsymbol{k}''')\\&\quad+\langle E_iE_j\rangle_{\omega,\boldsymbol{k}}\langle E_sE_r\rangle_{\omega',\boldsymbol{k}'}\,\delta(\omega+\omega'')\,\delta(\boldsymbol{k}+\boldsymbol{k}'')\,\delta(\omega'+\omega''')\,\delta(\boldsymbol{k}'+\boldsymbol{k}''')\\&\quad+\langle E_iE_r\rangle_{\omega,\boldsymbol{k}}\langle E_sE_j\rangle_{\omega',\boldsymbol{k}'}\,\delta(\omega+\omega''')\,\delta(\boldsymbol{k}+\boldsymbol{k}''')\,\delta(\omega'+\omega'')\,\delta(\boldsymbol{k}'+\boldsymbol{k}'')\,.\end{aligned}\tag{12.1.15}$$

Ternary correlators are nonzero and of the order of E^4, therefore high precision is necessary in their calculation. Let us write (12.1.16) up to the cubic terms in the field

$$\begin{aligned}M_{ij}(\omega,\boldsymbol{k})\,E_j(\omega,\boldsymbol{k})&=-\int d\omega_1\,d\boldsymbol{k}_1\,\varepsilon_{ijs}(\omega,\boldsymbol{k},\omega_1,\boldsymbol{k}_1)\\&\quad\times E_j(\omega-\omega_1,\boldsymbol{k}-\boldsymbol{k}_1)\,E_s(\omega_1,\boldsymbol{k}_1)\,.\end{aligned}\tag{12.1.16}$$

When solving this equation by means of successive approximations, we obtain

$$\begin{aligned}E_i(\omega,\boldsymbol{k})&=E_i^{(0)}(\omega,\boldsymbol{k})-A_{ir}(\omega,\boldsymbol{k})\int d\omega_1\,d\boldsymbol{k}_1\,\varepsilon_{rjs}(\omega,\boldsymbol{k};\omega_1,\boldsymbol{k}_1)\\&\quad\times E_j^{(0)}(\omega-\omega_1,\boldsymbol{k}-\boldsymbol{k}_1)\,E_s^{(0)}(\omega_1,\boldsymbol{k}_1)\,,\end{aligned}\tag{12.1.17}$$

where $E_i^{(0)}(\omega,\boldsymbol{k})$ is the solution of the linear equation (12.1.16), i.e., without the right-hand side, and

$$A_{ij}(\omega,\boldsymbol{k})=M_{ij}^{-1}(\omega,\boldsymbol{k})=\left[E_{ij}(\omega,\boldsymbol{k})-\frac{k^2c^2}{\omega^2}\left(\delta_{ij}-\frac{k_ik_j}{k^2}\right)\right]^{-1}.\tag{12.1.18}$$

Now we can write down the ternary correlators if we take into account that the fields $E_i^0(\omega,\boldsymbol{k})$ are mutually uncorrelated; thus

$$\begin{aligned}&\langle E_i(\omega,\boldsymbol{k})\,E_s(\omega',\boldsymbol{k}')\,E_j(\omega'',\boldsymbol{k}'')\rangle=-\int d\omega_1\,d\boldsymbol{k}_1\,[A_{ia}(\omega,\boldsymbol{k})\,\varepsilon_{abc}(\omega,\boldsymbol{k};\omega_1,\boldsymbol{k}_1)\\&\quad\times\langle E_b^{(0)}(\omega-\omega_1,\boldsymbol{k}-\boldsymbol{k}_1)\,E_c^{(0)}(\omega_1,\boldsymbol{k}_1)\,E_s^{(0)}(\omega',\boldsymbol{k}')\,E_j^{(0)}(\omega'',\boldsymbol{k}'')\rangle\\&\quad+A_{sc}(\omega',\boldsymbol{k}')\,\varepsilon_{cba}(\omega',\boldsymbol{k}';\omega_1,\boldsymbol{k}_1)\langle E_b^{(0)}(\omega'-\omega_1,\boldsymbol{k}'-\boldsymbol{k}_1)\,E_a^{(0)}(\omega_1,\boldsymbol{k}_1)\\&\quad\times E_i^{(0)}(\omega,\boldsymbol{k})\,E_j^{(0)}(\omega'',\boldsymbol{k}'')\rangle+A_{jb}(\omega'',\boldsymbol{k}'')\,\varepsilon_{bca}(\omega'',\boldsymbol{k}'';\omega_1,\boldsymbol{k}_1)\\&\quad\times\langle E_b^{(0)}(\omega''-\omega_1,\boldsymbol{k}''-\boldsymbol{k}_1)\,E_a^{(0)}(\omega_1,\boldsymbol{k}_1)\,E_j^{(0)}(\omega,\boldsymbol{k})\,E_s^{(0)}(\omega',\boldsymbol{k}')\rangle]\,.\end{aligned}\tag{12.1.19}$$

Substituting (12.1.12, 15 and 19) into the averaged equation (12.1.11), accounting for terms up to the fourth order, we finally obtain the equation of nonlinear electrodynamics for waves with chaotic phases:

$$\frac{\partial}{\partial t}\left[\frac{1}{\omega}\frac{\partial}{\partial \omega}\omega M_{ij}^{\mathrm{H}}(\omega,\boldsymbol{k})\langle E_j E_i\rangle_{\omega,\boldsymbol{k}}\right]-\frac{\partial}{\partial \boldsymbol{r}}\left[\frac{1}{\omega}\frac{\partial}{\partial \boldsymbol{k}}\omega M_{ij}^{\mathrm{H}}(\omega,\boldsymbol{k})\langle E_j E_i\rangle_{\omega,\boldsymbol{k}}\right]$$
$$=2i\varepsilon_{ij}^{\mathrm{a}}(\omega,\boldsymbol{k})\langle E_j E_i\rangle_{\omega,\boldsymbol{k}}+\operatorname{Im}\{\textstyle\int d\omega'\,d\boldsymbol{k}'$$
$$\times\,[A_{ia}^{*}(\omega,\boldsymbol{k})\,S_{ijs}(\omega,\boldsymbol{k};\omega',\boldsymbol{k}')\,S_{abc}^{*}(\omega,\boldsymbol{k};\omega',\boldsymbol{k}')\langle E_s E_c\rangle_{\omega',\boldsymbol{k}'}$$
$$\times\,\langle E_j E_b\rangle_{\omega-\omega',\boldsymbol{k}-\boldsymbol{k}'}+2A_{jb}(\omega-\omega',\boldsymbol{k}-\boldsymbol{k}')\,S_{ijs}(\omega,\boldsymbol{k};\omega',\boldsymbol{k}')$$
$$\times\,S_{bca}(\omega-\omega',\boldsymbol{k}-\boldsymbol{k}',\omega,\boldsymbol{k})\langle E_s E_c\rangle_{\omega',\boldsymbol{k}'}\langle E_a E_i\rangle_{\omega,\boldsymbol{k}}$$
$$-2V_{isac}(\omega,\boldsymbol{k};\omega',\boldsymbol{k}')\langle E_a E_i\rangle_{\omega,\boldsymbol{k}}\langle E_s E_c\rangle_{\omega',\boldsymbol{k}'}\}\,. \qquad (12.1.20)$$

Here, the terms containing $A_{ij}(0,0)$ are omitted as they give no contribution to the problem of nonlinear wave interaction, since ω and $\boldsymbol{k}$ cannot be simultaneously equal to zero for waves, and V_{ijab} and S_{ijs} are, respectively:

$$V_{ijab}(\omega,\boldsymbol{k},\omega',\boldsymbol{k}')\equiv\varepsilon_{ijab}(\omega,\boldsymbol{k};\omega+\omega',\boldsymbol{k}+\boldsymbol{k}';\omega',\boldsymbol{k}')$$
$$+\,\varepsilon_{ijba}(\omega,\boldsymbol{k};\omega+\omega',\boldsymbol{k}+\boldsymbol{k}';\omega',\boldsymbol{k}')\,, \qquad (12.1.21)$$

$$S_{ijs}(\omega,\boldsymbol{k};\omega',\boldsymbol{k}')\equiv\varepsilon_{ijs}(\omega,\boldsymbol{k},\omega',\boldsymbol{k}')+\varepsilon_{isj}(\omega,\boldsymbol{k};\omega-\omega',\boldsymbol{k}-\boldsymbol{k}')\,. \qquad (12.1.22)$$

In the linear approximation, i.e., when the fourth-order terms in field are negligible, the averaging (12.1.20) reduces to (12.1.11).

Thus, the composition of a nonlinear equation for the wave interaction is reduced to calculating the tensors of higher ranks for the medium. We do this for the plasma in the subsequent sections, where (12.1.11, 20) will be simplified. Without using a particular model of the medium, these equations can be simplified for either purely longitudinal or purely transverse waves:

$$\boldsymbol{E}^{\mathrm{lo}}(\omega,\boldsymbol{k})=E^{\mathrm{lo}}(\omega,\boldsymbol{k})\frac{\boldsymbol{k}}{k}\,,\qquad \boldsymbol{E}^{\mathrm{tr}}(\omega,\boldsymbol{k})=E^{\mathrm{tr}}(\omega,\boldsymbol{k})\frac{\boldsymbol{h}}{h}\,, \qquad (12.1.23)$$

where $\boldsymbol{h}\cdot\boldsymbol{k}=0$.

Strictly speaking, (12.1.11, 20) are valid only to describe the interaction of waves, the amplitudes of which greatly exceed the thermal noise level, since the effects of spontaneous radiation and wave absorption by plasma particles (i.e., interparticle collisions in the plasma) are left out. In the following discussion, we deal with these cases.

12.2 Multi-Index Dielectric Tensors of Homogeneous Plasmas

Before starting the analysis of the particular nonlinear effects of the wave interaction in plasma, the multi-index dielectric tensors in (12.1.11, 20) should be calculated. In this section, we calculate these tensors for a collisionless spatially homogeneous plasma obeying Vlasov's equation:

$$\frac{\partial f_a}{\partial t} + \boldsymbol{v}\frac{\partial f_a}{\partial \boldsymbol{r}} + e_a\left\{\boldsymbol{E} + \frac{1}{c}[\boldsymbol{v},\boldsymbol{B}]\right\}\frac{\partial f_a}{\partial \boldsymbol{p}} = 0 \,. \tag{12.2.1}$$

The equilibrium distribution function of particles of the type α in the nondegenerate plasma is assumed to be Maxwellian

$$f_{0\alpha} = \frac{N_a}{(2\pi m_a T_a)^{3/2}} \exp\left(-\frac{m_a v^2}{2T_a}\right) \tag{12.2.2}$$

and in the degenerate one to be Fermian

$$f_{0\alpha} = \begin{cases} 0 & \text{if } \; p > p_{F\alpha} = (3\pi^2)^{1/3}\hbar N_\alpha^{1/3} \,, \\ \dfrac{2}{(2\pi\hbar)^3} & \text{if } \; p < p_{F\alpha} \,. \end{cases} \tag{12.2.2 a}$$

12.2.1 Solution of the Vlasov Equation

Let us expand the deviation of the distribution function from equilibrium in powers of perturbation fields (the subscript α can be omitted)

$$\delta f = f_1 + f_2 + \ldots + f_n + \ldots, \qquad f_n \sim E^n \,. \tag{12.2.3}$$

Substituting this expansion into (12.2.1) and equating the terms of the same order yields

$$\frac{\partial f_n}{\partial t} + \boldsymbol{v}\frac{\partial f_n}{\partial \boldsymbol{r}} = -e\left\{\boldsymbol{E} + \frac{1}{c}[\boldsymbol{v},\boldsymbol{B}]\right\}\frac{\partial f_{n-1}}{\partial \boldsymbol{p}} \tag{12.2.4}$$

for the plasma in the absence of external fields. After passing over to the Fourier components

$$f_n(\boldsymbol{p},t,\boldsymbol{r}) = \int d\omega \, d\boldsymbol{k} \, \mathrm{e}^{-\mathrm{i}\omega t + \mathrm{i}\boldsymbol{k}\cdot\boldsymbol{r}} f_n(\boldsymbol{p},\omega,\boldsymbol{k}) \tag{12.2.5}$$

and using the field equation $\boldsymbol{B} = [\boldsymbol{k},\boldsymbol{E}]c/\omega$ for a separate component $f_n(\boldsymbol{p},\omega,\boldsymbol{k})$, we have

$$f_n(\boldsymbol{p},\omega,\boldsymbol{k}) = -\,\mathrm{i}e\int d\omega' d\boldsymbol{k}' \frac{\alpha_{ij}(\boldsymbol{v},\omega',\boldsymbol{k}')}{\omega-\boldsymbol{k}\cdot\boldsymbol{v}}\,E_j(\omega',\boldsymbol{k}')\,\frac{\partial f_{n-1}(\boldsymbol{p},\omega-\omega',\boldsymbol{k}-\boldsymbol{k}')}{\partial p_i}$$

$$= -\,\mathrm{i}e\int d\omega' d\boldsymbol{k}' d\omega'' d\boldsymbol{k}'' \frac{\alpha_{ij}(\boldsymbol{v},\omega',\boldsymbol{k}')}{\omega-\boldsymbol{k}\cdot\boldsymbol{v}}\,E_j(\omega',\boldsymbol{k}') \tag{12.2.6}$$

$$\times \frac{\partial f_{n-1}(\boldsymbol{p},\omega'',\boldsymbol{k}'')}{\partial p_i}\,\delta(\omega-\omega'-\omega'')\,\delta(\boldsymbol{k}-\boldsymbol{k}'-\boldsymbol{k}'')\,.$$

Here the notation

$$\alpha_{ij}(\boldsymbol{v},\omega,\boldsymbol{k}) = \frac{1}{\omega}\left[k_i v_j + \delta_{ij}(\omega-\boldsymbol{k}\cdot\boldsymbol{v})\right] \tag{12.2.7}$$

is introduced, and the assumption of adiabatic switching-on of the field in the infinite past

$$f_n(\boldsymbol{p},t\to-\infty,\boldsymbol{r})\to 0 \tag{12.2.8}$$

is applied.

Note that we obtain the expression in the linear theory for $n = 1$, provided that $f_0(\boldsymbol{p},\omega,\boldsymbol{k}) = f_0(\boldsymbol{p})\,\delta(\omega)\,\delta(\boldsymbol{k})$ (Chap. 4):

$$f_1(\boldsymbol{p},\omega,\boldsymbol{k}) = -\,\mathrm{i}e\,\frac{\alpha_{ij}(\boldsymbol{v},\omega,\boldsymbol{k})}{\omega-\boldsymbol{k}\cdot\boldsymbol{v}}\,E_j(\omega,\boldsymbol{k})\,\frac{\partial f_0(\boldsymbol{p})}{\partial p_i}\,. \tag{12.2.9}$$

Substituting it into (12.2.6) for $n = 2$ results in

$$f_2(\boldsymbol{p},\omega,\boldsymbol{k}) = (-\mathrm{i}e)^2\int d\omega_1 d\boldsymbol{k}_1 d\omega_2 d\boldsymbol{k}_2$$

$$\times \frac{\alpha_{i_1j_1}(\boldsymbol{v},\omega-\omega_1,\boldsymbol{k}-\boldsymbol{k}_1)}{\omega-\boldsymbol{k}\cdot\boldsymbol{v}}\,\frac{\partial}{\partial p_{i_1}}\,\frac{\alpha_{i_2j_2}(\boldsymbol{v},\omega_1-\omega_2,\boldsymbol{k}_1-\boldsymbol{k}_2)}{\omega-\boldsymbol{k}_1\cdot\boldsymbol{v}}$$

$$\times \frac{\partial f_0(\boldsymbol{p},\omega_2,\boldsymbol{k}_2)}{\partial p_{i_1}}\,E_{j_1}(\omega-\omega_1,\boldsymbol{k}-\boldsymbol{k}_1)\,E_{j_2}(\omega_1-\omega_2;\boldsymbol{k}_1-\boldsymbol{k}_2)\,. \tag{12.2.10}$$

Thus, all terms of the series (12.2.3) can be obtained by means of successive substitutions. It is easily seen that

$$f_n(\boldsymbol{p},\omega,\boldsymbol{k}) = (-\mathrm{i}e)^n\int d\omega_1 d\boldsymbol{k}_1 \dots d\omega_n d\boldsymbol{k}_n g\Gamma_{j_1} g_1 \Gamma_{j_2} \dots g_{n-1}\Gamma_{j_n} f_0(\boldsymbol{p},\omega_n,\boldsymbol{k}_n)$$

$$\times E_{j_1}(\omega-\omega_1,\boldsymbol{k}-\boldsymbol{k}_1)\dots E_{jn}(\omega_{n-1}-\omega_n,\boldsymbol{k}_{n-1}-\boldsymbol{k}_n)\,. \tag{12.2.11}$$

Here we introduce the following notations

$$g = \frac{1}{\omega - \boldsymbol{k}\cdot\boldsymbol{v}}\,, \qquad \Gamma_{j_1} = \alpha_{i_1 j_1}(\boldsymbol{v}, \omega-\omega_1, \boldsymbol{k}-\boldsymbol{k}_1)\,\frac{\partial}{\partial p_{i_1}}\,,$$
$$g_n = \frac{1}{\omega_n - \boldsymbol{k}_n\cdot\boldsymbol{v}}\,, \qquad \Gamma_{jn} = \alpha_{i_n j_n}(\boldsymbol{v}, \omega_{n-1}-\omega_n, \boldsymbol{k}_{n-1}-\boldsymbol{k}_n)\,\frac{\partial}{\partial p_{i_n}}\,. \tag{12.2.12}$$

12.2.2 Three- and Four-Index Tensors of the Isotropic Plasma

When the solution of the kinetic equation is obtained in the form of an expansion in powers of the field, one can proceed to calculating multi-index dielectric tensors of the isotropic plasma. To do this, one must calculate the density of the current induced in the plasma by particles of the type α:

$$\boldsymbol{j}(\omega, \boldsymbol{k}) = e\int \boldsymbol{v} f\, d\boldsymbol{p} = \boldsymbol{j}_1 + \boldsymbol{j}_2 + \ldots + \boldsymbol{j}_n + \ldots \tag{12.2.13}$$

(in the equilibrium state the current is assumed to be absent in the plasma). On the basis of (12.2.12), it is easy to show that

$$\begin{aligned} \boldsymbol{j}_n(\omega, \boldsymbol{k}) &= e(-\mathrm{i}e)^n \int d\boldsymbol{p}\, \boldsymbol{v} \int d\omega_1 d\boldsymbol{k}_1 \ldots d\omega_n d\boldsymbol{k}_n g\Gamma_{j_1} g_1 \Gamma_{j_2} \ldots g_{n-1}\Gamma_{j_n} \\ &\times f_0(\boldsymbol{p}, \omega_n, \boldsymbol{k}_n)\, E_{j_1}(\omega-\omega_1, \boldsymbol{k}-\boldsymbol{k}_1) \ldots E_{j_n}(\omega_{n-1}-\omega_n, \boldsymbol{k}_{n-1}-\boldsymbol{k}_n)\,, \\ \varrho_n(\omega, \boldsymbol{k}) &= e(-\mathrm{i}e)^n \int d\boldsymbol{p} \int d\omega_1 d\boldsymbol{k}_1 \ldots d\omega_n d\boldsymbol{k}_n g\Gamma_{j_1} g_1 \Gamma_{j_2} \ldots g_{n-1}\Gamma_{j_n} \\ &\times f_0(\boldsymbol{p}, \omega_n, \boldsymbol{k}_n)\, E_{j_1}(\omega-\omega_1, \boldsymbol{k}-\boldsymbol{k}_1) \ldots E_{j_n}(\omega_{n-1}-\omega_n, \boldsymbol{k}_{n-1}-\boldsymbol{k}_n)\,. \end{aligned} \tag{12.2.14}$$

From this equation the *multi-index tensor of the isotropic plasma* is obtained for the current density ($n \geq 1$):

$$\begin{aligned} \varepsilon_{ij_1\ldots j_n}(\omega, \boldsymbol{k}; \omega_1\boldsymbol{k}_1; \ldots; \omega_{n-1}, \boldsymbol{k}_{n-1}) &= \delta_{n1}\delta_{ij_1} \\ -4\pi(-\mathrm{i}e)^{n+1}\int d\boldsymbol{p}\frac{v_i}{\omega}\, g\Gamma_{j_1} g_1\Gamma_{j_2} \ldots g_{n-1}\Gamma_{j_n} f_0(\boldsymbol{p})\,. \end{aligned} \tag{12.2.15}$$

Hence, for $n = 1$ the dielectric tensor of the linear theory is

$$\begin{aligned} \varepsilon_{ij_1}(\omega, \boldsymbol{k}) &= \delta_{ij_1} + \frac{4\pi e^2}{\omega}\int d\boldsymbol{p}\, v_i g\Gamma_{j_1} f_0(\boldsymbol{p}) \\ &= \delta_{ij_1} + \frac{4\pi e^2}{\omega}\int d\boldsymbol{p}\, v_i\, \frac{\alpha_{sj_1}(\boldsymbol{v}, \omega, \boldsymbol{k})}{\omega - \boldsymbol{k}\cdot\boldsymbol{v}}\,\frac{\partial f_0}{\partial p_s} \\ &= \left(\delta_{ij_1} - \frac{k_i k_{j_1}}{k^2}\right)\varepsilon^{\mathrm{tr}}(\omega, k) + \frac{k_i k_{j_1}}{k^2}\,\varepsilon^{\mathrm{lo}}(\omega, k) \end{aligned} \tag{12.2.16}$$

and for $n = 2$ the three-index tensor is

$$\varepsilon_{ijs}(\omega, \boldsymbol{k}, \omega', \boldsymbol{k}') = -\mathrm{i}\frac{4\pi e^3}{\omega}\int d\boldsymbol{p}\,\frac{v_i}{\omega - \boldsymbol{k}\cdot\boldsymbol{v}}$$

$$\times\,\alpha_{mj}(\boldsymbol{v}, \omega'', \boldsymbol{k}'')\,\frac{\partial}{\partial p_m}\,\frac{\alpha_{ns}(\boldsymbol{v}, \omega', \boldsymbol{k}')}{\omega' - \boldsymbol{k}'\cdot\boldsymbol{v}}\,\frac{\partial f_0}{\partial p_n}\,, \tag{12.2.17}$$

where $\omega'' = \omega - \omega'$; $\boldsymbol{k} - \boldsymbol{k}' = \boldsymbol{k}''$.

Along with ε_{ijs} we should write out the tensor S_{ijs} appearing in the nonlinear equation (12.1.20). We obtain by using (12.1.21) after integration by parts and symmetrization

$$S_{ijs}(\omega, \boldsymbol{k}, \omega', \boldsymbol{k}') = \frac{4\pi \mathrm{i} e^3}{m^4\omega\omega'\omega''}\int d\boldsymbol{p}\,\beta_{ni}$$

$$\times\,(\beta'_{ns}\beta''_{mj} + \beta'_{ms}\beta_{nj} - \beta'_{ns}\beta_{mj})\frac{\partial f_0}{\partial p_m}\,, \tag{12.2.18}$$

where $\beta_{ij}(v) = \omega\alpha_{ij}(\boldsymbol{v}, \omega, \boldsymbol{k})/(\omega - \boldsymbol{k}\cdot\boldsymbol{v})$ and the primes at β_{ij} correspond to the primed arguments ω and $\boldsymbol{k}$.

Equation (12.1.18) can be conveniently presented in the form

$$S_{ijs}(\omega, \boldsymbol{k}, \omega', \boldsymbol{k}') = -\frac{4\pi \mathrm{i} e^3}{m^5\omega\omega'\omega''}\int d\boldsymbol{p}\, f_0\beta_{ai}$$

$$\times\,\beta''_{bj}\beta'_{cs}\left(\frac{k_a}{\omega_0}\delta_{bc} + \frac{k''_b}{\omega_2}\delta_{ac} + \frac{k'_c}{\omega_1}\delta_{ab}\right), \tag{12.2.19}$$

where $\omega_0 = \omega - \boldsymbol{k}\cdot\boldsymbol{v}$, $\omega_1 = \omega' - \boldsymbol{k}'\cdot\boldsymbol{v}$, $\omega_2 = \omega'' - \boldsymbol{k}''\cdot\boldsymbol{v}$. This expression is derived from (12.2.18) by means of an integration by parts: it is entirely symmetric over the arguments and indices. In the limit of the cold plasma, $\boldsymbol{k}\cdot\boldsymbol{v}/\omega \to 0$, (12.2.18) yields

$$S_{ijs}(\omega, \boldsymbol{k}, \omega', \boldsymbol{k}') = -\frac{\mathrm{i} e\omega_p^2}{m\omega\omega'\omega''}\left(\frac{k_i}{\omega}\delta_{js} + \frac{k'_s}{\omega'}\delta_{ij} + \frac{k''_j}{\omega''}\delta_{is}\right). \tag{12.2.20}$$

The three-index tensor S_{ijs} completely determines the nonlinear decay interaction of waves (also wave coalescence). In order to analyze the induced wave scattering on particles one must also consider a four-index tensor V_{ijab}, which by (12.1.21) is a sum of two four-index dielectric tensors ε_{iajb}. On the basis of (12.2.15) it can be written for $n = 3$ as

$$V_{iajb}(\omega, \boldsymbol{k}, \omega', \boldsymbol{k}') = -\frac{4\pi e^4}{\omega}\int d\boldsymbol{p}\, v_i\left(\frac{\alpha_{ma}(\boldsymbol{v}, \omega, \boldsymbol{k})}{\omega - \boldsymbol{k}\cdot\boldsymbol{v}}\frac{\partial}{\partial p_m}\right.$$

$$\times\,\frac{\alpha_{nj}(\boldsymbol{v}, \omega, \boldsymbol{k})}{\omega + \omega' - (\boldsymbol{k} + \boldsymbol{k}')\cdot\boldsymbol{v}}\,\frac{\partial}{\partial p_n}\,\frac{\alpha_{lb}(\boldsymbol{v}, \omega', \boldsymbol{k}')}{\omega' - \boldsymbol{k}'\cdot\boldsymbol{v}} + \frac{\alpha_{ma}(\boldsymbol{v}, \omega', \boldsymbol{k}')}{\omega' - \boldsymbol{k}'\cdot\boldsymbol{v}}\,\frac{\partial}{\partial p_m}$$

$$\times\left.\frac{\alpha_{nb}(\boldsymbol{v}, \omega', \boldsymbol{k}')}{\omega' + \omega - (\boldsymbol{k} + \boldsymbol{k}')\cdot\boldsymbol{v}}\,\frac{\partial}{\partial p_n}\,\frac{\alpha_{lj}(\boldsymbol{v}, \omega, \boldsymbol{k})}{\omega - \boldsymbol{k}\cdot\boldsymbol{v}}\right)\frac{\partial f_0}{\partial p_l}\,. \tag{12.2.21}$$

Equation (12.2.21) does not possess such a high degree of symmetry as the tensor S_{ijb}, therefore it is not simplified in a general form. However, while investigating nonlinear processes in the plasma, some simplifications will be made. Here only the expression for the tensor V_{iajb} in the case of the cold plasma is given. By taking into account the integration over ω' and $\boldsymbol{k}'$ in (12.1.20), and parity of the binary field correlators over ω' and $\boldsymbol{k}'$, we can immediately write

$$V_{iajb}(\omega, \boldsymbol{k}, \omega', \boldsymbol{k}') + V_{iajb}(\omega, \boldsymbol{k}; -\omega', -\boldsymbol{k}') = 2\,\frac{e^2\omega_p^2}{m^2\omega^2(\omega^2-\omega'^2)}$$

$$\times \left\{ \delta_{ia}\left(\frac{k_j'k_b'}{\omega'^2} - \frac{k_jk_b}{\omega^2}\right) + \delta_{jb}\left(\frac{k_i'k_a'}{\omega'^2} - \frac{k_ik_a}{\omega^2}\right) \right.$$

$$\left. - \,\delta_{ia}\delta_{jb}\left[\left(\frac{\boldsymbol{k}}{\omega} - \frac{\boldsymbol{k}'}{\omega'}\right)^2 + 2\,\frac{k^2+k'^2}{\omega^2-\omega'^2} - 2\,\frac{\boldsymbol{k}\cdot\boldsymbol{k}'}{\omega\omega'}\,\frac{\omega^2+\omega'^2}{\omega^2-\omega'^2}\right]\right\}. \tag{12.2.22}$$

Here the frequencies ω and ω' are assumed to be positive.

Finally, let us give the expression for the tensor $A_{ij}(\omega, \boldsymbol{k})$, see (12.1.18), appearing in (12.1.20) for the isotropic plasma:

$$\begin{aligned} A_{ij}(\omega, \boldsymbol{k}) &= \frac{k_ik_j}{k^2}\,A^{\mathrm{lo}}(\omega, \boldsymbol{k}) + \left(\delta_{ij} - \frac{k_ik_j}{k^2}\right)A^{\mathrm{tr}}(\omega, \boldsymbol{k}) \\ &= \frac{k_ik_j}{k^2}\,\frac{1}{\varepsilon^{\mathrm{lo}}(\omega, \boldsymbol{k})} + \left(\delta_{ij} - \frac{k_ik_j}{k^2}\right)\frac{1}{\varepsilon^{\mathrm{tr}}(\omega, \boldsymbol{k}) - k^2c^2/\omega^2}\,. \end{aligned} \tag{12.2.23}$$

12.2.3 Nonlinear Solution of the Vlasov Equation for the Magneto-Active Plasma

Now let us consider multi-index dielectric tensors for the magneto-active plasma. As before, we proceed from Vlasov's kinetic equation for the collisionless plasma

$$\frac{\partial f}{\partial t} + \boldsymbol{v}\frac{\partial f}{\partial \boldsymbol{r}} + e\left\{\boldsymbol{E} + \frac{1}{c}\,[\boldsymbol{v}, \boldsymbol{B} + \boldsymbol{B}_0]\right\}\frac{\partial f}{\partial \boldsymbol{p}} = 0\,, \tag{12.2.24}$$

$\boldsymbol{B}_0$ being the external magnetic field. By presenting the function f in the form of (12.2.3) we derive the following system of coupled equations for non-equilibrium corrections to the unperturbed function

$$\frac{\partial f_n}{\partial t} + \boldsymbol{v}\frac{\partial f_n}{\partial \boldsymbol{r}} + \frac{e}{c}[\boldsymbol{v}, \boldsymbol{B}_0]\frac{\partial f_n}{\partial \boldsymbol{p}} = -\left\{\boldsymbol{E} + \frac{1}{c}[\boldsymbol{v}, \boldsymbol{B}]\right\}\frac{\partial f_{n-1}}{\partial \boldsymbol{p}}, \tag{12.2.25}$$

where $n = 1, 2, 3 \ldots$. The equation for $n = 1$ is the same as (5.1.1) of the linear theory.

As before, either the nonrelativistic Maxwellian distribution function (12.2.2) or the Fermian (12.2.2 a) is used as f_0. After introducing the notation $\tau = (\phi - \phi')/\Omega$ and Fourier transforming (12.2.5), we can write the solution of (12.2.25) for $n = 1$:

$$f_1(\boldsymbol{p}, \omega, \boldsymbol{k}) = -\frac{e}{m}\int_{-\infty}^{0} d\tau \exp\{-\mathrm{i}\omega\tau + \mathrm{i}\boldsymbol{k}\cdot\delta\boldsymbol{R}[\tau, \boldsymbol{v}(\tau)]\}$$
$$\times\, \alpha_{ij}(\boldsymbol{v}, \omega, \boldsymbol{k})\, E_j(\omega, \boldsymbol{k})\frac{\partial f_0}{\partial p_i(\tau)}. \tag{12.2.26}$$

Here

$$\boldsymbol{v}(\tau) = v_z\boldsymbol{e}_z + [\boldsymbol{v}\cdot\boldsymbol{e}_z]\sin\Omega\tau + [\boldsymbol{e}_z[\boldsymbol{v}, \boldsymbol{e}_z]]\cos\Omega\tau\,, \tag{12.2.27}$$

$$\delta\boldsymbol{R}[\tau, \boldsymbol{v}(\tau)] = \int_0^{\tau} d\tau'\boldsymbol{v}(\tau')\,.$$

The term $\alpha_{ij}(\boldsymbol{v}, \omega, \boldsymbol{k})$ is determined by (12.2.7), $\boldsymbol{v}(\tau_0)$ being the quantity appearing in the argument of the function f_1.

Analogously, (12.2.25) gives for an arbitrary value of n

$$f_n(\boldsymbol{p}, \omega, \boldsymbol{k}) = -e\int d\omega' d\boldsymbol{k}' d\omega'' d\boldsymbol{k}'' \int_{-\infty}^{0} d\tau \exp\{-\mathrm{i}\omega\tau + \mathrm{i}\boldsymbol{k}\cdot\delta\boldsymbol{R}[\tau, \boldsymbol{v}(\tau)]\}$$
$$\times\, \alpha_{ij}[\boldsymbol{v}(\tau, \boldsymbol{v}), \omega', \boldsymbol{k}']\, E_j(\omega', \boldsymbol{k}')\frac{\partial f_{n-1}[\boldsymbol{p}(\tau, \boldsymbol{v}), \omega'', \boldsymbol{k}'']}{\partial p_i(\tau)}. \tag{12.2.28}$$

By expressing successively f_n in terms of f_{n-1}, etc., we finally obtain

$$f(\boldsymbol{p}, t, \boldsymbol{r}) = f_0(\boldsymbol{p})\sum_{n=1}^{\infty}(-e)^n\int d\omega\, d\boldsymbol{k} \exp(-\mathrm{i}\omega t + \mathrm{i}\boldsymbol{k}\cdot\boldsymbol{r})$$
$$\times\int d\omega_1 d\boldsymbol{k}_1 \ldots d\omega_n d\boldsymbol{k}_n \int_{-\infty}^{0} d\tau_0 \int_{-\infty}^{0} d\tau_{n-1} G\Gamma_{j_1}(\tau_0)\, G_1\Gamma_{j_2}(\tau_0 + \tau_1) + \ldots$$
$$\ldots G_{n-1}\Gamma_{jn}(\tau_0 + \tau_1 + \ldots + \tau_{n-1})\, E_{j_1}(\omega - \omega_1, \boldsymbol{k} - \boldsymbol{k}_1)\ldots$$
$$\ldots E_{j_n}(\omega_{n-1} - \omega_n, \boldsymbol{k}_{n-1} - \boldsymbol{k}_n)\, f_0[\boldsymbol{p}(\tau_0 + \ldots + \tau_{n-1}, \boldsymbol{v}), \omega_n, \boldsymbol{k}_n]\,. \tag{12.2.29}$$

Here we introduced the notations ($n = 1, 2, 3 \ldots$):

$$G = \exp[-\mathrm{i}\omega\tau_0 + \mathrm{i}\boldsymbol{k} \cdot \delta\boldsymbol{R}(\tau_0, \boldsymbol{v})] \, ,$$

$$G_n = \exp\left\{-\mathrm{i}\omega_n\tau_n + \mathrm{i}\left[\boldsymbol{k}_n \cdot \delta\boldsymbol{R}\left(\sum_{l=0}^{n} \tau_l, \boldsymbol{v}\right) - \boldsymbol{k} \cdot \delta\boldsymbol{R}\left(\sum_{l=0}^{n-1} \tau_l, \boldsymbol{v}\right)\right]\right\} ,$$

$$\Gamma_{j_1} = \alpha_{i_1 j_1}[\boldsymbol{v}(\tau_0, \boldsymbol{v}), \omega - \omega_1, \boldsymbol{k} - \boldsymbol{k}_1] \frac{\partial}{\partial p_{i_1}(\tau_0, \boldsymbol{v})} ,$$

$$\Gamma_{j_n}\left(\sum_{l=0}^{n} \tau_l\right) = \alpha_{i_n j_n}\left[\boldsymbol{v}\left(\sum_{l=0}^{n-1} \tau_l, \boldsymbol{v}\right), \omega_{n-1} - \omega_n, \boldsymbol{k}_{n-1} - \boldsymbol{k}_n\right] \times \frac{\partial}{\partial p_{i_n}\left(\sum_{l=0}^{n-1} \tau_l, \boldsymbol{v}\right)} , \tag{12.2.30}$$

$$\boldsymbol{v} = [\tau_1, \boldsymbol{v}(\tau_0, \boldsymbol{v})] \equiv \boldsymbol{v}(\tau_0 + \tau_1, \boldsymbol{v}) \, ,$$

$$\delta\boldsymbol{R}[\tau_1, \boldsymbol{v}(\tau_0, \boldsymbol{v})] = \delta\boldsymbol{R}(\tau_0 + \tau_1, \boldsymbol{v}) - \delta\boldsymbol{R}(\tau_0, \boldsymbol{v}) \, .$$

12.2.4 Three- and Four-Index Tensors of the Magneto-Active Plasma

The derived solution of the kinetic equation (12.2.29) permits us to determine the induced current and all *multi-index dielectric tensors of the magneto-active plasma* with any degree of accuracy:

$$\varepsilon_{ij_1 \ldots j_n}(\omega, \boldsymbol{k}, \omega_1, \boldsymbol{k}_1, \ldots, \omega_{n-1}, \boldsymbol{k}_{n-1}) = \delta_{n_1}\delta_{ij_1} - 4\pi\mathrm{i}(-e)^{n+1} \int d\boldsymbol{p} \frac{v_i}{\omega} \times \int_{-\infty}^{0} d\tau_0 \ldots \int_{-\infty}^{0} d\tau_{n-1} G\Gamma_{j_1}(\tau_0)\, G_1\Gamma_{j_2}(\tau_0 + \tau_1) \ldots \times G_{n-1}\Gamma_{j_n}\left(\sum_{l=0}^{n-1} \tau_l\right) f_0\left[\boldsymbol{p}\left(\sum_{l=0}^{n-1} \tau_l, \boldsymbol{v}\right)\right] . \tag{12.2.31}$$

It is easy to show that in the limit $\boldsymbol{B}_0 \to 0$ this expression leads to (12.2.15).

From (12.2.31) for $n = 1$, we obtain the two-index dielectric tensor of linear theory (Chap. 5)

$$\begin{aligned} \varepsilon_{ij}(\omega, \boldsymbol{k}) &= \delta_{ij} - \frac{4\pi\mathrm{i}e^2}{\omega} \int d\boldsymbol{p}\, v_i \int_{-\infty}^{0} d\tau_0 G\Gamma_j(\tau_0)\, f[\boldsymbol{p}(\tau_0, \boldsymbol{v})] \\ &= \delta_{ij} - \frac{4\pi\mathrm{i}e^2}{\omega^2} \int d\boldsymbol{p}\, v_i \int_{-\infty}^{0} d\tau_0 \exp[-\mathrm{i}\omega\tau_0 \\ &\quad + \mathrm{i}\boldsymbol{k} \cdot \delta\boldsymbol{R}(\tau_0, \boldsymbol{v})]\, \alpha_{nj}[\boldsymbol{v}(\tau_0, \boldsymbol{v}), \omega, \boldsymbol{k}] \frac{\partial f_0}{\partial p_n(\tau_0, \boldsymbol{v})} . \end{aligned} \tag{12.2.32}$$

For $n = 2$, the three-index dielectric tensor of the magneto-active plasma is

$$\varepsilon_{ijs}(\omega, \boldsymbol{k}, \omega', \boldsymbol{k}') = \frac{4\pi i e^3}{\omega} \int d\boldsymbol{p}\, v_i \int_{-\infty}^{0} d\tau_0 \int_{-\infty}^{0} d\tau_1 \exp[-i\omega\tau_0 + i\boldsymbol{k} \cdot \delta\boldsymbol{R}(\tau_0, \boldsymbol{v})]$$

$$\times\, \alpha_{nj}[\boldsymbol{v}(\tau_0, \boldsymbol{v}), \omega'', \boldsymbol{k}''] \frac{\partial}{\partial p_n(\tau_0, \boldsymbol{v})} \exp[-i\omega'\tau_1 + i\boldsymbol{k}' \cdot \delta\boldsymbol{R}(\tau_0 + \tau_1, \boldsymbol{v})$$

$$- i\boldsymbol{k}' \cdot \delta R(\tau_0, \boldsymbol{v})]\, \alpha_{ms}[\boldsymbol{v}(\tau_0 + \tau_1, \boldsymbol{v}), \omega', \boldsymbol{k}'] \frac{\partial f_0}{\partial p_m(\tau_0 + \tau_1, \boldsymbol{v})}, \quad (12.2.33)$$

where $\omega = \omega' + \omega''$, $\boldsymbol{k} = \boldsymbol{k}' + \boldsymbol{k}''$.

Finally, for $n = 3$, there is a four-index tensor

$$\varepsilon_{iajb}(\omega, \boldsymbol{k}, \omega' + \omega, \boldsymbol{k} + \boldsymbol{k}', \omega', \boldsymbol{k}') = -\frac{4\pi i e^4}{\omega} \int d\boldsymbol{p}\, v_i \int_{-\infty}^{0} d\tau_0 \int_{-\infty}^{0} d\tau_1 \int_{-\infty}^{0} d\tau_2$$

$$\times \exp[-i\omega\tau_0 + i\boldsymbol{k} \cdot \delta\boldsymbol{R}(\tau_0, \boldsymbol{v})]\, \alpha_{nj}[\boldsymbol{v}(\tau_0, \boldsymbol{v}), \omega', \boldsymbol{k}'] \frac{\partial}{\partial p_n(\tau_0, \boldsymbol{v})}$$

$$\times \exp\{-i(\omega + \omega')\tau_1 + i(\boldsymbol{k} + \boldsymbol{k}')[\delta\boldsymbol{R}(\tau_0 + \tau_1, \boldsymbol{v}) - \delta\boldsymbol{R}(\tau_0, \boldsymbol{v})]\}$$

$$\times\, \alpha_{mj}[\boldsymbol{v}(\tau_0 + \tau_1, \boldsymbol{v}), \omega, \boldsymbol{k}] \frac{\partial}{\partial p_m(\tau_0 + \tau_1, \boldsymbol{v})}$$

$$\times \exp\{-i\omega'\tau_2 + i\boldsymbol{k}' \cdot [\delta\boldsymbol{R}(\tau_0 + \tau_1 + \tau_2, \boldsymbol{v}) - \delta\boldsymbol{R}(\tau_0 + \tau_1, \boldsymbol{v}]\}$$

$$\times\, \alpha_{lb}[\boldsymbol{v}(\tau_0 + \tau_1 + \tau_2, \boldsymbol{v}), \omega', \boldsymbol{k}'] \frac{\partial f_0}{\partial p_l(\tau_0 + \tau_1 + \tau_2, \boldsymbol{v})}. \quad (12.2.34)$$

The tensors S_{ijs} and V_{isab}, appearing in the averaged equation (12.2.20), are calculated by (12.1.21, 22). These calculations, however, are so complicated that we do not consider them here.

12.3 Nonlinear Interaction of Waves in Isotropic Plasmas

From a variety of cases of nonlinear wave interactions in the plasma, only the interaction of longitudinal waves in the isotropic plasma is chosen for consideration here. When we take into account the fact that plasma instabilities of all types are generally excited by longitudinal waves, the choice of these waves may seem quite justified and sufficient. The isotropic plasma is chosen only for simplicity. Besides, we proceed from the averaged equation (12.1.20), and assume the wave phases to be chaotic.

First of all, note that the following relations can be written for the isotropic plasma:

$$\langle E_i E_j \rangle_{\omega, k} = \langle (E^{lo})^2 \rangle_{\omega, k} \frac{k_i k_j}{k^2} + \frac{1}{2} \langle (E^{tr})^2 \rangle_{\omega, k} \left(\delta_{ij} - \frac{k_i k_j}{k^2} \right) ,$$

$$M_{ij}(\omega, \boldsymbol{k}) = \varepsilon^{lo}(\omega, k) \frac{k_i k_j}{k^2} + \left[\varepsilon^{tr}(\omega, k) - \frac{k^2 c^2}{\omega^2} \right] \left(\delta_{ij} - \frac{k_i k_j}{k^2} \right) ,$$

$$\varepsilon_{ij}(\omega, \boldsymbol{k}) = \varepsilon^{lo}(\omega, k) \frac{k_i k_j}{k^2} + \varepsilon^{tr}(\omega, k) \left(\delta_{ij} - \frac{k_i k_j}{k^2} \right) , \tag{12.3.1}$$

$$\begin{aligned} A_{ij}(\omega, \boldsymbol{k}) &= M_{ij}^{-1}(\omega, \boldsymbol{k}) \\ &= \frac{1}{\varepsilon^{lo}(\omega, k)} \frac{k_i k_j}{k^2} + \left[\varepsilon^{tr}(\omega, k) - \frac{k^2 c^2}{\omega^2} \right]^{-1} \left(\delta_{ij} - \frac{k_i k_j}{k^2} \right) . \end{aligned}$$

By substituting them into (12.1.20) and regarding only the longitudinal field $\langle (E^{lo})^2 \rangle_{\omega, k}$ as nonzero, we obtain

$$\begin{aligned} &\frac{\partial \mathrm{Re}\{\varepsilon^{lo}(\omega, k)\}}{\partial \omega} \frac{\partial \langle (E^{lo})^2 \rangle_{\omega, k}}{\partial t} - \frac{\partial \mathrm{Re}\{\varepsilon^{lo}(\omega, k)\}}{\partial \boldsymbol{k}} \frac{\partial \langle (E^{lo})^2 \rangle_{\omega, k}}{\partial \boldsymbol{r}} \\ &= -2\, \mathrm{Im}\{\varepsilon^{lo}(\omega, k)\} \langle (E^{lo})^2 \rangle_{\omega, k} + 2 \langle (E^{lo})^2 \rangle_{\omega, k} \int d\omega' d\boldsymbol{k}' \frac{k_i k_a}{k^2} \frac{k'_c k'_s}{k'^2} \\ &\times \langle (E^{lo})^2 \rangle_{\omega', k'} \mathrm{Im} \left\{ \int d\omega'' d\boldsymbol{k}'' \delta(\omega - \omega' - \omega'') \, \delta(\boldsymbol{k} - \boldsymbol{k}' - \boldsymbol{k}'') \, S_{ijs}(\omega, \boldsymbol{k}, \omega', \boldsymbol{k}') \right. \\ &\times \left. S_{bca}(\omega'', \boldsymbol{k}'', \omega, \boldsymbol{k}) \frac{k''_j k''_b}{k''^2 \varepsilon^{lo}(\omega'', k'')} - V_{icas}(\omega, \boldsymbol{k}, \omega', \boldsymbol{k}') \right\} \\ &+ 2 \langle (E^{lo})^2 \rangle_{\omega, k} \int d\omega' d\boldsymbol{k}' d\omega'' d\boldsymbol{k}'' \delta(\omega - \omega' - \omega'') \, \delta(\boldsymbol{k} - \boldsymbol{k}' - \boldsymbol{k}'') \langle (E^{lo})^2 \rangle_{\omega', k'} \\ &\times \mathrm{Im} \left\{ \left(\delta_{bj} - \frac{k''_j k''_b}{k''^2} \right) \left[\varepsilon^{tr}(\omega'', k'') - \frac{k''^2 c^2}{\omega''^2} \right]^{-1} \frac{k_i k_a}{k^2} \frac{k'_s k'_c}{k'^2} S_{ijs}(\omega, \boldsymbol{k}, \omega', \boldsymbol{k}') \right. \\ &\times \left. S_{bca}(\omega'', \boldsymbol{k}'', \omega, \boldsymbol{k}) \right\} - \mathrm{Im} \left\{ \frac{1}{\varepsilon^{lo}(\omega, k)} \right\} \int d\omega' d\boldsymbol{k}' d\omega'' d\boldsymbol{k}'' \\ &\times \delta(\omega - \omega' - \omega'') \, \delta(\boldsymbol{k} - \boldsymbol{k}' - \boldsymbol{k}'') \langle (E^{lo})^2 \rangle_{\omega', k'} \langle (E^{lo})^2 \rangle_{\omega'', k''} \\ &\times \left| S_{ijs}(\omega, \boldsymbol{k}, \omega', \boldsymbol{k}') \frac{k_i k''_j k'_s}{k k' k''} \right|^2 . \end{aligned} \tag{12.3.2}$$

The first term on the right-hand side of (12.3.2) describes the linear damping of longitudinal waves; the second and the third ones govern the *induced scattering of longitudinal waves* in the plasma via intermediate longitudinal and transverse virtual waves, respectively. The last term describes the coalescence of two longitudinal waves into the third one. Scattering through a longitudinal virtual wave is often called nonlinear *Coulomb scattering* and through a transverse wave, *delayed scattering*. Usually, all these different nonlinear processes are studied separately.

12.3.1 Induced Scattering of Plasma Waves in the Isotropic Plasma

We start the analysis of concrete nonlinear effects with the Coulomb scattering of electron Langmuir oscillations in the nondegenerate plasma. Only the second term on the right-hand side of (12.3.2) must be accounted for, and the limit

$$k^2 r_{\mathrm{De}}^2 \ll 1 , \quad k'^2 r_{\mathrm{De}}^2 \ll 1 , \quad k''^2 r_{\mathrm{De}}^2 \ll 1 \tag{12.3.3}$$

shall be taken. On introducing the spectral energy density of longitudinal waves $W^{\mathrm{lo}}(\omega, \boldsymbol{k})$, see (12.1.14):

$$W^{\mathrm{lo}}(\omega, \boldsymbol{k}) = \frac{1}{4\pi} \frac{\partial[\omega \operatorname{Re}\{\varepsilon^{\mathrm{lo}}(\omega, k)\}]}{\partial \omega} \langle (E^{\mathrm{lo}})^2 \rangle_{\omega, \boldsymbol{k}} , \tag{12.3.4}$$

after tedious but straightforward calculations, we obtain

$$\begin{aligned} \left(\frac{\partial W^{\mathrm{lo}}(\boldsymbol{k})}{\partial t}\right)_{\mathrm{coul}} &= \frac{\partial W^{\mathrm{lo}}(\boldsymbol{k})}{\partial t} - \frac{\partial \operatorname{Re}\{\varepsilon^{\mathrm{lo}}(\omega, k)\}}{\partial \boldsymbol{k}} \left(\frac{\partial \operatorname{Re}\{\varepsilon^{\mathrm{lo}}(\omega, k)\}}{\partial \omega}\right)^{-1} \frac{\partial}{\partial \boldsymbol{r}} W^{\mathrm{lo}}(\boldsymbol{k}) \\ &= \frac{\partial W^{\mathrm{lo}}(\boldsymbol{k})}{\partial t} + 3 v_{\mathrm{Te}} r_{\mathrm{De}} \boldsymbol{k} \frac{\partial W^{\mathrm{lo}}(\boldsymbol{k})}{\partial \boldsymbol{r}} = W^{\mathrm{lo}}(\boldsymbol{k}) \int d\boldsymbol{k}' Q(\boldsymbol{k}, \boldsymbol{k}') \, W^{\mathrm{lo}}(\boldsymbol{k}') . \end{aligned} \tag{12.3.5}$$

Here, we introduce the following notation

$$\begin{aligned} Q(\boldsymbol{k}, \boldsymbol{k}') &= -\frac{\omega_{\mathrm{pe}}}{16\pi^3 N_{\mathrm{e}} T_{\mathrm{e}}} \frac{(\boldsymbol{k} \cdot \boldsymbol{k}')}{k^2 k'^2} \frac{(k'' r_{\mathrm{De}})^2}{|\varepsilon^{\mathrm{lo}}(\omega'', k'')|^2} \Big[\operatorname{Im}\{\delta\varepsilon_{\mathrm{e}}^{\mathrm{lo}}(\omega'', k'')\} \\ &\quad \times \frac{4[\boldsymbol{k}, \boldsymbol{k}']^2 r_{\mathrm{De}}^2}{k''^2} \big(2 \operatorname{Re}\{\delta\varepsilon_{\mathrm{e}}^{\mathrm{lo}}(\omega'', k'')\} + 2 \operatorname{Re}\{\delta\varepsilon_{\mathrm{i}}^{\mathrm{lo}}(\omega'', k'')\} \\ &\quad \times \operatorname{Re}\{\delta\varepsilon_{\mathrm{e}}^{\mathrm{lo}}(\omega'', k'')\} + |\delta\varepsilon_{\mathrm{e}}^{\mathrm{lo}}(\omega'', k'')|^2\big) + (1 + \delta\varepsilon_{\mathrm{i}}^{\mathrm{lo}}(\omega'', k''))^2\Big] \\ &\quad + \operatorname{Im}\{\delta\varepsilon_{\mathrm{i}}^{\mathrm{lo}}(\omega'', k'')\} \, |\delta\varepsilon_{\mathrm{e}}^{\mathrm{lo}}(\omega'', k'')|^2\Big] , \end{aligned}$$

where $\omega'' = \omega - \omega'$, $\boldsymbol{k}'' = \boldsymbol{k} - \boldsymbol{k}'$ and $\delta\varepsilon_{\mathrm{e}}^{\mathrm{lo}}(\omega, \boldsymbol{k})$ and $\delta\varepsilon_{\mathrm{i}}^{\mathrm{lo}}(\omega, \boldsymbol{k})$ are the electron and ion contributions to the longitudinal plasma permittivity, respectively. It is easily shown that

$$\omega'' = \frac{3}{2}(k^2 - k'^2)\frac{v_{\mathrm{Te}}^2}{\omega_{\mathrm{pe}}} . \tag{12.3.6}$$

The kernel $Q(\boldsymbol{k}, \boldsymbol{k}')$ of the integral equation (12.3.5) is antisymmetric with respect to the substitution $\boldsymbol{k} \rightleftarrows \boldsymbol{k}'$. Thus its integration gives

$$W^{\mathrm{lo}} = \int d\boldsymbol{k}\, W^{\mathrm{lo}}(\boldsymbol{k}) = \mathrm{const} , \tag{12.3.7}$$

which means the energy conservation of longitudinal waves in the process of the nonlinear Coulomb scattering. Here the energy transfer from one spectrum range to the other is possible without any variation of the total field energy. In order to find out the transfer direction, we consider two narrow packets of longitudinal waves and their evolution with time under the assumption of their spatial homogeneity. Let $W_1(0)$ and $W_2(0)$ be known at $t = 0$. Then

$$W^{\mathrm{lo}}(\boldsymbol{k}) = W_1(t)\,\delta(\boldsymbol{k} - \boldsymbol{k}_1) + W_2(t)\,\delta(\boldsymbol{k} - \boldsymbol{k}_2) . \tag{12.3.8}$$

Substituting this expression into (12.3.5) yields

$$\frac{\partial W_1}{\partial t} = Q(\boldsymbol{k}_1, \boldsymbol{k}_2)\, W_1 W_2 , \quad \frac{\partial W_2}{\partial t} = -Q(\boldsymbol{k}_1, \boldsymbol{k}_2)\, W_1, W_2 . \tag{12.3.9}$$

Hence $W_1 + W_2 = \int d\boldsymbol{k}\, W^{\mathrm{lo}}(t) = W_0 = \mathrm{const}$, and

$$\frac{W_1(t)}{W_2(t)} = \frac{W_1(0)}{W_2(0)} \exp[Q(\boldsymbol{k}_1, \boldsymbol{k}_2)\, W_0 t] , \quad \text{or} \tag{12.3.10}$$

$$\frac{W_1(t)}{W_2(t)} = W_0\{W_1(0) + W_2(0)\exp[-Q(\boldsymbol{k}_1, \boldsymbol{k}_2)\, W_0 t]^{-1}\} . \tag{12.3.11}$$

Since $Q(\boldsymbol{k}_1, \boldsymbol{k}_2) < 0$ for $k_1 > k_2$, then in the case of Coulomb scattering, energy transfer from short wavelengths to long ones occurs. The characteristic time of this process estimated by (12.3.11) is

$$t_{\mathrm{coul}} = -\frac{1}{W_0 Q(\boldsymbol{k}_1, \boldsymbol{k}_2)} \quad \text{for} \quad k_1 > k_2 . \tag{12.3.12}$$

To illustrate these results, let us consider the limit $T_{\mathrm{i}} \to 0$, $M \to \infty$, when the ion contribution to $Q(\boldsymbol{k}_1, \boldsymbol{k}_2)$ can be neglected. Then the character of the

Coulomb scattering is entirely determined by the electrons, and $Q(\boldsymbol{k}_1, \boldsymbol{k}_2)$ takes the form

$$Q(\boldsymbol{k}_1, \boldsymbol{k}_2) = -\frac{3}{2} \frac{\omega_{\mathrm{pe}} r_{\mathrm{De}}^3}{(2\pi)^{5/2} N_{\mathrm{e}} T_{\mathrm{e}}} \frac{k_1^2 - k_2^2}{|\boldsymbol{k}_1 - \boldsymbol{k}_2|^3} \frac{[\boldsymbol{k}_1, \boldsymbol{k}_2]^2 (\boldsymbol{k}_1 \cdot \boldsymbol{k}_2)^2}{k_1^2 k_2^2} . \tag{12.3.13}$$

Hence, $Q(\boldsymbol{k}_1, \boldsymbol{k}_2) < 0$ for $k_1 > k_2$. Besides, provided that $\boldsymbol{k}_1 \perp \boldsymbol{k}_2$ or $\boldsymbol{k}_1 \parallel \boldsymbol{k}_2$, the value of $Q(\boldsymbol{k}_1, \boldsymbol{k}_2) = 0$, i.e., in the approximation considered, the nonlinear Coulomb scattering for either strictly perpendicular or strictly parallel waves is absent, and therefore $t_{\mathrm{coul}} \to \infty$. The order of magnitude is

$$t_{\mathrm{coul}} \approx 30 \frac{N r_{\mathrm{De}}^3}{\omega_{\mathrm{pe}}} \frac{T_{\mathrm{e}}}{W^{\mathrm{lo}}(k)} \cdot \frac{1}{k^5 \Delta k r_{\mathrm{De}}^6} , \tag{12.3.14}$$

where $\Delta k = |\boldsymbol{k}_1 - \boldsymbol{k}_2|$ and $W_0 = W^{\mathrm{lo}}(k)\,(\Delta k)^3$.

Note that Coulomb scattering on the electrons prevails over scattering on the ions under the condition $T_{\mathrm{e}}/T_{\mathrm{i}} > (m/M)\,(mc^2/T_{\mathrm{e}}) \cdot \ln(M T_{\mathrm{e}}^3/m T_{\mathrm{i}}^3)$. This is valid for the hydrogen plasma when $T_{\mathrm{e}}/T_{\mathrm{i}} > 10^6$ K/T_{e} and for the plasma with heavy ions ($A \approx 100$) when $T_{\mathrm{e}}/T_{\mathrm{i}} > 10^4$ K/T_{e}, i.e., practically always.

Let us substitute (12.3.13) into (12.3.12) and compare the obtained result with the absorption time of longitudinal waves due to electron-ion collisions

$$t_{\mathrm{col}} = 1/\nu_{\mathrm{ei}} = [(4/3)\,\sqrt{2\pi/m}\; e^4\, L\, N_{\mathrm{e}}/T_{\mathrm{e}}^{3/2}]^{-1} .$$

Here the condition for neglecting interparticle collisions is reduced to the requirement $\nu_{\mathrm{ei}} t_{\mathrm{col}} \gg 1$, or

$$\frac{W^{\mathrm{lo}}}{T_{\mathrm{e}}} \gg \frac{L}{k^5 \Delta k r_{\mathrm{De}}^6} \gg 1 . \tag{12.3.15}$$

This inequality justifies neglecting the spontaneous radiation of longitudinal waves by plasma electrons compared to the induced scattering considered above. For $W^{\mathrm{lo}} \gg T_{\mathrm{e}}$ the spontaneous radiation is small in comparison with the collisionless Cherenkov absorption of longitudinal waves by the electrons. The latter leads to the exponential decrease of W^{lo} with the characteristic time, equal to the inverse Landau damping decrement $t_{\mathrm{L}} \approx 1/\delta$, where δ is defined by (4.2.6). Comparing it with (12.3.14) yields the following condition for neglecting the linear collisionless absorption of longitudinal waves, compared to the nonlinear Coulomb scattering:

$$2 k^2 r_{\mathrm{De}}^2 \ln\left(\frac{T_{\mathrm{e}}}{W^{\mathrm{lo}} N_{\mathrm{e}} r_{\mathrm{De}}^3}\right) \ll 1 . \tag{12.3.16}$$

Since the quasilinear relaxation time of plasma waves always exceeds the time of their linear damping, (12.3.16) is also the condition for neglecting quasilinear relaxation of the equilibrium electron distribution in the process of nonlinear Coulomb wave scattering.

12.3.2 Nonlinear Coalescence of Plasma Waves

Let us now consider the influence of a transverse virtual field on the nonlinear interaction of longitudinal waves in the plasma. It is described by the third term of (12.3.2). The concrete effect to be studied involves the coalescence of two longitudinal waves into a transverse one. This process is of great importance for applied problems. The point is that longitudinal waves can be easily excited in the plasma, for instance, due to the plasma-beam interaction, but they are trapped in the plasma and are not radiated outside. Transverse waves, on the contrary, can easily leave the plasma. Therefore the coalescence of longitudinal waves into transverse ones is the radiation channel for longitudinal waves excited in the plasma.

A transverse wave with the frequency $\omega'' \approx 2\omega_{pe}$ and the wave vector $k'' \approx \sqrt{3}\,\omega_{pe}/c$ is created due to the interaction of two longitudinal waves with the frequencies $\omega \approx \omega' \approx \omega_{pe}$. This process is governed by the contribution to (12.3.2) from the pole

$$\varepsilon^{tr}(\omega'', k'') - \frac{k''^2 c^2}{\omega''^2} = 0\,. \tag{12.3.17}$$

Neglecting the thermal motion of the particles, the rate of the coalescence process can be obtained from (12.3.2). One finds

$$\left(\frac{\partial W^{lo}}{\partial t}\right)_{coal} = -\frac{\omega_{pe} r_{De}^4 c^2}{12(2\pi)^2 N_e v_{Te}^2}\,\frac{W^{lo}(\boldsymbol{k})}{T_e}\int d\boldsymbol{k}'\, W^{lo}(\boldsymbol{k})$$
$$\delta\left(3 - \frac{(\boldsymbol{k}+\boldsymbol{k}')^2 c^2}{\omega_{pe}^2}\right)\frac{[\boldsymbol{k},\boldsymbol{k}']^2}{k^2 k'^2}\,(k^2 - k'^2)^2\,, \tag{12.3.18}$$

where we assumed $\omega'' = \omega + \omega' \approx 2\omega_{pe}$, $\boldsymbol{k}'' = \boldsymbol{k}' + \boldsymbol{k}$, and

$$\left[\varepsilon^{tr}(\omega'', k'') - \frac{k''^2 c^2}{\omega''^2}\right]^{-1} \approx -i\pi\delta\left[\varepsilon^{tr}(\omega'', k'') - \frac{k''^2 c^2}{\omega''^2}\right] = -4\pi i\delta\left(3 - \frac{k''^2 c^2}{\omega_{pe}^2}\right).$$

The characteristic time of the coalescence for the longitudinal waves with wavelengths $\lambda_0 \approx c/\omega_{pe}$ is easily estimated by means of (12.3.18):

$$t_{coal} \simeq 10^3\,\frac{N_e r_{De}^3}{\omega_{pe}}\,\frac{T_e}{W^{lo}}\left(\frac{c}{v_{Te}}\right)^3. \tag{12.3.19}$$

For $\lambda_0 \sim 1/k \sim c/\omega_{pe}$ and $\Delta k \sim \omega_{pe}/c$, the comparison of the last equation with (12.3.14) reveals that $t_{coal}/t_{coul} \approx 30\ v_{Te}/c < 1$. In other words, the transfer over a spectrum goes more slowly than the wave coalescence. However, for such waves the condition $t_{conf}\,\nu_{ei} < 1$ is difficult to be realized because of the inequality $W^{lo}/T_e \gg (c/v_{Te})^3 \gg 1$.

12.3.3 Electromagnetic Scattering of Plasma Waves

Besides, the process of the longitudinal-wave coalescence, the scattering of a longitudinal wave off an intermediate transverse virtual wave, is described with the aid of the third term of (12.3.2). This process is governed by the contribution outside the pole (12.3.17), i.e., from ω'' and $\boldsymbol{k}''$ for which the quantity (12.3.17) is nonzero. Then, (12.3.2) yields

$$\left(\frac{\partial W^{lo}}{\partial t}\right)_{del} = -\frac{\omega_{pe}}{6\,(2\pi)^{5/2}N_e}\,\frac{W^{lo}(\boldsymbol{k})}{T_e r_{De}}\int d\boldsymbol{k}'\,W^{lo}(\boldsymbol{k})\,\frac{[\boldsymbol{k},\boldsymbol{k}']^2}{k^2k'^2}$$
$$\times\left(1+4\,\frac{(\boldsymbol{k}\cdot\boldsymbol{k}')\,c^2}{\omega_{pe}^2}\right)\frac{|\boldsymbol{k}-\boldsymbol{k}'|}{k^2-k'^2}$$
$$\times\left(\frac{4}{9}\,\frac{c^4\,(\boldsymbol{k}-\boldsymbol{k}')^2}{v_{Te}^4\,(k^2-k'^2)}+2\frac{c^2}{v_{Te}^2}+\frac{\pi}{2}\,\frac{1}{r_{De}^2\,(\boldsymbol{k}-\boldsymbol{k}')^2}\right)^{-1}. \tag{12.3.20}$$

For short-wavelength oscillations ($k, k' > \omega_{pe}/c$), the wave transfer over a spectrum due to the scattering off an intermediate transverse field is always smaller than that due to the Coulomb scattering. However, for long-wavelength oscillations ($k, k' < \omega_{pe}/c$), this is not the case. In this limit (more exactly, for waves with $k, k' \ll \omega_{pe}v_{Te}/c^2$), (12.3.20) reduces to

$$\left(\frac{\partial W^{lo}}{\partial t}\right)_{del} = -\frac{2\,r_{De}}{3\,(2\pi)^{7/2}}\,\frac{\omega_{pe}W^{lo}(\boldsymbol{k})}{N_eT_e}\int d\boldsymbol{k}'\,W^{lo}(\boldsymbol{k},\boldsymbol{k}')$$
$$\times\,\frac{[\boldsymbol{k},\boldsymbol{k}']^2}{k^2k'^2}\,\frac{|\boldsymbol{k}-\boldsymbol{k}'|^3}{k^2-k'^2}\,. \tag{12.3.21}$$

Hence, due to the delayed as well as the Coulomb interaction the transfer of waves over the spectrum occurs from short to long wavelengths. The transfer time

$$t_{del} \sim 10^3\,\frac{N_e r_{De}^3}{\omega_{pe}}\,\frac{T_e}{W^{lo}}\,\frac{1}{k^2\Delta k r_{De}^3} \tag{12.3.22}$$

in some cases turns out to be shorter than the time (12.3.14) associated with the Coulomb scattering.

Finally, note that the last term in (12.3.2) turns out to be zero for a problem involving nonlinear interaction of the Langmuir oscillations. The reason is that it describes the coalescence of two longitudinal waves into one longitudinal wave. For Langmuir waves such a process is impossible. The last summand in (12.3.2) is manifested in such processes as coalescence of two ion-acoustic waves into a third one, or of a longitudinal electron plasma wave with an ion-acoustic one. However, we do not deal with a great variety of nonlinear interactions of longitudinal and transverse waves in an isotropic plasma, and in particular in a magnetoplasma, because the analysis of these effects will be quite lengthy and the results would not be much different than those mentioned above. There exists a sufficient amount of literature dealing with these problems.

In conclusion, we present one more example of a nonlinear equation which describes the wave interaction in a plasma. Equations (12.3.5, 18, 20) demonstrate that the nonlinear equation governing the interaction of longitudinal waves in a plasma can be written as

$$\frac{dW^{\text{lo}}(\boldsymbol{k})}{dt} = W^{\text{lo}}(\boldsymbol{k}) \int d\boldsymbol{k}\, Q(\boldsymbol{k}, \boldsymbol{k}')\, W^{\text{lo}}(\boldsymbol{k}') \,. \tag{12.3.23}$$

The number of oscillations *(the number of plasmons)* is usually introduced instead of the energy spectral density

$$N_{\boldsymbol{k}} = \frac{W^{\text{lo}}(\boldsymbol{k})}{\hbar\omega(\boldsymbol{k})} \,, \tag{12.3.24}$$

where $\omega(\boldsymbol{k})$ is the frequency of longitudinal waves. It is easy to verify that $N_{\boldsymbol{k}}$ satisfies

$$\frac{dN_{\boldsymbol{k}}}{dt} = \hbar N_{\boldsymbol{k}} \int d\boldsymbol{k}'\, Q(\boldsymbol{k}, \boldsymbol{k}')\, N_{\boldsymbol{k}'}\omega(\boldsymbol{k}') \,. \tag{12.3.25}$$

Analogously, the appropriate numbers of plasmons are introduced to describe the interaction of waves of different kinds (longitudinal and transverse) in a plasma.

12.4 Nonlinear Three-Wave Interaction in a Plasma in the Field of Strong Electromagnetic Waves

In the previous section, a general theory of the wave interaction in a plasma has been applied to the analysis of nonlinear phenomena of lower orders accurately up to quadratic terms in the electromagnetic field energy. In this approximation, nonlinear processes involve three-wave interactions in the form of a decay, a coalescence and a scattering on plasma oscillations. The equations describing these processes are rather complex, therefore we have analyzed them only qualitatively. However, the analysis of the three-wave interaction in a plasma can be greatly simplified provided that the amplitude of one of these waves is considered to be given and to be sufficiently large. Due to the latter assumptions only nonlinear terms in the field with respect to the wave with the assigned amplitude should be taken into account. Such an analysis can be carried out analogously to the parametric interaction of large amplitude high-frequency fields with plasma (Chap. 7).

12.4.1 Equilibrium Distribution Function in the Field of Strong Electromagnetic Waves

Let the isotropic plasma be imbedded in the field of a monochromatic electromagnetic pump wave

$$\begin{aligned} \boldsymbol{E}_0(\boldsymbol{r},t) &= \boldsymbol{E}_0 \cos(\omega_0 t - \boldsymbol{k}_0 \cdot \boldsymbol{r}) , \\ \boldsymbol{B}_0(\boldsymbol{r},t) &= \frac{c}{\omega_0} [\boldsymbol{k}_0, \boldsymbol{E}_0] \cos(\omega_0 t - \boldsymbol{k}_0 \cdot \boldsymbol{r}) . \end{aligned} \tag{12.4.1}$$

Neglecting the interparticle collisions for the distribution function of particles in the ground state, we can write

$$\frac{\partial f}{\partial t} + \boldsymbol{v}\frac{\partial f}{\partial \boldsymbol{r}} + \mathrm{e}\left\{\boldsymbol{E}_0(\boldsymbol{r},t) + \frac{1}{c}[\boldsymbol{v}, \boldsymbol{B}_0(\boldsymbol{r},t)]\right\}\frac{\partial f}{\partial \boldsymbol{p}} = 0 . \tag{12.4.2}$$

As before, we solve this equation by means of successive approximations, assuming that the wave causes a weak deviation of the distribution function of particles $f(\boldsymbol{p},\boldsymbol{r},t)$ from the equilibrium distribution $f_0(\boldsymbol{p})$ which is of Maxwellian (12.2.2) or Fermian (12.2.3) form. Then, we obtain accurately up to the second-order terms in the field (12.4.1)

$$\begin{aligned} f(\boldsymbol{p},\boldsymbol{r},t) &= f_0(p) + \Delta f(\boldsymbol{p}) \\ &\quad - \frac{e\alpha_{ij}(\omega_0,\boldsymbol{k}_0,\boldsymbol{v})}{\omega_0 - \boldsymbol{k}_0\cdot\boldsymbol{v}} E_{0j}\frac{\partial f_0}{\partial p_i}\sin(\omega_0 t - \boldsymbol{k}_0\cdot\boldsymbol{r}) , \end{aligned} \tag{12.4.3}$$

where

$$\begin{aligned} \Delta f(\boldsymbol{p}) = \frac{e^2}{4\omega_0} E_{0j}E_{0\mu}\Bigg[&\frac{\partial}{\partial p_\mu}\left(\frac{\alpha_{ij}(\omega_0,\boldsymbol{k}_0,\boldsymbol{v})}{\omega_0 - \boldsymbol{k}_0\cdot\boldsymbol{v}}\frac{\partial f_0}{\partial p_i}\right) \\ &+ \frac{\partial}{\partial p_i}\left(\frac{\alpha_{ij}(\omega_0,\boldsymbol{k}_0,\boldsymbol{v})}{(\omega_0 - \boldsymbol{k}_0\cdot\boldsymbol{v})^2} k_{0\nu}v_\mu\frac{\partial f_0}{\partial p_\nu}\right)\Bigg] . \end{aligned} \tag{12.4.4}$$

Here, we omit the time-varying double-frequency corrections quadratic in the field (12.4.1), since in the dispersion equation they give rise to corrections of higher orders compared to the terms which are kept in (12.4.3).

In order to determine the equilibrium plasma state, the solution (12.4.3) must be combined with the Maxwell equations. Since the pump wave is considered to be weak, the linear approximation is adequate. As a result, the relationship between the frequency ω_0 and the wave vector $\boldsymbol{k}_0$ is obtained

$$\varepsilon^{\mathrm{lo}}(\omega_0, \boldsymbol{k}_0) = 0 \tag{12.4.5}$$

for the longitudinal field $\boldsymbol{E}_0$, i.e., $\boldsymbol{E}_0 \| \boldsymbol{k}_0$, and

$$k_0^2 c^2 - \omega_0^2 \varepsilon^{\mathrm{tr}}(\omega_0, \boldsymbol{k}_0) = 0 \tag{12.4.6}$$

for the transverse field $\boldsymbol{E}_0$, i.e., $\boldsymbol{E}_0 \perp \boldsymbol{k}_0$. Here $\varepsilon^{\mathrm{lo}}(\omega_0, \boldsymbol{k}_0)$ and $\varepsilon^{\mathrm{tr}}(\omega_0, \boldsymbol{k}_0)$ are the longitudinal and transverse permittivities of the isotropic plasma, see (4.1.14).

12.4.2 Dispersion Equation for Small Oscillations

We may now study the stability of the equilibrium by allowing a small deviation $\delta f(\boldsymbol{p}, \boldsymbol{r}, t)$ which is caused by the perturbed fields $\delta \boldsymbol{E}(\boldsymbol{r}, t)$ and $\delta \boldsymbol{B}(\boldsymbol{r}, t)$. The function $\delta f(\boldsymbol{p}, \boldsymbol{r}, t)$ satisfies

$$\frac{\partial \delta f}{\partial t} + \boldsymbol{v} \frac{\partial \delta f}{\partial \boldsymbol{r}} + e\alpha_{ij}(\omega_0, \boldsymbol{k}_0, \boldsymbol{v}) E_{0j} \frac{\partial f}{\partial p_i} \cos(\omega_0 t - \boldsymbol{k}_0 \cdot \boldsymbol{r})$$
$$+ e\left\{\delta \boldsymbol{E} + \frac{1}{c}[\boldsymbol{v}, \delta \boldsymbol{B}]\right\} \frac{\partial f}{\partial \boldsymbol{p}} = 0 , \tag{12.4.7}$$

where $f(\boldsymbol{p}, \boldsymbol{r}, t)$ is given by (12.4.3).

Representing all the perturbed quantities in the form of the Fourier expansion

$$A(\boldsymbol{r}, t) = \sum_h A(\omega + n\omega_0, \boldsymbol{k} + n\boldsymbol{k}_0) \exp[-\mathrm{i}(\omega + n\omega_0)t + \mathrm{i}(\boldsymbol{k} + n\boldsymbol{k}_0) \cdot \boldsymbol{r}] ,$$

we obtain from (12.4.3)

$$(\omega - \boldsymbol{k} \cdot \boldsymbol{v})\, \delta f(\omega, \boldsymbol{k}, \boldsymbol{p}) + \frac{\mathrm{i}e}{2} \alpha_{ij}(\omega_0, \boldsymbol{k}_0, \boldsymbol{p}) E_{0j} \frac{\partial}{\partial p_i} [\delta f(\omega + \omega_0, \boldsymbol{k} + \boldsymbol{k}_0, \boldsymbol{p})$$
$$+ \delta f(\omega - \omega_0, \boldsymbol{k} - \boldsymbol{k}_0, \boldsymbol{p})] + \mathrm{i}e\alpha_{ij}(\omega, \boldsymbol{k}, \boldsymbol{v})\, \delta E_j(\omega, \boldsymbol{k}) \frac{\partial}{\partial p_j} [f_0(p) + \Delta f(\boldsymbol{p})]$$
$$- \frac{e^2}{2} [\alpha_{\mu l}(\omega + \omega_0, \boldsymbol{k} + \boldsymbol{k}_0, \boldsymbol{v})\, \delta E_l(\omega + \omega_0, \boldsymbol{k} + \boldsymbol{k}_0) - \alpha_{\mu l}(\omega - \omega_0, \boldsymbol{k} - \boldsymbol{k}_0, \boldsymbol{v})$$
$$\times \delta E_l(\omega - \omega_0, \boldsymbol{k} - \boldsymbol{k}_0)] E_{0j} \frac{\partial}{\partial p_\mu} \left(\frac{\alpha_{ij}(\omega_0, \boldsymbol{k}_0, \boldsymbol{v})}{\omega - \boldsymbol{k}_0 \cdot \boldsymbol{v}} \frac{\partial f_0}{\partial p_i} \right) = 0 . \tag{12.4.8}$$

Equation (12.4.8) connects the harmonic $(\omega, \boldsymbol{k})$ of the perturbed distribution function with the harmonics $(\omega \pm \omega_0,\ \boldsymbol{k} \pm \boldsymbol{k}_0)$ of perturbations of the distribution function and the electromagnetic fields. This represents a system of coupled equations. Assuming the field $\boldsymbol{E}_0$ to be weak, we can easily solve this system and find $\delta f(\omega, \boldsymbol{k})$ accurately up to quadratic terms in the field and linear in the perturbation $\delta\boldsymbol{E}$. Since the function $f(\omega, \boldsymbol{k})$ is lengthy, we do not present it here. Making use of this function, an expression for the nonequilibrium current density can be written

$$\begin{aligned} j_i(\omega, \boldsymbol{k}) &= e \int d\boldsymbol{p}\, v_i \delta f(\omega, \boldsymbol{k}) = -\frac{i\omega}{4\pi} \{[\varepsilon_{ij}(\omega, \boldsymbol{k}) + \Delta\varepsilon_{ij}(\omega, \boldsymbol{k}) - \delta_{ij}] \\ &\times \delta E_j(\omega, \boldsymbol{k}) + \tfrac{1}{2} E_{0j} [S_{ilj}(\omega, \boldsymbol{k}; \omega_0, \boldsymbol{k}_0)\, \delta E_l(\omega - \omega_0, \boldsymbol{k} - \boldsymbol{k}_0) \\ &+ S_{ilj}(\omega, \boldsymbol{k}, -\omega_0, -\boldsymbol{k}_0)\, \delta E_l(\omega + \omega_0, \boldsymbol{k} + \boldsymbol{k}_0)] + \tfrac{1}{4} E_{0s} E_{0j} \delta E_l(\omega, \boldsymbol{k}) \\ &\times [V_{islj}(\omega, \boldsymbol{k}; \omega_0, \boldsymbol{k}_0) + V_{islj}(\omega, \boldsymbol{k}; -\omega_0, -\boldsymbol{k}_0)]\} \,. \end{aligned} \tag{12.4.9}$$

The quantities $S_{ijl}(\omega, \boldsymbol{k}; \omega_0, \boldsymbol{k}_0)$ and $V_{ijls}(\omega, \boldsymbol{k}; \omega_0, \boldsymbol{k}_0)$ are defined by (12.1.21 and 22), respectively; $\varepsilon_{ij}(\omega, \boldsymbol{k})$ is the dielectric tensor of the isotropic plasma (4.1.9), and

$$\Delta\varepsilon_{ij}(\omega, \boldsymbol{k}) = \sum \frac{4\pi e^2}{\omega} \int d\boldsymbol{p}\, \frac{v_i a_{lj}(\omega, \boldsymbol{k}, \boldsymbol{v})}{\omega - \boldsymbol{k}\cdot\boldsymbol{v}} \frac{\partial \Delta f(\boldsymbol{p})}{\partial p_l} \,. \tag{12.4.10}$$

Inserting (12.4.9) into the Maxwell equations, we find a system of homogeneous coupled equations for the field perturbation $\delta\boldsymbol{E}$:

$$\begin{aligned} T_{ij}(\omega, \boldsymbol{k})\, \delta E_j(\omega, \boldsymbol{k}) &= \tfrac{1}{2} E_{0j} [S_{ijl}(\omega, \boldsymbol{k}; -\omega_0, -\boldsymbol{k}_0)\, \delta E_l(\omega + \omega_0, \boldsymbol{k} + \boldsymbol{k}_0) \\ &+ S_{ijl}(\omega, \boldsymbol{k}; \omega_0, \boldsymbol{k}_0)\, \delta E_l(\omega - \omega_0, \boldsymbol{k} - \boldsymbol{k}_0)] \,, \end{aligned} \tag{12.4.11}$$

where

$$\begin{aligned} T_{ij}(\omega, \boldsymbol{k}) &= \varepsilon_{ij}(\omega, \boldsymbol{k}) + \Delta\varepsilon_{ij}(\omega, \boldsymbol{k}) - \frac{c^2k^2}{\omega^2}\left(\delta_{ij} - \frac{k_i k_j}{k^2}\right) \\ &+ \tfrac{1}{4} E_{0s} E_{0l} [V_{islj}(\omega, \boldsymbol{k}; \omega_0, \boldsymbol{k}_0) + V_{islj}(\omega, \boldsymbol{k}, -\omega_0, -\boldsymbol{k}_0)] \,. \end{aligned} \tag{12.4.12}$$

The solvability condition for the homogeneous system (12.4.11), accurate up to the second-order terms in the field $\boldsymbol{E}_0$, is the unknown dispersion equation for small oscillations of the isotropic plasma in the field of the electromagnetic pump wave (12.4.1)

$$\begin{aligned} &|T_{ij}(\omega, \boldsymbol{k}) - \tfrac{1}{4} E_{0\nu} E_{0s} [S_{i\mu\nu}(\omega, \boldsymbol{k}; \omega_0, \boldsymbol{k}_0)\, T^{-1}_{\mu l}(\omega - \omega_0, \boldsymbol{k} - \boldsymbol{k}_0) \\ &\times S_{ljs}(\omega - \omega_0, \boldsymbol{k} - \boldsymbol{k}_0; -\omega_0, -\boldsymbol{k}_0) + S_{i\mu\nu}(\omega, \boldsymbol{k}; -\omega_0, -\boldsymbol{k}_0) \\ &\times T^{-1}_{\mu l}(\omega + \omega_0, \boldsymbol{k} + \boldsymbol{k}_0)\, S_{ljs}(\omega + \omega_0, \boldsymbol{k} + \boldsymbol{k}_0; \omega_0, \boldsymbol{k}_0)]| = 0 \,. \end{aligned} \tag{12.4.13}$$

For low-frequency perturbations ($\omega \ll \omega_0$), the last equation is significantly simplified. If the phase velocity of the pump wave is assumed to be larger than the velocities of plasma electrons, i.e., $\omega_0 \gg k_0 v_{\mathrm{Te}}, k v_{\mathrm{Te}}$, then (12.4.13) splits up into two equations. The first equation describes transverse low-frequency field perturbations independent of the pump wave; the second one describes longitudinal (potential) low-frequency perturbations interacting with the pump wave. The latter equation is of the form:

$$\varepsilon^{\mathrm{lo}}(\omega, \boldsymbol{k}) + \frac{1}{4}\,\delta\varepsilon_{\mathrm{e}}^{\mathrm{lo}}(\omega, \boldsymbol{k})\left[1 + \delta\varepsilon_{\mathrm{i}}^{\mathrm{lo}}(\omega, \boldsymbol{k})\right]\frac{k^2}{\omega_0^2}\left\{\frac{((\boldsymbol{k}-\boldsymbol{k}_0)\cdot\boldsymbol{v}_{\mathrm{E}})^2}{(\boldsymbol{k}-\boldsymbol{k}_0)^2}\right.$$

$$\times\frac{1}{\varepsilon^{\mathrm{lo}}(\omega-\omega_0, \boldsymbol{k}-\boldsymbol{k}_0)} + \frac{((\boldsymbol{k}+\boldsymbol{k}_0)\cdot\boldsymbol{v}_{\mathrm{E}})^2}{(\boldsymbol{k}+\boldsymbol{k}_0)^2}\cdot\frac{1}{\varepsilon^{\mathrm{lo}}(\omega+\omega_0, \boldsymbol{k}+\boldsymbol{k}_0)}$$

$$+\frac{[(\boldsymbol{k}-\boldsymbol{k}_0), \boldsymbol{v}_{\mathrm{E}}]^2}{(\boldsymbol{k}-\boldsymbol{k}_0)^2}\left[\varepsilon^{\mathrm{tr}}(\omega-\omega_0, \boldsymbol{k}-\boldsymbol{k}_0) - \frac{c^2(\boldsymbol{k}-\boldsymbol{k}_0)^2}{(\omega-\omega_0)^2}\right]^{-1}$$

$$\left.+\frac{((\boldsymbol{k}+\boldsymbol{k}_0)\cdot\boldsymbol{v}_{\mathrm{E}})^2}{(\boldsymbol{k}+\boldsymbol{k}_0)^2}\left[\varepsilon^{\mathrm{tr}}(\omega+\omega_0, \boldsymbol{k}+\boldsymbol{k}_0) - \frac{c^2(\boldsymbol{k}+\boldsymbol{k}_0)^2}{(\omega+\omega_0)^2}\right]^{-1}\right\} = 0\,. \tag{12.4.14}$$

Here $\boldsymbol{v}_{\mathrm{E}} = e\boldsymbol{E}_0/m\omega_0$ is the electron quiver velocity.

We can easily derive (12.4.14) if we assume the low-frequency field perturbations to be potential. Moreover, this equation is also valid for a magneto-active plasma provided that a high-frequency pump wave satisfies $\omega_0 \gg (\Omega_{\mathrm{e}}, \omega, k_0 v_{\mathrm{Te}}, k v_{\mathrm{Te}})$. Here, however, $\varepsilon^{\mathrm{lo}}(\omega, \boldsymbol{k})$ and $\delta\varepsilon_{\mathrm{e,i}}^{\mathrm{lo}}(\omega, \boldsymbol{k})$ represent the corresponding longitudinal permittivities $k_i k_j \varepsilon_{ij}(\omega, \boldsymbol{k})/k^2$, $k_i k_j \delta\varepsilon_{\mathrm{i,e}}(\omega, \boldsymbol{k})/k^2$.

Note that the first (last) two terms describe the interaction of the longitudinal (transverse) pump wave with the plasma. Therefore, here only the first or the second pair of terms is significant; this depends on the pump wave polarization.

Finally, for a dipole external field (i.e., for $k_0 \to 0$) which consists of potential perturbations (i.e., for $\omega^2 \ll k^2c^2$), (12.4.4) goes over to (7.5.22). The latter describes the parametric interaction of the high-frequency electric field with an isotropic plasma.

In order to demonstrate the application of the dispersion equation (12.4.14), let us consider a problem of high-frequency transverse wave scattering ($\omega_0 \gg \omega_{\mathrm{pe}}, \boldsymbol{k}_0 \cdot \boldsymbol{E}_0 = 0$) in an isotropic plasma. As shown in the previous section, for the scattering process the matching conditions

$$\omega_0 = \omega_{\mathrm{s}} \pm \omega\,, \qquad \boldsymbol{k}_0 = \boldsymbol{k}_{\mathrm{s}} \pm \boldsymbol{k} \tag{12.4.15}$$

have to be satisfied. Here, ω_{s} and $\boldsymbol{k}_{\mathrm{s}}$ are the frequency and the wave vector of the scattered wave. By the applicability condition (12.4.14) we have

$\omega_s = \omega_0 \pm \omega \gg \omega$, and hence the scattered wave is also transverse. The scattered wave with the frequency $\omega_0 - \omega\,(\omega_0 + \omega)$ is called a "red" ("blue") satellite, or a Stokes (anti-Stokes) scattering line.

12.4.3 Induced Raman Scattering of Electromagnetic Waves in the Isotropic Plasma

Let us first consider the case when $\omega \gtrsim \omega_{pe} \gg kv_{Te}$, and the contribution from the ion terms in (12.4.14) can be neglected. Here, the electromagnetic wave scatters off the high-frequency plasma oscillations, namely, the stimulated Compton or the stimulated Raman scattering. For such a process with the "red" satellite, (12.4.14) can be written as[1]

$$1 + \delta\varepsilon_e^{lo}(\omega, k) + \frac{\delta\varepsilon_e^{lo}(\omega, k)}{4} \times \frac{k^2(\omega-\omega_0)^2\,[\boldsymbol{k}-\boldsymbol{k}_0, \boldsymbol{v}_E]^2}{\omega_0^2(\boldsymbol{k}-\boldsymbol{k}_0)^2\,[(\omega-\omega_0)^2\,\varepsilon^{tr}(\omega-\omega_0) - c^2(\boldsymbol{k}-\boldsymbol{k}_0)^2]} = 0\,, \qquad (12.4.16)$$

where

$$\delta\varepsilon_e^{lo}(\omega, k) = -\frac{\omega_{pe}^2}{\omega^2} + \mathrm{i}\sqrt{\frac{\pi}{2}}\,\frac{\omega_{pe}^2\omega}{k^3 v_{Te}^3}\exp\left(-\frac{\omega^2}{2k^2 v_{Te}^2}\right),$$
$$\varepsilon^{tr}(\omega-\omega_0) = 1 - \frac{\omega_{pe}^2}{(\omega-\omega_0)^2} \approx 1\,. \qquad (12.4.17)$$

Equation (12.4.16) shows the importance of high-frequency longitudinal plasma oscillations obeying the equation $1 + \delta\varepsilon_e^{lo}(\omega, k) = 0$ and possessing their natural frequency ω_{pe}. The scattering significantly differs according to the location of the frequency ω whether it is near or far from the resonance frequency ω_{pe}. Near the resonance frequency, i.e., for $\omega = \omega_{pe} + \mathrm{i}\delta$, δ being a small correction, (12.4.16) gives

$$\delta = \left(\frac{k^2 v_E^2}{16}\,\frac{\omega_b}{\omega_0}\right)^{1/2}. \qquad (12.4.18)$$

Here, the exponentially small Cherenkov dissipation is neglected. This equation implies that both the longitudinal plasma oscillations and the scattered light wave fields grow exponentially with time. The increment (12.4.18)

1 Due to the complete symmetry the facts mentioned for the Stokes scattering line with the "red" satellite are also valid for the anti-Stokes scattering line.

attains a maximum for the backscattering ($k \approx 2k_0 \approx 2\omega_0/c$) and therefore $\delta_{max} \approx \sqrt{\omega_0\omega_{pe}}v_E/(2c)$. This resonance scattering is called the induced *Raman scattering*.

The resonance Raman scattering occurs in relatively weak fields under the conditions $\omega_{pe} \gg \delta \gg kv_{Te}$ or, which is the same[2],

$$\sqrt{\frac{\omega_{pe}}{\omega_0}} \gg \frac{v_E}{c} \gg \frac{v_{Te}}{c}\sqrt{\frac{\omega_0}{\omega_{pe}}} \tag{12.4.19}$$

are satisfied. In strong fields, the first inequality in (12.4.19) is violated. For $v_E \ll c$, the resonance Raman scattering is replaced by a modified decay. In this case, $\omega \gg \omega_{pe}$ and

$$\omega = \frac{-1 + i\sqrt{3}}{2}\left(\frac{k^2\omega_{pe}^2 v_E^2}{8\omega_0}\right)^{1/3}. \tag{12.4.20}$$

Note that the instability acquires an aperiodic character. Here, the maximum increment also occurs for the backscattering ($k \approx 2k_0 \approx 2\omega_0/c$) and $\delta_{max} = \operatorname{Im}\{\omega_{max}\} = (\sqrt{3}/2)\,(\omega_0 v_E^2\omega_{pe}^2/2c^2)^{1/3}$.

12.4.4 Mandelstam-Brillouin Scattering in the Nonisothermal Plasma

Now let us analyze scattering of a transverse electromagnetic wave in a nonisothermal plasma ($T_e \gg T_i$) when $\omega \ll \omega_{pi}$, kv_{Ti} (very small frequency variations). Then the resonance scattering on ion-acoustic plasma oscillations, known as the Mandelstam-Brillouin scattering, becomes essential. Ion terms cannot be neglected and (12.4.14) takes the form

$$k^2r_{De}^2 - \frac{\omega^2}{\omega_{pi}^2} - i\sqrt{\frac{\pi}{2}}k^2r_{De}^2\frac{\omega}{kv_{Te}} = \frac{k^2}{4}\left(\frac{[\boldsymbol{k}-\boldsymbol{k}_0, \boldsymbol{v}_E]^2}{(\boldsymbol{k}-\boldsymbol{k}_0)^2}\right.$$
$$\left.\times\frac{1}{2\omega\omega_0 + c^2(k^2 - 2\boldsymbol{k}\cdot\boldsymbol{k}_0)} - \frac{[\boldsymbol{k}+\boldsymbol{k}_0, \boldsymbol{v}_E]^2}{(\boldsymbol{k}+\boldsymbol{k}_0)^2}\,\frac{1}{2\omega\omega_0 - c^2(k^2 + 2\boldsymbol{k}\cdot\boldsymbol{k}_0)}\right). \tag{12.4.21}$$

Here for $\omega \gg kv_{Ti}$ we completely neglect the exponentially small terms specified by the Cherenkov wave absorption on ions. The terms which are

2 In weaker fields, the last inequality (12.4.19) is violated, and the considered hydrodynamic instability transfers into the kinetic one. Here, the Cherenkov wave dissipation becomes significant. The maximum value of the kinetic instability increment is of the order $\delta \approx \sqrt{\pi/8} \times v_E^2\omega_b^2/(\omega_0 v_{Te}^2)$ (Exercise 12.5.2).

included on the right-hand side of (12.4.21) correspond to the terms with resonance denominators in (12.4.14)

$$(\omega \pm \omega_0)^2 \varepsilon^{\mathrm{tr}} (\omega \pm \omega_0) - c^2 (\boldsymbol{k} \pm \boldsymbol{k}_0)^2 \approx 0 \qquad (12.4.22)$$

corresponding to the "red" (Stokes, $\omega_0 - \omega$) and "blue" (anti-Stokes, $\omega_0 + \omega$) satellites in the transverse scattered wave. In this case of relatively small pump wave fields, these satellites are shifted with respect to the frequency ω_0 at the value of the ion-acoustic oscillation frequency, i.e., $\omega = kv_{\mathrm{s}}$, where $v_{\mathrm{s}} = \sqrt{T_{\mathrm{e}}/M}$. Accounting for the small quantity of the imaginary term in (12.4.21) which is specified by the Cherenkov wave dissipation of electrons, we easily obtain the increment of ion-acoustic oscillations, $\omega \to \omega + \mathrm{i}\delta$, and thus of the scattered wave. When $k^2 \approx \pm 2\boldsymbol{k} \cdot \boldsymbol{k}_0$, this increment achieves a maximum for "red" and "blue" satellites, respectively, and one obtains

$$\delta_{\max} \approx \frac{1}{2}\left[\sqrt{\frac{\pi}{8} k^2 v_{\mathrm{s}}^2 + \frac{k v_{\mathrm{s}} \omega_{\mathrm{pi}}}{4 v_{\mathrm{s}}^2}\left(v_{\mathrm{E}}^2 - \frac{(\boldsymbol{k} \cdot \boldsymbol{v}_{\mathrm{E}})^2}{k_0^2}\right)} - \sqrt{\frac{\pi}{8}}\, k v_{\mathrm{s}}\right]. \qquad (12.4.23)$$

Finally, note that if $\varepsilon^{\mathrm{lo}}(\omega, \boldsymbol{k})$ and $\delta\varepsilon^{\mathrm{lo}}_{\mathrm{e,i}}(\omega, \boldsymbol{k})$ represent the corresponding longitudinal dielectric permittivities of the magneto-active plasma, then due to (12.4.14) we can also study three-wave processes in a magneto-active plasma.

12.5 Exercises

12.5.1 Derive the expression for two- and three-index dielectric permittivities of the cold isotropic electron plasma in the model of independent particles.

Solution. Let us proceed from the system of continuity and Euler equations

$$\frac{\partial N}{\partial t} + \operatorname{div} N\boldsymbol{V}, \qquad \frac{\partial \boldsymbol{V}}{\partial t} + (\boldsymbol{V} \cdot \nabla)\boldsymbol{V} = \frac{e}{m}\left\{\boldsymbol{E} + \frac{1}{c}[\boldsymbol{V}, \boldsymbol{B}]\right\}. \qquad (12.5.1)$$

The fields $\boldsymbol{E}$ and $\boldsymbol{B}$ satisfy the Maxwell equation

$$\begin{aligned} \operatorname{curl} \boldsymbol{B} &= \frac{1}{c}\frac{\partial \boldsymbol{E}}{\partial t} + \frac{4\pi}{c}\boldsymbol{j}, \qquad & \operatorname{div} \boldsymbol{B} &= 0, \\ \operatorname{curl} \boldsymbol{E} &= -\frac{1}{c}\frac{\partial \boldsymbol{B}}{\partial t}, \qquad & \operatorname{div} \boldsymbol{E} &= 4\pi\varrho, \end{aligned} \qquad (12.5.2)$$

where

$$\varrho = eN\,, \quad j = eNV\,. \tag{12.5.3}$$

In the absence of the fields $\boldsymbol{E}$ and $\boldsymbol{B}$ we assume

$$N_0 = \text{const}\,, \quad V_0 = \text{const.} \tag{12.5.4}$$

The changes $\delta N(t, \boldsymbol{r})$ and $\delta V(t, \boldsymbol{r})$ appear due to the perturbation fields. For small fields E and B we can write

$$N = N_0 + N_1 + N_2 + \dots\,, \quad V = V_0 + V_1 + V_2 + \dots\,, \tag{12.5.5}$$

where

$$N_1 \sim \boldsymbol{E}\,, \quad N_2 \sim \boldsymbol{E}^2, \dots, \quad V_1 \sim \boldsymbol{E}\,, \quad V_2 \sim \boldsymbol{E}^2, \dots\,. \tag{12.5.6}$$

In the linear approximation we have

$$\frac{\partial N_1}{\partial t} + \operatorname{div} N_0 V_1\,, \quad \frac{\partial V_1}{\partial t} = \frac{e}{m}\boldsymbol{E}\,. \tag{12.5.7}$$

Hence, after the Fourier transformation we obtain

$$N_1(\omega, \boldsymbol{k}) = \frac{ieN_0}{m\omega^2}\boldsymbol{k}\cdot\boldsymbol{E}(\omega, \boldsymbol{k})\,, \quad V_1 = \frac{ie}{m\omega}\boldsymbol{E}(\omega, \boldsymbol{k})\,. \tag{12.5.8}$$

In the linear approximation, we find

$$\begin{aligned} j_i(\omega, \boldsymbol{k}) &= eN_0 V_{1i}(\omega, \boldsymbol{k}) = \sigma_{ij}(\omega, \boldsymbol{k})\, E_j(\omega, \boldsymbol{k})\,, \\ \sigma_{ij}(\omega, \boldsymbol{k}) &= \mathrm{i}\frac{e^2 N_0}{m\omega}\,, \quad \text{or} \end{aligned} \tag{12.5.9}$$

$$\varepsilon_{ij}(\omega, \boldsymbol{k}) = \delta_{ij} + \frac{4\pi \mathrm{i}}{\omega}\sigma_{ij}(\omega, \boldsymbol{k}) = \left(1 - \frac{\omega_{\mathrm{pe}}^2}{\omega^2}\right)\delta_{ij}\,, \tag{12.5.10}$$

where $\varepsilon_{ij}(\omega, \boldsymbol{k})$ is the two-index dielectric tensor of the cold isotropic electron plasma.

To derive the three-index tensor, the current density must be written up to second-order terms in the field

$$\boldsymbol{j}(\boldsymbol{r}, t) = eN_0 V_1(\boldsymbol{r}, t) + eN_1(\boldsymbol{r}, t)\, V_1(\boldsymbol{r}, t) + eN_0 V_2(\boldsymbol{r}, t)\,. \tag{12.5.11}$$

Here $N_1(\boldsymbol{r}, t)$ and $\boldsymbol{V}_1(\boldsymbol{r}, t)$ are defined by (12.5.7) and

$$\frac{\partial \boldsymbol{V}_2}{\partial t} + (\boldsymbol{V}_1 \cdot \nabla)\boldsymbol{V}_1 = \frac{e}{mc}[\boldsymbol{V}_1, \boldsymbol{B}] . \tag{12.5.12}$$

By eliminating the field $\boldsymbol{B}$ by means of the Maxwell equation and applying the Fourier transformation we obtain

$$V_{2i}(\omega, \boldsymbol{k}) = \frac{1}{\omega} \int d\omega' d\boldsymbol{k}' d\omega'' d\boldsymbol{k}'' \delta(\omega - \omega' - \omega'')\, \delta(\boldsymbol{k} - \boldsymbol{k}' - \boldsymbol{k}'')$$
$$\times \left[\frac{\mathrm{i}e}{m} \alpha_{ijs}(\omega'', \boldsymbol{k}'') E_j(\omega'', \boldsymbol{k}'') + k_s'' V_{1i}(\omega'', \boldsymbol{k}'')\right] V_{1s}(\omega', \boldsymbol{k}') , \tag{12.5.13}$$

$$\alpha_{ijs}(\omega, \boldsymbol{k}) = \frac{1}{\omega}(k_i \delta_{js} - k_s \delta_{ij}) .$$

Substituting $\boldsymbol{V}_1(\omega, \boldsymbol{k})$ from (12.5.8) gives

$$V_{2i}(\omega, \boldsymbol{k}) = -\frac{e^2}{m^2} \int d\omega' d\boldsymbol{k}' \frac{\delta_{js}}{\omega\omega'} \frac{k_i - k_i'}{\omega - \omega'} E_s(\omega', \boldsymbol{k}') E_j(\omega'', \boldsymbol{k}'') , \tag{12.5.14}$$

where $\omega'' = \omega - \omega'$ and $\boldsymbol{k}'' = \boldsymbol{k} - \boldsymbol{k}'$.

Finally, from (12.5.11) we obtain the Fourier component of the current density up to the second-order terms

$$j_i(\omega, \boldsymbol{k}) = \sigma_{ij}(\omega, \boldsymbol{k}) E_j(\omega, \boldsymbol{k})$$
$$+ \int d\omega' d\boldsymbol{k}' \sigma_{ijs}(\omega, \boldsymbol{k}; \omega', \boldsymbol{k}') E_j(\omega'', \boldsymbol{k}'') E_s(\omega', \boldsymbol{k}') , \tag{12.5.15}$$

$$\sigma_{ijs}(\omega, \boldsymbol{k}; \omega', \boldsymbol{k}') = -\frac{e^3 N_0}{m^2} \frac{1}{\omega\omega''} \left(\frac{k_i''}{\omega}\delta_{js} + \frac{k_s'}{\omega'}\delta_{ij}\right) \quad \text{or}$$

$$D_i(\omega, \boldsymbol{k}) = \varepsilon_{ij}(\omega, \boldsymbol{k}) E_j(\omega, \boldsymbol{k})$$
$$+ \int d\omega' d\boldsymbol{k}' \varepsilon_{ijs}(\omega, \boldsymbol{k}; \omega', \boldsymbol{k}') E_j(\omega'', \boldsymbol{k}'') E_s(\omega', \boldsymbol{k}') , \tag{12.5.16}$$

$$\varepsilon_{ijs}(\omega, \boldsymbol{k}; \omega', \boldsymbol{k}') = -\frac{\mathrm{i}e}{m} \frac{\omega_{\mathrm{pe}}^2}{\omega\omega'\omega''} \left(\frac{k_i''}{\omega}\delta_{js} + \frac{k_s'}{\omega'}\delta_{ij}\right) .$$

The tensor $\varepsilon_{ijs}(\omega, \boldsymbol{k}; \omega'', \boldsymbol{k}'')$ is obtained from $\varepsilon_{ijs}(\omega, \boldsymbol{k}; \omega', \boldsymbol{k}')$ by means of the substitution $\omega' \rightleftarrows \omega''$, $\boldsymbol{k}' \rightleftarrows \boldsymbol{k}''$, $s \rightleftarrows j$. Finally, for their symmetrized combination we have, see (12.2.20),

$$S_{ijs}(\omega, \boldsymbol{k}; \omega', \boldsymbol{k}') = -\frac{\mathrm{i}e}{m} \frac{\omega_{\mathrm{pe}}^2}{\omega\omega'\omega''} \left(\frac{k_i}{\omega}\delta_{js} + \frac{k_j''}{\omega''}\delta_{is} + \frac{k_s'}{\omega'}\delta_{ij}\right) . \tag{12.5.17}$$

The four-index tensors $\varepsilon_{iajb}(\omega, \boldsymbol{k}; \omega', \boldsymbol{k}'; \omega'', \boldsymbol{k}'')$ and $V_{iajb}(\omega, \boldsymbol{k}; \omega', \boldsymbol{k}')$ are obtained analogously, see (12.2.21, 22).

12.5.2 Study the process of the *induced Compton scattering* of a transverse wave in an isotropic plasma (kinetic instability) on the basis of (12.1.20).

Solution. Let us account for the Coulomb scattering on longitudinal fluctuations and assume the phases of incident and scattered waves to be random. Substituting the field correlator (12.3.1) into (12.1.20) and assuming

$$W^{\mathrm{tr}}(\boldsymbol{k}) = \frac{\partial}{\partial\omega}\left\{\omega\left[\mathrm{Re}\,\{\varepsilon^{\mathrm{tr}}(\omega, \boldsymbol{k})\} - \frac{k^2c^2}{\omega^2}\right]\right\}\langle(E^{\mathrm{tr}})^2\rangle_{\omega,\boldsymbol{k}} \tag{12.5.18}$$

to be nonzero for the induced Compton scattering of transverse waves, we obtain, cf. (12.3.5):

$$\left\{\frac{\partial W^{\mathrm{tr}}}{\partial t}\right\}_{\mathrm{coul}} = W^{\mathrm{tr}}(\boldsymbol{k}) \int d\boldsymbol{k}'\, Q(\boldsymbol{k}, \boldsymbol{k}')\, W^{\mathrm{tr}}(\boldsymbol{k}')\,, \tag{12.5.19}$$

where

$$\begin{aligned} Q(\boldsymbol{k}, \boldsymbol{k}') = &-\frac{\omega_{\mathrm{pe}}^4 (k'' r_{\mathrm{De}})^2}{8(2\pi)^3 N_{\mathrm{e}} T_{\mathrm{e}} \omega^3} \frac{1}{|\varepsilon^{\mathrm{lo}}(\omega'', \boldsymbol{k}'')|^2} \left\{\mathrm{Im}\,\{\delta\varepsilon_{\mathrm{e}}^{\mathrm{lo}}(\omega'', \boldsymbol{k}'')\}\right. \\ &\times \left[\frac{2v_{\mathrm{Te}}^2}{\omega^2} \frac{[\boldsymbol{k}, \boldsymbol{k}']^2}{k^2k'^2} (k'^2 + k^2 - \boldsymbol{k}\cdot\boldsymbol{k}')(2\,\mathrm{Re}\,\{\delta\varepsilon_{\mathrm{e}}^{\mathrm{lo}}(\omega'', \boldsymbol{k}'')\}\right. \\ &+ 2\,\mathrm{Re}\,\{\delta\varepsilon_{\mathrm{e}}^{\mathrm{lo}}(\omega'', \boldsymbol{k}'')\} \times \mathrm{Re}\,\{\delta\varepsilon_{\mathrm{i}}^{\mathrm{lo}}(\omega'', \boldsymbol{k}'')\} + |\delta\varepsilon_{\mathrm{e}}^{\mathrm{lo}}(\omega'', \boldsymbol{k}'')|^2) \\ &\left.+ \left(1 + \frac{(\boldsymbol{k}\cdot\boldsymbol{k}')^2}{k^2k'^2}\right)|1 + \delta\varepsilon_{\mathrm{i}}^{\mathrm{lo}}(\omega'', \boldsymbol{k}'')|^2\right] \\ &\left.+ \left(1 + \frac{(\boldsymbol{k}\cdot\boldsymbol{k}')^2}{k^2k'^2}\right) \mathrm{Im}\,\{\delta\varepsilon_{\mathrm{i}}^{\mathrm{lo}}(\omega'', \boldsymbol{k}'')\}|\delta\varepsilon_{\mathrm{e}}^{\mathrm{lo}}(\omega'', \boldsymbol{k}'')|^2\right\}. \end{aligned} \tag{12.5.20}$$

Here $\omega'' = \omega - \omega'$, $\boldsymbol{k}'' = \boldsymbol{k} - \boldsymbol{k}'$ and $\omega^2 = k^2c^2 + \omega_{\mathrm{pe}}^2$.

The antisymmetry of the kernel of (12.5.20) with respect to the substitution $\boldsymbol{k} \rightleftarrows \boldsymbol{k}'$ immediately gives the conservation law for the total energy of the transverse electromagnetic waves

$$\int d\boldsymbol{k}\, W^{\mathrm{tr}}(\boldsymbol{k}) = W^{\mathrm{tr}} = \mathrm{const.} \tag{12.5.21}$$

Thus scattering has an elastic character and the energy transfer from the one spectrum to the other is possible only when the total field energy is constant.

When the wave scattering is off the purely electron plasma, then the ion contribution into (12.5.20) is negligible:

$$Q(\boldsymbol{k},\boldsymbol{k}') = -\frac{\omega_{\mathrm{pe}}^4 (k'' r_{\mathrm{De}})^2}{8(2\pi)^3 N_{\mathrm{e}} T_{\mathrm{e}} \omega^3} \frac{1}{|\varepsilon^{\mathrm{lo}}(\omega'',\boldsymbol{k}'')|} \left\{ \mathrm{Im}\,\{\delta\varepsilon_{\mathrm{e}}^{\mathrm{lo}}(\omega'',\boldsymbol{k}'')\} \right.$$

$$\times \left[\frac{2\,v_{\mathrm{Te}}}{\omega^2} \frac{[\boldsymbol{k},\boldsymbol{k}']^2}{k^2 k'^2} (k'^2 + k^2 - \boldsymbol{k}\cdot\boldsymbol{k}')(2\,\mathrm{Re}\,\{\delta\varepsilon_{\mathrm{e}}^{\mathrm{lo}}(\omega'',\boldsymbol{k}'')\} \right.$$

$$\left.\left. + |\delta\varepsilon_{\mathrm{e}}^{\mathrm{lo}}(\omega'',\boldsymbol{k}'')|^2) + \left(1 + \frac{(\boldsymbol{k}\cdot\boldsymbol{k}')^2}{k^2 k'^2}\right)\right]\right\}. \tag{12.5.22}$$

Let us estimate this expression for the backscattering of high-frequency transverse waves ω, $\omega' \gg \omega_{\mathrm{pe}}\, c/v_{\mathrm{Te}}$, $\boldsymbol{k}' = -\boldsymbol{k}$. Thus

$$Q(\boldsymbol{k},\boldsymbol{k}') \approx -\frac{1}{4(2\pi)^3} \frac{\omega_{\mathrm{pe}}^4}{\omega^3} \frac{1}{N_{\mathrm{e}} T_{\mathrm{e}}}. \tag{12.5.23}$$

Hence the characteristic growth time of the process under consideration is obtained from (12.5.19):

$$\tau \approx 10^3 \frac{\omega^3}{\omega_{\mathrm{pe}}^4} \frac{N_{\mathrm{e}} T_{\mathrm{e}}}{W_0^{\mathrm{tr}}} \approx 5 \frac{\omega^3 N_{\mathrm{e}} T_{\mathrm{e}}}{\omega_{\mathrm{pe}}^4 E_0^{\mathrm{tr}2}}, \tag{12.5.24}$$

where $W_0^{\mathrm{tr}} = (2\pi)^3 (E_0^{\mathrm{tr}})^2/8\pi$ is the intensity of the incident transverse wave. Correspondingly, the amplitude of the scattered wave grows with time:

$$W_{\mathrm{sc}}^{\mathrm{tr}} \sim \exp(t/\tau). \tag{12.5.25}$$

12.5.3 Analyze the induced scattering of a high-frequency transverse electromagnetic wave on a magnetized monoenergetic rectilinear electron beam (*free-electron laser*).

Solution. In the beam frame the scattering process is a decay of an incident transverse wave with the frequency ω_0' into a scattered wave with the frequency ω_{s}' and a space-charge wave (beam longitudinal oscillations) with the frequency ω' (all the values in the intrinsic beam system are marked by primary indices), i.e.,

$$\omega_0' = \omega_{\mathrm{s}}' + \omega', \qquad \boldsymbol{k}_0' = \boldsymbol{k}_{\mathrm{s}}' + \boldsymbol{k}'. \tag{12.5.26}$$

Let us apply the Lorentz transformation formulas (Sect. 6.1) and take into account that for both incident and scattered high-frequency transverse waves the equations

$$k_{0,s}'^2 = \frac{\omega_{0,s}'^2}{c^2}, \quad k_{0,s}^2 = \frac{\omega_{0,s}^2}{c^2} \tag{12.5.27}$$

are valid. Here for the wave vector component oriented along the magnetic field $\boldsymbol{B}_0 \parallel 0z$ we have

$$k_{z0,s} = \frac{\omega_{0,s}}{c}\cos\phi_{0,s} = \pm\frac{\omega_{0,s}}{c}\sqrt{1-\frac{\omega_{\mathrm{cr}0,s}^2}{\omega_{0,s}^2}}, \tag{12.5.28}$$

where $\omega_{\mathrm{cr}0,s}^2 = k_{\perp 0,s}^2 c^2$ and $\phi_{0,s}$ is the angle included between the wave vector $\boldsymbol{k}_{0,s}$ and the $0z$-axis. Waves with $k_z > 0$ are passing waves and those with $k_z < 0$ are counterwaves.

Equations (12.5.26–28) give the scattered wave frequency at the given scattering angle

$$\omega_s = \omega_1\gamma^2\left(1 \pm \frac{u}{c}\sqrt{1-\frac{\omega_{\mathrm{cr},s}^2}{\omega_1^2\gamma^2}}\right), \quad \omega_1 = \omega_0\left(1+\frac{u}{c}\cos\phi_0 - \frac{\omega'}{\gamma\omega_0}\right). \tag{12.5.29}$$

Here two signs correspond to the backward and forward scattering with formation of passing and counterwaves, respectively. Then the frequency ω' of the beam longitudinal wave is determined from (12.4.16). In the beam frame the latter is

$$1 + \delta\varepsilon_e^{\mathrm{lo}}(\omega', k') + \frac{k'^2\delta\varepsilon_e^{\mathrm{lo}}(\omega', k')(\omega_0'-\omega')^2\,[\boldsymbol{k}_0'-\boldsymbol{k}', \boldsymbol{v}_E']^2}{4\,\omega_0'^2(\boldsymbol{k}_0'-\boldsymbol{k}')^2[(\omega_0'-\omega')^2\varepsilon^{\mathrm{tr}}(\omega_0'-\omega') - c^2(\boldsymbol{k}_0'-\boldsymbol{k}')^2]} = 0\,. \tag{12.5.30}$$

Here $\boldsymbol{v}_E' = e\boldsymbol{E}_0'/(m\omega_0')$ is the electron oscillation velocity in the moving frame and

$$\delta\varepsilon_e^{\mathrm{lo}}(\omega', k') \approx -\frac{\omega_b^2}{\omega'^2\gamma}\frac{k_z'^2}{k'^2}, \quad \varepsilon^{\mathrm{tr}}(\omega_0'-\omega') = 1 - \frac{\omega_b^2}{\gamma(\omega_0'-\omega')^2} \approx 1\,. \tag{12.5.31}$$

The analysis of (12.5.30) is completely analogous to that given in Sect. 12.4, in particular for the Raman scattering $\omega' = \omega_b'|k_z'|/(\sqrt{\gamma}\,k') + \mathrm{i}\delta'$, where

$$\delta' = \sqrt{\frac{\omega_0' v_E'^2\omega_b}{8c^2\sqrt{\gamma}}}\,. \tag{12.5.32}$$

For the Compton or the modified decay scattering, we have $\omega' \gg \omega_b k_z'/(\sqrt{\gamma}\,k')$ and

$$\omega' = \frac{-1+\mathrm{i}\sqrt{3}}{2}\left(\frac{k_z'^2\omega_b^2 v_E'^2}{8\gamma\omega_0}\right)^{1/3}. \tag{12.5.33}$$

The Raman scattering occurs in relatively weak fields when

$$\frac{v_E'^2}{c^2} \approx \frac{v_E^2}{c^2} \ll \frac{\omega_b}{\sqrt{\gamma}\,\omega_0'} \approx \frac{\omega_b}{\gamma^{3/2}\,\omega_0}\,, \tag{12.5.34}$$

and the Compton scattering prevails in the opposite limit.

Note that the incident wave field must be sufficiently large, so that the thermal spread of beam electrons over velocities may be neglected. This requirement for $\omega' \gg k_z' v_T$ is reduced to

$$1 - \frac{\omega_b}{\gamma^{3/2}\omega_0} \gg \frac{v_T}{c}\,, \tag{12.5.35}$$

where v_T is the thermal spread of electrons in the beam frame.

Knowing the temporal increment of oscillations we may easily obtain the spatial one ($k' \to k' + i\delta k'$):

$$\delta k' = \frac{\delta'}{v_{gr}'} \approx \frac{\delta'}{c}\,. \tag{12.5.36}$$

Here v_{gr}' is the group velocity of the scattered wave in the beam frame. In this case the group velocity is close to the velocity of light. Now we can write the temporal and spatial increments in the laboratory frame

$$\delta = \gamma(\delta' + u\delta k')\,, \quad \delta k = \frac{\delta}{v_{gr}} \approx \frac{\delta}{c}\,. \tag{12.5.37}$$

Finally, note that the discussed process of the induced scattering on a relativistic beam presents some interest due to the transformation of the incident wave frequency ω_0 into a higher frequency ω_s of the scattered wave. A significant frequency growth is possible for the backscattering of the incident counterwave $\omega_s \sim \gamma^2\omega_0$, i.e., for $\gamma \gg 1$. Thus, such an induced scattering process is called a free-electron laser.

All formulas mentioned above are also valid for the scattering of an incident high-frequency E-wave on the electron beam moving in a round metallic waveguide. Here, $\omega_{cr0,s}$ means the waveguide cut-off frequencies for incident and scattered waves, i.e., $\omega_{cr0,s} \equiv \mu_{0,s} c/R$, where R is a waveguide radius and $\mu_{0,s}$ are the roots of the Bessel functions $J_l(\mu_{0,s}) = 0$, which characterize radial oscillation modes of incident and scattered waves.

12.5.4 Let the harmonic perturbations of the field $E_1(z,t) = E_1 \sin(k_1 z)\,\delta(t)$ and $E_2(z,t) = E_2 \sin(k_2 z)\,\delta(t)$ be consistently excited in the isotropic plasma by an external source at the time moments $t = 0$ and $t = \tau$. Their wavelengths $\lambda_1 = 1/k_1$ and $\lambda_2 = 1/k_2$ are smaller than the Larmor radius of electrons and thus such perturbations quickly damp in the plasma. The time interval τ greatly exceeds the damping period of the perturbation fields. Find the response of the system after such perturbations damp.

Solution. The response of the system to the first signal can be found from

$$\frac{\partial f_1}{\partial t} \pm ik_1 v f_1 = \frac{e}{m} E_1 \frac{\partial f_0}{\partial v} e^{\pm ik_1 z} \delta(t) , \tag{12.5.38}$$

where f_0 is the unperturbed distribution function of electrons. The solution has the form

$$f_1(t) = \theta(\tau) \frac{e}{m} E_1 \frac{\partial f_0}{\partial v} \exp[\pm ik_1 (z - vt)] , \tag{12.5.39}$$

where $\theta(t)$ is the unit step function, i.e., $\theta(t) = 1$ when $t > 0$ and $\theta(t) = 0$ when $t < 0$. Due to fast oscillations of function f_2 in time, the integral of f_1 vanishes for large t. Physically, the modulated electron beam produced by source 1 after a long period of time t does not perturb the electric field in the plasma, though the distribution function is perturbed.

Source 2 affects the electron distribution function already perturbed by the first signal, i.e.,

$$\frac{\partial f_2}{\partial t} \pm ik_2 v f_2 = \frac{e}{m} E_2 \frac{\partial (f_0 + f_1)}{\partial v} e^{\pm ik_2 z} \delta(t - \tau) . \tag{12.5.40}$$

Hence

$$\begin{aligned} f_2 = {} & \theta(t-\tau) \frac{e}{m} E_2 \frac{\partial f_0}{\partial v} \exp\{\pm ik_2 [z - v(t-\tau)]\} + \theta(t)\,\theta(t-\tau) \\ & \times \frac{e^2}{m^2} E_1 E_2 \exp\{\pm ik_2 [z - v(t-\tau)]\} \frac{\partial}{\partial v} \exp[\pm ik_1 (z - vt)] \frac{\partial f_0}{\partial v} . \end{aligned} \tag{12.5.41}$$

The perturbation arising from the first term in (12.5.41) does not perturb the field in the plasma after a long period of time, i.e., the field damps. The second term in (12.5.41) is velocity independent if

$$t = \frac{k_2}{k_2 \pm k_1} \tau . \tag{12.5.42}$$

Thus it is free of fast oscillations. Finally, the perturbations of the charge and current and therefore of the field appear nonvanishing in the system.

Hence at the time moment (12.5.42) macroscopic oscillations of the density of electrons and the electric field arise in the plasma. This effect is usually called the *plasma echo*.

Appendix

A. The Main Operators of Field Theory in Orthogonal Curvilinear Coordinate Systems

The main operators of field theory are the gradient, the divergence, the rotation and the Laplacian. In Cartesian coordinates where

$$r = ix + jy + kz \ ,$$

i, j, k being the unit vectors along the coordinate axes, we have

$$\begin{aligned}
\operatorname{grad} \Psi(x, y, z) &= \nabla \Psi = i \frac{\partial \Psi}{\partial x} + j \frac{\partial \Psi}{\partial y} + k \frac{\partial \Psi}{\partial z} \ , \\
\operatorname{div} A(x, y, z) &= \nabla \cdot A = \frac{\partial A_x}{\partial x} + \frac{\partial A_y}{\partial y} + \frac{\partial A_z}{\partial z} \\
\operatorname{curl} A(x, y, z) &= [\nabla, A] = \begin{vmatrix} i & j & k \\ \dfrac{\partial}{\partial x} & \dfrac{\partial}{\partial y} & \dfrac{\partial}{\partial z} \\ A_x & A_y & A_z \end{vmatrix} \\
&= i\left(\frac{\partial A_z}{\partial y} - \frac{\partial A_y}{\partial z}\right) + j\left(\frac{\partial A_x}{\partial z} - \frac{\partial A_z}{\partial x}\right) + k\left(\frac{\partial A_y}{\partial x} - \frac{\partial A_x}{\partial y}\right) , \\
\Delta \Psi(x, y, z) &= \frac{\partial^2 \Psi}{\partial x^2} + \frac{\partial^2 \Psi}{\partial y^2} + \frac{\partial^2 \Psi}{\partial z^2} \ .
\end{aligned} \tag{A.1}$$

Thus

$$\nabla = i\frac{\partial}{\partial x} + j\frac{\partial}{\partial y} + k\frac{\partial}{\partial z} \ , \qquad \Delta = \frac{\partial^2}{\partial x^2} + \frac{\partial^2}{\partial y^2} + \frac{\partial^2}{\partial z^2} \ . \tag{A.2}$$

It is not difficult to write these operators in another curvilinear coordinate system (q_1, q_2, q_3), related to the Cartesian system by the transformation formulas

$$x = x(q_1, q_2, q_3), \quad y = y(q_1, q_2, q_3), \quad z = z(q_1, q_2, q_3) . \tag{A.3}$$

If the transformation determinant is nonzero:

$$D = \left|\frac{\partial x_i}{\partial q_j}\right| = \begin{vmatrix} \frac{\partial x}{\partial q_1} & \frac{\partial x}{\partial q_2} & \frac{\partial x}{\partial q_3} \\ \frac{\partial y}{\partial q_1} & \frac{\partial y}{\partial q_2} & \frac{\partial y}{\partial q_3} \\ \frac{\partial z}{\partial q_1} & \frac{\partial z}{\partial q_2} & \frac{\partial z}{\partial q_3} \end{vmatrix} \neq 0 , \tag{A.4}$$

the following relations are valid:

$$\operatorname{grad} \Psi(q_1, q_2, q_3) = \frac{\boldsymbol{e}_1}{H_1}\frac{\partial \Psi}{\partial q_1} + \frac{\boldsymbol{e}_2}{H_2}\frac{\partial \Psi}{\partial q_2} + \frac{\boldsymbol{e}_3}{H_3}\frac{\partial \Psi}{\partial q_3} ,$$

$$\operatorname{div} \boldsymbol{A}(q_1, q_2, q_3) = \frac{1}{H_1 H_2 H_3}\left[\frac{\partial}{\partial q_1}(A_1 H_2 H_3) + \frac{\partial}{\partial q_2}(A_2 H_1 H_3) + \frac{\partial}{\partial q_3}(A_3 H_1 H_2)\right] ,$$

$$\operatorname{curl} \boldsymbol{A}(q_1, q_2, q_3) = \frac{\boldsymbol{e}_1}{H_2 H_3}\left[\frac{\partial}{\partial q_2}(A_3 H_3) - \frac{\partial}{\partial q_3}(A_2 H_2)\right] + \frac{\boldsymbol{e}_2}{H_1 H_3} \times\left[\frac{\partial}{\partial q_3}(A_1 H_1) - \frac{\partial}{\partial q_1}(A_3 H_3)\right] + \frac{\boldsymbol{e}_3}{H_1 H_2}\left[\frac{\partial}{\partial q_1}(A_2 H_2) - \frac{\partial}{\partial q_2}(A_1 H_1)\right] , \tag{A.5}$$

$$\Delta \Psi(q_1, q_2, q_3) = \frac{1}{H_1 H_2 H_3}\left[H_2 H_3 \frac{\partial}{\partial q_1}\left(\frac{1}{H_1}\frac{\partial \Psi}{\partial q_1}\right) + H_1 H_3 \frac{\partial}{\partial q_2}\left(\frac{1}{H_2}\frac{\partial \Psi}{\partial q_2}\right) + H_1 H_2 \frac{\partial}{\partial q_3}\left(\frac{1}{H_3}\frac{\partial \Psi}{\partial q_3}\right)\right] .$$

Here $\boldsymbol{e}_1, \boldsymbol{e}_2, \boldsymbol{e}_3$ are the unit vectors of the curvilinear coordinate system and H_1, H_2, H_3 the so-called Lamé coefficients determined by

$$H_i = \sqrt{\left(\frac{\partial x}{\partial q_i}\right)^2 + \left(\frac{\partial y}{\partial q_i}\right)^2 + \left(\frac{\partial z}{\partial q_i}\right)^2} , \tag{A.6}$$

for $i = 1, 2, 3$.

Throughout the book we have frequently used the cylindrical coordinate system where

$$x = r\cos\phi\ , \quad y = r\sin\phi\ , \quad z = z \tag{A.7}$$

and $e_1 = e_r$, $e_2 = e_\phi$, $e_3 = e_z$ are the unit vectors and $q_1 = r$, $q_2 = \phi$, $q_3 = z$. Then we have

$$H_1 = 1\ , \quad H_2 = r\ , \quad H_3 = 1\ . \tag{A.8}$$

Thus, in the cylindrical coordinate system we obtain

$$\begin{aligned}
\operatorname{grad}\Psi(r,\phi,z) &= e_r\frac{\partial\Psi}{\partial r} + \frac{e_\phi}{r}\frac{\partial\Psi}{\partial\phi} + e_z\frac{\partial\Psi}{\partial z}\ ;\\
\operatorname{div}A\,(r,\phi,z) &= \frac{1}{r}\left[\frac{\partial}{\partial r}(rA_r) + \frac{\partial A_\phi}{\partial\phi} + r\frac{\partial A_z}{\partial z}\right]\\
&= \frac{1}{r}\frac{\partial}{\partial r}(rA_r) + \frac{1}{r}\frac{\partial A_\phi}{\partial\phi} + \frac{\partial A_z}{\partial z}\ ;\\
\operatorname{curl}A\,(r,\phi,z) &= e_r\left[\frac{1}{r}\frac{\partial A_z}{\partial\phi} - \frac{\partial A_\phi}{\partial z}\right] + e_\phi\left[\frac{\partial A_r}{\partial z} - \frac{\partial A_z}{\partial r}\right]\\
&\quad + e_z\left[\frac{1}{r}\frac{\partial}{\partial r}(rA_\phi) - \frac{1}{r}\frac{\partial A_r}{\partial\phi}\right]\ ;\\
\Delta\Psi(r,\phi,z) &= \frac{\partial^2\Psi}{\partial r^2} + \frac{1}{r}\frac{\partial^2\Psi}{\partial\phi^2} + \frac{1}{r}\frac{\partial\Psi}{\partial r} + \frac{\partial^2\Psi}{\partial z^2}\ .
\end{aligned} \tag{A.9}$$

A.1 Exercises

A.1.1 Calculate the Lamé coefficients in the spherical coordinate system, where

$$x = r\sin\theta\cos\phi\ , \quad y = r\sin\theta\sin\phi\ , \quad z = r\cos\theta\ .$$

Solution. We obtain

$$\begin{aligned}
&q_1 = \mathrm{r}\ , \quad q_2 = \phi\ , \quad q_3 = \theta\ ,\\
&H_1 = 1\ , \quad H_2 = r\sin\theta\ , \quad H_3 = r\ .
\end{aligned}$$

A.1.2 Assuming the $0z$-axis to be oriented along the vector $\boldsymbol{B}$, calculate the operator $[\boldsymbol{v}, \boldsymbol{B}]\, \partial/\partial\boldsymbol{v}$ in the cylindrical coordinates of the velocity space.

Solution. Taking into account that

$$\frac{\partial}{\partial v_x} = \frac{\partial v_\perp}{\partial v_x}\frac{\partial}{\partial v_\perp} + \frac{\partial \phi}{\partial v_x}\frac{\partial}{\partial \phi} = \cos\phi\frac{\partial}{\partial v_\perp} - \frac{\sin\phi}{v_\perp}\frac{\partial}{\partial \phi},$$

$$\frac{\partial}{\partial v_y} = \frac{\partial v_\perp}{\partial v_y}\frac{\partial}{\partial v_\perp} + \frac{\partial \phi}{\partial v_y}\frac{\partial}{\partial \phi} = \sin\phi\frac{\partial}{\partial v_\perp} + \frac{\cos\phi}{v_\perp}\frac{\partial}{\partial \phi}$$

we easily obtain

$$[\boldsymbol{v}, \boldsymbol{B}]\frac{\partial}{\partial \boldsymbol{v}} = B\left(v_y\frac{\partial}{\partial v_x} - v_x\frac{\partial}{\partial v_y}\right) = -B\frac{\partial}{\partial \phi}.$$

A.1.3 Write the vector $(\boldsymbol{v}\cdot\nabla)\,\boldsymbol{v}$ in cylindrical coordinates.

Solution. Using the identity

$$(\boldsymbol{v}\cdot\nabla)\,\boldsymbol{v} = \tfrac{1}{2}\,\mathrm{grad}\, v^2 - [\boldsymbol{v}, \mathrm{rot}\,\boldsymbol{v}]$$

and the expressions for $\mathrm{grad}\, v^2$ and $\mathrm{rot}\,\boldsymbol{v}$ in the cylindrical coordinates (A.9), we easily obtain

$$\{(\boldsymbol{v}\cdot\nabla)\,\boldsymbol{v}\}_r = v_r\frac{\partial v_r}{\partial r} + \frac{v_\phi}{r}\frac{\partial v_r}{\partial \phi} + v_z\frac{\partial v_r}{\partial z} - \frac{v_\phi^2}{r},$$

$$\{(\boldsymbol{v}\cdot\nabla)\,\boldsymbol{v}\}_\phi = v_r\frac{\partial v_\phi}{\partial r} + \frac{v_\phi}{r}\frac{\partial v_\phi}{\partial \phi} + v_z\frac{\partial v_\phi}{\partial z} + \frac{v_r v_\phi}{r},$$

$$\{(\boldsymbol{v}\cdot\nabla)\,\boldsymbol{v}\}_z = v_r\frac{\partial v_z}{\partial r} + \frac{v_\phi}{r}\frac{\partial v_z}{\partial \phi} + v_z\frac{\partial v_z}{\partial z}.$$

B. Elements of Tensor Calculus

The concept of a tensor is closely related to the transformations of coordinate systems. Above we have primarily used the three-dimensional orthogonal Cartesian coordinate system $0xyz$, which will be written in the following in the symmetrical form $0x_1x_2x_3$. We consider two Cartesian systems $0x_1x_2x_3$ and $0x_1'x_2'x_3'$ with the common origin 0. Then the coordinates of a point M in the primed and the primeless system are related by

$$x_k = e_{kj} x_j' , \quad x_k' = e_{kj}^{-1} x_j = e_{jk} x_j , \tag{B.1}$$

where e_{ij} are the cosines of the angles between the axes of the primed and primeless system:

	x_1'	x_2'	x_3'
x_1	e_{11}	e_{12}	e_{13}
x_2	e_{21}	e_{22}	e_{23}
x_3	e_{31}	e_{32}	e_{33}

. (B.2)

To simplify the notation, the summation from 1 to 3 is assumed to be carried out over repeated (dummy) indices. It is easy to show the relation

$$e_{ik} e_{jk} = e_{ki} e_{kj} = \delta_{ij} = \begin{cases} 1 & \text{for} \quad i = j , \\ 0 & \text{for} \quad i \neq j . \end{cases} \tag{B.3}$$

The transformation (B.1) is called an orthogonal affine transformation (a rotation of the coordinate system) and the matrix

$$e_{ij} = \begin{pmatrix} e_{11} & e_{12} & e_{13} \\ e_{21} & e_{22} & e_{23} \\ e_{31} & e_{32} & e_{33} \end{pmatrix} \tag{B.4}$$

is known as the transformation matrix.

Since the vector $\boldsymbol{x}$ with the components x_1, x_2, x_3 or $\boldsymbol{x}'$ with x_1', x_2', x_3' in the primeless and primed coordinates, respectively, represent the point M, (B.1) constitutes the transformation law of vectors when the coordinate system is rotated. Moreover, a set of three quantities transformed by (B.1) is a vector.

If two vectors $\boldsymbol{a} = (a_1, a_2, a_3)$ and $\boldsymbol{b} = (b_1, b_2, b_3)$ are given, then their scalar product defined by

$$\boldsymbol{a} \cdot \boldsymbol{b} = a_i b_i = a_1 b_1 + a_2 b_2 + a_3 b_3 \tag{B.5}$$

is invariant under the transformations of the coordinate system (B.1), i.e.,

$$a_i b_i = a_i' b_i' . \tag{B.6}$$

Indeed, due to (B.3) we have $a_i' b_i' = e_{ki} a_k e_{si} b_s = \delta_{ks} a_k b_s = a_s b_s$, which was to be proven.

The invariance of the scalar product (B.6) is often used as the definition of a vector. Thus, if the vector $\boldsymbol{x} = (x_1, x_2, x_3)$ and a set of three quantities $a_i = (a_1, a_2, a_3)$ are given and the linear form

$$F_1 = a_i x_i \tag{B.7}$$

is invariant with respect to transformations of the coordinate system (B.1), then the set forms a vector $(a_i) = \boldsymbol{a} = (a_1, a_2, a_3)$.

A second-rank tensor is defined analogously: if two vectors $\boldsymbol{x} = (x_1, x_2, x_3)$ and $\boldsymbol{y} = (y_1, y_2, y_3)$ are given and the quadratic form

$$F_2 = d_{ij} x_i y_j \tag{B.8}$$

is invariant with respect to transformations of the coordinate system (B.1), then the set of the nine quantities d_{ij} is called a second-rank tensor.

The set δ_{ij} defined by (B.3) is also a tensor. Actually,

$$\delta_{ij} x_i y_i = x_j y_j = \boldsymbol{x} \cdot \boldsymbol{y} = \text{const.} \tag{B.9}$$

From (B.8) there follows the transformation law of second-rank tensors

$$d_{ij} x_i x_j = d'_{ij} x'_i y'_i = d'_{ij} e_{mi} x_m e_{nj} y_n = d'_{sk} e_{is} e_{jk} x_i y_j \,, \quad \text{or} \tag{B.10}$$

$$d_{ks} = e_{ki} e_{sj} d'_{ij} \,. \tag{B.11}$$

Analogously we obtain

$$d'_{ks} = e_{ik} e_{js} d_{ij} \,. \tag{B.12}$$

Thus, a second-rank tensor is transformed like the outer product of two vectors $a_i b_j$. Therefore, one can define such a tensor as a set of nine quantities being transformed like the outer product of two vectors.

Tensors of higher rank are defined analogously. Thus, a third-rank tensor β_{ijk} is a set of 27 quantities, leaving the cubic form

$$F_3 = \beta_{ijk} x_i y_j z_k \tag{B.13}$$

invariant with respect to the transformation (B.1), when $\boldsymbol{x}, \boldsymbol{y}, \boldsymbol{z}$ are vectors. Equivalently, the set of the quantities β_{ijk} defines a tensor when it is transformed like the outer product of three vectors $a_i b_j c_k$. A scalar quantity can be regarded as a zero-rank tensor and a vector as a first-rank tensor.

The components of a tensor can be both real and complex. Therefore, in general, we have to deal with complex tensors. Then, the concept of the Hermiticity of a tensor is important. A second-rank tensor is called Hermitian if ("*" means complex conjugation)

$$\alpha_{ij}^{*\mathrm{H}} = \alpha_{ji}^{\mathrm{H}} , \tag{B.14}$$

but if

$$\alpha_{ij}^{*\mathrm{a}} = -\alpha_{ji}^{\mathrm{a}} \tag{B.15}$$

the tensor is called anti-Hermitian. Any tensor can be decomposed into a Hermitian and an anti-Hermitian part.

In the above, tensors have been referred to as a set of complex quantities. The tensor components, however, can be functions of both a scalar [e.g., of the time, $\alpha_{ij}(t)$] and a vector [e.g., of the coordinate, $\alpha_{ij}(\boldsymbol{r})$]. Therefore, in general, we must write:

$\phi(t, \boldsymbol{r})$	for a scalar (zero-rank tensor)
$a_i(t, \boldsymbol{r})$	for a vector (first-rank tensor)
$\alpha_{ij}(t, \boldsymbol{r})$	for a second-rank tensor
$\beta_{ijk}(t, \boldsymbol{r})$	for a third-rank tensor, etc.

Tensors as functions of more than one scalar or vector or even tensor variable are defined analogously.

Differentiating a tensor with respect to a scalar its rank is not altered; however, the differentiation with respect to a vector enlarges the tensor rank. Hence,

$\dfrac{\partial \phi(t, \boldsymbol{r})}{\partial r_i}$ is a vector (a first-rank tensor),

$\dfrac{\partial a_i(t, \boldsymbol{r})}{\partial r_j}$ is a second-rank tensor,

$\dfrac{\partial \alpha_{ij}(t, \boldsymbol{r})}{\partial r_k}$ is a third-rank tensor, etc.

We must take account of this fact when expanding a tensor in powers of a vector quantity:

$$\alpha_{ij}(t, \boldsymbol{r} + \Delta\boldsymbol{r}) = \alpha_{ij}(t, \boldsymbol{r}) + \frac{\partial \alpha_{ij}(t, \boldsymbol{r})}{\partial r_k} \Delta r_k + \frac{\partial^2 \alpha_{ij}(t, \boldsymbol{r})}{\partial r_k \partial r_s} \Delta r_k \Delta r_s . \tag{B.16}$$

The expansion of a scalar quantity in a power series is performed in the usual way.

Up to now we have dealt with a rotation of the coordinate system only, and the tensor quantities have been defined by a certain symmetry property with respect to the rotation transformation (B.1). We now consider the mirror reflection of the symmetry axes of the coordinates. With respect to these transformations the tensors can be subdivided into real and pseudotensors. A scalar quantity, which is invariant not only with respect to the rotation of the coordinate system but also with respect to the mirror reflection, is called a real scalar. If it is not modified by a rotation but reverses its sign under a mirror reflection it is called a pseudoscalar. Real and pseudotensors of any rank are defined analogously. A real tensor of an even rank does not reverse its sign under the transformation of the mirror reflection. A pseudotensor, however, does reverse it. A real tensor of an odd rank reverses its sign under the transformation of the mirror reflection and a pseudotensor of this rank conserves it.

Real vectors in the three-dimensional space are, e.g., the radius vector $\boldsymbol{r}$, the velocity $\boldsymbol{v}$ and the momentum $\boldsymbol{p}$ vectors, the wave vector $\boldsymbol{k}$, the vectors of the electric field strength $\boldsymbol{E}$ and the electric induction $\boldsymbol{D}$, the vector of the current density $\boldsymbol{j}$, etc. Real scalars are the time t, the charge density ϱ, the particle energy $\mathscr{E}(\boldsymbol{p})$ and the frequency $\omega(\boldsymbol{k})$. We have

$$\boldsymbol{v}=\frac{d\boldsymbol{r}}{dt}\,,\qquad \boldsymbol{v}=\frac{\partial\mathscr{E}(\boldsymbol{p})}{\partial\boldsymbol{p}}\,,\qquad v_{\mathrm{gr}}=\frac{\partial\omega(\boldsymbol{k})}{\partial\boldsymbol{k}}\,. \tag{B.17}$$

A pseudovector can be obtained as the vector product of two real vectors $\boldsymbol{a}$ and $\boldsymbol{b}$, i.e., $[\boldsymbol{a},\boldsymbol{b}]$. On the other hand, the vector product of a real vector $\boldsymbol{a}$ and a pseudovector $\boldsymbol{d}$ is a real vector $[\boldsymbol{a},\boldsymbol{d}]$. Thus, the magnetic field $\boldsymbol{B}$ is a pseudovector, since its vector product with the velocity vector $\boldsymbol{v}$ constitutes the real vector of the force $\boldsymbol{F}\sim[\boldsymbol{v},\boldsymbol{B}]$.

In electrodynamics of material media, the completely antisymmetric third-rank unit tensor e_{ijk}, determined by

$$e_{ijk}\begin{cases}0 & \text{if two of the indices } i,\,j,\,k \text{ coincide,}\\ 1 & \text{if the indices } i,\,j,\,k \text{ form a regular succession of the numbers } 1,2,3,\\ -1 & \text{if the indices } i,\,j,\,k \text{ form an irregular succession of the numbers } 1,2,3,\end{cases} \tag{B.18}$$

is of special importance. A cyclic succession of the numbers 1, 2, 3 is called regular and a noncyclic one irregular.

Using the unit tensor e_{ijk}, the product of two vectors $\boldsymbol{a}$ and $\boldsymbol{b}$ can be written as

$$[\boldsymbol{a},\boldsymbol{b}]_i=e_{ijk}a_jb_k\,. \tag{B.19}$$

The scalar product of a real vector $\boldsymbol{a}$ and a pseudovector $\boldsymbol{d}$, in contrast to the scalar product of two real vectors, is a pseudoscalar:

$$\boldsymbol{a}\cdot\boldsymbol{d}=a_i d_i=a_i e_{ijk} b_j c_k\ . \tag{B.20}$$

Here $d_i = e_{ijk} b_j c_k$ is a pseudovector, $\boldsymbol{b}$ and $\boldsymbol{c}$ are real vectors.

The aforementioned properties are given not only for vector fields $a_i(\boldsymbol{r})$ and tensor fields $\alpha_{ij}(\boldsymbol{r})$, $\beta_{ijk}(\boldsymbol{r})$ but also for vector and tensor operators. As stated above, the differentiation with respect to a vector argument increases the rank of the matrix. We can introduce the differentiation operator as the vector $\partial/\partial r_i = \partial/\partial\boldsymbol{r} = \nabla_r$ and define the differentiation process by a vector or scalar product:

$$\begin{aligned}
&\frac{\partial}{\partial r_i}\phi(\boldsymbol{r}) = \nabla_r\phi(\boldsymbol{r}) = \operatorname{grad}\phi(\boldsymbol{r})\ ,\\
&\frac{\partial}{\partial r_i}a_i(\boldsymbol{r}) = \nabla_r\boldsymbol{a}(\boldsymbol{r}) = \operatorname{div}\boldsymbol{a}(\boldsymbol{r})\ ,\\
&e_{ijk}\frac{\partial}{\partial r_j}a_k(\boldsymbol{r}) = [\nabla_r,\boldsymbol{a}(\boldsymbol{r})]_i = \operatorname{curl}\boldsymbol{a}(\boldsymbol{r})\ldots\ .
\end{aligned} \tag{B.21}$$

If $\phi(\boldsymbol{r})$ is a real scalar and $\boldsymbol{r}$ is a real vector, then the first quantity in (B.1) is a real vector. If $\phi(\boldsymbol{r})$ is a pseudoscalar and $\boldsymbol{r}$ a real vector, then it is a pseudovector. It will also be a pseudovector if $\phi(\boldsymbol{r})$ is a real scalar and $\boldsymbol{r}$ a pseudovector. The other quantities appearing in (B.1) and also the variables and operators of a higher rank can be interpreted analogously. For instance,

$$\begin{aligned}
\operatorname{div}\operatorname{grad}\phi(\boldsymbol{r}) &= \frac{\partial}{\partial r_i}\frac{\partial}{\partial r_i}\phi(\boldsymbol{r}) = \Delta\phi(\boldsymbol{r}) = \left(\frac{\partial^2}{\partial x^2}+\frac{\partial^2}{\partial y^2}+\frac{\partial^2}{\partial z^2}\right)\phi(\boldsymbol{r})\ ,\\
\operatorname{curl}\operatorname{curl}\boldsymbol{a}(\boldsymbol{r}) &= [\nabla_r[\nabla_r,\boldsymbol{a}(\boldsymbol{r})]]_i = e_{imn}e_{nkj}\frac{\partial}{\partial r_j}\frac{\partial}{\partial r_m}a_k\\
&= \frac{\partial}{\partial r_i}\frac{\partial}{\partial r_j}a_j(\boldsymbol{r}) - \frac{\partial^2}{\partial r_j\partial r_j}a_i(\boldsymbol{r}) = \operatorname{grad}\operatorname{div}\boldsymbol{a}(\boldsymbol{r}) - \Delta\boldsymbol{a}(\boldsymbol{r})\ldots\ .
\end{aligned} \tag{B.22}$$

The described theory of three-dimensional tensors can be easily generalized to the four-dimensional case. In the four-dimensional space of the time and space coordinates $(t,\boldsymbol{r})$ the Lorentz transformations are rotation transformations and form the basis for the definition of four-dimensional vectors and tensors. Besides, $(t,\boldsymbol{r})$, the current and charge densities $(\varrho,\boldsymbol{j})$, the wave vector and frequency $(\omega,\boldsymbol{k})$, etc., are four-dimensional vectors. However, we do not expand on the theory of four-dimensional tensors, since, in fact, they have not been used in the book.

B.1 Exercises

B.1.1 Using the identity

$$e_{ikl}e_{mnl} = \delta_{im}\delta_{kn} - \delta_{in}\delta_{km} ,$$

verify the equality

$$[\boldsymbol{A}, [\boldsymbol{B}, \boldsymbol{C}]] = \boldsymbol{B}(\boldsymbol{A}\cdot\boldsymbol{C}) - \boldsymbol{C}(\boldsymbol{A}\cdot\boldsymbol{B}) .$$

Solution. We apply the tensor notation

$$\begin{aligned}[\boldsymbol{A}, [\boldsymbol{B}, \boldsymbol{C}]] &= e_{ikl}A_kB_mC_ne_{lmn} = A_kB_mC_n(e_{ikl}e_{lmn}) \\ &= A_kB_mC_n(\delta_{im}\delta_{kn} - \delta_{in}\delta_{km}) = B_i(A_nC_n) - C_i(A_nB_n) ,\end{aligned}$$

which was to be proven.

B.1.2 Compose the general second-rank tensor ε_{ij} with $\varepsilon_{ij}(\boldsymbol{k}) = \varepsilon_{ji}(-\boldsymbol{k})$ for real $\boldsymbol{k}$. For $\boldsymbol{k} = 0$ reduce ε_{ij} over the indices, i.e., take the sum ε_{ii}.

Solution.

$$\varepsilon_{ij}(\boldsymbol{k}) = \alpha_1\delta_{ij} + \alpha_2k_ik_j = \left(\delta_{ij} - \frac{k_ik_j}{k^2}\right)\varepsilon^{\text{tr}} + \frac{k_ik_j}{k^2}\varepsilon^{\text{lo}} ,$$

i.e.,

$$\alpha_1 = \varepsilon^{\text{tr}} , \qquad \alpha_2 = \frac{\varepsilon^{\text{lo}} - \varepsilon^{\text{tr}}}{k^2} .$$

In the limit $\boldsymbol{k} \to 0$ we have $\varepsilon_{ij}(0) = \alpha_1\delta_{ij}$ and $\varepsilon^{\text{lo}} = \varepsilon^{\text{tr}} = \varepsilon$. For vanishing $\boldsymbol{k}$, $\varepsilon_{ij}(0) = \varepsilon\delta_{ij}$ is the general second-rank tensor. The reduction over the indices reads

$$\varepsilon_{ij}(\boldsymbol{k}) = (3-1)\,\varepsilon^{\text{tr}} + \varepsilon^{\text{lo}} = 2\,\varepsilon^{\text{tr}} + \varepsilon^{\text{lo}}, \qquad \varepsilon_{ii}^{\text{lo}}(0) = 3\,\varepsilon .$$

B.1.3 Compose the second-rank tensor $\varepsilon_{ij}(\boldsymbol{B}) = \varepsilon_{ij}(-\boldsymbol{B})$ from a pseudovector $\boldsymbol{B}$ and reduce it over the indices.

Solution.

$$\varepsilon_{ij}(\boldsymbol{B}) = \alpha_1\delta_{ij} + \alpha_2b_ib_j + \alpha_3e_{ijk}b_k = \varepsilon_\perp\delta_{ij} + (\varepsilon_\parallel - \varepsilon_\perp)\,b_ib_j + ige_{ijk}b_k ,$$

where $\boldsymbol{b} = \boldsymbol{B}/B$. This tensor has the following matrix form

$$\varepsilon_{ij}(\boldsymbol{B}) = \begin{pmatrix} \varepsilon_\perp & ig & 0 \\ -ig & \varepsilon_\perp & 0 \\ 0 & 0 & \varepsilon_\parallel \end{pmatrix} .$$

Here the $0x_3$-axis is oriented along the vector $\boldsymbol{B}$. For vanishing B we have

$$\varepsilon_{ij}(0) = \varepsilon\delta_{ij}\,, \quad \varepsilon_\perp = \varepsilon_\parallel = \varepsilon\,, \quad g = 0\,.$$

Finally, the reduction over the indices is

$$\varepsilon_{ii}(\boldsymbol{B}) = 2\,\varepsilon_\perp + \varepsilon_\parallel\,, \quad \varepsilon_{ii}(0) = 3\,\varepsilon\,.$$

B.1.4 Compose the second-rank tensor $\varepsilon_{ij}(\boldsymbol{k},\boldsymbol{B}) = \varepsilon_{ji}(-\boldsymbol{k},-\boldsymbol{B})$ from a real vector $\boldsymbol{k}$ and a pseudovector $\boldsymbol{B}$ and reduce it over the indices.

Solution.

$$\begin{aligned}\varepsilon_{ij}(\boldsymbol{k},\boldsymbol{B}) = {} & \alpha_1\delta_{ij} + \alpha_2 k_i k_j + \alpha_3 b_i b_j + \alpha_4\, e_{ijm} b_m \\ & + \alpha_5 e_{imn} e_{jrs} k_m b_n k_r b_s + \alpha_6 (e_{imn} k_m b_n k_j - e_{jmn} k_m b_n k_i)\,,\end{aligned}$$

where $\boldsymbol{b} = \boldsymbol{B}/B$. Orienting the $0x_3 = 0z$-axis along the vector $\boldsymbol{b}$ and the $0x_1 = 0x$-axis so that the vector $\boldsymbol{k}$ takes the form $\boldsymbol{k} = (k_\perp, 0, k_\parallel)$, we obtain

$$\varepsilon_{ij}(\boldsymbol{k},\boldsymbol{B}) = \begin{pmatrix} \varepsilon_{11} & \varepsilon_{12} & \varepsilon_{13} \\ -\varepsilon_{12} & \varepsilon_{22} & \varepsilon_{23} \\ \varepsilon_{13} & -\varepsilon_{23} & \varepsilon_{33} \end{pmatrix}, \quad \text{with}$$

$$\begin{aligned} \varepsilon_\parallel &= \alpha_1 + \alpha_2 k_\perp^2\,, & \varepsilon_{22} &= \alpha_1 + \alpha_5 k_\perp^5\,, \\ \varepsilon_{12} &= \alpha_4 + \alpha_6 k_\perp^5\,, & \varepsilon_{23} &= -\,\alpha_6 k_\perp k_\parallel\,, \\ \varepsilon_{13} &= \alpha_2 k_\perp k_\parallel\,, & \varepsilon_{33} &= \alpha_1 + \alpha_2 k_\parallel^2 + \alpha_3\,. \end{aligned}$$

Reducing the derived tensor over the indices yields

$$\varepsilon_{ii}(k,B) = \varepsilon_{11} + \varepsilon_{12} + \varepsilon_{33} = 3\,\alpha_1 + \alpha_2 k^2 + \alpha_3 + \alpha_5 k_\perp^2\,.$$

B.1.5 Write down the general relation $B_i = \mu_{ij}(\boldsymbol{H})\,H_j$ between the pseudovectors $\boldsymbol{B}$ and $\boldsymbol{H}$.

Solution. Evidently, $\mu_{ij}(\boldsymbol{H})$ must be a real tensor. Therefore

$$\mu_{ij}(\boldsymbol{H}) = \alpha_1\delta_{ij} + \alpha_2 h_i h_j + \alpha_3 e_{ijk} h_k\,,$$

where $\boldsymbol{h} = \boldsymbol{H}/H$. The general relation under discussion thus obtains the form

$$\boldsymbol{B} = (\alpha_1 + \alpha_2)\boldsymbol{H} = \mu(\boldsymbol{H})\boldsymbol{H}\,.$$

References

Chapter 1

Balescu, R.: *Statistical Mechanics of Charged Particles* (Wiley-Interscience, New York 1963)
Ecker, G.: *Theory of Fully Ionized Plasmas* (Academic, New York 1972)
Frank-Kamenetskij, D.A.: *Lektsii po Fizike Plazmy* [*Lectures on Plasma Physics*] (Gosatomizdat, Moscow 1964)
Golant, V.E., Zhilinskij, A.P., Sakharov, A.P.: *Osnovy Fiziki Plazmy* [*Principles of Plasma Physics*] (Gosatomizdat, Moscow 1974)
Klimontovich, Yu.L.: *Kinetic Theory of Non-Ideal Gases and Non-Ideal Plasmas*, Int. Ser. Nat. Phil., Vol.105 (Pergamon, New York 1982)
Krall, N.A., Trivelpiece, A.W.: *Principles of Plasma Physics* (McGraw-Hill, New York 1973)
Landau, L.D., Lifshitz, E.M.: *Physical Kinetics*, Course of Theor. Phys., Vol.10 (Pergamon, New York 1981)
Mikhailovskij, A.B.: *Teorija Plazmennykh Neustojchivostej* [*Theory of Plasma Instabilities*] Vol.1-2 (Gosatomizdat, Moscow 1975)

Chapter 2

Akhiezer, A.I., Akhiezer, I.A., Polovin, R.V., Sitenko, A.G., Stepanov, N.P.: *Plasma Electrodynamics* (Pergamon, London 1975)
Ginzburg, V.L., Rukhadze, A.A.:"Waves and Resonances in Magneto-Active Plasma": In *Geophysics III*, Encyclopedia of Physics, Vol.XLIX/4 (Springer, Berlin, Heidelberg 1972)
Gurevich, A.V., Shvartsburg, A.B.: *Nelinejnaja Teorija Rasprostranenija Radiovoln v Ionosfere* [*Nonlinear Theory of Radiowave Propagation in Ionosphere*] (Nauka, Moscow 1973)
Landau, L.D., Lifshitz, E.M.: *Electrodynamics of Continuous Media*, Course of Theor. Phys., Vol.8 (Pergamon, New York 1960)
Landau, L.D., Lifshitz, E.M.: *Physical Kinetics*, Course of Theor. Phys., Vol.10 (Pergamon, New York 1981)
Lawson, J.D.: *The Physics of Charged Particle Beams* (Clarendon, Oxford 1977)
Silin, V.P., Rukhadze, A.A.: *Elektromagnitnye Svojstva Plazmy i Plazmopodobnykh Sred* [*Electromagnetic Properties of Plasma and Similar Media*] (Gosatomizdat, Moscow 1961)

Chapter 3

Akhiezer, A.I., Akhiezer, I.A., Polovin, R.V., Sitenko, A.G., Stepanov, N.P.: *Plasma Electrodynamics* (Pergamon, London 1975)
Balescu, R.: *Statistical Mechanics of Charged Particles* (Wiley-Interscience, New York 1963)
Bogolyubov, N.N.: *Dinamicheskie Problemy Statisticheskoj Fiziki* [*Dynamic Problems of Statistical Physics*] (Gostekhizdat, Moscow 1946)
Ecker, G.: *Theory of Fully Ionized Plasmas* (Academic, New York 1972)
Galeev, A.A., Sudan, R. (eds.): *The Principles of Plasma Physics*, Vol.1 (North-Holland, Amsterdam 1983)
Ginzburg, V.L.: *The Propagation of Electromagnetic Waves in Plasmas* (Pergamon, New York 1971)
Ginzburg, V.L., Rukhadze, A.A.:"Waves and Resonances in Magneto-Active Plasma":In *Geophysics III*, Encyclopedia of Physics, Vol.XLIX/4 (Springer, Berlin, Heidelberg 1972)

Ichimaru, S.: *Basic Principles of Plasma Physics. A Statistical Approach* (Pergamon, New York 1973)
Klimontovich, Yu.L.: *Kinetic Theory of Non-Ideal Gases and Non-Ideal Plasmas*, Int. Ser. Nat. Phil., Vol.105 (Pergamon, New York 1982)
Klimontovich, Yu.L.: *The Statistical Theory of Non-Equilibrium Processes in a Plasma*, Int. Ser. Nat. Phil., Vol.9 (Pergamon, New York 1967)
Krall, N.A., Trivelpiece, A.W.: *Principles of Plasma Physics* (McGraw-Hill, New York 1973)
Landau, L.D., Lifshitz, E.M.: *Physical Kinetics*, Course of Theor. Phys., Vol.10 (Pergamon, New York 1981)
Pozhela, J.: *Plasma and Current Instabilities in Semiconductors* (Pergamon, New York 1981)
Risken, H.: *The Fokker-Planck Equation*, Springer Ser. Syn., Vol.18 (Springer, Berlin, Heidelberg 1983)
Silin, V.P.: *Vvedenie v Kineticheskuju Teoriju Gasov* [*Introduction to the Kinetic Gas Theory*] (Nauka, Moscow 1971)

Chapter 4

Akhiezer, A.I., Akhiezer, I.A., Polovin, R.V., Sitenko, A.G., Stepanov, N.P.: *Plasma Electrodynamics* (Pergamon, London 1975)
Artsimovich, L.A., Sagdeev, R.Z.: *Fizika Plazmy dlia Fizikov* [*Plasma Physics for Physicists*] (Gosatomizdat, Moscow 1974)
Galeev, A.A., Sudan, R.(eds.): *The Principles of Plasma Physics*, Vol.1 (North-Holland, Amsterdam 1983)
Ginzburg, V.L.: *The Propagation of Electromagnetic Waves in Plasmas* (Pergamon, New York, 1971)
Ginzburg, V.L., Rukhadze, A.A.: "Waves and Resonances in Magneto-Active Plasma": In *Geophysics III*, Encyclopedia of Physics, Vol.XLIX/4 (Springer, Berlin, Heidelberg 1972)
Golant, V.E., Zhilinskij, A.P., Sakharov, A.P.: *Osnovy Fiziki Plazmy* [*Principles of Plasma Physics*] (Gosatomizdat, Moscow 1974)
Ichimaru, S.: *Basic Principles of Plasma Physics. A Statistical Approach* (Pergamon, New York 1973)
Kondratenko, A.N.: *Proniknovenie Polja v Plazmu* [*Field Penetration into the Plasma*] (Nauka, Moscow 1979)
Krall, N.A., Trivelpiece, A.W.: *Principles of Plasma Physics* (McGraw-Hill, New York 1973)
Lawson, J.D.: *The Physics of Charged-Particle Beams* (Clarendon, Oxford 1977)
Pozhela, J.: *Plasma and Current Instabilities in Semiconductors* (Pergamon, New York 1981)
Risken, H.: *The Fokker-Planck Equation*, Springer Ser. Syn., Vol.18 (Springer, Berlin, Heidelberg 1983)
Silin, V.P.: *Vvedenie v Kineticheskuju Teoriju Gasov* [*Introduction to the Kinetic Gas Theory*] (Nauka, Moscow 1971)

Chapter 5

Akhiezer, A.I., Akhiezer, I.A., Polovin, R.V., Sitenko, A.G., Stepanov, N.P.: *Plasma Electrodynamics* (Pergamon, London 1975)
Allis, W.P., Buchsbaum, S.G., Bers, A.: *Waves in Anisotropic Plasmas* (MIT Press, Cambridge, MA 1963)
Artsimovich, L.A., Sagdeev, R.Z.: *Fizika Plazmy dlia Fizikov* [*Plasma Physics for Physicists*] (Gosatomizdat, Moscow 1974)
Frank-Kamenetskij, D.A.: *Lektsii po Fizike Plazmy* [*Lectures on Plasma Physics*] (Gosatomizdat, Moscow 1964)
Galeev, A.A., Sudan, R. (eds.): *The Principles of Plasma Physics*, Vol.1 (North-Holland, Amsterdam 1983)
Ginzburg, V.L.: *The Propagation of Electromagnetic Waves in Plasmas* (Pergamon, New York, 1971)
Ginzburg, V.L., Rukhadze, A.A.: "Waves and Resonances in Magneto-Active Plasma": In *Geophysics III*, Encyclopedia of Physics, Vol.XLIX/4 (Springer, Berlin, Heidelberg 1972)

Golant, V.E., Zhilinskij, A.P., Sakharov, A.P.: *Osnovy Fiziki Plazmy* [*Principles of Plasma Physics*] (Gosatomizdat, Moscow 1974)
Ichimaru, S.: *Basic Principles of Plasma Physics. A Statistical Approach* (Pergamon, New York 1973)
Kondratenko, A.N.: *Proniknovenie Polja v Plazmu* [*Field Penetration into the Plasma*] (Nauka, Moscow 1979)
Krall, N.A., Trivelpiece, A.W.: *Principles of Plasma Physics* (McGraw-Hill, New York 1973)
Landau, L.D., Lifshitz, E.M: *Electrodynamics of Continuous Media*, Course of Theor. Phys., Vol.8 (Pergamon, New York 1960)
Lawson, J.D.: *The Physics of Charged-Particle Beams* (Clarendon, Oxford 1977)
Lominadze, D.G.: *Cyclotron Waves in Plasma* (Pergamon, New York 1981)
Platzman, P.M., Wolff, P.A.: *Waves and Interactions in Solid-State Plasmas* (Academic, New York 1973)
Pozhela, J.: *Plasma and Current Instabilities in Semiconductors* (Pergamon, New York 1981)
Silin, V.P., Rukhadze, A.A.: *Electromagnitnye Svojstva Plazmy i Plazmopodobnykh Sred* [*Electromagnetic Properties of Plasma and Similar Media*] (Gosatomizdat, Moscow 1961)
Stix, T.H.: *The Theory of Plasma Waves* (McGraw-Hill, New York 1962)
Vinogradova, M.B., Rudenko, O.V. Sukhorukov, A.P.: *Teorija Voln* [*The Theory of Waves*] (Nauka, Moscow 1979)
Vladimirov, V.V., Volkov, A.F., Mejlikov, E.Z.: *Plazma Poluprovodnikov* [*Plasma of Semiconductors*] (Gosatomizdat, Moscow 1979)

Chapter 6

Akhiezer, A.I., Akhiezer, I.A., Polovin, R.V., Sitenko, A.G., Stepanov, N.P.: *Plasma Electrodynamics* (Pergamon, London 1975)
Allis, W.P. Buchsbaum, S.G., Bers, A.: *Waves in Anisotropic Plasmas* (MIT Press, Cambridge, MA 1963)
Davidson, R.C.: *Theory of Nonneutral Plasmas* (Benjamin, Reading, MA 1974)
Golant, V.E., Zhilinskij, A.P., Sakharov, A.P.: *Osnovy Fiziki Plazmy* [*Principles of Plasma Physics*] (Gosatomizdat, Moscow 1974)
Gurevich, A.V., Shvartsburg, A.B.:*Nelinejnaja Teorija Rasprostranenija Radiovoln v Ionosfere* [*Nonlinear Theory of Radiowave Propagation in Ionosphere*] (Nauka, Moscow 1973)
Ichimaru, S.: *Basic Principles of Plasma Physics. A Statistical Approach* (Pergamon, New York 1973)
Ivanov, A.A.: *Fizika Sil'no Neravnovesnoj Plazmy* [*Physics of Strongly Nonequilibrium Plasma*] (Gosatomizdat, Moscow 1977)
Krall, N.A., Trivelpiece, A.W.: *Principles of Plasma Physics* (McGraw-Hill, New York 1973)
Landau, L.D., Lifshitz, E.M.: *Physical Kinetics*, Course of Theor. Phys., Vol.10 (Pergamon, New York 1981)
Longmire, C.L.: *Elementary Plasma Physics* (Wiley-Interscience, New York 1981)
Mikhailovskij, A.B.: *Teorija Plazmennykh Neustojchivostej* [*Theory of Plasma Instabilities*] Vol.1,2 (Gosatomizdat, Moscow 1975)
Platzman, P.M., Wolff, P.A.: *Waves and Interactions in Solid-State Plasmas* (Academic, New York 1973)
Silin, V.P.: *Parametricheskoe Vozdejstvie Izluchenija Bol'shoj Moshnosti na Plazmu* [*Parametric Effect of High-Power Radiation on Plasma*] (Nauka, Moscow 1973)

Chapter 7

Akhiezer, A.I., Akhiezer, I.A., Polovin, R.V., Sitenko, A.G., Stepanov, N.P.: *Plasma Electrodynamics* (Pergamon, London 1975)
Allies, W.P., Buchsbaum, S.G., Bers, A.: *Waves in Anisotropic Plasmas* (MIT Press, Cambridge, MA 1963)
Bas, F.G., Gurevich, Yu.G.: *Gorjachie Elektrony i Sil'nye Elektromagnitnye Volny v Plazme Poluprovodnikov i Gazovogo Razrjada* [*Hot Electrons and Strong Electromagnetic Waves in Plasmas of Semiconductors and Gas Discharge*] (Nauka, Moscow 1975)
Ginzburg, V.L., Rukhadze, A.A.: "Waves and Resonances in Magneto-Active Plasma": In *Geophysics III*, Encyclopedia of Physics, Vol. XLIX/4 (Springer, Berlin, Heidelberg 1972)

Gurevich, A.V., Shvartsburg, A.B.: *Nelinejnana Teorija Rasprostranenija Radiovoln v Ionosfere* [*Nonlinear Theory of Radiowave Propagation in Ionosphere*] (Nauka, Moscow 1973)
Ivanov, A.A.: *Fizika Sil'no Neravnovesnoj Plazmy* [*Physics of Strongly Nonequilibrium Plasma*] (Gosatomizdat, Moscow 1977)
Krall, N.A., Trivelpiece, A.W.: *Principles of Plasma Physics* (McGraw-Hill, New York 1973)
Landau, L.D., Lifschitz, E.M.: *Physical Kinetics*, Course of Theor. Phys., Vol.10 (Pergamon, New York 1981)
Longmire, C.L.: *Elementary Plasma Physics* (Wiley-Interscience, New York 1981)
Mikhailovskij, A.B.: *Teorija Plazmennykh Neustojchivostej* [*Theory of Plasma Instabilities*] (Gosatomizdat, Moscow 1975)
Nezlin, M.V.: *Dinamika Puchkov v Plazme* [*Dynamics of Beams in Plasma*] (Gosatomizdat, Moscow 1982)
Platzman, P.M., Wolff, P.A.: *Waves and Interactions in Solid-State Plasmas* (Academic, New York 1973)
Silin, V.P.: *Parametricheskoe Vozdejstvie Izluchenija Bol'shoj Moshnosti na Plazmu* [*Parametric Effect of High-Power Radiation on Plasma*] (Nauka, Moscow 1973)
Vinogradova, M.B., Rudenko, O.V., Sukhorukov, A.P.: *Teorija Voln* [*The Theory of Waves*] (Nauka, Moscow 1979)
Vladimirov, V.V., Volkov, A.F., Mejlikov, E.Z.: *Plazma Poluprovodnikov* [*Plasma of Semiconductors*] (Gosatomizdat, Moscow 1979)

Chapter 8

Artsimovich, L.A., Sagdeev, R.Z.: *Fizika plazmy dlia Fizikov* [*Plasma Physics for Physicists*] (Gosatomizdat, Moscow 1974)
Galeev, A.A., Sudan, R. (eds.): *The Principles of Plasma Physics*, Vol.1 (North-Holland, Amsterdam 1983)
Ginzburg, V.L., Rukhadze, A.A.: "Waves and Resonances in Magneto-Active Plasma": In *Geophysics III*, Encyclopedia of Physics, Vol. XLIX/4 (Springer, Berlin, Heidelberg 1972)
Golant, V.E., Zhilinskij, A.P., Sakharov, A.P.: *Osnovy Fiziki Plazmy* [*Principles of Plasma Physics*] (Gosatomizdat, Moscow 1974)
Ichimaru, S.: *Basic Principles of Plasma Physics. A Statistical Approach* (Pergamon, New York 1973)
Kadomtsev, B.B.: *Kollektivnye Javlenija v Plazme* [*Collective Phenomena in Plasma*] (Nauka, Moscow 1976)
Krall, N.A., Trivelpiece, A.W.: *Principles of Plasma Physics* (McGraw-Hill, New York 1973)
Landau, L.D., Lifshitz, E.M.: *Physical Kinetics*, Course of Theor. Phys., Vol.10 (Pergamon, New York 1981)
Longmire, C.L.: *Elementary Plasma Physics* (Wiley-Interscience, New York 1981)
Mikhailovskij, A.B.: *Teorija Plazmennykh Neustojchivostej* [*Theory of Plasma Instabilities*] Vol.1,2 (Gosatomizdat, Moscow 1975)
Platzman, P.M., Wolff, P.A.: *Waves and Interactions in Solid-State Plasmas* (Academic, New York 1973)
Vladimirov, V.V., Volkov, A.F., Mejlikov, E.Z.: *Plasma Poluprovodnikov* [*Plasma of Semiconductors*] (Gosatomizdat, Moscow 1979)

Chapter 9

Ichimaru, S.: *Basic Principles of Plasma Physics. A Statistical Approach* (Pergamon, New York 1973)
Kondratenko, A.N.: *Plazmennye Volnovody* [*Plasma Waveguides*] (Gosatomizdat, Moscow 1976)
Kondratenko, A.N.: *Proniknovenie Polja v Plazmu* [*Field Penetration into the Plasma*] (Nauka, Moscow 1979)
Krall, N.A., Trivelpiece, A.W.: *Principles of Plasma Physics* (McGraw-Hill, New York 1973)
Nezlin, M.V.: *Dinamika Puchkov v Plazme* [*Dynamics of Beams in Plasma*] (Gosatomizdat, Moscow 1982)
Pozhela, J.: *Plasma and Current Instabilities in Semiconductors* (Pergamon, New York 1981)

Stix, T.H.: *The Theory of Plasma Waves* (McGraw-Hill, New York 1962)
Vinogradova, M.B., Rudenko, O.V., Sukhorukov, A.P.: *Teorija Voln* [*The Theory of Waves*] (Nauka, Moscow 1979)

Chapter 10

Akhiezer, A.I., Akhiezer, I.A., Polovin, R.V., Sitenko, A.G., Stepanov, N.P.: *Plasma Electrodynamics* (Pergamon, London 1975)
Ecker, G.: *Theory of Fully Ionized Plasmas* (Academic, New York 1972)
Galeev, A.A., Sudan. R. (eds.): *The Principles of Plasma Physics*, Vol.1 (North-Holland, Amsterdam 1983)
Klimontovich, Yu.L.: *The Statistical Theory of Non-Equilibrium Processes in a Plasma*, Internat. Ser. Nat. Phil., Vol.9 (Pergamon, New York, 1967)
Krall, N.A., Trivelpiece, A.W.: *Principles of Plasma Physics* McGraw-Hill, New York 1973)
Landau, L.D., Lifshitz, E.M.: *Electrodynamics of Continuous Media*, Course of Theor. Phys., Vol.8 (Pergamon, New York 1960)
Landau, L.D., Lifshitz, E.M.: *Physical Kinetics*, Course of Theor. Phys., Vol.10 (Pergamon, New York 1981)
Lawson, J.D.: *The Physics of Charged-Particle Beams* (Clarendon, Oxford 1977)
Pustovalov, V.V., Silin, V.P.: *Nelinejnoe Vzaimodejstvie Voln v Plazme* [*Nonlinear Wave Interaction in Plasma*] Trudy FIAN, Proceedings of P.N. Lebedev Physical Institute, Vol.61 (Moscow 1972)
Risken, H.: *The Fokker-Planck Equation*, Springer Ser. Syn., Vol.18 (Springer, Berlin, Heidelberg 1983)
Sheffield, I.: *Plasma Scattering of Electromagnetic Radiation* (Academic, New York 1975)
Sitenko, A.G.: *Fluctuations and Non-Linear Wave Interactions in Plasmas* (Pergamon, New York 1981)
Tsytovich, V.N.: *An Introduction to the Theory of Plasma Turbulence* (Pergamon, New York 1972)

Chapter 11

Akhiezer, A.I., Akhiezer, I.A., Polovin, R.V., Sitenko, A.G., Stepanov, N.P.: *Plasma Electrodynamics* (Pergamon, London 1975)
Balescu, R.: *Statistical Mechanics of Charged Particles* (Wiley-Interscience, New York 1963)
Ivanov, A.A.: *Fizika Sil'no Neravnovesnoj Plazmy* [*Physics of Strongly Nonequilibrium Plasma*] (Gosatomizdat, Moscow 1977)
Kadomtsev, B.B.: *Kollektivnye Javlenija v Plazme* [*Collective Phenomena in Plasma*] (Nauka, Moscow 1976)
Klimontovich, Yu.L.: *Kinetic Theory of Non-Ideal Gases and Non-Ideal Plasmas*, Internat. Ser. Nat. Phil., Vol.105 (Pergamon, New York 1982)
Landau, L.D., Lifshitz, E.M.: *Physical Kinetics*, Course of Theor. Phys., Vol.10 (Pergamon, New York 1981)
Lawson, J.D.: *The Physics of Charged-Particle Beams* (Clarendon, Oxford 1977)
Nezlin, M.V.: *Dinamika Puchkov v Plazme* [*Dynamics of Beams in Plasma*] (Gosatomizdat, Moscow 1982)
Pustovalov, V.V., Silin, V.P.: *Nelinejnoe Vzaimodejstvie Voln v Plazme* [*Nonlinear Wave Interaction in Plasma*] Trudy FIAN, Proceedings of P.N. Lebedev Physical Institute, Vol.61 (Moscow 1972)
Tsytovich, V.N.: *Nelinejnye Effekty v Plazme* [*Nonlinear Effects in Plasmas*] (Nauka, Moscow 1967)

Chapter 12

Akhiezer, A.I., Akhiezer, I.A., Polovin, R.V., Sitenko, A.G., Stepanov, N.P.: *Plasma Electrodynamics* (Pergamon, London 1975)
Balescu, R.: *Statistical Mechanics of Charged Particles* (Wiley-Interscience, New York 1963)
Ecker, G.: *Theory of Fully Ionized Plasma* (Academic, New York 1972)
Galeev, A.A., Sudan, R. (eds.): *The Principles of Plasma Physics*, Vol.1 (North Holland, Amsterdam 1983)

Golant, V.W., Zhilinskij, A.P., Sakharov, A.P.: *Osnovy Fiziki Plazmy* [*Principles of Plasma Physics*] (Gosatomizdat, Moscow 1974)
Gurevich, A.V., Shvartsburg, A.B.: *Nelinejnaja Teorija Rasprostranenija Radiovoln y Ionosfere* [*Nonlinear Theory of Radiowave Propagation in Ionosphere*] (Nauka, Moscow 1973)
Ichimaru, S.: *Basic Principles of Plasma Physics. A Statistical Approach* (Pergamon, New York 1973)
Kadomtsev B.B.: *Kollektivnye Javlenija v Plazme* [*Collective Phenomena in Plasma*] (Nauka, Moscow 1976)
Klimontovich, Yu.L.: *The Statistical Theory of Non-Equilibrium Processes in a Plasma,* Internat. Ser. Nat. Phil., Vol.9 (Pergamon, New York 1967)
Landau, L.D., Lifshitz, E.M.: *Physical Kinetics*, Course of Theor. Phys., Vol.10 (Pergamon, New York 1981)
Lawson, J.D.: *The Physics of Charged-Particle Beams* (Clarendon, Oxford 1977)
Pustovalov, V.V., Silin, V.P.: *Nelinejnoe Vzaimodejstvie Voln v Plazme* [*Nonlinear Wave Interaction in Plasma*] Trudy FIAN, Proceedings of P.N. Lebedev Physical Institute, Vol.61 (Moscow 1972)
Sheffield, I.: *Plasma Scattering of Electromagnetic Radiation* (Academic, New York 1975)
Silin, V.P.: *Parametricheskoe Vozdejstvie Izluchenija Bol'shoj Moshnosti na Plazmu* [*Parametric Effect of High-Power Radiation on Plasma*] (Nauka, Moscow 1973)
Tsytovich, V.N.: *Nelinejnye Effekty v Plazme* [*Nonlinear Effects in Plasma*] (Nauka, Moscow 1967)
Vinogradova, M.B., Rudenko, O.V., Sukhorukov, A.P.: *Teorija Voln* [*The Theory of Waves*] (Nauka, Moscow 1979)

Subject Index

I. I. Sobelman

Atomic Spectra and Radiative Transitions

1979. 21 figures, 46 tables. XII, 306 pages (Springer Series in Chemical Physics, Volume 1)
ISBN 3-540-09082-7

Contents: Elementary Information on Atomic Spectra: The Hydrogen Spectrum. Systematics of the Spectra of Multielectron Atoms. Spectra of Multielectron Atoms. - Theory of Atomic Spectra: Angular Momenta. Systematics of the Levels of Multielectron Atoms. Hyperfine Structure of Spectral Lines. The Atom in an External Electric Field. The Atom in an External Magnetic Field. Radiative Transitions. - References. - List of Symbols. - Subject Index.

This volume is continued by *Excitation of Atoms and Broadening of Spectral Lines* published as Volume 7 in this series.

I. I. Sobelman, L. A. Vainshtein, E. A. Yukov

Excitation of Atoms and Broadening of Spectral Lines

1981. 34 figures, 40 tables. X, 315 pages (Springer Series in Chemical Physics, Volume 7)
ISBN 3-540-09890 9

Contents: Elementary Processes Giving Rise to Spectra. - Theory of Atomic Collisions. Approximate Methods for Calculating Cross Sections. - Collisions Between Heavy Particles. - Some Problems of Excitation Kinetics. - Tables and Formulas for the Estimation of Effective Cross Sections. - Broadening of Spectral Lines. - References. - List of Symbols. - Subject Index. - Errata for volume 1 of this series.

S. Büttgenbach

Hyperfine Structure in 4d-and 5d-Shell Atoms

1982. 14 figures. VIII, 97 pages (Springer Tracts in Modern Physics, Volume 96)
ISBN 3-540-11740-7

Contents: Introduction. - Theoretical Considerations. - Experimental Methods. - Experimental Hyperfine Interaction Constants for 4d- and 5d-Shell Atoms. - Intermediate Coupling Wavefunctions, Atomic g_J Values and Parametric Interpretation of Level Isotope Shifts. - Effective Radial Parameters of the Magnetic Dipole and Electric Quadrupole Interaction. - Nuclear Moments and Hyperfine Anomalies. - Concluding Remarks. - References. - Abbreviations in this Work. - Subject Index.

I. Lindgren, J. Morrison

Atomic Many-Body Theory

1982. 96 figures. XIII, 469 pages (Springer Series in Chemical Physics, Volume 13)
ISBN 3-540-10504-2

Contents: Angular-Momentum Theory and the Independent-Particle Model: Introduction. Angular-Momentum and Spherical Tensor Operators. Angular-Momentum Graphs. Further Developments of Angular-Momentum Graphs. Applications to Physical Problems. The Independent-Particle Model. The Central-Field Model. The Hartree-Fock Model. Many-Electron Wave Functions. - Perturbation Theory and the Treatment of Atomic Many-Body Effects: Perturbation Theory. First-Order Perturbation for Closed-Shell Atoms. Second Quantization and the Particle-Hole Formalism. Application of Perturbation Theory of Closed-Shell Systems: Application of Perturbation Theory to Open-Shell Systems. The Hyperfine Interaction. The Pair-Correlation Problem and the Coupled-Cluster Approach. - Appendices A-D. - References. - Author Index. - Subject Index.

Springer-Verlag
Berlin
Heidelberg
New York
Tokyo

The manufacturer's authorised representative in the EU is Springer Nature Customer Service Centre GmbH, Europaplatz 3, 69115 Heidelberg, Germany. If you have any concerns regarding our products, please contact ProductSafety@springernature.com

Printed and bound by CPI Group (UK) Ltd, Croydon, CR0 4YY

07/01/2026

02030924-0013